Springer Complexity

Springer Complexity is an interdisciplinary program publishing the best research and academic-level teaching on both fundamental and applied aspects of complex systems-cutting across all traditional disciplines of the natural and life sciences, engineering, economics, medicine, neuroscience, social and computer science.

Complex Systems are systems that comprise many interacting parts with the ability to generate a new quality of macroscopic collective behavior, the manifestations of which are the spontaneous formation of distinctive temporal, spatial, or functional structures. Models of such systems can be successfully mapped onto quite diverse "real-life" situations like the climate, the coherent emission of light from lasers, chemical reaction-diffusion systems, biological cellular networks, the dynamics of stock markets and of the Internet, earthquake statistics and prediction, freeway traffic, the human brain, or the formation of opinions in social systems, to name just some of the popular applications.

Although their scope and methodologies overlap somewhat, one can distinguish the following main concepts and tools: self-organization, nonlinear dynamics, synergetics, turbulence, dynamical systems, catastrophes, instabilities, stochastic processes, chaos, graphs and networks, cellular automata, adaptive systems, genetic algorithms, and computational intelligence.

The three major book publication platforms of the Springer Complexity program are the monograph series "Understanding Complex Systems" focusing on the various applications of complexity, the "Springer Series in Synergetics," which is devoted to the quantitative theoretical and methodological foundations, and the "Springer Briefs in Complexity" which are concise and topical working reports, case studies, surveys, essays, and lecture notes of relevance to the field. In addition to the books in these two core series, the program also incorporates individual titles ranging from textbooks to major reference works.

Editorial and Programme Advisory Board

Understanding Complex Systems

Founding Editor: S. Kelso

Future scientific and technological developments in many fields will necessarily depend upon coming to grips with complex systems. Such systems are complex in both their composition – typically many different kinds of components interacting simultaneously and nonlinearly with each other and their environments on multiple levels – and in the rich diversity of behavior of which they are capable.

The Springer Series in Understanding Complex Systems (UCS) promotes new strategies and paradigms for understanding and realizing applications of complex systems research in a wide variety of fields and endeavors. UCS is explicitly transdisciplinary. It has three main goals: first, to elaborate the concepts, methods, and tools of complex systems at all levels of description and in all scientific fields, especially newly emerging areas within the life, social, behavioral, economic, and neuro- and cognitive sciences (and derivatives thereof); second, to encourage novel applications of these ideas in various fields of engineering and computation such as robotics, nano-technology, and informatics; and third, to provide a single forum within which commonalities and differences in the workings of complex systems may be discerned, hence leading to deeper insight and understanding.

UCS will publish monographs, lecture notes, and selected edited contributions aimed at communicating new findings to a large multidisciplinary audience.

More information about this series at http://www.springer.com/series/5394

George E. Mobus • Michael C. Kalton

Principles of Systems Science

 Springer

George E. Mobus
Associate Professor
Faculty in Computer Science & Systems,
  Computer Engineering & Systems
Institute of Technology
University of Washington Tacoma
Tacoma, WA, USA

Michael C. Kalton
Professor Emeritus
Faculty in Interdisciplinary Arts & Sciences
University of Washington Tacoma
Tacoma, WA, USA

ISSN 1860-0832   ISSN 1860-0840 (electronic)
ISBN 978-1-4939-1919-2   ISBN 978-1-4939-1920-8 (eBook)
DOI 10.1007/978-1-4939-1920-8
Springer New York Heidelberg Dordrecht London

Library of Congress Control Number: 2014951029

Printed on acid-free paper

Springer is part of Springer Science+Business Media (www.springer.com)

About the Authors

George E. Mobus is an Associate Professor of Computer Science and Systems and Computer Engineering and Systems in the Institute of Technology at the University of Washington Tacoma. In addition to teaching computer science and engineering courses, he teaches courses in systems science to a broad array of students from across the campus. He received his Ph.D. in computer science from the University of North Texas in 1994. His dissertation, and subsequent research program at Western Washington University, involved developing autonomous robot agents by emulating natural intelligence as opposed to using some form of artificial intelligence. He is reviving this research agenda now that hardware elements have caught up with the processing requirements for simulating real biological neurons. He also received an MBA from San Diego State University in 1983, doing a thesis on the modeling of decision support systems based on the hierarchical cybernetic principles presented in this volume. He did this while actually managing an embedded systems manufacturing and engineering company in Southern California. His baccalaureate degree was earned at the University of Washington (Seattle) in 1973, in zoology. He studied the energetics of living systems and the interplay between information, evolution, and complexity. By using some control algorithms that he had developed in both his undergraduate and MBA degrees in programming embedded control systems, he solved some interesting problems that led to promotion from a software engineer (without a degree) to the top spot in the company. All because of systems science!

Michael C. Kalton is Professor Emeritus of Interdisciplinary Arts and Sciences at the University of Washington Tacoma. He came to systems science through the study of how cultures arise from and reinforce different ways of thinking about and interacting with the world. After receiving a Bachelor's degree in Philosophy and Letters, a Master's degree in Greek, and a Licentiate in Philosophy from St. Louis University, he went to Harvard University where in 1977 he received a joint Ph.D. degree in East Asian Languages and Civilizations, and Comparative Religion. He has done extensive research and publication on the Neo-Confucian tradition, the dominant intellectual and spiritual tradition throughout East Asia prior to the twentieth century. Environmental themes of self-organizing relational interdependence and the need to fit in the patterned systemic flow of life drew his attention due to their resonance with East Asian assumptions about the world. Ecosystems joined social systems in his research and teaching, sharing a common matrix in the study of complex systems, emergence, and evolution. The interdisciplinary character of his program allowed this integral expansion of his work; systems thinking became the thread of continuity in courses ranging from the world's great social, religious, and intellectual traditions to environmental ethics and the systems dynamics of contemporary society. He sees a deep and creative synergy between pre-modern Neo-Confucian thought and contemporary systems science; investigating this potential cross-fertilization is now his major research focus.

Preface

Understanding

This book is about *understanding*. When can a person say that they understand something? Is understanding different from what we normally call "knowledge?" Do we actually understand a phenomenon when we can make predictions about its behavior? Perhaps an example of the latter question will serve as a key to the concept of understanding.

Consider the law of gravity. We all know what gravity is; who hasn't experienced its insistence on one being pulled toward the Earth, sometimes painfully? Yet it is the case that we actually still do not really understand gravity in the sense of what causes this force to act upon mass. Sir Isaac Newton formulated the laws of motion and particularly the mechanics of planetary motions from Johannes Kepler's planetary "laws." Kepler, in turn, had derived his laws from discovering the patterns contained in Tycho Brahe's astronomical observations of the planets' motions. Newton invented a descriptive language, the calculus, and advanced the universal laws of gravitation as a formula[1] that would predict with reasonable accuracy (even today) how bodies behave when acted upon by its force (one of four fundamental forces of nature[2]). NASA engineers can predict with tremendous accuracy just how much time and with what force a small rocket engine should fire to maintain a trajectory of a space probe millions of miles from Earth so that it neatly passes by a moon of Saturn to get pictures and data.

Albert Einstein "improved" our ability to predict such behavior, indeed for all objects of all masses and all distances in the universe, with his theory of General Relativity. Rather than describe this behavior as resulting from a mysterious force, Einstein converted the language of gravitation to geometry, explaining how the behavior of objects, such as planets orbiting the Sun, is a consequence of the distortions in space (and for really fast objects, time).

Both theories provide adequate predictions for celestial mechanics. We can say we humans understand the behavior from the outside. That is, we can, given the initial conditions of any two bodies of known masses at time 0, predict with great accuracy and very impressive precision what will happen in the future. But, and this is a crucial "but," we don't know why gravity works the way it does. For example, just saying that space is curved in the region of a massive object doesn't begin to say why. Physics is still actively seeking that kind of understanding. Our knowledge includes the formulas needed to predict planetary and satellite motions, which we routinely use, but it does not include the internal workings of nature sufficient to explain why those formulas work.

And this condition, what we must call "partial understanding," is often more true of much of our knowledge than we might like to acknowledge. Systems science is ultimately about gaining more complete understanding. Notice we said "more complete" rather than merely "complete." Understanding comes in degrees. As far as anyone knows, there is no such thing as absolute (complete) understanding or knowledge (see our discussion of knowledge in Chap. 7). Rather there are approaches to understanding more about phenomena by gaining knowledge of their inner mechanics. All of the sciences work at this.

In this regard systems science can be considered the universal science. All sciences seek to gain and organize knowledge systematically. They all use methodologies that, while geared to the specific domain of interest (say physics or psychology), nevertheless are variations on concepts you will find in this volume. They all seek to establish organizations of knowledge (invariably hierarchical in nature) that expose patterns of relations, for example, Dmitri Mendeleev's Periodic Table for chemistry (and its many improvements since then) or Carolus Linnaeus' classification hierarchy for species that helped lead to the Theory of Evolution proposed by Charles Darwin. As you will see in this text, organization, structure, and many other aspects of knowledge form the kernel of systems science.

---

[1] $F = G(m_1 m_2 / r^2)$. F is the force due to gravitational attraction. m_1 and m_2 are the masses of the two bodies (it takes two!) and r is the distance between the centers of the two bodies.

[2] The other three being electromagnetic, weak, and strong forces. The first of these describes how elementary particles behave due to attraction and repulsion. The latter two apply to interactions between components of atomic nuclei.

What systems science does, above and beyond the efforts of any of the domain-oriented sciences, is to make the whole enterprise of gaining better understanding explicit. All scientists (in the broadest interpretation of that word) are systems scientists to one degree or another, even when they don't know that.

Mental Models of the World: Cognitive Understanding

Whenever you think about what may happen during an upcoming day in your life you are accessing what we call a *mental model* of your world. As will be described in several sections of this book, our brains construct these models based on our experiences as we grow up and age. Most of our knowledge is tucked away in what cognitive scientists call implicit form. This could be "procedural" knowledge, such as how to ride a bicycle or drive a car, or it could be more general knowledge that isn't automatically accessible to conscious thinking; you need to expend some mental effort to do so. Your ability to live in a society with a culture and to go about daily life all depends on your having built up a large repertoire of mental models about how things work. When you enter a restaurant, for example, you know basically what to do without even thinking about it. You know how to wait to be seated, how to examine a menu and decide your order, how to give your order to a server, etc. You have done this so often that it is like second nature. The places and people and menus may change, but you know the general script for how to behave and accomplish your goal (getting fed!). Perhaps as much as 80–90 % of your daily interactions with things and people are the result of processing these mental models subconsciously!

Models are manipulatable representations of things (especially people), relations of things, and how they behave in the world. Mental models are those we build up in our neural network systems in, especially, our neocortex. Our understanding of the world depends on us being able to learn what to expect from the things and people we interact with into the possible future.

We will have much more to say about mental models in Chaps. 7–9 (Part III). What we intend for this book to accomplish is to help you organize your mental models, to make connections between aspects of the world you may not have explicitly recognized. We believe that systems science is capable of helping people make more sense of their mental models—to help them better understand the world.

Formal Models of the World: The Extension of Cognitive Understanding

One of the great achievements of the human mind has been to develop abstract, external representations of the world. This started with the evolution of language (maybe 150–200 thousand years ago) as a way to communicate complex mental models. It later gave rise to the development of *signs* and *symbols* marked on a medium, the beginning of written language and mathematics. Humans have, since then, developed extremely sophisticated ways to use those signs and symbols to construct models of the world externally to their own minds. Mathematics is such a way to compactly express a "formal" model of, for example, the attributes of things (measurements with numbers), relations of things (algebraic and geometric), and behaviors (dynamics).

Formal models have extended the human ability to much better understand the world and communicate that understanding to others. Today we have computer-based models of incredibly complex phenomena (e.g., the climate and weather) that allow us to make more detailed predictions about the future than could be done with mental models alone. Part III also will cover aspects of formal modeling, and Chap. 13 will explain how modern models are built and used in the sciences and engineering.

Unfortunately the very power to build formal models has contributed to what we feel is a negative side effect, which is a major motivation for this book. We describe the tendency for the disciplines to become more isolated from one another below. In part, this tendency is enhanced by the very nature of formal modeling in the sense that each discipline has developed its own specialized language of signs, symbols, syntax, and semantics. In essence, as the models get better at helping experts understand their small piece of the world, they start to hide the connections between those pieces. Our sincere hope is that a more explicit education in systems science will help correct this situation. The left hand absolutely needs to know what the right hand is doing, and vice versa.

Why an Education in Systems Science?

A quick word of explanation for those who would equate the terms "systems science" strictly with computers and communications systems; while those are examples of human-built systems (see Chap. 14), systems science is not just about technology. Indeed the latter is just a tiny part of systems science (and engineering). In today's jargon, the word "system" has come to be dominantly associated with computational technology. This is another consequence of our education system's propensity to work against integrative understanding in preference for specialization.

Both authors have taught courses that are either explicitly about systems science or stealthfully bring systems science into the curriculum. In every case, the students' general responses invariably show surprise to learn that the world can be understood as a system of systems and that they had never been exposed to this perspective in their education previously. Moreover, they express deep gratitude for being shown how the world can be interpreted in a more holistic fashion and one that they can readily grasp. Why do they react this way?

The modern American and many other countries educational systems, in our opinion, have devolved into promoting and serving silo-based thinking. By that we mean domain-specific subjects and majors are the norm. Increasingly this tendency also squeezes out the traditional liberal studies courses that were considered essential for students to develop broad knowledge and develop critical thinking skills outside the context of just one domain. Until recently, systems science, in the form we present in this book, has not been a discipline per se. Parts and pieces of systems science have been pursued for their own sakes, but the integrated whole subject did not have an integrated whole body of knowledge that could be explored and improved in its own way. The needs of the marketplace have dominated the methods and approaches of education in such a way that an educated person, today, is expected to find a job in industry or government, in which the skills they acquired in school can be put to immediate productive use. And up until very recently, the perception of society has been that those skills were domain-specific.

But something interesting has been developing in the worlds of commerce, government, and, indeed, all fields. People are beginning to recognize that the kinds of problems we seek to solve no longer involve single domains of knowledge and skills. Rather every field is experiencing the need to involve other fields and do integrative (what has been called cross-disciplinary) work. This has led to a new problem for scientist, engineers, business people, and others. Essentially, what is the common language we can all speak that will allow us to integrate our different domains? And they are finding, increasingly, that a systems approach provides that common language. It is a kind of Rosetta stone for systemic work.

As we watch the world developing a much stronger need for systems science and systems development, we anticipate the need for more explicit education in systems science in the near future. Students who grasp systemness (that term is introduced in Chap. 1 and more thoroughly defined in Chap. 3) are better able to understand complex systems and how the various disciplinary languages can be ameliorated in a common view of those systems. We assert that a basic education in systems science will better prepare any student for any major in which they are expected to tackle and solve complex problems. Even if it is just a two-course sequence based on this book, they will emerge with a much greater understanding of what it means to understand and how to gain that understanding in whatever field they decide to specialize.

Why a Textbook on Systems Science?

It seems strange to say that there have not been any introductory textbooks in systems science[3] if this subject is "meta" to all other sciences. But that seems to be the situation. There are general books that are about systems science or systems thinking, but they do not attempt to systematically outline the subtopics and then provide an integrated perspective of the whole subject. They are excellent for motivating

---

[3] To be fair there have been many books that introduce the ideas of systems theory, even with titles purporting to be introductions to systems science. Many of these will be found in the bibliography. However, our assessment of the books that we have surveyed of this kind is that they are not really comprehensive attempts to lay out all of the modern topics in systems science in the form of a textbook suitable for pedagogical uses. This assessment can arguably be contested. But of all the books on the subject we assert this is the most balanced and integrative volume of its kind.

students in the idea of systems thinking but do not expose the body of systems subjects with pedagogy in mind.

We think the subject of systems science will begin to take a front seat in education because the grasp of systemness is a powerful mental framework for thinking about literally everything in the world.

There is a truism held by almost all students that no textbook is ever written well. Trade books and story books, on the other hand, are written so the average reader can understand what the author is saying. The worst textbooks are in technical and science fields where the writing is dry and, well, technical. They are hard to read.

So why write a textbook about systems science that students are going to find hard to understand? Well, our answer is that we have not written a textbook that is hard to understand because we are telling a story—a very big story.

This is an introductory textbook in a subject that is universal to many other subjects in which the reader might decide to major. We claim the reader will be able to understand, but that doesn't mean they will not have to put some effort into it. The book covers a broad array of subjects with many examples from various disciplines. We recognize that not all students will have had courses in some of these subjects. But we also don't think coursework in these subjects is actually prerequisite to grasping the main ideas in this book. In all cases where we have used explicit examples, say from biology, we also have provided reference links to articles in Wikipedia[4] that we think do a good job of explaining details. We encourage readers to use these links and get a passing familiarity with the subjects or, at least, get good definitions of terms that might be foreign to the reader.

Why Is This Textbook the First of Its Kind?

As you are about to find out, systems science is a huge subject. That is because the concepts covered here are actually found in all of the other sciences (so-called natural and social) in one form or another. Systems science is a universal science. It is therefore surprising (at least it was to us when we started researching for the kind of textbook we had in mind for our teaching) that no general, introductory textbook seemed to exist.

The idea of a "general systems theory" and several related ideas were first put forth in a formal way in the late 1940s and through the next decade. For example, general systems theory was developed by Ludwig von Bertalanffy (September 19,

---

[4]Wikipedia, if you don't already know, is a FREE online, crowd-sourced encyclopedia containing pages on just about everything that anyone knows (or believes they do). But, there is no such thing as free, as many of these chapters will show. Wikipedia is supported by a foundation, the Wikimedia Foundation (http://en.wikipedia.org/wiki/Wikipedia:Wikimedia_Foundation) that accepts charitable contributions to keep Wikipedia going. We use Wikipedia links extensively throughout the book where we think the information is good and could provide readers with additional links to a vast warehouse of information. Please consider a donation to the Foundation if you find yourself taking advantage of this rich resource.

1901–June 12, 1972) during the 1950s and published in English in the 1968.[5] von Bertalanffy was a biologist, and many think of him as the father of systems biology. What he sought were the principles of organization and dynamics, the spatiotemporal patterns that were common across all kinds of systems. He felt these could be captured in universal laws that would apply to all systems and could be codified in mathematics.

But the emphasis on mathematics (or at least the appearance of the emphasis) kept the concepts from gaining broad acceptance, let alone understanding. Many researchers already possessing the mathematical skills, of course, jumped onto the various aspects of systems science and have done tremendous work in those areas. But the overall subject has remained invisible to the average educated person.

Several other fields of research coming out of efforts made during WWII, such as cybernetics, information theory, operations research, computation, etc., were also mathematical in their origins and so remained inaccessible but to a few mathematicians who could grapple with the equations. And those who studied these fields, even while extolling the notion that they were all deeply related in the nature of "systems," found it easier to isolate themselves into their respective subdomains, driving deeper into those domains and creating an invisible boundary between them. As a consequence, the idea of general systems got more and more difficult to envision from a higher perspective. And the underlying interrelations gave way to increasingly real language barriers. Ironically, what started out as a truly integrative idea ended up in the same kinds of disciplinary silos into which all the other academic subjects had fallen.

And the general public, even those with higher education degrees, partly because of the continuing emphasis on more sophisticated mathematics and partly because the systems scientist themselves encouraged increasing insulation, grew ever more ignorant of the concept of general systems theory even while using the word "system" in increasing frequency. Everyone knows (or "feels" they know) what you mean by a "computer system." They know what you mean by the "educational system." But all they really know is that somehow the parts of those "systems" are related and the whole "system" is supposed to perform a function. Beyond that, the deeper principles remain in shadows, not even hinted at by the cyberneticists, the communications theorists, and the evolutionary and systems biologists.

The areas that actually had much better success in recognizing systemness and the importance of general systems theory has been business management and military science. Much of the seminal thoughts had come from efforts by mathematicians to discover principles of control and command, both organizational and mechanical. Communications, especially encrypted during WWII, gave rise to information theory. Thermodynamics was an old science in physics, but there were new surprises there as well. But after WWII, in the west, business management theorists started applying concepts from cybernetics and information theory in a framework of systemic organization and process management. Other organization theorists developed languages to describe models of organizations

---

[5] See: von Bertalanffy (1968) and http://en.wikipedia.org/wiki/Ludwig_von_Bertalanffy.

and, eventually, computer simulations of those models that demonstrated systems dynamics of system behaviors.

With all of this foment and active research into systems-related subjects, the question remains. Why are there no general textbooks that introduce the broad range of sub-subjects in an integrated way and make the concepts accessible to, say, lower division baccalaureate students? To be fair, there are a number of books with titles like *Introduction to Systems Science*[6] and *An Introduction to General Systems Thinking*.[7] And these books do attempt to explore systems science and systems thinking, but, truthfully, they are not very comprehensive. This is because their authors harken back to a time when there was very little knowledge about some subjects that would, more recently, change many perspectives on what general systems theory might encompass. Most of the authors who have written introductory books have taken a more philosophical approach to describing systems science. They sought generalizations but were less concerned with fundamentally tying the pieces together. They were not writing textbooks but summaries of every insight they had gained up till the time of writing.

And insightful they were. We hope many of those insights have been captured in these pages.

But an introductory textbook to any subject has to explore the breadth of it and dip into some depths when it is appropriate to show how the whole fabric is stitched together. In this book we have attempted to do three basic things. The first is to outline what we think are the fundamental principles of systems science and show how they apply across a wide array of systems examples. Second, we are attempting to demonstrate some depth in the sub-subjects so that you get a better understanding of them and what kinds of work go on when digging deeper within each. The third objective is to show how all of these different sub-subjects relate to one another, *strongly*.

Unlike most other subjects where subfields tend to become more specialized and distant from one another, we claim that systems science has strong interrelations between the sub-subjects at all levels of study. You cannot really isolate, for example, internal dynamics from network theory. Dynamics work themselves out in networks of relations. Someone studying overt dynamics (external behavior) might be able to ignore some details of the network organization of the parts of the system, but ultimately, in order to fully understand that system, they will need to show how the dynamical properties and behaviors are partly a consequence of the network structure.

The same can be said for complexity theory and, for example, emergence and evolution. All of these principle-based sub-subjects have to be understood in light of all the others. We attempt to show this in Chap. 1.

So the answer to the question is that this may be a unique confluence in time of several "systemic" factors that allow an approach such as we have taken. First the existence today of accessible high-speed computers makes a kind of experimental systems science feasible, but more than that, the way in which computers work and

---

[6] See Warfield (2006).

[7] See Weinberg (2001).

are organized wholes has provided an intellectual scaffold for grappling with systems principles. Second, many new areas not well understood by earlier thinkers have developed in the last two to three decades. One in particular, the exponential growth in understanding of the brains of animals and man has forced many systems thinkers to reconsider ideas about complexity. Along those lines, new understandings of emergent phenomena and evolution have added another dimension to systems thinking that was not well understood even into the current century. The capabilities to sequence and catalog the genomes of many species, especially us, and the ability to map those genes and their developmental control programs have changed the way we understand information and knowledge.

Third, there is a social problem that systems science might be able to help with. The modern specialist education was seen in the mid- and late twentieth century as the route to a more effective and efficient economy. Liberal studies took a back seat to silo-based and professional degrees. This worked in the early part of the so-called Information Economy, but as the kinds of endeavors humans have been undertaking keep getting more and more complex, with components needed from multiple disciplines, the need for a higher-level viewpoint and an ability to grapple with complex patterns has emerged as a new capability needed by society. Generalists are hard to come by because most people think, and rightly so as far as it goes, that you can't know everything and you can't be a specialist in everything. As everyone knows you can be a jack of all trades but will not be a master of any.

Except that, general systems science and systems thinking apply everywhere. And a deep knowledge of systems science may yet prove to be the twenty-first century equivalent to liberal studies in that it promotes generalist understandings along with real critical thinking and integrative thinking. The world needs many more systems scientists to help integrate the work of specialists. Systems scientists have a basic vocabulary and semantics that can readily fit into any discipline, and they are thus positioned to grasp what the specialists are talking about. They can provide translation services when two different disciplines have to work together.

The need for broad systems thinking and the tools of systems science are needed more than ever today owing to some of the planet-wide systemic problems that are facing humanity. Our hope in writing this book, and telling the story, is that introducing more students to the concepts and the way of thinking will induce them to pursue whatever majors they choose from the perspective of systems.

About the Math

We mean for this book to be accessible to a very broad audience. The reason is straightforward. We feel that knowledge of systems science is something that every thinking person could benefit from. And we recognize, even while there is a current panic in our society that students aren't learning enough math (or math well enough), not every person will be comfortable believing that they will never understand something if they don't understand the math. Our feeling is that the fundamentals and principles of systems science are completely understandable without necessary recourse to mathematics. And so we have minimized the use of mathematics in the book in the hopes that non-math-oriented readers will not be intimidated into feeling they cannot understand the principles.

We do assume that readers will have had at least a course in algebra since algebraic expressions can often convey the kinds of relations we present. Even when this much math is included in the text, it is possible for readers to extract the relational information from the verbiage, but just with a little more effort on their part. We are not advocating that it is OK for people to avoid mathematics. Rather we are trying to show that these ideas can be expressed in English but could generally be expressed more compactly mathematically. Perhaps some people who were math-phobic at the start of this book will start to see the benefit of using math to express these ideas by the time they reach the end.

But, if the reader is a math-phile, we have included special boxes (*Quant Boxes*) that illustrate the kinds of problems encountered in various sub-subject domains and the kind of math that is used to solve those problems. Or they give examples of how special topics are defined mathematically. Many of the reference works in the chapter bibliographies could take the reader much deeper into the mathematical side of the subjects.

About a Central Theme: The Brain as a Complex Adaptive System

Starting in Chap. 3 we have constructed a set of boxes (*Think Boxes*) that carry a theme throughout all of the chapters thereafter. That theme is about the human brain as a complex adaptive system. The purpose of these focused boxes is to show that the brain is best understood as a complex system that demonstrates all of the principles discussed throughout the book. Thus in each chapter we introduce aspects of the brain that can be understood from the perspective of the principle discussed in that chapter. An obvious example is the fact that the brain is composed of high-level organized networks of neurons (Chap. 4) and the principle of network representation is nowhere better seen than in the way the brain encodes and stores memories of concepts.

Our hope is that these Think Boxes will not only be interesting for what they may reveal regarding how the brain works, but they will help students pause to consider something we find remarkable (and mind boggling; no pun intended). The brain is a system that is capable of understanding itself. It is a complex system capable of modeling its own complexity (see Chap. 13 and Principles 9 and 10). It is our belief that an understanding of the brain as a system will help students think about some of the most important and difficult existential questions that regularly invade intellectual life: What am I? How does my mind work? What is my place in the universe?

About the Pedagogy

Textbooks generally have questions and problems at the end of every chapter to exercise the student's learning of the material covered in that chapter. But those are not textbooks about systems science! This book does not take that route.

Systems science, as we argued above, is integrative in a way that almost no other subject is. Even though we have broken this subject into chapters that each focus on an aspect of systems science, as you will soon experience, these sub-subjects cannot really be taken in effective isolation such that little pieces of one can be memorized without reference to the rest. In every chapter you will find forward and backward references to other chapter contents as we try to establish how all of the aspects interrelate. The book reflects the holism that systems science is about.

Instead, throughout the book we have positioned *Question Boxes* near subjects that we want to get readers actively engaged in thinking about. Often those questions ask the reader to consider what the current subject means in relation to subjects that have been covered previously. And the questions are open ended. That is, there is no necessary single right answer. Rather the questions act as probes to elicit critical thinking on the part of the student.

We envision this book being used in a course that is conducted more along the lines of a seminar, that is, a general discussion around the current topic, but with the freedom to explore its relations with other topics. To that end, the Question Boxes can be used to spur discussions in class. The teacher can act more as a facilitator than an instructor. We have conducted several such classes at both undergraduate and graduate levels in this manner and have routinely found that student learning is much greater when the student is actively engaged in thinking and expressing their thoughts than when they are motivated by the need to pass a test.

Teachers can always construct various means of assessments, of course, to see if learning is taking place.

About the Use of the Book

In truth it is hard to suggest how the book "should" be used because there are not many courses devoted to the way in which this book integrates sub-subjects within systems science. In other words, there is no "norm" to point to and to which to map the contents of the book.

For Students

There are probably many different ways to approach this book based on your background, previous coursework, and interests. Our main objective is to promote critical and holistic thinking about the world.

Given the way we have organized the book, in chapters like typical textbooks, might imply you should read straight through from Chap. 1 to the end. But that linear approach would not be as productive as actually tracking the forward and backward references when given in the text (e.g., when you see a parenthetical "see: Chap. 4, Sect. 4.2.3"). The subject of systems science is so integrated that it really is not possible to think of one sub-subject without reference to many or all of the others.

Nevertheless, the subjects do build as the book progresses. So even though you could start reading a later chapter covering a topic you might be attracted to, the reading would eventually point you back to something in an earlier chapter (or several).

About the Think Boxes

The name we chose for these focus boxes has a double meaning. They are meant to get you thinking, of course. But they are also *about thinking*. That is, they reflect on how your brain actually does what it does. In some cases the Think Box will come toward the end of the chapter where they will attempt to show how the subject of the chapter applies to the study of brains. In these cases they can act as a review of the subjects in the chapter. In other cases they can act as previews of what is to come. Think about it!

About the Quant Boxes

As indicated above we intend this book to be read by a very diverse audience. Some chapters, such as Chaps. 7 and 8, have mathematics throughout the text, but it is relatively low level and is needed to explain the content. Elsewhere we rely on qualitative descriptions and reserve the math for the Quant Boxes. And those are only meant to be illustrative of the kind of math that is routinely used in the sub-subject. Occasionally we ask a question that would require some exercise of that particular math as a challenge to your thinking (and understanding). We assert that at this stage of your learning in systems science you do not need to get caught up in mathematical details in order to understand the subjects. If you stick with systems science, you will take courses in each of these subjects where the math will be made more explicit and you will have to exercise your skills in solving problems relevant to the domain.

About the Question Boxes

More important than the Quant Boxes, in our view, are the Question Boxes that pose open-ended questions that we hope will push you to think holistically, integratively, and critically. There are no right or wrong answers to most of these questions. They are not meant to show up in an exam, but rather to drive the tone of a discussion.

For Teachers

Both authors have taught courses that drew on materials found here. And we have discussed how such a book "could" be used in courses. For an undergraduate program, we've envisioned the book being used in a two-semester sequence at a sophomore level. The first third of the book (Parts I and II) and Chaps. 7 and 8 could be covered in one semester as the foundations needed. Chapter 9, Cybernetics, and the rest of the book could be covered in the second semester. Chapters 9–11 use the principles and fundamental ideas developed in the first chapters and are fairly heavy in terms of intellectual load. The final part is all about methodologies, somewhat similar to many technical subjects, but they are more like surveys of the subjects rather than instructive in details. Chapters 7 and 8 should probably be reviewed at the start of the second semester. Or some of the material might be moved into the second semester to lighten the load in the first semester. But these are just suggestions.

It is possible that upper-division courses (junior and senior) might be able to cover the entire book (or large sections of it), especially in programs that have bits and pieces of systems science already in their other offerings. For example, a junior in a biology program will already have a lot of background knowledge that will allow them to move through the book more quickly.

We also suggest that the book would make a good basis for a graduate course in any of the sciences (social and natural) as a way to broaden the students' perspective to see how their chosen field can be seen as systemic as well as related to other fields. The potential for encouraging interdisciplinary studies can be enhanced.

At this stage of the maturation of the subject and with no feedback from practitioners who have taught courses like this, we prefer to let others develop their courses (and pedagogy) in ways that seem good to them. We would appreciate hearing of their experiences.

Tacoma, WA, USA

George E. Mobus
Michael C. Kalton

Bibliography

von Bertalanffy L (1968) General system theory. George Braziller, New York
Wolfram S (2002) A new kind of science. Wolfram Media Inc., Champaign, IL
Warfield JN (2006) An introduction to systems science. World Scientific, New Jersey

Acknowledgements

Over the years of development of this book, numerous people have contributed ideas and critiques of various parts. Many students have used some of the materials in here and provided invaluable feedback. Unfortunately the numbers of people who have helped in these casual and semi-casual ways are too many to list (even if we could recall every one!). But we do want to thank a specific few who contributed more than just a little to improving the work. We'd like to thank Joseph Simpson for his careful review of a number of chapters and very useful suggestions for making them stronger. Wayne Wakeland provided early consultations that helped shape the work. His experience in teaching systems courses in the Portland State University Ph.D. program was invaluable. We are especially grateful to Don McLane who helped proofread the Quant Boxes. Arabie Jaloway was a former student of both authors who has continued helping us see the material from a student's perspective. Ugo Bardi, Carey King, and Barbara Endicott-Popovsky provided encouragement that helped keep us going. Scott Hansen of the Puget Creek Restoration Society provided us with one very compelling story about auto-organization and emergence of a social organization (Chap. 10). We especially want to thank our Executive Editor, David Packer, for steering us in the right directions when we otherwise might have gone off course. Lastly we owe gratitude to our wives, Janet Mobus and Margaret Kalton, who have exercised the epitome of patience while we labored many a weekend on this work.

Contents

Part I
Introduction to Systems Science

1.1 Getting Perspective and Orientation

Chapters 1 and 2 will introduce the reader to the general concepts of systems and how they can be formulated as a set of principles. In Chap. 1 we define "system-ness," which might be thought of as the properties or attributes that make something a system. We list the principles that systems adhere to. We also discuss the nature of the science of systems. Unlike many other disciplines in the sciences, systems science is more like a metascience. That is, its body of knowledge is actually that which is common to all of the sciences.

In Chap. 2, we provide an example of systems science at work. We use the principles outlined in the first chapter to elucidate a large complex problem, the evolution of drug-resistant tuberculosis (TB). We show how looking at the problem from the perspective of a system provides insights that go beyond the epidemiology of the problem, possibly suggesting ways to tackle the problem.

These chapters will provide the reader with the first hints of a new perspective on the world around them. They will introduce the reader to a different way of thinking about the structures and functions of the objects and phenomena they see in the world. This way of thinking is called systems thinking, and it is quite different from the more predominant reductionist thinking that most of the sciences have tended to favor in the past. Systems thinking is about wholeness, completeness, function, and purpose.

Chapter 1
A Helicopter View

Surveying the evolution of modern science, we encounter a surprising phenomenon. Independently of each other, similar problems and conceptions have evolved in widely different fields.

Ludwig von Bertalanffy, 1969, 30

A new view of the world is taking shape in the minds of advanced scientific thinkers the world over, and it offers the best hope of understanding and controlling the processes that affect the lives of us all. Let us not delay, then, in doing our best to come to a clear understanding of it.

Ervin Laszlo, 1996, viii

Abstract Systems science provides a somewhat unique mode of inquiry in revealing not just how one kind of system, say a biological system, works, but rather how all kinds of systems work. That is, it looks at what is common across all kinds of systems in terms of form and function. In this sense, it is a metascience, something that informs all other sciences that deal with particular kinds of systems. In this chapter, we describe the attributes that all systems share in common. We identify 12 non-exclusive principles that apply to all or most systems of significant interest. These principles provide the guidance for the rest of the book.

1.1 Why Systems Science: The State of Knowledge and Understanding

As of the writing of this chapter, there are estimated to be over 100,000,000 nonfiction books and perhaps 20 times that number of journal articles. On top of this stack of knowledge objects, there are uncounted millions of newspapers and magazines with daily and weekly publications containing between 30 and 100 articles. And that only covers the print medium. When you include the databases, computer programs, papers-in-progress, and dozens of other forms of digital text media, you could easily multiply the number of words written that contain what we typically call *information* by several orders of magnitude!

© Springer Science+Business Media New York 2015
G.E. Mobus, M.C. Kalton, *Principles of Systems Science*, Understanding Complex Systems, DOI 10.1007/978-1-4939-1920-8_1

Then consider the graphic, video, audio, and any other form of symbolic representation of human knowledge and you begin to see that we humans have produced and continue to produce unimaginably large volumes of information. It is, in fact, more than anyone can really account for.

How much of this symbolic representation would actually be considered true knowledge? That is, how much of what we have recorded in sharable media could we rely on to inform us of how the world works and how we humans subsist and thrive? We know a priori that much of what has been produced is in the form of opinions and mistaken observations or conclusions that would not stand up to some kind of rigorous test for usable (or, as is said, actionable) knowledge.

But even after you account for all of the gibberish and false knowledge and boil it down to that which we could agree is real knowledge, there remains a problem. If we humans know so much, then why does our modern world face so many seemingly intractable problems? In our globalized world, with a world population headed toward 9+ billion persons, we face threats to food and water supplies; from soil erosion, climate change, and dwindling natural resources; and especially from depleted storages of fossil fuel energy. If we really understood how the world works, would this be the case?

Clearly there is some disconnect between the state of our understanding and the way the world system really works or we would be well on our way to a bright and sustainable future for humans and the rest of nature. We use the term "world system" to name the world as a whole, with the thought that humans are a part of the natural world, even if a special case.

Could it be that our knowledge of these fundamental requirements for life and civilization is that incomplete? Or is it possible that we have not succeeded in integrating the knowledge that we have in such a way that larger but more subtle patterns that have more importance for the whole world system become visible and understandable? Are we still infants when it comes to understanding the whole, even while we have extensive knowledge about the parts? A world system in which all of the parts interoperate in some kind of harmony has always been part of the human dream of utopia. Yet we never can seem to use our knowledge to manage or even nudge the world in this direction.

Science is held out as the epitome of gaining usable knowledge, and with good cause, we think. But science, as it has been practiced historically (see below), has been more concerned with the parts and generally not the whole. There are perfectly good reasons for this from a historical perspective. But as the problems that we humans and our cohabitants of the spaceship Earth face demonstrate, we who by our very nature have such an impact on the world system have not actually been very good at putting the pieces together. Lack of attention to the complex, multidimensional relationships that organize the subsystems into a dynamic whole leads to our being constantly blindsided by unanticipated consequences. We are expert in fine-grained analysis that enables us to maximize particular desirable functions, but we are ignorant regarding the impact of the new functionality on the relational dynamics of the whole. It is relatively easy, for example, to measure the improved function of our transportation and communication systems, but the consequent globalization of virtually every aspect of daily life is a mighty transformation rife with mighty consequences that beg for understanding. Not everything about a complex system can be predicted, and hindsight is often our final teacher. But there are good systemic reasons for unpredictability, and it behooves us to at least understand and anticipate when we are making major moves with unpredictable consequences.

In sum, we need a complement to the kind of science that goes deeper and deeper into ever more refined and limited components of a system. We need also a science that can see borders as places of meeting and transition, a science that can follow the complex dynamics of how components function together in terms of one another with a meaning not captured in any individual component. Fields of science are designated in terms of the borders that define the particular aspects of the world they study. Thus, we have major fields such as physics, chemistry, biology, etc., and each of these is continually subdivided into finer and finer specializations as the analysis unfolds. In contrast to this, a science that crosses borders to address whole relational systems would be a metascience, "meta" being Greek for "beyond." When we call systems science a metascience, the idea is not that systems science is literally beyond science but that it deliberately goes beyond the boundaries of any particular science to include them all. Such systems science would not know physics or chemistry better than physicists and chemists, but since it studies the complex systemic relationships that are broken down into diverse fields of study, systems science should have more to say than physicists and chemists regarding the relationship of physics to chemistry, or of both to biology, etc.

Over the last 75 years, or so, many scientists and philosophers have been exploring the meaning of whole systems thinking. They have discovered a variety of tools and techniques for discovering connectedness between seemingly separate components of the world system (and, indeed the universe). These scientists have developed a conceptual and formal framework for thinking about wholeness and interrelatedness. They have discovered ways to understand the interrelations between previously isolated subjects and found ways to go beyond the boundaries of academic disciplines to develop truer integration across our entire domain of knowledge.

This book, then, is about a metascience. Systems science is a way to look at all parts of the world in a way that is unifying and explanatory. Insofar as it provides a way to integrate the knowledge produced by the other sciences, it can also provide useful guidance to those other sciences.

Virtually everything in the universe, including the universe itself, is a system! And systems are composed of systems. How this system unfolds and organizes in the ramifying complexity of ever richer subsystems is the story of our own emergence. Understanding the organizing dynamics and principles of the world system, the life system, and our socioeconomic systems is both fascinating in itself and critical to our own life project, for it is as participants in the complex dynamics of this layered system that we organize our own lives and find our fit.

1.2   The Distinctive Potential of Systems Science

Systems science is a *universal science* in the sense that it does apply to literally everything in the universe. But that does not mean that systems science subsumes all traditional scientific disciplines. Traditional sciences, as they have evolved over time into specialties, are not geared to cross boundaries. Their typical movement of thought moves to more and more intense specialization within their disciplinary boundaries, reducing and analyzing their subjects into more refined objects and mechanisms. When they address whole systems, it is within the bounded disciplinary perspective and generally takes the form of a reconstruction of the whole from the analyzed parts. This can be very useful and is a means of solving certain types of problems that have components and factors that remain in the disciplinary specialist's purview. But systems science, by contrast, typically follows the ramifying network of relationships outward, becoming more and more inclusive rather than more and more exclusive. It sees boundaries as relational transitions en route to a more inclusive systemic level. This is also useful and complements reductionist approaches. Humans have made extraordinary technological advances through specialization. However, we are increasingly facing intractable multi-causal environmental and social problems that seem less amenable to solutions of this kind.

Every scientific discipline investigates and seeks to understand a network of relationships that characterize its subject matter. But it studies those relationships precisely as part of a particular subject matter. The general subject matter of systems science is systems, and systems are comprised of relational organization. So even while looking at one sort of system or another, systems inquiry probes and pushes to understand principles or dynamics that go with relational systems as such rather than with a particular kind of system, which would be the purview of some specialized area of science. Because specialized disciplines and systems science share the study of the complex relational nature of reality, but with different lenses, there should be a lively interaction of systems science with other areas of study. Systems thinkers often have a background in some particular disciplinary area, and the rich detail of relationship revealed by disciplinary investigation is an invaluable resource, for it constitutes manifold streams feeding systems understanding.

What systems science contributes to this dialogue with the disciplines stems largely from its boundary-crossing, inclusive nature. All too often, disciplines are so specialized they do not know how to talk to one another. Systems science, by contrast, is more like the linguistic study of syntax, a feature shared by all diverse and mutually unintelligible languages. Systems study can reveal shared structure and dynamics; it provides scaffolding for thinking and inquiry shared by and bridging the diverse subject areas of natural science, social science, and the humanities. If syntax itself could speak, languages foreign to one another could communicate. That is not the case; but systems science can indeed speak, and insofar as it finds its voice and becomes articulate, it can serve as a critical disciplinary bridge for a society that increasingly recognizes the need for interdisciplinary understanding to grapple with daunting and complex systemic problems.

Systems science is not one monolithic field of study. Rather, it is a large collection of conceptual frameworks that interrelate and share. This textbook strives to consolidate the various frameworks to collectively form a general theory of systems and to provide scientific tools for studying objects in nature as systems.

1.2.1   What Is a Science?

When we think of biology or chemistry, we think of a field of inquiry in which related objects are studied, dissected, manipulated, and otherwise brought into the realm of human understanding. We have been taught since an early age that there is a method, the scientific method, that allows us to formulate hypotheses (speculations of an informed sort), design and conduct tests involving controls and measurements, and perform mathematical analysis on the results to form our understanding of what the phenomenon of interest is all about, how it works, etc. While many of the social sciences may employ qualitative rather than quantitative methods, there has long been an expectation that the sciences dealing with nonhuman aspects of the natural world[1] use mathematically based analysis. This method, we are told, is what scientists do as their jobs. Few people actually experience much more than a cursory exercise in "doing an experiment" in a lab in high school or as a freshman in college. So most people in our western societies have a very limited understanding of what we will call the "scientific process."

But science is so much more than just this common account of the scientific method. It is a way of thinking about some part of the universe and trying to find ways to understand how things work. It certainly involves using the scientific method once we have been able to produce some preliminary ideas about how our subject of interest works. But getting to that point involves a larger context, including considerable effort put into observing, being curious when your observations do something unexpected, and trying to use what we already know about the subject to speculate about what might account for the unexpected behavior. And this personal inquiry will be deeply informed by what prior scientists in our fields have done, what they have discovered, and what the majority have come to claim as the basic cause-effect relationships within the subject area. In other words, it takes a lot of education to get down to the basics and have a broad understanding of the field in order to do science.

---

[1] Social sciences are those typically seeking to understand human interactions and organizations; e.g., political science, sociology, cultural anthropology, and economics are often categorized this way. Natural sciences are the ones generally associated with how the rest of nature works. Typical examples are physics, chemistry, geology, and biology. The latter group often uses qualitative methods to isolate or identify phenomena of interest and quantitative methods to further explicate the behavior and dynamics of the phenomena. Some of the social sciences tend to use qualitative methods more extensively and perhaps resort to statistical analysis to refine their results. And more frequently now, the social sciences are turning to systems thinking and quantitative methods such as modeling.

Sciences have tended to focus on a few broad modes of organization and then to spawn a wide array of subdisciplines as they progress. Physics addresses the general organization of matter/energy from the cosmic to the subatomic and quantum levels. Chemistry attends to the mechanical interactions between atoms and molecules, a form of interaction that emerged as fusion processes in stars began filling the universe with more complex heavy elements. From chemical complexity arises the even more complex dynamic interactions of living organisms, the object of biology and related life sciences. Ecology inquires into the complex organization of the community of living organisms and the physical flows which sustain it, while the social sciences investigate the parallel but even more complex phenomena of organization among humans. Systems science, focusing on relationship and organization, sees each of these levels as distinctive emergences that have taken place within a single dynamic contextual process, that of increasing and evolving complexity.

In general then, science is a process for discovering and codifying our understanding of how objects in our universe work and interact with one another. Those objects must have some real embodiment (e.g., atoms, rocks, people, etc.) even if the embodiment is hidden or seemingly ephemeral. Our thoughts, for example, are still electrical stimulations coursing through a neural circuit in a brain. Generally we can find some way to observe and measure some aspects of their behaviors. The great advantage of measurement is that it sets up a shared framework where other observers can repeat the experiment, check data, and contribute to the inquiry.

For a good reason, physics is the king of the sciences when it comes to using mathematics to describe its object of investigation. But, as we shall see, emergent systemic complexity (chemistry, biology, ecology, etc.) is accompanied by new dynamic characteristics which call for appropriately proportioned new approaches. It is therefore a mistake to take physics, or the approach of any other scientific discipline, as paradigmatic of what "science" should be.

1.2.2 What Is Systems Science?

There are physical systems, bio-systems, ecosystems, social systems, economic systems, global systems, local systems, gambling systems, computer systems—and the list goes on and on. Further, all these systems involve subsystems, which in turn may be reduced to and analyzed into yet more subsystems. Within the academic world, every discipline devotes itself to a specialized systemic understanding, and the object of systemic investigation is itself a system.

Systems science undertakes the understanding of systems as such, i.e., not this kind of system or that kind of system (physics, chemistry, biology, sociology, etc.), but the investigation of general and useful attributes, dynamics, characteristics, and behaviors of systems as systems—including key differences among subclasses of systems such as linear, nonlinear, closed, open, complex, etc. (for discussion of the nature and properties of systems, see Chap. 3). When one understands systems or "systemness" this way, one becomes aware of features that run through every area of study. These features manifest differently in different systemic contexts, so systems science prepares us to see continuities-with-a-difference across boundaries where disciplines tend to see only differences.

Arising from the convergence of cybernetics, computers, and information theory on the one hand and ecology and its acolyte physical and life sciences on the other, such systems science has made major strides in the past 40 years. A Google™ search on "whole systems" yields 210 million entries, and a quick look at even a few will reveal an extensive, diverse, and lively literature exploring a wide range of applications. Scholars come to this study from the most diverse areas: one can find rich mathematical explorations, a wide range of exploratory computer simulations, side by side with theological musings on the implications of Steven Hawking's cosmology, holistic, and naturalistic philosophers engaged with Earth as a whole (living) system and ecological thinkers exploring the applicability of their natural systems understanding to culture and society.

Systems theory has been multiform, with varied expressions and currents of development finding powerful application in physical, biological, ecological, social, economic, cultural, and psychological realms. Its shared core is the exploration of virtually any phenomenon as a web of relationships among elements; it looks to cross-referenced and interdependent causal networks rather than assuming chains of deterministic mechanical causes as in the analytic model inherited from the Enlightenment. The mechanistic model thrived especially on the promise of prediction and control, and some forms of systems theory extend the traditional emphasis on prediction and control into new realms of complex organization. But from the 1960s and 1970s, a new kind of system science has emerged, stimulated by chaos theory and our ability to use computers to disclose and model patterned emergence within nonlinear process. The core inquiry has thus shifted to process and dynamics, with special attention to the emergence of the new, the unexpected, and the unpredictable.

Such features are especially pronounced in systems that have the capacity to learn from experience. This has given rise to a rich development called complex adaptive systems (CAS). The paradigm for this kind of systems thinking is not the machine, but evolution. While CAS is a fertile, cutting-edge area of computer-related technological development, it also encompasses the most spectacular and complex system of systems, the web of life. The dynamics of ever-evolving ecosystems and ever-changing cultures and societies are equally its purview.

Basic CAS understanding is relevant for students pursuing a number of majors. Therefore, we have endeavored to make this textbook useful as an adjunct to an array of courses with a variety of disciplinary focus as well as to courses in systems science as such. Those concerned primarily with the natural world will find broad applicability in areas of environmental science, and those concerned with the systemic interface of humans and nature will find systems science fundamental for engaging the question of sustainability and related issues. In general there is wide applicability for the methods and insights of CAS throughout the social sciences and humanities (see Bourke 2011), and this dimension of systems science receives constant attention throughout the text.

In fact, systems science engages the real and necessary intersection of natural science, social science, and the humanities, areas that have traditionally been separated as discrete academic disciplines. In reality the world of human affairs treated in social sciences and humanities is always contextualized in the systemic physical and living global web in which we are enmeshed. And the world investigated by physics, chemistry, and life sciences is critically bound up with human motivations, visions, and projects. These come together most urgently in the emergence of sustainability as the major question and challenge of the twenty-first century. For too long, our failure to grasp the dynamic interrelationships of this interdependent whole has spawned a myriad unanticipated and unintended consequences that we must now confront. This leads many to view a new paradigm of systems-informed thinking as not only desirable but imperative for the twenty-first century.

1.3 Systems Science as a Mode of Inquiry

Systems inquiry views phenomena as a web of relationships among elements that thereby constitute a system. Systemic phenomena of all sorts are scrutinized especially for common patterns, properties, or behaviors that might be relevant to the understanding of systems as such. Since virtually all disciplines study some sort of complex system, any grasp of principles that belong to all such systems would be an important contribution to a more unified framework for what now appear simply as disparate areas of academic or scientific investigation.

1.3.1 The Heritage of Atomism

Our assumptions about the world shape the kind of questions we ask of it and the way we pursue the answers. Most often those assumptions are the unexamined shared views of entire societies and their traditions. The Western world within which science emerged holds many shared assumptions about things existing as inherently individual units, and this assumption has deeply influenced belief about what makes things happen. Confronted with an unwelcome event in human affairs, for example, "Who's to blame?" seems like an obvious question, setting in motion inquiries seeking to identify the responsible individuals or groups (conceived of as individuals). One sees this pattern at work, for example, after 9/11 or Hurricane Katrina. Deep-seated cultural values such as individual responsibility, autonomy, and even human rights draw on these conceptual wellsprings. And beyond society, we have employed similar analyses to the world of nature, as when "pest" species are identified and targeted for their depredations on our herds or crops.

One could trace the roots of such thinking back to ancient Greece, but it received fresh impetus with the emergence of modern science in the seventeenth century. In classical Newtonian physics, the entire cosmos was imagined as a vast machine.

The machine became a general model or paradigm for complex function, thereby spawning an array of scientific disciplines to investigate the "mechanisms" of the various subsystems, be they psychological, social, biological, chemical, or atomic, each considered a unit in itself, open to understanding independently from other units. The whole, it was assumed, could be understood by adding together our understanding of the several parts. However, specialization soon became so intense, profound, and idiosyncratic that any hope of putting Humpty Dumpty back together again has long been abandoned.

1.3.2   Holism

Holism has arisen in self-conscious contradistinction to the atomism implicit in many of our common ways of thinking. As the term itself indicates, confronted with wholes made up of parts, it gives priority to the whole as the necessary framework for understanding the parts. Holism explicitly denies the atomistic proposition that a whole is nothing but the sum of its parts, and the corollary that by understanding the parts one may understand the whole. A systems thinker would not deny the importance of analyzing parts or that significant questions may be answered in terms of the behavior of parts. But at the same time, one who thinks in terms of systems recognizes the critical limitations and frequent shortcomings of the divide-and-conquer approach. The elements in a system do not simply pool their functions to add up to the behavior of the whole; they rather perform in synergy with one another so that the behavior of each is critically shaped and informed by its relation to the whole. From our atomistic heritage, we have tended to think of the world in terms of self-subsisting "things," to which relationships are somehow added to make more complex "things." In a systems view, the web of relationships we refer to as context or environment are themselves constitutive aspects of each thing or activity, so there are no individual objects that can just be understood in isolation. To neglect this relational web reduces understanding and increases the risk of misunderstanding.

Consider smoking, for example—the relationship of cigarettes, persons, and lungs. The personal decision of John Smith puts the cigarette in his mouth, and smoke fills his lungs, producing pleasure, relaxation, and eventually perhaps cancer. Even with this very simple account, investigation and understanding would require a psychologist and perhaps an ethicist for the decision aspects and a biologist or team of biological sub-specializations (addiction, cancer). That is, in fact, more or less how the matter stood for some decades after the Surgeon General of the United States announced that smoking is dangerous to your health. Makers of personal decisions were henceforth forewarned of this danger by notification on every pack of cigarettes, and the notion of purely personal freedom and responsibility insulated tobacco companies from lawsuits brought by unhappy users of their product. One sees here the social and policy consequences of a relatively simple and linear cause-and-effect analysis—an analysis that was held in place with considerable assistance from the tobacco companies.

A systems-oriented analysis would be attuned to scoping out the entire web of relationships, producing a multi-causal and very different account of John Smith's decision. Start inside the brain of John Smith himself, assuming his initial decision regarding smoking was made when he was a teenager, the most common case. Recent research reveals that the portion of the brain that processes serious consideration of consequences is not fully wired until one is in the twenties, making teens the ideal target for marketing a dangerous, addictive product. How "personal" was his decision? Was it shaped and moved by his peer group? And how was peer opinion shaped by advertising, media, and adult behavior? And the sponsors of the advertising, the tobacco companies, were participating in typical marketing behavior of competitive, profit-driven corporate institutions in a capitalist society. Except marketing an addictive product has obvious advantages and disadvantages. On the side of advantages, how did they deal with the possibility of increasing their competitive edge by manipulating the addictive nicotine content? Among the disadvantages is the government's inclination to regulate addictive substances in response to political pressures. A lobbying industry addresses that in part, along with subtle processes like addicting state legislatures to the revenues generated by "sin taxes." And who grows the tobacco, and where? Livelihoods are involved here; are equally profitable alternatives available? State legislatures are invested in supporting existing ways of making a living, which produce happy voters. And what are the effects of this crop on soil fertility? How do chemical companies figure in producing the right fertilizers, pesticides, and most desirable seeds? Into what rivers does the runoff from these crops flow, with what consequences? And circling back to John Smith, how does the insurance industry manage his risk, and what are the medical expenses, and how are they taken care of for the uninsured?

This web could be both filled in and expanded almost indefinitely, but even this sketch is enough to illustrate the transformation of a question when it is entertained in a systemic perspective. There is great practical import here. From the simple linear cause-and-effect account, one might think the central issue is John's decision to start smoking, with the obvious solution being a strategy to convince him that it is a bad idea. But positioning his decision in its broad systemic context, we see that the question is far more complex, even if we keep our focus on how to prevent John Smith from becoming a smoker. Every relational line we explore suggests further questions, ramifications, consequences, and points for leverage and intervention.

1.3.3   System Causal Dynamics

When the complex web of systemic interdependent causality is fully in view, gridlock seems the inevitable consequence, since every part is interlocked and supported by every other part. Indeed, an analysis of systemic relations at a given moment tends to sound like a traditional linear causal system, but with many hopelessly intertwined lines of causality. Especially when dealing with deeply entwined social phenomena, this perception of gridlock is amply supported by the experience of reformers of every type. On a positive note, this interlocked support for the status quo also maintains stability and contributes to systemic resilience; as we shall see, the question of resilience will be an important one for systems science.

Short-term gridlock notwithstanding, in a larger time frame, all bets are off. Complex systems can change in ways that completely belie our ordinary expectations of steady, incremental modifications. From the ubiquity and acceptance of smoking in American society of the mid-twentieth century, one could never have predicted the huge lawsuit rewards and extensive restrictions that encumber the practice in the early twenty-first century. The industry has been forced to mount advertising campaigns designed to discourage smoking, celebrities no longer appear in the media with cigarettes in their mouths, smoking is banned by many states even in bars, and the notion of big important deals being cut in "smoke-filled rooms" has become a quaint anachronism.

How did this happen? Simple linear causal analysis offers little clue, and even the extended, more systemic framing of the situation pointed more toward stability than change. The problem here is that taken at a given moment, even a good analysis of complex relationships follows the implicit determinism of cause and effect: there is this vast interwoven complex of physical, biological, social, psychological, etc., causes, each producing their necessary effect, with the cumulative result that everything is the way it has to be. This offers no clue to the system's potential for sudden and unpredictable change; taken only this far, we still have not gotten beyond the perennial model of mechanistic causal thinking.

Systems inquiry indeed automatically looks to the entire web of all-too-often neglected relationships and is useful in thus identifying a more complex relational structure. But taking on the related critical question of *process* is what has moved systems thinking definitively beyond the frame of classical cause-and-effect analysis. Causality in reality takes place only in time, so causality is necessarily a temporal process. Cybernetics emerged about the time of World War II as an attempt to program the control of complex temporal causal processes, ushering the world of manufacturing into what became the computer age. The challenge of automating complex processes was far more daunting than simply lining up a series of actions to be performed, for in many cases the timing of the action is critically linked to a precise condition of the system that cannot itself be timed with precision. Every cook knows that when the instructions say, "bake the cake for 45 min at 350°," after about 40 min you go and test the cake, taking it out when done, knowing that may be at 40 or 50 min depending on conditions. Testing the cake provides information feedback to guide the next move—putting the cake back in the oven or taking it out. Automation is made possible only by such feedback, where instrumentation sends information back to the controller, closing a causal loop in which the effects launched earlier register and circle back to modify the directing source to produce a new move. Thus, cybernetics gave to systems thinking one of its most basic tools, the concept of *feedback loops*.

Cybernetics was largely concerned with information feedback loops, for information is the key to control. But in the larger context of systems thinking, the feedback loop proved to be the paradigm for the transformation in virtually all causal thinking. Even Newtonian physics alerted us to the "equal and opposite" action of forces, but this two-way street of causality was simplified in practical linear cause-and-effect considerations. Carpet manufacturers, for example, would want to know how many footfalls it would take to wear out their carpet. But how fast the carpet was wearing out the soles of the falling footwear could be taken as a totally separate issue, if anyone wanted to think of it at all. Systems thinkers, however, inquiring into the behavior of complex networked relational webs over time, could no longer ignore the reality of the fact that causes are themselves continually conditioned and changed by the system which they modify by their causality. The essence of this idea is captured beautifully in Eric Sevareid's famous remark, "The chief cause of problems is solutions." The image of a loop seems simple, but unlike classical causal thinking, it catches the necessary conditioning of every cause by the system to which it is introducing causal change. This introduces the prospect of a new level of inquiry into the dynamic reality of process and change.

1.3.4 Nonlinearity

System dynamics are of major concern in systems inquiry. The actual intersecting paths of transformative interaction that effects follow en route to reinforming their cause provide a map of the dynamic structure of the system. But the dynamics of systems go beyond the predictable stability of water flowing through a network of pipes, as the notion of "map" might suggest. Surprising things may happen, and the character of causal relationships can shift in unsuspected ways. In mechanical relationships, the effects of "more" are generally predictable: more pressure on the gas pedal should make the car go faster. But in other sorts of systems, this is not necessarily the case. Doubling the dose of a medication is not necessarily twice as effective. At some point, cooling water by one more degree produces not cooler water, but ice, which is quite a different matter. Such nonlinear phenomena give rise to the concept of systemic "thresholds," tipping points where the dynamic behavior of the system suddenly modifies in a manner out of all seeming proportion to the incremental change. Thus, the present fixation with the prospect of climate change: If the global climate warms by 4°, systems models suggest we will have to deal not with summers that are simply 4° warmer but rather with a threshold crossing beyond which lies an unpredictably different weather pattern for the Earth.

In systems which are close to what we regard as mechanical, such as warming and cooling, we are ill-prepared for such major transformations wrought by seemingly minor inputs. In our social systems, however, we are only mildly surprised by revolutions or bursting economic bubbles or other sweeping changes over relatively short time periods. To go from a society that advertised the brand of cigarettes "most smoked by doctors" (Camels) to smokers exiled to huddled groups puffing in the rain outside workplaces in a few decades is surprising in some ways, but not at all beyond the familiar range of change in society. Who knows what things will be like in 50 years? In the world of nature as well, we are now getting used to the idea of change cascading through ecosystems, and we have become wary of the potentially far-reaching consequences of our various interventions into our living system. What patterns run through all this? Can we decipher principles of systemic dynamics that would shed light on the emergence of the complex and seemingly volatile dynamics of living systems from the seeming predictability of the nonliving physical world?

Some linkages have indeed emerged. Systems dynamics investigates the behavior through time of various sorts of systems. In the relative simplicity of humanly engineered systems such as the world of computers, the careful construction of pathways for information processes reproduces the kind of predictability and control we have come to expect due to our long industrial experience with constructing complex mechanisms. But even there we have discovered that looping feedback can surprise us: the field of *chaos theory* emerged with the discovery that there are simple nonlinear equations which, when reiterated thousands of times by computers, can produce random looking behavior patterns of great beauty and virtually limitless complexity.[2] In fact, it was discovered that some of those equations could produce images eerily similar to the self-similar but never exactly repeated patterns of the natural world, such as tree leaves, ferns, and flowers. While this discovery immediately spawned a new generation of computer screen savers, it also opened up an entirely new version of Galileo's observation that the language used in the creation of the universe was mathematics. If so, God was particularly fond of nonlinear mathematics, the mathematics of turbulent flows and self-similar but never repeated living organisms.

In the world of linear mathematics and mechanical systems, turbulence had long been an aggravation, demanding tricks to transform it into some kind of mathematically tractable regularity. The new tools and perspective of systems have transformed *turbulence* into an exciting area of investigation. Watching the smooth laminar flow of water from a faucet suddenly transform into a turbulent writhing as the flow rate crosses a particular threshold initially seems to be the loss of order; but with eyes attuned to questions of dynamic relationships, this is actually a sudden shift from relative simplicity to great complexity. How does a smoothly flowing system suddenly ratchet itself to entirely new levels of complexity, in seeming contradiction to the normal order-dissolving law of entropy? This question, and the ensuing investigation of "far-from-equilibrium dissipative systems," led to a Nobel Prize for Illya Prigogene in 1977. His work brought major new insights into how systems of flowing energy can *self-organize*[3] themselves, yielding what complex systems theorist Stuart Kauffman has termed, "order-for-free" (Kauffman 1995, p. 71).

---

[2] For the classic introduction to chaos theory, see Gleick (1987).

[3] The term "self-organizing" is the most common reference to processes whereby components in a system tend to form stable linkages (see Chap. 10). We left this term usage here because that is what Prigogine and most other authors have used. The use of the word "self," however, may carry a little too much emotive baggage and possibly convey a denotation of mental intention. For example, one could easily and innocently attach the notion of components "wanting" to interlink in such-and-such a manner, leading to a stable configuration. Throughout the rest of the book, we use an alternate term "auto-organization" to mean what most researchers mean by self-organization. This term does not seem to carry any sense of a mental process and more correctly, in our view, labels the nature of the process of organization without outside manipulations taking place.

Here we start to see how systems inquiry spans conventional disciplinary boundaries, not with the specialized insights of the disciplines themselves but with concepts that can move with significant questions from topic to topic because as relational webs these topics participate in common-yet-different systemic dynamics. Prigogene's work starts in physics, then chemistry, and then into considering biology, because every living organism is a "far-from-equilibrium dissipative system," continually taking in, transforming, and dissipating energy (See Prigogene and Stengers 1984). The entire globe, bathed in a continual stream of solar energy, is such a system, as are the innumerable ecosystems which constitute the subnarratives of global self-organization.

We have here a foretaste of how the dynamics of evolving systems, manifested in the many forms of auto-organization found within physics, chemistry, biology, ecology, sociology, psychology, and all the other "-logys," may indeed form a single overarching principle. Many researchers pursue one aspect or another of this many-faceted story; specialization has not disappeared. But further hope lies in the prospect of communication and sharing, where insights arising in one area lead to fruitful inquiry in others. Boundaries are no longer boundaries but fascinating intersections and thresholds of emergent complexity: how does a world of basic physical processes become chemical, then biological, then social? Far from the classical anticipation that all of existence could be reduced to a suitably complex form of physics, systems inquiry expects this narrative to be a story of the ongoing emergence of novel complexity, a story in which each new chapter occupies a unique place that cannot be swallowed up in the chapter before. However complex and seemingly disparate the areas of systems inquiry may be, the thrust of the inquiry is inherently holistic, fully admitting irreducible differences, yet hoping to understand the wholeness of this entire ongoing emergent process.

While dynamics describe the interacting flows of an entire system, feedback loops equip systems thinkers to deal with fine-grained process and change, critical features of the real world that have been hard to capture in the deterministic framework of linear causality. The popular adage, "What goes around comes around," is only one possibility in the interwoven, looping network of systemic causality. Familiar behaviors disappear, and new patterns can emerge as the status quo gives rise to increasing complexity or decays to a lower level of organization. As the tide turned against smoking, for example, whole new levels of regulation and enforcers of regulations emerged in local and national government. In the new organizational complexity, a significant segment of the vending machine industry melted down, but the work and arrangements for counter service in grocery and convenience stores became more complex as tobacco products demanded new handling. Even in carefully studied systems and relatively well-understood complex systems, one can always expect the unexpected. Note the long list of side effects on packages of medication; these lists reflect the inherently unpredictable consequences of taking a given medication, especially in light of the almost limitless variety of people and their circumstances.

The power of linear causality is its promise of predictability and hence of control: If we have this causal arrangement, we will get these consequences. Within that framework, if we know what we're doing, we can tune an automobile engine, for example, to finer and finer degrees to maximize performance. "Knowing what we're doing" is here virtually synonymous with understanding the networked linear causality of a system, and under this canopy our advanced technological civilization has arisen.

At the same time, when we introduce a significant time dimension, we know that sooner or later, the engine will break down, an unpredictable and often unexpected event that seems to belong to a different causal system. The breakdown in fact belongs to the same real causal system, but not to the dimensions abstracted out and entertained by design engineers. Design engineers can and do factor in rates of wear and tear on engine components. But then, every automobile has a driver, and the vehicle is not only driven differently but in different conditions. What kind of driving will the vehicle's performance evoke in a particular sort of driver, and how will the driving affect the performance? And then consider the intersection of a driver's response to various sorts of road conditions, mediated by his or her expectations of the vehicle's capacities. Analytically these can be broken down into distinct loops, but systemically they intersect; the high number of SUV rollover accidents certainly has something to do not only with their high center of gravity, but drivers' expectations of these "rugged" vehicles (advertised for effortless navigation of challenging terrain) also contribute to miscalculated pushing of the vehicles beyond their limits of balance.

Our ability to analyze and predict how looping causal processes will respond is inversely proportional to their complexity. While the instinct of systems inquiry to be inclusive in searching out these networked relationships exacts a price in terms of predictability—long the forte and dominant value of western science—it leads to the asking of better and more relevant questions. Sometimes this leads to preventive measures, sometimes to walling off legal responsibility for what cannot be controlled. In any case, for those charged with responsibility for complex processes, the attitude counseled by China's ancient Taoist classic, the *Tao te ching*, seems appropriate: "The sage is cautious and alert, like one crossing a river in winter" (Chap. 15).

1.4 The Principles of Systems Science

1.4.1 Principles as a Framework

As we have seen, systems embrace all areas and objects of study. Ever since this was understood, there has been a felt need for some kind of scientific framework for systems that might serve as a metascience or umbrella conceptual guide for all fields of knowledge. A notable step in this direction in the mid-twentieth century was Ludwig von Bertalanffy's development of what became known as general systems theory (see von Bertalanffy 1969). General systems theory made important contributions and was influential in the social sciences, but it emerged before some critical well-recognized phenomena such as complexity and network theories were understood in the systems context.[4]

Since then the understanding of systems has enjoyed lively development fed from a number of sources and perspectives which have emerged as an array of stand-alone disciplines within the field of systems study. In particular, complexity theory is studied as a discipline that is often considered virtually synonymous with systems science. Indeed, many of the central and distinctive characteristics of systems such as nonlinear dynamics and adaptive learning are intimately linked with complexity. But complexity is still one aspect of a system, a characteristic of critical importance but not a total window on the subject of systems. Table 1.1 provides a short summary of the various subject areas that have emerged as stand-alone discipline areas within system studies, giving some of their internal interests (principles) and indications of where they interrelate with the other subjects. Clearly there is a tremendous amount of overlap or interrelation among these disciplines.

For more than a decade, we[5] have been surveying the subjects that are recognized to be related to systems theory, and we have attempted to consolidate the principles we find running through these areas. We offer the following set of 12 principles that seem in some degree to apply to all complex adaptive systems. Many apply to all systems. They are by no means exhaustive, but they provide a coherent and sufficiently inclusive framework for systems science. Typical of frameworks, the principles interdepend, overlap, and cross-refer with one another. Some of these principles will be the subject of extended discussion, even whole chapters, and all of them will emerge repeatedly as we take up the critical characteristics of systems.

The principles include:

1. *Systemness*: Bounded networks of relations among parts constitute a holistic unit. Systems *interact* with other systems, forming yet larger systems. The universe is composed of systems of systems.
2. Systems are *processes* organized in *structural and functional hierarchies.*
3. Systems are themselves, and can be represented abstractly as, *networks of relations* between components.
4. Systems are *dynamic* on multiple time scales.
5. Systems exhibit various kinds and levels of *complexity.*
6. Systems *evolve.*

---

[4] For further development, see, for example, Klir and Elias (1969, 1985). Klir served as president of the Society for General Systems Research (now International Society for the Systems Sciences), which was founded by von Bertalanffy and others in 1954.

[5] For much of that time we (Kalton and Mobus) worked independently and only discovered our mutual understanding of principles a few years before agreeing to tackle this textbook project. The principles outlined here are the result of integrating our convergent ideas.

Table 1.1 Some example subjects with interrelations that make up a *system* of system science

Subject area	Examples of principles enunciated	Relations with other subject areas
Complexity theory	Relations between components Structure, dynamics, functions, adaptive systems	Assumes structure and networks of relations. Borrows aspects of emergence from evolution. Provides an interesting relation with information theory
Network theory	Network connectivity, topology Evolving networks	Is used in one form or another by all of the others since systems are fundamentally networks of relations among component parts. Provides mathematical and modeling tools for all the others
Information theory	Effective messages, knowledge encoding Communications	Found at the core of all the others in seemingly different guises
Cybernetics	Error feedback for regulation Decision science	Applicable as a useful model mostly to more complex dynamic systems. However, wherever information theory is applicable, some form of cybernetic control (i.e., feedback) may be discovered
Evolution	Change in system structure/ function over time	Though this has mostly been used in biological systems science, it is now shown to be applicable in all systems when studied over very long time scales
Systems dynamics	Modeling complex systems	At some level of resolution, all systems (even rocks) are dynamic, meaning that the relations vary in some kind of time-dependent way
Psychology/ neurobiology	Brain architecture Behavior Adaptation/learning	Though a specialization that is still in its infancy with respect to application or relations with other systems subjects, it draws on all of the principles and provides other subjects with insights about things like adaptive control and dynamics of system behaviors
Systems engineering	Analysis and design Relational networks Dynamic systems	Another seemingly narrow field, but if one includes the practice of "reverse" engineering, the tools and techniques of this discipline become more broadly useful in all other scientific fields
Social systems	Politics, governance, cultures	Still developing as a contributor to the other subjects, but one very fruitful area is agent-based modeling. This approach adds agent intentions and behaviors to complex, networked, and adaptive systems thinking

7. Systems encode *knowledge* and receive and send *information.*
8. Systems have *regulation* subsystems to achieve stability.
9. Systems contain *models* of other systems (e.g., protocols for interaction up to anticipatory models).
10. Sufficiently *complex, adaptive* systems can contain *models of themselves* (e.g., brains and mental models).
11. Systems can be *understood* (a corollary of #9)—science.
12. Systems can be *improved* (a corollary of #6)—engineering.

We will explain these in brief below. Chapter 2 will exemplify their application as perspectives for analyzing a complex systemic problem.

1.4.2   Principle 1: Systemness

The observable universe is a system containing systems. Every system contains subsystems, which are, in turn, systems. Chapter 3 will investigate the nature and properties of systems, including the difficult question of boundaries and what may constitute "a system." Here we call attention to the intertwined wholeness as one moves up or down a scale of many levels and types of systems which both encompass and are encompassed by systems.

Systems of all kinds are found to be composed of components, which may themselves be subsystems (i.e., systems in their own rights).[6] Systemness is a recursive property in which, starting at some mid-level, one can go upward (seeing the initial system of interest as a subsystem of a larger system) or downward (analyzing the components and their interrelations).[7] This principle guides both analysis, in which systems are understood by breaking them down into components and the functions of components, and synthesis, in which the functioning of systems is understood in terms of its place in a larger relational web.

Systemness means that as new levels of complex organization emerge, the new levels intertwine in dynamic systemic relationships with other levels. The emergence of sentient life is one such new level, symbolic language with its capacity for self-reference is another, and computer software systems may have the potential to become yet another. This gives us three (four) levels of interacting systems to consider:

- Systems in the world—the ontological aspect.
- Systems in the mind—the epistemological aspect.

---

[6] Arthur Koestler (1905–1983) used the term "holarchy" to describe this fundamental structuring of systems. See Koestler (1967).

[7] One might think there is a potential problem with infinite regress in this definition. We address this in Principle 2 and later in the book. There are reasonable stopping conditions in both directions of analysis. Practically speaking, however, most systems of interest will not require approaching those conditions.

- Systems in the abstract—the mathematical/symbolic language aspect.
- Systems in software—a new addition to this category that seems to take on a life of its own and is becoming affective in its own right (see below).

Systems exist in the world (*world* is a shorthand for the observable universe) in an objective sense. That is, everything outside of our mental concepts, and independent of a human observer, is a system and a subsystem. The structures and behaviors of all sorts of systems in the world have been the subject of the sciences, physical as well as social. The sciences all study and improve our understanding of systems in the world (see Principle 11). Science has even turned to studying systems in the mind as also actual system in the world. Modern neuroscience is making considerable progress in understanding how systems in the world are reflected in the mind as systems of malleable neuronal networks.

Systems in the mind, percepts, and concepts formed in brains (especially of humans) are constructed models of systems in the world (see Principle 9). These models are formed by observation, associative learning, and generating questions to be tested. These objects are less distinct in terms of boundaries and very much imprecise and inaccurate as mappings from the systems in the world. They work well enough when the world of interest to humans is fairly simple. Imprecise models worked well enough for predicting the future to get us to the point of generating complex cultures. But they become unreliable as advances in complexity and precision reduce the tolerable margin of error.

Humans express these less precise models in natural language. This works well enough for ordinary social interaction. But something a bit more rigorous was needed in terms of building models that had reliable and more precise predictive capabilities. A further recursive reflection on the systemic relationships of propositions yielded what we have called "systems in the abstract," the precise yet intangible order of logic and mathematics. Science uses measurement to bridge systems in the world with this abstract realm, making it possible to express the patterned regularities of systems in the world as abstract laws of nature. Figure 1.1 shows the relations between these three "kinds" of systems.

The figure also shows software as a fourth and potentially new kind of system.[8] It is still probably too early in the development of systems in software, which literally are an amalgam of the other three systems, to say that they are fully a fourth kind of system equivalent in stature to the first three. However, one can make an argument that systems in software represent an entirely new level of organization that has "emerged" from the matrix of the first three kinds. This form of system is an extension of systems in the abstract, but one that is taking on new capabilities.

---

[8] Software, as in computer programs and accouterments such as relational data sets, are an offshoot of systems in the abstract, but with a more direct causal relation with systems in the world. For example, a robot can have causal influence over objects in physical reality. Or a program can influence how a user reacts to situations.

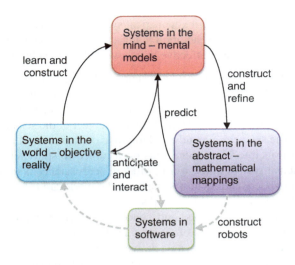

Fig. 1.1 Systems can be classified in these three domains. The *arrows* show the operational relations between the three types. Note that the interactions between systems in the world and systems in the abstract must necessarily be through systems in the mind. With the advent of computer-driven robots, however, we are seeing systems in software interacting directly with systems in the world. This may be considered a new kind of system, but it is derived from systems in the abstract and is still in its infancy

1.4.3   Principle 2: Systems Are Processes Organized in Structural and Functional Hierarchies

Since all components and their interactions exist only as processes unfolding in time, the word "system" and the word "process" are essentially synonymous. We often use the word when wishing to denote a holistic reference to an object considered as an organized relational structure. When we use the term, we are usually denoting the internal workings of an object that take inputs and produce outputs. Even systems that seem inert on the time scales of human perception, e.g., a rock, are still *processes*. It is a somewhat different way to look at things to think of rocks as processes, but at the atomic/molecular scale inputs like water seepage, thermal variations, etc. cause the component molecules and crystals to change. The rock's output, while it still exists, is the shedding of flakes (e.g., of silica) that end up as sands and clays in other parts of the environment. So in order to understand the organized structure of the Earth, the geologist must study it as process, not just structure!

The hierarchical nature of system structures has long been recognized. As process, functional hierarchies correspond with the structural hierarchical architecture of systems. Hierarchies are recognized as the means by which systems naturally organize the work that they do. Analytical tools that decompose systems based on these hierarchies are well known, especially in reductionist science. But also when we attempt to construct a system that will perform some overall function for us, we find it is best to design it as a hierarchy of components integrated into working modules, which, in turn, are integrated into meta-modules. The notion of hierarchy will become especially important when we take up the question of coordination and control in our discussion of cybernetics.

These first two principles will be the subject of Chap. 3, Organized Wholes.

1.4.4   Principle 3: Systems Are Networks of Relations Among Components and Can Be Represented Abstractly as Such Networks of Relations

Systems are networks of components tied together via links representing different kinds of relations and flows. This principle ties several other principles together. Namely, Principles 9 and 11 have to do with how we can create models of systems in the world with systems in the mind, or systems in the abstract. The emerging network science (Barabási 2002) provides us with a range of formal tools for understanding systems. For example, graph theory, in mathematics, provides some powerful tools for examining the properties of networks that might otherwise be hidden from casual observations.

For example, Fig. 1.2 shows the existence of a node type, the hub, that was not understood until the application of network theory to several example networks. A "hub" is a node that is strongly connected to many other nodes in such a way that it provides a kind of bridge to many other nodes (depending on the direction of connectivity—the two-way connections in this figure represent the more general case).

Another powerful way to use graph and network theories, very closely related to one another, is the "flow" graph. In a standard graph, the links represent relations and directions of influence. In a flow graph, the links show a single direction of influence, but the influence is carried by a flow of a real substance, i.e., matter,

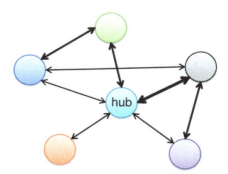

Fig. 1.2 A network of components (nodes) can abstractly represent interrelations by links (edges) in a graph structure. Interaction strengths are represented by *arrow thickness*, but could be represented by numerical labels. This is a bidirectional graph meaning that the relation goes both ways, e.g., like electromagnetic force. In this graph, the node labeled "hub" is connected to all other nodes, so it would be expected to play a key role in the functional dynamics of this network

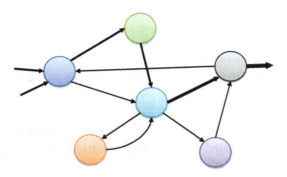

Fig. 1.3 The same network as represented in Fig. 1.2 is here represented as a "flow network." The *arrows* are unidirectional indicating that the net flow is toward the node with the *arrow point*. Flow networks provide additional mathematical tools for analyzing the dynamics of a system. Here we have added inputs and outputs in conformance with the idea that such a network is a process with an overall function (outputs given inputs). Again the "volume" of a flow is indicated by the thickness of the *arrow* for simplicity

energy, or informational messages. In these cases, the rate and magnitude of the flow are considerations and need to be represented in some fashion. Typically we use numeric and textual labels to identify those flows. More abstractly, as in Fig. 1.3, they can be represented by the thickness of the arrows showing direction.

These kinds of graphs and the networks represented have been used to analyze so many kinds of systems to date that they have become an essential tool for the pursuit of systems science. A grounding in network and graph theoretical methods is thus very helpful. Even if the quantitative methods of graph theory are not fully made explicit, it is still an invaluable conceptual tool to know how to qualitatively characterize systems as networks of interacting components and to provide detailed descriptions of the nature of the links involved in order to provide a "map" of the inner workings of a system.[9]

1.4.5   *Principle 4: Systems Are Dynamic over Multiple Spatial and Time Scales*

Dynamics refers to how the processes operate or change inputs into outputs over time. In the most general sense, the lower the level of resolution in space dimensions, the smaller the resolution in time scales relevant to dynamics. At very small

---

[9]The relationship between a network and a map should be really clear. The word "map" is used generically to refer to any graphic representation of relations between identified components. A map of a state or country is just one example of such a network representation, as the network of roads that connect cities, etc.

spatial scales (e.g., molecular), such processing proceeds in the micro- and millisec-
ond time scales. At somewhat larger spatial scales, say at the level of whole cells,
the time constants might be given in deci-seconds (1/10th of a second). On still
larger spatial scales, processes might be measured in seconds and minutes. On geological spatial scales, geophysical processes might be measured in centuries or even
millennia. What about the universe as a whole?

We sometimes find that critical processes that operate over sufficiently different
time scales can have hidden negative consequences for the system as a whole.
Systems constantly adjust themselves by feedback loops, but when interdependent
components operate with feedback loops of different temporal scales, the system
may become unstable. [10] In understanding what goes wrong and leads to disruption
and collapse of function in natural and human built systems, we generally find
dynamical mismatches at the root. For example, the fast economic payoff for
clear-cutting forests or harvesting fish by factory trawlers is not in itself scaled to
match the reproductive cycles of trees or fish. Systems science explicitly calls for
attention to dynamics at all time scales in which conflicts could threaten the sustainability of the system. In those cases where sustainability is desirable, we look
for ways to find "harmony" among the different levels of system composition (the
hierarchy).

1.4.6   Principle 5: Systems Exhibit Various Kinds and Levels of Complexity

Complexity, like network science, is really one characteristic of *systemness*. But
since the complexity of systems is a critical attribute in understanding why a system
might behave as it does or fail to behave as might be expected, complexity science
has emerged as a subject standing on its own (see Mitchell 2009). Chapter 5 will
discuss the nature of complexity in more detail. And when we take up the transformation and evolution of systems in Part IV, we will see that as systems become
more complex, new functionality and unexpected potentials may emerge. But complexity also carries a price: some of the more important findings in complexity science, such as deterministic chaos, self-organized criticality, and catastrophe theory,
have shown us that complexity and nonlinearity can, themselves, be sources of disruption or failure.

Human societies and institutions present one of the most trenchant examples of
the trade-offs of complexity. Joseph Tainter (1988; also Tainter and Patzek 2011)
has put forth a very credible argument that as societies become increasingly complex as a result of trying to solve local problems, only to create bigger problems, the
marginal return (e.g., in stability) decreases and even goes negative. This phenom-

---

[10] For an extensive analysis of the dynamics of development and collapse in human and natural
systems, see Gunderson and Holling (2002).

enon is linked with the fall of many historical civilizations, such as the Roman Empire, and causes some social scientists today to voice concerns regarding the trajectory of our modern technological civilization.

Note for principles 6–12: these principles apply mainly to more complex systems, especially those described as "complex adaptive systems" (CAS). We highlight them as such in this book because such systems occupy a major place in systems science—so much so that, as mentioned above, the study of complexity is sometimes even identified as the main subject of systems study.

1.4.7   Principle 6: Systems Evolve

In many ways, this principle, itself composed of several subprinciples, is the most overarching of them all. Indeed, it can be reasonably argued that the complexity of the systemness we find in the universe is an outcome of evolution. All systems can be in one of three situations. They can be evolving toward higher organization, maintaining a steady-state dynamics, or decaying. The principle that systems evolve is based on the systemic effects of energy flows. If there is an abundance of inflowing *free* energy, that which is available to do useful work, then systems (as a general rule) will tend toward higher levels of organization and complexity (see Principle 8 below). Real work is needed to maintain structures and to create new, compound structures. When the energy flow is diminished, the second law of thermodynamics[11] (entropy) rules, and instead of the uphill climb to higher order and complexity or the energy-demanding maintenance of complex order, a process of decay sets in and systemic order deteriorates toward random disorder.

1.4.8   Principle 7: Systems Encode Knowledge and Receive and Send Information

Information and knowledge are most often thought of as pertaining to systems in the mind, a subset of systems. Another way of looking at it, however, finds them in the operation of all systems as they move into a future with possibilities already shaped by the present state of the system. This approach usefully grounds knowledge and information in systemic structure, which not only *is* as it is but *means something* for any possible reaction to events as they unfold. That is, the system by

---

[11] The second law will show up many times throughout this book so it would be worthwhile for the reader to take some time to study its physical basis. The second law describes the way in which energy has a tendency to "diffuse" throughout a system or degrade to low temperature heat from which no additional work can be obtained. See http://en.wikipedia.org/wik/Laws_of_ thermodynamics for a general overview of the laws of thermodynamics.

its very structure "knows" how to react. From this foundation in physics, we will be able to more clearly trace the emergent systemic differences in modalities of knowledge and information as biological and psychological life evolves from the original matrix of physical and chemical systems. This will allow a more careful differentiation of the mental way of possessing knowledge and processing information from the physical way, making it clear that the way living organisms hold knowledge and process information is a more complex, evolved form of doing something every system does.

1.4.9   Principle 8: Systems Have Regulatory Subsystems to Achieve Stability

As systems evolve toward greater complexity, the interactions between different levels of subsystems require coordination. At a low level of complexity, cooperation between subsystems may emerge as a matter of chance synergies, but more complex systems need more reliable mechanisms of control to coordinate the activities of multiple components. Thus, control, typically exercised through feedback processes linked with specialized subsystems, becomes an important issue in any discussion of the function of both fabricated and evolved systems. When we take up cybernetics in Chap. 8, we will see how complex logistical coordination is achieved through the development of control hierarchies (multiple controllers require another layer of coordination among themselves!). And then the question will reemerge in an even more challenging form when we discuss the reproductive ability that marks the emergence of life, where not just coordination but accurate copying of the entire system pushes the control question to new levels.

1.4.10   Principle 9: Systems Can Contain Models of Other Systems

We are all aware of the function of mental models, how the image of how someone looks aids in meeting with them, how the map modeling the street layout enables us to navigate the city, or how the blueprint guides the construction of the building. But modeling occurs not just with minds but in all sorts of systemic relations where one system or subsystem somehow expects another. Thus, a piece of a puzzle models inversely the shape of the piece that will fit with it, and in a similar way molecules by their shape and distribution of charges model the molecules with which they might interact. In general, systems encode in some form models of the environment or aspects of the environment with which they interact, though this modeling element of functional relationships is realized in many different ways and levels in different sorts of systems.

functional relationships

1.4.11   Principle 10: Sufficiently Complex, Adaptive Systems Can Contain Models of Themselves

Adaptive systems such as living organisms can modify their models of an environment to adapt to changes, or simply for greater accuracy (i.e., learning). Creatures capable of having mentally mediated roles and identities include models of themselves, and these likewise may involve greater or lesser accuracy. For humans, as we shall see, the intertwined models of the world and of themselves become structured into their societies and inform the way societies interact with the environment. Systems science reveals the dynamics by which such models are shaped and supplies a framework within which to critique their validity. Insofar as inaccurate models contribute to dysfunctional interaction between society and the environment, systems science thus offers an especially valuable window on the question of sustainability.

1.4.12   Principle 11: Systems Can Be Understood (A Corollary of #9)

As discussed above, science is a process for explicating the workings of systems in the world, and it has very recently been turned to a better understanding of systems in the mind as well. It has moved our understanding of these systems to new levels by employing formal systems in the abstract. As these formal systems mature and are fed back into mental models arising from experience, we humans can develop better understanding of how things work both in the world and in our own minds. We will never reach an end to this process of understanding systems, and some levels of systems may continue to elude us, but in principle systems function in terms of relational dynamics, and this is an appropriate object for human understanding.

The reason we call this principle a corollary of Principle 9 is that the understanding comes from the efficacy of the models we hold of the systems we study. Science is the paradigmatic example. As a social process, science seeks to characterize and model natural phenomena by a piecewise approximation process. The models are improved in terms of accuracy and precision as well as predictive capacity over time. Models are sometimes found to be incorrect and so are abandoned in pursuit of better models. Alchemy evaporated as chemistry arose. In the end, efficacy is the test of the explanatory power of the models. This is what is meant by "understanding" something. When you understand, you can make predictions, or at least project scenarios that can then be tested. Then, the accuracy, precision, and explanatory power of the models can all be assessed, and, according to Principle 8, using information feedback for self-regulation, the models can be further improved (or found wanting and abandoned).

The human capacity to learn, especially in abstract conceptual models, is an individualistic form of Principle 11. Whereas science builds formal models in the abstract (and increasingly in software), we individuals construct knowledge in our neural networks. Knowledge is just another word for model in this sense. Our brains are capable of building dynamic models of systems in the world and using those models to predict or spin scenarios of the future state of the world given current and possible conditions. We have a strong tendency to anticipate the future, and in so doing we are relying on subconscious models of how things work in order to generate plausible explanations, both of what has happened and what might happen in the future. When we say we learn from our mistakes, we are, just as in the case of science, correcting our models based on errors fed back to our subconscious minds where the construction takes place.

When we say that systems can be understood, then we are referring to our ability to function successfully through the guidance of models in the mind that correlate with relevant features of systems in the world. A model is not identical with the object it models, so our understanding of a system is not identical with the system itself and therefore is never final. A given model can always be carried further, and another perspective yielding an alternative model is always possible. And this takes us to our twelfth and final principle.

1.4.13   Principle 12: Systems Can Be Improved (A Corollary of #6)

If one has the boldness to assert that something is an improvement, they are likely to meet the familiar counter, "Who's to say?" If I say it's great to have a new highway, someone else can always bring up sacrifice of land, air quality, noise, or others of a virtually unlimited (and equally systemic) number of ways in which the improvement might also be considered a degradation. Systems science will furnish a framework for thinking through these issues. Principle 6 notes that with available free energy, systems can evolve to higher complexity with emergent new properties. But this is not to say the dynamics that ratchet up complexity automatically lead to improvement. Quite the contrary, increased complexity can also lead to instability and collapse. And then again, who's to say that in the big picture stability is better than collapse?!

We will frame the systemic question of improvement in terms of function. Dynamic systems in their operation necessarily produce consequences. This is their functioning. And when we study auto-organization and evolution, we will find that functioning changes as complexity increases. But unless this causal functioning somehow *aims* at some result, all results are equal, and the notion of improvement has no basis, no metric. So a systems account of evolution will have to take up the question of when and how causal functioning gets to the condition where the operation of the system can be observed *to aim* selectively at some kind of result. Although aim is hard to identify in the prelife universe, the world of life is full of such processes. How then does evolution ramp up to start working in terms of *improved* function, the selection of the fittest?

We, our metabolisms, and the entire world of life operate and organize in an ongoing process of looking for what fits and better fits our varied aims and purposes. And of course all those aims and purposes are not perfectly harmonized, as both individual lives and whole systemic sectors intersect with vectors that include tensions, competitions, and sometimes outright contradiction. The success of the predator is the failure of the prey. So with a narrow focus, one can improve either the hunting of the predator or the elusiveness of the prey. At a more inclusive systemic level, improvement would have to look for the best dynamic balance, since too much success on either side, while good for the individuals involved, would have negative consequences at the level of the species well-being. And improving an integral ecosystem in which myriad intersecting competitions are woven into a dynamically shifting mutual fit would be a yet more daunting challenge. And the same goes for social systems, where clarity regarding individual lives or narrowly focused issues is much easier than the endlessly contested visions of what constitutes an improved society.

1.5   The Exposition of Systems Science

As even our rough sketch of the fundamental principles makes clear, almost anything one might say about systems is involved with and dependent on a lot of other things that also must be in place for understanding the topic at hand. Systems resist the clear, step-by-step linear organization of textbook presentation, so as we discuss them, frequent cross-referencing and pointing to topics to be developed in upcoming sections or chapters is unavoidable.

Figure 1.4 illustrates major systems topics and their intersections. The central oval in the figure is labeled "conceptualization" to indicate that ultimately all of systems science is a way to organize thinking, or how to conceptualize things in the world. Our ability to conceptualize a system is thought to be built right into the human brain. We automatically (subconsciously) categorize, note differences and similarities, find patterns, detect interconnections and patterns, and grasp changes over time (dynamics).

Concepts are inherently relational and hierarchical so in a very real sense, the concepts we hold in our minds (encoded in real physical neural networks) are truly systems in all of the senses of the above twelve principles. This should not be surprising since our brains are physical objects that evolved to interact successfully with the rest of the world. It seems appropriate that the organization of concepts and thoughts using those concepts are reflections of all that they seek to represent.

The outer rectangle in the figure includes the tools used to study and improve our understanding of systems: mathematics, relational thinking, computation, logic, and modeling. These tools employed by systems science include both qualitative and quantitative aspects, which mutually inform and complement one another. We recognize that the mathematical background of our readers will vary

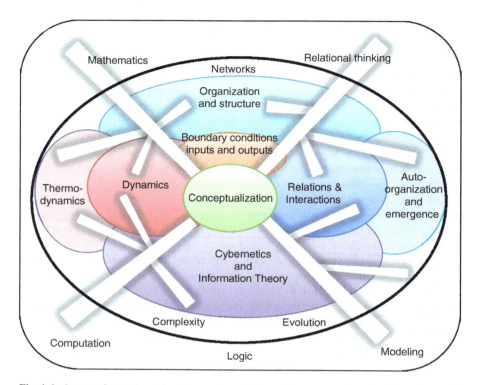

Fig. 1.4 A general overview of systems science topics with a sense on how they interrelate with one another and relate to the principles is discussed above. The overlapping ovals indicate roughly the near relations with other topics. The rays indicate that conceptualization binds the whole together. The ovals capture the topical areas to be covered, while the outer framework indicates some of the tools that help us think about systems. The large white oval contains subjects that are highly related and collectively are related to all of the others

considerably. Some may have only a moderate knowledge of mathematics, while others will have more advanced mathematical skills and/or computer programming skills. What you bring to this endeavor in the way of mathematics is less important than your ability, desire, and discipline to organize concepts in meaningful ways. Math, logic, and programming skills are handy, to be sure. Such skills facilitate the representation and manipulation of the more complex concepts by using well-established rule systems. Mathematics is not the only way of doing this, but it offers the advantage of efficiency.

But, far more important than being efficient is being able to correctly express the relations. This is not a matter of mathematics as much as it is visualizing relations between entities, of seeing behaviors over time in your mind. Formal mathematics and logic (and computer programs) can only help you manipulate the concepts

AFTER you have envisioned them. In this book, in order to make the key concepts accessible to those who prefer qualitative approaches, we strive to present the key concepts in ways that allow anyone with a bent for relational thinking to understand the important structures.

For those with more mathematics and computational background, we have organized the chapters so that you will be able to dig a little deeper into the quantitative tools needed to approach the subject from this perspective. We also cultivate a middle ground where we hope that some of the quantitative parts of this development are still accessible by those who are more qualitative in orientation, because we believe it is important to be able to appreciate how the qualitative (relational thinking) and the quantitative (mathematics and computation) interact.

Our next chapter will complete Part I, the introductory overview, with an extended example of how the fundamental principles of systems science can be used to better understand aspects of the complex web of social, environmental, economic, and intellectual systems within which we exist.

Part II and Part III of this book are closely linked. As we observed above, structure and process are complementary ways of viewing a system and can hardly be separated as topics. Thus, Part II will discuss the structural point of view, but we include dynamics and function since otherwise one can hardly see the point of relations among components. And Part III will look at how systemic structure maintains and adapts itself through time, but structures of control emerge as a critical element in considering how those processes can be sustained. Part IV will build upon this understanding of structure and process to investigate the short- and long-term processes by which systems self-organize (auto-organization) and evolve higher structural complexity and new functionality. Part V will draw this to a conclusion with a discussion of its practical application for the modeling, analysis, and engineering of systems.

1.6   An Outline History of Systems Science

Contemporary systems science and systems thinking is the product of a number of important areas of creative development over the last century. Typical of systems work, boundaries of these areas overlap, and extensive cross-fertilization and concept sharing has taken place. Insights that furnished the core of these semi-distinct movements remain key focal points in the study of systems science and appear as the subjects of many of the chapters of this book. In our presentation of systems science, however, we are concerned mainly with the coherence and articulation of this material and so do not use historical development to structure the chapters. The following brief outline history is intended as an overview of the emergence of the strands now woven together as facets of systems science. This is just a sketch, presenting enough of the key developments and names of major contributors that readers can use as a take-off point for further investigation. For readers interested in a more in-depth development of this history, we highly recommend Fritjof Capra's book,

The Web of Life, and Melanie Mitchell's *Complexity: a Guided Tour.* Both are quite accessible, but rich in detail and exposition, and they have served as major sources for the much simplified outline presented here.

1.6.1   Early Twentieth Century

In the early years of the twentieth century, the predominant paradigm in the sciences was mechanistic reductionism, the expectation that all phenomena could be finally reduced to the interactions of components at the level of chemistry and physics. But the roots of what has now become systems science were already forming in currents of dissent and dissatisfaction with the reigning pattern. In philosophy Henri Bergson advocated a vitalistic, dynamic, continually creative and unpredictable reality in opposition to the unchanging interacting units of mechanistic thinkers. This was advanced in the widely influential process philosophy of Alfred North Whitehead, which analyzed reality as a fabric of events and relations among events. Such philosophies fit well with the broader current of organicism, a concern for dynamic pattern and relational wholes gaining increasing currency in various areas of life science. In the 1920s Walter Cannon came up with the term "homeostasis" to describe what he saw as the organized self-regulation by which living bodies maintain a stable equilibrium among their complex interrelated components and processes. Gestalt theory with its emphasis on the priority of wholes in our perception influenced both neurological and psychological studies of perception. In 1905 Frederic Clements, in the first American ecology book, *Animal Ecology*, described plant communities as so intensely interrelated that they constituted a kind of superorganism. And in 1927 Charles Elton introduced the concept of ecological systems comprised of food chains or "food cycles," an idea soon refined into the more complex interdependency of "food webs."

These lively new explorations of relational wholes were focused especially on the dynamics of living organisms, so it is not surprising that the first more comprehensive theories of systems should come from thinkers with a background in organismic biology.

1.6.2   Von Bertalanffy's General Systems Theory

An Austrian biologist with organicist leanings and a strong interest in philosophy, Ludwig von Bertalanffy, is generally credited with first introducing systems as such to the world of serious scientific investigation. From the late 1920s, he published papers and lectured as a strong advocate of organismic biology, stressing that living organisms could not be reduced to a machinelike interaction of their parts. Something more, the relational whole and its dynamic organization was required to explain fundamental characteristics of life such as metabolism, growth, development, self-regulation, response to stimuli, spontaneous activity, etc. He originated the influential description of organisms as "open systems," distinguished from closed systems by an organization arising through and maintained by a constant flow of energy. Accordingly he stressed that metabolism achieved a dynamic steady state quite unlike the lowest energy state entropic equilibrium condition produced by the second law of thermodynamics in closed systems.

Von Bertalanffy saw how his insight into open systems provided new understanding for the patterned dynamics of ecosystems, social systems, and a wide range of other fields of inquiry. This led him to advocate and develop what he called "general systems theory":

> The theory of open systems is part of a *general system theory*. This doctrine is concerned with principles that apply to systems in general, irrespective of the nature of their components and the forces governing them. With general system theory we reach a level where we no longer talk about physical and chemical entities, but discuss wholes of a completely general nature. Yet, certain principles of open systems still hold true and may be applied successfully to wider fields, from ecology, the competition and equilibrium among species, to human economy and other sociological fields. (von Bertalanffy 1969, p. 149)

Von Bertalanffy was ready to publish these ideas at the end of the 1930s, but was interrupted by the war. In the two decades following WW II, systems theory entered the mainstream of ecology, social sciences, and business management. The spread of the systems concept was evidenced in the emergence of new fields such as systems design, systems engineering, and systems analysis. Immediately following the war systems, ideas developed in strong synergy with the emergence of cybernetics and information theory.

1.6.3 Cybernetics (See Chap. 9)

Norbert Wiener, one of the founding figures and leading developers of the new field, borrowed the Greek term for a ship's helmsman to create the term cybernetics, which he defined as the science of communication and control in both machines and animals. The roots of cybernetics go back to World War II when researchers including Wiener, John von Neumann, Claude Shannon, and Warren McCulloch tackled the problem of creating automatic tracking systems to guide antiaircraft guns. Out of this work came the critical concepts of information and feedback loops, the essential basis for understanding all sorts of systemic regulation and control. The notion of feedback was so broadly applicable for the analysis of mechanical and biological regulation and control mechanisms, and automated systems quickly proved so useful that cybernetic ideas quickly spread to every area of life. Indeed, by the late 1960s von Bertalanffy found it necessary to protest against a tendency to identify systems theory as such with cybernetics (von Bertalanffy 1969, p. 17).

The early cybernetics movement cross-fertilized with the postwar emergence of computers (von Neumann), and information theory (Claude Shannon, Gregory Bateson) and cybernetic regulation in machines became a common model for investigating neural function in the brain (Ross Ashby, Heinz von Foerester).

1.6.4   Information (See Chaps. 7 and 9)

Claude Shannon, working for Bell Labs, laid the foundation for modern information and communication theory in a two part article, first published as "A Mathematical Theory of Communication" in *Bell System Technical Journal, 1948,* and then expanded and popularized in a book with Warren Weaver, *The Mathematical Theory of Communication.* His intent was to find how much information could be transmitted through a given channel even with errors caused by noise in the channel. His work emulated Ludwig Boltzmann's statistical approach to entropy. Shannon analyzed information as a message sent from a source to a receiver. His intent was to find how much information could be transmitted through a given channel even with errors caused by noise in the channel. For this he needed first of all some *measure* of information, and his stroke of genius was to define this as the amount of uncertainty *removed* by the message. The minimal move from uncertainty to certainty is the binary either/or situation, as in the flip of a coin. This yielded the fundamental unit of information we know as a "bit," incorporated in the binary code processed by on/off switches in computers—a strategy Shannon had already introduced several years earlier. "Shannon information" has given rise to the coding techniques and powerful statistical manipulations of information in the complex field of information technology and also has played a critical role in the development of molecular biology and its seven of genes.

Shannon quantified information without really discussing what it is. Gregory Bateson, taking cybernetic insights into understanding control systems in psychology, ecology, and sociology, defined information as "a difference that makes a difference" (Bateson 1972). The binary either/or might still express the minimal informational difference, but Bateson makes clear that what distinguishes information is not the physically embodied code or message but the receipt of the message and its translation into some response, some difference related to the content or meaning of the message. This qualitative definition has proved especially fertile in the life and social sciences, where information feedback processes of all sorts are selectively shaped and structured precisely in terms of what sort of difference they make.

1.6.5   Computation (See Chaps. 8 and 9)

The basic idea of the modern programmable computer was conceived by Alan Turing in 1935 some ten years before such machines were actually built. The first machine implementations were produced by John Mauchly and J. Presper Eckert,[12] but the first truly practical design incorporating the switching theory (binary) developed by Shannon was accomplished by John von Neumann, who worked with

---

[12] See http://en.wikipedia.org/wiki/ENIAC.

Mauchly and Eckert at the University of Pennsylvania's Moore School of Electrical Engineering. As a mathematician von Neumann realized that base two numbers would make the design of computers much simpler since the simple position of a switch (ON or OFF) could be used to represent bits. Joined with the emergence of cybernetics and its closely related areas of information and communication theory, computers became for the next decades the paradigm of mechanical and biological information processing. Computation used to be thought of as simply a mathematical process. Now, on the other side of the computer-driven digital revolution, we find that virtually every form of information can be translated into and out of a digital code. So maybe "information processing" and "computation" are simply different terms for the same thing. At least that is the background of thinking as we now pursue research on how information is processed/computed in cells, in ecosystems, in stock markets, and in social systems.

As computers advanced in both memory and processing speed, they opened up applications of nonlinear mathematics that were critical for the emergence of a new approach to modeling and to the understanding of complex systems. In particular, computers opened up the world of nonlinear math, as equations could be repeated over and over with the product of the prior becoming the basis for the next iteration. With the digital ability to render such computation processes as visual information, an unsuspected world of nonlinear pattern and organization was revealed, and chaos theory became a major gateway in the 1960s and 1970s to the investigation of complex and complex adaptive systems.

1.6.6   Complex Systems (See Chap. 5)

Multiple areas of research have fed into the burgeoning study of complex systems, and there are likewise a variety of descriptions or ways of defining and measuring complexity. One of the early revelations of the distinctive character of complex systems occurred when Edward Lorenz discovered in 1963 that in running his computerized weather model, even the tiniest differences in starting variables could cause unpredictably large differences in the predicted weather patterns. This "sensitivity to initial conditions" was found to be a common trait of many complex dynamic systems, a major shock to the scientific assumption common at the time that mathematically determined systems were necessarily predictable.

"Deterministic chaos," as it was called because of this unpredictability, was found to harbor unsuspected forms of regularity and order which could be explored with computers. Iterated equations could be tracked as trajectories in multidimensional phase space, and a new field, dynamic systems theory, developed a mathematical language that could describe this behavior in terms of bifurcations (sudden shifts), attractors (patterns to which trajectories would be "attracted"), and other qualitative regularities. Linear systems would result in trajectories concluding in a single point (a "point attractor"), like a pendulum slowly winding down, or in trajectories that finally repeat a former point and so cyclically repeat the whole process (a "periodic attractor"). Nonlinear equations on the other hand might produce a smooth trajectory for a time and then at some value shift dramatically (a bifurcation) to a new configuration; this patterned yet not wholly predictable behavior is classed as a "strange attractor." Bifurcations also became the focus of a sub-specialization, catastrophe theory, a mathematical investigation by Rene Thom[13] with attractive application to the ways complex systems hit thresholds at which behavior changes drastically, including such phenomena as the sudden onset of cascading positive feedback (more leading to more) evidenced in the collapse of hillsides in landslides or of ecosystems as a food web disintegrates.

The natural world produces many phenomena with regular but always varied patterns such as we observe in the shapes of clouds and plants. Some of these also involve self-similarity across different scales or degrees of magnification: the jagged patterns that outline of coastlines of entire countries reappear even on the level of clods and bits of dirt, or the branching patterns of tree limbs is replicated as one moves level by level to smaller branches and twigs. In the 1960s Benoit Mandelbrot developed a new kind of geometry to describe and investigate patterned phenomena that exhibit self-similar patterning across a range of scales of magnification. He called these "fractals." He discovered that some nonlinear equations produce out of apparent randomness emergent fractal patterns of immense complexity which are characterized by self-similarity at whatever scale. In the mid-1970s, it was found that the attractors that describe the patterns of chaotic trajectories also have this fractal geometry, and fractals became another area in the exploration of chaos theory.

While chaos theory, dynamic systems, and fractals offered new ways and new tools to describe and investigate the nonlinear patterning process of complex systems, complexity is also marked by structural characteristics. In the early 1960s, Herbert A. Simon typified that structure as a hierarchical organization of "nearly decomposable" modules or components. Hierarchy points to the range of more and more inclusive or more and more fine-grained levels at which a complex system may be analyzed: components have components which have components, a nested structural hierarchy. Components are modular insofar as they interact within themselves more strongly than with exterior components, but they cannot be completely decomposed as self-contained modules insofar as some of their behavior is caused by their external relations with other components. Simon thus describes a hierarchically structured systemic whole which is made up of many components, but with an interdependence among them that makes the whole more than just the sum of the parts. His description suggests strategies useful for the structural analysis of complexity and also for measuring the relative complexity of systems (see Chap. 5).

---

[13] See http://en.wikipedia.org/wiki/Ren%C3%A9_Thom.

1.6.7   Modeling Complex Systems (See Chap. 13)

John von Neumann in the 1940s had come up with the idea of cellular automata to investigate the logic of self-reproduction in machines. In 1970 John Conway adopted cellular automata in a simple form he called the Game of Life, an emulation of the relational dynamics of some kinds of complex organization. Social insects such as bees, ants, and termites evidence a high degree of differentiated but coordinated behavior with no evident central control. Cellular automata are composed of a large number of individual units or cells which can be programmed to act (turn on or off) in reaction to the state (on or off) of cells in their immediate vicinity, every cell doing this simultaneously and then repeating it following the same rule but based on the new on/off array. Rules can be varied in a number of ways to explore different modalities or the consequences of shifting parameters. But it was immediately evident that a few simple relational rules iterated many times could give rise to a range of surprisingly complex and patterned dynamic behaviors, with changing groups clustering and transforming, or producing "gliders" that sail like flocks of birds across the computer screen. Conway's game brought cellular automata to the attention of both a wide popular audience and inspired serious scientific investigation as well. Stephen Wolfram, who began studying cellular automata in the 1980s, came out in 2002 with an influential book, *A New Kind of Science*, proposing that something like cellular automata rules may govern the dynamics of the universe.

Attractors, fractals, and cellular automata represented a new way to graphically model the unpredictable behavior of complex systems. The iterative math and graphics capability of computers has in fact introduced a new dimension to scientific method. Computer simulations of complex systems now offer an avenue to investigate the way varied parameters may affect systems such as global weather, economics, and social dynamics that operate at scales not open direct experiment. The results of course are only as good as the necessarily limited models, and models themselves must be continually critiqued with feedback from real-world observation. The process of constructing models and running them to see what will happen will never supplant the need for traditional experiment and verification, but it expands the reach of science into dimensions hitherto inaccessible.

1.6.8   Networks (See Chap. 4)

Networks are another mathematically based model for exploring the connectivity structure of systemic organization. A network model portrays a system in terms of a web of nodes and links. We are surrounded by networks and accustomed to hear electric, neural, social, computer, communications, and virtually any other system nowadays discussed in terms of networks—the Internet having perhaps now overshadowed all others. In math, networks are studied in graph theory, a discipline going back at least to the famed mathematician Leonhard Euler in the eighteenth century, and many fields of application have independently developed their own forms of network theory. It became especially prominent in social studies in the 1970s and now finds application in everything from the study of the spread of epidemics to understanding the networked control mechanism for the expression of genes. With the growing awareness of how the phenomenon of network structure seems to transcend any particular system, it is now being suggested that identifying and understanding principles that apply to networks as such may provide a common way of thinking about all systemic organization (Barabási 2002).

1.6.9 Self-Organization and Evolution (See Chaps. 10 and 11)

Von Bertalanffy's general systems theory made a major contribution by introducing the notion of organisms as dynamic open systems maintaining themselves in a far-from-equilibrium stability, and he called for a new kind of thermodynamic theory to account for such dynamics. In the 1970s, open systems were carried a decisive step further by Nobel Laureate Ilya Prigogene's work on "far-from-equilibrium dissipative systems." Such systems, like von Bertalanffy's open systems, exist far from equilibrium in a context of dependence on constant energy input and output. But unlike the earlier focus on living systems, Prigogene's work was rooted in physics and chemistry, and thermodynamics was thoroughly integrated into the discussion. Open systems were concerned mainly with explaining metabolic homeostasis, but Prigogine was concerned with showing how, given a suitable flow of available energy, systems could actually ratchet themselves up to a new level of complexity. About the same time, Harold Morowitz provided the detailed vision of exactly how energy flow produced the increases in organization and later explained how new levels of organization (complexity) emerged from lower levels (Morowitz 1968).

This new understanding of the process of mounting systemic organization reframes evolution. Darwinian natural selection remains critical in understanding the ongoing process of increasing complexity and diversity in the community of life, but the newer understanding of self-organization roots bio-evolution more deeply by exploring the rise of the physical and chemical complexity that takes a system to the threshold of life. In sum, a full systems account should now be able to look at the junctures where chemistry emerges from physics, biology from chemistry, and sociology and ecology from biology. Being able to address the rise of auto-organizing complexity at the level of physics and chemistry moves decisively beyond former notions of organization by statistically improbable random chance, much as the understanding of open systems put to rest vitalist theorizing that demanded some ethereal animating principle to account for life functions not adequately accounted for by reductionist mechanism. Systems thinkers such as Stuart Kauffman and Terrence Deacon now actively explore the organizational threshold of life (see Kauffman 1995; Deacon 2012). As Kauffman puts it, "If I am right, the motto of life is not We the improbable, but We the expected" (Kauffman 1995, p. 45).

1.6.10 Autopoiesis (See Chaps. 10 and 11)

In the 1970s, the Chilean biologists Humberto Maturana and Francisco Varela introduced "autopoiesis," from the Greek terms for "self" and "making." They were particularly intent to describe the distinctive self-referential feedback loops by which elements of cells continually construct themselves and maintain a shared, bounded interdependent context which they themselves create. The term autopoiesis came to be used broadly as virtually synonymous with "self-organization" by many who do not adhere closely to the strict context of the original work. The major impact of autopoiesis was to draw wide attention to the systemic dynamics of self-generation in a variety of areas including social, economic, legal, and even textual systems, making the observation that participants in a system create the very system in which they participate a relatively commonplace observation.

1.6.11 Systems Dynamics (See Chaps. 6 and 13)

Systems dynamics has become one of the most widely used systems tools for many sorts of policy analysis. Its main concepts have to do with feedback loops, stocks, and flows. The creator of systems dynamics was Jay Forrester, an MIT-trained electrical engineer who in 1956 became a professor in the MIT Sloan School of Management. His systems dynamics creatively apply engineering concepts of the regulated storage and flow of energy to the understanding of the functioning of complex industrial, business, and social organization. The timing which regulates the coordination of stocks and flows among the components in a complex system is a particularly critical concern. His ideas first found application in the organization of complex manufacturing processes but then jumped to the arena of urban planning and policy, and thence, via contact with the Club of Rome, to application to the global issue of sustainability. Systems dynamics thinking was the underpinning of the famous 1972 *Limits of Growth* book (Meadows et al. 1972) which launched the international movement for sustainability. Computerized simulations based on systems dynamics principles graphically portray the consequences of a range of variables subject to managerial decision, so they continue to enjoy wide application in the consideration of policy in business, society, and government.

Bibliography and Further Reading

Barabási AL (2002) Linked: the new science of networks. Perseus, Cambridge, MA
Bateson G (1972) Steps to an ecology of mind: collected essays in anthropology, psychiatry, evolution, and epistemology. University of Chicago Press, Chicago, IL
Bourke AFG (2011) Principles of social evolution. Oxford University Press, Oxford
Csikszentmihalyi M (1996) Creativity: flow and the psychology of discovery and invention. HarperCollins Publishers, New York, NY

Capra F (1996) The web of life. Anchor Books, New York, NY

Deacon TW (1997) The symbolic species: the co-evolution of language and the brain. Norton, New York, NY

Deacon TW (2012) Incomplete nature: how mind emerged from matter. W. W. Norton and Company, New York, NY

Forrester J (1968) Principles of systems. Pegasus Communications, Waltham, MA

Geary DC (2005) The origin of mind: evolution of brain, cognition, and general intelligence. American Psychological Association, Washington, DC

Gilovich T, Griffin D, Kahneman D (2002) Heuristics and biases: the psychology of intuitive judgment. Cambridge University Press, Cambridge

Gleick J (1987) Chaos: making a new science. Penguin, New York, NY

Gunderson LH, Holling CS (eds) (2002) Panarchy: understanding transformations in human and natural systems. Island, Washingdon, DC

Johnson-Laird P (2006) How we reason. Oxford University Press, Oxford

Kauffman S (1995) At home in the universe: the search for the lows of self-organization and complexity. Oxford University Press, New York, NY

Klir G, Elias D (1969) An approach to general systems theory. Van Nostrand Reinhold, New York, NY

Klir G, Elias D (1985) Architecture of systems problem solving. Plenum, New York, NY

Koestler A (1967) The ghost in the machine. Macmillan Publishing, New York, NY

Laszlo E (1996) The systems view of the world. Hampton, Cresskill, NJ

Mandelbrot B (1982) The fractal geometry of nature. W. H. Freeman and Co., New York, NY

Maturana H, Varela F (1980) Autopoiesis and cognition: the realization of the living. D. Reidel Publishing Co., Dordecht

Meadows DH et al (1972) Limits of growth. Universe Books, New York, NY

Mitchell M (2009) Complexity: a guided tour. Oxford University Press, New York, NY

Mobus GE (1994) Toward a theory of learning and representing causal inferences in neural networks. In: Levine DS, Aparicio M (eds) Neural networks for knowledge representation and inference. Lawrence Erlbaum Associates, Hillsdale, NJ

Morowitz H (1968) Energy flow in biology. Academic, Waltham, MA

Prigogene I, Stengers I (1984) Order out of chaos: man's new dialogue with nature. Bantam Books, New York, NY

Shannon C, Warren W (1949) The mathematical theory of communication. University of Illinois Press, Champaign, IL

Simon HA (1996) The sciences of the artificial: third edition. MIT, Cambridge, MA

Tainter JA (1988) The collapse of complex societies. Cambridge University Press, Cambridge MA

Tainter JA, Patzek TW (2011) Drilling down: the gulf oil debacle and our energy dilemma. Springer, New York, NY

von Bertalanffy L (1969) General systems theory: foundations, development, applications. George Braziller, New York, NY

Wiener N (1948) Cybernetics. MIT, Cambridge, MA

Wolfram S (2002) A new kind of science. Wolfram Media Inc., Champaign, IL

Chapter 2
Systems Principles in the Real World: Understanding Drug-Resistant TB

If a problematic situation is to be resolved, the variety available to the designer of a means of resolving the situation must have controlling access to the same variety as that found in the situation.

John N. Warfield, 2006

In complex situations, decision makers are not presented with problems and alternative solutions. Decision makers must search for problems, as well as solutions...

Kaye Remington and Julien Pollack, 2012

Abstract An example of how a complex modern problem for humankind can be considered in terms of systems science should help in understanding how the principles introduced in Chap. 1 can be applied. Drug-resistant tuberculosis has become a threat brought on by our very use of antibiotics and the power of evolution to select more fit bacteria strain—fit that is to not be affected adversely by antibiotics originally developed by humans to kill them and prevent disease. This chapter lays out the complexity of the problem and examines its facets through the lenses of the principles.

2.1 Introduction

Both of the above statements are interpretations of what is known as Ashby's Law of Requisite Variety. Ross Ashby, a seminal thinker in the then emergent field of cybernetics, came up with his mathematically formulated variety thesis in the late 1950s (see Ashby 1958). His thesis brings the control theory of cybernetics to bear on the world of complex problems. As Warfield's paraphrase brings out, the basic idea is that a complex, multi-faceted problem can be controlled only by means that have as much complexity (variety) as the problem being addressed. Sometimes referred to as "the first law of cybernetics," Ashby's variety law is the antithesis of the always attractive search for a "silver bullet" that will somehow make a complex problem go away. Although the notion that a control must be complex enough to address all the dimensions requiring control seems almost self-evident, we have all

© Springer Science+Business Media New York 2015
G.E. Mobus, M.C. Kalton, *Principles of Systems Science*, Understanding Complex Systems, DOI 10.1007/978-1-4939-1920-8_2

too many high-profile cases where it is ignored—with sad consequences. One has only to think of our erstwhile "war on drugs" or the once popular "three strikes and you're out" approach to control crime in order to see the appeal and damage wrought by simple solutions to complex problems.

Remington and Pollack, specialists in the management of complex projects, draw the corollary for real-life situations: the first order of business is to get the complex, interconnected dimensions of the problem in view. What sounds at first like a single problem often turns out to be a multi-level interactive web of problems (Chaps. 4 and 5). Simple solutions often snarl the web further in the effort to address a single element as if it could be isolated. The systems challenge and then extends beyond identifying the many aspects of a complex situation as if they are self-enclosed components, each a solvable problem: rather they must also be seen in their dynamic interaction, for that intertwining is often enough the most critical and challenging dimension of a complex problem.

This chapter will give an extended example that will further elucidate the meaning of each of the 12 principles of systems science introduced in the first chapter. But in addition, it is intended to illustrate how these common features of complex systems can serve as windows through which we may see and address the web of interdependent issues that must be identified if we are to make headway as we address complex questions.

Analytic thinking in science and the social sciences typically defines the border of a problem or question to be considered and then proceeds by inspecting the internal structures and dynamics within those borders. We follow a somewhat similar approach when we decompose a complex system by analyzing component subsystems. But the distinctive mark of a whole systems approach is that it can then follow relational lines across would-be topical and disciplinary boundaries to see how the whole fabric of the problematic situation hangs together. These approaches are complementary. We need analysis to identify the many parts or aspects of a problem (Chap. 12). But without seeing the interrelated dependencies and dynamics, we risk dealing with facets of the problem in a counterproductive way. The requisite cognitive variety for dealing with complex problems requires both the manyness arrived at by analysis and the integral understanding of the whole that must inform action on any facet of the problem. We have chosen drug-resistant TB as our example, for it typifies the levels of interwoven clarity and disagreement, solubility and intractability, which are common to many of the problems that confront society.

2.2 Drug-Resistant TB

It is now a well-known and alarming fact that the protective walls of antibiotic drugs we have come to take for granted are crumbling, their defenses circumvented by the evolution of "super bugs." Of course there is nothing unique or "super" about these new strains of bacteria except that they are no longer vulnerable to elements in their environments that in earlier generations were lethal. As we shall see when we discuss evolution (Chap. 11), this is a characteristic of evolution in action. Among the new drug-resistant bugs, the TB bacterium has attracted a lot of attention, and for a good reason, TB was once one of the deadliest scourges faced by humans, especially those living in dense urban populations. TB is most often a lung infection, and because it is easily transmitted by air, a cough could put all the people in a room, a market place, a workshop, or a factory at risk of infection. In the 1800s TB was responsible for almost 25 % of deaths in Europe. Over the next 150 years, improvements in public health based on better understanding of transmission (e.g., warnings against spitting in public and pasteurizing to prevent the transmission of TB through infected cow's milk) did much to improve the situation by the mid-twentieth century. The real corner was turned with the development of the antibiotic, streptomycin, in 1946, the first really effective treatment and cure. Since that time TB has largely receded from public consciousness among the more affluent, antibiotic protected communities of the developed world, but has remained a killer among impoverished populations.[1]

Drug-resistant TB is a many-sided issue, as is evident in the wide array of experts who address the subject. Of course there are medical researchers arrayed in their numerous subdisciplines and public health agencies, epidemiologists, the media, governments, the UN, sociologists, and economists to study its spread. In fact, systemic sectors from the microscopic to global organizations all are involved. Each has a piece of the action, but what do the pieces look like as a whole?

2.2.1 Systemness: *Bounded Networks of Relations Among Parts Constitute a Holistic Unit. Systems* Interact *with Other Systems. The Universe Is Composed of Systems of Systems*

Our first principle, systemness, tells us that any of these aspects may be considered as a system but that each of these systems can also be considered as a component of a larger system. The component view invites crossing boundaries to inquire about relations and dynamics among the components vis-à-vis a larger whole. Growing scales of scope, size, and complexity often signal a "nested" systemic structure, a system of systems in which each successive level enfolds the previous level as an environment. TB bacilli typically are housed in the lungs, the lungs that belong to persons, who are members of families, social groups, the wider community, regions, nations, and the whole globally organized human race. We move from the organization of metabolisms to interpersonal relations, thence to social, political, and economic organization on incremental scales from the local environment to the entire globe (Chaps. 10 and 11). This brief outline of the systemness of the subject at hand furnishes a useful and workable map of relevant questions to be asked as we investigate the relational network that is the matrix for the emergence and spread of drug-resistant TB.

---

[1] See Tuberculosis: Society and Culture. wikipedia.org/wiki/Tuberculosis#Society_and_culture.

2.2.2 Systems Are Processes Organized in Structural and Functional Hierarchies

System as process (Chap. 6) invites us to look at the dynamic interaction among components of the system. We might start at the level of bacteria and metabolic processes. Bacteria are not necessarily the enemy. The body hosts from 500 to 1,000 species of bacteria internally and about the same on the surface of the skin. Most of these are either expected cooperators in our health maintenance processes or neutral; only a small subset is problematic. Without the bacteria in our digestive tracts, for example, we cannot break down the food we eat into the nutrients we actually absorb. As a result digestive problems are often side effects as antibiotics take out these bacteria essential to our digestive co-op along with the enemies that were targeted. Our life processes are flows combining the input of many streams, some of which, like the bacteria in our intestines, originate somewhere else but nonetheless are expected members of the structured metabolic system. For every cell in our body that "belongs" to us, there are about ten resident visiting microbes (Wenner 2007). The common focus on bacteria as agents of disease, such as TB, needs to be reframed.

This systemic reframing raises new questions, with consequences for practice. If our life system expects and needs so many foreign transients, how does it identify and deal with the bacteria dangerous to our well-being? The immune system, like any defense force confronted with a continual influx of visitors only some of whom are "invaders," has to be pretty sharp to respond with the necessary and proportional discernment (Chap. 9). Is this discernment and resistance totally inborn and automatic, or partially a feature of training? And are there ways of neutralizing invaders in a process less drastic than total destruction? Such questions lead to understanding how we build up "resistance," a natural process that indeed trains the immune system, and one that can be artificially mimicked by the development of vaccines.[2] However, becoming aware of the process of developing resistance also calls into question an overenthusiasm for making household environments as sterile as possible. If some degree of exposure helps train resistance to bacteria, overly sterile environments can create people who, like an isolated community, can lose the knowledge for dealing with outsiders. Indeed, this trained resistance and its absence has had major historical ramifications. A systems thinker such as Jared Diamond can observe how centuries of living at close quarters with their livestock made European explorers, merchants, and missionaries highly resistant to their microbes. At the same time, they became unwitting carriers of invading microbial armies which decimated the residents of the Americas who had not grown up with domestic livestock and so had no resistance to their microbes (Diamond 1999).

At the next level of process, we have not microbes and metabolisms but whole persons and their dynamic interactions with each other. Here reflection turns not just to the disease process of the individual but to the lived experience of being diseased

---

[2] Katsnelson (2011).

as a member of a family, a wage earner, and a participant in the community. And we must also include the experience of families and communities in dealing with diseased members. A highly communicable disease like TB has very different ramifications from a health problem which is not "catching," for it impacts all the routine forms of close contact which sustain our daily life. Adding to this is the high fear factor that goes with the words drug resistant and feelings of dashed hopes, desperation and despair, that accompany an often prolonged and expensive search for alternatives, and one can see how this disease can pose a special sort of challenge for networks of interpersonal relationships.

Individuals and their families are also enmeshed in an interwoven system with dynamics that span levels from the familial and local to the regional, national, and even global (Chaps. 3 and 4). TB flourishes in crowded conditions and especially among populations where poor nutrition or other factors such as HIV weaken immune systems. It becomes drug resistant due to repeated partial treatments which stop prematurely when symptoms disappear, but the stronger bugs have not yet been wiped out. Patients may stop taking their meds, or low-quality drugs may have been provided; health agencies may not be careful enough with instruction and follow up, or prison inmates may be released before their course of medication has been completed, or some combination of such factors often results in incomplete treatment. Such conditions are especially associated with poverty, so it comes as no surprise that the incidence of TB is highest where the world's poor are crowded together in the rural villages or urban slums of the Third World. Weakened immune systems and crowded, unsanitary conditions work in synergy with low education and inadequate public health and medical facilities, which results in the high incidence of inadequately treated TB that in turn can lead to the emergence of resistant strains.

Since poverty, malnutrition, crowding, lack of education, and weak public health infrastructure empower the TB bacterium, a broad array of systemic processes at every level of local and regional social organization are entangled in the issue. But global dynamics also play a role. Poor rural villages have been a hotbed for TB because of the lack of access to information and treatment services. And now the global market structure is transforming traditional sustenance farming by peasants into commercial agribusiness, displacing peasant farmers and resulting in a wave of migration from the countryside to burgeoning urban slums. Regional infrastructure for sanitation, education, public health, and employment, inadequate to begin with, is overwhelmed by the influx. But cheap labor and unrestricted labor conditions are enticements for foreign investment, so national governments may tolerate conditions, however regrettable, which they regard as a necessary ramp for economic growth and development.[3]

---

[3] On the intersection of global and national economics on the health of the poor, see especially Kim et al. (2000).

2.2.3   Systems Are Themselves and Can Be Represented Abstractly as Networks of Relations Between Components

Process brings out the complex dynamics of a system, while networks look to the more or less stable relational web within which the process unfolds. With networks, then, our attention focuses on structural linkages. This perspective is useful for considering causality. We often think in terms of chains of causality where A produces B which causes C, etc. But networks call attention to the fact that in complex systems a given effect is commonly the product of multiple causes and a given cause has multiple effects within a system. Or to put it another way, linkages in systems are complex, so real consequences in a network are always more than the single result that is too often our sole focus. Careful consideration of the networked linkage of a system would forewarn us of "side" effects and reduce the frequency of unintended consequences.

The human body is an incredibly complex network. In that interconnected environment, any drug has multiple effects—hence the long list of side effects (actually they are just effects) we hear in drug advertisements. And the list of predicted effects quickly veers to unpredictability for any given patient when the interaction of multiple drugs not only introduces new effects but modifies each other's effects in unexpected ways.

The biological network is complemented by the many-layered social, economic, and political networks we have discussed. Each of these can be analyzed both in terms of their inner systemic linkage and in terms of their linkage to one another, which is the boundary crossing that takes us to the whole system. A system of systems is, in this way, also a network of networks (Chap. 4).

Not only are there many kinds of organizational linkage, there are many degrees of linkage strength as well. Understanding the texture of relative strengths is often critical for understanding both what happens and, equally important, what does not happen in a network. In the network of nations, drug companies and their research facilities tend to be located in wealthy nations, far from the TB-infested rural villages and urban slums of Asia and Africa. Distance weakens the linkage to local problems by the "not my problem" factor. And profit, the guiding link in corporate behavior, is also a weak link when it comes to addressing the diseases associated with poverty. Drug-resistance, however, strengthens the linkage weakened by distance, for global travel now ensures that TB has begun circulating among us in the resistant form that even our most advanced medical facilities cannot cure.

Consequently, it becomes our problem, and so has already received enough media attention to result in public outcry and congressional hearings about the notable lag in antibiotic research.

Between 1945 and 1968, 13 new categories of antibiotics were invented; these for a time saturated the market. Since 1968 just two have been added, even though we have known about growing resistance for decades. After more than two decades of outcry from public health agencies about the emergence of drug resistance, only four of the twelve major pharmaceutical companies are engaged in the research. This research is very expensive and the payoffs are much more modest than for any number of other kinds of drugs such as statin drugs, sleeping pills, or diet pills, to mention a few. Moreover, the FDA has become more reluctant to approve new antibiotics after a scandal in 2007 regarding fraud and safety issues in connection with Ketek (Telithromycin), an antibiotic introduced in 2001. In effect the effort to protect consumers by tighter regulation has contributed to endangering them in new ways, as drug companies become even more reluctant to engage in expensive research with even higher barriers to bringing a new antibiotic to market. The linkage between consumers, drug companies, and the FDA now actively also includes the media, Congress, and the voters. As drug resistance became a high-profile issue, Congress in 2012 enacted provisions in an FDA authorization bill to grant drug companies engaging in antibiotic research an additional 5 years of patent protection (i.e., no generics), thus readjusting the strength of the profit linkage with the reasonable expectation; this will prove motivational for the pharmaceutical companies (Vastag 2012).

> **Question Box 2.2**
> Drug companies, the FDA, medical clinics, TB patients, and TB bacteria are all networked components. What are the linkages among them?

2.2.4 Systems Are Dynamic on Multiple Time Scales

The time scale difference of most immediate interest for drug-resistant bacteria and antibiotics is that between rapid microbe reproduction rates and the slow pace of human social change. Yet the adaptive dynamics of human societies are lightning fast compared with the adaptive standards of the systems of most larger organisms. The pace at which we are able to conspire and introduce new strategies far exceeds the rate at which most of the larger life forms can adapt. For good reason even in the case of stocks of wildlife, we now use the term "harvest" when managing our fishing and hunting activities.

If we reduce the reproductive scale to that of insects and even more to microbes, however, their basic adaptive dynamics make our cultural adaptation seem glacial by comparison. By "basic" in this case we mean their rate and quantity of reproduction.

We have for decades been waging all-out chemical and biological warfare against the insects and microbes we define as the enemy. Victory seems rapid and impressive enough to sell a lot of pesticides and antibacterial drugs, but as they are widely used, resistant "super bugs" are bound to emerge. Because of the rapid rate and scale of their reproduction, insects and microbes manage to problem solve by evolution, which can keep pace with and eventually outrun the calculated strategies with which we attack them. If there is any variation in the gene pool that happens to render its carrier more resistant to the current wave of attack, it is likely to survive long enough to produce offspring in numbers, who will in turn produce yet more progeny endowed with fortunate resistance in a geometrically escalating population wave (Chap. 11).

Within the human social system itself, the varying time scales of different systemic levels is a constant source of friction and frustration. Businesses try to be light and lively to take advantage of any opportunity. Bureaucratic agencies need defined rules and procedures to maximize regularity and predictability (Chap. 9). So the regulation of drug companies by the FDA is naturally an area of tension. Predictably, from the point of view of commerce, regulation seems most often to be somewhat out of date and counterproductive, in short "behind the curve," thus hindering rapid response to new opportunities or needs. Consequently the weight of FDA regulation in the area of antibiotics currently seems to make the process of bringing a new antibiotic to market too long and too uncertain to turn a profit on a time scale acceptable to business interests. [4]

Improvements in public health and the availability of antibiotics and other drugs has dramatically reduced mortality rates and extended human life. This advance contributes to another time scale problem. In just 40 years after 1950, the human population doubled from 2.5 billion to 5 billion; by 2050 we expect about 9 billion people to inhabit the Earth. At the time of Christ, world population was only about 300 million. Accelerating growth (a time scale factor) in one segment of a system typically intersects a limit imposed by other interdependent sectors which move at a slower time scale, producing cycles of boom and bust (Chap. 6). In the absence of predation, for example, herbivores such as deer are likely to multiply rapidly until they consume more forage than a growing season can produce, after which they may undergo a population crash of 90 % or more. The acceleration of our population has been matched by technologies that accelerate the extraction processes not only of gas and oil but also of food, and, increasingly, of water upon which we are all dependent. But such acceleration has only been making the pipeline bigger, not increasing the size of the well. Some of our resources, such as oil, replenish only in geological time, a much slower time scale which we cannot manipulate. For trees and plants and the livestock that depends upon them, we can accelerate growth rates to some extent, but the deep processes of fertility have their own timeline that intersects and finally reverses booming populations. Getting the

---

[4] For an excellent overview of the situation and suggested measures to encourage renewed efforts in research and development of antibacterials from the point of view of the biomedical industry, see Gollaher and Milner (2012).

time scales of human reproduction and resource consumption in sync with the other systemic time scales is one definition of sustainability, a reality so vital to this world upon which we depend.

2.2.5   *Systems Exhibit Various Kinds and Levels of* Complexity

An easy way to get a start on kinds of complexity is to ask what kind of problem something is. To say drug-resistant TB is a health problem does not narrow the field much. As we have seen, the health problem includes not only the complex metabolisms of biology but also social, psychological, educational, economic, and governmental dimensions, to name only the most evident. Describing any one of these immediately gets us into distinctive types of layered complexity. The TB bacterium is a complex organism which includes many interacting components, each with its own complexity. How it sustains itself and reproduces in the environment of human lungs both relates to its internal organization and to its relation to that environment. And what do TB bacteria do to that environment as they make a living and multiply? The typical course of a disease is a process with its own kind of complexity, and the issue of contagion and transmission intertwines with the complexity of social contacts. All of these considerations belong just to the physiology of the disease, so even this single area harbors diverse kinds of complexity.

We move into a whole different order of complexity when we look at the relationships among humans, who have conscious and subconscious, psychological and social, economic, political, and religious dimensions which are all manifested in the complex systems and subsystems that order their shared lives. Particular kinds of complexity develop in each of these areas, plus a different order and type of complexity unfolds at the holistic level where all these areas interact and shape one another (Chap. 5). This difference is reflected in academic studies: natural and social sciences separate these areas as different kinds of specialization, each delving with special training and jargon into a particular kind of systemic complexity. The humanities, in contrast, typically engage the complex dynamics of the whole, for the mutual interaction of all these areas is quite different from anything that could be understood by studying each of them individually and then attempting to add the results together.

Drug-resistant TB involves all these kinds and levels of complexity. It calls for many specialized kinds of study and intervention, everything from research laboratories to government agencies and UN-sponsored educational outreach. The treatment of patients in the actual practice of medicine stands at a particularly complex intersection. Treatment needs to be informed by all the specializations: the ideal clinic would be up-to-date on medical science and best practices, with good community relations, efficient agency procedures, skilled doctors, etc. But if it is just processing patients as diseased bodies to be tallied as caseload turnaround, something vital will be missing. What every patient wants, in the midst of all that expertise, is to be treated as a human being. Thus, the doctors held in highest esteem, in addition to their specialized skills, will be good at the holistic humanities side of practice, communicating not just expertise but human concern and care. Indeed, part of the complexity of a medical clinic is the motivation of the personnel who undertake such work, and further complexity arises when the clinic is dealing with a highly dangerous communicable disease festering in impoverished areas.

Question Box 2.3
Areas such as medicine, economics, politics, and religion are each so complex that they are often broken into sub-specializations for study by experts. Yet at another level, they all intersect with complex dynamics and linkages as components of a larger system. What are some of the issues you might not see or predict by just becoming expert in one or even several of these areas?

2.2.6   Systems Evolve

Given time, everything changes. But some change is directional over more or less long periods of time. Evolving systems get on a vector that heads somewhere because of some kind of selective pressure that keeps building on a characteristic as it is transmitted over and over again. Drug resistance is a sort of poster child for biological evolutionary process. You take an organism with inheritable variations, put it in an environment where certain variations allow for a good life while alternatives generally perish, and then watch as reproduction increasingly fills subsequent generations with the favorable variation . Partially effective or partially completed courses of antibiotics create exactly such a selective environment, allowing successive generations to become more and more characterized by the recipe that allows the bacteria a good life even in an antibiotic environment.

Society underwent a long evolution to get to the point of producing antibiotics in the first place. Reflecting over decades, centuries, even thousands of years, we can identify the selective pressures that have given our world a social shape in which the conditions are right for the emergence of drug-resistant TB bacteria. From scattered hunting and gathering tribes, we have increased our numbers, invented entirely new ways of making a living, and eventually organized ourselves into a globe-encompassing market economy. The advanced medical research facilities, giant drug companies, and global marketing which produce both conventional and new antibiotics, as well as the conditions of urban poverty in which TB thrives, are themselves the contemporary manifestation of this long and ongoing evolution (Chap. 11).

Hindsight on evolving systems is 100 %—or at least pretty good—but what the future holds involves considerable unpredictability. Where will the emergence of drug-resistant TB and other antibiotic-resistant infections lead? The outcome of intersecting selective pressures is uncertain, especially in the case of humans who can shift priorities so quickly. There are reputations and careers to be made in advancing medical research, money to be made from drug sales, and lives to be prolonged by defeating infectious diseases. Everything seems to point the same way. But these selective pressures do not at present line up quite so nicely. Researchers now understand that the less an antibiotic is used, the less chance there is that it will encounter and launch some randomly resistant strain which will then undermine its own effectiveness. Thus, in order to preserve their effectiveness, doctors should resort to new antibiotics only as a last resort, when all others have failed (Gill 2008). But such wisdom works against the current in a world shaped by the profit motive. Not much money can be made in drugs that are seldom used. And when it does come time to use them, their rare usage also guarantees the initial price will be sky-high, making them beyond the means of many families. Added to that, the communities in greatest need have the least means to pay. So at this time, society has not evolved much of an effective response to being outflanked by these rapidly evolving TB bacteria.

At some point, however, the drug resistance and contagion base will reach a tipping point which is likely to cause significant social reorganization. But the nature of the reorganization will be highly dependent on timing and circumstances. A celebrity could contract the disease and spur an early response. Or political gridlock about spending could paralyze a government intervention that might otherwise have funded the research activity the market alone cannot. Or fact-finding committees might figure out who to blame, with variable consequences. The Department of Health might grow a new agency to make sure this never happens again. The threat of contagion could reshape housing, workplaces, and schools. In any case, some change or changes will emerge and affect the shape of all the other possibilities and probabilities for evolution as society moves onward into our collective future.

If we focus narrowly and separately on matters such as the evolution of drug resistance in bacteria, likely business response to profit incentives, or human desire to maintain health, their trajectories are all fairly predictable, and such information can be used to anticipate the future. But on a more complex level, the predictable systemic trajectories that evolve under these diverse selective pressures actually intersect *un*predictably, giving us a future that in hindsight always seems as though it should have been foreseeable, although in fact it can never be securely foreseen.

Question Box 2.4

What systemic factors contribute to the selective pressures that drive the evolution of drug-resistant TB bacteria? What sorts of changes might lessen or even remove the pressure?

2.2.7 Systems Encode Knowledge and Receive and Send Information

Knowledge and information are critical and demanding topics that will call for considerable discussion and development. But for the moment, let's take two basic propositions and see where they lead in considering drug-resistant TB. The first is that a system knows how to act. This kind of knowledge does not require a brain; all it demands is structure. System structure itself encodes ways of acting in relation to an array of shifting circumstances of the level and type appropriate to the structure. For example, there's something about mass that draws stuff to it; so it is in the structure of things that our Earth revolves around the Sun, that our Moon is attracted to the Earth, and that if we throw a ball into the air, it falls back to Earth, time after time—the system does know how to act, how to function. Second, insofar as systems exist in a universe of process and change, their structurally encoded knowledge of how to act is modified moment to moment by an information flow, information being the news of difference which arrives as some level of structural modification. This structure-encoded knowledge-behavior linkage, mediated by the continual flow of information, becomes the changing world of process. Process is quite deterministic and predictable at the level of physics, but becomes open in new ways as we move through levels of further systemic organization and complexity (Chap. 6). While physics may be basic, that does not mean sociology is just a complex form of physics!

Viewed through this systemic lens, what can we see about drug-resistant TB? On the level of physics and chemistry, the behavior of every atom or molecule is encoded in its structure and informed by modifications in its relational matrix. On quite a different level from simple physics, the TB bacterium is a complex biological structure that encodes the knowledge of how to keep all those molecules hanging together in a very particular way, and it must likewise manage energy flows to maintain and repair a complex order that would otherwise fall apart. Further, the bacterium exists in a hostile environment where its presence sends information activating the destructive agents of the host's immune system. One component of the immune system includes the macrophages ("big eaters"), which are structurally encoded to react to the information of this sort of bacterial presence by engulfing it, putting it in an environment where the encoded response of critical molecules is to disassemble, i.e., to be digested. Even before evolving drug resistance, TB bacteria evolved defenses against this, restructuring in ways that no longer encoded a disassemble response when ensconced in a macrophage. Rather it substituted the equivalent of "go to sleep," in effect turning the macrophage into a bedroom from which it might awaken and emerge when circumstances were more hospitable.[5] Thus, while as much as one-third of the world population is thought to have dormant (sleeping!),

---

[5] For a graphic series portraying this process, see Rockefeller University's *TB Infection Timeline*. rockefeller.edu/pubinfo/tbanim.swf.

asymptomatic, and noncommunicable TB, only about 10 % of these cases will ever become active (or about 30 % for those with HIV compromised immune systems).

Knowledge comprises the "how-to" of relational responsiveness within and among systems; such knowledge is encoded in all sorts of organizational structures, from microbes to our most complex social institutions. Just as a social organization can flex and change, so new things can happen in the bacterial world, as when the attacking macrophage is turned into a protective dormitory for the bacterium. Insofar as a given organization endures, a degree of predictability exists: it is the very consistency of the macrophage response to TB that made it a consistent selective pressure toward an evolved reorganization of TB bacteria. Because organization encodes the knowledge that shapes response, we expect personalities, institutions, and organizations to behave certain ways. It is no surprise that drug companies are motivated by profit, that impoverished, densely populated, poorly educated populations are exploited for cheap sweatshop labor, that malnourished bodies are vulnerable to TB, or that some governments are responsive to the needs and well-being of citizens while others are corrupt and ineffective. In each case the knowledge of how to act is built right into the structural organization.

The corollary follows that changing knowledge means changing organization. We process continually changing information from the environment, and as we do so we also engage in a dynamic and restless flow of thought. The knowledge organized into the pattern of our lives is harder to change than the fluctuating stream of our mental life, but some thoughts or new ideas may modify mental models in ways that are literally life-changing (Chap. 13).

If the thought is some insight about significant change, such as combatting the spread of drug resistant TB, one soon encounters the hills and valleys of the relevant kinds of knowledge structured into successive layers of social organization. "How can we help?" "Sorry, it's not our job." "What are your credentials?" "We've always done it this way." "Do you have a permit?" "How much can you pay?" "Let's apply for a grant to fund the research." Because they typify the sort of responsiveness or knowledge structured into the organizations, we can guess likely organizations to match each such response. If we wish to change the status quo, it is critical to understand the way the status quo is programmed into the knowledge encoded into the organizational structure at relevant levels. TB bacteria have the structural knowledge to evolve around antibiotics. But whether or not it does so depends largely on the shape of the knowledge structured into our varied and multilayered social organization.

Knowledge structured into large-scale organization is much more resistant to change than the knowledge of individuals for good reason. A society composed of such organizations results in a relatively stable and predictable world even though in principle everything can change. But it also gives us the all-too-common experience of seeing clearly that such conditions as oppressive poverty, malnutrition, and a high incidence of TB could change and need to change, but somehow nonetheless endure for decade after decade. In our enthusiasms, we often feel "We can change the world," all too commonly followed with frustrating experiences and the observation that things just are as they are and we can't do anything about it.

The former is naïve concerning the structural depth and linkages of knowledge inherent in our socioeconomic system, while the latter mistakes short-term rigidity for a kind of absolute invulnerability to change which in fact is not possible for any complex organization. Understanding the nature of structural knowledge and its power in guiding organizational behavior in a given situation is critical for mounting effective and robust strategies to bring about necessary change. While it gives us reason to hope, it also counsels for a patient- and system-wise strategy.

> **Question Box 2.5**
> Systemic knowledge keeps organizations performing in a similar way even as the personnel change. The larger and more complex the organization, the more knowledge is embedded in the structure, so it is very difficult, for example, to change a government bureaucracy. The knowledge in a one-person business, in contrast, exists mainly in the mind of the one-person, or at least that is likely to outweigh what is embedded in the structure of the business. At what size do you think organizations start to become "impersonal," where structurally embedded knowledge is the main thing governing responses?

2.2.8 Systems Have Regulation Subsystems to Achieve Stability

Stability means maintaining system integrity and function over time. Simple systems take care of themselves. But in proportion as a system becomes complex, there are also more ways it can breakdown or malfunction. In fact, a narrow range exists of ways of everything going right, compared with the very wide range of ways things can go wrong. Consequently as systems evolve to greater complexity, they also spawn subsystems for the kind of monitoring and correcting needed to keep things on track.

Regulation necessarily involves some kind of expectation and some kind of feedback of information that registers deviation from the expected state of affairs (Chap. 9). Living organisms do this with metabolisms that regulate maintenance and repair through myriads of intertwined feedback subsystems that monitor and shape flows of energy and nutrition.

The question of drug-resistant TB takes us to the heart of a particularly complex area of regulatory feedback subsystems, the problem of maintaining and controlling defense systems. Much as social systems, at the bacterial level the information feedback challenge for appropriate regulation revolves around a dynamic game of detection and eluding detection. Immune systems must be triggered by real invaders, yet remain calm in reacting to the unavoidable host of casual visitors. Allergies represent a familiar failure of the appropriate regulatory response, resulting in one's own defense system becoming a threat as it mounts a violent response to the presence of ordinarily benign agents. Symptoms from seasonal allergies can be indistinguishable from an upper respiratory infection for many people allergic to tree pollen. Even our next level of defense, the medical community, can inadvertently do grave harm by wrong discernment. Thus, the practice of medicine is heavily layered with subsystems of rules and regulations that cover every aspect of training, practice, and the array of technologies used to supplement our onboard defense system. The stability of the medical system depends upon the predictability and reliability conferred by this complex regulatory infrastructure, even though the weight of regulation has the side effect of sometimes slowing the speed and flexibility of response.

Public health agencies are governmental subsystems for regulating the community conditions and habits that concern threats to the health of the general public. The role of these agencies is especially important for a highly contagious airborne disease such as TB. Stable, healthy governments commonly have good, effective public health agencies, much as stable, healthy bodies have sound immune systems. And the converse is also true: governments in turmoil commonly have impaired public health defenses that are ineffectual in regulating and remedying the conditions that promote the spread of TB or the emergence of drug-resistant strains.

Because complex systems rarely can regulate for just one thing, a further complication arises. Governments function at a high level of systemic complexity; consequently they regulate for multiple outcomes which may work together or at cross purposes. Setting priorities is thus a major function of high-level social regulation. In developed nations, public health most often takes priority, though not without economic tensions. Regulatory agencies such as OSHA may be unpopular with business interests, or the FDA with the pharmaceutical companies, and yet they are essential for worker safety and consumer protection. Impoverished nations often feel economic development is the necessary priority, perhaps even the chief method to attain good public health. But the health and well-being of the workers is often in tension with strategies for rapid economic gain. In countries like India and China, when public policy opts to maximize economic growth, masses of farm workers move to urban slums and into the unhealthy and crowded conditions that often characterize concentrations of cheap labor. Wealth and poverty may work in an unhealthy synergy. The governments of wealthier nations rarely prioritize the health of workers in other countries, certainly not to the point of preventing the exploitation of cheap labor in foreign countries. And cheap labor, health issues notwithstanding, can serve as a major advantage for have-not economies to bootstrap themselves up the development ladder.[6] As for the health of the workers, the main way that priority is reasserted and given regulatory teeth is through the corrective force of consumer public outcry in wealthier countries stoked by a sense of common humanity. This constitutes a transnational kind of regulatory system which is facilitated by the global information feedback through the Internet and other media sources.

---

[6] For an example of these dynamics close to home, see the case discussion of NAFTA and the factory system established on the US-Mexico border in Brenner et al. (2000).

Question Box 2.6
Even criminal systems of any complexity such as gangs or drug cartels have
regulatory subsystems to stabilize their operations. What in fact does "regulate" mean in a system context?

2.2.9   Systems Contain Models of Other Systems (e.g., Protocols for Interaction up to Anticipatory Models)

Systems interact with an appropriate degree of consistency. That is, reactions are
not random but are to some extent prefigured in the structure of the systems. The
protocol that guides this interaction amounts to a model of the other system—not
a complete model-->, but one that specifies what is relevant for the interaction
(Chap. 13). The array of signals that set immune systems into action, for example,
amounts to a model of the enemy that must be defended against. In the interaction
of living systems, such models are often critical strategic factors. An insect, such as
a walking stick that looks like a twig, can often elude a bird's model of what lunch
looks like; a fisherman can exploit this situation but in reverse, by fashioning cork,
paint, and feathers into a lure that fulfills a fish's model of lunch, even though to our
eyes the similarity may be far-fetched. Since models guide systemic interaction,
every living organism has a structural model of the environment in which it survives
and makes a living. For example, our lungs model an atmosphere rich in oxygen.

The creation of an antibiotic is in some ways similar. It begins with a model of
the life maintenance system of the target bacteria and then seeks strategies to
disrupt some necessary condition. Evolution selectively remodels the offspring of
a species to fit changed conditions. When an antibiotic becomes a sufficiently frequent disruptive factor in the environment of a bacterium, the evolutionary process
will select any available alternatives which are not disrupted, i.e., those bacteria
which survive to reproduce. So the antibiotic-resistant TB in effect has evolved into
a form that structurally models an environment in which the presence of the antibiotic is expected but no longer disruptive (Chap. 11). And researchers, in turn, have
a new model of the bacterial life maintenance system to figure out and strive to
disrupt anew.

The TB bacterium has a model of its world that is continually shaped and
reshaped by the selective hand of evolution. But as we move to social, political,
economic, and cultural levels, we find that they too each enshrine multiple models,
and these models guide all sorts of interactions from our personal daily routines all
the way up to the dynamics of global markets. As in the case of predator–prey relationships, or the TB bacterium modeling its environment and the researchers in turn
modeling TB's survival in its environment, the organized world is full of fluctuating,
cross-referenced models carried by different but related systems and used in all
sorts of strategic competitions and cooperations. In this living dance of models, the
bottom line for any model is what works for the aims of the system. In the shifting

circumstances of life, almost nothing works permanently, so as the world changes, systemic models either change in response to the new conditions, or they become dysfunctional.

We build models from common or repeated experiences, but often need to tweak them with education. A common model for the use of medication is to feel that when you feel better, you can stop taking it, because you think its job is done. This intuitive but misinformed model of effective drug use gives a big boost to bacterial resistance to antibiotics. Unfortunately the more resistant bacteria are still hanging on even when the symptoms seem to have disappeared. Stopping the medication or missing scheduled doses allows resistant bacteria to survive and pass their resistance along to a next generation. When the patient is educated in this new model of what is happening, it shows them the necessity of completing the full course of medication even after they begin to feel better.

The model is not in fact the reality, but a simplified version of reality that typically focuses on one or a few elements designed to address the question immediately at hand. Because of this selective simplification, no model can be pursued single-mindedly without producing unexpected and problematic "side effects." Pharmaceutical companies, for example, "just doing business" in line with a the common profit-maximizing model, have given rise to a conundrum: there's an overabundance of vigorous research on diet pills, statin drugs, and sleep aids for which there is a massive market in wealthier nations, and paralysis when it comes to meeting a growing but not particularly lucrative health challenge posed by increasing resistance to presently available antibiotics.

This exemplifies the difficulties that arise when the market model is applied too exclusively to the health-care system. It begs to be complemented by the common ethical model of humans forming communities in which they take care of one another. Frequently this model is invoked as a perspective quite critical of the capitalist dynamics of just doing business in the area of health. Business is quick to respond that if we single-mindedly pursue such health and ethical models, commerce will suffer, prices rise, jobs disappear, and everyone will be the worse off—an anticipatory model from a business point of view of how linkages will work in our society. Socialism is one alternative model that attempts to synthesize markets and an ethics of communal care, and frequently those who criticize the capitalist model are simply labeled as "socialists." In fact, much political controversy pivots around competing (and often poorly understood) models.

Institutions, professionals, and everyday people alike engage the world through models fitted to their particular situations; frequently these partial models are mistakenly thought of by those who constantly use them as "the way the world works." Such simplistic models might suggest seemingly obvious paths to optimization— the wishful "if only" thinking. It might go something like this: the emergence of drug-resistant strains of TB could be curbed if only people stayed on the farm, if only exploitation ceased and poverty and ignorance were alleviated, if only health agencies had good funding and regulatory power, if only governments behaved responsibly, if only we could find the right market mechanisms, if only...

We have seen that models can be effective protocols for areas or types of systemic interaction, but they are partial, so none can be maximized without disruption. Nor can they simply be added together to constitute an adequate model of a whole, for on closer examination they most often involve too many contradictory dynamics and trade-offs for any additive process or formula to work. Being aware of the role of models as protocols for system interactions alerts us to both their inevitability and their unavoidable partiality. Being forewarned about their partial nature does not allow us to somehow magically find non-partial alternatives. It does, however, introduce a very useful measure of caution and prudence as we follow our necessarily partial and limited courses of action in a reality always spilling beyond our models.

Question Box 2.7
One of the common "if only" propositions is, "If only people behaved the way they should!" But there are very different (and importantly different) models of what constitutes proper behavior. What are some of the competing models? Can such a model be dismissed just because few people live up to it? What is the role or function of such ideal models?

2.2.10 Sufficiently Complex, Adaptive Systems Can Contain Models of Themselves (e.g., Brains and Mental Models)

All the humans and human institutions or organizations involved in the drug-resistant TB situation carry models of themselves, self-conscious images of who and what they are that inform and guide their activity. Less evident is that the TB bacterium also carries a model of itself—not in consciousness but in its DNA. The DNA, as a model of the whole organism, serves as the critical protocol for subsystem interactions that function to produce another copy of the original, which will in turn be complete with its own onboard model of itself. In the process of evolution, variations of this DNA model are continually sifted, screened for how well the resulting organism fares in comparison with other models in the challenge of fitting the current environment. The evolution of drug resistance in an environment laced with antibiotics exemplifies a search for models of itself that still work even when antibiotics are present.

Models, as we have seen, are essentially functional, serving as protocols that guide system interactions. The example of DNA shows the distinctive functionality of a self-model, functioning both for integral maintenance and for the reproduction of a complex system. Similarly, the self-models carried in consciousness maintain and reproduce personal psychological, social, and institutional identities. While engaged in the continual flux of experience, these self-models give a relatively consistent identity through time to the fluid world of consciousness and the forms of organization created and maintained by consciousness (Chap. 13).

Insofar as it is the source of continuity and identity, this onboard model of self easily becomes a bulwark against change, even when change is needed. The conditions that help spread TB are to some extent produced by external forces, but they also become internalized in the habits which become part of the identities of individuals and communities: "This is just who we are and how we do things." Governments and their agencies may resist changes to their mission or routines, especially if they wish to preserve the appearance of a good self-image. And crusaders for change often take on that crusading self-image and have a hard time knowing when to back off or compromise.

The contents of a system's model of itself include not only relations among its internal components but also its relations with its external environment. DNA lays out an organism, but the way in which it lays it out has been selectively shaped already by the environment in which it must fit. Similarly self-models at the personal level are shaped selectively through family within a culture; cultures are shaped in dynamic interaction with other cultures. Institutions anticipate a fit with other institutions, often across cultures. Thus, self-models are not only a protocol for inner organization but also for interaction with other systems. The dynamics that shape any system's self-model or identity necessarily extend far beyond the perceived confines of that self. Nothing exists in itself alone.

Thus, systems composed of multiple layers of systems which have their own self-model possess, a particular kind of complexity. These layered systems of self-models are continually being shaped in a restless dance of internal-external definition and redefinition which includes influence from self-models up and down the hierarchy. Our families make us what we are even as we make them what they are, and the same goes for all the other layers of organization that emerge. However, models of hierarchical control (Chap. 9) need to make room for the dynamics of self-definition among components, and models of individual freedom need to be complemented with the necessary relational fit into the larger environing system. This causal loop between layers means that at any level, self-maximizing strategies may become short-sighted and eventually counterproductive. While neither family nor company, community or government can totally define who we are, neither can we ourselves. Yet the self-model of every community, company, or government does inform its component units, i.e., family members, employees, citizens, etc., and our own self-models critically include some sort of fit, comfortable or not, within these various and layered contexts.

The mutuality of this many-sided dance of self-modeling ensures that systemic social evolution is open-ended: there is no self-enclosed, self-defining, and unchangeable systemic identity. Thus, a situation such as the emergence of antibiotic-resistant TB and the conditions of poverty and ignorance in which it thrives can be challenged and changed. But an understanding of this dance of self-definition must necessarily underlie any effective strategy for change. Self-models also include individuals' roles in the dance, i.e., to whom one should listen, and who has the authority to change things. Such understanding helps identify points of leverage for intervention and constructive change.

In the corporate world, the guiding role of the profit motive can be difficult to change (Chap. 11). Except in extreme cases such as war or economic collapse, democratic governments do not ordinarily legislate what should be manufactured. In light of this, the self-models of capitalism and democracy may trap their cultures into a short-term profit-oriented status quo which may need to be modified or replaced by more far-sighted systems in order to address problems not immediately resolved with profit-oriented thinking. Thus, in order to solve perceived problems, revolutionary thinking may challenge the basic self-model of capitalism with a line of argument at odds with its core beliefs. Frustration with the way in which capitalism can ignore glaring problems sometimes leads to extreme proposals: if the capitalist system were overturned, we could do away with poverty and ignorance along with the urban slums which lay populations open to TB and other epidemics.

In our democratic model, we attempt to use less drastic approaches to systemic change while trying to respect existing self-models. Taxes and tax breaks, surcharges and subsidies, consumer protection and patent rights, etc. are routine tools to enable government to shape the terrain of profit for the greater good. In the world of drug development, a carrot of profitability can be extended by these means to drug companies by governments, and business can be expected to react to the opportunity with research on new antibiotics. Congress, for its part, keeps a self-interested ear open for voter sentiment, which in turn is often shaped by the media, which may quote experts, some spinning facts to further a political agenda. This dance of ideas, alternatives, and compromises often results in watered-down baby steps in the right direction, but steps nonetheless. Our intertwined self-models at every level suggest ways of acting, sometimes in mutually reinforcing synergies, sometimes in corrective tensions. Understanding the layers and interrelation of these self-models becomes a map of points for strategic intervention and leverage in a complex system in which we are both components of the system and agents for change.

Question Box 2.8
Self-identities are formed and maintained in a constant negotiation between inner and outer and up and down the systemic layers of identity (individuals, families, communities, businesses, regions, governments, etc.). What happens in systems dynamics when any self-layer behaves as if it is truly self-enclosed, ignoring the claims of other levels/selves (e.g., individuals focused entirely on themselves, totalitarian governments, businesses that exploit their workforce)?

2.2.11 Systems Can Be Understood (A Corollary of #9): Science

The systemness of the universe forms the basis for relational patterning that makes the world comparatively predictable and hence understandable. The relational patterning of the system of current interest enters the realm of one's conscious understanding as a mental model of causal relationships: we understand how it works well enough to have expectations regarding it that can guide our interaction with it. Actual interactions may further fill out the model, either by simply reaffirming it or challenging it as expectations prove wrong. In this ongoing process of modeling and experiential feedback, understanding is never full, perfect, or complete, though continually open to revision (Chap. 13).

Our living world eludes complete understanding not only because of the limited and selective nature of models, but because, unlike some passive object, it becomes different as it is understood differently. Understanding guides the ways we act, and systems are reshaped in response, actively as well as passively. For example, land, soil, and communities in rural societies become different when agriculture is understood as another industrial process, with productivity subject to efficiencies of scale as in any other industry. Similarly microorganisms and immune systems both become different when people mistakenly regard every environment as a health risk to be "improved" by spraying disinfectants around. The discipline of cultural anthropology is replete with examples of cultures becoming virtually different worlds through their radically different understandings and expectations. Sometimes unexpected side effects persuade us to revise previously held understandings in light of bitter experience; for example, water engineers concerned with flood control have spent the last several decades reintroducing twists and turns to the very waterways they spent earlier decades straightening, with the unintended consequences of devastating floods downstream. But often understanding, especially of people and social relationships, creates the very thing they expect, making the phrase "self-fulfilling prophecy" a commonplace. Children expected to do well in school often do so, and vice versa. An economy with low consumer confidence is likely to perform badly.

Cognition, the ability to know as a dynamic function of a system, as opposed to knowledge imbedded in structure, emerged in the course of evolution as a more effective way of guiding an increasingly broad and flexible range of an organism's life-sustaining interactions with its environment. This kind of knowing, of understanding, has a pragmatic base: the models it employ must correlate sufficiently with the relevant aspects of its world to support functional interaction as it pursues its well-being. However, human cognitive faculties have reached such breadth and flexibility that what they expect and look for as functional in interactions can vary widely. And in the absence of a common reference point, the feedback of functionality does not necessarily shape understandings in any one direction. An example most of us experience daily is lunch, most often a light meal taken in the middle of the day. This thing we all call lunch, we actually think of and experience in many ways. Some people focus on nutrition, while those in a hurry settle for a fast and easy snack; some look for the least expensive option, while others insist on a fine dining experience, or what might be uniquely tasty, or just uniquely unique. All of these different ways of approaching lunch concern the function of lunch, and they feedback into different personal understandings. Thus, in matters of lunch, as well as in many areas of life, the really important shared understanding is the one honed by the experience of diversity: we have learned to tolerate a wide variety of understandings.

As we collectively organize, ever more complex technologies and ways of interacting with the world, however, shared understanding becomes critical. Science and its measurement-based methodology, the process by which both observations and expectations are translated into numbers, arose as a method for grounding the all-too-flexible feedback between understanding and experience in a common framework of agreed numerical processing. Such measurement has become an effective way to cut through the myriad personal, organizational, ideological, and cultural differences of understanding and perception. All the different approaches to lunch in our example above, for instance, could be brought together in a comparative statistical review of preparation and consumption time, number of calories, percentages of daily nutrition requirements, etc. With the advent of this kind of understanding, the world of personal preferences can be subjected to an "objective" critique with a claim to general validity, possibly leading some individuals to a better understanding of what is in fact in line or out of line with their well-being.

A particular advantage of the introduction of scientific measurement was that it opened the prospect of an especially powerful way to improve systems. The ability to track by means of measurements what differences various kinds or quantities of intervention make in systemic functions promises insight into strategies to improve those functions. The effectiveness of approaches to coping with disease or improving public health, for example, could be tracked and continually improved. The disunity of an array of personal interests and understandings is replaced with a precise and agreed-upon standard of functionality and an objective way of measuring it. Ideally this approach opens the prospect of continual improvement of both understanding and function.

Such enthusiastic expectations concerning the power of scientific method gave rise in the nineteenth century to hopes for universal human progress. Such a dream persisted among some well into the twentieth century. Experience, however, has shown that only some areas of life submit to the precision of measurement in a way that allows calculated and steady improvement. We have come to take such improvement for granted in technology and industrial systems. But the anticipated convergence of understanding in the arenas of politics, religion, and social mores has not occurred. Measurement as applied to these areas has limited effect; it tends to measure, and therefore emphasize, the differences rather than transcend those differences to converge on some single model of functionality.

With regard to the case of drug-resistant TB, scientific understanding and the technology of intervention have improved greatly. Mistaken ideas about the causes of the disease, such as spirit possession or as divine punishment, could not (in the larger picture) withstand the power of measurable causes and effects; as the world's outlying populations becomes better educated, these ways of understanding the cause of disease are fading. But other causes of the emergence of drug-resistant TB have thus far proved intractable. After years of research, reams of statistics are available on income disparities, levels of education, access to clean water, square feet of household living space per person, daily caloric intake of food, and other measures concerning living conditions on all levels—local, regional, national, and international. By correlating all this information with statistics on the rates of TB and the emergence of resistant TB, we are able to understand the many layers of systemic causality involved and even predict levels of incidence and where antibiotic resistance is likely to emerge and spread. The human community already possesses the wealth, material, and technological resources to address every one of the problem factors statistically correlated with TB.

But social problems are inherently multi-causal, and there is no convergence of understanding and motivation to improve any given function as the key to remedying the situation. While the dynamic interplay of the complex factors involved in the system of impoverished urban slums and the way it feeds into the rise of disease and drug-resistant bacteria can be understood, the multiple causes are assessed differently. Political and social groups, and even whole nations, have different interests, and from their different perspectives, they tend to focus on one or another of the causes, but protect the status quo in other areas. Statistics are often used to support opposing positions: some groups will use them to prove the inevitable inequality and exploitation that feed the market system, while others will use the same statistics to show how poverty follows from the restriction of a fully free market. Corruption is often supported in systemic local practice even as it is bemoaned publicly and outside the local system. Religion may be used both to support the status quo and as its fiercest critic. While understanding abounds, those understandings pull in different directions rather than converge on any one or several solutions.

This should come as no surprise. Cognition, as you may recall, arose to guide activity effectively for maintaining well-being. Insofar as well-being is a many-faceted condition involving tensions and tradeoffs among individual units at any given level and across systemic levels (individual vs. community etc.), we should expect that the understanding proportioned to all these different systemic locations/ perspectives likewise will involve tensions and oppositions.

> **Question Box 2.9**
> What do you make of the multiple ways of understanding just about anything? Are they all equal? What would constitute a mistaken understanding?

2.2.12   Systems Can Be Improved (A Corollary of #6): Engineering

We improve systems all the time, or so we think. However, viewed from a systemic perspective, the notion of improvement is not so simple. Virtually any improvement from one point of view can be found to be problematic from some other point of view. In fact, it is hopeless to disentangle improvement from a particular point of view, representing some individuals or levels, but necessarily in tension with the points of view of other individuals, groups, or levels. Improvement must really be thought of in terms of trade-offs. In the maintenance of human health, for instance, let's consider our case concerning TB, which can be cured with the use of antibiotics. Curing the disease can contribute to both better reproductive success and higher life expectancy. These desirable consequences, good for virtually all individuals and their families, at another level also contribute to the hugely problematic systemic challenge of population growth.

When we try to improve a given function, the question of from what particular points of view it is an improvement cannot be neglected. A particular function that fulfills a particular interest can always be improved. In fact, our conscious life is so filled with arranging and rearranging the contents of our multiple mental models precisely because the systems so modeled are responsive to our interested intervention. However, backed up by ample experience, one caveat here is that in the network of complex systems, any change or "improvement" has not only the intended effect but also side effects which may or may not be an improvement, even from the limited point of view of our own interests. In the wider world, and on a larger scale, this becomes even more obvious. In China, for example, increased subsidies to enable health agencies to treat more poor TB patients not only increased the number of patients helped, but it also became an incentive for clinics to keep patients longer than necessary and to avoid referring them to more accessible dispensaries or care units for follow-up care (Tobe et al. 2011).

As we play with our mental models, systems seem easy to improve: there is hardly any function in life that we cannot think of ratcheting up (Chap. 14). In part that is because we can mentally abstract a function of interest from its real-life embeddedness in the complex matrix of related and competing interests and functions. We can easily think, for instance, of educating and treating more impoverished at-risk slum dwellers; but in real life, such ideal intervention will require facilities, staffs, and supporting agencies, all likely to require funding, often from some government with a limited supply of tax dollars—time, effort, and money—for which fierce competition abounds. And as in the case of the Chinese clinics mentioned above, the desire for expansion of these facilities, and the trajectory of staff careers, etc. may themselves become goals that compete with the implementation of the specified functional improvement they were originally intended to bring about.

Even when resources are in theory available, how they might be used in the process of systemic improvement can be surprising, as illustrated in the following table from a 2004 Worldwatch report (Table 2.1) (Gardiner et al. 2004).

This table reflects how some kinds of improvements (to consumer goods) are inherently easier to make than others (social improvements). It also demonstrates how different systems are better, or worse, at working out a given sort of improvement. Luxury goods are produced and marketed through complex systems involving many component technologies, skills, and techniques; all of these are honed continually for productivity, efficiency, competitive marketability, etc. From the consumer's point of view, greater functionality takes the form of satisfaction of an interest and for the producer, profitability. The market serves to link these two in a feedback loop that in theory generates continual improvements—although we also have found that short-term profit can be a misleading guide for systemic improvement!

Table 2.1 Annual expenditure on luxury items compared with funding needed to meet selected basic needs

Annual expenditure on luxury items	Annual expenditure	Social or economic goal	Additional annual investment needed to achieve goal
Makeup	$18 billion	Reproductive health-care for all women	$12 billion
Pet food in Europe and United States	$17 billion	Elimination of hunger and malnutrition	$19 billion
Perfumes	$15 billion	Universal literacy	$5 billion
Ocean cruises	$14 billion	Clean drinking water for all	$10 billion
Ice cream in Europe	$11 billion	Immunizing every child	$1.3 billion

Luxury goods reflect human desires rather than needs, while social goals are concerned with deep needs inherent in human well-being. Market forces take care of deep needs for populations with enough money, but are less effective where satisfying those needs is not allied with the prospect for profit. Poverty is also often accompanied by weak social agencies and corrupt government (a nonproductive profit motive!). Unfortunately this combination often spawns fertile conditions for situations like epidemic TB morphing into drug-resistant strains because of incomplete treatment. In such conditions, local feedback through voting or even revolutions may easily be infected with the very systemic dysfunction it was meant to improve. The deeper systemic problem here may be the local and global distribution of resources, which is inevitability tied with diverse and competing interests with their disparate views on functionality and improvement. Consequently the negotiation of social change for the improvement of society is perhaps the human community's most complex challenge.

The social and economic philosophies that divide the international community are grand mental models for improvement, and implementing any of them in reality runs into the complexity, the constraints, and the unanticipated consequences we have described above. Yet even at this complex, social level systems clearly can and do improve; the process, however, becomes difficult and uncertain in proportion as multiple interests and intersecting dynamics on various scales make definition of function increasingly contested and arbitrary.

And we can also see why, conversely, technology and all sorts of engineering mediated by science and with agreed-upon definitions of measurable function have shown spectacular improvements. Of course this is the easy framing of the notion of improvement, achieved by limiting consideration to a narrowly defined function. Pursuing the question of new and improved technologies in terms of their functionality in a larger social or environmental system again raises the questions about relative trade-offs among differing interests with differing perspectives on function. But there is also hope for improvement at this more complex level. As the problematic social and environmental consequences of our technology are identified, this understanding can circle in a feedback loop between levels to furnish new definitions of technological functionality. Power plants and automobiles, for example, can be engineered for superior function in terms of carbon emissions if that function is given priority. Or on a more complex systemic level, the regulations governing drug companies can be modified so that companies have an adequate incentive to do research and produce the next generation of antibiotics.

Having spent much time and money undoing the large-scale dysfunctionality brought on by short-sighted or insufficiently informed engineering (such as changing, and re-changing, the courses of water ways), we have learned to intervene in the function of complex systems such as the environment with greater caution. In framing our consideration of functionality, competing interests, priorities, and even the boundaries of the system to be considered are all often contested and negotiated at length. But once such hurdles are overcome, system function itself can be modified and improved. In the case of drug-resistant TB, the technological improvements should be the easy part. As we have seen however, that is, contingent upon complex social, economic, and political systemic factors. Improved function at these complex levels is a formidable challenge, but fortunately improvement is not an all-or-nothing proposition, and these systems are in principle open to improvement.

> **Question Box 2.10**
> Building on our technological success, the notion of "social engineering" became popular, though it is now in less favor. Technology relies on accurate measurements applied to implementing and improving functionality. To what extent can we use the statistics of social and political science in a similar manner to improve our social and political systems?

2.3 Conclusion

Running throughout our illustration of these twelve principles of systems has been the very first, the principle of systemness. The systemness of the universe means that a double perspective is constantly in play: at any level we can consider a system and its environment and then at another consider the system and environment as itself a system and inspect its interchanges with a yet wider environment. Thus, every aspect of our discussion is marked by movement through multiple levels of consideration and changing dynamics. We move from bacteria to hosts to families, communities, nations, and the world. We see productive and reproductive processes that intersect and interdepend on micro- and macro-scales of extent and time. We see cooperative networks framed in competitive environments that are in turn cooperative networks on other levels.

The principles we have introduced are tools to explore the modalities and dynamics of systemic organization. Systemness articulates relationship among all these dimensions, so their connectedness is explicit, or at least the implicit subtext of any investigation. As we move from the organization of metabolisms to households, daily routines, social dynamics, or global economics, it is simply exploring the systemic topography of a single topic, and any topic is in fact enmeshed in a similar topography.

Becoming skilled navigators of this interconnected skein of systemness has important advantages. Our conscious lives are bubbles of anticipation, where the awareness of what is enables our adjustment for what will be. In so living, an awareness of this web of multi-leveled systemic relations is critical both for better anticipating the range of consequences from any given action and for understanding points for strategic intervention. On all scales and levels of human life, understanding the fact and functioning of this connectedness enhances the strategies by which we live and minimizes the unintended side effects of our actions.

Bibliography and Further Reading

Almeida D et al (2003) Incidence of multidrug-resistant tuberculosis in urban and rural india and implications for prevention. Clin Infect Dis 36(12):e152–e154, http://cid.oxfordjournals.org/content/36/12/e152.long. Accessed 14 Sept 2013

Ashby WR (1958) Requisite variety and its implications for the control of complex systems. Cybernetica 1(2):83–99, Available online: http://medicinaycomplejidad.org/pdf/soporte/ashbyreqvar.pdf

Brenner J et al (2000) Neoliberal trade and investment and the health of *Maquiladora* workers on the U.S.-Mexico border. In: Kim JY et al (eds) Dying for growth: global inequality and the health of the poor. Common Courage, Monroe, ME, pp 261–290

Cooper R (2012) The battle to discover new antibiotics. The Telegraph,12 Jan 2012. http://www.telegraph.co.uk/finance/newsbysector/pharmaceuticalsandchemicals/9010738/The-battle-to-discover-new-antibiotics.html. Accessed 13 Sept 2013

European Center for Disease Prevention and Control. TB in vulnerable populations. c.europa.eu/en/activities/diseaseprogrammes/programme_tuberculosis/pages/tuberculosis_vulnerable_populations.aspx?MasterPage=1. Accessed 14 May 2014

Farmer P (2005) Pathologies of power: health, human rights, and the new war on the poor. University of California Press, Berkeley, CA

Gardiner G et al (2004) The state of consumption today. In: Worldwatch (ed) The state of the world 2004. W. W. Norton and Co., New York, NY

Gill V (2008) The trouble with antibiotics. Chemistry World, March 2008. http://www.rsc.org/chemistryworld/Issues/2008/March/TheTroubleWithAntibiotics.asp. Accessed 15 Sept 2013

Gollaher DL, Milner PG (2012) Promoting antibiotic discovery and development. California Healthcare Institute. chi.org/uploadedFiles/Industry_at_a_glance/CHI Antibiotic White Paper_FINAL.pdf. Accessed 20 May 2014

Katsnelson A (2011) How microbes train our immune system. Nature, 21 September 2011. nature.com/news/2011/110921/full/news.2011.550.html. Accessed 5 Oct 2012

Kim JY et al (eds) (2000) Dying for growth: global inequality and the health of the poor. Common Courage, Monroe, ME

Remington K, Pollack J (2012) Complexity, decision-making and requisite variety. http://www.elefsis.org/Complexity_Decisionmaking_and_Requisite_Variety_Remington. Accessed 17 May 2014

Tobe RG et al (2011) The rural-to-urban migrant population in China: gloomy prospects for tuberculosis control. BioSci Trends 5(6):226–230

Vastag B (2012) NIH superbug outbreak highlights lack of new antibiotics. The Washington Post, 24 Aug 2012. washingtonpost.com/national/health-science/nih-superbug-outbreak-highlights-lack-of-new-antibiotics/2012/08/24/ec33d0c8-ee24-11e1-b0eb-dac6b50187ad_story.htm. Accessed 28 Sept 2012

Warfield JN (2006) An introduction to systems science. World Scientific, London

Part II
Structural and Functional Aspects

1.1 Properties of Systems

The next four chapters address the various properties that make systems what they are. This includes the structures, organization, interactions, and behavior of systems. Chapter 3 will start with descriptions of structure. Structure refers primarily to how the components interact with one another and how boundaries impose an inside and outside character to what might be called "entity-hood." We will show how systems are identifiable entities and that those entities have interactions with other entities. Moreover, we will show how systems of subsystems form structural hierarchies of "levels of organization."

In Chap. 4 we demonstrate a powerful way to consider system structure as a relational network of components and their interactions. Networked structures can be represented abstractly, for example, as a graph, which then means powerful analytic methods can be brought to bear to identify and explicate subtleties within the system structure. Systems are networks of component relations that are physical connections, such as flows of material or energy. But they can be represented in these abstract networks as nodes and links, which are, in turn, representable in computer data structures and subjected to powerful computational techniques like graph search.

Large and complicated networks lead us naturally into the subject of Chap. 5, complexity. To a large degree, the coming to understanding of systems is a process of managing complexity. We start by recognizing the difficulties in coming up with a universally accepted definition of complexity, partly we think because there are so many seemingly different examples of complex structures or functions. We survey some common approaches to defining complexity but commit to a specific one that we think ties many other ideas in systems science together.

Systems have behavior even when we can't see it directly. Chapter 6 takes up the issues of system behavior or dynamics; what happens when systems move and interact with other systems? We look at both the outward behavior of systems and the inner behavior of components that give rise to the outward behavior.

Chapter 3
Organized Wholes

"A system is a set of things—people, cells, molecules, or whatever—interconnected in such a way that they produce their own pattern of behavior over time. The system may be buffeted, constricted, triggered, or driven by outside forces. But the system's response to these forces is characteristic of itself, and that response is seldom simple in the real world."

Donella Meadows, 2008 (Meadows 2008, p. 2).

Abstract We start with an overview of the main attributes of systems in general. These common attributes are found in all systems. They can be examined in the abstract as concepts or concretely in actual example systems. The overall concept of a system begins with the concept of an organized whole entity that has connections to other such entities as well as exists in an environmental milieu. We have shamelessly invented the word "systemness" to encompass these general attributes. Our first Quant Box provides a starting place for a formal definition of systemness, which will apply as we construct models of systems for formal analysis. But the rest of the chapter provides descriptions that expand on that formality in terms easy to understand without the math. The first Think Box introduces a theme that will run through the rest of the chapters—that of how the brain is a wonderful model of a complex adaptive system and how the principles and subjects of the chapters apply to understanding this remarkable system.

3.1 Introduction: Systems, Obvious and Not So Obvious

The principle of "systemness" is a starting place for understanding systems science. It is, in effect, a statement of universality of the following principles, introduced in Chap. 1, the root of a branching tree of those interrelated principles.

The essence of systemness is the existence of a "something," a bounded object. Systems are studied in many ways and for diverse purposes, and this gives rise to a corresponding diversity in ways of defining them. For a book such as this, we need a broad and flexible descriptive definition. For our purposes, a *system* is a whole of some sort made up of interacting or interdependent elements or components

© Springer Science+Business Media New York 2015
G.E. Mobus, M.C. Kalton, *Principles of Systems Science*, Understanding
Complex Systems, DOI 10.1007/978-1-4939-1920-8_3

integrally related among themselves in a way that differs from the relationships they may have with other elements. This difference is what allows us to recognize a boundary that defines the system as such. The boundaries of a system, however, are not as easily determined as it might seem they should be. In some senses, the boundaries are natural, observable, physical containers, such as a cell membrane or your skin. In others, they are less observable, determined as much by choices of the observer as by any natural "skin." Indeed, in the latter case, as we learn more about a system we study, it is possible that the choices of boundaries will change. As we will show shortly, the wholeness and integrity of a system are still very real, even in the case where an observer has chosen what is to be considered as within the system and what is to be outside, in the environment of the system. In all cases, systems are objects of interest with boundaries, and we identify them through boundaries.

At the same time, a system will be found to be composed of parts. And those parts have the same quality of objecthood themselves. Take a human body as an example. It is a whole object bounded by skin. It is composed of many subsystems, some of which are clearly objects themselves (e.g., the liver). And there are other components that have a more distributed form, such as the circulatory system. Even the latter, however, is an object in that it has a natural boundary, the tissues that form the tubes, arteries, veins, and capillaries.

The last example is also a good one to note the problem with boundaries and their relation to what exactly we count as the system of interest. The tubes of the circulatory system (along with the heart) do constitute an easily identified object/ system. As long as we are only interested in the circulation of blood, this might suffice. But blood is a complex tissue in its own right and not all of it stays inside the tubes! Except for where the tubes route through the brain, where a blood–brain barrier prevents anything other than oxygen, carbon dioxide, and low-weight amino acids and glucose molecules to transport across the membrane, the fluid known as plasma (itself complex) can ooze out between the cells forming the membrane and collect in the other tissues of the body. White blood cells, too, can get out. This complicates the issue of what exactly is the circulatory system. If our interest is only in how blood circulates through the tubes, then we can consider the tubes as boundaries. On the other hand, if we want to understand how components of blood that cross over that boundary circulate through the body tissues as well, then we have to include the lymphatic (sub)system as part of the whole circulatory system since it collects and returns plasma and white blood cells to the main tubular system. This kind of example comes up repeatedly when we study systems. It points to the role played by the questions we are trying to answer in determining what is in the system and what is out and what the boundaries are. This is the reason systems being investigated are referred to as "the system of interest," for the questions asked, i.e., our interest, enter into defining the boundaries of the system.

This caveat is not to negate the notion of systemness. It actually underscores the principle. Systems can have both real physical boundaries and seemingly arbitrary (though not really arbitrary) boundaries because all systems are always part of a larger system. Systems interact with one another in regular ways and in so doing act as components in a yet larger unit of organization. The key to understanding systemness is in the interactions between and among subsystems. Depending on what kind of understanding an observer seeks, she chooses a boundary such that some interactions are treated as inputs and outputs to the system of interest, while others, those among the components of the system, are treated as internal flows.

This may be one of the hardest intellectual challenges in comprehending systemness: boundaries can be physical on the one hand and conceptual, yet real, on the other hand. In this chapter we will try to explicate this seeming duality and show how it is a matter of perspective and choice of questions to be answered.

Physical
conceptual

Quant Box 3.1 Formal Definition of System

There have been a number of approaches to defining a system in a formal mathematical way so that formal analysis can be done. Donella Meadows' definition that led this chapter is informal but conveys some clues to what a formal definition might look like. In this Quant Box, we will describe a formal definition and show some ways in which it can be used to relate the various principles outlined in Chaps. 1 and 2 and more deeply described throughout the book. We will use *set theoretical* formalisms to define a system and then explain how we would use the notation to analyze the system more deeply. The rest of this book attempts to explain all of these elements in a somewhat less formal framework.

A *system* S_l is a 6-tuple:

$$S_l = \{C, N, I, B, K, H\}_l, \, l = 0, 1, 2 \dots m$$

where:

l is an index related to the level of complexity (Chap. 5). Level 0 is the top for the system of interest ($l = 0$). When C is decomposed and each c_i is treated as its own distinct system, $S_{l,i}$. The definition recurses with $l = 1$, then 2, and so on. See Chap. 12 regarding system decomposition.

C is a multiset[1] of component subsystems, $\{\{c_1, n_1\}, \{c_2, n_2\}, \{c_3, n_3\}, \{c_i, n_i\}, \dots \{c_k, n_k\}\}$, where each c_i is a *type of component* and may also be a subsystem but at level $l + 1$. That is, each c_i may itself be defined by a similar 6-tuple in a recursive fashion. Below, we explain how the recursion is prevented from being infinite (at least in a practical sense). Each n_i in the component tuples is an integer enumeration of the number of components of that particular type. These tuples, $\{c_i, n_i\}$, are multisets. As written here, their cardinality is one. However, in very complex systems, we can aggregate any number of variations of components of the same basic type (i.e., a category) into a multiset whose cardinality is the number of distinct variations identified, for example, the multiset of all alleles of a specific gene in a genome in a population. There can be many viable variations in the gene but then many occurrences of any one allele in the population.

(continued)

Quant Box 3.1 (continued)

N is a network description in graph theoretical terms. A graph, $G = \{V, E\}$, is a tuple containing two sets, V, the vertices or nodes, which in this case correspond to the components in C, and E, the edges or connections between the nodes (see Quant Box 4.1 and Wikipedia—Graph theory: http://en.wikipedia.org/wiki/Graph_theory). N provides a map of what components are connected to what other components at this level of complexity.

I is also a graph in which half of the nodes are not inside the system per se but are either sources or sinks or other entity types in the environment. That is, I contains elements of the environment that are relevant to the system. The other nodes are component subsystems within the system. I describes the connections a system has with objects in its environment using the same format as N (which is strictly internal connections).

B is a complex object that will vary based on details. Fundamentally, it is a description of the set of boundary conditions that maintain the system identity (see Sect. 3.3.1.1 re: boundaries).

K is also a complex object. This object contains the descriptions of how the various components interact with one another and with the environment (Chap. 6 covers system internal and external behavior, or dynamics). That is, it contains the state transition rules for all interactions. It is an augmentation to the N and I graphs. For instance, for each edge $e_i \in N$, there is an object k_i containing the parameters associated with e_i. For example, k_i could contain a tuple representing the capacity of a flow, the substance flowing, simply the strength of the connection, etc. A similar representation is provided for the graph in I where augmentation of the connections between external (environmental) elements and internal components is needed. K is called the system knowledge. It is both self-knowledge (i.e., internal structural and functional knowledge) and other knowledge to the extent it details the external connections. Note that the external objects are not, in this definition, modeled. That is, the system contains no explicit knowledge of the internals of external objects, only how they behave. See Chaps. 7–9—Part III to see how this can be extended in complex adaptive systems to include knowledge of the environmental objects.

Finally, H is a super complex object that records the history of the system, or its record of state transitions, especially as it develops or evolves. For example, brains learn from experience, and as such their internal microstructures change over time. This is called *memory* and the current state of K is based on all previous states. Some simple systems, like atoms for example, may have a NULL H; that is, there is no memory of past states. As just mentioned, on the other hand, brains (and indeed all biological systems) have very rich memories. H is an augment for K.

Each one of these elements is a description of some aspect of the system and all are needed to provide a complete description. Many of them, such as K, can be extremely complex. For example, in the case of the brain mentioned

(continued)

Quant Box 3.1 (continued)

above, K is actually distributed between the genetic inheritance of the individual, which prescribes the major wiring diagram and the major processing modules, and any culturally determined factors that push the direction of development (learning). This is why K must also contain augments for I. It is also why H may become extremely complex.

In Chap. 12, Sect.12. 3, we show how this definition, albeit in a less formal form, is used to identify and enumerate the systems, subsystems, and low-level components in a system. Essentially, the process of system decomposition fills in the details in each one of these objects. In that section, we also explain how to keep the decomposition from getting into an infinite recursion. In essence, a decomposition stops at a level $l = L$ when the components in S_L are all defined as *atomic* for the particular kind of system being analyzed. Atomic, here, simply means the simplest component, not needing further decomposition because its functions, etc. are already known.

By having a formal definition such as this, it is now possible to apply a great many mathematical tools to analyze many aspects of the system. The rest of the book will contain examples of some of these tools and how they are used.

1. A multiset is similar to a regular set but we relax the condition that all items in the set are unique. As used here and in Chap. 5, a system is comprised of a set of component types, but there can be many instances of each component type. See Wikipedia—Multiset, http://en.wikipedia. org/wiki/Multiset.

3.1.1 Systems from the Outside

It is often easy to perceive a system when there is an obvious physical boundary. We will say that everything inside the boundary is "in" the system, or a part of the system, while everything else is outside of the system, or in the environment. Physical boundaries simplify our ability to conceive of the system. We can look for instances of materials or energy or messages that flow from the outside environment into the system, across the boundary. And we look for flows that come from inside the system and move to the environment. This perspective is looking at a system from the outside and discerning its wholeness.

When perceived from the outside, a system or object is said to be a "black box." A better term would probably be an "opaque box" since this is really the essence of what we are trying to describe.[1] The formal approach to analyzing such opaque

---

[1] The term black box is a holdover from the reverse engineering field where engineers have to deduce the inner workings of a machine (the box) without taking it apart for fear that doing so would destroy its functionality. More accurately, we should call systems that we can only observe from the outside as opaque processes.

systems is to construct a monitoring boundary around the system. This monitor measures all inputs and outputs over time, and with the data, we try to infer several things about the system.

First, we try to infer if the system is behaving stably over time. Does it produce the same basic outputs given the same inputs over extended time? In Chap. 5, Dynamics, we will see how this works. Our ability to infer systemness from the outside comes from seeing that the object of interest produces consistency in outputs given inputs. In other words, the system performs a function in the sense that the output is a function of the input and is consistent (nonrandom). As we will see later in this chapter, consistency in functional relations implies internal organization; only organized systems can consistently transform inputs into consistent outputs. Conversely, finding random outputs given the pattern of inputs implies an internal lack of organization that can give rise to that randomness, this despite the existence of a physical boundary.

> **Question Box 3.1**
> When a system such as math has largely conceptual boundaries, do the ideas of "inside" and "outside" apply? Can math be a "black box" system?

There is another category of output behaviors, however, that on first appearances look like randomness but are actually not random. This phenomenon is due to nonlinear interactions between components inside the system. The components are organized in the conventional sense, but their interactions take on a characteristic random-like appearance called *chaos*.[2] The friction between the walls of a water pipe and the flowing water, for example, will at unpredictable points produce feedback dynamics that take over the water column so that the stream of water from a faucet suddenly changes from smooth regularity to chaotic turbulent flow.

In the case that a system of interest observed from the outside shows nonrandom behavior that persists over time, we can also start to infer some things about the internals even though we can't see them. Functions, the production of given outputs for given inputs, might be accomplished in many ways, but sometimes the physics of the system suggests that there might be a small range of possible mechanisms operating inside to produce the specific functions observed. Thus, we can "speculate" about how a specific system might work internally just by knowing something about its function(s). Of course if the system is chaotic, the process of inference involves some clever mathematics. But if it is statistically predictable, then it is easier to infer internals. In either case, we can then build *models* of the system's internals and use the model to explore other aspects of the system, such as how it

---

[2] Chaos in the sense used here is not the conventional purely randomness that most people think about when using that term. Chaos theory in mathematics and physics deals with a form of nonpredictability in systems' behavior that is not amenable to regular statistical methods. More in Chap. 5, but see http://en.wikipedia.org/wiki/Chaos_theory.

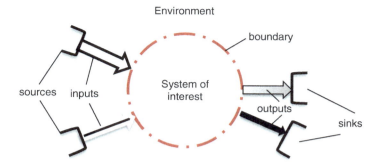

Fig. 3.1 The "system of interest" is situated in an environment where it interacts with other entities. The system is impacted by various inputs of material, energy, and information (*arrows*). The system also produces outputs of matter, energy, and information. The boundary of the object delineates what is inside and what is outside the system. Boundaries may be porous to allow entry of inputs and exits of outputs. The open rectangular objects represent the environmental entities, undefined sources (of inputs), and sinks (for outputs) that interact with the object

might interact with its environment under other than normal conditions. In Chap. 11, we will provide a formal approach to the various kinds of models that scientists use to make inferences about systems, such as the climate models used to predict how the climate might be changing under conditions of global warming.

Throughout the book, however, we use the notion of models extensively. Models are just abstract representations of systems that capture the essential features of that system. They can be as simple as diagrams or as abstract as mathematical formulas. Figure 3.1 shows a diagrammatic model that captures the essential features of the relationship between a system of interest and its environment. It is only meant to show the nature of a bounded entity, the flows of inputs through the boundary, and the flows of outputs through the boundary outward. The immediate environment consists of other un-modeled entities that are sources of the input flows and sinks for the output flows. The different kinds of arrows represent different kinds of material or energy flows.

In Chap. 6, Dynamics, we will introduce a more formal visual "language" or set of diagramming primitives that can be used to capture many more of the essential features of systems and their interactions. Until then we will use a less formal approach to diagramming that will suffice to provide the essence of the concepts we cover.

3.1.2 Systems from the Inside

There is a different and rather difficult situation when the observer of a system is actually *inside* the system. Consider the case of a human or any other complex adaptive system (CAS, covered in Chap. 5) that is also an observer trying to understand its environment. It observes the behaviors of other systems, treating them as black boxes, and begins to suspect that there is some kind of grander organization in which it participates—it suspects it is part of a larger system!

How does this observer formulate a model of the larger system in which the observer is enveloped? That is, of course, the problem that we humans have faced in understanding how our societies, our world, and the universe work. The problem of coming to understand the larger system is complicated by the fact that we must also come to understand ourselves as components of that larger system. As components, we modify the system of which we are part, and since we ourselves are adaptive systems, we also continually modify ourselves—and hence the system to which we belong—as a consequence of our observations. And the problem is made even more complex by the fact that we are a multi-agent entity (our species), each agent of extraordinary complexity and unique in many aspects.

Such considerations have given rise to the celebrated "postmodern" debate regarding objectivity and subjectivity. Modern science modeled its procedures on the outside observation of systems described above. This, however, ignores the very real fact that every observed system *includes* the observer, for without the observer, it would not be an observed system! The consequences of this more systematic understanding of the nature of observation are to relativize what once seemed a simple and clear distinction between outside and inside, objective versus subjective observations. There are differences, but they vary case by case and application by application.

Question Box 3.2
When observing one's own family, one must clearly include oneself as a component in the system of interest. Observing someone else's family, one is no longer part of the system of interest in the same obvious way. But does that make one an "objective" observer? Why not? If one cannot describe a completely objective observer, can you describe what might go to make some observers relatively more or less objective?

Science introduced the math-based experimental method as a way to correct for the differences of individual observers. The major efforts of the sciences have been to study other systems, first as black boxes and then as "white boxes," where they begin to produce models of the internal workings of those systems in an attempt to further understand the general phenomena in which those systems participate. This has been successful in proportion as the systems of interest have been subject to generally agreed upon standards and procedures of measurement, as in physics. The social sciences, for the reasons discussed above, have more reason to be wary of an oversimplified notion of objectivity. Even in the natural sciences, as the fierce debates surrounding global warming remind us, we pay a price if we do not take time to sort out the ways in which good "objective" science still must factor in the presence of observers in the system in which they have an interest. The fact that science unavoidably is done by interested observers does not destroy the difference between reasonably objective versus manipulated results. On the other hand, it is also useful to be always aware of the inevitable imprint of *human* interest: we get the versions of the system in which we participate of interest to humans, even when we do our best to remove ourselves and our projects from the picture.

> **Question Box 3.3**
> Granted the observer is always present in an observed system, this fact is more important in some cases than in others. Give examples of a few kinds of cases where you consider the observer's personal characteristics—apart from expertise (always relevant)—an important element. What are examples where the convention of "objective" observation works? What kinds of features make these cases different? Can we generalize about when we can ignore the observer, or is it always case by case or somewhere in between?

The perplexity of observing a system from the inside, then, comes ultimately from the need to include ourselves as components. And the role of the human component itself calls for a sort of outsider, black box to white box sort of analysis. As we seek to observe and understand ourselves, we, the ultimate insiders, face the conundrum of our own opacity to ourselves as we use our subjective powers to observe ourselves as objectively as possible.

3.1.3 Systems Thinking

The approach of applying principles of systems science has been dubbed "systems thinking." It entails conceptualizing the objects we encounter as systems and subsystems that are parts of larger systems. It turns out that systems thinking can be explained in terms of systems science because those entities of the mind we call concepts are actually systems of neuronal networks which have learned to represent the systems in the world that we encounter. We will return to this toward the end of the chapter.

For now, we want to provide a broad outline of systemness and show how all of the other principles interrelate to one another within that overarching concept. To do so, we start by discussing the key properties of systems in general and show the "hooks" on which the principles hang within that framework.

3.2   Philosophical Background

The roots of systems thinking in human history are deep. We provide a number of references for works that provide a history and overview of systems thinking so we will not dig too deeply here. However, some words to situate the field of systems science in the context of human thinking in general are in order. In this section, we provide a brief look at the philosophical background issues that are pertinent to systems science, particularly with respect to organizational aspects.

3.2.1   Ontological Status: Parts and Wholes

Ontology is the study of the nature of reality. More precisely, it is the study of what exists, what it is that makes up reality. As we have seen above, we have a strong tendency to model all aspects of the real world—activities, events, ideas, etc.—after the material "things" or objects that we perceive directly through our senses. Philosophically, this has given rise to a long and rich tradition of thinking in terms of substances, entities that stand on their own, existing in themselves. Early on, this led to the idea that more complex objects must be composed of more simple units, the "real" thing that exists in itself. The Greek philosopher Democritus (ca. 460 BCE to ca. 370 BC) followed this line of thought to suggest the smallest unit of reality must be the *atomos*, a Greek word meaning "not divisible" (a-, "not" temnō, "I cut"). Thus was born atomism, and with it a fascinating but never-ending discussion concerning wholes and parts and the "reality" status of each.

We will not take up that perennial debate. But several important and useful insights emerge from reflecting on this tradition. The first is the familiar notion that complex wholes are composed of parts, which in turn may be composed of further parts. And a follow-on idea may be even more useful: wholes at one level may consist of components that are in turn wholes on their own level (Fig. 3.2).

The question of a self-contained "real" unit turns out to be highly problematic, but it prepares us to think about different kinds of reality at different levels and in different modes of organization. Is a red blood cell—a bounded entity moving freely through ones arteries—a self-existing reality or a part of a larger self-existing reality, i.e., the person? Is a baseball team a reality or just a fictional whole made up of an aggregate of real individual players? Systems science will not argue the "which is really real" question in either terms but prefers to address instead questions such as the nature of boundaries, properties, and internal/external relations that constitute distinct but interwoven levels. When these features are adequately described, we find that wholeness takes on different meanings at different levels and with different sorts of boundaries and that any of these sorts of levels or wholes are real at least insofar as they entail consequences of a sort consonant with the system being described. Observable differences in organization make each of these kinds or levels of wholes fitting "objects" for systems inquiry.

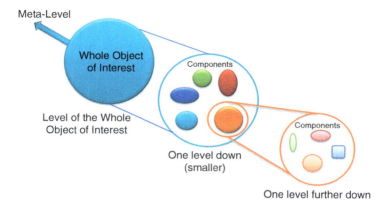

Fig. 3.2 A whole object is comprised of multiple components, which, in turn, can be seen as whole objects. At an even smaller scale, these may have internal components. Ad infinitum? In turn the object of interest can be found to be a component of a larger system at a "meta" level

---

Question Box 3.4

A living person is a whole object. McDonald's Corporation is another whole, as is the local McDonald's franchise at the shopping center. In what way are these all wholes? What do you see as important differences? How do these differences relate to their mode of organization?

It is important to note the statement that wholes are composed of parts needs to be paired with the proposition that wholes also may be more than the sum of their parts. A reductionistic analysis starts with a whole and breaks it down into successively finer levels of parts. But as we will see when we take up the phenomenon of the emergence of new properties or characteristics as systems become more complex (see Chap. 10), wholes on their own level have properties that are more than could be predicted simply by understanding the properties of each of the constituent parts or the sum of such properties. We can understand much about systems by breaking them down and studying the behavior of components, but then the wholes must also be understood on their distinctive level. One needs chemistry to understand biology, for example, but one will never understand what is going on in organisms just by studying chemistry.

The central questions of the ontology of objects (parts and wholes) involve the nature of matter and energy. But modern science, and particularly systems science, has "discovered" another aspect of reality which must be granted an ontological status, and that is information (see below). Information, and its scientific theory, provides a link to another major philosophical concern, the nature of knowledge, or *epistemology*.

3.2.2   Epistemological Status: Knowledge and Information

Epistemology is the philosophical inquiry into knowledge, what it is, and how we get it. In the present case, we are interested in what we can know about objects (and thus systems) and how we acquire the said knowledge. In Chap. 1, Principle 7, we mentioned the importance of distinguishing between *information* and *knowledge*. Here, we will introduce the ideas behind that distinction, and in Chap. 7, we will provide a much more complete view. Our purpose here is to clarify enough of the distinction so that much of what we discuss later in this chapter and in subsequent chapters will be clear.

3.2.2.1   Information

The world is full of data, correlations that can become information if (and only if) an appropriate interpreting element is available. Causal interactions in the world of matter and energy continually create correlations, raw data that can become information. A footprint or fingerprint is the physical effect brought about by a cause of particular shape operating with particular force at a particular time; the physical consequence is there, but *data about the cause* is likewise present in the new physical configuration. That is, the cause and its effect are correlated—the effect has potential information about the cause. But data requires interpretation to decipher it and turn it into real information; fingerprints and footprints carry information about their causes only to one equipped to interpret that data.

Since everything in the world is involved in this dance of mutual causal modification, the world is full of potential information; in fact, what we ordinarily call "world" is really information—visual, tactile, auditory, taste information, etc.—that allows us to interact with the world; because our world of sensory information is correlated with the world which it is about, it can guide our activity fittingly.

From the very origins of life, evolutionary process already strongly selected organisms for successful information processing (interpreting). Single-celled organisms read chemical signals that lead them to nutrients; DNA is a code that is read as instructions to build complex proteins; metabolic processes of all sorts are initiated or curtailed by chemical signals. Our eyes and ears are highly evolved organs that transmit modifications in light and sound patterns to an interpreting brain/mind that reads them as meaningful patterns carrying information about their sources.

In all of this, what is being read or interpreted are *differences* in the environmental medium that correlate with some item of interest, use, or need. Gregory Bateson,[3] one of the founders of cybernetics, thus described information as "a difference that makes a difference" (Bateson 1972). Not every difference available is read as meaningful in a given interpretive scheme. The letter "a," for example, may be written in quite different typefaces, but the lexical information content stays the same. On the

---

[3] See http://en.wikipedia.org/wiki/Gregory_Bateson.

other hand, if the discussion is the esthetic appeal of different typefaces, the different ways of shaping the "a" become the critical information content, and its lexical meaning is irrelevant. So, only differences that "make a difference" for the interpretive scheme count as information.

Qualitatively, then, information may be described as "news of difference." It is the quality of a message that a receiver gets that was not expected. The repetition of a message adds no new information because it is not different after the first receipt. On the other hand, pure difference does not become information because it does not have the patterning element necessary for interpretation to take place. One could not write (or read) with an unlimited alphabet. Interpretation is always a matter of reading differences against a background of expected patterning. We have the expected pattern for our own language, so the varied words and sentences we process are meaningful differences, i.e., they carry information. In contrast, listening to a foreign language we do not know can leave us somewhat amazed that anyone can get meaningful information out of that stream of noise. The difference between noise and information is having the pattern necessary to allow differences to make a difference. And this brings us to knowledge.

3.2.2.2 Knowledge

Knowledge is the base of patterned expectation against which incoming information can be interpreted. Such knowledge may be innate, that is, "hardwired," or it may be acquired through a learning process. Beetles may be born ready to deal with a beetle world, while humans survive only by learning and retaining true or veridical[4] knowledge. Our ongoing life experience is a continuous stream of manifold forms of information which in turn patterns and repatterns configurations and processes within the brain (see below). This patterning is the physical embodiment of knowledge, which is the ongoing and ever-adapting patterned expectation against which we interpret the world of difference continuously thrown up by sense experience and mental manipulation.

The question of possessing the interpretive patterns or expectations that allow differences to make a difference routinely leads to discussion of hardwired versus soft-wired ("learning") processes. Organisms that live a short time in relatively predictable conditions and reproduce in great numbers can do well with hardwired interpretive schemes; these are relentlessly checked for accuracy in a changing world by the process of natural selection. Those with longer lives dealing with more complex and variable circumstances and having fewer offspring emerge on an evolutionary trajectory involving more and more ability to learn and adapt according to

---

[4]Truth and the verification of truth are much disputed questions in philosophy. A common philosophical definition of knowledge is "justified true belief" or a belief one holds that corresponds with reality to a reasonable degree. For the purposes of this book, a reasonable correspondence is one that is sufficient to guide activity appropriately for the situation.

experience. Humans, at the cutting edge of this evolutionary development, possess brains hardwired for a remarkable range of further, experience-based patterning.

In creatures that live adaptively by accumulating knowledge, the flow of information continually shapes the more enduring, patterned knowledge, which itself in turn serves as the basis for interpreting the information. This feedback loop is the essential dynamic that keeps knowledge sufficiently correlated with the real world. The repetition of information (my living room furniture is kept being in the familiar place) reinforces knowledge and expectations, keeping a given model of the world in place. New and unexpected information proportionately modifies or may even challenge whole areas of knowledge.

Our subjective experience of surprise furnishes a rough measure of the degree of newness or difference carried by information. The greater the surprise factor, the more potential it carries for modifying the knowledge base. Really surprising news about the conduct of a friend, for example, is likely to challenge my former opinion and cause me to modify my knowledge base regarding that person. Surprise should occasion new learning, but the absence of surprise is quite ambiguous: it can indicate that one already knows a lot, so there is not much room for surprise (new information), or it can mean that one does not know enough to have any expectation—i.e., there is insufficient basis to interpret the new data.

Question Box 3.5
When we consider the difference between information and knowledge and the nature of the feedback loop between them, it reveals a lot about the flexibility and adaptability available with different modes of organization. Whether new information immediately modifies a knowledge base or whether the base is so deeply ingrained that it disallows any or most new information is a critically different organizational strategy. What sorts of natural or cultural environments foster one or the other of these tendencies? Is it always a strength to be highly adapted to a set of circumstances? Always a weakness? Is there a predictable sweet spot between flexibility and rigidity?

3.2.2.2.1   The Brain's Natural Tendency

We have a basic template in our brains for identifying patterns when we see them, even from the sketchiest exposure, say when we run across something for the first time. We can see the template in action reflected in our use of language. The word "thing" functions to mark a pattern as being something without us actually having much detailed knowledge of what it is. As we get more exposure and have an opportunity to observe more aspects of a thing, we may either recognize it as being like something else we already know about or start forming a new concept about something we didn't previously know about. Eventually, we find a better name (noun) to denote the thing and generally only revert to using the word "thing" in the shorthand way.

We are so adapted to pattern recognition that for us absence of pattern is described in terms that have a strong negative connotation: chaos (in the vernacular sense—there is a formal version of chaos to be dealt with in another section of the book), randomness, disorder, and disorganization. There is no identifiable system, and our minds do not deal with that state of affairs very well. So strongly are we compelled to see the world as organized that people see images in a scatter diagram of randomly placed dots and see in clouds the shapes of familiar objects.

3.2.2.2.2 Subjectivity: Conception Driving Perception

We have seen how our knowledge base serves as an interpretive framework for the ongoing stream of information and how the new information acts as a continual check, correction, and amplifier of the knowledge base. The section immediately above describes how the brain is actually patterned and repatterned or patterned more deeply with this knowledge. This constant feedback loop keeps our knowledge functionally in sufficient correlation with the character of the world about us to guide our activity. There is considerable leeway in the correlation between patterns and systems in the mind and in the world: we routinely learn to live surrounded by a wide diversity of opinion and, as controversy surrounding issues such as climate change makes evident, even scientific methods do not result in absolute agreement on truth. The same information inevitably is open to multiple interpretations as viewed from different knowledge bases. Interpretation is an inherently subjective element of the knowing process, based as it is on the knowledge patterned into our individual minds.

This subjectivity can be a strength, offering multiple viewpoints that can mutually inform and correct one another. We recognize that "group think"—whereby everyone gets coordinated into a single shared interpretive framework—is a weakness. But the term "subjective," as used in ordinary conversation, most often carries a negative connotation. The negative usage reflects a problem inherent in the circularity of the knowledge-information feedback loop: the knowledge already patterned into my mind and thinking is the necessary frame for interpreting and understanding incoming information. We know "open-minded" people who continually readjust their thinking in terms of new information, and we also are familiar with the phenomenon of "closed minds," which rigidly interprets the flow of information in a way that only reinforces the already existing ways of thought.

The closed mind is clearly vulnerable to getting locked into its own patterning, becoming out of touch with the world. The open mind is more in line with the inherent dynamic we have seen in the knowledge-information feedback, but such openness also requires something more. Giving too much weight to new information or the potential for new information can lead to such weak conviction concerning knowledge that timely action is replaced by unending hesitation. How can we find our way beyond the situation of either thinking we know too much or of not recognizing the solidity of what we do know?

3.2.2.2.3 Objectivity: Perception Based on Standards

The word "objectivity" indicates the notion of correct and accurate correspondence with the object being apprehended. In a naïve form, the idea is that the object itself totally determines the content of its apprehension, with no contribution from the knower. But all perception involves our interpretive knowledge base, so there is no such thing as objectivity that does not include this subjective component. This loop of information and knowledge is to a certain extent self-correcting. Functionality is its own standard: the world exacts a price from too deviant interpreters. But there is huge latitude beyond the requirements of survival, and by different ways of standardizing the perceptive-interpretive process, humans have historically created very different societies and ways of life. Scientific method has become one of the most widely accepted and consequential of these standardizations, and it distinguishes modern societies from all earlier forms.

"Scientific objectivity" has become a byword for duly critiqued and rectified knowing processes. Scientists do not simply magically put their own knowledge base aside, but they must find a way to both ensure its solidity and keep it open to continual reformation and growth through incoming information. They do this by standardizing processes in a way that enhances a central feature of all information processes: redundancy.

Redundancy means repetition; it comes in many forms and is our intuitive recourse for checking our accuracy and understanding. Repetition of the same event, the same experience, or even the same opinion is a powerful confirmation. Random variation confirms nothing, but repetition is a key indicator of some ordering at work. The cycle of the seasons, stages of life, the shapes of maple leaves never exactly duplicate, but each occurs with sufficient self-similarity that we recognize "another" one and are confirmed in our conviction they reveal something of the nature of the world, not just a particular condition of our local mental weather.

The methods of confirmation through redundancy are manifold, but in general, experience that cannot be repeated is suspect. We respect individual experience, but reliability is established through repetition. Even the report of unique individual experience is filtered for believability through repeated experiences of the veracity and accuracy of that person: an unknown stranger has a built-in credibility problem.

Scientific method self-consciously refines and standardizes the rather vague intuitive criteria of repetition. Standardized units of measure and standardized processes of measurement furnish a foundation for the precision of experimental methodology. The power of measurement to capture experiential process in numbers gives new exactness and rigor to the all-important confirmation by repetition. And the redundancy of the shared experimental process itself becomes a matrix for introducing deliberate variation, which can lead to a wide variety of differing (but repeatable!) outcomes representing new information. Once there is pattern (redundancy), variation becomes meaningful.

Science is very good at giving us a reliable account of constant features of a system/object, i.e., those features that can be confirmed through some form of repetition. Information is substantiated not only by repetition of experiments but by its conformity with a larger web or pattern of accepted knowledge. Falling in with a pattern is yet another form of redundancy. But then what do we do about dealing with novelty and newness? What do we do with the evolving emergence of the unprecedented? The common response to novelty is first to try to fit it in with familiar patterns, for the failure to repeat and confirm those patterns may either disconfirm the pattern or disconfirm the new discovery. Thus, scientific advance into new dimensions or new interpretations is routinely accompanied by a buzz of activity to introduce some form of replication, confirming the new by repeated observations and/or by finding a fit for it by tweaking accepted patterns.

With these considerations in mind, we can tackle the ways in which we human beings, as cognitive, sentient beings, understand systemness in the world. Systems in the world somehow have a correlated existence as systems in the mind. There are two aspects of this phenomenon to be investigated. The first is identifying actual properties of systems that we observe. And the second, and more subtle aspect, is seeing how our brains are also systems that have to construct the knowledge of other systems. In other words, we have to conceptualize both systemness and objects/actions that mentally embody that state. In the next two sections, we will take up each of these two aspects in turn.

> **Question Box 3.6**
> How would you respond to the assertion that science is subjective because it is a matter of interpretation and there are always competing interpretations, so it comes down to a matter of opinion?

3.3 Properties of Systems

We start our exploration of the principles by considering some general properties of systems that are universal. Explicating these properties will also give us an opportunity to see how the various principles outlined in Chap. 1 are associated with the nature of systemness.

3.3.1 Wholeness: Boundedness

The property of "wholeness" reflects our intuition that there are identifiable objects or things that are composed of parts that interrelate to one another in some kind of organized fashion. Let us first consider the property that lends weight to our perception of wholeness—the boundary of a system.

3.3.1.1 Boundaries

One way we know we are looking at (perceiving) an object is that we see a boundary that differentiates the object from its environment, giving perception of a foreground/background structure. Perceptually, we are able to differentiate between an object and its surroundings by the special properties of a boundary condition. Sometimes, boundaries are easily identified upon inspection—the role of a boundary "component" (like skin) is clearly ascertained as part of the object and, yet, itself being somewhat different in composition from the "inside" components. At other times, we can see an object as having an <u>inner cohesion between elements</u> that serves to delineate the boundaries of the object from its surroundings. A flock of geese, for example, is held together by internal forces or couplings that are not visible by inspection but seems to strongly compel the component parts, the geese, to keep together in flight formation. In this latter case, the coupling turns out to be informational and knowledge based.

3.3.1.1.1 Concrete Boundaries

A boundary is said to be concrete if its demarcation between the internals of an object and the surrounding environment is localized in space and time. Usually, we would expect to see a physical substance standing between the inside and outside. And, in general, there is a set of coordinates in space-time that mark the transition between outside and inside such that if we pass those coordinates, we "know" definitively that we are inside or outside of the object (Fig. 3.3).

Being inside or outside of a system is said to be a discrete event (of course passage through the boundary might take time and energy if it is "thick"). The concreteness of a boundary may arise from the outer component pieces being tightly coupled to each other and/or tightly coupled to something further inside the system. For example, a cell membrane is formed from complex molecules comprised of lipids (cholesterols) and phospholipids. Lipids and phospholipids strongly attracted to

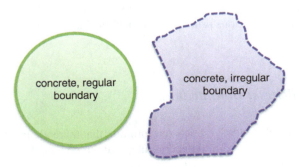

Fig. 3.3 Concrete boundaries are easily identifiable, even if the object is of an "irregular" shape. The boundary on the left is meant to convey a discrete enclosing substance (like a rubber balloon enclosing air under pressure). The boundary of the right-hand object is discrete, but porous (like a living cell membrane), allowing other substances to enter or leave the system

each other, but the lipids are water averse, so they are repelled by both aqueous external and aqueous internal environments. That is, some of the components of the cell membrane "prefer" to not interact with the water outside the cell and equally prefer to not interact with the water inside the cell. But they "like" to interact with one another (see Fig. 3.4 below). The result of this combination of attraction and repulsion is a fairly tightly held sheet of molecules that form quite naturally into spheres, the configuration that minimizes their interactions with water. This arrangement benefits the formation of living cells and helps to give them their identity as distinct objects. The terms "hydrophilic" and "hydrophobic" mean "water loving" and "water hating," respectively. They describe the way in which the molecules seem to act as a result of the forces. The terms "prefers," "like," and "hates" (see the caption below) are, of course, metaphors that relate to the idea that repelling and attracting forces are acting on the molecules involved.

More generally, boundaries can be formed at the periphery of a unitary system when there are components that have a natural ability to interact (say through attractive forces) preferentially with each other and/or with more inward positioned components. Figure 3.4 shows this diagrammatically. When the coupling strengths between components of a boundary exceed those with other components or external entities, then a physical boundary will form and maintain itself under the right conditions (see also Fig. 3.5).

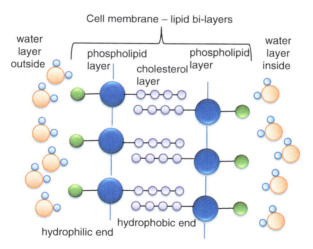

Fig. 3.4 A cell membrane demonstrates how repulsive and attractive forces can act to organize a system, in this case the boundary of a cell. The blue circles connected chemically to the green ones represent phospholipids (a combination of phosphate molecules and lipid molecules). The phosphate molecules are happy to be interacting with water molecules (water layers on the inside and outside of the membrane). The lipids (fatty molecules) "hate" water, which is why the phosphates are all turned into the surrounding waters. The purple tails attached to the blue lipids are another kind of lipid called cholesterol. The geometrical arrangement allows the tails to intermingle and be protected from the hated water molecules. Thus, the membrane is formed and remains stable in an aqueous environment. It should be noted that these kinds of membranes can be formed spontaneously when these kinds of molecules are introduced into a water bath. It does not require a living cell. This is an example of auto-organization

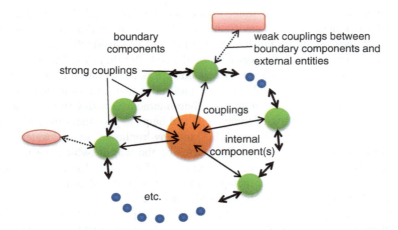

Fig. 3.5 Concrete boundaries are formed by the actions of strong coupling forces between boundary components and internal components. The couplings with external entities need to be weaker than the ones between boundary components

3.3.1.1.2   Porous Boundaries

Any real physical boundary can also have some aspects of porosity, allowing the import and export of materials, energy, and messages. The cell membrane mentioned above is an excellent example of this type. In living cells, there are large protein complexes that provide special "pores" or tunnels between the outside and inside of the cell. These pores, which are embedded in the lipid bilayer and penetrate through the membrane, have the ability to pass specific kinds of molecules, either into or out of the cell. This allows the cell to obtain necessary resources while preventing the incursion of harmful or useless molecules. These pores are active in that they open and close depending on the needs of the cell.

Doors and windows, as well as incoming water and gas pipes, outgoing sewage pipes, and incoming electric wires, are all examples of the porous boundary of a house.

3.3.1.1.3   Fuzzy Boundaries

Boundaries may not have the features of a discrete transition from outside to inside. Consider the atmosphere of our planet. As one ascends through the air column above a point on the surface of the Earth, the air gets progressively thinner until at some point it is so thin that we claim we have reached "outer" space where the environment is a nearly hard vacuum. At what point did we leave the Earth system and enter outer space? Such a transition is what we call fuzzy. At one point, say on the surface of the planet, we say we are definitely "in" and then at some higher elevation we are definitely "out." But the transition from in to out was gradual and difficult to specify with fixed coordinates. Fuzziness can be described mathematically (see the Quantitative section) but does not give us, as observers, a clear and clean definition of when we are in or out.

Fuzzy boundaries (depicted in Fig. 3.6) arise as a result of their substance not being sufficiently differentiable from the surroundings of the system. Often, as in the case of the flock of geese, the boundary is formed by some internal attractive force working on the component parts. The parts toward the outside are more diffusely connected to the center. In the case of Earth's atmosphere, molecules of gasses that make up the composition of air are held by gravitational force to the planet. However, gravity (and the electric force) falls off in attraction as the square of the distance from the center of gravity. Thus, molecules higher in the atmosphere, and hence farther from the center of the Earth, are more loosely coupled to the planet. In addition, these gasses tend not to chemically bond with one another or generally have strong electronic force attractions that might give them cohesion in spite of the weaker gravitational pull. Hence, the upper atmosphere is much more diffuse. Indeed, the concentration of gasses (number of molecules per unit volume) gets less as we ascend, and this defines what we call the fuzzy membership function. As we go up and the concentration of air gets less, we are becoming less "in" and more "out".

Fuzzy boundaries are abundant in biological, ecological, and social systems. In fact, in very complex systems, boundaries may be comprised of multiple "sub-boundaries." Take an organization like a company, for example. Every employee could be considered as "in" the company even when not physically inside the building envelope. Being an employee defines a kind of coupling between the employee and the company that extends over time. But is an employee an employee while on vacation? Or, for that matter, should we call someone an employee when they are at home, where they are clearly part of another organizational system?

As you can see, the concept of a boundary can get quite difficult and complex. Quite often it is necessary to selectively define boundaries of such complex systems for the benefit of defining the system of interest. For example, if we were to attempt

Fig. 3.6 Fuzzy boundaries are hard to pinpoint. Even though they can be described mathematically, they are hard to "measure" in any conventional sense

fuzzy, irregular boundary

to build a computer model of a company, we would probably ignore the fact that employees leave the building completely. Our model would probably be concerned with the financial performance of the company and treat labor as a large undifferentiated pool of workforce without concern for when someone is actually in the building working and when they are not. On the other hand, if we are looking at productivity issues, then consideration for when an employee is actually working versus in some non-company-related activity would be an issue.

Complex systems may have multiple boundaries, some of which are concrete, if porous, and some of which are fuzzy. Nevertheless, it is feasible to analyze the details of a system's relation with its environment in such a way as to discern (perceive) what belongs to the inside of the system and what does not and when.

3.3.1.1.4   Conceptual Boundaries

As described above, it is sometimes necessary to make decisions about what to call a boundary and when it might be functioning as a boundary for the system of interest. Some systems thinkers prefer to refer to such boundaries as conceptual. That is, they think of these boundaries as being truly arbitrarily selected for a matter of convenience, for the purposes of the specific kind of analysis that they wish to pursue. While it is useful to recognize the role of the mind in identifying and to some extent creating boundaries, we believe it is also important not to lose sight of the fact that even our conceptual boundaries are generally somehow anchored in the reality of the system of interest. Just as it is naïve to identify one-on-one the features of conceptual systems in the mind with features of systems in the world, it is also an oversimplification to treat them as entirely unrelated and independent. As we have indicated, and will describe in greater detail later in the book, concepts are themselves systems or bounded objects encoded in brain neural networks. They are, however, fuzzy in that the neurons that tend to fire together in the activation of a specific thought of a specific concept also may weakly activate related neural networks encoding similar or related concepts. This is why some thoughts automatically lead to other related thoughts. Each named concept has its own set of neurons that are co-activated when that concept is active, especially in conscious awareness. Thus, there is a correspondence between neural net activations in the brain and the nature of the object of the concept in the physical world. There is some subset of those neurons that correspond to the boundary conditions we are interested in from any given perspective on the system of interest.

Hence, the term conceptual boundary is apt in recognizing that a particular perspective on the boundedness of a system determines what counts (at that moment) as a boundary. But that doesn't mean that related boundaries from different perspectives are any less real at that same moment. A company employee may be bounded within the company when working in the building (or, these days, telecommuting) but also bounded by the strength of the employer-employee relationship (a contract) when at home or at the grocery store.

In the figure above, a conceptual boundary is created by treating the former sources and sinks (from Fig. 3.1) as modeled subsystems linked to the original

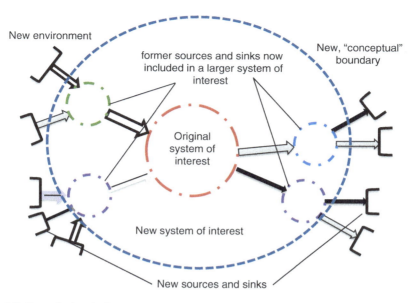

Fig. 3.7 By analyzing the former sources and sinks from Fig. 3.1 and treating them as subsystems, a new system of interest with a new boundary is created

system of interest. We have expanded the boundary (in Fig. 3.7) to include these subsystems within a larger system of interest based on our need to understand more about the relations between those sources and sinks, our original system of interest (the flows they provide or accept), and the larger environment. In the latter are more sources and sinks that provide inputs and outputs, respectively, to our old sources and sinks.

A crucial point to recognize is that although there is no physical boundary that is crisp and delineated in this figure, the blue dashed circle constitutes a *real* boundary by virtue of the strong links of inputs and outputs from the larger environment. These act effectively in the same way as the links between components in Fig. 3.4. The five circles within the blue circle have an ongoing, long-term functional relation with one another that we can now treat as a whole. Thus, we talk about the blue circle as if it is a boundary in the physical sense even though it is one that we "created" by including the four smaller subsystems into the picture. Thus, though this boundary is conceptual, it is also very real and has a physical embodiment in the form of the flow links from/to the environment.

Boundaries, as we see, can be simple enclosures or they can be multifaceted and difficult to pin down in space and time. Nevertheless, boundedness is an essential property of all systems. The relational organization we have called "systemness" can take many forms, including those that are fuzzy and hard to identify as one relational mode shades off into another. But the differences are real, even when difficult to specify.

Boundedness

3.3.1.1.5   Boundary Conditions

Once we have identified and/or defined a boundary for a system, we must then consider what conditions pertain to both the inside and outside of that boundary. We will have a great deal more to say about this subject when we discuss processes and dynamics (Chap. 6). Boundary conditions are those properties that determine, largely, what gets into or out of a system, that is, what goes through the boundary. The pores penetrating a boundary, as mentioned above, are examples of controlled entrance and exit of such stuff. Systems in general would be pretty dull if nothing ever got in or stuff inside never got out. Systems exchange material, energy, and messages with other systems. This is how couplings occur in the first place. This is how transactions are possible.

The nature of the boundary, for example, its porosity, will determine the flows of these three substances into and out of the system of interest. Thus, much of systems analysis begins with teasing out an understanding of what happens at that boundary. Whole systems are often described or named by the nature of the boundary conditions. For example, a "red" apple is an apple-shaped object that is of color red. This actually describes boundary conditions: the color of an object depends on what frequencies of visible light are reflected versus absorbed by the outer surface (the boundary) and shaped boundary of the surface emerges where those conditions change. There will be much more said throughout the book about boundaries and boundary conditions.

Question Box 3.7
Persons have boundaries described as "privacy." What sorts of flows are controlled at these boundaries? These boundaries are negotiated differently in different societies. What impact does this have on social organization?

3.3.2   Composition

A key property of any system is its composition. Systems are composed of subsystems and their interactions with one another, so describing these is critical for understanding the system. Subsystems, as we have hinted, are then composed of yet smaller subsystems. Below, we will take up the hierarchical structure of systems obtaining from this fact. Here, we will focus on the nature of the parts that comprise the subsystems and determine their interaction. We do this from a very general perspective, that is, from the perspective of what properties are relevant to all systems. But we will also provide some examples of these properties in real systems.

3.3.2.1 Components and Their "Personalities"

As we have seen, systemness gives us systems composed of systems, meaning that the notions of whole and part change in step with the level of our consideration. The term "component" can take on many different meanings depending on the resolution of observations made inside a system. Here, we will allow that at any resolution we will be able to see component subsystems as simply whole entities in their own right. That is, components, when we consider them "in themselves," are whole entities that have interaction potentials expressed in their boundary conditions.[5] We can treat each as a black box and need not necessarily know much about them as subsystems, i.e., decompose them into their components.

What we note, however, is that interesting systems, i.e., complex systems, are found to be composed of numerous kinds of components, each kind with its own unique *personality*. By personality we mean that components expose various kinds of interaction potentials at their boundaries. For example, atoms have valence electron shells; different atoms (elements) have different numbers of electrons in their outermost shells and have different levels of shells out from the nucleus. These differences give the elements different potentials for forming bonds with one another, that is, for making molecules.

Individual humans certainly display personalities that are an integral part of their interactions with one another. They also have talents and skills that allow them to interact with other entities in the environment. In a similar vein, organizations have distinctive personalities that come into play when they interact with, for example, customers, or other so-called stakeholders.

The core idea in describing the personalities of components is that the system is composed of various component subsystems, all of which have different capacities to interact with other types of components. The more types of components we find in a system, the richer the possibilities for interactions.

The shape features in the figure below illustrate a sort of relational geometry which is often a major factor in organization. For example, component D in the figure is a circle with three different arrow types. Its circularity implies that the arrows could point out in any radial direction from the center. On the other hand,

---

[5] Actually, we know that components have internal structure and complexity themselves precisely because they have different personalities. The different interaction potential types tell us that something is going on internally within the components, which means they have subcomponents that are interacting in different ways. A beautiful example of this is that atoms, which we once thought were indivisible, have component particles (protons, electrons, and neutrons) that interact in combinations that give rise to the atomic types. But it goes further. Each component of an atom, itself, is comprised of yet smaller components called quarks. Since there are multiple kinds of quarks, many physicists believe that even they have internal structures. No one knows where or if this "matryoshka doll" (also called babushka dolls or Russian nesting dolls) phenomenon bottoms out, though string theory might provide an answer. But since the pattern appears to be consistent in spite of what level we examine, we are content to start, abstractly, somewhere in the middle as if it were the bottom.

component B is a rectangle which may only show its interaction arrows at 90°
angles. Such geometrical considerations start to show up in molecular structures in
chemistry and play an absolutely critical role in chemical reactions of biological
interest.

Once we have categorized and inventoried the components in a system, we are
ready to analyze the structures that these components can form as networks.

3.3.2.1.1   Modularity Versus Overlap

A "module" generally refers to a component subsystem that has a fairly concrete
boundary and stable set of interaction potentials. The module may function as a
process, changing inputs into outputs, or as a structural element, holding multiple
other modules together to form a functional structure in the larger system.

In contrast to modularity, systemic overlap is a compositional strategy that
utilizes component subsystems that have more fuzzy boundaries and less
defined or stable interaction potentials. At times, these components may appear
to merge together, while at other times, they may appear as more distinct, individuated forms. A small office or factory where "everyone does everything"
exemplifies this kind of overlap. Systems frequently use overlap to provide a
fail-safe mechanism for especially critical functions, so if one component fails,
another takes over.

An alternative compositional strategy is to overlap the functions of a single component, often a way of enhancing efficiency or economy. A single module (subsystem) may serve multiple purposes at different times, as do stem cells, for example,
which are able to differentiate at different times and places into any sort of tissue
cell. Or a single module may perform the same function but in service to different
other "client" modules at different times. This is a type of overlap often used in the
writing of computer software, where along with modules which serve a single purpose in the system are others which perform routine subtasks that many larger
modules need services from. For example, a common need of many larger modules
is to output characters or graphics to the screen. Rather than each module doing its
own output (which would be unmanageable in a multiprocessing system), they send
a request to the operating system output module to do the task at the appropriate
time. All modules can share the output function and the output module can more
readily manage the process of serving so many requests.

3.3.2.1.2   Boundary Conditions: External Personalities

Under ordinary circumstances the inner workings of a subsystem component are
generally inaccessible to the other subsystems. More often than not we do have to
judge books by their covers. The boundaries are opaque, so to speak, making components black boxes to one another. All that any one subsystem can observe of
another subsystem is the input and output or force interactions displayed at the
boundary. For example, one person is not privy to what is going on inside the head

of another person. All they have to use to infer what *might* be going on is what the other person says or does in some particular set of circumstances.

The more complex the subsystem components are, the more degrees of freedom they have in terms of what they take as input or produce as output at any given instant. As human society richly exemplifies, the more freedom system components have in choosing actions to take in response to stimuli, the harder it gets to obtain a high level of regularity in how those subsystems interact. As we will see in Chaps. 7–9, and again in Chap. 11, very specific kinds of whole system-level internal control mechanisms need to emerge in order to coordinate and regulate complex component interactions.

> **Question Box 3.8**
> We frequently hear that as countries develop economically, the new middle class becomes a source of political unrest. How would you explain that in terms of the above discussion?

3.3.2.1.3 Homogeneity Versus Heterogeneity

Chapter 5 is dedicated to the study of *complexity* as a subject within systems science. There we will revisit the issues of system composition that contributes to complexity. Here, we simply note that one important attribute of a system, in terms of its composition, is the degree of homogeneity versus heterogeneity, that is, how many different kinds of components make up the system. A more homogeneous composition implies a simpler internal structure and function. Conversely, the higher the number of different kinds of components, with many more interaction potentials available, implies greater complexity in internal structure and function.

3.3.3 Internal Organization and Structure

Our current concern is simply to describe the relevant aspects of organization and structure that constitute systemness. The processes by which structures originate or change through time will be addressed later in Chaps. 6 (Behavior) and 10/11 (Emergence and Evolution).

Above, we covered characteristics of composition relating to the multiplicity of system components. In this section we focus on how those components are linked together into structures and interactions that produce the functional aspects of systems. *Structure* refers to the way in which components are stably linked to one another to form what we call *patterns* that persist over time. *Organization* is a somewhat higher-level concept that concerns not just the internal structure of the system but also how that structure leads to functions or processes. Here, we will introduce the relation that structure has with the internal dynamics of a system, a subject we will delve into much more in Chap. 6.

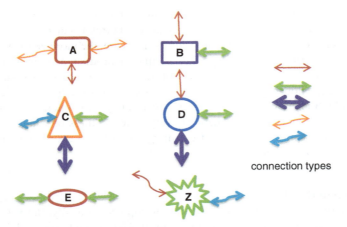

components and their "personalities"

Fig. 3.8 This is a highly abstract representation of the idea of component "personalities." Different component subsystems can have different personalities. Here, we represent six different component types using shape and color to differentiate the types. Connection potentials (either forces or flows) are schematically represented by *bidirectional arrows* of different *shapes* and *colors* too. Components can interact with other types of components if they share a similar *arrow* (*shape* and *color*) projecting from them as their personalities. Each component has different connection potential characteristics represented by the *arrows*. Component A could link with another A by virtue of either the *curvy orange arrows* or the *straight thin red arrow*. It could link with either B or D by the *straight thin red arrow*. Component E can only link with other components that have straight, medium thickness, *green arrows*, B, C, D, and other Es. Thus, the complexity of a system having many different component types with many different combinations of connection potentials can get to be quite huge. See Chap. 5 for more details about complexity

3.3.3.1   Connectivity

Figure 3.9 shows some of the components from Fig. 3.8 now connected to form a stable configuration. This is just a schematic representation of the idea that certain components can make effective (strong) connections with some, usually limited, number of other components.

This figure is highly schematic (abstract), but it captures the main features of what constitutes structure. Note that there are "unused" interaction potentials. The structure as a whole now has a composite personality derived from these various potentials left available. This structure, thus, may become connected to other structures to form larger-scale structures. We will see this aspect of connectivity developed further in future chapters.

3.3.3.1.1   Coupling Strength

A very important feature of connectivity involves the strength of the connection between any two components. The strength of coupling refers to the amount of change that will be effected in one component as a result of changes in another component.

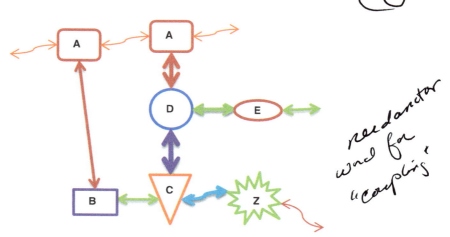

*redundant
word for
"coupling"*

Fig. 3.9 Components connect with each other according to their mutual interconnection potentials. Each component displays its individual personality and some aspects couple with other similar components

*forces
flows*

Coupling strength manifests in a number of ways (e.g., see the below discussion of forces and flows). Two components that are directly connected will typically display the greatest degree of coupling strength through their direct connection. However, there are variations that are important to understand.

First, if one component affects another component directly, the second component will affect the third component as a result. This is the chaining effect or a *transitive relation* among components. Since all components are, by definition, connected to all other components in a system, a change in one will propagate to all others to one degree or another, even if the change is vanishingly small or immeasurable. The strength between any two components can be said to be weighted by the number of intervening connections along with some factor for attenuation (or amplification). Figure 3.10 gives some sense of this.

In the figure, component A is not as affected by what happens to component B (input at 1) due to attenuation as the effect propagates through the connections and components. It also shows another case where the effect of D on A is actually boosted by the effect of C on D (input at 4 through effect 5). Such composite effects can be simple linear (additive or multiplicative) or nonlinear depending on the specific nature of the components involved. In Chap. 7 (Information), we will discuss several types of connectivity that can lead to increases or decreases in the strength of coupling, e.g., amplification.

Couplings may be transient and weak, being made and broken intermittently depending on the internal dynamics of the system. Or they may be quite strong and persistent. The overall organization of a system depends on the extent and strength of couplings between the component subsystems. In general, the stronger the coupling between components, the more stable and persistent the system will be. In Chap. 6, we will examine more closely the close relation of coupling strength and systems dynamics.

amplification

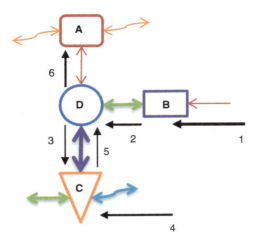

Fig. 3.10 The effects of changes in several components, caused by external factors, propagate through the connections between components. (*1*) A large change imparted to B; (*2*) the change in B is attenuated as the effect is passed through the *green* connection to D; (*3*) the effect is further attenuated from D to C through the *purple* connection; (*4*) another input to C; (*5*) an attenuated effect propagates to D through the bidirectional connection; (*6*) D combines the effects at *2* and *5* to produce a stronger effect on A. The latter case shows that effects can combine in complicated ways in such networks

3.3.3.1.2 Forces

Components may be connected by forces.[6] In nature we observe two fundamental kinds of forces, one of attraction between component objects and one of repulsion between component objects. We use attraction and repulsion to understand many sorts of organization in space and time. Why are components close together or far apart? Why do they move toward or away from one another? Or how do they maintain a constant distance in some cases? When these arrangements are not accidental, but due to connections, we usually appeal to forces of attraction and repulsion to explain the organization.

Physics discusses the organization of the material universe in terms of four forces that operate with different strengths and at different scales. The universe is literally held together by the attractive force of gravity.[7] The strong force binds protons and neutrons together or, at another scale, binds the quarks that comprise protons,

---

[6] Contemporary physics now theorizes forces in terms of exchanges (flows) of special kinds of particles, which would reduce the second form of connection to the first. But "force" is a well-established category in a broad variety of usages and we will retain it here.

[7] Thus far, the evidence that there might be a repulsive analog to gravitational attraction, called "dark energy," is sketchy, but it does make sense that there might be such. See http://en.wikipedia.org/wiki/Dark_energy.

neutrons, and other particles, collectively called "hadrons."[8] The weak force is associated with certain kinds of radioactive decay. And the electromagnetic force figures in the attraction and repulsion of positively and negatively charged particles, the general rule being that unlike charges attract and like charges repel.[9]

The physics describing the organization of our complex universe can be discussed in terms of these four forces. This illustrates another systems principle: very complex phenomena can arise through the operation of a few simple rules. Part of the secret here is what can emerge not from a single rule but from the interaction of the rules together. The nuclei of atoms are composed of subatomic particles called protons and neutrons that are held together by the strong force. Protons, which are positively charged particles, would ordinarily repulse each other (opposite charges attract) by the electromagnetic force, but the strong force overpowers the electromagnetic force to keep atomic nuclei from blowing apart (in fact occasionally some types of nuclei do blow apart in a nuclear fission reaction!). The electromagnetic force acts through two types of "charges" associated with different subatomic particles. Electrons have a negative charge, while protons have positive charges. In a normal atom of some element, there is a balance of protons in the nucleus and electrons in a cloud buzzing around at a respectful distance. For example, in the model of the simplest atom, hydrogen, the attraction of the electromagnetic force between a single positively charged proton and a negatively charged electron is just right to keep the energetic electron buzzing about the proton at an average distance, as illustrated in Fig. 3.11.[10]

The electromagnetic force mediates the attraction between these two particles and creates a condition by which, under ordinary circumstances, the system is exceptionally stable. Indeed, the universe as we know it could not exist if this weren't the case.

But how do we get heavier elements with multiple protons and neutrons? If left only to the electromagnetic force, the rule of like charges repelling would forbid any stable combination of protons. The answer is nuclear fusion, in which nuclei overcome repulsion when sufficiently compressed by the attractive force of gravity. The proportion between the repulsive electromagnetic force and gravity's attractive force in turn dictates the gravitational mass a star must achieve to begin the process of fusion burning. In some cases, the end of a star's fusion process results in a supernova, an explosion which scatters the newly formed heavy elements into the dust from which can form planets such as ours, made of complex particles such as carbon and oxygen and ready for interesting chemistry.

But how big is big enough to start fusion? Change the attraction-repulsion proportion in one direction and you get a very small, fast universe; change it the other

---

[8] See http://en.wikipedia.org/wiki/Hadrons for a basic explanation.

[9] For a basic description of "force" and the four forces of nature, see http://en.wikipedia.org/wiki/Forces.

[10] The real physics of atoms is much more involved than we can go into in this book. If the reader has not had a course in physics, a good general book on particle physics, the fundamental forces of nature, and the cosmological origins of atoms is Primack and Abrams (2006).

Fig. 3.11 A hydrogen atom is perhaps one of the simplest systems in the cosmos. It is comprised of two components, one proton and one electron connected by the electromagnetic force. The proton carries a positive "charge" while the electron carries a negative one. In electromagnetic force fields, opposite charges attract whereas similarly charged particles repel one another

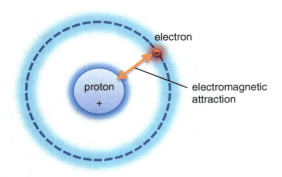

way, and it becomes so large and slow nothing of interest might have yet happened. The possibility of a universe of sufficient duration and complexity to give rise to life and intelligence hinges on some very fine tuning in the proportions of these forces.[11] Rules of forces that have to do with size and duration, by their proportioned interaction, turn in another systemic step into rules that determine the size of cosmic bodies and the duration of their processes!

On a seemingly quite different level of organization, we might ask, what keeps a family together? What causes some members of a family to, on occasion, want to get as far apart from one another as possible? Here again the same ideas of attraction and repulsion enter in to explain organization. The field of personality psychology deals with characteristics of individual people that give rise to attraction and repulsion. Everyone has had the experience of meeting someone whom they immediately find attractive. It could be sexual attraction or just a case of charisma. Whatever the reason, we feel as if some kind of force was compelling us to want to be closer to the other individual. And then there are those even more complex cases of combined attraction and repulsion, and questions of which overcomes which in what circumstances and what kinds of social clumping occur through shared attractions, shared repulsions, etc.

At first it seems ridiculous to be using similar categories to analyze interpersonal relationships, atomic structure, and the organization of the universe. But as we range through very different levels and kinds of systems, shared patterns organized in terms of attraction and repulsion emerge that inspire a broadly applied concept such as "force." When we speak of gravity, electromagnetism, love, and hatred as "forces" that attract or repel, there is no implication that they are the same thing.

---

[11] Fusion, supernovas, the emergence of heavy elements, and complex chemistry are common topics in descriptions of cosmic evolution. The fine tuning that allows a universe with the possible emergence of intelligent life, such as ours, is a special topic generally discussed as the "anthropic principle."

The sameness rather is in the patterned phenomena of directed motion and consequent forms of clumping and dispersion that we observe. We directly experience pushing and pulling, both physically and emotionally ("motivation" literally moves us!) in bringing about organization. However, the causal explanation of those forces is a different question, for it relates not to patterns of organization as such but to the particular level or nature of the system being observed.

Forces of attraction and repulsion may involve different and only poorly understood forms of causality, but they are often measurable and predictable and so serve an important role in understanding organization. It turns out, for example, that we really don't know why electrons and protons are attracted to one another. The assignment of a property we call charge to the particles is really a way of describing simply what they do. By careful experimental measurements of patterns of behavior, we have determined that the force of attraction or repulsion is reliable and consistently predictable, and we use the language of positive and negative charges as ways to account for these observed and predictable behaviors. In fact we really don't know what causes this phenomenon to occur! And among systems, there is a sliding scale of measurability and predictability, but that does not preclude the emergence of similar and consistent patterning. For example, even though the attraction between individual humans is much fuzzier, the outcome is proportionately as predictable as between oppositely charged particles. In fact, many of the online dating or matching services that are operating today rely on this predictability.

So whatever may be the cause or "force" that brings about the directed motion signified by attraction and repulsion, attraction and repulsion represent important organizing connections between components in all sorts of systems based on forces. The structural organization of cells is in part determined by the fact that lipids (fats) are repelled by water (Fig. 3.4). The foraging behavior of ant colonies is organized by the laying down of attractive scent trails to food sources. And humans build heavily on hillsides with attractive views and avoid areas with repulsive odors. As various as these phenomena may be, we can find in them a similar organizing principle that explains the observed connection among components.

Forces are real even if they operate at levels that seem esoteric. The lesson here is that attraction and repulsion, however implemented in the substrates, are to be attended to in understanding systems organization.

3.3.3.1.3 Flows

Flows of matter, energy, and information between or among components of a system constitute the second form of organizing connection. The organizational consequences of flows in, out of, and through a system are a complex topic that will be considered several times more as we proceed. Here, we will give it a preliminary look.

Flows are special cases of forces in which something (a material, or energy, or message) moves from one place to another and, specifically, causes changes in

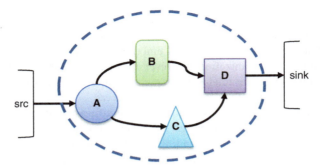

Fig. 3.12 Flows (of material, energy, or messages) between components constitute another kind of connectivity. "Src" refers to an un-modeled "source" of the flow; the open rectangle means we don't know how the source works, we only know that it supplies the flow. The "sink" is the receiver of an output flow. Again, it is un-modeled but should reliably receive the output flow

target systems.[12] Figure 3.12 shows a basic concept of flows between components in a system. Flows are characteristically unidirectional and are represented by single-headed arrows as shown in the figure. Counterflows are also possible and would be represented by a second arrow pointing in the opposite direction rather than a double-headed arrow. This is because the flow rates in either direction need to be specified separately.

In Chap. 6 (Behavior), we will explore the concept of a *process* as a fundamental way to think about systems in terms of how they process inputs to produce outputs. Here, we introduce the basic nature of flows in systems as a form of connectivity. Figure 3.12 shows a representation of a system receiving input flows from various undefined sources and producing outputs that flow to undefined sinks.

The system in Fig. 3.12 is a process that converts inputs into outputs. Component subsystems, of course, have this same structure where many of the sources and sinks are actually other subsystems. The strength of coupling in the case of flows is usually measured in terms of the "units of stuff" per unit of time, such as when we measure electric current as the flow rate of electrons in a conductor. But strength is a multidimensional concept; it can be gauged in a variety of ways, such as by the necessity of the flow, that is, how important the flow is to the normal function of the system of interest, or even by the reliability of the flow. Other, more complex characterizations are possible and we will see examples throughout the book (Fig. 3.13).

In this figure, we show some basic conventions that will be used to describe complex systems. Essentially, all complex (and dynamic) systems receive inputs of

---

[12] Physics treats forces as moving (accelerating) material. We used force above in a relational context with explicit reference to its vector, introducing separation or convergence, repulsion or attraction. In the case of flows, we have a specific interest in the change they bring about in the recipient or object system.

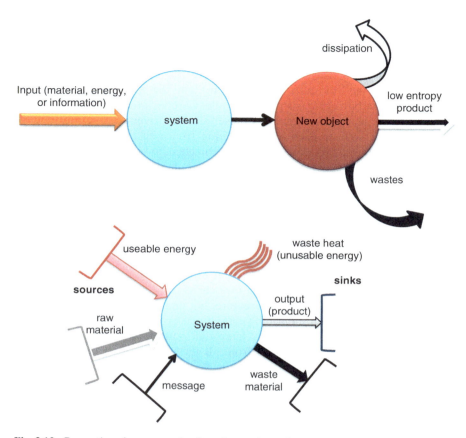

Fig. 3.13 Conventions for representing flows into and out of a system recognize physical aspects. Sources and sinks are shown as *open rectangles* that are not modeled but known to produce or receive flows, respectively. Generally, the convention calls for showing flows from left to right when the system of interest is the focus of analysis as shown here. *Colors* and *arrow* shapes are somewhat representative of the kinds of flows. For example, "raw material" might represent medium entropy matter, such as a metal ore. The output "product" represents much lower entropy material (e.g., purified metal ingots), whereas "waste material" (garbage, e.g., slag from refining ores) is a very-high-entropy matter in that it cannot be used for anything directly by the system. Similarly, energy is shown entering a system as a "useable" form and exiting as high-entropy waste heat. These conventions will be refined in Chap. 6

"usable energy" and what we call "raw material." They also receive inputs of messages which may or may not be informational (covered in Chap. 7, Information, etc.). Useable energy is that energy from which useful work can be obtained. Internally, the system is using energy to do work on the raw materials, thereby producing products (very-low-entropy material) along with waste heat and waste material. The production of waste heat is a natural consequence of the second law of thermodynamics, which we will come back to in later chapters.

In a semi-closed macro-system like the surface of the Earth, essentially, all outflows from systems become inflows to other systems, even if they pass through long periods of storage. On the scale of the whole Earth, we consider interactions or flows between the atmosphere (air system), the hydrosphere (water system), the lithosphere (rock system), and the biosphere (life system). The so-called carbon cycle is a good example. Carbon, which is the most essential atom in life, cycles through all of these "spheres" over a span of millions of years, spending time in each before flowing from one into another. Being a cycle, we can start at any point. Carbon permeates the biosphere, cycling back and forth into organisms from air and water and moving in organic compounds from body to body up the food chain. It is so shared around that it is possible that you have at least an atom or two of carbon in your body that once was in the body of Julius Caesar! That atom may get combined with oxygen to form CO_2 that you breathe out into the atmosphere. Plant life is a biological sink that reabsorbs CO_2, but if you live near the seashore, it is possible that within your lifetime that particular molecule may end up dissolved in the ocean, another major sink for CO_2. In the hydrosphere, it might undergo a chemical change where the carbon atom becomes part of a carbonate ion and then reenters the biosphere in the shell of some sea creature that upon death sinks to the depths of the ocean. Over several thousand years, it gets covered over by silt and then mud and then sand. A million years later, that portion of the sea bed, having been subjected to higher pressures from all the weight of sediments atop it, turns to rock. Ten million years later, the veritable crashing together of two tectonic plates causes that area to be raised into dry land and a mountain range. Then after another several million years, rain finally erodes the rock, and the carbon atom that was once in you and in Julius Caesar a few centuries before that is freed from the rock and ends up in some silt deposited by a flood in a fertile valley where a grass plant absorbs the carbon in its mineral form, thus bringing it back to the biosphere.

Essentially all of the molecules and atoms needed by living systems flow through cycles such as this, some much shorter and some even longer. In a systems perspective, we can recognize the great reservoirs of water, air, biomass, and rock as storage areas, systems that are themselves affected by the presence of their constitutive matter and at the same time systemically interrelated by the flows among them.

Throughout both the natural world and the human-built world at all scales, we find flows that connect together subsystems as components of the larger system.

3.3.3.2 Systems Within Systems

Principle 1 (Chap. 1) says that systems are composed of subsystems at all scales. That is, a system can be decomposed (in principle) to expose multiple subsystems. Those, in turn, could be decomposed into their subsystems.

The above figure represents this composition aspect. A larger system is composed of a number of smaller subsystems. The system as a whole has organization based on interconnections between the various subsystems. The coupling strengths

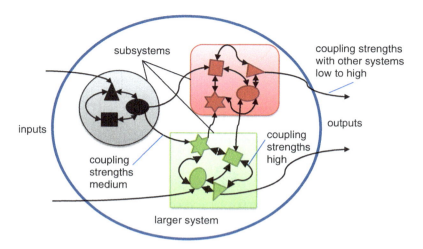

Fig. 3.14 Systems are composed of subsystems. Here, a larger system receives inputs from the external environment (other systems) and produces outputs that go to other entities in the environment. However, the larger system contains several subsystems that exchange flows. Those subsystems, in turn, have components (sub-subsystems) that are strongly coupled

of these connections are, in general, stronger than those between these subsystems and external entities (not shown). As discussed previously, this relative difference in coupling strength accounts for why this object is a system, even if it is a subsystem within a yet larger system by virtue of the inputs and outputs shown.

Each of these subsystems is composed of yet smaller components. These, in turn, are subsystems that are very strongly coupled with one another, thus creating the systemness of the member subsystems of the "larger system (Fig. 3.14)."

For example, suppose the larger system is society. Then the component subsystems include people, organizations, artifacts, etc. We can further decompose a person (at least in principle!) into organs and tissues. These are composed from cells and those are composed of molecules. Molecules are composed of atoms, and so on it goes down to components as yet to be understood. Although comparisons can be difficult (e.g., the relative forces of love and gravity), in general, as noted in Principle 4, the dynamics vary with differences in spatial and temporal scale. In general, the forces and flows connecting components are stronger than at the next higher scale, and this gradation across scales gives integrity and "wholeness" to systems and subsystems.

3.3.3.3 Hierarchical Organization

The fact of systems being composed of subsystems gives rise naturally to the notion of *hierarchy*, as expressed in Principle 2. Using the system components presented above in Fig. 3.14, the component hierarchy is represented in Fig. 3.15 below (as an inverted

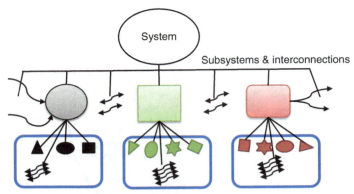

Fig. 3.15 Any complex system can be represented by a hierarchical structure, what is technically called a "tree." The root (*top*) of this inverted tree is the whole system (as seen in Fig. 3.13 above). As already shown, this system can be decomposed into component subsystems (*gray*, *green*, and *red* shapes) along with their interconnections with one another as well as the other entities outside the system. Those subsystems, in turn, are decomposed into their components and the interconnections between them. Tree abstractions will be revisited in the next chapter on network structures

Fig. 3.16 Less modular (more overlapping) subsystems can still be represented in a tree structure

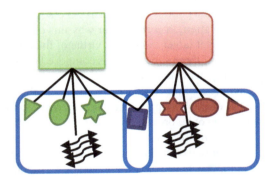

"tree" structure. A hierarchy such as this is an abstraction of the system, but it has its basis in the reality of composition. The system we show here is modular (as defined above). That is, all of the subsystems are unitary, having reasonably well-defined boundaries. Overlapping subsystems could be shown as sharing one or more components at a lower level. Figure 3.16 shows a similar set of subsystems in an overlapping composition in which one component is shared. This is the case when the coupling strengths between other components in each subsystem are relatively strong so that the shared component would have to be considered part of the two subsystems.

3.3.3.4 Complexity (A Preview)

Hierarchical structures are tied intimately to complexity (see Chap. 5). Our intuitions about complexity come from the observation of systemness already discussed, namely, the hierarchical structure of the system of interest. Recall Figs. 3.13, 3.14, and 3.15 above. These figures show the concept of hierarchies of subsystems within a larger system. Using even just our intuitive notion of complexity, we can see readily that the more levels in the structure (height of the tree), the more complex the system becomes. Combining this structural organization with the heterogeneity of the subsystems (and their sub-subsystems, etc.) gives a good sense of realized complexity.

The subject of complexity is actually far from settled in terms of definitions accepted by everyone working in the various subfields of systems science. There is, in fact, a whole subfield titled complexity science that engages in a lively investigation of the concept and its relevance to other fields of knowledge.[13] Principle 5 calls attention to the variety of types and levels of complexity. We will be particularly interested on how systems grow in complexity, maintain it, or lose it, for complexity exists in a temporal flux with these parameters. Potential and its realization are ways of assessing the stages of this temporal process, for the realized condition of any system also contains its potential for the future. So for our purposes, it is useful to define two basic categories of complexity: potential complexity and realized complexity.

3.3.3.4.1 Potential Complexity

"Potential" complexity[14] is an indexical function of a bounded system which takes into account the number of kinds of components, the number of kinds of *possible* interactions between all combinations of component types, and the number of objects of each type of component contained within the boundary. Imagine a huge sphere containing dozens of chemical elements (like hydrogen or oxygen), each element being represented by billions and billions of atoms per type, and assume for the moment that this mixture has been thoroughly mixed up. Technically, such a system, if at thermal equilibrium, would be characterized as maximally entropic (maximum disorder). This would give you an idea of what we mean by potential complexity. We could derive a single index number that took all of these combinations into account and gave us a measure of complexity from this standpoint.

However, there is an important consideration to understand before we could allow such a system as being potentially complex. At equilibrium, there is no energy differential among the components, whatever their theoretical connectivity, to do

---

[13] See Mitchell 2009. Chapter 7 deals with this definitional problem (p. 94).

[14] This concept is borrowed directly from Thermodynamics and consideration of systems in the equilibrium state or maximum entropy.

the work required for connections to happen. So there must be a source of the correct kind of energy that *could*, at least in principle, be used to push the system far from equilibrium. We return to this in the chapter on complexity and again in the chapters on emergence and evolution.

Question Box 3.9
Which has more potential complexity, a lego set of 20 pieces or a picture puzzle of 1,000 pieces? Why?

3.3.3.4.2   Realized Complexity

"Realized" complexity comes closer to what most people probably consider the word to mean.[15] We readily recognize the complexity of systems in which numerous components are interconnected and function together. For example, if you found a bag containing a bunch of pocket watch parts all jumbled up (potential complexity), you would not consider this to be a very complex "thing" because it is highly disordered. Whereas if you pulled a fully assembled and working watch out of the bag, you might be inclined to think of it as a complex object.

The structural complexity of the assembled watch was actually inherent in the parts. There are only so many (actually very few) ways in which the parts could go together to produce a working watch. But somehow, the pile of parts does not seem truly complex. This is why we need to differentiate between these two kinds of complexity. In the first instance, all that needs to be added to the mix is the skill and time for a watch maker to put the parts together in the right fashion to produce the watch. That is, the watchmaker has to do work on the parts in order to obtain realized complex organization from the potential complexity of the parts.

The movement from potential to realized complexity will require careful and extensive consideration. It not only highlights the thermodynamic framework of systems science, it also introduces us into the nature of process, emergence, and evolution. In some respects, we constantly assess potential connectivity as we deal with every situation (always processes!), from crossing a street to eating a sandwich. In other ways, potential always moves forward beyond our grasp as realized connections make entirely new forms of connectivity possible. With hindsight, it is evident that students in a university taking a course on systems science were one of the potentials, now a realized complexity, of the universe. It is a challenge for systems science to figure out, with hindsight, the process of transforming, evolving connectivity that has brought this realization. For an observer studying systemic potential just after the big bang, such a potential would be deeply buried in layers of

---

[15] Again we borrow from physics. Here we are referring to systems far from thermal equilibrium, where energy is being supplied to do the useful work of constructing linkages between components to form meta-components.

modes of potential connectivity (chemistry, biology, sociology, economics, etc.) that would evolve over billions of years. From the array of potential complexity, some realized complexity emerges, with its own array of further potentials and some realization with yet further potentials.

> **Question Box 3.10**
> So where does our very complex world system go from here? What are some of the factors you would consider in assessing the potentials and their relative likelihood?

3.3.3.5   Networks (Another Preview)

The reader will no doubt have noticed that many of the diagrams presented so far consisted of discrete objects (shapes) connected by arrows or lines. These are representations of systems comprised of components and their interconnections. It turns out that all systems may be described as organizations based on networks of components and their various interconnections/interactions through forces and flows (Principle 3). Thus, we consider network organization as a key concept in systems science. Systemness means that everything is ultimately connected to everything else, even if the coupling strength is infinitesimal. Finding the nontrivial connections, including their characteristics, and construction network maps that represent these (components and connections) is a major first step in understanding systems. Throughout the rest of the book, we will be working with network representations of systems as conceptual maps of those systems in the real world. And the next chapter will explore the nature of networks and their use in understanding systems structures more thoroughly.

3.3.3.5.1   Function and Purpose

Let us focus for a time on flow interconnections and more specifically on the behavior of systems as components in larger systems. Typically, systems receive flows of material, energy, and information, as already described, and produce output flows of wastes and products (or movement).

The transformation of inputs into outputs is a key characteristic of a system and a function of its organization. One way we know we are observing a system is by noting a consistent production of predictable outputs given the observed inputs. This works for material systems (matter and energy) as well as conceptual systems (mathematics).

A function is a process that transforms inputs into outputs. Every process performs a function. Another way to view the organization of a system is to look at the structure of functions. This too will look like a tree. Here, however, we must keep track of the functional relations, namely, inputs and outputs. The hierarchy of functions is very

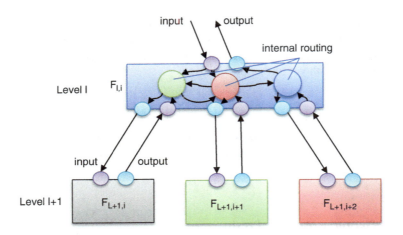

Fig. 3.17 Functions can be organized in a hierarchical manner. Here, the function $F_{1,i}$ at level 1 requires the services of functions at level $1+1$ (i through $i+2$). Internally, the function uses routing mechanisms to control the sequence of functions used. Depending on the exact inputs at some particular time to $F_{1,i}$, the routing can be different at different times. The *green, red,* and *blue circles* inside the function perform logical operations rather than transformative ones. However, the effect of routing its inputs to different functions at level $1+1$ in different ways results in $F_{1,i}$ being responsible for providing functionality to one or more functions in level $1-1$. The smaller circles at the edges represent input receivers and output senders

similar to the structural hierarchy because functions are, after all, performed by processes. Figure 3.17, below, shows a function hierarchical tree. Here, we show the inputs and outputs in a generic form rather than as strength-weighted flows. Of course the arrows in this view correspond with the flow arrows in the previous figures. What is different in this view is the explicit notion of modifiable routing of flows to perform different functions at different times depending on the exact form of input. Recall that in Fig. 3.16, we showed the existence of a component that was shared by two subsystems. The functional view allows us to represent this kind of sharing more explicitly. For example, in the figure below, we could show two input arrows to the $F_{1,i}$ (top) function and two outputs, both going to different other functions in the next higher level. So this one function could serve different higher functions at different times and even perform different versions of its function(s) based on those differences. Or it could provide the same function for multiple higher functions.

Functional decompositions and resulting tree organization diagrams like these have found their greatest uses in the computer industry in both hardware and software design. However, the procedures for decomposing functions in this manner are also used in other forms of engineering design (see Chaps. 13 and 14 for more details on functional decomposition).

Purpose is a word that is related to function in systems thinking but is generally reserved for those systems where we can speak of something being *aimed at*. The aim of human engineers is evident in the systems they design, and in such systems and "purpose" are used almost interchangeably. For example, an automobile serves the purpose of transporting people and goods, and a computer has the purpose of computing for a wide range of applications. Although function and purpose in such cases seem almost identical, the difference introduced by an aim at some functionality emerges when we think of improving a system (Principle number 12): something must be *aimed at*, or all we have is variation, not improvement. Over the years, we have greatly modified the functions of our automobiles and computers. These modifications are definite improvements for us, where the aim resides, but mere functional variations for the machines, which (we presume) are not aiming at anything in particular.

Aim or purposefulness is a characteristic of our consciously directed activity, but consciousness is not required for aim. On a more basic level, life is a self-maintaining process: conscious or not, the complex metabolic functions of organisms would be difficult to comprehend apart from their purposefulness in maintaining life. And metabolisms functionally interdepend in ecosystems and over time evolve improved functionality; thus, life evolution and ecology are further areas where function is conjoined with purpose. For example, in ecology we might ask questions like "what is the purpose of a keystone species [16] in maintaining the stability of a particular system?" Here, the keystone species plays a functional role by balancing some resource aspects of the ecological system. Or we can ask "what is the purpose of the giant, colorful feathers on a peacock?" Why did these adornments, which certainly make the peacock more vulnerable to predation, evolve? Evolutionary biologists contend that these feathers are used to attract mates. Male peacocks are in competition for mates, and females choose a mate based on the show he can put on, so fancier feathers serve a purpose in giving their owners a functional advantage that is transmitted to their offspring in proportion to its success.

As systems evolve across the threshold from nonliving to living, we are confronted with the emergence of systemic functions and processes which apparently function with aim or purpose. How (or even whether!) this happens has been a matter of intense philosophical controversy. From a systems perspective, purposeful function is a phenomenon that emerges, at the very least, at the threshold of life, and it is a matter for systems science to try to understand and describe how such a critical systemic emergence can take place. We will discuss this question in greater detail in the third section of this book when we take up evolutionary process.

---

[16] A keystone species is any single species in a particular ecosystem that has a large impact on the stability of the ecosystem. Take the species out of the system and it usually crashes or radically remodels to find a new mix of species and relations. See http://en.wikipedia.org/wiki/Keystone_species.

3.3.4   External Organization: System and Environment

We have considered at length the internal organization of systems and the types of connections that organize their constituent systems. Insofar as virtually every system is a component of a yet larger system, much of what we have said about relations among components applies in general to relations among systems. But new considerations arise when we look at any given system as a system in itself rather than as a system in a larger system. In that case, the most relevant parameters become characteristics inside the system and related characteristics outside the system, or, in other words, the system and its environment. The question is the nature of the relationship between these two.

3.3.4.1   Meaning of Environment

One level of analysis considers systems in terms of their internal organization. Looked at only in this way, one might get the impression that systemic wholes somehow exist in themselves. But the laws of thermodynamics are such that systemic order arises and is maintained against spontaneous decay (entropy) only by work, which means some sort of flow of energy, matter, or information from outside the system is required, and input is always accompanied by output. Input comes from somewhere and output goes somewhere, so sources and sinks are fundamental terms for understanding the meaning of environment. Figure 3.13 above depicts the universal systemic condition of existing context within flows from environmental sources with eventual output that goes to some sort of environmental sink.

Systems that maintain themselves by a constant input–output flow are called "dissipative systems," because they continually take in and dispel matter, energy, and information. Because dissipative systems are essentially patterns that emerge and subsist within a flow, the interdependent linkage of inside and outside is especially clear in such systems and corrects the common perception of environment as simply the surrounding context. The whirlpool that forms as water flows from the bathtub is a common example. But so are all living organisms and all the subsystems (economies) they create in support of their life. All life takes in matter and energy—"food"—from its environment. This is used for systemic self-maintenance in the metabolic process and output in the form of various activities and waste products. If sources dry up or the capacity to take in the waste or by-products of activity is exhausted, the life organized in the flow from source to sink cannot subsist. Because it must work in terms of available sources and sinks, just as whirlpools must fit and feed off of the properties of water, these environmental features are reflected in the internal organization of the organism. Lungs arise only with free oxygen in the atmosphere, gills with oxygen available in the water, and sulfur-based metabolisms point to an environment of subsea volcanic vents.

Environmental flows into and out of all sorts of systems are key to understanding the behavior of the system. Even systems we do not think of as dissipative are elucidated by these considerations. For example, it takes an input of work (energy), matter, and information to get matter into the thermodynamically unlikely (i.e., nonrandom) shape of an automobile. The systemic pattern of your car does not take in a flow of energy to maintain itself, so random flows eventually break down the precise order in the far more likely random direction. Your car will seem to disintegrate all by itself but will never reassemble itself without the external hand of a mechanic introducing more matter, energy, and information input. No system simply "sits there," independent of its environs, and the attempt to understand any system therefore leads beyond the boundaries of that system to the larger relational matrix, the "environment," within which it subsists for a time.

The boundaries that distinguish an object are the meeting place of that object and an impinging, encompassing environment. And since they mediate all exchange and relationship between inside and outside, boundary conditions are critical to understanding a system's relation to its environment. The internal organization of objects can be maintained only by excluding what would break down the organization and by permitting flows that support and foster it. This is another way of saying that systemic organization involves the emergence in space and time of constraints that allow some sorts of flows from the environment but not others. The variable nature of these connections with their environment is thus reflected in the variety of boundary conditions that determine various sorts of objects.

3.3.4.1.1 Environmental Flows and Messaging

We have seen that, with a change of perspective, components become whole systems in an environment—an environment which also hosts other systems (fellow components, at another level of consideration). Interactions among all these fellow participants in an environment are of such major importance that when we speak of "the environment," it is often shorthand for this web of interaction, especially when we are speaking of living systems. We have covered the interaction aspects between components and systems in terms of physical flows and forces. But to have a full appreciation of the nature of the environment, we need to cover critical issues in the way information gets around, for this is the critical guidance function that shapes all interaction.

Information flows are always mediated, carried by some kind of flow of matter or energy. For instance, visual information arises only in environments of reflected light and sound only when there is air to carry sound waves. The environmental conditions that constrain and shape flows of matter and energy thus critically shape the informational flows that connect the subsystems that share the environment.

The laws of physics are the most basic of such conditions. Most often we look simply at how the physical conditions of a given environment limit the possibility of relevant physical flows, as when a cave blocks light or a vacuum prevents the propagation of sound waves. But there are additional ways physics conditions the flows that carry information. The most important of these may be the principle of diffusion, which gives rise the phenomenon of gradients. Gradients, as we shall see, form a kind of environmental information topology in the world about us.

3.3.4.1.2   Diffusion and Gradients

Diffusion is the second law of thermodynamics at work, in the sense that substances in a concentration will always attempt to disperse or equilibrate within their boundaries. In essence, a product emitted by a system into its environment will dissipate outward over time and become more dilute the farther it removes from the source. The principle of diffusion is very general and applies particularly to physical and chemical processes. We see it at work when we put a lump of sugar in a glass of water and let it sit or when drop of ink enters a glass of water. But it also applies to analogous particulate behaviors at other systemic levels as well. For example, any idea that is spread in a society by word of mouth usually undergoes some weakening effect as it spreads in the population. Similarly, gene mutations that "drift" through a population tend to become more diffuse with distance from the origin.

Any sort of life system (including subsystems or components) output chemicals into their environments, and these emissions also carry messages for a receptor suitably equipped to interpret it. But as these chemical messages become dispersed and get weaker in concentration as they get further from the source, it sets up a gradient from highly concentrated at the source to more and more dispersed as distance from the source increases.

Gradients add a new form of difference to an environment, a difference that can make a difference, i.e., become information, for an interpreter. Diffusion, being a differentiating (spreading out, thinning) process in both space and time, fills an environment with spatial and temporal information tracks. Information serves a purpose. Many (perhaps all) organisms that are motile (move under muscular control), for example, are able to follow gradient signals. They will follow a gradient toward an increasing concentration if the scent represents something they seek or away from the source if the scent is something they wish to avoid, like a poison or a predator. A dispersed scent may also be interpreted as distance in time: the rabbit was here yesterday, don't bother following that scent! Or on a completely different scale, scientists by measuring the degree of the dispersed background radiation are able to calculate the age of the universe.

The intensity of various sorts of chemical gradients is also a crucial signaling message for metabolic processes. One of the most impressive examples has to do with embryogenesis, the critical process in which an organism over time develops its differentiated spatial organization. How do brain cells develop in the brain and muscle cells in the heart, when the ancestor stem cells could go either way? For an embryo to go through the familiar but incredible process of developing over 9 months from a single fertilized cell to becoming a spread out, organized and highly differentiated system of perhaps three trillion cells requires an incredible feat of coordinating timing and location! It turns out that key cells in the body of a developing embryo emit certain signaling molecules that tell other cells where they are relatively. Those other cells, the ones that will later differentiate into specific tissue type cells as the embryo matures, can receive signals from a variety of key cells. Depending on the concentration of the signal molecules, the developing cell can tell exactly where it is in the presumptive body plan and commence to differentiate into the appropriate tissue type depending on that location.

> **Question Box 3.11**
> When laying out an organism, why is it important for cells to "know" where they are relative to other cells?

3.3.5 System Organization Summary

Systems that persist in time and interact with their environments achieve this by virtue of maintaining an internal organization that is stable due to the strong coupling between components that make it up. There are other factors as well, which we will get into in Chap. 9, in particular. But there we will see that the issues are still based on the structural and functional attributes of the system.

Systems that have transient and weak interconnections between their internal components will generally come apart and not endure for long. They will not last in environments that are competing with the internal structure, for example, in trying to attract some internal component(s) to some different systems. When a lion eats a zebra, the parts of the zebra are confiscated into the parts of the lion.

Organization is a real phenomenon in that it occurs regardless of whether someone sees it or not. Systems science is concerned with discovering and describing those organizational features of systems in order to understand them better. In Fig. 1.11, in Chap. 1, we showed a Venn diagram of the main areas of interest in systems science. At the core of that diagram was a single oval called "Conception." Without the human mind, there would be no systems science, so we need to understand how it is that we have the ability to conceive of systemness and perceive systems in the world. This is doubly intriguing since we can come to understand how we come to understand anything by recognizing our own brains/minds as systems. That is our next topic.

3.4 Conception of Systems

Observing systems from the outside, as discussed above, we typically conceive of them as organized wholes. What does it mean to say "we conceive of" something? In Chap. 1, we introduced a fundamental principle of systems science (#9) which holds that systems can contain within them models of other systems. Our brains are such systems. We introduce the idea of a *concept* as a model operating inside the brain(s) of a human(s) and many other vertebrate animals. Concepts, then, are neural-based models of those other systems we observe.

Below we will get into some details of how systems are perceived/conceived,[17] but here we first have to consider the concept of *concepts* in order to fully appreciate

---

[17] Percepts are low-level patterns that are integrated in the sensory cortex of higher vertebrates, for example, the perception of texture on a surface or the shape of an object. Concepts are higher-order integrations of perceptions that are triggered not only by bottom-up perception but can be independently manipulated mentally. See the later text for more details.

what conception is about. In order to comprehend how systems in the mind can usefully conceptualize systems in the world, we need at least a preliminary exploration of how the brain represents concepts in neuronal networks.[18] Indeed, neuroscientists are really just starting to get a handle on how these networks are formed and modified over time as we learn. Several authors have attempted to formulate the "rules" by which these networks must work in order to observe the mental behavior we see in our ability to work with concepts (see Johnson-Laird 2006 and Deacon 1997). Much work is yet needed to see precisely how the networks implement these rules, but the basic mechanisms for wiring and encoding memory traces in neural tissues are now fairly well known.[19] We will revisit this topic in Chap. 8 where we demonstrate what we call "neuronal computations."

The concept of concepts has long intrigued philosophers and now neuroscientists have become involved. It is quite clear from subjective experience that there are objects and actions in our minds that relate strongly to things we witness (sense) in the world "out there." In terms of content, these range from the highly concrete and individual, like the memory of a particular relative, to the highly abstract and general, like the idea of a generic "person," or the even more abstract realm of mathematics. Such degrees of abstraction constitute hierarchical structures of more particular and more general concepts, so "my brother" belongs to the more general categories of "male" and "person" or a specific math function may belong to calculus and even more generally is a mathematical object.

Clearly, a good deal of mental activity goes into forming such conceptual structures. The term "construct" has been used by a number of philosophers and learning theorists to describe a mental object as an "ideal" (i.e., "idea like") or mentally created representation of things in the world. This term emphasizes the idea that a concept or construct needs to be "constructed," put together from some sort of component parts. In other words, more complex concepts are derived from simpler concepts. For example, the concept "person" is constructed from all of the various canonical attributes and features possessed by all persons. An individual develops this concept by having experiences with many people (e.g., inductive learning) and constructing the ideal person form in their minds.

This constructive activity is more than just abstract logic, however. We now realize that this construction takes place by linking neurons from many different locations in the brain so that they tend to fire in synchrony when the concept is active in the mind. The concepts of more concrete objects, for example, your brother, appear

---

[18] These refer to networks of neurons in various brain regions, in particular the cerebral cortex. Neurons are the cells that encode memory traces of experiences generated both from perceptual processing and from conceptual processing. These networks are able to encode features of objects like boundaries or shape. Conceptual neuronal networks integrate all of the features along with the behavioral aspects of a system.

[19] See Alkon (1988) and LeDoux (2002). At the time of this writing, a new paper has been published showing how neurons throughout the cerebral cortex form "semantic maps" of concepts. See Huth et al. (2012).

to be combinations of the ideal (canonical) representation and a set of specialized attributes and features—some of which may even be in the form of "deviations" from the canonical norms.

Concepts may interweave things and activities in almost any combination. The inherently relational structure of neural networks grounds this relational nature of concepts and forms the basis of the noun-verb relation in language. The concept of a man flying by flapping his arms, while never directly experienced, combines a range of concepts, relating all at once a generic "man," the action of flapping his arms, and the notion of arms acting as wings (like a bird's). This construction is made possible by the way the different conceptual representations encoded in neural networks can be synchronized to form a higher-level concept.

The central point is that concepts, even though we tend to experience and think of them as abstract, elusive, and even nonmaterial, are actually also real objects composed of neurons connected and activated. Thus, what we experience is based on a real, if constantly transforming, systemic configuration of neural activity. Concepts are systems that can be activated and have a causal effect on working memory in the brain and affect our conscious states. Concepts are systems.

We are used to the idea that experienced categories of sensation such as sight and touch are configured or hardwired in our physical system. Now we are pushing to understand how conceptual structuring and processes may have a similar grounding. It is not unreasonable to suspect that our brains have something like a generic system of neural connections that are used to represent a *general system* (i.e., any entity or object). This might be somewhat like the diagram in Fig. 3.1. The system in Fig. 3.1 is a *canonical* system; that is, it represents features found in all systems. The hypothesis is that there exists a neuronal network that forms under genetic control during development of the brain and thus is hardwired into the brain. That network acts as a template for forming our concepts of specific systems. Upon first seeing a new object, our brains apply the template (probably making a copy of the template network into a working memory location where it can be modified) and give it the abstract verbal tag of "thing" or "object." As we observe the particulars of the newly instantiated systems, our brain transforms this template pattern by adding links to the already learned patterns. As an example, neuroscientists have determined that there is a collection of neurons in monkeys' brains that are already tuned to recognize face-like features at birth.[20] These neuronal networks are hardwired to recognize a "face system," as it were, and to then innervate other neuronal networks that begin to encode the differences between faces of individuals. From work tracking the amount of time a newborn human infant spends attending to its mother's eyes, it is thought that a similar pre-recognition function is taking place in humans as well. The infant's point of focus can be tracked as it interacts with its mother. Its gaze will wander around the mother's face but always return to her eyes as if it is trying to attach other features to those eye shapes in order to learn to recognize the mother's face specifically.

---

[20]Perrett et al. (1982), Rolls (1984), and Yamane et al. (1988).

3.4.1   Conceptual Frameworks

As suggested above, neural configurations can amount to inborn templates or frameworks for conceptual processes. In this section, we will outline a number of seemingly hardwired features that collectively provide us with ways of thinking and perceiving. As already noted, our brains have evolved to automatically adopt these frameworks when we observe the world, which, not coincidentally, includes ourselves as actors in the world.

3.4.1.1   Patterns

A *pattern* is any set of components that stand in an organized relationship with one another from one instance of a system to another. Patterns exist in both spatial and temporal domains. It is the regularity of the organization of components in time and space that qualifies an object as having a pattern. What we think of as "the world," including both physical and temporal objects (e.g., events, activities), are matters of pattern, and the brains that guide our function in the world are fundamentally pattern recognizers.[21]

Repeated experience furnishes our fundamental means of recognizing the regularity of relationship that constitutes pattern or organization. The word "thing" allows us to mark a pattern as being something without us actually having much detailed knowledge of what the thing is. As we get more exposure and have an opportunity to observe more aspects of a thing, we may either recognize it as being like something else we already know about or start forming a new concept about something we didn't previously know about. Eventually, we find a better name (noun) to denote the thing and generally only revert to using the word "thing" in the shorthand way.

3.4.1.1.1   Spatial Patterns

Fundamentally, a spatial pattern exists when there is a consistent (statistically significant) set of spatial relations or connections between a set of components in a system (see Fig. 3.18). In other words, all of the pieces have an ongoing relation to all the other pieces that persists from one sample to another. The ability to recognize faces is one of the most complex feats of spatial pattern recognition, but we are so primed for it we seem to do it effortlessly. There is a very rich literature on the subject of patterns and pattern recognition that provides a matured perspective on systemness (some references here).

---

[21] The neurobiologist, Elkhonon Goldberg, among others, gives an extensive account of the perceptual mechanisms in the brain as pattern recognizing devices. See Goldberg (2001).

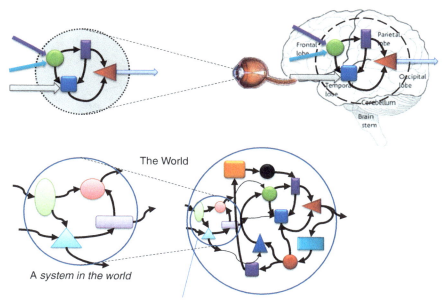

Fig. 3.18 Principle 9 states that complex systems can have subsystems that are models of other systems in the world. This graphic represents this situation. The system on the right contains a model of the system on the left which exists independently in the world of both systems. In the case of vertebrate brains, such as ours, such a model is represented in neuronal networks as described later. These are what we refer to as *concepts*

3.4.1.1.2 Temporal Patterns

Our ability to comprehend process depends upon recognizing patterns of relationship extending in time as well as in space. One especially important aspect of some kinds of temporal patterns is their repetition at regular time intervals, the phenomenon we call "cycles."

Consider the pattern of seasonal cycles. Here, we see the same kinds of relations arising over a period of time (a year). In the spring, we see the renewal of life in the budding of leaves of deciduous plants. In summer, we see the maturing and growth of these plants. In fall, we see the dropping of leaves and the animals preparing for winter. And in winter, we see the quiet sleep of nature as it protects against the cold. The cycling of the seasons is a form of temporal pattern that allows us to expect and prepare for the recurrence of states of the world from year to year, and in general, our ability to apprehend temporal patterns is critical to our ability to anticipate and deal with the future.

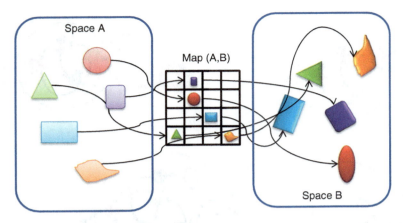

Fig. 3.19 A map is any device that translates the position of distinct objects in one space to objects in another space. Suppose one has objects in space A (perhaps these could be names of objects in space B). The mapping is comprised of links (*arrows* from space A to the map (A,B)) from that object to specific locations in the map. At those locations, one finds a link to the object in space B. A telephone book is a map from "name_of_person" space to "address_and_phone_number" space. The links from A to the map, in this case, is by virtue of the names being listed in alphabetical order, thereby rendering a simple lookup procedure effective in finding the location in the map (book) where the sought information is stored that will then provide a link to the actual physical and telephonic location of the person

3.4.1.1.3   Maps

A map, in a much generalized sense, is an abstract representation providing a linkage between a pattern in one "space" and that in another "space." There is more to this than one might think. First, by space we do not just mean the three dimensions of physical space alone. Rather, we include the temporal dimension and, indeed, any number of dimensions beyond space-time if needed. Mathematically, there is no limit to the number of dimensions that a "space" might contain. However, here, we will just work with the ordinary four-dimensional space-time of our ordinary experience.

The linkage between two spaces refers to the principle that a point in one space (a specific coordinate measured in an appropriate metric) can be mapped or linked to a point in another space, with different coordinates and metrics (see Fig. 3.19). We say that objects in one space are isomorphic with objects in another space. Let's look at some examples.

As a first example, take an ordinary street map. What this "device" does is to provide a visual layout on the two-dimensional surface of a sheet of paper of the pattern of streets by proportionately showing their relative coordinates in the real space of geography. But more subtle, yet, is the real purpose of such a map: the map is used by your brain to translate these coordinates into physical actions that result in your getting where you want to go from where you are. Thus, we say the paper map mediates another mapping from the physical location of features in geographic space to the behavioral sequences in mental space that provide for actions and results. In reality then, the street map is just the beginning of a series of mappings from physical space to the final behaviors you exercise to get to where you want to go. There are many intermediate maps in the brain to translate the represented physical relations in the paper map into actual behavior. We will have more to say about the brain as a pattern recognizer and mapping system later.

A final example of patterns that act as maps comes from the world of mathematics, relations, and functions. Remember from algebra?

Quant Box 3.2 Mathematical Maps

Relation

A relation is a mapping from one set of objects (say numbers) to another (though possibly the same) set of objects such that any object in the first set is connected with zero or more objects in the second set. As an example, consider the "binary" relation between integers, "$\leq$." We would write something like $a \leq b$, meaning that the value of a is constrained to either be exactly the same as the value of b or it could be any value from the set of integers strictly less than the value of b.

Function

A function is a mapping from one object in a set to exactly one other object in another set (again, both sets may be the same, e.g., integers). Functions are often written in the form $b = f(a)$. Here, $f()$ stands for the function that produces the conversion of the argument, a, into the result, b. For example, a specific function such as $b = a^2 + 3a$ would set b to the value of a squared plus three times a's value.

3.4.1.2 Properties and Their Measurement

The specific "objects" that constitute a pattern are the properties and features of the components. There is not a lot of consensus in the literature of systems and patterns about the meaning of these terms. In general, a *characteristic* is a property of an object if it is always present in that *kind* of object. In physics, we find one of the more consistent definitions of properties as they apply to physical objects. For example, physical objects can have properties such as mass, volume, reflectivity, and so on. Note that color, an effect produced in our senses, is not a property in the sense meant in physics. Rather, the object reflects light only in certain bandwidths of the visible light spectrum, and the brain interprets that reflection as color. The property that belongs to the object, then, technically is its reflectivity, whereas the color is called a phenomenal experience of the observer.

This concept of physical properties embodies an important principle from systems theory. Namely, the properties of an object are actually based on properties of the system itself or of the component objects (subsystems) making up that object. So, for example, the reflectivity of the surface of an object, as mentioned above, is a result of the interactions between light rays (photons) and the particular atoms and molecules that make up the surface of that object. We can think of not just a fixed thing called a "property" but rather a hierarchy of properties that may differ depending on the level of observation. Water, for example, is a liquid at certain temperatures, but individual molecules of H_2O are not liquid and have no temperature. A mouse may be alive, and life processes can still be identified at the level of various component organs, but it would be fruitless to search for the distinctive properties of life in the individual atoms and molecules that together constitute a living mouse.

Question Box 3.12
How does the hierarchy of properties discussed above relate to the proposition that "the whole is more than the sum of its parts"?

Properties, in general, are that aspect of systems that we can measure via some form of sensing device. Measurement is itself an interesting systems concept. We normally think of measuring something (a property like mass) as finding a numerical value to assign to the object of interest. But the act of taking a measurement fundamentally involves comparing one system to another system through the forces of interaction between them, where one of the systems is considered to be the referent (the measuring device) and the other the subject. In this respect, all of our senses are in fact measuring devices that compare and calculate differences. The color that we see reflected from an object, for example, is our brain's way of measuring differences in the light spectrum being reflected (via our eyes). Although our senses do not use numbers as their metric, the differences they register as colors can also be expressed numerically. Indeed, all our clever measuring devices are strategies to extend this measuring activity of our senses. The numerical assignments are totally arbitrary but become fixed by a consensus agreement (various scales) for purposes of comparing different objects. For example, we measure mass by reference to the gravitational force inherent between the Earth and an object. We put a spring-loaded device between the Earth and the object and read off a numerical amount that shows how much the spring has been deformed in comparison with its condition before the object was put on it. We call this weight, which is an indirect measure of mass. All measurements are the result of such interactions between an agreed upon referent system and the subject system. Thanks to the laws of nature being consistent, the measures will provide us with a consistent comparison among objects, their properties, and the measures themselves.

3.4.1.3 Features

Features can generally be thought of as the arrangements of component parts of a system, where such arrangements have properties that can be discriminated by measuring devices. Features are also referred to as the characteristics of a particular object or kind of object. This is because we can use features to differentiate among individuals that comprise a class of similar objects. Thus, our most basic measuring instruments, our senses, register features when we identify individual persons or pick out our car in the parking lot.

Features, like properties, come in hierarchies. Features that distinguish species, for example, differ from the features that distinguish individual members of a species. But unlike properties, it is generally possible to directly observe micro-features that make up a larger-scale feature. This is important, because similar large-scale features can be differentiated in terms of the precise arrangement or composition of micro-features. Take, for example, the features of a human face. At a higher level, we would list things like eyes, a nose, a mouth, eyebrows, etc. as being features of a face. Being common to all people, this helps us differentiate people from trees, but not from each other. But then we would note that each of these features consists of features such as outlines, colors, textures, and so forth, and these micro-feature components may vary from person to person.

At the finest level of visual feature detection, for example, we can identify aspects such as lines, ovals, and other primitive shapes. Figure 3.20 shows a profile image of a face. The outline shows smooth curves (black line) but these can be broken down into a series of straight line segments (green lines overlaying black).

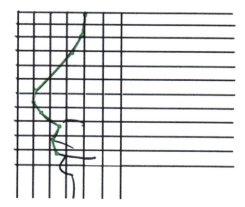

Fig. 3.20 A face profile can be decomposed into a set of line segments that our eyes and brains can detect as micro-features. Taken together, these comprise a larger-scale feature (nose and upper lip). The grid represents a coordinate space for assigning numerical values to the locations of the end points of the line segments. No scale is given. Our capacity for feature detection is at a very much finer scale than depicted

The curved lines in this case are comprised of a set of micro-features (straight lines). The uniqueness of this particular curve, then, is not the uniqueness of the micro-features, but the particular way the micro-features are joined to make up the curve. Of course, in our vision system, the detection of lines is at a much, much finer granularity than shown in this figure. But the idea is the same.

The relative position and properties of features constitute their relations to one another. If we were to detail out the positions of the end points of all line segments in the coordinate space, as shown in Fig. 3.20, we could then specify the relations between all of the segments in that coordinate space and thereby create a specification for that particular nose! It would then be the particulars of that specification, essentially a list of the end points for the sequence of line segments that would provide the definition of the nose feature for that particular individual.

A pattern can now be defined as the set of all relations that organize the set of features at any given level in a hierarchy of features, into a map. The map is a representation of the actual object in an abstract, but usable form. Pattern recognition involves the detection of features that work through the map to point to another representation, namely, that of the object being recognized.

So, in the above example, we have a map that translates a set of line segments into a meta-feature we call a nose. In the example, we have differentiated the junction of the lines only enough to indicate a nose rather than an ear or some other general form, but more exacting differentiation will further individualize the nose. We then can have a map that translates a specific set of relations of those line segments into a specific person's nose. The person is identified, in part, by the specific form of the nose! This process of pattern recognition and identification takes place in the brain, but it can also take place in a computer, at least to some degree.

3.4.1.4 Classification

We have seen how a pattern of features (with their specific properties) gives rise to meta-features and a pattern of such meta-features gives rise to objects. But there is one more aspect of this kind of feature hierarchy that we need to consider. Some kinds of features can be grouped in such a way as to provide a generalization that we can call a "class."

All the people we know can be grouped into a class we call human beings. In a similar way, all human beings can be grouped into a class we call mammals. This ability to recognize certain features that are found in many different specific representatives of objects is the basis for what we call classification, grouping representatives under a single abstract heading based on their possession of specific general characteristics. All mammals have hair and the females have mammary glands (males having essentially placeholders for the same!). These generalizations can be applied across a wide array of animals that possess those specific features and we can aggregate them as a group (one common name).

There is a deeper significance to this kind of aggregation. The fact that so many different animals that we encounter have these features tells us something important about the world we live in, namely, that there is a systemic relation between these animals. Our knowledge of how the world works begins with an understanding of what is related to what. And that depends, critically, on our ability to group like things together.

This is what we mean by categorization. Aggregates of specific (and important) features suggest that objects possessing them are somehow related. Our whole understanding of the universe begins with an ability to categorize objects based on these common, important features. Planets have common features. So do mammals. So do oceans.

Our conceptual grasp of the universe begins with our ability to create categories based on common features and differences between categories based on different features. This is no small feat. At the same time, it is not beyond understanding. Once we grasp how the world is parsed into different systems with unique sets of features, yet having common features that unify our perceptions, we are in a position to categorize and name these variations. All systems have common features (to be covered below and throughout the rest of this book). At the same time, all systems have unique features that differentiate them and allow us (as human perceivers) to recognize individual systems (like friends and acquaintances) as distinct.

Classification of groups of objects with similar features is a means of increasing the efficiency of mapping and, hence, pattern recognition and selection. In this sense, maps do not have to provide the details of common group features for every member of the group. Rather, those common features can be stored as single concepts (one per feature and also one per common aggregates of features in a hierarchy). Then, when learning a mapping, it isn't necessary to provide links from every "atomic" (that is micro-) feature to every instance of every member of the class. Instead, a simple link from the feature concept (stored once and shared by all instances of that class) to the instance object is needed. Since most dogs have visible fur, it is possible to represent a canonical dog with a single link to fur (which it shares with most other mammals) rather than store a "fur" feature with every single instance of dog. By providing a linkage from your specific dog back to the canonical representation, you will still have a pattern mapping including fur. Of course, there may be many different kinds of fur types. But this still represents an enormous saving in memory space to not have to particularize every single dog you ever met with its own stored pattern of fur.

As we shall see, these efficiencies of pattern mapping and memory are advantages that over evolutionary time have found their way into the brain's neural architecture and way of recognizing, processing, and storing patterns.

3.4.2 Pattern Recognition

The process of decomposing an image into features, checking the specific features and their relations, and finding a mapping between the set of features and an identifiable object is called pattern recognition. It is what our brains do magnificently and what we are starting to teach computers how to do, even if primitively. In what follows, we are going to cover briefly some interesting points about each. We should note that while

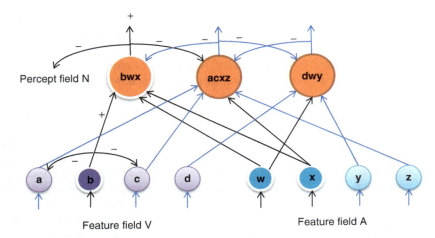

Fig. 3.21 Clusters of neurons are tuned to detect specific features in the early sensory processing fields of the sensory cortex. Percepts are mappings of specific sets of features that consistently go together in experience. Whenever this specific combination of features is detected in the sensory field, the active set of feature detectors (indicated by *circles with white outlines*) activates the neural network designated as the percept. The activated percept is forwarded to conceptual processing in higher-order processing in the brain, e.g., association cortex areas. *Plus signs* indicate excitation and *negative signs* indicate inhibition. Cross inhibition helps strengthen the feature or percept activation as it is passed on to the next higher level. See Chap. 8 for a more in-depth description of neuronal computations for perception and conception

the mechanisms by which brains and computers achieve pattern recognition are quite different, the underlying descriptive mathematics are the same. This is why it is possible to program computers to have some level of competence in doing pattern recognition as a precursor to artificial intelligence. Here, we explore, briefly, the differences between human perception (of patterns) and computer-based pattern recognition to give you a sense of those underlying (systems-based) similarities.

3.4.2.1   Perception in the Human Brain

In Chap. 8, Computation, we will take up in greater detail the way in which natural neural networks (i.e., brains) process data to produce recognition of patterns that persist in the world over time. Here, we offer just a taste of what that will involve.

The brain, and specifically the cerebral cortex, is organized to process the hierarchy of features (line segments) and properties (colors representing visible light reflections) that give rise to higher-level percepts (noses and faces). These, in turn, are components in the recognition of concepts at a still higher level, such as the recognition of a specific person's face that goes along with their name and personality (as well as one's whole history of interactions with that person!). Here, we will only be concerned with the nature of perceptual recognition of patterns and leave the discussion of conceptual processing to later discussions.

The nature of perceptual processing in the brain can be demonstrated by what is currently known about visual processing of patterns by the early visual cortex (the occipital lobes of the mammalian brain). This will necessarily be a very cursory explanation of what goes on in the brain. The subject is extraordinarily complex. What we hope to show with this simplified explanation is the power of the systems approach to understanding otherwise very complex processes.

Figure 3.21 is a schematic representation of the mapping from feature detectors to perceptual objects. We do not explain how these mappings came to be; that is the subject of learning and adaptation covered in Chap. 8. Here, all we want to show is that the mapping concept is involved in how the brain translates a feature set into a percept (a pattern). Signals from the sensory relays in the primitive parts of the mammalian brain are relayed to the feature detection areas in the cerebral cortex. There, clusters of neurons analyze the inputs, and when a specific feature is detected, that cluster is activated and sends a signal upward toward the higher levels of percept analysis. It also sends inhibitory signals to other feature detectors to help increase the influence of the strongest feature detected in the next level of analysis.

Many important feature detection and processing circuits in the mammalian brain have been found and examined. The way in which the cerebral cortex is able to process this information and pass it on to higher levels of perceptual and conceptual processing has now been elucidated in nonhuman and human brains as well.

3.4.2.2 Machine Pattern Recognition

In addition to evidence from animal brain studies, there are now numerous examples of pattern recognition processes operating in computers that emulate the processes that take place in living brains. That these computer emulations are successful in detecting and processing patterns that can be used to affect, say, the behavior of robots, gives us a great deal of support in thinking that we have properly understood the nature of patterns and perceptions in living systems.

We will only mention a single approach to pattern recognition in machines (computers as an example of how a systems approach to understanding objects and their meanings can be implemented by nonliving as well as living systems). Over the last several decades, a number of approaches have been tried in programming computers to recognize patterns and objects as a preliminary step in having these machines effectively interact with their environments. This is the general field of intelligent, autonomous robotics.

As mentioned above, there is a clear, mathematical way to specify patterns such that they are amenable to machine recognition. As it turns out, however, the variability in patterns within a particular class of pattern types (e.g., meta-features like noses) is so high that it is impossible, even in principle, to specify all of the possibilities. As a result, computer scientists have attempted to emulate the approaches taken by brains in the form of what are called artificial neural networks (ANN). These are programs that do not attempt to do pattern recognition directly but rather attempt to simulate neural circuits that perform this function in the brain. There are very many approaches to doing these simulations. Some are more successful than others, and all of them have strengths and weaknesses. To date, no single computerized approach can match the flexibility and comprehensiveness of the brain in performing pattern recognition and mapping to discrete identifications of objects or situations. But several have been quite successful in performing subsets of these functions. Certain kinds of ANNs, for example, have been used in very successful performances of autonomous vehicles that are able to drive themselves around natural and city terrains.[22]

We are still at the very beginnings of producing human-level pattern recognition and perceptual processing in machines, but the application of the systems approach to understanding these processes in natural brains is expected to provide guidance to achieving that goal.

3.4.2.3 Learning or Encoding Pattern Mappings

One very important question for both brain-based and machine-based pattern recognition is how do the mappings get created in the first place? As noted above, the variations in features and feature combinations across all representatives of any given group of similar objects are generally very high. It would be impossible, even in principle, to preprogram all of the possibilities into a computer program, and, in a similar vein, it would be impossible for there to be a genetic program that prespecified cortical circuits that would do the work.[23] Brains, and now ANNs, are designed to modify circuit connections as a result of experience. This is done so as to capture the statistically significant properties of features and their relations in the real world.

In other words, the mappings from one space to another space (the links to the map and the links from the map to the target space) are developed as the brain/ANN experiences associations in real life. This is accomplished by some sort of strengthening the preexisting connections (the arrows) when experience shows the objects from the two worlds are correlated in space-time. In other words, if object ω always is found present at the same time that object α is present, then the system will strengthen the mapping between them. At the same time, it will weaken the links between ω and other objects that are NOT found co-occurring. After some number of trials where ω and α co-occur a significant number of times, it will become possible to "predict" the occurrence of α simply by the occurrence of ω in space-time. Thus, the mapping is the result of an adaptive process based on actual experience. Basically this is what neurons do in learning.

---

[22] See Autonomous vehicles—http://en.wikipedia.org/wiki/Autonomous_vehicle.

[23] This latter fact became painfully clear when the completion of the Human Genome Project showed that there are probably around 30,000 genes specifying the human body and brain plan. Clearly, the brain's detailed wiring cannot be a result of genetic control. But we already knew that most of what we carry around in our heads is the result of learning!

Think Box. Concepts are Systems

What is a concept? Is it a thought? Where do concepts go when you are not thinking of them consciously?

Is it possible to think about a concept as a system? For example, would it be appropriate to think of a concept as having a boundary? In fact, this is almost necessary in order for us to say we have a specific thought. Consider a simple example: the concept of a dog. Here, we don't mean a specific dog, say your pet Fido. Rather we mean the idea of dog-ness—the qualities of a dog. Ahead, in Chap. 4 (e.g., see Fig. 4.3), we consider something called a "concept map" related to dog-ness. Such a map can be formulated in this abstract manner because each concept can be treated as an object with definable qualities and capabilities to be linked through relations with other concepts.

Recent research in neuroscience is showing that concepts such as a dog or horse, or a face, are present in consciousness when specific patches of neocortex (the outer layer of the cerebrum) light up in imaging investigations. The growing picture is that clusters of neurons tend to encode concepts. The neurons in a cluster fire synchronously when the concept they represent is active in working memory. But the concept may have sensory and/or motor associations, e.g., a face has a number of visual features common to all faces. These are held in lower-level sensory cortex, essentially sub-concepts. It appears that when you think of a general dog (or horse or face), the cluster of higher-level neurons excites the lower-level component sub-concept clusters. Moreover, when you think of a specific dog (Fido), all of those clusters, plus another cluster that represents Fido specifically, fire in synchrony. It may be the case that the brain forms a general concept from experience with multiple instances of objects with many more degrees of similarity than differences. Then in every new encounter with something that excites the general concept, the new thing is immediately recognized as belonging to that general category.

These clusters have porous and fuzzy boundaries since each one can have slightly different associations with other similar clusters. They receive inputs in the form of excitations from other clusters to get them excited and active in working memory. They produce outputs as excitation signals to other clusters. For example, you see a dog on the sidewalk. Your visual processing system gets the clusters of features excited and they communicate upward in the concept hierarchy to your "dog" cluster. You recognize the dog, and it is in your working memory as a conscious idea. But then that cluster sends a message to another cluster that is a memory of your dog Fido and he comes to mind.

We now have methods for modeling neural networks as concept encoding clusters and can watch them operate, taking inputs from other areas of the brain and sending messages out to other areas. The brain is a large complex system of many complex subsystems. Concepts are just one aspect of systems operating in the brain.

3.5   Chapter Summary

We have covered a lot of ground in this chapter. But we will find that this broad coverage of the notion of an object and the various characteristics of objects that give rise to what we have called systemness become the basis of all else this book covers. All systems are observed as some form of object to the human mind. We have seen that the human brain perceives objects based on their boundaries and their behaviors in interactions with their environments. That means the human mind must also perceive the nature of an environment in the background as well as a system or object in the foreground. In that vein, we have seen how environments themselves are organized and, indeed, can be thought of as a meta-system, a larger system with its own level of internal dynamics.

The internal dynamics of systems consist of flows and forces that allow components of a system to interact with one another in a wide variety of ways. Sometimes components attract, at other times they repel, and still at other times they may exchange matter, or energy, or information, or all of these, through flows. These exchanges can happen only if energy is available, so active systems have boundary conditions that permit flows from sources and back into sinks in the environment.

Organization arises over time by the ongoing interactions of different components or subsystems, each with their unique boundary attributes. Complexity arises from the potential of a bounded system based on its components and the possible interactions between them. Realized complexity is seen in systems that have had time to allow their components to accomplish the work to realize those connective interactions. Complexity and form emerge from these temporal evolutions. The more potential complexity a system possesses, the more varied the possible outcomes of system evolution and dynamics become, and growth of complexity brings yet further potential.

Now, we will dive into the particulars. These have been the basic principles. The particulars are more complex! What we hope to show you is that wherever you look, you will find these principles at work in a huge variety of arenas where it was once thought that only special laws operated. Everything in the universe is a system of one kind or another. Therefore, everything should obey system principles. Let's see if that works out.

Bibliography and Further Reading

Alkon DL (1988) Memory traces in the brain. Cambridge University Press, Cambridge, UK

Bateson G (1972) Steps to an ecology of mind: collected essays in anthropology, psychiatry, evolution, and epistemology. University of Chicago Press, Chicago, IL

Capra F (2000) The tao of physics. Shambhala Publications, Boston, MA

Capra F (2002) The hidden connections. Doubleday, New York, NY

Deacon TW (1997) The symbolic species. W.W. Norton & Co., New York, NY

Goldberg E (2001) The executive brain: frontal lobes and the civilized mind. Oxford University Press, New York, NY

Huth AG et al (2012) A continuous semantic space describes the representation of thousands of object and action categories across the human brain, *Neuron* 76: 1210–1224. Elsevier, Inc., New York, NY

Johnson-Laird P (2006) How we reason. Oxford University Press, Inc., New York, NY

LeDoux J (2002) Synaptic self: how our brains become who we are. Viking Penguin, New York, NY

Meadows DH (2008) Thinking in systems. Chelsea Green Publishing, White River Junction, VT

Mitchell M (2009) Complexity: a guided tour. Oxford University Press, New York, NY

Perrett DI, Rolls ET, Caan W (1982) Visual neurons responsive to faces in the monkey temporal cortex. Exp Brain Res 47:329–342

Primack JR, Abrams NE (2006) The view from the center of the universe. Riverhead Books, New York, NY

Rolls ET (1984) Neurons in the cortex of the temporal lobe and in the amygdala of the monkey with responses selective for faces. Human Neurobiol 3:209–222

Smith BC (1996) On the origin of objects. MIT, Cambridge MA

Yamane S et al (1988) What facial features activate face neurons in the inferotemporal cortex of the monkey? Exp Brain Res 73:209–214

flows + Forces

The internal dynamics of systems consist of flows + forces —

See Graduate Medical Education as a system

Chapter 4
Networks: Connections Within and Without

"...the web of life consists of networks within networks. At each scale, under closer scrutiny, the nodes of the network reveal themselves as smaller networks."

Fritjof Capra, The Web of Life, 1996

"The mystery of life begins with the intricate web of interactions, integrating the millions of molecules within each organism. The enigma of the society starts with the convoluted structure of the social network... Therefore, networks are the prerequisite for describing any complex system..."

Albert-László Barabási, Linked, 2002

Abstract A key attribute of systems is that internally the components are connected in various relations. That is, the physical system is a network of relations between components. It is also possible to "represent" a system as an abstract network of nodes and links. The science and mathematics of networks can be brought to bear on the analysis of these representations, and characteristics of network topologies can be used to help understand structures, functions, and overall dynamics.

4.1 Introduction: Everything Is Connected to Everything Else

As our third principle of systems science states: systems are themselves and can be represented abstractly as networks of relations. In order to understand the nature of systems in terms of structure, organization, and function (dynamics), we need to understand networks.

When someone says that "everything is connected to everything else," it is easy to brush this statement off as trivially true. Without actually knowing what the connections are that supposedly make this statement true, we intuitively realize that "something" connects all things in the universe together, even if it is only gravity.

© Springer Science+Business Media New York 2015
G.E. Mobus, M.C. Kalton, *Principles of Systems Science*, Understanding Complex Systems, DOI 10.1007/978-1-4939-1920-8_4

What makes us consider the statement to be trivial is that we also intuitively realize that these connections are rarely direct, from one object to another. Rather, one object may connect to another object, which in turn connects to a third object, and so on down some chain of connections. It is through this chain of connections that we say that the first object is connected to something much further down the chain. But some kinds of connections seem much more significant than others. If the connections between a set of objects are some kind of solid form, say a non-stretchable string, then pulling on one end at the first object will transmit a force down the chain of objects, through the string, such that the distant object is as affected as is the first, even after a short time delay. In this situation the statement of connectivity is not very trivial at all. What happens to the first object will happen to the distant object.

But in other cases causal consequences seem weakened almost to the vanishing point as they travel through a web of connections. To take another example from physics, say there is a room full of white spheres laying around in the dark (say they are white billiard balls randomly scattered). Now we shine a flashlight on one of the spheres from the side. The light will reflect off the sphere and strike one or more other spheres. It will reflect off of them and, thus, strike yet other spheres. In this case the spheres are all connected, at least potentially, by light beams bouncing between them. But distant spheres will receive much less light than those nearest the original sphere. This is because the light rays attenuate (grow dimmer) as they scatter among the spheres. There will most likely be many spheres that will not be illuminated well enough to detect or simply fall into a shadow from some other sphere. All will receive a miniscule amount of light that has simply scattered to fill the space at a very low intensity, but it would take a more sensitive instrument than an eye to register it. Here we have a case of all things being more or less connected, but by significantly different degrees, such that a small change in the intensity of the original light beam may have no discernible impact of most of the other spheres.

Universal connectivity of all things is a truism. It is a postulate or axiom from which we can build a large array of derived truths about the universe. But all connections are not equal, and we need a deeper analysis of connectivity to derive useful tools for describing particularly meaningful (nontrivial) senses of how things being connected to one another is important. That is the purpose of this chapter.

In Chap. 3 we began the discussion about networks, the structures of systems, and the functions those structures perform. In this chapter we will delve much more deeply into the subjects introduced in the last chapter. Specifically, we look at how the organization, or pattern, of a system can be described as networks of components and relations. We will be concerned with how different structures arise within networks to produce different functions. Moreover, we will draw from real-world systems examples in greater detail than before. We will explore examples from biology, sociology, and economics to demonstrate these principles in action.

4.2   The Fundamentals of Networks

Let us start with a rather abstract view of networks. Later we will apply this view to real systems to see how the principle of network organization is realized.

A network can be described easily enough as a set of objects, or nodes, and a set of connections, or links, between the objects. Below we will introduce a powerful formalism called graph theory that is used to analyze networks in terms of nodes and connections. The seemingly simple technique of displaying networks as nodes linked by lines, as in Fig. 4.1, turns out to be a sophisticated tool for disclosing the properties of transforming and ramifying networks.

Recall that black boxes are units whose inner connectivity and functionality are not analyzed, so only input and output to and from the unit are considered. In Fig. 4.1, the whole system is analyzed in terms of the linkages of its component subsystems, but the subsystems are all black boxes. In Fig. 4.2, we use the same graph formalism to carry the analysis further, converting some of the black boxes to analyzed subsystems, or white boxes. Of course these new sub-nodes are more black boxes that could be analyzed into yet another layer of nodes and linkage. Here we see graphically revealed not only the principles of systemness and hierarchical organization but the meaning of systemic levels and the methodology of layered analysis necessary to disclose the connective functionality of a complex system. As we will see shortly, this architecture of networks leads to many opportunities to work with systems as networks.

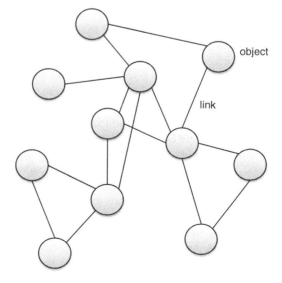

Fig. 4.1 A network is a set of distinct objects and a set of connections (links) between these objects. This figure is a highly abstract and stereotypical kind of network, called a basic undirected graph (see section below on mathematics of networks). Notice that some objects, also called nodes, have many linkages, while others have few. The objects are subsystems. The links define the connectivity of subsystems within the larger system

Fig. 4.2 Some of the black box objects in Fig. 4.1 are shown to be composed of internal networks of objects and links

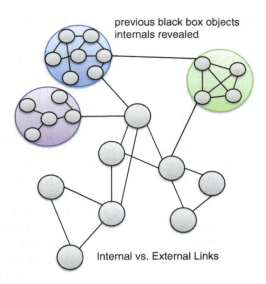

previous black box objects internals revealed

Internal vs. External Links

4.2.1   Various Kinds of Networks

We can start with a classification of networks based on some very general properties that differentiate them. Here we focus on just a few of the most basic.

4.2.1.1   Physical Versus Logical

If you recall from Chap. 3, we identified the problem of boundaries having different qualities depending on how the system was observed and what kinds of questions you might be asking about it. There we focused mainly on systems with obvious physical, if fuzzy, boundaries like organisms or corporations. We noted, however, that having an obvious physical boundary was not the main determinant of what we could choose to be the boundary of a system. Different questions about the nature of the system of interest might indeed require changing the boundary scope, such as taking into account some resource sources or sinks in order to better understand the behavior of the "bounded" system. For example, we might need to consider the nutritional quality of foods we eat in an attempt to better understand health issues. Food is technically coming from the outside of the "skin" boundary. But as a resource that will cross the boundary and eventually provide material and energy to the body, we might be curious to understand where it comes from and how it was produced.

Logical boundaries are set by the nature of the questions being asked or problems we attempt to solve. They are no less real in the sense that even if there is no actual physical boundary, they still represent a conceptual grouping, a boundary which includes some things (and their connections) and excludes others.

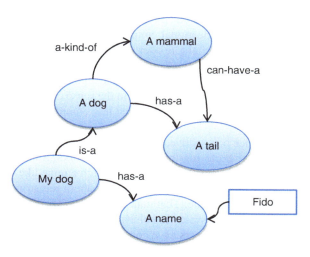

Fig. 4.3 A conceptual map is a network of concepts linked by relations. Here "My dog" is related to the category concept of "A dog" by an "is-a" relation, meaning that my pet is a dog. He has a name, Fido. He also has a tail, which is an attribute that many mammals can have. Finally, a dog is "a-kind-of" mammal with all of the same basic attributes that all mammals share (e.g., hair, mammary glands, etc.). Note that this kind of concept map can be realized in the neural networks of a human brain

Approaching systems as networks makes this clear, since concepts and conceptual connections qualify as objects and links. A similar network analysis will thus help us comprehend systems whether they are physical, conceptual, or a combination of both. A physical network, as its name implies, involves actual physical connections, like roads, between physical nodes, like cities. A logical network is one where the connections are purely ideational and the nodes are conceptual. For example, Fig. 4.3 shows what we call a conceptual map, a network of relations between conceptual objects (recall our discussion of *maps* from Chap. 3). A paper map of, say, a state or a country, showing lines that represent the roads that connect dots, the cities, is an example of a logical network that corresponds with a physical network. Such maps attempt to replicate the actual physical layout of the network but at a much smaller scale. Other kinds of maps, say a subway connection map, do not show the actual geometry of what they represent, but rather the logical nature of the connections themselves.

Another example of a network that is both physical and logical is the Internet and the overlaid World Wide Web. The Internet is comprised of nodes, including routers, switches, and computers (including mobile devices of numerous kinds). It has links that include wires, radio waves, and coaxial and optical cables. There are many different services such as e-mail and the web that are logically superimposed on this physical network. The complexity of the web services involves a mutual feedback loop between conceptual and physical connectivity (e.g., online shopping) so that both the physical and ideational connectivities of the globe have grown at something like a super-exponential rate over the last two decades.

The web is obviously a real network. But it is established by logical design. All of the software, the computer programs that implement the web (e.g., HTTP servers and browsers), are part of the logical mapping, in the sense covered in Chap. 3, from one very complex logical space to the somewhat less complex physical space—the web to the Internet.

From a systems science perspective, the Internet and web have provided a seemingly ideal laboratory for exploring the nature of networks. Since it is possible to program a computer to explore the web, just as a search engine like Google™ does, it is possible to collect tremendous volumes of data about the structure, functions, and evolution of the web over time. That data can then be analyzed (we call it data mining) for patterns that show systemic behaviors. Most of the major advances in the study of networks in recent years have come from insights gained from analysis of the World Wide Web. The principles derived from these insights have been used to view many other kinds of systems in new light.

4.2.1.2 Fixed Versus Changing

Fixed networks can be represented or modeled by arranging a given number of nodes and links in a pattern that conforms to the object system. Although any real system will be involved in change and flux, it can be useful to abstract from these considerations in order to get at certain structural issues relevant at a given time. Thus, a well-made graph may disclose the surprising hub-like connectedness of a given unit, or the critical interdependence among clustered subunits, without reference to how these features change over time.

While the analysis of a network as fixed at a given time serves many purposes, it leaves unaddressed an important question: how did the networked structure emerge in the first place? Are there any general rules or patterns in the way networks develop? One of the most exciting insights to have emerged from studying the way the World Wide Web has been changing over the years is to recognize patterns of network evolution that can be applied to other systems.

Fixed networks are still worth studying for interesting structural properties, but changing networks are proving to be of significant interest. A major form of change, which we will take up below, is the abovementioned evolution, which involves both dynamic and structural changes (i.e., addition of nodes and/or links). Another form of change has to do with just the dynamic behavior of networks in which materials, energy, and messages flow (see next section). The structures of such networks can be fixed over the time of study, while the flows themselves may change for various reasons. The graphing of these so-called flow networks gives us an invaluable tool for representing and analyzing this dimension of network change.

4.2.1.3   Flow Networks

Question Box 4.1
Once upon a time the American West was relatively empty insofar as cities or towns were concerned. Spurred largely by the gold rush and a restless population looking for land, settlers started invading the Western lands and setting up first villages and small settlements. Later as the railroads developed, some of these grew to become towns and even cities (transportation, as well as local resources, was the major key to community growth). In other words, the West became a network of population centers with roads and rail lines connecting them that evolved in complexity and power over time. Similarly, the Internet and World Wide Web started out empty and has evolved structure, population density, and connectivity over the years. What patterns can you spot in the development of the West that might also apply to the way the web has developed? Do you see anything in the dynamics of web development that might offer an insight into the evolving topography of population centers?

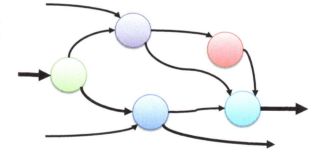

Fig. 4.4 A flow network is used to represent connections where stuff flows into a node, is presumably processed in some fashion, and then flows out of the node

Since real systems involve flows of materials, energies, and messages, it is useful to represent networks in which flows of substance can be tracked through the various nodes. A very abstract representation of such a network is seen in Fig. 4.4 above. The connections between nodes are now represented by directional arrows, which indicate flows. In this figure the larger volume flows are represented by thicker arrows, but in a more useful representation, the arrows would be labeled with flow rates (volume per unit time). In dynamic systems these flow rates could vary over longer time scales.

Flow networks are an ideal way to model real systems which, as will be seen in Chap. 6, are comprised of subsystems that act as nodes in a network. As with other types of networks, flow networks have particular properties that enter into how we develop and use these models. For example, flow networks have to obey the conservation principles for matter and energy, which means that the sum of outflows from a node can neither exceed nor be less than the sum of the inflows for any type of flow.

Nodes can be represented simply as seen here. If we are interested in the details of processing of inputs to outputs, we simply add another layer of analysis, transforming these black box subsystems into analyzed representations of nodes and their linking flows. Nevertheless, a lot of information about the internal dynamics of a system can be explored just by capturing the primary flows.

Question Box 4.2
What information can you get about the system from Fig. 4.4? Can you give an example of an actual system and a situation in which even this basic information would be important?

4.2.2   Attributes of Networks

Here we will examine a few of the more important attributes of networks that are used to characterize network organization in systems in general. These attributes can be found in any network analysis of any kind of system, from simple chemical systems to complex ecosystems.

4.2.2.1   Size and Composition

There are several different metrics that can be used in describing the *size* of a network. Which metric is apropos is somewhat dependent on another aspect of the network, its composition. By this we mean how many different kinds of nodes and links are involved in the network. We can talk about networks as being homogeneous in either node or link composition, or both. For example, the node types and types of links in the Internet are relatively few in number. This network is relatively homogeneous in composition so that its metric of size might simply be the number of nodes and links as a rough measure.

For heterogeneous networks we usually are interested in size metrics related to the number of different kinds of nodes and kinds of links. As networks include more of such differences in their composition, they become more complex, a kind of metric we will take up in the next chapter. Network structures/functions and complexity measures are directly tied to one another to the extent that talking about one aspect is impossible without mention of the other. In addition to complexity (Chap. 5), the issue of heterogeneous network structures and functions will also factor into our discussions of dynamics (Chap. 6), emergence (Chap. 10), and evolution (Chap. 11). This many-faceted relevance of composition is another reflection of the interrelatedness of the principles of systems science.

4.2.2.2 Density and Coupling Strength

Another size-like metric that relates to network structure is what we call *density*, meaning how many nodes are connected to how many other nodes. In networks where the links are physical flows of stuff, density relates to how many different pathways there are to get from one node to another without necessarily having to pass through an intermediate node. For example, in the case of cities and roads connecting them, the geographic limitations of distance means that cities tend to be linked only to cities that are nearby in physical space. In order to get to more distant cities, say via interstate highways, you have to go through other cities; you cannot just go from city A to city Z directly because the cities are nodes in a relatively low-density network. But when you include the airplane routes as links, it becomes possible (at least in principle) to provide direct links from every city to every other city. Of course the cost of doing so is prohibitive, but at least some cities, such as Los Angeles and Chicago, become densely connected hubs with numerous national and international links. This varying density of linkages within a network is an important structural feature we will revisit below when we consider hubs.

Coupling strength means how strongly two nodes are linked when the linkages can have different levels of strength. For example, a freeway between two cities can carry far more traffic than a two-lane road. A fiber optic cable can carry far more many messages than a copper wire cable. And a rope tying two horses together is more likely to keep them working together than if they were tied together with a thread! Strength here may mean the volume of flow of some stuff (material, energy, and messages),[1] or the level of a force intermediating between nodes, such as the difference between the rope and the thread.

As we will see several times in this book, coupling strengths play an important role in understanding things like stability, persistence, redundancy, and sustainability, all aspects of dynamics. This is due in part to the way density and coupling strength are interrelated. High density of linkages typically points to many alternative paths if one should fail, so the coupling strength of a particular link may be less critical. On the other hand when nodes are more remotely linked through other nodes, then the coupling strength between them is subject to the "weakest link" syndrome, and so coupling strength becomes a critical factor.

4.2.2.3 Dynamics (Yet Another Preview)

Dynamics, which we will take up in Chap. 6, describe the way in which the flows through a network, the coupling strengths of links, the relative geometrical or logical positions of nodes, and the overall function of a network may change in time. If

---

[1] It turns out that flows and forces are actually the same thing where a flow imparts a force that was imparted to it at its source. The four forces of nature, in quantum theory, are often described as the exchange of particles, e.g., gluons for the strong nuclear force, between node particles, e.g., protons and neutrons.

you have ever seen a video[2] of an amoeba (a single-celled organism) crawling along under a microscope, you have a sense of the impact of dynamics on all of these dimensions. The amoeba shifts its shape around, sending out pseudopodia (false feet) as it tests the environment in multiple directions at once. It flows along the direction that suits it (perhaps tasting the molecules in the water for signals that food lies in a particular direction). It accomplishes its flowing transit via an internal web of protein molecules that are very much like those in muscle cells in multicellular animals, except that these molecules form in a mesh structure and form nodes where several cross over one another. What looks like a mess under the microscope, however, is an intricate network of molecules that allow the cell to crawl around over a surface. The system is extraordinarily dynamic, yet it is the same network of components merely changing relations in an ongoing manner.

The dynamics of the Internet are not nearly as complicated. Messages from one terminal node to another are broken into small packets and put into what amounts to digital envelopes with addresses (to and return) and sent into the network links. Routers are devices that receive packets and have some internal guidance as to which out-link to send the packet on its journey to its destination. Typically packets pass through many hundreds, even thousands, of routers on their way. Moreover, packets from the same message may take slightly different paths through the network as a result of lighter or heavier loads on particular routers. The routers keep track of how busy their neighboring nodes are and reroute packets to avoid heavy loads. The packets are just digitized electronic or optical signals moving under the laws of electronics or optical physics (including radio waves). When one looks at the traffic load at a single node, deep in the network, one sees some high and low levels fluctuating over time but a more or less steady flow. This is because the routing algorithms used in the routers attempt to optimize the load balance to prevent packets from being lost due to buffer overflows. Of course if there is some event, like a natural disaster, taking place and many, many people are trying to get news or send e-mails (and now Tweets!) at once, the system may still get clogged. The Internet transmission protocols provide for repeating the sending of packets that may have gotten dropped, but this just slows down the overall progress of getting a message through.

Studies of the dynamic attributes of the Internet have led to exciting discoveries about dynamics in all kinds of networks involving flows of stuff. Even some transportation networks (e.g., long-haul semitrucks) have dynamic properties similar to those in communications networks.

Question Box 4.3

Locate a video of a chaotic pendulum on the web. What is the essential linkage change that turns a clockwork pendulum, the essence of predictable regularity, into a chaotic system of unpredictable dynamic irregularity?

---

[2] Just Google "amoeba movement."

4.2.3 Organizing Principles

The above attributes can be measured in one way or another. When we do so and analyze over whole networks, we often see patterns of organization that were not obvious just by looking at a map of the network. This has become especially clear as network theorists have examined those networks that grow and evolve over time. That is, they are interested in the class of networks that are not just dynamic but show organizing behaviors that are a result of growth and/or change in composition and/or density. Such networks are in the class of systems we call complex adaptive systems (CAS), as they are lifelike, if not actually living, and show a regularity of responses to changes in their environments. Network theorists have thus far identified a number of important organizing principles that, as we will see in Chap. 10, Emergence, help us understand how these complex adaptive systems come into being and evolve over time.

4.2.3.1 Networks That Grow and/or Evolve[3]

A living multicellular organism, whether at a molecular, cellular, tissue, or organ level of organization, can be described as a network of those entities (nodes) linked by relevant connections (e.g., bonds, signaling, structural links, functional links, respectively). All such organisms start out as single-celled entities that must grow, divide, and differentiate through a development phase to become a reproducing adult.[4] That is, the networks, starting with the single cell, grow and develop new nodes by ingesting material and energy; they reorganize and form new linkages internally; and as the networks thus transforms themselves and the next new level of organization emerges and develops, they change their character and become a new level of network structure.

How this embryological unfolding works had long been a mystery in biology, but in recent years the ability to describe organisms as networks of interacting elements has allowed biologists to better understand the mechanisms. They now know that it is an intricate network of control elements encoded in the DNA, along with equally intricate epigenetic[5] mechanisms, that regulate the unfolding of form and function.

---

[3] We highly recommend a very readable book by Albert-László Barabási (2002) called *Linked: How Everything is Connected to Everything Else and What It Means for Business, Science, and Everyday Life*, Penguin Group, New York.

[4] Some exceptions to adulthood meaning reproduction exist but they are evolved to support the ultimate success of the species. For example, the worker ants or bees in colonies do not reproduce, but support the queen who serves as the main reproductive organ for the colony as a whole.

[5] Epigenetics refers to molecular processes that take place outside of the DNA molecules themselves but that regulate the transcription of DNA into RNA and eventually determine which proteins get manufactured in different cell types, e.g., insulin produced in specialized pancreatic cells. Some of these mechanisms can actually be inherited by offspring, so they are a nongenetic form of inheritance. See Jablonka and Lamb (2005).

The closely observed growth and evolution of the World Wide Web has made it a paradigm case for analyzing network change and evolution. Its vitality and adaptivity make it seem almost like a living entity. The web grows new nodes by the accretion of more servers and user computers. As it grows it also evolves by changing the composition and even the detailed nature of its linkages, so we find not just quantitative growth but new levels of organization emerging.

The unfolding transformations of biological growth represent perhaps the most complex and challenging case of network evolution. A more tractable case for study has been the growth of the web. Less complex than organic growth but one of life's most complex products, its growth has been carefully tracked from the beginning. When the web was first invented by Sir Tim Berners-Lee,[6] it had only static links between web pages and static pages. A page was edited in a special language called hypertext markup language (HTML) and stored in a server database. When a user clicked on a link, that sent a message from their browser to the server which then "served" up the page by sending the text of it back to the browser. The latter then rendered the document based on the embedded formatting tags in HTML.

That was a remarkable and worthy achievement. But it was only the beginning. As more and more computer scientists grasped the significance of the protocol, they saw how it could be used to move all kinds of content, not just text, and the nature of web pages and the way they were linked began to change. HTML itself was extended to accommodate the demand for a more dynamic, fluid linkage. Instead of static pages, dynamic HTML allows users to request content that changes dynamically such as the daily news or stock reports. The web of today becomes continually more complex, with an ever-expanding array of emergent functionality. It is true that the detailed changes to the protocol and language(s) used to deliver content today were made by engineers trying to improve the functionality of the web. But those changes were not as much part of an overall design as they were part of a process of auto-organization and emergence (Chap. 10). The web can be said to adapt to the many demands placed on it from its environment (Chap. 11), even as its new functions continually inspire new demands.

Now a new kind of network has emerged on top of the web in the form of *social networks*. These are not just connections between pages but between people communicating by both synchronous (chat rooms) and asynchronous methods (writing on a "friend's" *wall* in Facebook™). The end nodes here are really people, and they are forming highly dynamic linkages via formatted (like a protocol) dynamic web pages that include input editing and file sharing. These networks cannot be described solely in terms of the web, just as the latter could not be described solely in terms of the Internet underlying it. A new field of study is emerging just to analyze and understand the dynamics and evolution of social networks. It is still a very young

---

[6] Sir Berners-Lee, now Director of the World Wide Web Consortium (W3C), was a computer scientist working at the European Organization for Nuclear Research (CERN) when he devised the communications protocol (HTTP) and the architecture of hypertext linked web pages. For his incredible innovation and continued work on the evolution of the WWW, he was benighted by Queen Elizabeth II of England. See http://en.wikipedia.org/wiki/Tim_Berners-Lee

field, but as with its predecessor disciplines, it is ultimately based on understanding the systemness of its subject.

Social networks have long existed outside of the World Wide Web of course. Societal organizations such as businesses, religious communities, neighborhoods, and the like are describable as growing or evolving subnetworks that emerge within and continue to transform as part of the evolving human social network. What constitutes the links in such networks is extraordinarily varied in nature and strength. People can simply like each other, or they may share a common belief, or a common purpose. This is the realm of social psychology. The nodes (the individuals) are extraordinarily complex agents who are themselves growing and adapting (evolving in thoughts and ideas).[7] Systems sociology is starting to make headway in terms of analyzing and understanding social networks of all kinds. It is possible, though extremely difficult, to map out the nodes and links in a social network, but such maps must themselves be highly dynamic and evolvable. In Chap. 12 (Models) we will discuss a particular modeling methodology called agent-based modeling in which networks of interacting complex adaptive systems, which can make decisions under uncertainty (agents), can be used to demonstrate the evolution of a social system over time.

4.2.3.2 Small World Model

There are several patterns of network connectivity that seem to consistently emerge in growing/evolving networks. The first is called the *small world* or the "degrees of separation" model.[8] Most people have heard of the "six degrees of separation," or its popular online application, the Kevin Bacon Game. That game actually beats the six linkage steps that in theory can connect any two people on Earth. Working in the more limited community of Hollywood, the claim is that every actor in Hollywood can be linked either directly to Kevin Bacon (the actor) in a film, or by, on average, three links (some more, some less). In principle you pick any actor in Hollywood at random, and using a filmography database you can find within three films (on average), where the films are nodes and the actors are the links, one that Kevin Bacon has been in. Thus, out of the thousands (counting minor role actors, maybe hundreds of thousands) of actors in Hollywood-produced films, no one of them is separated far from Kevin Bacon. And on average no further than three degrees!

Analysis of many different kinds of systems networks shows this same kind of structural feature. It is a property that derives from the density attribute discussed above. The notion of degrees of separation was the subject of a famous experiment

---

[7] See Ogle (2007) for an insightful extended analysis of the emergence of a global information network he characterizes as an "extended mind." His book offers an excellent description of the attributes and dynamics of networks and their application and potentials for contemporary social, economic, and cultural systems.

[8] See Barabási (2002), ch. 4. We borrow his term, "small world," to describe this characteristic of networks.

by American psychologist Stanley Milgram. Milgram sent a kind of pyramid letter out to a fixed number of randomly chosen people nearby geographically. In the letter he asked the receiver who knew a certain person in a distant city to sign the copy and mail it to them or, if they did not know them personally, send it to someone they thought might know that person. The receiver also sent a message to Milgram so he could track the progress. The letter did get to the intended recipient; in fact several copies did, by different intermediaries. On average it only took six such intermediaries, hence the phrase, six degrees of separation.

The success of this experiment can be attributed to both the density of the network of people (due to the way people were spreading out, moving to new cities, etc., many people got to know many other people) and to the coupling strength of the connections (the mail system and its ability to get letters from place to place.) Social scientists would probably want to add the ingenuity of the agents (intermediaries) along the way. They probably thought hard about who they might know that would likely know the intended target. But really this is not too much different from what Internet routers do when deciding which next router they should send a packet to! The lower complexity of the Internet, compared with a human social network, makes it easier to solve the problem by algorithmic means (see Chap. 8, Computation, for a description of algorithmic problem solving).

Now it seems that wherever researchers look at networks that grow and evolve, they find this same small world phenomenon taking shape. It turns out that as networks grow they tend to do so in ways that take advantage of existing nodes and links, a dynamic Barabási (2002) characterizes as, "the rich get richer." This phenomenon is so consistent that it can be described as a mathematical relation: as a network grows linearly in the number of nodes added per unit of time, the separation between any two nodes grows logarithmically (Barabási 2002 p. 35). So on a base 10 (every node carrying 10 linkages), the log of 10 nodes is 1, of 100 2, of 1000 3, of a million only 6, and a billion 9. So any two people in a society of a billion would be separated on average by 9 degrees if each knew 10 other people. Clearly as networks become populated with more nodes, the connectivity of paths linking any two remains very short—the "small world" phenomenon. That is an amazing result and it seems robust across many different kinds of systems. So this is a case where the network principle can be used to generate hypotheses about completely newly described systems and provide guidance in what to look for in the structure/function/dynamics and evolution of that system.

4.2.3.3 Hubs

In a random network, all nodes are assumed to be equal. But many real networks are composed of heterogeneous nodes, which give rise to the phenomenon of hubs of connectivity within networks. This pattern emerges rather consistently when heterogeneous node networks grow, for not every node is equally attractive in forming new connections, so certain nodes acquire many connections and become what are called "hubs" (with the connections fanning out like spokes in a bicycle wheel—see Fig. 4.5 below). In the Internet this phenomenon can be seen when highly useful services are brought into the network. For example, search engine sites and social media sites are found to have many nodes connected to them because they provide a service that many people want.

A somewhat similar pattern exists in biological systems. We can construct a logical network comprised of the atoms that constitute living systems. Different atomic structures accommodate a greater or lesser variety of linkages. Accordingly we find that the carbon atom links to just about everything else when different kinds of

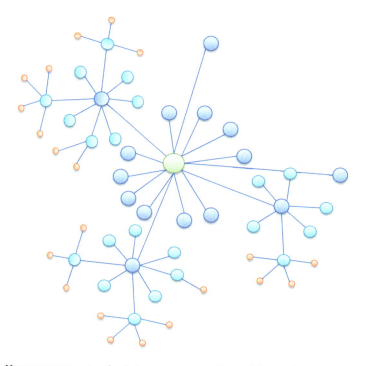

Fig. 4.5 Heterogeneous networks that are growing and/or evolving tend to organize in patterns that are consistent. Here a network is organizing as a few hubs with many connections, more with fewer connections, and many with very few or only one connection. Internet search engines and social network servers follow this pattern. See also Power Laws below

Question Box 4.4

Figure 4.5 could well be an aerial view of the layout one sees flying into any large city. What are some of the connective/differentiating factors that create urban, suburban, neighborhood, and rural hubs of varying density? How would changes in a selection of those factors modify the hub layout (e.g., new roads, cyber connections, WalMart, etc.)?

bonds are allowed. Hydrogen is linked to many other atoms. Oxygen and nitrogen are linked to a few others. Sulfur is linked to fewer still, and so on. This is the reason we talk about life on Earth being carbon based: it is the hub of all biochemical activities and biological structures.

4.2.3.4   Power Laws

Figure 4.5 also shows another pattern that is found in networks that grow and evolve with heterogeneous nodes (and links). The distribution of link counts for hubs follows an inverse power law. A power law is so called because frequency distribution is linked as a power of some attribute, giving a continuous distribution curve from high to low, which is very different from the sharp peak or bell curve associated with more random systems. The inverse power law means that there will tend to be very few hubs with the high-end number of links, a moderate number of hubs with a moderate number of links, and a very large number of hubs with very few (or only one) links.

Power laws show up in a large variety of heterogeneous networks and in connection with different attributes. We have described a power law in terms of the probability of occurrence of a hub of a particular size (number of links). But another example could be the phenomenon of clustering, which is also a common attribute of evolving networks. In clustering, groups form with stronger interlinkage and then link more weakly with other groups. Figure 4.6 shows a network in which clusters of densely connected nodes form. The density of their inner linkage follows a power law: a few will have many nodes with high density, a greater number will have intermediate numbers of nodes and intermediate density, and a majority will have low numbers of nodes and low density. This pattern emerges in social networks, for example, where friends and acquaintances tend to form clusters with occasional out-links to members of other clusters.

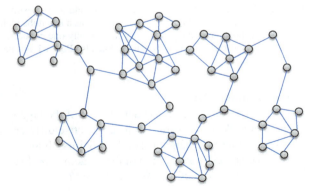

Fig. 4.6 Clusters can form in growing/evolving networks in a manner similar to the development of hubs in Fig. 4.5. Hubs and clusters can co-exist, e.g., a cluster may grow around a hub, but we show them separately to clarify the points

The dynamics of forming hubs and clusters is a function of the mechanisms underlying the formation of the links. These mechanisms vary with different kinds of networks, so networks will be found to have different patterns of clustering or hub formation. But it appears to be a general rule that evolving heterogeneous networks will develop these kinds of patterns and that the sizes/densities of hubs and clusters will follow a power law distribution.

An interesting aspect of power law distributions such as these is that they can be found to hold over a wide range of scales. For example, suppose the network shown in Fig. 4.6 is just a small portion of a much bigger network, say one that is 100,000 times larger (considering the number of users on Facebook, this is not much of an exaggeration). If we were to zoom out to take in, say, ten times as many nodes, we would find that the clustering pattern not only holds but we would likely find a cluster even bigger than the one at the top center of the figure and a fair number of those of that same basic size. Indeed, were we to zoom out again, to take in ten times more nodes, we would likely find a still larger cluster and many, many like the smaller clusters in the figure. In other words the size distributions of the clusters would hold to the power law at any scale (within boundaries). Somewhere in the world of Facebook™, there is likely to be a gigantic cluster of friends (probably around some popular rock star or Apple™ computers!)

We will be discussing power law functions in Chap. 6, Dynamics, as they are a particularly dominant and important feature of the dynamic characteristics of many natural processes.

Question Box 4.5
We form friends for many reasons, sometimes just because of who we like, but often combined with strategic, connective factors (e.g., who knows who). Say Fig. 4.6 is a social network of clustered friends, and you are a newcomer with this godlike view of everyone's relationship. Strategically, what are the implications (for various purposes) of where you begin making friends?

4.2.3.5   Aggregation of Power

The above kinds of patterns and the underlying mechanisms that give rise to them suggest a generalization that we might almost treat as a law of network dynamics and evolution. Namely, as the network grows over time, nodes like hubs and clusters that have more connections tend to attract more new connections. The way a network of friends in Facebook™ grows around a rock star may provide an example: the more the fans, the more likely it is to attract new fans. How often have you heard it said that the powerful get more powerful, or the rich get richer (see Barabási 2000, ch. 7)? These hub nodes and clusters attract "joiners" as the network grows, and in doing so become more attractive. This positive feedback of network aggregation seems much like the law of gravitation that works to aggregate stars and planets!

4.3   The Math of Networks

Representing the structure of a system as a network of components and their interactions not only has a correspondence with physical reality, but also it offers an opportunity to apply some very powerful mathematical tools to the analysis of systems. A field of mathematics called graph theory has developed methods for analyzing structures, functions, dynamics, and evolution in networks. Much of our above discussion of attributes and organizing principles of networks in fact derives largely from the outcomes of the application of graph theory to real systems such as the Internet.

4.3.1   Graphs as Representations of Networks

Graph theory uses a special language to construct purely abstract representations of networks. The representations are called "graphs," but these are nothing like the bar or line graphs and the like used to show data relations graphically. The graphs of graph theory rather are similar to the figures we have already been using to represent networks. Nodes are called *vertices* (as in a point where lines intersect) and links are called *edges* (for obscure reasons!). The vertices can be labeled with *identities* and edges can be labeled with *weightings* appropriate to the nature of the network and the questions of interest. The weightings are numerical measures of things like costs involved in traversing that particular edge (e.g., how much gas will it take to get from city A to city B).

Algorithms have been developed which identify pathways through the graph, starting at a particular vertex and ending in another vertex, by showing which edges are traversed in the path. Depending on the identities of the objects (vertices) and the weightings of the edges, a wide variety of questions may be addressed. For literal travel between locations, for example, when the graphs may have many vertices and many edges so that multiple possible paths are possible, this provides a method to determine what is the shortest distance (or least cost) path from vertex A to vertex Z? Other questions may involve finding the kinds of clusters and hubs shown above. Marketers and advertisers, for example, may be interested in finding subgraphs in the larger graph that have higher density (where the vertices are densely linked). See the Quant Box below for an example of graph theory at work.

Quant Box 4.1   Quantitative Issues

A mathematical tool that allows us to analyze all of the characteristics of networks of components and their interconnections is graph theory. In its abstract form a graph is a set, G, comprised of two subsets, V and E. V is a set of vertices (also called nodes in network applications) and E is a set of edges (links or connections). Thus, $G = \{V, E\}$ formally.

(continued)

Quant Box 4.1 (continued)

Vertices are point objects where edges intersect. Each vertex has a unique identifier (such as an integer) in order to fulfill the definition of a set. However, a vertex can be labeled in a number of different ways such that one can designate types of vertices, contents (see below), and coordinates in some appropriate space and time. Similarly, edges can be labeled with, for example, weights that represent costs associated with traversing from a source vertex to the destination vertex. Edges are designated as a binary tuple, $\{Vi, Vj\}$, where each vertex is at one end of the edge.

Graph objects (what we would call a system) can be easily represented in computer formats, and there are powerful algorithms that allow one to examine properties of specific instances of such objects. These algorithms start with efficient search techniques that allow a computer program to work through a graph, following edges and determining pathways from vertex to vertex, in terms of the list of edges that one must traverse to get from vertex i to vertex j. A list with only one edge in it means that the two vertices are directly connected. From these efficient search methods, it is also possible to ask questions like: What is the least cost pathway from Vi to Vj (where i is the start and j is the endpoint)?

Figure QB 4.1.1 shows a simple directed graph. A directed graph is one in which traversal from one vertex to a connected vertex is unidirectional. Undirected graphs assume that a traversal between two connected vertices can go both ways. In this graph the set V would be: $\{1, 2, 3, 4, \ldots 9\}$. The set E would be: $\{\{1,3\}, \{2,3\}, \{3,4\}, \{3,6\}, \{6,7\}, \{7,2\}, \{7,8\}, \{6,8\}, \{6,5\}, \{5,4\}, \{8,9\}, \{9,5\}\}$. It should be easy to see that a path from vertex 1 to vertex 2 requires traversing first to vertex 3, then to 6, then to 7, and finally to 2. Thus, path $P1,2 = \{1, 3, 6,7, 2\}$ and cost $C\{P1,2\} = 30 + 21 + 3 + 15 = 69$ (in whatever units are being used).

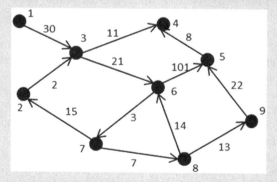

Fig. QB 4.1.1 A simple graph with 9 vertices and 12 edges. Vertices are labeled with unique identifiers and edges carry weights which could represent costs of traversal. This graph is called a "directed" graph since the *arrows* only point in one direction

Question: What is the lowest cost path from vertex 2 to vertex 5 in the graph in Q1?

Another version of objects in graph theory is called network flow graphs. In these graphs edges carry some kind of flow (material, energy, information) that is quantifiable. Vertices represent convergence or divergence points in the flows. Flows only go in one direction (a directed graph) but a two-way flow can be accommodated as needed by simply having two arrows going opposite directions.

It should now be clear that graph theory holds a tremendous amount of analytical power for working with networks (e.g., the flow of packets through routers in a packet-switched digital communications network), or essentially any system in which components are connected. Force connections, for example, can be represented in a kind of undirected graph where two-way arrows are used. Arrows pointing toward vertices could represent mutual repulsion, while edges without any arrow points would represent mutual attraction. The weight value would represent the coupling strength, and the addition of a time index would allow analyzing changes in these strengths over time. Similarly, flow networks provide a powerful way to represent and analyze flow-based connectivity in system components. See the references at the end of the chapter for some very good books on graph theory.

4.3.2 Networks and the Structure of Systems

In Chap. 3 we explored the idea of observing a system from outside of its boundary, viewing it as a whole. This perspective allows us to treat a system as an *object*. Objects are what we most readily perceive in nature and so it seemed prudent to start from this obvious fact. However, objects (and objecthood) are really only the superficial perception of systemness. In Chap. 3 we introduced a number of examples of sets of objects that were connected or linked to one another. Different kinds of linkage among objects create different sorts of systems.

Objects may be linked together by *physical* connections. For example, as we saw in Chap. 3, neurons are linked together by axons, acting something like wiring between components. Electrochemical impulses travel along these links, propagating from one neuron to the next. As we will see later, however, the physical linkage is just part of the picture. We need to also understand the strength of what we might call "affective" linkage, or how strong the dynamic properties of the link are, and how such factors cause changes to occur in the downstream object.

But we also saw in Chap. 3 the nature of conceptual linkages, or networks of *relations* between objects. There can be many kinds of relational links in a network. The connections do not represent physical linkage per se, only logical linkages. For example, say one object is Person A with a link to Person B. The link might be labeled "knows" meaning that Person A knows Person B and vice versa. Or Person A might be linked to Person B as "boss of," meaning that Person A is Person B's supervisor at work. These relational linkages are every bit as "real" as are the physical linkages. Relational networks are the basis of conceptual maps, or the ways that we have of diagramming how various components of conceptual systems relate to one another.

Linkages, physical or logical, determine the structure of a system internally. Physical links determine physical structures and relational links determine logical structures. What is remarkable is that both kinds of systems can be described and analyzed using the same basic language, network theory using the mathematics of graph theory.[9]

4.4 Networks and Complexity

The concept of complexity comes up again and again in various contexts. We will devote a whole chapter (Chap. 5) to the concept (or perhaps we should say concepts since there are many ways to look at complexity in systems).

This word, complexity, has already shown up as a topic in the last chapter where we introduced the notions of potential and realized complexity, along with a hint of the difference between a focus on components and on organization. We will revisit this word and its significance in a much more holistic fashion in Chap. 7. But in this chapter we need to broaden the basic understanding of complexity since it now starts to dominate the semantics of many other principles we will be covering. Immediately, this involves the concept of networks in systemness.

4.5 Real-World Examples

Study of the World Wide Web has yielded valuable insights into network organization, growth, and dynamics, making it a favored network example. But networks have far-reaching importance in the function of all sorts of complex systems, with particular relevance to virtually any living or life-produced system. We will turn, then, to the world of life and conclude this chapter, with examples drawn from a progression of systemic levels, from the metabolic processes of organisms, to eco-systems flows that enable life, to the food webs that support it, and finally conclude with the manufacturing organizations that typify the way humans build out their environment to enhance their lives.

---

[9] Graph theory is a rich mathematical formalism for working with graphs, network objects such as shown in Fig. 4.1. For background see http://en.wikipedia.org/wiki/Graph_theory

4.5.1 Biological: A Cellular Network in the Body

Biological systems provide some of the very best examples of network organization leading to functional stability. Bio-systems, which range from bacteria to populations of mammalian species (such as humans), are characterized by much richer interconnections between components at a given level, many levels of interconnected organization (e.g., from the biochemical, through the tissues, to the whole body), and highly probabilistic interconnections. This latter point means that balance and proportionality play an important role in biological systems, offering both flexibility and ways of going wrong not shared by machines or machinelike mechanisms.

In Fig. 4.7 we depict three cells in the body: two of them are insulin-producing cells from the islets of Langerhans in the pancreas, and the other is a single muscle cell. The two pancreatic cells are bound together by an adhesion protein that makes them stick together to form a tissue. These cells secrete insulin when blood sugar levels are high, causing the muscle cells (and all other tissues) to absorb the blood sugars for metabolism. The insulin is shown as narrow single-directional arrows,

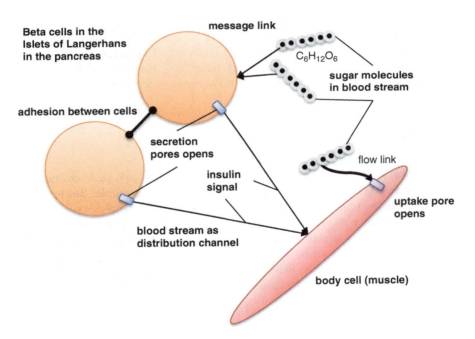

Fig. 4.7 Beta cells in the pancreas, islets of Langerhans, can detect the presence of sugar molecules in the blood and then release insulin to signal muscle cells to absorb the sugar for metabolism. This is only a small part of a complex network of communications (signals or messages) that regulate blood sugar (glucose) and help the muscle cells take up the energy packets of sugar. The two beta cells secrete through pore molecules (as discussed above). The muscle cell similarly has a pore structure that opens to allow the uptake of sugar. We also see an example of a strong physical link between the cells in the form of adhesion that keeps them together as a tissue

whereas the adhesion is shown as a thick two-way arrow. A sugar molecule near one of the pancreatic cells provides a signal to the cell to start releasing insulin. The muscle cell then captures any sugar molecules in its vicinity (curved narrow arrow).

Suppose we take a look at all of these network connections. The connection between two pancreatic cells (simplified in this example—there are more connections in the real cells) is by way of a shared molecule that binds onto the surface of the cell's membrane. This binding is what keeps the cells together working as a unit or tissue. The adhesion molecule may be produced by entirely different cells, and it ends up available in the intercellular medium where it attaches to the pancreatic cells automatically. This is an example of a force connection, in this case, an attractive force.

The pancreatic cells are sensitive to the presence of sugar. The details are not important (and they are complicated), but suffice it to say that the presence of a high concentration of sugar molecules will transmit a message to the cell telling it to start producing insulin. The message is conveyed in the mere presence of matter (the sugar molecule). The pancreatic cells then secrete insulin molecules into the blood stream to signal all other cells that sugar is available in the blood and needs to be absorbed. The reasons for having special cells in the pancreas that do this, and not simply having every cell signal itself when sugar is present, is that sugar regulation is very finely tuned in the body and must be controlled carefully and in a coordinated way so that cells do not get either too much or too little. The pancreas thus serves as the coordinator and controller of this vital process. When the insulin system is compromised, it can result in a disease such as Type II diabetes.

In this case insulin molecules from the pancreas convey a message; the concentration of insulin is what affects the other body cells to cause them to absorb and use the sugar. We indicate the absorption connection between sugar and muscle by a curved one-way arrow. Here this indicates the flow of material into a component (as a subsystem) that also conveys a message.

Thus, we see how a combination of force connectivity and message flows controls sugar uptake. This is even more complex if we consider that the process here is not just a matter of molecules but moves from a molecular level to the functions of whole organs, back to cells but then to the function of various sorts of tissue, and finally to the healthy state of the whole body.

4.5.2 The Earth Ecosystem as a Network of Flows

Systems ecology studies ecological processes as a network of components (e.g., organisms and physical forces) and their relations. It is particularly concerned with the flows of energy and materials through the networks that produce the complexity and sustainable biomass of, for example, a climax ecosystem.[10]

---

[10] A "climax" system is distinguished by a relatively sustainable equilibrium, in that the varieties of plants cohabit and may even mutually benefit one another, versus systems in transition as some forms of plants drive out and replace others.

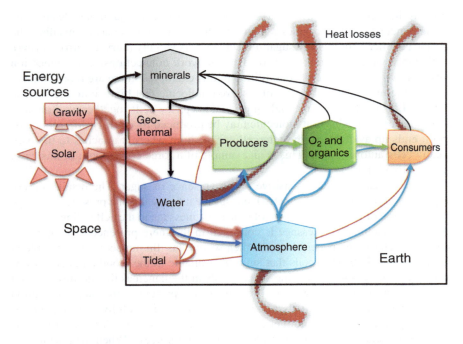

Fig. 4.8 A high-level representation of the Earth ecosystem shows the complex network of relations between the major components of the Earth system. This is a systems ecological diagram that maps the flows of energy and materials from and to reservoirs (e.g., water, atmosphere, producer, and consumer biomass). Primary energy comes from the Sun in the form of light and gravitational influences (along with the effect from the Moon). All energy eventually flows out of the system as waste heat dumped to outer space

Here, in Fig. 4.8 above, the network is represented by what we call a "stock-and-flow" model, which we will investigate in Chap. 12 in greater detail. The various shaped nodes represent physical stocks of substances (both matter and energy) as well as the internal processes or transitions that take place. For example, the half-box/half-circle shape labeled "Producers" is the stock of all photosynthetic plants that are converting solar energy into organic molecules (food for consumers) and the oxygen that enters the atmosphere.

The arrows represent the flows of energy and matter between the stocks. Solar energy enters the systems affecting the photosynthetic plants, the hydrosphere (e.g., evaporation) and the atmosphere (e.g., generating winds). Gravity (from the Sun and the Moon) also has energetic impacts on most aspects of the system (though most are not shown since they tend to be minor). Matter (minerals) already in the Earth system is recycled in many different ways.

Figure 4.8 provides a highly *macroscopic* view of the entire Earth ecosystem. All more local or regional ecosystems have all of the same components and flows as shown in that figure. All that varies is specific details of configurations (e.g., mountains and valleys versus plains). Here we see the value of capturing the relational transformative network of primary flows, which is valuable information in itself and also provides a structure that both suggests and frames questions for further levels of analysis as we look into the internal processes of the various nodes and their external networked relations to the rest of the system. In the next section we provide another level of analysis to investigate a more local example of an ecosystem, and in particular the flow of material and energy in what is called a "food web."

Question 4.6
One use of a diagram of primary flows is the way it indicates how questions in any given node are intertwined with what goes on in other nodes. Global warming is an atmospheric phenomenon, but following arrows into and out of the atmosphere, what related issues and questions immediately present themselves?

4.5.3   Food Webs in a Local Ecosystem

Compared to the tight and regulated organization within an organism, ecosystems composed of multiple species of whole organisms are much more open to change. But they too are woven of interdependencies in which needs for habitat, nutrition, and reproduction are met by the dynamic connectivity of the multiple inhabitants of a given environment. The selective pressures, which maintain a certain equilibrium or weave new versions of the system after critical changes are introduced, reside in the interwoven needs and opportunities provided by a system in which every creature is simultaneously part of an environment for every other creature. The selective key here is mutual fit. As Darwin observed, the fit survive; that is, their needs are satisfied by their environment, and their activities in meeting those needs do not destroy the environment upon which they depend. In fact, there is even a selective reward for hitting upon strategies that enhance the network of life upon which a species depends, giving rise to cooperative symbiotic strategies along with the predatory relationships that first come to mind when we think of the great nutritional energy flow called a food web.

The dynamics that realize the ongoing, more or less balanced, life of an ecosystem sometimes become clear when the balance is disturbed: our attention is often drawn to how something worked when it ceases to work that way. Our example here will be a simplified version of the ecosystem of the Bering Sea kelp forest habitat in the region of the Aleutian Islands off the coast of Alaska. The kelp forests provide habitat for a wide range of marine life. Here we will simplify our systemic consideration to eight players caught in a recent dynamic of potential systemic unraveling or collapse.

We begin our analysis with a typical scientific reduction of the system we wish to consider to its components: orcas, sea otters, sea urchins, seals, sea lions, large fish, small fish, and kelp. Describing the components would include considering for each, minimally, the two most basic facts of life, feeding and reproduction. Consideration of feeding, i.e., what eats what, takes us to the level of synergistic analysis, the relational energy flow dynamic that moves us to analyzing the system as a system rather than just a collocation of components. This dynamic relational flow sustains the system as a system. Orcas eat seals and sea lions, which eat large fish, which eat smaller fish that hide out in the kelp. This is intertwined with another web, sea otters eating sea urchins which feed on the kelp. Notice the role of kelp as both food source and as habitat, a dual role that will become critical.

But maintaining the system as food web critically depends upon the components' reproduction as well, for only by each population reproducing itself in more or less expected proportions can the food web be maintained. Recall here our comments above about probabilistic interconnections and balance. The size of any population is constrained by the size of its food supply and clearly the food supply must reproduce at a rate and quantity to maintain itself in the face of the rate of predation. Thus, the size of the component populations and relative rates of consumption and reproduction are interwoven in a network of dynamic interdependencies. One can anticipate here both the systemic need for flexible give-and-take to withstand fluctuations and the presence of limits that, if transgressed, might degrade or even crash the system.

The full significance of any component can be understood only in terms of the whole system to which it belongs, for the dynamic consequences of changes to a given component varies with its place in the systemic structure. And because of networked interdependencies, the consequences of a given change can be hard to anticipate. Ecologists speak of "keystone species," species whose change or absence can unleash a cascade of change that unravels a system. Students new to ecological systemic interdependence sometimes get the feeling that every species must play a keystone role, but systems would be far too brittle if such were the case. But systemic structure tends to include points of convergent dependencies, just as the forces of an arch converge on the single keystone.

The structure of the Aleutian Island kelp forest community (see Fig. 4.9) has been highlighted by a crisis in the sea otter population, which has crashed dramatically, from a high of around 53,000 to about 6,000 in 1998. The reasons could be many, but scholars have noted repeated observations of orcas consuming sea otters, a phenomenon first observed in 1991. The ordinary diet of orcas in this system is seals and sea lions, which in turn feed on large fish. But heavy commercial fishing has greatly reduced stocks of large fish, which has in turn impacted the seal and sea lion communities. Orcas on short rations have options, including just moving on, so it was a surprise that they turned instead to feeding on sea otters, which at 50–70 lb are a mere snack compared to thousand-pound sea lions. Scientists calculate that consuming five otters apiece each day, a single pod of four orcas could account for the total decrease in the otter population.

The Aleutian kelp forest system may have a hard time surviving without the sea otters, which are a keystone species in the system. The problem is that the population of prickly sea urchins which feed ceaselessly upon kelp goes into a

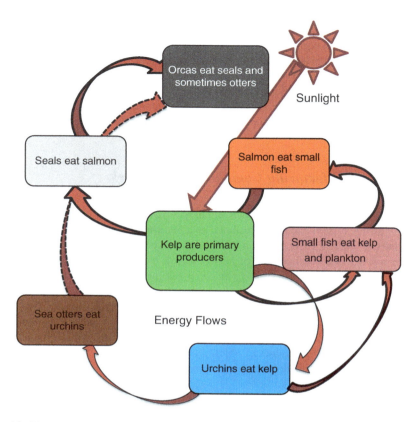

Fig. 4.9 The Bering Sea food web is typical of ecosystems. The kelp beds are the primary producers, converting sunlight into biomass which can then be consumed by a whole host of animal organisms, including the sea urchins. In addition, the beds provide a breeding ground and nursery for a wide variety of smaller fish that are then consumed by larger fish, such as the salmon. Sea lions and seals eat these larger fish and, in turn, may be eaten by orca whales. The latter are at the top of the trophic (energy flow) hierarchy (see Chap. 11). Ordinarily orcas would not eat sea otters since they are small. The sea otters provide an essential service to the ecosystem in that they eat urchins, helping to protect the kelp from overgrazing

reproductive positive feedback loop unless kept in check by the otters, for which they are a favorite menu item. More urchins will produce more urchins, and the kelp cannot adapt its growth rate to keep up. If the kelp habitat is overgrazed by a burgeoning population of urchins, the nurseries and protective cover for small fish and the hunting grounds for all the other members of the system disappear, replaced by a much more impoverished marine system sometimes referred to as "an urchin desert."

One sees here the critical interdependence of population sizes, reproduction rates, and feeding relationships in maintaining a stable system. When such a system is in balance, we can see the parts and relationships clearly, but the consequences of changes follow such complex and indirect relational paths they are often difficult to predict. Experts in various sorts of complex systems often see potential problems that seem as counterintuitive as the notion that overfishing could hurt the kelp. This facility for seeing possibilities in complex networks that skeptics see as groundless doom and gloom sometimes gives ecologists a bad name, but this is exactly the kind of understanding needed as we impact and alter complex living systems.

4.5.4 A Manufacturing Company as a Network

As with the examples above, the flows of energy, materials, and messages through a manufacturing system can be represented as a heterogeneous network of nodes and flow links. Figure 4.10 provides a fairly simple version of this idea. In the next chapter we will see another kind of network view of an organization based on the management hierarchy. For now we can see the company as in the figure.

The nodes in this network are the operational units and the management units. Material management keeps track of the availability of parts and orders from vendors when the inventory is low. Parts are received into the inventory that acts as a coordinating buffer for manufacturing. The latter uses energy inputs to transform the parts into finished products, which flow into a product inventory awaiting purchase by customers. The management nodes (purple) process and communicate

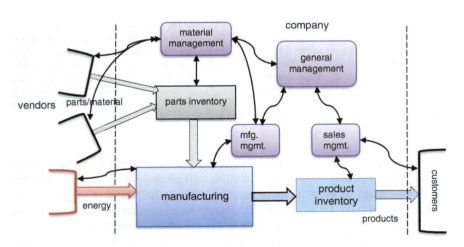

Fig. 4.10 A manufacturing company provides yet another kind of network representation of a system. Here the flows of matter, energy, and messages provide the networking. The *two-headed arrows* represent a two-way communication. The *dashed lines* show the fuzzy and conceptual boundary of the company. Here the flow of energy into the company can be thought to also represent labor (workers coming to do their jobs!) as well as electricity to run the machines. Quite a few other details are not shown just to keep this example simple

messages which coordinate and control phases of the process. Note that arrows to management nodes are double-headed, indicating the information feedback loop structure essential to the control function in complex systems.

This is a highly simplified view of a firm. Each of the nodes could be decomposed to find very complex subsystems within. Even a small manufacturing company has a complex network of flows and communications that constitute the structure of the company. Organization charts are another kind of network representation of a company based on the management structure, but that network (a top-down tree) can be mapped to something like the above representation.

The real beauty of being able to represent and think about systems like a manufacturing company (or any organizations really) as networks is the fact that representations of the network can be reduced to more abstract forms that are amenable to computational analysis and thus even to the designing of control models for the organization (this subject will be more fully explored in Chaps. 7–9). In Fig. 4.11, below, we show an alternative representation of the manufacturer in Fig. 4.10 but now with simple nodes and directed arrows replicating the more maplike version seen above. The nodes are labeled with the same name of subsystem as in Fig. 4.10.

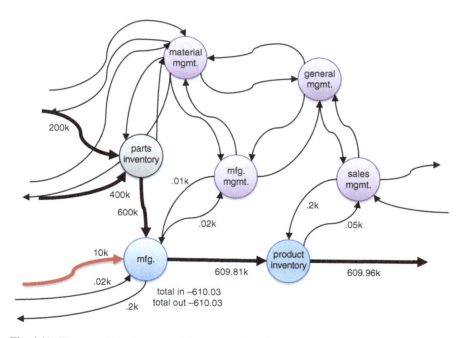

Fig. 4.11 The manufacturing network has been reduced to an abstract version in which the flow of a single resource (monetary cost) flows are indicated as we analyze this network for a specific purpose (see the text). Here the *arrow colors* represent the original type of substance in the flow (e.g., materials, energy, and messages). The nodes are reduced to simple labeled circles because the details are not important for this analysis. The *double-headed arrows* of Fig. 4.10 have been replaced with *labeled single-headed arrows* to show that different values may flow in different directions. Only a few arrows have been labeled for simplicity

But the links are now weighted with numerical values representing a volume per unit time flow translated into its monetary value. This allows us to analyze how *money flows through the system as costs* move from one node to another.

In this example, then, we are only looking at the dollar values associated with the flow of materials, energy, and messages. The red arrow accounts for the use of energy in the manufacturing process. The dollar values of message flows represent the cost of producing the information in the message. From left to right if we track the costs of basic materials (parts, say) coming into the system, along with cost of energy and the administrative costs associated with message flows, we can potentially spot problems. Note that the total flow of costs into the node labeled "mfg" equals the total cost flow out of the node. This follows from the simple fact that (in theory) there is a conservation of money value rule that has to be upheld. The total flows out of a node must equal the total flows into the node in the same time frames. There are flow network algorithms that can be used in a computer to analyze the situation for the whole system to detect discrepancies if any exist. A standard cost accounting system has been used for ages to do this using bookkeeping methods and human calculations to detect problems.

The computer's ability to analyze networks and network flows introduces a broad range of applications for network models of business and economics. For example, beyond accounting, this kind of cash-flow network model has applications in various sorts of planning. Suppose management wants to increase manufacturing throughput by some dollar amount, for example. With a computer model implementation of the system, it would be easy to test the implications of this kind of change. Assuming one would want to keep the same relative proportions of inputs to maintain dynamic balance, if they set a dollar amount, say exiting the product inventory, the program could backtrack through the flows to determine what upstream adjustments would be needed to maintain a balance of inputs and outputs from every node affected. Not only cash flows but the effects of changes in any sorts of complex but interdependent network flows can be modeled and studied this way.

Think Box Abstract Network Models as Systems
Now we have looked into the black box and find that systems are internally organized in complex networks of components whose interactions work through the network links (flows and forces). Abstract network models can have additional link types. For example, in Think Box 3 we mentioned concept maps and showed such a map above in Fig. 4.3. The links there are labeled with terms like IS-A, A-KIND-OF, and HAS-A. These are relational links rather than flows in the ordinary sense. They show the relationships between concept nodes in terms of spatial, temporal, membership, etc. much like a prepositional phrase.

(continued)

Think Box (continued)

As we mentioned in Think Box 3, networks of neurons in the brain are real physical instantiations of concepts, and it turns out that the kinds of links we see in the concept map are realized in neuronal links (through synapses) so that we could say there really is a flow of messages between concept clusters. In other words, the brain actually forms concrete networks that resemble what we ordinarily think of as abstract networks. The brain may just be the most ideal modeling platform of all!

Working in the other direction, though, that is, from real physical systems to abstractions, is one of our most powerful tools for thinking about those systems. Any of the real-world examples of networks (Sect. 4.5) can be represented as an abstract graph using nodes and links as shown in Sect. 4.2. Such graphs can then be analyzed for their structural properties. Using the graph methods described in Quant Box 4.1, one can determine if a network has, for example, a small world structure where any two nodes are separated in the number of jumps (intermediate nodes) by a small number, e.g., degrees of separation. Or one can discover hubs in a complex food web (Sect. 4.5.3) that represent keystone species, those that have the greatest influence on the entire web.

Another example of using network modeling is in analyzing the flow of packet traffic in the Internet communications system. Packets are the units of flow, containing just pieces of an entire data stream. The original data is broken up into these packets and sent into the network to be routed to their destination (each packet contains the destination IP address and a sequence number in case they arrive out of order). Network traffic can sometimes cause delays in packet deliveries, and the Internet routing protocols are designed to try to balance flows through the various routing and switching nodes. Once again, the physical details of the network are unimportant in analyzing the system since just a few laws apply to the behavior of the flows.

Figure 4.11 shows the manufacturing system as an abstract network and provides some clues about how one kind of flow (that of money or value) through the system can be analyzed to determine how resources need to be channeled.

Graphs like this are very likely just external (on paper, as it were) representations of exactly what is going on in our brains. The abstract graphs reduce representation of the objects to mere nodes and their relations to labeled arrows. All of the features of the nodes have been stripped away so as to not clutter the picture and that allows us to focus on the major interrelations that make the whole a system. In your brain your concept maps are graph-like, but you have to concentrate consciously to subdue the features and focus on the main components.

(continued)

Think Box (continued)

And that brings us to what language is! Words, particularly nouns and verbs (along with their modifiers), are just the names of nodes and links in the graph. When we express a concept in words (sentences), we are producing a very abstract version of the graph/map. The name Fido can be invoked in a sentence representing a larger dynamic concept (i.e., Fido did something significant!) without flooding your mind with a complete vision of Fido and his dog-ness.

When you diagram a sentence, are you not producing a kind of graph based on the relations the words have to one another?

Bibliography and Further Reading

Barabási AL (2002) Linked: how everything is connected to everything else and what it means for business, science, and everyday life. Penguin Group, New York

Anderson WT (2004) All connected now: life in the first global civilization. Westview Press, Boulder, CO

Capra F (1996) The Web of life. Anchor Books, New York

Capra F (2002) The hidden connections. Doubleday, New York

Jablonka E, Lamb M (2005) Evolution in four dimensions. The MIT Press, Cambridge, MA

Ogle R (2007) Smart world: breakthrough creativity and the new science of ideas. Harvard Business School Press, Boston

Chapter 5
Complexity

"I shall not today attempt further to define the kinds of material I understand to be embraced ... [b]ut I know it when I see it ..."
United States Supreme Court Justice Potter Stewart, 1964
"...complexity frequently takes the form of hierarchy and... hierarchic systems have some common properties independent of their specific content."

Herbert A. Simon (1996. The science of the artificial,
The MIT Press, Cambridge MA, p 184)

Abstract Complexity is another key concept in understanding systems, but it is not an easy concept to define. There are many approaches to understanding complexity and we will review several representatives. However, we make a commitment to a definition that we feel is most compatible with the breadth of systems science, Herb Simon's (The science of the artificial. The MIT Press, Cambridge, MA, 1996) concept of a decomposable hierarchy (as explained in Chap. 3). Systems that have many levels of organization are, generally speaking, more complex. This definition will come into play in later chapters, especially Chaps. 10 and 11 where we look at how complexity increases over time. Toward the end of the chapter, we examine some of the downside of higher complexity, especially as it affects modern civilization.

5.1 Introduction: A Concept in Flux

Justice Potter was having a hard time trying to obtain a precise definition for the controversial "entertainment" (hard-core pornography). The term "complexity" shares this attribute with that form of entertainment; it is extremely hard to define in a precise way. Yet define it we must if we are to make any great use of it as a scientific principle. As of this writing there are still several disparate approaches to producing a workable definition. In spite of this, there is sufficiently widespread agreement that there is a quality about systems that is embodied within our intuitive notion of complexity (or its opposite—simplicity). We know a complex system (or think we do) when we see it.

© Springer Science+Business Media New York 2015
G.E. Mobus, M.C. Kalton, *Principles of Systems Science*, Understanding Complex Systems, DOI 10.1007/978-1-4939-1920-8_5

The concept of a complex system or the notion of complexity itself has garnered a great deal of attention in the academic research world of late. The intuition of what is meant by complexity has a long history in philosophy and science. Indeed up until the early part of the twentieth century, science was restrained in its ability to tackle really interesting phenomena in the macro world by the fact that it did not have the mathematical tools to deal with nonlinear systems readily. With the discovery of deterministic chaos[1] and associated phenomena, made possible by electronic computation, interest in the concept of complexity as a formal property of many (perhaps most) natural phenomena took off. Today there are many research centers devoted to the study of complexity and a fair number of journals, textbooks, and papers covering the mathematical/computational investigations of complexity.

In this chapter we are interested in a systems perspective that seeks a more holistic notion of complexity than is often found in the detailed investigations covered in the literature. What we hope to accomplish is the synthesis of a general concept of complexity based on the integration of several foundational approaches. We will leave the surveys of the investigative tools used to grapple with complexity to the many fine books listed in the bibliography at the end of this chapter.

The approach we adopt for the exploration of complexity was developed by Herbert Simon, that of structural and functional hierarchy of nearly decomposable systems.[2] This follows from what we covered in Chap. 3 on organization. We will develop this view as it seems the most compatible with the breadth of systems science principles (Chap. 1). For example, much of the descriptive work on complex systems involves networks (as in the last chapter) since complexity involves not just the components of a system but also, and especially, the connections between those components.

Later in the chapter we will introduce some other perspectives on complexity. But we will see that seemingly different views of the subject can be related to the notion of complex hierarchies.

5.2 What Is Complexity?

In Chap. 3, we gave a brief introduction to the notion of complexity as a concept relevant to the organization and structure of a system. In this chapter we will focus on the concept in much greater detail and develop basic principles that will be used in subsequent chapters, especially Chaps. 10 and 11, Emergence and Evolution.

---

[1] The subject of chaos theory will be explored in Chap. 6, Dynamics. We mention it here because of its triggering effect in kicking off the quest for understanding the nature of complexity.

[2] Simon (1996). Especially Chap. 8, "The Architecture of Complexity: Hierarchic Systems," p 183.

From here on we will speak of "complex systems" rather than treat complexity as a stand-alone concept (see below). Such systems have properties we can generally agree upon:

- Complex systems often display behaviors that surprise us. We cannot easily predict what a complex active system will do next, even when the external conditions are seemingly the same.
- Complex systems require that a considerable amount of work be done in order to "understand" them.
- Complex systems cannot be easily described, and certainly not by simply listing their parts (obtained from the previous concern).

When we say that a system's behavior is surprising, we mean that the observer had some kind of a priori expectation of what that system would do next and the system didn't do it. In Chap. 7 we will delve into the deeper nature of what is called *information*, but for now we claim that information, as a property of messages, corresponds with the amount of surprise or unpredictability that comes with a particular message received by an observer. What we will see in this chapter is how the complexity character of a system can give rise to surprising, and we will argue, interesting, behaviors relative to the observer.

Understanding a system is actually a deep philosophical (epistemological) issue; there are various levels of understanding we need to consider. It is one thing to "know" what the parts of a system are, another to "know" how the parts interact with one another (or can interact), and yet another to know enough about how those parts do interact in a specific form of the whole system. In a very real sense, understanding is somehow the opposite of surprise. The more we understand a system, its composition, its structure(s), its intrinsic behaviors, and its reactions to environmental contingencies, the less often it will surprise us. But as we will see, understanding really means that we have incorporated some of the system's complexity into our own model of that system, i.e., our model (understanding) has become more complex! Chapter 7 will also introduce the notion of *knowledge* as the dual of information.

One possible measure of complexity might come from a ratio of our amount of surprise to our amount of understanding. In other words, when we think we have a high level of understanding and still a system surprises us with its actual behavior (at times), then it is because the system is more complex than we realized. Unfortunately, since the notions of amount of surprise and amount of understanding are still vague, this measure is at best conceptual. Nevertheless, it seems to fit within our intuitive understanding of complexity.

Building an understanding of a system takes work, as we discussed in the last chapter. We have to analyze the system, we have to analyze the components, we have to test behaviors, and so on. All of this takes effort on our part. So understanding comes with a cost. This cost, say in energy terms, is another kind of indexical measure for complexity on its own, though only a rough one. Nevertheless, it gives us a sense of how complex something is when we have to work hard to grasp it (e.g., think about how hard it is to learn math).

We understand a system when we can make more reasonable estimates of what it is likely to do (behavior) under observed conditions of the environment. But additionally, we, as abstract concept builders, can say we understand it when we can describe it in some model language that can be interpreted (understood) by others. The amount of work, again, needed to *record* a full description of a system is also a kind of indexical measure of its complexity.

These kinds of attributes provide the underpinnings for the several approaches to define complexity that have been developed so far. We will spend some time reviewing them below as background for what we propose in this chapter as a working definition of complexity, i.e., one that can be applied to systems in general.

5.2.1   *Intuitions About Complexity*

Let's make a start by first reviewing the kinds of common intuitions that most people share when they think of complexity. If asked to list off some attributes, many people might say things like this.

A complex system is one that has a *large* number of:

- Different kinds of parts (components)
- Each kind of part
- Kinds of interactions between different kinds of parts
- Part aggregations
- Behaviors of parts and aggregations of parts (tightly coupled in a network)

In this approach the phrase "large number of…" is common in that most people do think that complexity relates in some way to the size of numbers used to measure these attributes. It turns out that in real physical systems[3] the first characteristic, large number of kinds of parts, is often correlated with complexity, especially of behavior. But the largeness of the numbers of each kind, or kinds of interactions, is not necessarily a prerequisite. We can think of all kinds of systems that have complex behaviors with very few numbers of each kind or kinds of interactions. It is also possible to find examples of complex behaviors coming out of systems with few kinds of parts but large numbers of each kind.[4]

Part of the problem with these intuitions is that the term "kind" is not really that well defined. For example, consider a rushing mountain stream cascading over rocks. Such a stream displays extraordinarily complex behavior as it rushes down the mountain over rocks and tree stumps. The swirls and waterfall mists perform all sorts of chaotic dances. We abstractly call the stream turbulent.

---

[3] Remember, for our way of thinking, this is actually a redundant phrase since we claim all systems, even so-called abstract ones, have physical embodiment in one form or another.

[4] This is the case for cellular automata (see http://en.wikipedia.org/wiki/Cellular_automaton). Wolframn (2002) suggests cellular automata rules may be the law of the universe. We will say more about cellular automata and variants in the realm of "artificial life" later in the chapter.

It is relatively easy to define one "kind" of component in the stream system as the molecules of water that make up the bulk of the stream. But, what about the rocks and stumps? Should we lump them together and call the "kind" debris? Or should we categorize rocks by size, differentiating little rocks from medium sized, etc. After all the size of the rock will influence how the water flow behaves in its vicinity! This is where interaction kinds come into play.

You should be able to see the problems we would have being very precise about just "how" complex we think the stream system is (compared with other systems). Our intuitions about complexity have served us reasonably well, especially primitive humans dealing with naturally occurring systems in the wild, like the stream. For example, when crossing a raging stream, it is possible to notice regularities occurring in the turbulence patterns that indicate the presence of stable rocks that can be footholds for the crossing. Our ancestors learned how to spot these regularities in seemingly random natural systems and exploit them, otherwise we wouldn't be here to muse over it.

Intuitions may be a good starting point in trying to systematically understand complexity. But it is far from sufficient for our present purposes. Humans today deal with all kinds of very complex systems that demand our much better grasp of what complexity means in order to deal effectively. Systems science attempts to provide such a grasp.

> **Question Box 5.1**
> Everyone has some intuition about the nature of complexity. What is yours? Can you write down five attributes of a system that make it a complex one?

5.2.2   A Systems Definition of Complexity

There are many ways to define and describe complexity. The concept itself has been approached starting from many different perspectives (see below for a review of some of these).[5] What follows is motivated by the desire to integrate these formal approaches along with some of our intuitions to derive a systems science definition. The key concept here is that of a structural and functional hierarchical model, following Simon's work. In the end, the degree or measurement of complexity derives from the depth of the hierarchy; the deeper a hierarchy, the more complex the system. Let us examine this idea from two perspectives, a structural hierarchy and then a functional hierarchy. Both are absolutely related, but it might be easier to see this relation after examining each separately.

---

[5] See Mitchell (2009). Chapter 7 provides a good review of several of these perspectives, including Simon's.

5.2.2.1 Structural Hierarchy

We can actually start from the intuitions discussed above, and using the language of network theory developed in the previous chapter, let's examine some basic ideas that will be involved in developing a systems definition. In Fig. 5.1 we see four different categories of complexity that could be described in terms of the network (graph) properties involved. The categories range from what we can call "simplicity" up to what we can call "structural complexity."

Figure 5.1.A is a fully unconnected graph in which all of the nodes are of the exact same type. Most people would agree there is nothing very complex about this arrangement. So we have called it "simplicity." It acts as a kind of baseline for comparing with other systems.[6]

Figure 5.1b includes links (edges) that constitute relations or connections between certain nodes. This is what we will call the beginnings of *relational* complexity . The graph is still not densely connected and there is only the one kind of node with only one kind of link connecting different nodes. This network is only slightly more complex than the simplicity figure.

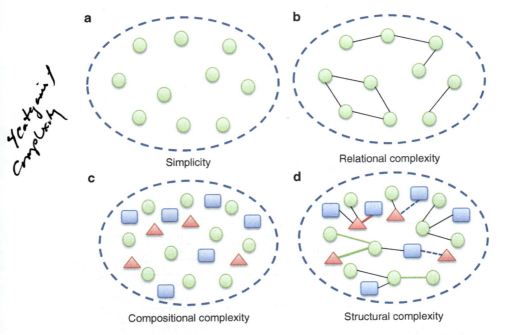

Fig. 5.1 We can use the language of networks (Chap. 4) to develop our concepts of complexity. Here we view three kinds of complexity defined by graph theoretic attributes

---

[6] Technically we would not be able to call this figure a "system." The blue dashed oval outline could represent a boundary, in which case this might be a potential system, as covered below.

In Fig. 5.1c we revert to a non-relational graph but include multiple *types* of nodes. This is an example of a *compositionally* complex system which we will describe more fully later.

Finally, in Fig. 5.1d we start to see a network that suggests what most people would agree is somewhat more complex. We call this structural complexity in that it combines compositional complexity with relational complexity and includes the possibility for multiple kinds of links as well as types of nodes. Every kind of system that we call complex will have these kinds of features, rich in node types, linkage types, and actual linkages made. But in addition to these features, we will see that some kinds of structural complexity can lead to functional complexity as well, under the right conditions.

Figure 5.1d demonstrates what we mean by a heterogeneous system, whereas a and b show homogeneous systems. c is heterogeneous only in terms of composition, whereas d shows heterogeneity in linkages as well.

The systems in Fig. 5.1 are only concerning organization and we have not introduced dynamic considerations. Functions result from organizations in which subsystems interact with one another by virtue of exchanges of information, matter, and energy (flows) in which subsystems accept or receive inputs from other subsystems and convert them in some manner to outputs (recall Fig. 3.1).

In Chap. 3 we developed the concept of systems being composed of subsystems. When we treated subsystems as black boxes (unitary entities with "personalities"), we described the system of interest as a gray box; we possessed partial knowledge of how the system worked by virtue of knowing how the various subsystems interacted with one another and how the whole system interacted with its environment. We called the subsystems "components" and assumed that they all had their own internal structures that self-similarly resembled that of the parent system, namely, being composed of their own "simpler" components. We also hinted that each of those subsystems could be treated as systems in their own right, treating the larger parent system as the environment for purposes of a reductionist-style decomposition of the component(s).

We demonstrated that this structure constitutes a hierarchy of increasingly simple components (going downward) or increasingly complex subsystems (going upward). Recall the simple view presented in Fig. 3.1. Later in that chapter we made this more explicit (see Figs. 3.14 and 3.15) showing another property of such structures, near decomposability. Subsystems (components) are identifiable because the internal links between their components are stronger than the links the subsystems have between them in the larger parent system. This is a robust property of complex systems (Simon 1996). Dynamically the near decomposability property translates into the fact that the behaviors of subsystems are more strongly determined by the internal connections than by the behaviors of other subsystems.

5.2.2.1.1   Hierarchy Revisited

Figures 5.2 and 5.3 show yet another look at a hierarchical organization. Figure 5.2 shows schematically the same kinds of lowest-level components that were shown in Fig. 3.8.

atoms with interaction potentials

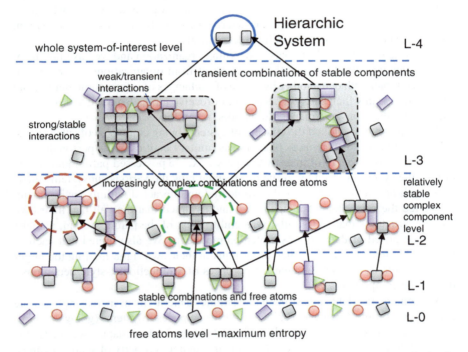

Fig. 5.2 The components from Fig. 3.8 are here treated as non-decomposable "atoms" that display "personalities" in the form of interaction potentials (*arrows*) as discussed in Chap. 3. These atoms are "conceptual" rather than literally atoms in chemistry, but the same principles of interaction apply

Fig. 5.3 A system (*blue circle*) consists of subsystems that, in turn, are composed of sub-subsystems. At the bottom of the hierarchy, the lowest level of decomposition, we find atoms, i.e., non-decomposable entities. See text for details

We call these lowest-level components *atoms* in order to assert a rule for stopping decomposition of the system. For example, if our system of interest is a social network (e.g., Facebook), then the atoms could be people with the links being the various kinds of messages (and pictures) they post on each other's "walls." As with conceptual boundaries (in Chap. 3), the determination of atomic elements in a particular systems analysis is largely determined by the kinds of questions we are asking about the system.

Of course all real systems ultimately decompose down to elementary particles as discussed previously. Real element atoms (e.g., hydrogen or carbon) provide a useful way to think about conceptual atoms. Every element's personality is determined by its atomic weight and the quantum nature of the electron "shells"[7] (remember every proton in the nucleus is matched in the electrically neutral atom by an electron orbiting about). The kinds of chemical bonds (covalent, ionic, etc.) that the atom can make with other atoms within the considerations of the larger environment in which they exist determine how more complex molecules or crystals obtain. The world of chemistry, and especially organic and biochemistries, provides marvelous examples of combinatorial relations that result in hierarchies of complexity. Readers who have not taken a basic chemistry course should consider doing so.[8]

Note that we are not just using chemistry as some sort of analogy. The correspondence between how molecules are structured and come into existence from basic atoms and how societies are structured and come into being is fundamentally the same, as discussed in Chap. 3. What we are going to look at is the pattern of formation and structural characteristics in complex systems in general.

Figure 5.3 shows this pattern and provides a basic schema for what we will call "levels of organization."

Figure 5.3 expands on Fig. 3.15. Here we start at the lowest level of the hierarchy which we labeled "L-0" (level zero), denoting that there are no lower levels of decomposition as far as we are concerned. At this level we find the conceptual atoms of the system, the smallest elements that will combine to create the complex system (level 4). Each atom in the figure has its own personality as shown in Fig. 5.2. We did not include the arrows (interaction potentials) here to help manage the amount of detail.

At L-0 the atoms are shown as individual entities. We also claim that this level represents the level of maximum entropy or disorganization in the physical sense (particularly in the case of physical atoms). Later, when we integrate the energetics and emergence aspects, we will see this correspondence more clearly.

The next level in the hierarchy, L-1 shows the beginning of actualized interactions between atoms that form combinations and give rise to stable entities given the constraints in the environment (in Chap. 10, Emergence, we will provide a much more rigorous explanation of the role of the environmental conditions in defining stability). Similarly in level L-2 the combination of atomic combinations continues to produce larger, and more complex, entities (the black arrows indicate which entities from the lower level combine in the higher level; note that free atoms are generally assumed to be available in all environments). A very important aspect enters into the situation as we move from L-1 to L-2 and that is the consideration for geometrical constraints. Note that the combinations of combinations rely on the fact that there are "unused" interaction potentials protruding, as it were, from the outfacing atoms. For example, the combination in the red dashed circle (in L-2) depends on the fact that a black rectangle can interact with a red circle and these two atoms are

---

[7] In elemental atoms the electrons distribute in shell-like orbitals around the nucleus. These shells are governed by quantum laws of energy. The outermost shell of any atom may contain fewer electrons than quantum laws allow and that is the basis for chemical interactions such as covalent or ionic bonding. See http://en.wikipedia.org/wiki/Electron_configuration.

[8] Or at least consider Moore (2003).

exposed in their respective combinations in L-1. But as atoms form interactions with one another, they also become less able to form some specific combinations. Though not shown here, but hinted at in Chap. 3, the various kinds of interaction potentials can also have variable strength potentials that depend on which other atoms they actually combine with. Moreover, the remaining free interaction potentials can be modified in potential strength by the kinds of interactions already obtained. Thus, we see multiple sources of both more complexity in what obtains and a new factor in constraining exactly what form the more complex entities can take.

The green dashed circle in L-2 calls attention to another aspect that cannot be easily shown but follows from the variable strength and geometry arguments. The entity in the circle (looking somewhat like a little person!) represents a particularly stable combination as compared with others at that level. In L-3 we see that this entity has become the basis for several different combinations. This result will be further explained in Chaps. 10 and 11 as the result of selection and emergence.

In level L-3 we show some consequences of variable strength interactions (especially weakening) and geometry in terms of building even more complex combinations. Here we show two of the subsystems that are part of what makes up the final system. We have not shown interactions between these two. That will be addressed later. What we do show is that these entities have a potentially transitory status. That is, the interactions between the entity shown in the green dashed circle in L-2 and the other two different entities are weaker and thus subject to possible breaking. This means there are more possible combinations that could arise if conditions in the milieu of the system change. These are semi-stable arrangements and will become relevant in real systems that are not only complex, but able to adapt to changing environments, our so-called complex adaptive systems (CAS). Moreover, this situation demonstrates the near decomposability property from a structural perspective.

Functionally entities that have transitory relations with one another can develop significant complexity in terms of patterns of those relations. Under the conditions we will introduce, shortly dynamic systems can develop cycles of interactions that recur regularly in time.

In the above generic system of interest, we identified five levels in a hierarchy, from independent atoms at the lowest level to a unified system at the highest. Of course this is somewhat arbitrary since we pre-identified the atoms (of interest) as well as the top-level system of interest. In reality the atoms may very well be subsystems in their own rights. We simply stipulated the objects as atomic for convenience rather than some absolute physical reason. For example, real elemental atoms, as we have pointed out, are composed of subatomic particles. And even some of those are composed of yet smaller particles (quarks) which take the role of "atoms" with respect to the elemental atom. Subatomic particles take the role of stable combinations that, in turn, combine to form whole atoms (nuclei anyway).

So the levels as shown in Fig. 5.3 are schematic only. They show a pattern of combination and some rules that apply to how combination possibilities change as systems become more complex. We could have just as easily started, for example, with elemental atoms as the "atoms" and ended with unicellular organisms as the system of interest (see below). This would lead to a hierarchy with five levels as shown above but where structural and functional combinations are different. Had we decided to end with a multicellular organism as the system of interest but started with the chemical atoms as before, the hierarchy would have to show at least several more levels, e.g., the tissue level and the organ level before coming to the whole organism.

Question Box 5.2
Consider a typical university as a system of interest. What might you consider the lowest level in a hierarchy of organization? What are the "atoms" (hint: consider just the kinds of people involved)? What are the kinds of connections between these that give rise to the L-1 level of organization? What might you estimate the number of levels of organization to be (hint: think of departments and classes)?

And that isn't the end of it. If our system of interest is the species, or a population, or an ecosystem in which many organisms (unicellular and multicellular) participate while our starting atoms remain the chemical ones, then you should be able to see our hierarchy has many more levels. This of course is the situation for the whole field of biology. And it is essentially a measure of complexity—that is, *the number of levels between the lowest level of atoms and the highest level of the biosphere as a whole provides an index of complexity*. This is why biology is such a complex subject by comparison to many of the natural sciences. And it is why biology is divided into so many subdisciplines and sub-subdisciplines! In truth the whole subject of biology is so complex because it reflects the complexity of real biological systems. Even for all that biologists have discovered and documented, the collection of knowledge about biology, many experts consider that we humans have only a very small fraction of the knowledge that might be available in that field of study collectively. No one could probably venture a statement about how many levels there are in the whole of biological phenomena if we take the biosphere as our system of interest, though we might make an educated guess based on this systems principle. Much work will need to be done to make such a guess.[9]

---

[9] It gets even worse if you consider the field of exobiology! This is the study of life not on this Earth. Biologists are now convinced that life is not a purely chance phenomenon on this one little planet, but a natural consequence of the properties of the planet. Recently astronomers have discovered many planets orbiting many different stars, including some coming very close to Earthlike properties such as distance from their star, mass, etc. The current thinking is that life will arise naturally on any Earthlike planet, but it might reflect very different variations based on random conditions. Ergo, biology becomes an even more complex subject. If we find life on Mars, it will get very interesting. See Chaps. 10 and 11 for more about life emerging from chemical bases.

Quant Box 5.1 Complexity, Hierarchy, and Abstraction
Figure 5.3 gives a hint of the nature of complexity wherein the number of possible combinations of "atoms" explodes with the number of atom types and the possible ways in which they are *allowed* to combine. To start to get a formal handle on the nature of complexity in hierarchical depth, let's look at the concepts from an abstract viewpoint. We will do this from what might be considered a "worst-case" perspective. Let us assume that there is a set of atoms (where atom as used here means the lowest level of organization in which we are interested), which is to say there is only one atom of each kind in the set. But let us also assume that any atom can combine with any other atom. For example, suppose we have a set $S = \{A, B, C, D\}$ (Fig. QB 5.1.1). We can start with all combinations of elements from the set taken two at a time. This is a new set, $P = \{\{A,B\}. \{A,C\}, \ldots \{C,D\}\}$.

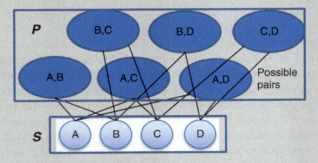

Fig. QB 5.1.1 Combinations of elements from one set in possible pairs

Set P represents *potential* complexity derived from S. However, *realized* complexity will be something less since if A pairs with B, say, then only C and D are left to pair up. Therefore, there are only two realizable elements in R regardless of which two pair first (Fig. QB 5.1.2).

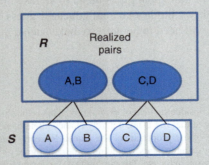

Fig. QB 5.1.2 Realized set of pairs is much smaller than the potential number of pairs

(continued)

Quant Box 5.1 (continued)

There could be further restrictions on the realized complexity. For example, suppose that the atomic personalities of the elements in S only allow certain couplings. Suppose A can combine only with B, but B can combine with C also, which can combine with D. Then the potential complexity $P = \{\{A,B\}, \{B,C\}, \{C,D\}\}$ and the realized complexity R would be either $\{\{A,B\}, \{C,D\}\}$, or $\{\{B,C\}\}$, a greatly reduced complexity.

When considering an upper bound on potential complexity under the conditions shown in Fig. QB 5.1.1, we can compute the value using the formula:

$$C = \frac{n!}{k!(k-1)!} \qquad (QB\ 5.1.1)$$

where n is the total number of elements (atoms) and k is the number of them taken at a time, i.e., 2, 3, etc. So $C_2^4 = 6$, as shown in Fig. QB 5.1.1.

Matters get more difficult when restrictions on combinations are taken into account. C represents an upper bound if no combination restrictions apply—a worst-case situation for potential complexity. By Eq. (QB 5.1.1), if n were, say, 20 and k were 10, we would have a potential complexity of 1,847,560 combinations.

Realized complexity, the number of actual combinations, cannot be obtained directly from knowing the potential complexity. Realization of various combinations is gotten from a process extended over time. This is the nature of auto-organization and emergence, the subjects of Chap. 10. In the first time units, pairs of atoms form, then triples, then quadruples, etc. Each time a bonding occurs, there are fewer atoms left in the pool and thus fewer realizable combinations.

And, when we examine the situation for a multiset, that is, a set-like object that can contain multiple numbers of each atom type, the situation becomes even more murky analytically speaking. Moreover, when we also consider some dynamic aspects, such as strengths of affinities between pairs, triples, etc., we discover that formations of realized complexity units can take multiple history-dependent pathways. Again we will see this aspect in Chap. 10 on the phenomenon of emergence.

Now let us look at these combinations from a slightly different perspective, that of a temporal sequence of realizations. Figure QB 5.1.3 shows a diagram of combinations taking place in a tree diagram. First, the dark blue atoms combine pairwise to form the light blue objects. Now, if we consider such objects to be able to combine, also pairwise, and their combined objects to do likewise, we get this binary tree representation of realized complexity over time. Such a tree has analytical properties that can be brought to bear on describing the process of combination.

(continued)

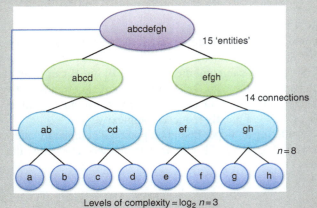

Fig. QB 5.1.3 Assuming the first pairwise combinations of atoms (a, b, c, etc.) giving rise to the light blue ovals, and then their pairwise combinations to form the light green ovals, etc., we obtain a tree structure that is an organizational hierarchy. The blue lines indicate levels of complexity according to Simon's definition

In this implementation we only allowed pairwise combinations at each level to form a regular hierarchy. Note that in this case the number of levels can easily be derived from the number of original atoms as $\log_2 n$. The base of the logarithm is two since the combinations are made two at a time. For eight atoms, in the figure, the height of the tree is $\log_2 n + 1$, or four, and the levels of complexity is three (blue lines). The latter number can also be used as an index of the complexity of a system.

But hold on! We restricted the combinations at each level to be only pairwise. Unfortunately real life doesn't work that way. One kind of additional complication is represented in Fig. QB 5.1.4. Here we allow ternary combinations at the first level, but with only eight atoms we run into a problem immediately.

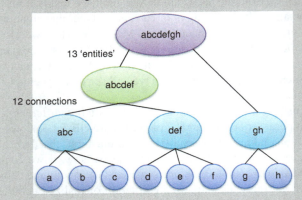

Fig. QB 5.1.4 With ternary combinations at the atomic level and an even number of atoms in the set, somebody at the first level gets shortchanged

(continued)

Quant Box 5.1 (continued)

In the figure we show the case where we allow less than ternary combinations (i.e., pairwise), but it leads to an "unbalanced" tree. The left side, the path through (abcdef), has three levels of complexity again, but the right-hand side suggests only two levels.

Challenge: *Can you think of a way to mathematically describe this condition?*

Figure QB 5.1.5 provides a hint.

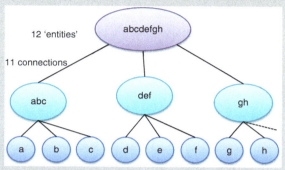

12 'entities'

11 connections

Levels of complexity $= \log_3 N = 1.9 \tilde{\ } 2$

Fig. QB 5.1.5 The *dotted line* represents a ternary combination that is not completely realized with only eight atoms. What would be a mathematical formula for deriving the number of levels of complexity for this object?

5.2.2.2 Real Hierarchies

Let's take a look at a few examples of some real-life hierarchies that will solidify these concepts. The first two examples are directly from biology, from where our most fundamental understanding of complexity comes. The third example comes from social science, in this case the study of formal organizations like a business. The final example comes from the man-made machine world—the computer, being a good example of a less heterogeneous system whose complexity derives from the sheer number of parts.

5.2.2.2.1 Complex Hierarchy in a Living Cell

Figure 5.4 demonstrates the organizational hierarchy in a living cell. This one, of course, starts with real physical atoms, carbon, hydrogen, etc. and shows the progressive construction of more complex molecules that eventually interact to form functional units that contribute to the metabolism and activities of the cell.

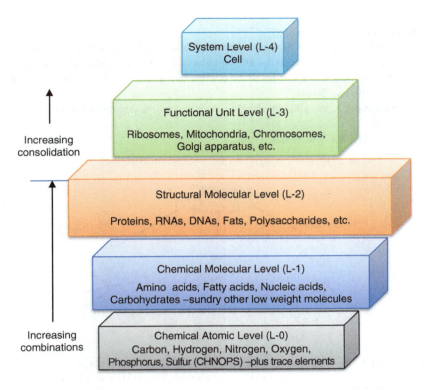

Fig. 5.4 Living cells demonstrate a natural hierarchy of component structures from an unorganized atomic level (L-0), through an emerging, semiorganized, molecular level (L-1), through a highly organized structural molecular level (L-2), through a functional unit level (L-3), and finally to a system level of a whole cell (L-4)

In Fig. 5.4 we choose a single living eukaryotic cell as the system of interest and start at the level of elemental atoms (carbon, oxygen, etc.). This allows us to demonstrate the levels corresponding with the above generic hierarchical complex model.

The atoms of life (carbon, hydrogen, nitrogen, phosphorus, and sulfur—or CHNOPS—along with many trace elements in the figure) combine in particular ways to form basic building blocks of the more complex molecules involved in metabolic processes. Amino acids, for example, are polymerized (formed into long chains like beads on a string) to form proteins, which, in turn, act as enzymes or structural elements. Many of these, in turn, interact by forming higher-level structures like ribosomes, the organelles that are responsible for manufacturing the proteins from the basic amino acids.

Finally, all of these complexes are encapsulated in a cell membrane (which is itself one of the structural components) to form a single living cell.

Question Box 5.3

In Fig. 5.4 it shows something called "consolidation" where the sizes of the boxes get smaller instead of larger, going up the hierarchy. Can you explain this? Is complexity still increasing as one goes upward beyond L-2?If you say yes, what is the nature of the complexity and how is it different from that in L-2 and below?

The basic atomic level involves not that many different kinds of atoms, compared with the number of naturally occurring atoms on the Earth's surface (atomic numbers 1, hydrogen, through 98, californium). So the lowest level in the figure is shown smaller in size than two of the levels above, L-1 and L-2. This is because these few atoms have a tremendous number of interaction potentials between them (in terms of covalent bonds alone). Moreover, carbon (atomic number 6) is capable of forming very long polymeric chains that can have many different side chains. It can form loops and other geometrical shapes in L-1 that produce new interaction potentials resulting in forms operating in L-2. In L-2 we find low and intermediate weight molecules that are then incorporated into a few biologically important structural elements in L-3, the organelles and various matrices that provide functions and structures for the organization of the cell. One such matrix that we saw in Fig. 3.4 is the cell membrane which encapsulates the organelles and mediates material transports into and out of the cell. With this encapsulating membrane and the functional organization of organelles, we arrive at L-4, the whole system of the living cell.

In level L-3 we see another aspect of complexity that we will go into in Chaps. 7 through 9. In this figure we see that the number of components in L-3 actually shrinks after the expansion of components seen in L-2. What accounts for this shrinkage are two aspects. One is that the components in L-2 form a much smaller number of functional subsystems in L-3. Cellular metabolism depends on a few kinds of structures that collectively form L-4, the cell system as a whole. The other aspect is the result of employing special kinds of subsystems of internal control needed to manage complexity. The functional subsystems, working together, cover the full requirements for a living system, but they need to be coordinated in order to do so. As we will cover in these future chapters, this requires special information flows and control decision processing components. In essence, complexity has been compartmentalized (modularized) and regulated as a set of massively parallel processes.

5.2.2.2.2   Complex Hierarchy in the Tree of Life

The evolution of life on the Earth is often represented by tree diagrams in which the earliest living things are at the base and form the "root" of the tree. The branching represents divergences of major groupings. Following the origin of cellular life (life) in what is thought to be the universal common ancestor (Morowitz 1992), we

Fig. 5.5 The tree of life is depicted as having a root starting with the earliest common ancestor, a primitive bacterial-like cell that incorporated all of the major biochemical features shared among all life forms on the planet today. The points of branching are not meant to represent actual time of divergence in this figure. There is a correspondence between the branches and complexity

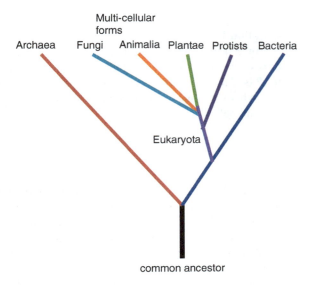

find branching starting in what is now called the "domain" of life into three major categories: Archaea, Bacteria, and Eukaryota. The first two comprise the non-nucleated cells such as extremophiles (lovers of extreme environments such as hot water or highly salty water) and the ubiquitous bacteria. These are considered the most primitive types of life on the planet. The eukaryotes (true nucleus) are thought to have derived from earlier bacteria that were symbiotic. Bacteria possessing different characteristics were able to cooperate with one another for their mutual benefit to produce the first eukaryotes.[10] Later, eukaryotic life forms developed into multicellular forms that became plants, animals, and fungi that we can see in our macroscopic world. Figure 5.5 provides a simplified schematic of this process of divergence or radiation from a central trunk.

The phylogenetic tree implies an increase in complexity of life forms as one proceeds upward and notes the points of branching. The top of the tree is the present and the root is in the dim past, some 3.5 billion years ago. The tree in Fig. 5.5 is just schematic and not scaled to actual time in any way. What it shows is that there was an early branching of bacteria and archaea and then a later branching of eukaryotes. The latter eventually gave rise to multicellular forms of plants, animals, and fungi. All of these forms exist today in many different phyla, classes, orders, families, genus, and species (in that order of refinement).[11]

---

[10] Margulis and Sagan (2000, chapter 5). Also see the Theory of Endosymbiosis: http://en.wikipedia. org/wiki/Endosymbiotic_theory. The creation of eukaryotic cells in this fashion is actually a reflection of the point made in the section above on the need to "reduce" complexity by introducing a coordination control system.

[11] See Taxonomic rank: http://en.wikipedia.org/wiki/Taxonomic_rank.

There is no clear relationship between complexity and these rankings as one goes higher in them. That is to say, it becomes increasingly difficult to claim that a later species of any particular animal is more complex than an older species from which it diverged. This is because speciation is based on environmental fitness, and two different environments may simply be different and not have any particular complexity relation with one another. Complexity has more to do with the number of parts and the relations between them that give rise to more complex behaviors. So it is easy to say that a gazelle is more complex, say, than a crab but only by carefully noting the numbers of different kinds of responses to different stimuli that are found in each of their behaviors.

> **Question Box 5.4**
> In Fig. 5.5 the hierarchy of organization and increasing complexity are demonstrated in the branching points for various phyla. The claim is that organisms that branch off higher in the tree are more complex. What is it that makes these organisms more complex than those lower in the branching?

The safest way to make claims about relative species complexities, at least among animals, is to look at their brains. Brain size alone does not provide a clue as to how complex the behavior of the animal will be. Brain size and weight proportional to body size and weight seems to be a somewhat better indicator of what is commonly called *intelligence*; and that is another way of talking about complexity of behavior. Smarter animals (those that can learn more options and/or show an ability to solve problems) are generally deemed "higher" on the phylogenetic tree, and there does seem to be a correlation between position in that sense and the complexity of the brains. This is an area of continuing investigation—a potentially fruitful realm of scientific inquiry in the future.

Brain size, the ratio of weight to body weight, for example, is only a rough index of complexity. To really see that brains get more complex with evolutionary time, we have to compare the details of brain morphology and cytology, or, in other words, the architectural features of brains as we go from "primitive" animals to "advanced" ones. The evolution of brain complexity shows direct correlation with complexity of behavior (Striedter 2005).

5.2.2.2.3   Complex Hierarchy in Human Organizations

Who is not familiar with the hierarchical "organization chart" that shows who is the boss of whom in organizations? Or who hasn't had some complaint about the hierarchical bureaucracy in government agencies that seem to isolate citizens from the "decision makers" who could help them?

Fig. 5.6 An organization
displays the same kind of
hierarchy. All of the physical
elements of production, the
workers, their specific jobs,
capital equipment, supplies,
etc. might be considered
as the "atomic" level (L-0).
The organizational structure
basically follows the same
pattern as in Fig. 5.3

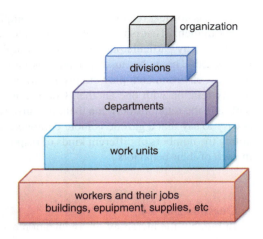

Organizations are designed to accomplish some goal or goals. They have an overall purpose as envisioned by their founders. They may produce products or services, sometimes for profit, sometimes for the good of society. In all cases, when organizations grow to a large size and there is much work to be done internally, they become sufficiently complex that the management must be divided among many people. Moreover, organizational management involves decision making that involves different scopes and time horizons. Typically organization management is layered such that the scope of concerns and the time scales for decisions get larger from the bottom layer to the top, where the scope is the whole organization and the time scale may involve the distant future. In Chap. 9 we will examine the management hierarchy in detail. Here we consider the issue of complexity as it relates to the structure of management. Figure 5.6 depicts the levels of organization with workers and their tools, workspace, etc. taken as the atoms. Workers are organized into work units, or jobs. These are generally organized into departments, like accounting or inventory or production. The latter are then placed into divisions like finance and operations. The organization as a whole is the system of interest.

The lowest level of an organization is where the work gets done to produce the products or services that constitute the purpose of the organization. This level involves *real-time management* to keep the work going and maintaining quality. But for large organizations, this work is distributed among many subprocesses that must be coordinated. Also, the obtaining of resources and the output of final products/ services must be managed in coordination with the external environment (e.g., suppliers and customers). So the level above real-time management handles *coordination* among all of the subprocesses and between the real-time work and the environment. Coordination is typically divided between *tactical* (interactions with the environment) and *logistical* (interactions among subprocesses) coordination. The managers at this level have wider scopes for decisions and make those decisions over longer time scales than real time. For example, the manager of purchasing (tactical) in a manufacturing organization must keep track of parts inventory levels and production schedules to make sure the parts needed are on-site when they are needed.

When the work processes are complex in their own right, then managers at any level will need staff to which they will delegate responsibilities and authority. No single human being can take care of every detail, so the organization expands horizontally at each level of management. The organizational chart (the tree) reflects this increase in complexity. Other factors can increase complexity. Organizations can diversify their products and services.

Roughly speaking, the above figure also corresponds with the typical organizational chart used to depict the management hierarchy. Managers at higher levels in the figure have broader responsibilities.

By some definitions of (or perspectives on) complexity, the measure of complexity is gotten from the lowest level in a hierarchy such as an organization. That is where you count the number of components and the number of types of components to determine how complex an operation is. But the reconciliation between the depth of the hierarchy versus the raw counts of components views is in recognizing that the higher layers in a hierarchy not only add to the numbers and types of components (in that higher layer) but also represent an abstraction of all that comes below. The depth of a hierarchy subsumes both the counts and the amount of organization (interactions between components) in that a very deep hierarchy correlates with count-based measures when systems are nearly decomposable.

5.2.2.2.4   Complex Hierarchy in a Computing Machine

Computers are often described as extremely complex systems. As with living systems the basic components (logic circuits) are built from just a very few "atomic" components—transistors, resisters, capacitors, and wires for the most part. The logic gates (see Chap. 8) are generally considered the fundamental components for designing almost all computational elements. Aside from the wires and other electronic components needed to combine logic gates into circuits, the gates themselves determine the kind of computation performed by the circuits. Just using a few gates computer designers have devised a relatively small number of functional circuits such as registers, adders, memory cells, etc. (Fig. 5.7). In turn these circuits are combined in fairly standard ways to produce a computing device.[12]

---

[12] Here a computing device is any combination of circuits that accomplish a computational function. For example, the central processing unit (CPU) is a computational device even though it is only useful when combined with other devices like memory and input/output devices to produce a working computer.

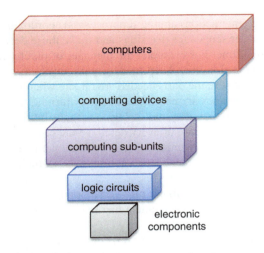

Fig. 5.7 The complexity of a computer systems and networks of computer systems increases as combinations of electronic components and circuits. This complexity is inverted from that in Fig. 5.3 since very few components are used in combinatorial fashion to produce many computational devices. Those in turn are combined in seemingly unlimited ways to produce a myriad of computers

Computing devices can be combined in many different ways to produce working computers of many different kinds. And it doesn't stop there. Computers of many different kinds can be combined via communications channels to produce extremely complex networks able to do extraordinarily complex computations.

Thus, we see that unlike the hierarchy of an organization or cellular life after the initial broadening of components through L-2 where as the hierarchy deepens, the number of kinds of combinations declines, for computing systems, the opposite seems to be the case. That is, as we go up the hierarchy, we see a combinational explosion in many different aspects such as applications and multi-computing complexes. There seems to be no end to the recombination of computers (and their interconnectivity) to form new systems. Is there something logically different about computation that gives rise to this? No, not really. It depends on what we take as the system of interest. In both the case of living cells and organizations, we could have expanded our boundaries and taken a larger meta-system as our system of interest. For example, had we considered multicellular life forms than the "cell" level in our diagram (L-4) would have been much broader since there are many different kinds of cells (not just one) in a single body. But then we would have probably started L-0 somewhere higher up for the analysis of complexity. These are not simple relations and this is an example of why there remain some disagreements over what is meant by complexity. Still, a systems perspective on the issue, such as recognizing the boundary choice problem, might help alleviate some of this dispute.

5.2.2.3   Functional Hierarchy

We now bring dynamics back into the picture. In Chap. 6 we will explore some of the underlying details of dynamics and develop further some of the concepts we will introduce here.

Components and subsystems don't generally just sit there in a static structure. They more often have behaviors. In the section above, on structural hierarchies, we described the notion of interaction potentials possessed by atoms and subsystems. If you will remember from the previous chapter, connections between components (nodes in a network) are not just limited to static links but can be flows as well. Connections can be persistent or intermittent. And they can have various strengths at different times depending on the context (this was also the case in structural hierarchy giving rise to the property of near decomposability). In other words, connections are generally dynamic.

Connections may either involve forces or flows (or mixtures of these). Forces are either attractive or repulsive and carry variable strength levels depending on the type of interaction potential (between two components). Both forces and flows may also vary in time as a function of environmental factors influencing one or both components. Depending also on the internal complexity of the component, forces and flows can be intermittent, sporadic, and episodic.[13] They can be regular or what we call periodic, like a clock tick or sinusoidal rhythm.

Importantly, components make connections because their interaction potentials preordain them to. For example, in the case of a system whose components are processes such as in Fig. 5.8, the output flows called products are necessary inputs to other processes in the system. Thus, systems and subsystems are said to perform *functions*.

Complex systems produce more than a single product output as a rule. As seen in the figure, and explained in Chap. 6, aside from product(s) output(s), real dynamic systems produce waste materials and waste heat. Accounting for these requires something more than a typical algebraic function, e.g.,

$$y = f\left(x, z\right)$$

which says that the value of y (a measure of the product) is a function of variables x and z (inputs).

A functional hierarchy is basically the same as a structural hierarchy but puts more emphasis on flows between components, subsystems, and into/out of the system as a whole. Components (atoms to subsystems) tend to have a few flow interconnections at a given level as compared with the flows within the subsystem to

---

[13] Intermittent means they come and go. Sporadic implies they do not occur regularly. And episodic implies that when they occur, they last for a while and may have varying amplitudes of strength during an episode. What all of this implies is that many kinds of connections introduce a great deal of uncertainty into system structures.

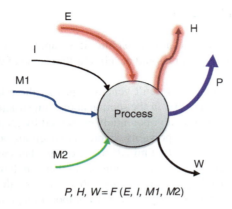

$$P, H, W = F(E, I, M1, M2)$$

Fig. 5.8 A subsystem (or component) can be a process that acts on inputs to produce outputs (flows). Following conventions to be further explained in Chap. 6, Dynamics, this shows that a process serves a function in that it converts the inputs of energy (E), information (I), and materials (M1 and M2) into a product (P) with some waste materials (W) and waste heat (H). The formula shows the designation of the function (F)

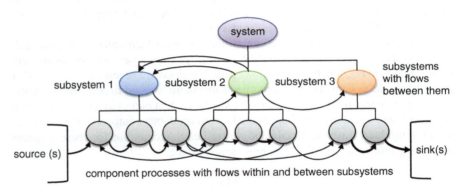

Fig. 5.9 Flows between components within a subsystem are greater in number and/or "volume" or strength than between subsystems. The *heavier black arrows* between components are meant to convey this much tighter coupling within a subsystem. The flows between subsystems, however, are sparse in number or weak in volume

which they belong. Figure 5.9 shows how the structural hierarchy of system relates to the functional hierarchy in terms of input and output flows between components within the subsystems and between subsystems. Once again this is a reflection of the nearly decomposable property we saw up above.

Another way to characterize subsystems is to note that the components of a subsystem are tightly integrated with one another as compared with components in one subsystem being so tightly integrated with components in another subsystem. Unremarkably this is yet another way of recognizing subsystems as such. In Fig. 5.9 if there were an equal number of flow arrows of equal size going between all of the processes, then we would have to treat it as one large subsystem (or system), and it would be much more complex at one level.

Question Box 5.5

Functions or purposes seem to be a necessary attribute of all systems. Is this true? Can you think of any kind of system that does not actually have a function? If so, what does the system do over time? If not, can you explain why systems seem to always have functions?

As we will see in Chap. 9, systems that are overly complex at a given level of organization tend to have numerous stability problems due to an adequate method of coordinating the behaviors (functions) of numerous interacting components. Rather complex dynamic systems tend to modularize into subsystems in which the component count can be kept lower and the coordination between subsystems is more manageable. This is, in fact, why large organizations tend to develop more levels in their hierarchy (e.g., middle management) as they grow. The same phenomenon will be seen in complex biological systems as well (see Chap. 11).

5.2.2.4   Complexity as Depth of a Hierarchical Tree

Structural and functional hierarchies are merely different perspectives on the same basic phenomenon. The depth of a hierarchical structure is a response to increasing complexity of forms and functions. Thus, in Simon's proposal we have a natural link between a concept of complexity and a physical manifestation. More complex entities tend to have more levels in a hierarchy of structure and function.

This view of complexity seems most suited for a full systems science approach. After establishing the boundary of a system of interest and specifying the atomic components, the number of levels in the structural/functional hierarchy provides a reasonable index of complexity for comparison between similar systems. It also provides a measure of comparative complexity between systems. For example, if we were to find bacterial life on Mars (or some other planet), we would intuitively think of Mars as a less complex planet compared with Earth. And by the hierarchical depth of the Earth ecosystem compared with that of Mars, we would have a more objective basis for supporting that intuition.

Below we will outline a few other perspectives or theories of complexity that have gained recognition. We will try to show how each could be related to Simon's hierarchic complexity theory. That might open the door to a consilience of views in the future. Time will tell.

Think Box The Complexity of Concepts and Their Relations

In this chapter we have focused on the concept of complexity as represented by hierarchical depth. You may have noticed that we often show this concept in the form of a tree representation which is merely a kind of graph (network!) As it turns out, concepts themselves are organized in your brain in exactly this fashion. That is, higher-order concepts are composed of lower-order concepts with the depth of the concept hierarchy providing a measure of your brain's capacity to mentally handle complexity in the world. In a very real sense, the competency of a brain to deal with the real world depends on just how high a concept level it can represent.

Figure 5.3 showed the sense of how hierarchies form through combinations of low-level "atoms" or components, and those combinations then are able to form yet more complex combinations on up the hierarchy of possibilities. In this structural emergence (see Chap. 10 for more on the process of combinations), once atoms are committed to a combination, they are no longer available to form different combinations at the same level of complexity, except for their ability to bring their current combination to form another combination at a higher level.

The cerebral cortex, as it happens, has the same kind of capability to form more complex concepts out of what we call lower-level concepts. But in the case of concept formation, all of the lower-level concepts are never completely committed to just one higher-level concept. Instead, all concepts (and this also goes for really low-level percepts) at any given level of complexity are available to form higher-level concepts. That is, they are all reusable in infinite combinations! (Fig. TB 5.1)

Figure TB 5.2 shows a layout of both the perceptual/conceptual and behavioral hierarchical mappings in the cerebral cortex (see Fuster 1995, esp. chapter 5).

The small boxes in the primary sensory areas represent the learned and stable features that the organism's brain has encountered over and over again in its environment. As the brain learns (during juvenile development), these features combine in nearly infinite ways to form percepts and those then form concepts. As learning concepts proceeds (from the back part of the brain, perception, to the front, decision processing), the combinations of lower-level concepts get much more complex. A concept such as a baseball pitcher requires concepts of person, baseball (the ball and the game), the act of throwing, etc. A more complex concept would be the baseball team, what comprises the team, what it means to be a team, the business of baseball, and perhaps thousands of subsidiary and similar concepts.

Neuron clusters learn by association between subclusters that is reinforced over time; that is, one must encounter the same patterns of association in perception over and over again to form stable relations.

(continued)

Think Box 5.1 (continued)

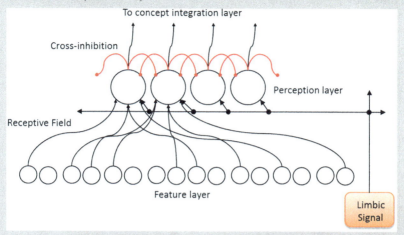

Fig. TB 5.1 The neuron clusters represented by *small white circles* at the bottom recognize various low-level features in an image (e.g., a visual field image). These features are the regularities that can be found in any image, and they tend to co-occur in various combinations to form patterns. The *arrows* from these feature detectors are the excitatory connections to the *larger white circles* in the perception layer. Those neuron clusters learn which feature combinations represent different percepts. After many repetitions of experience with specific combinations of features belonging to a single percept, each cluster becomes an "expert" at recognizing that particular percept when it is present. For an example, go back and look at Fig. 3.20, which shows how line segments at different angles (features) combine to make up a contour. The nose is a percept; all noses have very similar features. "Nose" is also a low-level concept that composes with other percepts/concepts to form the concept of a face. "Face" is a higher-level concept that composes with other body parts to form the concept of a person, and so on

One thing to note is that while this picture of increasing complexity in a hierarchical fashion as we move from the rear of the brain to the front is essentially correct, it is not quite that simple. For one thing concepts at any level of complexity can communicate with other similar concepts so that the concept hierarchy is not strictly a tree structure. Rather it is a very complex directed graph containing not only feed-forward signaling but also feedback (or perhaps we should call it feed-down) from higher levels down to lower ones. For an example of the latter, we know that if a person is thinking about a particular concept, say a friend, that signals from the high-order concept of that friend will signal down to activate the low-level concepts and percepts that go to make up the friend concept. The friend's image is brought to mind. Also, the higher-level concepts being thought about can prime the lower-level ones to "expect" to get sensory inputs related to the higher concept. You could be thinking about your friend because you are anticipating their arrival. If a stranger that has some of the same features as your friend appears in the distance, your brain can actually mistake the visual image for that of your friend because you were expecting to see her.

(continued)

Think Box 5.1 (continued)

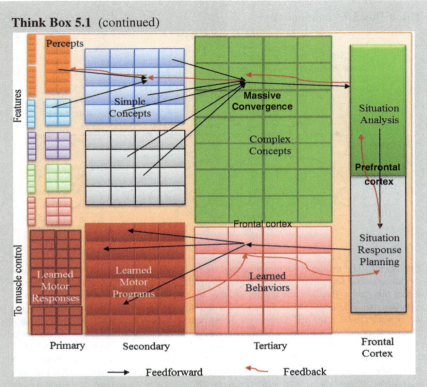

Fig. TB 5.2 A rough layout of the cerebral cortex, running from primary sensory cortex (*upper left corner*), through the concept hierarchy (*across the top*), into the "thinking" part of the brain—the frontal cortex—where decisions for action based on the concepts being processed in consciousness are made. Then commands are issued to behave (hopefully appropriately), which are communicated down an action hierarchy all the way to motor responses. The *black arrows* in the sensory-to-associative (*blue, gray, and green*) show the way in which sub-concepts converge to form higher-level concepts. The *blue arrows*, on the other hand, show how one sub-concept might contribute to many higher-level concepts through divergent pathways. The *red arrows* show that there are recurrent connections from higher to lower levels of cortex. These may provide feedback or be used in top-down commanding (see Think Box 6)

Question Box 5.6

Can you construct a "concept map" of a system? Consider the concept of a dog. Draw a circle roughly in the middle of a piece of notebook paper. Label it dog. Now, below that circle, consider all of the things that a dog has. Consider its body, its legs, its fur, etc. Draw circles for each kind of component that you think a dog has and draw a line from the dog circle to each of these lower circles. Then repeat the process for each component. For example, legs have paws, calves, thighs, etc. Fur is comprised of different kinds of hairs and colors. Can you see how a concept like a dog is composed of other component concepts? How do these concepts relate to the perceptions of shape and color?

5.3 Other Perspectives on Complexity

5.3.1 Algorithm-Based Complexity

The last example of complex systems in the last section, computers, provides a somewhat different way to talk about complexity, that is, how much time and space is required to compute the solution of a problem.[14] There are several ways to use computation as a basis for describing the level of complexity. The first of these is based on the amount of time it takes to solve a specific kind of problem. The other, algorithmic information complexity, asks questions about the minimum size of a program that would be needed to "produce" a computational object.

5.3.1.1 Time Complexity of Problems

In the realm of computational problem solving, it turns out there are classes of problems, some of which are "harder" to solve than others. The hardness of a problem is characterized by either the time it takes or the amount of memory that is required to solve varying "instances" of a problem. The same kind of problem, say sorting a list of names, has different instances meaning that different lists have different numbers of names depending on the application. The list of names belonging to a neighborhood association will be much less than the number of names in a city phone book. Computers are really good at sorting lists of names into alphabetical order, and there are many different ways or algorithms (see Chap. 8) for doing this. Some of those algorithms are inherently faster than others. But it is a little more sophisticated than simple speed.

In computational complexity we characterize the "time complexity" of an algorithm as the amount of time it takes to solve incrementally larger instances of a problem. So, in the case of sorting names, we consider the time it takes to sort lists of linearly increasing size, for example, starting with a list of 100 and increasing the size by 100 over many iterations. Suppose we are interested in a certain algorithm's ability to sort lists from 100 in length to 100,000 in length. We have mathematical tools for analyzing the algorithm's processing time on different instances (see Quant Box 5.1) of the problem. This allows us to compare algorithms being used on the same type (or class) of problem so that we can choose the most "efficient." For example, in Graph 5.1 we show three time-complexity curves derived from three different kinds of functions. A linear function produces the straight line increasing amount of time as the number of items in the list grows. The logarithmic function grows much more slowly (see the Quant Box for an explanation). But the exponential function literally explodes.

---

[14] Actually these days we also are concerned with energy consumed by computation, not just how big or how fast the computer is!

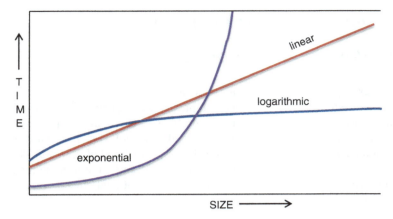

Graph 5.1 Problem types, and best algorithms available to solve them, come in varying time-complexity classes. Here we graph three different classes of problems in terms of the best known algorithms for solving each type. Note that the amount of time increases as the size of the instance of each problem class increases. The best known algorithms for problems such as sorting or searching a list increase in time logarithmically as the size of the list goes up. Other problems are inherently more difficult to solve, some showing increasing time in an exponential fashion. See text for details

Of the three the logarithmic algorithm would be the better choice. Indeed naïvely designed (i.e., simple) algorithms for sorting will show linear or possibly even slightly worse (e.g., polynomial) time. But efficient algorithms like the famous "quicksort" will often perform in logarithmic time. This is why even small computers can sort very long lists in reasonable real time, e.g., seconds to minutes.

So our analytic methods can help us choose between proposed algorithm designs to solve the same type of problem. But the method has also shown us that there are actually families of problem types for which the best known algorithms cannot be improved upon! The most general class of computational problems that can be solved efficiently are called polynomial-time (P) problems, meaning that their "worst-case" algorithms still perform in polynomial time. A polynomial function is one of the form:

$$y = a_i x^n + a_{i+1} x^{n-1} \ldots + a_n x^0$$

Here y is the time, a_i is a constant for each term, i being the index of the order of the term, and n is the starting exponent. An example would be: $y = .45x^2 - 12x + 2$.

Another class of problems doesn't seem to have algorithmic solutions that can run in polynomial time. In fact these problems are currently thought to require exponential time or worse (what could be worse? Try hyperbolic!). One famous problem is the traveling salesman problem (TSP—see Chap. 8). Essentially this problem asks: Is there a path through a network of cities in which no city is visited twice and the cost of travel is a global minimum? To date no polynomial-time algorithm has been devised to solve this problem, though many computer science students have tried. This class is called (for our purposes and for the sake of simplicity!) non-polynomial or NP,[15] meaning that no one has found an algorithm that works in polynomial time. Exponential time, as depicted in the graph above, means that the algorithm takes increasingly more time per unit of the size of the problem. Here the function takes the form $y=x^n$ where n is the size of items to be computed, such as the number of cities in the TSP.

It isn't entirely clear why this approach should be called a measure of complexity of systems. But if we consider computation as a real physical process (as we do in Chap. 8) and not just a mathematical exercise, then we can draw a relation between time complexity of computation and the amount of work that is necessary to complete the task. As we will see in Chaps. 8 and 9, computation is part of a larger system of control or management in real systems, so there are real issues with taking too much time to solve computational problems relative to the dynamic needs of the system. For example, suppose a shipping company such as FedEx were to actually try to compute the TSP for the thousands of cities between which it ships packages as a means for saving on fuel costs. Should they undertake that effort, they would have to have started the computation at the beginning of the universe and it would still be going long after our Sun is a red dwarf star!

NP class problems can be handled (occasionally) by employing a very different approach to computation than we find in our ordinary computers—serial steps taken in an algorithm. Massively parallel processing, as is done in the human brain, for example, can work on a special form of problem, such as pattern recognition, when the problem can be broken down into smaller chunks. Those can be distributed to many small processors to be worked on in parallel, thus cutting down the time needed to solve the whole problem. Indeed for very large problems that have the right structure, chunks can be broken down into yet smaller chunks so that we once again have something of a hierarchical structure. This then is a more direct relation between computational time complexity and Simon's hierarchical systems version. Alas, the TSP and similar forms of problems in the NP class do not have this structure, but very many practical problems do. For example, solving whole Earth climate models involves breaking the whole into chunks or grid sections and solving the equations for each grid and then reintegrating or putting the sections back together for a whole solution. Parallelization of computation definitely shows a structural resemblance to hierarchical measures of complexity.

---

[15] This isn't really what NP stands for, but the explanation would require far more space than we can afford to take. And, unless you plan on going into theoretical computer science, you wouldn't really appreciate it.

5.3.1.2 Algorithmic Information Complexity

This approach to complexity, while involving the ideas of algorithms and computation, is just a bit more subtle than time complexity. There are actually several versions of this form so we are only going to describe it briefly. The reader can find more resources in the bibliography at the end of the chapter.

What AIC entails is the idea that a computational object is produced by an algorithm (running in a program). For example, a string of random numbers could be generated by a fairly *simple* algorithm that just loops around so many times and each time calls a subroutine that provides the next random number.[16] Conversely, how complex would an algorithm have to be to generate the contents of the Library of Congress? In the latter case, the complete knowledge base of the LoC would have to be in the algorithm itself (seemingly). In other words, the algorithm would need to have the same measure of complexity as the object it attempts to generate.

More practically, if we look at many real computer applications, we find that they are always broken down into modules in the exact fashion of the Simon hierarchy. Many applications, such as a management information system in a company are, in essence, mirroring the systems in which they are embedded and thus are often almost as complex, in terms of the depth of the hierarchy of modules, as those systems.

5.3.2 Complexity of Behavior

Computers have been instrumental in exploring another domain of complexity. In this domain we examine the behavior of a system that seems to grow more complex over time but is based on fundamentally simple rules of interactions between the atoms. We will describe several versions of this here briefly.

5.3.2.1 Cellular Automata

The science of complex systems (sometimes called complexity science) started to emerge once cheap computing became available. One of the earlier explorations of complex behaviors involved constructing a regular lattice of cells (a matrix in a computer memory). Each cell would be found in a particular state. In the simplest version of this, each cell would be either "ON" or "OFF" (e.g., contain either a 1 or a 0), and these would be assigned in a random fashion initially. Next the computer

---

[16] Random number generators in computers are really pseudorandom number generators. That is, the sequence of numbers generated appears to be random with a uniform distribution. But in fact if you run the same program over and over, using the same "seed" value, you will get exactly the same sequence of pseudorandom numbers. Generating truly random numbers in computers requires special electronic circuits that produce white noise and that is used to generate random numbers. Even then there is some philosophical debate about what the meaning of this randomness really is; much more so than we care to get into here.

program iteratively runs through a set of simple rules applied to each cell in the lattice. These rules involved examining the state of surrounding or neighboring cells and then making a decision about what the state of the current cell should be. Based on the decision, the state is either changed or left alone. After all cells have been processed, the program starts the process over again. These outer loop iterations are effectively time steps. So the whole program represents a kind of evolution of the matrix over time. What is particularly astounding is the fact that a few relatively simple rules can sometimes lead to quite complex behaviors of the system. These models are called cellular automata.[17] Researchers have found an array of rule sets that produce many kinds of behaviors, some of which are very complex, including chaotic attractors (see next topic). The rules are indeed simple. For example, a rule could be that if three near neighbor cells are on (1), then set the center cell to on. Otherwise turn it off (0). A small number of rules of this kind are used to generate the changes that take place over many iterations. Some rule sets result in identifiable and repeating patterns that move across the grid, or "shoot" smaller self-propagating patterns that move away. Sometimes these patterns collide and generate still other patterns that have clear behaviors.

Cellular automata (CA) have been used to investigate a wide variety of phenomena. A whole field called "artificial life" derived from earlier work has shown that in combination with evolutionary programming, CAs demonstrate lifelike behaviors.[18] One author even goes so far as to suggest that the universe itself is a giant cellular automaton and has proposed "*A New Kind of Science*" (Wolfram 2002) in which all activities are explained by discrete dynamics (a claim that, if true, will make all students who hated calculus happy!). It would be hard to argue that the universe isn't very complex. But what is truly mind-blowing is the notion that all of that complexity emerges from some very simple rules of state transitions in a cellular matrix. In Chap. 10 we will return to the idea of how a complex system can emerge from a simple set of components and their interconnections that resembles the CA notion.

5.3.2.2 Fractals and Chaotic Systems

In a related vein, some simple mathematical rules can be iterated to produce amazingly complex objects that do not exist in an integer number of dimensions (1, 2, or 3), but in fractional dimensions, e.g., 1.26 dimensions! These objects are called fractals and they have some very peculiar properties such as self-similarity at many (or all) scales.[19] The complexity here is one of structure, although an alternative argument can be made that the resulting structures are not really complex in the intuitive sense.

---

[17] See Wikipedia—Cellular Automaton, http://en.wikipedia.org/wiki/Cellular_automaton.

[18] Langton CG et al. (eds) (1992. Also see Wikipedia—Artificial Life, http://en.wikipedia.org/wiki/Artificial_life. Additionally, John Conway's Game of Life, http://en.wikipedia.org/wiki/Conway%27s_Game_of_Life.

[19] Mandelbrot (1982). Also see Fractals, http://en.wikipedia.org/wiki/Fractal.

Once you have described the structure at one scale, it can be describe readily at any other scale. While true, if one examines some of the more elaborate versions, say in the Mandelbrot Set,[20] one is left with an immediate impression that these objects are complex in some sense of that word. The matter is unsettled and the number of examples of fractal-like systems in the real world seems small. Examples include river tributary patterns, tree and bush branching patterns, shorelines, and the circulatory system. While these example have a superficial resemblance to fractals in that they appear, at first, to be self-similar at multiple scales, the fact is that they have a limited range of scales, and close examination reveals that no single pattern description can be made to serve at all of the scales covered. Fractals are generated from simple rules applied iteratively, whereas branching patterns in living trees and circulatory systems have their roots in much more complex development rules which include reactions to environmental conditions. So the question of what role, if any, fractal generation applies to real system complexity is still open.

In the next chapter on dynamics, we will describe yet another related concept that involves complex behavior arising from simple mathematics, called deterministic chaos, or simply chaos.[21] A chaotic system is one whose behavior appears to follow a regular pattern but not entirely. These systems behave in a way that prevents one from making predictions about its status, especially further into the future. We will save discussion of these systems for the next chapter, but note here that chaotic behavior does lead to considerable complexity when such systems interact with other systems and especially other chaotic systems.

5.4 Additional Considerations on Complexity

There are two additional considerations on complexity that should be mentioned. These are ways to think about systems, especially from the standpoint of growth, development, and evolution. They concern the nature of complexity from two different states that systems can be in over what we would call their "life histories." These are:

- Disorganized versus organized complexity—Warren Weaver[22]
- Potential versus realized complexity

The first considers how much organization the system has at any given point in time, but especially after it has aged for a long time under the right conditions. It is essentially considering the amount of entropy in the system at a point in time.

---

[20] See Mandelbrot Set: http://en.wikipedia.org/wiki/Mandelbrot_set.

[21] See Chaos Theory: http://en.wikipedia.org/wiki/Chaos_theory.

[22] Weaver, W. (1948). "Science and complexity," in *American Scientist*, 36: 536–544. Accessed online at: http://philoscience.unibe.ch/documents/uk/weaver1948.pdf, Jan. 2013.

The second consideration is related but concerns a measure of complexity for disorganized systems as they initially form and begin to develop. A newly emerging system can be in a state of initial disorganization but has a high measure of potential complexity. As the system ages, and assuming appropriate flows of energy are driving the work of creating subsystems, as shown above, then a system with high potential complexity may develop realized organized complexity.

5.4.1   Unorganized Versus Organized

Warren Weaver described two kinds of complexity which we treat as exclusive extremes along a dimensional line. In Weaver's terminology, something could be a system in the sense of having many component parts (see below) that remain *disorganized*. That is to say, the parts do not have binding relations with one another that would give internal structure. A classic extreme example of such a system would be an inert gas contained in a fixed volume, at constant pressure and temperature. This might strain some people's credulity in calling it a system, but the fact is it does meet criteria as set out in the other considerations. For example, the system does have a physical boundary. This system, if aged long enough, would reach and stay at a maximum entropy level consistent with all of its parameters. Clearly, then, there is some relationship between the entropy of a system and its "complexity." Other kinds of systems can start out disorganized but actually have a substantial measure of potential complexity (below). Under the right conditions, which we will visit in Chaps. 6, 10, and 11, these systems will move from unorganized to organized along this dimension as a function of time. They will also move from potential to realized complexity as described next.

Organized complexity is more like what our intuitions from above suggest complexity to be. This refers to a system that not only has lots of parts but those parts have lots of connections or interactions with one another, as we saw in Chap. 4. This also corresponds with a system that has realized its complexity and is at least in the process of minimizing internal entropy. Or it could describe a fully realized complex system (having all the possible complexity it could ever muster) which corresponds with the minimum entropy for the system.

5.4.2   Potential Versus Realized Complexity Parameters

Potential complexity concerns arise when a system is not yet organized to its fullest, but the nature of the components along with a suitable source of energy flow sets the stage for developing or evolving organization. The system cannot be said to be complex in terms of its state at a fixed time. But it contains within its boundaries the potential to auto-organize with higher-level organizations emerging over time (Chap. 10) and the system ultimately evolving into its maximum organization (Chap. 11) which corresponds with its minimum entropy.

Potential Complexity. The potential complexity of a system depends upon:

- A sufficiently formed boundary that objectifies the system
- The number of *kinds of components* present within the boundary or that can pass through the boundary into the system
- The *number of components present of each kind*, the count of each kind present (note that by definition the kinds that can pass through the boundary but are not present cannot influence the complexity until they are present with respect to this parameter so it can change when time is taken into consideration)
- The *number of different pairwise connections* that *can* be made between all components in the set of kinds
- The *number of kinds of energies*, and their potential differences between sources and sinks, that can affect components in the set of kinds
- The geometry of the system with respect to its boundary conditions, i.e., what shape is it and where along the boundary are located the energies (sources and sinks)

In this dimension we seek to form a multicomponent index that tells us how much complexity *should* be possible if the system were to age a long time. This is clearly related to organized versus disorganized complexity but is not the same thing exactly. The two measures should converge at either end of their respective scales, i.e., at maximum organization realized complexity should be 1 (on a scale from 0 to 1 inclusive).

Realized complexity looks at the actual connections that have been made between all components in the system at any given point in time. But, in addition to just the raw number of connections, it looks at the actual structures within the system, identifying functional subunits or subprocesses that have obtained. We could even go so far as to quantify the degree of connectedness and sub-complexity of those subprocesses and add it to our measure. We've already seen that systems are defined recursively, so here is a point at which that definition becomes useful and operational.

5.5   Limits of Complexity

Up to this point we may have given the impression that complexity is a good thing, and you may have even thought that it is *always* a good thing. It turns out that this is not the case. It is possible for systems to be too complex and actually fail when certain events occur.

The term "resilience" refers to a system's ability to recover and restore its function subsequent to some disruptive event. We saw this in the last chapter. Some kinds of events may be possible but rare or unlikely over the life of a system of interest.[23]

---

[23] These kinds of events are called "black swans" by Nassim Nicholas Taleb. They are unexpected or of exceedingly low probability and so are not anticipated while the system is in development, i.e., no provision for resiliency is included in the structures and functions.

A system may have achieved a high level of complexity internally but with what we call "brittle" connectivity. That is, the connections between components cannot be altered without having dramatic effects on other parts of the system.

As we know from Chap. 4, in a network where components are connected with high coupling strength, even if sparsely so, any change in one component will propagate to every other part of the system with possibly minimal attenuation. In other words, a disruption will be felt everywhere within the system's boundary (and beyond due to that system's interconnections with other systems). But even more dangerous is the case where some of the components have nonlinear interactions as just discussed in the prior section. If positive feedback loops (amplifiers) are present in the connections between components, then the disruption will propagate with disproportionate and possibly growing strength.

Thus, systems can and do collapse or disintegrate from factors that stem from their own level of complexity. The external or internal event does not actually cause the collapse, it merely acts as a trigger. Rather, the collapse is intrinsic to the system itself.

We can say that such systems are at risk, not only from the occurrence of the triggering event but also from their inherent structure (see footnote 4, last page).

5.5.1 Component Failures

Component failure means that one or more components within the network structure suffer some kind of degradation in its personality, i.e., in its connectivity capacity. This may be due to many factors, but all will eventually boil down to our old friend the second law of thermodynamics. All components are assumed to have some internal structure of their own, recall. And entropy will always affect the order (organization) of those subsystems.

A very good example of this is the normal denaturing of protein molecules within a living cell. Proteins perform their functions primarily by virtue of their shape and the exposure of specific bonding sites due to that shape. Each protein molecule is like the clusters in the larger network in Fig. 4.6, and their bonding sites are like the sparse connections that link clusters together. As it turns out, under conditions of normal physiological conditions, many proteins are not super stable, they can lose their shape and, consequently, their function. This is called denaturing. When this happens the protein molecule actually becomes a liability to the cell's metabolism. Fortunately for all of us living beings, cells long ago came up with a way to sequester and destroy such proteins before they can do any harm. So here we have an example of a resilient system that has a negative feedback mechanism in place to deal with a disruption.

This evolved in living cells because protein denaturing is not a rare or unexpected event. But imagine if it was rare and cells were unprepared to handle the disruption. Every once in a while a cell would die because a rare event "poisoned" it.

As another example of a "brittle" system component failure leading to a catastrophic collapse, consider a single transistor in a computer's CPU (central processing unit—the main control for the entire computer). If even one tiny transistor fails, say due to heat exhaustion or gamma ray disruption, it will bring down the whole system. A single transistor is one of the literally billions of components, yet its failure is terminal.[24] The engineering and construction of computer components requires superlative care in order to assure a maximum of availability.

Finally, consider the hub and spoke topology discussed above. What happens if a hub component fails? That will really depend on how many other linkages the spoke nodes have to non-hubs or multiple hubs. Nevertheless, there are generally structural reasons why the hub arrangement came into being in the first place. Loss of such a hub will clearly have negative ripple effects throughout the whole system.

Question Box 5.7
Consider your home or automobile. Does either ever need repairs? What goes wrong and why? If your furnace fails and needs to be fixed, does the whole house fail? If your fuel injector fails, does the whole automobile fail? Can you develop a set of principles about whole system failures and complexity of the system?

5.5.2   Process Resource or Sink Failures

Since systems rely on their external environments for resources and to absorb their products and wastes, any failure of some other system in the chain of inputs or outputs to the system of interest will be disruptive. In Chap. 10 we will explore the various ways in which systems can be organized internally to handle disruptions of this sort. We will see that on the one hand, if such disruptions are relatively commonplace, systems will have a tendency to incorporate more complexity to deal with the situation (just as the cell has mechanisms to deal with protein denaturing).

The problem with too much complexity when the environment doesn't work as expected has more to do with critical subsystems that contain nonlinear processes or positive feedback loops necessary for stability. Below we will show how complex societies have collapsed due to restrictions in energy inputs needed to drive positive feedback loops!

---

[24] Today, modern computer designs are moving toward more redundancy in components, at least at a high level of organization. Such redundancy, such as what are known as multiple core CPUs, allows a computer to continue working even if one component fails at that level. Of course the system then operates at a reduced level of performance because the component cannot, at the present state of technology, be repaired in situ.

5.5.3 *Systemic Failures: Cascades*

Systems can be too complex for their own good! The more components, the more links, the more clusters and hubs in the network, the more things can go wrong. Moreover, the more complex a system is, the more likely that a small failure in one part of the network of components will lead to disruption and systemic failures throughout the network. Whole systems can collapse upon themselves under these circumstances. The rate of collapse can vary depending on the exact nature of the network structure and internal dynamics.

The cause of this collapse is the cascading effect of a failure at one node or cluster that propagates through a brittle network. The brittleness of the network links is actually a response to the complexity itself. We will see how this comes about in Chaps. 7 and 8. For now we will simply state that the number and coupling strengths of links in a complex network are determined in the long run by the need to maintain local stability. Unfortunately this very same condition can lead to total collapse when either something goes wrong internally or externally.

5.5.3.1 Aging

Biological systems go through distinct phases of complexity. Initially the system is simple (e.g., a single cell embryo). Over time it develops according to a master plan laid out in the genetic endowment modulated by environmental contingencies. That is, the complexity of the organism unfolds as it takes in material and energy resources and converts them to its own biomass. There follows a period of growth (and possibly more development). During this time the complex networks of cells, tissues, organs, etc. develop so that they are coordinated through numerous physiological feedback loops. At some point the growth ceases and the system matures, refining the functions to the best of its abilities. But then the inevitable second law of thermodynamics starts to win. In all somatic cells in almost all multicellular animals and plants, various little problems start to crop up. Imbalances in metabolites, genetic breakdowns, and several other age-related anomalies begin to accumulate. Eventually they dominate the tissues and bodies which go into what is called senescence or a kind of decay. Since living tissue has many redundant subsystems and reserves of structures, this decay does not result in death but a decline in functionality over time.

At some point the organism is overwhelmed by this decay and succumbs. This cycle of life seems to be inevitable.

5.5.3.2 Collapse of Complex Societies

One of the most enduring and fascinating themes taken up by many historians has been the collapses of numerous civilizations, societies, and empires. From the most recent examples such as the Chacoans of the southwestern United States, to the Mayans of the Yucatan peninsula, to the most ancient like the Mesopotamians of the

Middle East, and many others in between (most famously the Roman Empire), all social centers of power and wealth have either disintegrated or undergone rapid collapse .[25] We are all fascinated by social collapses because of the implication that our societies will one day collapse if there is some underlying natural cause for such. Ergo, historians, sociologists, archeologists, and anthropologists have sought causes of collapse from whatever historical (written or archeological) records can be pieced together. They have looked for both internal (e.g., cultural degradation) and external (e.g., climate change, invasions) causes, and several authors have tried to offer theories of main causes or common factors that seem to operate in all instances. They seek some universal principle that explains why civilizations eventually disintegrate or collapse.

One very compelling thesis about a common factor has been explored by Joseph A. Tainter (see footnote below). He describes how complexity of social institutions and commercial enterprises plays a role in collapse. Basically the thesis holds that human societies are always faced with various kinds of problems as the society grows larger. Getting more food and water for the populations, especially those living in the "cities" or centers, are certainly core issues, but so are waste disposal and protection of all overseen lands (e.g., protecting the farmers from marauders). As societies confront these problems, they find solutions that increase the complexity of the society as a whole. This is essentially the development of new nodes, new links, new clusters, and new hubs in the network of people, institutions, inventions, and every other subsystem of a society's culture. Once again we need to refer you to Chap. 8, Emergence and Evolution, where we will describe this process over longer time scales.

What Tainter has noted is that invariably many, maybe even most, solutions to problems generate new problems, at least over time. The law of unintended consequences usually attends most social solutions. The result is that now the society needs to find a solution to the new problem. So the problem-solution-problem cycle is a positive feedback loop or amplifier effect that forces increases in complexity. But, and this is the core of Tainter's thesis, the solution to problems must actually provide a benefit or payback. The solution must actually fix the problem even if it does cause new problems, or otherwise it is no solution and no one would implement it. Yet as societies become more complex, the payoff from solutions tends to diminish. The law of diminishing returns applies to increasing complexity. At some point the marginal returns do not exceed the marginal costs (in the form of operating costs but more importantly the creation of new problems).

Problems start to accumulate and societies only know how to respond by increasing complexity, which only compounds the problems. So at some point the institutions and other social mechanisms begin to fail (as described above) and disorder begins to creep in. In other words, societies essentially self-destruct by virtue of too much complexity. But we are left with a question about how this complexity could even be generated in the first place? What maintains whatever level of complexity obtains?

---

[25] See descriptions of these civilizations and their histories in Joseph A. Tainter's *The Collapse of Complex Societies*, Cambridge University Press, 1988.

More recently Tainter and other researchers have been looking at the role of energy flow through the societies. Remember it takes energy to do physical work and to keep people alive and functioning. Most civilizations have had to work off of real-time solar energy (food production) and short-term stored solar energy (water flow and wood). Homer-Dixon[26] has analyzed the energy situation for ancient Rome just as it was going into relatively steady decline (it was a bumpy ride down, but the trend was definitely downward until the empire finally collapsed in the West). He estimated the amount of energy from food (including food for work animals) and wood fuel that would be needed to support the population in Rome, the building of the Coliseum to assuage the growing restlessness, and the armies that were needed to keep the supply lines open. He found that to support the system ever-increasing areas of land were required since the flow of solar energy was essentially a constant for any unit of land. The problem turned out to be that as the empire expanded to meet the needs for more energy, the amount of energy needed to obtain and then govern the outer regions used up too much of the energy produced, so the marginal gain did not warrant the marginal costs. For example, since food had to be transported by horse and wagon, and horses need food to do their work, just the cost of feeding the horses to transport the remaining food back to Rome became unacceptable. The model of empire expansion to support the core civilization on real-time solar energy became unsustainable. Moreover, the energy needed to keep the core going was diminishing and that was a direct cause of the population's unrest (e.g., higher food prices).

In other words, by this model it seems that the collapse of Rome was partly the result of having too much complexity which required more energy flow, and eventually a declining energy flow due to lower net returns on energy invested in getting more energy. In the end, because the only response people had to increasing problems (from lack of sufficient energy flow) was to increase the complexity of the system through new laws, commercial enterprises, and institutions (the gladiator games to take the citizens' minds off their troubles), complexity itself became the problem.

The picture that has emerged from this work is that there are many possible internal failures and trigger events that lead to collapse. But all collapses (which have been studied with this new insight) have the common factors of over complexity and a decline in resources, especially energy, from their external environment. There are many pathways and mechanism failures that lead to collapse, but they all have their roots in the system's inability to sustain whatever level of complexity it has achieved.

The issue of increasing complexity as usable energy flows through a system is universal. Systems do evolve under these conditions (Chap. 11). But unless the energy flows increase to meet demand, at some point, the further increase in complexity cannot be sustained and the nature of the complexity turns it into a potential liability. If some necessary external factor is disrupted, it will result in the kind of cascade collapse described above.

---

[26] Thomas Homer-Dixon, "The Upside of Down: Catastrophe, Creativity, and the Renewal of Civilization," Island Press, 2006. Homer-Dixon was interested in not just collapse but also how such collapses freed up resources, especially human ingenuity, that then became the seeds for new civilizations.

Question Box 5.8

Based on your current understanding of the flow of energy in a system and the nature of complexity, can you describe the link between increasing energy flow and increasing complexity? Can you provide an example of increasing complexity in your experience and identify the energy that was needed to produce and support it?

Example Box 5.1 Solving a Social Problem, Increasing Complexity, and Unintended Consequences

Not long after northern industrializing societies discovered the power in fossil fuels, geologists and biologists figured out that these fuels were indeed laid down by the biological organisms of the past dying and some of their organic remains being covered by silt and sand, eventually being subjected to intense heat and pressure and cooking into the forms we find now, oil, coal, and natural gas. Most importantly we realized early on that these fuels were finite in quantity. In spite of this knowledge, industry started exploiting the power and consumers did too. Society kept inventing new ways to consume energy from these sources until presently they account for over 80 % of the world's energy consumption.

We've known for a long time that eventually these fuels would run out, but we didn't really act as if we knew it. Today we are actually facing the peak extraction rate for oil on a global basis. Moreover, most countries, like the United States, rely more heavily on imported oil from just a few (in some cases not terribly friendly) countries.

Problem to solve: *How can we keep our transportation capabilities, private and commercial, which are based on internal combustion engines (ICE), going into the indefinite future (or until we invent some wonderful replacement for the ICE)?*

Possible (partial) solution: *We can substitute some of our refined fuels, like gasoline, with ethanol, which is combustible with almost the same power as gasoline. We can produce this ethanol by fermenting corn mash and distilling the alcohol and mixing it directly into the gasoline prior to distribution.*

This seemed, in the early 1990s, like a great solution. Brazil had been producing ethanol as a fuel for years prior, but their feedstock was sugarcane. It is inherently easier (less energy consumed) to ferment sugar than starch, which has to first be broken down into its constituent sugars. So, for Brazil this looked like a great solution to feed their liquid fuel needs. Many of the policy makers in the United States asked why we couldn't do this in the United States. It turned out that corn growers were quite eager to develop a new market for their product and hoped that it would be one that would be sustained

(continued)

Example Box 5.1 (continued)

(at possibly higher prices) indefinitely into the future. So political forces conspired to solve part of the energy problem associated with the import of foreign-sourced oil with home-grown corn-based ethanol.

The policy makers took decisions that involved subsidizing the production and use of ethanol from corn, and in 2007 the Congress passed legislation, the Energy Independence and Security Act of 2007, to mandate the blending of ethanol into gasoline.

Unforeseen problems: *It takes almost as much external source energy (from fossil fuels!) to produce a unit of energy contained in ethanol!*

After many years of study, energy scientists have determined that by the time you end up adding in all of the energies used to plant, grow, fertilize, harvest, mash and ferment, distill, and finally deliver to market, you only get a net gain of one half unit (roughly) of energy. The *energy returned on the energy invested* (called the EROI or also EROEI) is only about 1.5 to 1 for corn. It is much higher for sugarcane because some of the input energies are significantly lower. These input energies do not even account for soil degradation or the runoff of excess nutrients that end up creating dead zones (eutrophication zones) due to intensive industrial agriculture.

In the meantime an extensive industry has developed for making corn ethanol. A new lobby organization is infiltrating the halls of the Congress, new organizations of corn growers have been formed, new legislation has been enacted, the gasoline distributers have to have facilities to blend in ethanol, and many other complex responses have been set in motion to accomplish this objective that turns out to have a very minimal impact on total net energy for society! This is the marginal return on increase in complexity that Tainter writes about.

Unforeseen consequences: *More corn going into ethanol means less available for human food and livestock feed, leading to higher prices and in some cases lower nutrition for poor people.*

The rush to substitute ethanol for gasoline has led to a new problem. Corn, or its derivatives such as corn sweeteners, is used in so many processed foods in the United States that much of that food supply has been affected by the higher prices. While there is a reasonable argument that these processed foods were not especially healthy in the first place, the fact is that much of our processed food industry upon which very many consumers depend is now starting to pass the increased costs on to those consumers. No one in the Congress who voted for the Energy Independence Act had any intention of causing price increases in foods just so other consumers could drive their vehicles as always.

What is wrong with this picture? How will the policy makers solve this new problem? There is a reasonable likelihood that it won't be by rescinding the act!

5.6   Summary of Complexity

We have tried to show in this chapter that complexity is a complex subject! The idea is intuitively attractive as a way to explain why we might not understand something as well as we think we should. A complex system surprises us often because by virtue of its complexity we fail to understand all of the parts, connections, and behaviors. Complex systems are sources of information (see Chap. 7 for an explanation) for this reason. They are also fascinating to observe.

Complexity remains an elusive concept in terms of easy definitions. Our approach in this book is to use the notion of hierarchical depth of a nearly decomposable system as put forth by Simon. We feel this idea captures most of what we generally mean by complexity and provides a reasonable quantitative measure that can be used as part of the description of systems. In the next set of chapters, we will start to explore the way in which complexity of organization and behaviors is actually managed in systems. Complexity could be a source of chaos, rather than the other way around, if it is not, in some way, controlled. The breakdown and collapse of complex societies shows us how that works when the governance processes are not properly tuned to the level of complexity of the society. This is true for all complex systems. Living organisms are complex systems that have evolved the capacity to self-regulate from the highest level of organization down to the lowest (within cells). So their example shows that complexity need not always lead to collapse per se. But we have to understand the principles that lead to organized and effective self-governance within systems.

Bibliography and Further Reading

Casti JL (1994) Complexification: explaining a paradoxical world through the science of surprise. Harper Collins, New York
Fuster JM (1995) Memory in the cerebral cortex. The MIT Press, Cambridge, MA
Gleick J (1987) Chaos: making a new science. Penguin, New York
Gribbin J (2004) Deep simplicity: bringing order out of chaos and complexity. Random House, New York
Langton CG et al (eds) (1992) Artificial life. Addison-Wesley Publishing Co., New York
Mandelbrot BB (1982) The fractal geometry of nature. W. H. Freeman and Co., New York
Margulis L, Sagan D (2000) What is life?. University of California Press, Los Angeles, CA
Mitchell M (2009) Complexity: a guided tour. Oxford University Press, New York
Moore J (2003) Chemistry for dummies. Wiley Publications, New York
Morowitz HJ (1992) Beginnings of cellular life: metabolism recapitulates biogenesis. Yale University Press, New Haven, CT
Nicolis G, Prigogine I (1989) Exploring complexity: an introduction. W.H. Freeman & Company, New York
Prigogine I, Stengers I (1984) Order out of chaos: man's new dialogue with nature. Bantam Books, New York
Simon HA (1996) The science of the artificial. The MIT Press, Cambridge, MA
Striedter GF (2005) Principles of brain evolution. Sinauer Associates, Inc., Sunderland, MA
Tainter JA (1988) The collapse of complex societies. Cambridge University Press, Cambridge, MA
Weaver W (1948) Science and complexity. Am Sci 36:536–544. http://philoscience.unibe.ch/documents/uk/weaver1948.pdf. Accessed Jan 2013
Wolfram S (2002) A new kind of science. Wolfram Media Inc., Champaign, IL

Chapter 6
Behavior: System Dynamics

"Prediction is very hard, especially when it's about the future."

Yogi Berra

"...I've got to keep on moving."

Ain't nothing going to hold me down, by Men at Work

Abstract Systems are never still. Even a rock weathers and often changes chemically over long enough time scales. Systems are dynamic, which means they have behavior. In this chapter we explore the dynamic properties of systems from a number of perspectives. Systems as a whole behave in their environments. But systems contain active components that also behave internally relative to one another. We look at a myriad of characteristics of system dynamics to understand this important principle. A key concept that pertains to system dynamics is that of energy flow and work. Every physical change involves the accomplishment of work, which requires the use of energy. The laws of thermodynamics come into play in a central way in systems science.

6.1 Introduction: Changes

To state the obvious: things change.

Look at the world around you. Your environment is always undergoing change. The obvious kind of change is the motion of objects—change in position relative to your position at any given time or more broadly. More broadly, change is inherent in the behavior of systems. On longer time scales objects change. Buildings are built or torn down. Roads are widened. On slightly longer time scales, the very styles of buildings and automobiles change. Altogether the amount of activity going on at any given instance in time itself is a process of change.

As it has been said, the only thing that is constant is that things change.

Consider yourself as a "system." You change over your lifetime. On a scale of seconds and minutes, your metabolism changes in accord with activity and environmental conditions. On the scale of hours and days, your brain is undergoing changes as you learn new things. Over years your body changes as you age, and changes accumulate with every attack by disease agents or accident.

© Springer Science+Business Media New York 2015
G.E. Mobus, M.C. Kalton, *Principles of Systems Science*, Understanding
Complex Systems, DOI 10.1007/978-1-4939-1920-8_6

Put simply, all systems undergo change and do so in different ways and on different scales of time and space. When change is itself systematic, it can manifest in patterns that we can analyze and work with. This chapter is about several specific kinds of changes that can be observed, measured, and analyzed so that we can find patterns in the change. Those patterns become the basis for developing expectations of how the systems of interest will behave in the future (e.g., predictions) or of how similar systems may behave in similar ways.

We call this study of how systems behave over time *dynamics*.[1] This chapter will explore several different kinds of changes that systems undergo as time progresses. First we will introduce these kinds of changes, and then we will show how they work in various kinds of systems. Ultimately we are concerned with understanding how and why systems behave as they do. Moreover, we will need to understand how that behavior itself may change as a result of certain kinds of changes within the system, a matter of particular importance for understanding adaptive systems. Adaptation brings us to another category of change that we will cover in great detail in the fourth section of this book, Evolution (Chaps. 10 and 11). Evolutionary changes are qualitatively different from dynamic ones in that the whole system of interest may change its very purpose (function) as a result of changes to its internal structure and processes.

Between the dynamics discussed in this chapter and evolutionary changes discussed in Part IV, there is a kind of intermediate form of change that does not easily fit into either category. That is the phenomenon of learning. Indeed this kind of change seems to be a combination of the other two, involving some aspects of dynamic changes and evolutionary changes.

Learning is a form of adaptivity (see below) that seems to involve some aspects of evolutionary process. We will introduce the dynamics of simple adaptivity in this chapter insofar as it can be treated as a dynamic process, and we will outline a more advanced kind of adaptivity that requires much more complex system capabilities. But the discussion of how this advanced form of adaptivity is achieved and managed must wait until we get to Chap. 7, *Information, Knowledge, and Cybernetics*. The discussion of learning will be covered in that section as well.

What qualifies as being treated as a dynamic process or phenomenon for discussion here depends on whether one can measure a specific parameter of the system that changes over time. Typically the observer takes a measurement of the parameter without disturbing the process (or such disturbance is miniscule and cannot, itself, be measured) at time intervals that are chosen such that the smallest detectable

---

[1] In this chapter we take a somewhat general approach to the concept of dynamics. In physics the change of position of material objects is covered in mechanics (kinetics). Changes in energy content, or form, are covered in thermodynamics. In chemistry these two kinds of dynamics seem to come together in the nature of molecular-level processes (i.e., reactions). Here we are describing all kinds of changes in system components which include matter, energy, and messages (covered later in information theory). Hence we do not go to great lengths to segregate the various approaches to dynamic systems descriptions unless it is cogent to the subject being discussed.

change from one *sample* to the next can be captured. This creates what we call a time series data set, where a list of measurements is kept in time order such as:

$$m_0, m_1, m_2, \ldots m_i, \ldots m_n$$

The first item is the measurement, m, taken at time t_0 or the first sample. The ith sample represents a typical measurement in the sequence, and the nth sample is at the end of the measurement period (the last time step in which a measurement was taken). Generally the intervals between measurements, called Δt, are equal with the duration, called a time constant, being short enough that the smallest interesting change can be measured.

As we will see in the descriptions below of different kinds of dynamics, they all share this property of having one or more measurable parameters (properties) the change of which is assumed to be important for understanding the system in question. This chapter will present many examples of this aspect. Graph 6.1 shows a generalized approach to understanding the dynamic behavior of a system. In this case we see a plot of data points representing measurements of population size taken at regular intervals, called time ticks. Overlaying the measurements (blue dots) is a smoothed curve that mathematically "fits" the data points. We can then derive a mathematical model (equation) that matches the curve, in this case an S-shaped curve called the "logistic function." The S-shape of the logistic function is generated when processes are characterized first by an exponential rise followed by an exponential deceleration to level off at a maximum. Population biologists encounter such processes continually and know a lot about why this model works the way it does. Knowing this overall system behavior, they assess the ecology of the population in terms of resources that support initial rapid population growth, which is followed by limits on key resources that cause growth to decline and stabilize. Field biologists can use such information to study the actual conditions in the real ecology.

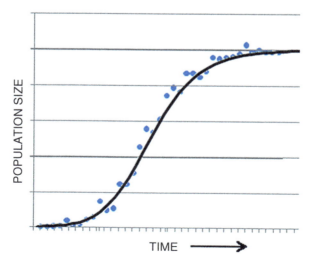

Graph 6.1 A time series of measurements of a population (*blue dots*) is fitted best by a logistic function

In most real-life systems, the measurements and the finding of a simple mathematical model are not so neat (see Quant Box 6.1 below). But the familiar S shape of the logistic function is often found in some form or another in population growth, and it shows how we can describe a complex dynamic process by transforming it into numbers through measurement and using relatively simple mathematics. As we will see below, however, this is a rare situation, and we do not always have access to a straightforward mathematical equation that can be used to completely describe the dynamic properties of a system. For most real-world systems, we need to use digital computers and build what are called *numerical models*, which we will introduce in Chap. 12. This chapter will provide a starting point for understanding how we can build such numerical models by introducing the conceptual framework and a graphical language that will help us think about system dynamics.

Quant Box 6.1 Functions, Curve Fitting, and Models
The simple form of the logistic function is given by

$$Y_{(t)} = \frac{1}{1 + e^{-t}}$$

where Y is the population number at t, the time index, and e is the base of the natural logarithm. This function produces the S-shaped curve shown above in the graph, one that rises at an accelerated rate to some midpoint where it then decelerates and levels off at some maximum value.

The logistical function presented in this section is very generic and not terribly useful in describing a specific system's growth. A more useful form of the equation is parameterized so as to produce a more "shaped" curve, appropriately scaled to the quantity that is said to be growing over time. Richards, in 1959, proposed a general formula that includes parameters that allow one to fit the S-shaped or sigmoid curve to a specific data set:

$$Y_{(t)} = A + \frac{K - A}{\left(1 + Qe^{-B(t-M)}\right)^{\frac{1}{v}}}$$

- A is the lower asymptote.
- K is the upper asymptote.
- B is the growth rate.
- $v > 0$ affects near which asymptote maximum growth occurs.
- Q depends on the value $Y(0)$, at the start of the growth.
- M is the time of maximum growth if $Q = v$

(see reference Wikipedia).

(continued)

Quant Box 6.1 (continued)

Application: Yeast growth study

Nick is an evolutionary biologist interested in the relative reproductive successes of several strains of a yeast species, *Saccharomyces cerevisiae*, an important contributor to human foods (breads) and drinks (beer and wine). This yeast metabolizes sugars into carbon dioxide and ethanol in a process called fermentation. But what interests Nick is that there are a few variants of this species that seem to have different rates of reproduction which he attributes to variations in one of the genes involved in fermentation. One variant seems to outreproduce the others, so Nick is going to do experiments to find out if this is the case. He will use population growth as a measure of reproductive success under varying conditions. To do this he grows a particular strain of yeast cells in a solution containing varying concentrations of sucrose, a sugar, and measures the population density in the test tubes every 2 h. Changes in population density over time can then be used to estimate the reproductive success of the strains. Here are the data he collected in one experiment on strain A (values obtained from a calibrated densitometer).

Data

0.002, 0.0035, 0.0065, 0.0089, 0.0400, 0.0131, 0.0220, 0.0451, 0.0601, 0.1502, 0.1001, 0.1100, 0.2501, 0.2503, 0.3103, 0.4580, 0.5605, 0.5410, 0.6120, 0.7450, 0.7891, 0.7662, 0.8701, 0.8705, 0.8502, 0.8804, 0.9603, 0.9511, 0.9592, 0.9630, 0.9802, 1.0301, 0.9902, 1.001, 0.9890, 0.9902, 0.9911, 0.9912, 0.9930, 0.9950, 0.9981, 0.9990, 0.9990, 0.9991, 0.9991

Here is a graph that shows the scatterplot of these data and a parameterized logistical function that "fits" the plot of the data (same as Graph 6.1 above) (Graph QB 6.1.1).

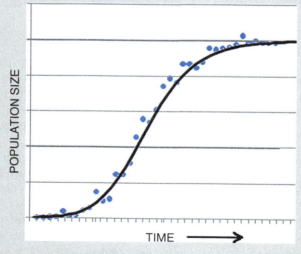

Graph QB 6.1.1 Same as Graph 6.1

(continued)

Quant Box 6.1 (continued)

Nick used a nonlinear curve fitting method to come up with the following parameters:

A	K	Q	B	M	v
0	1	0.5	0.5	1.6	0.5

Nick ran the same experiment on strain B that he suspected had a more efficient ability to ferment sugar. He got this scatterplot (Graph QB 6.1.2):

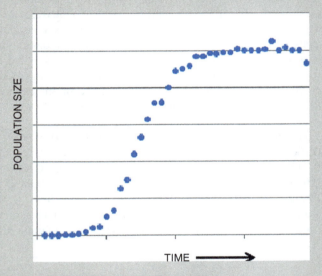

Graph QB 6.1.2 Plot of data from the second experiment

From this data:

0.00043, 0.00065, 0.00120, 0.00240, 0.00511, 0.00902, 0.02010, 0.03871, 0.04561, 0.10081, 0.13352, 0.25460, 0.30111, 0.44012, 0.53430, 0.63102, 0.72010, 0.72082, 0.80241, 0.89033, 0.90021, 0.91550, 0.97031, 0.97083, 0.98552, 0.98316, 0.98988, 0.98994, 1.00910, 1.00100, 0.99996, 0.99923, 1.00502, 1.05001, 0.99996, 1.01801, 0.99980, 0.99991, 0.92985, 0.99032, 1.00901, 0.99992, 0.98993, 1.01803, 0.99992

See if you can help Nick decide if this strain of yeast has greater reproductive success based on the comparison between the two. What parameters would you assign to get a good curve fit? Hint: B is growth rate and M is time of maximum growth! Also note that the density numbers are in a different range, so use $Q=0.25$. Using a spreadsheet program with graphing capabilities, plug in the formula above and the data. Use time steps of 1, 1.5, 2, 2.5, etc., up through

(continued)

Quant Box 6.1 (continued)
10 in the first column. Put the parameter names in the first row with values in the second, as shown above. You should be able to find a reasonable fit for this data set.
Can you explain why the data do not form a perfectly smooth curve?

References
1. Richards FJ (1959) A flexible growth function for empirical use. J Exp Bot 10: 290–300
2. Wikipedia. Generalized logistic function. http://en.wikipedia.org/wiki/Generalised_logistic_function

6.2 Kinds of Dynamics

In this section we briefly describe four of the most common categories of phenomena that can be understood in the framework of systems dynamics.

6.2.1 Motion and Interactions

Simple dynamics are covered in this category. That is, systems do not actually change structure or functions, but the components interact. Flows occur. Forces operate to cause components and whole systems to move in space. Chemical reactions may increase or decrease in rates but do not involve new elements.

Whole systems can move relative to other systems. They can be displaced in space over some relevant time scale. As observers of systems we can assign a frame of reference and scales of distance against which we measure motion as displacement along one or more dimensions per some unit of time, also scaled by the observer. This is the typical kind of dynamics studied by physics. The causes of motion are attributed to forces. Energy is used to do work in moving masses from one place to another. A ball is thrown. Or a cheetah runs at over 100 km per hour.

Of course component subsystems can also move relative to other component subsystems within a given system. Blood flows through the arteries, capillaries, and veins, for example. Traffic engineers study the timing and volume of vehicle flows, with special attention to those intersections which give rise to accidents or backups. So the dynamic of relative motion and interactions applies within systems as well as to systems acting in their environments.

Another kind of dynamic phenomenon is changes in *rates* of interactions between component subsystems or between systems. A common example of such changes is the way chemical reactions slow as the concentration of reactants thins as products are produced over time. Here we refer to cases where the reactants do not vary in kind, only concentrations.[2] Reactions can be speeded up by simply changing the temperature, another aspect of energy causing work to be done. But then a speed up of a chemical reaction without increasing the rate of resupply of the reactants will lead to reductions over time of the concentration of the reactants in the vicinity of the reaction. Here we see how several dynamic variables interdepend in producing a given rate of chemical interaction and production. Models of such processes are critical not only for every sort of industrial application but also for understanding and controlling the complex metabolic chemistry of living organisms.

Question Box 6.1
Motion among interacting systems or subsystems requires coordination, or dysfunction occurs. This applies to all sorts of systems, giving rise to shared descriptions across very different kinds of systems. How many applications can you think of for being "flooded"? What is the similarity that allows the term to apply to such different situations?

6.2.2 Growth or Shrinkage

Growth is generally considered as an increase in size (volume and/or mass) of a system per unit of time, while shrinkage is the opposite. One of these two or both parameters must be measured each unit of time in order to treat a growing system as a dynamic process. A very common example that everyone is familiar with is the growth of an economy as measured by an increase of the gross domestic product[3] (GDP) per time period (month to month, quarter to quarter, or year to year) where GDP is a mass measure of wealth. Another example is population growth. The dynamics of population growth and economic growth are intimately intertwined and carefully studied by governments as they try to assess the future economic trajectories of their countries.

---

[2] Chemical concentrations are measured in number of parts, e.g., molecules, per unit of some large summation of all types of chemicals present in a given volume or weight. For example, the concentration of hydrogen ions in a volume of water—the number of parts per liter, say—is a measure of the pH or acidity of the water. Similarly the number of CO_2 molecules in a liter of ordinary air is a meaningful measure that tells us something about the energy absorption capacity of the air, e.g., global warming effects.

[3] Gross domestic product is an attempt to measure the wealth of a nation in terms of monetary valued income. Briefly it is the sum of all income-producing transactions, such as wages and sales. See http://en.wikipedia.org/wiki/Gross_domestic_product

As a species that plans for the future, measurements of growth and shrinkage are of great practical import for humans, and the critical thing is that the type of measurement be maximally well suited to the use to which it is put—a matter of much controversy when it comes to policy debate! There are, for example, at least three conventional ways of measuring the GDP (production, income, and expenditure), each suited to different contexts. Even population growth is not a straightforward question: the populations of many species may be tallied by headcount, but sometimes aggregate biomass is a more meaningful measure. The context for using the measure is critical. If one wished to estimate nutritional flows necessary for the growing human populations, body size is such an important variable in consumption needs and is so varied among humans that biomass rather than headcount would be a far superior measurement.

Question Box 6.2
When Simon Kuznets first came up with how to measure the GDP in 1934 for the US Congress, in his first report he stated clearly, "The welfare of a nation can, therefore, scarcely be inferred from a measurement of national income as defined above." Nonetheless, the growth of the GDP is commonly identified with national well-being, an identification with extensive consequences in the world of policy and politics. In what ways is GDP inadequate and misleading as an indicator of well-being? What other sorts of growth might need to be included to make it a more adequate measure?

Growth dynamics are well understood in most systems (see Quant Box 6.1 above). Growth is sustained by a constantly increasing availability of multiple resources. We know that in physical systems infinite growth is not possible because many factors can limit the contribution to growth. For example, the growth of wealth needs continually increasing amounts of high-quality energy; the growth of populations depends on continually increasing food and water supplies. Understanding growth and the limits to growth in dynamic systems such as the economy or human populations has direct relevance to all of us.

6.2.3 Development or Decline

Development is the "programmed" change in complexity of a system as it matures, maturation being the realization of complexity over time as discussed in the prior chapter. For example, an embryo develops into a fully functional organism of a specific kind. The kind is completely encoded within the DNA of the fertilized egg so that the embryo cannot develop into just any kind of organism. The program for development is built into the DNA.

Development is, thus, based on inherent properties of the system. Developmental increases in complexity are quite often associated with growth of the organism[4] or organization (as is the case with embryonic or even child development), but development and growth are not quite the same. Growth is a quantitative measure, involving incorporating more resources into the structure, but development involves establishing new structures from existing ones. A pollywog grows larger and at some point develops into a frog.

The same dynamic can be seen in a growing business. As a business grows it often goes through some form of development that is, in essence, preordained by the very kind of business it is. If a subunit, like accounting, is growing to meet the needs of a growing business, then it will acquire new personnel who specialize in an already defined operation within the accounting function. Thus the accounting department is developing and has a preordained overall function, but is becoming both bigger and more complex over time. However, the question of development must be approached with caution since businesses are also capable of morphing into completely new arenas or markets in a non-programmed way. In this regard human organizations often fit more readily into evolutionary models rather than developmental models.

Question Box 6.3
As the counterpart of development, how would you describe decline?

6.2.4 Adaptivity

Adaptivity, in its basic form, involves change in the behavior of a system, but while using existing resources in response to environmental change.

When an environment changes in ways that have a bearing on the function of a given system, the system can persist and succeed only if it has a capacity to adapt to that change. Adaptation (as opposed to evolution) involves the ability to reallocate internal energy or material resources to existing sub-processes so that responding processes have more capacity to fit the conditions. Homeostasis, the process by which the body adjusts to maintain critical internal conditions such as temperature, is a good example of this. Expansion of muscle mass (a kind of growth of a subsystem) in response to external demands for the organism to do more physical work is another example. Nothing in the basic structure of muscle actually changes, it simply becomes more capable of responding to the demand for work.

---

[4] This is not actually the case in certain cases of metamorphosis, such as with a caterpillar turning into a butterfly. In this instance there is no real growth of the organism, merely a change in form. Nevertheless, it is a programmed change.

The dynamics of physical adaptation can be shown by measuring the relevant physical changes on some kind of time scale. Adaptation to increased demand on muscles, for example, can be tracked by charting how much more weight they can lift over a period of time. Or on a more complex level, the familiar treadmill stress test measures the rate of changes in the heart and circulation as the body adjusts homeostatically to conditions of increased exertion.

In addition to physiological adaptation, another kind of biological adaptation involves changes in behavior as a result of changing environmental conditions. For example, one could quantify the changes in foraging behavior of a predator as the prey species mix changed, say due to climate changes. And TV networks do an adaptive dance of program modification every season to respond to changes in viewer preferences and habits. Those changes can be measured and plotted just like other dynamic graphs.

Adaptivity involves the use of information and internal control systems to manage the adapting process. That is, the new conditions must somehow register on the system, and the system must have the inner capacity to respond to the change. We will cover this aspect in Part III of the book. Here we only discuss the outward observation and measurement of the dynamic process of a system responding to changes in its environment.

> **Question Box 6.4**
> Can you give an example of an adaptation that is also a development?

6.3 Perspectives on Behavior

There are two major perspectives for examining the dynamic behavior of systems. The first is the outside or black box perspective, which essentially involves looking at a whole system and measuring aspects of the external behavior. We can observe the nature of inputs and outputs, measure parameters of all, and describe the functional relations between outputs and inputs. For instance, in the above example of logistical population growth, we could measure the resource inputs to the population (food, water, etc.) as a function of time and then count the numbers of individuals as a function of that same time (as shown in the graph). We would find a high correlation between the two sets of measures. There might be a lead time on the resource curve or a lag on the growth curve, but the two would be highly correlated once adjusted for any time lags. The reason for this is well understood. Populations that have increasing access to necessary biological resources without any exogenous factors counteracting will tend to expand until at least one of the necessary resources starts to fall off in availability. That is, the population size is a direct function of the resource inputs. We can say a great deal about the system (the population) without actually knowing the details of the function(s). We can observe many different kinds of populations of plants and animals and find essentially similar behavior under similar conditions relative to the resources. Thus we can conclude that living populations will always grow in size as long as the requisite resources are available. It is a function of life itself.

However this doesn't really tell us much about how the function works within the living system. In order to understand the function more deeply, we need to open up the population and take a look at the behaviors of its components—the individuals that comprise it. The inside or white box perspective addresses subsystem behavior, and so, in the case of various sorts of population, for example, it can address questions such as just how the internal flows of particular resources lead to reproduction successes and thus population expansion for a species.

These are the two perspectives we use to discover and understand the dynamic behavior of systems, as wholes (from the outside) and as a system of subsystems (from the inside).

6.3.1   Whole System Behavior: Black Box Analysis

The cheetah (*Acinonyx jubatus*), an African cat, has been clocked running at up to 120 km/h (~75 mph) for short bursts. Biologists interested in knowing how the cheetah can achieve this world record speed have a problem in answering questions about how this magnificent animal can do this. In dealing with the behavior of a living creature, they are faced with a difficult conundrum. As engineers know, if you want to know how a machine works, you take it apart and look at the parts and how they fit together and how they work together. With a little luck you can even put the machine back together, and it will work again. Not so with a living creature, at least if you want to preserve its whole behavior.

The first thing biologists will do to understand a particular animal is to observe its total behavior over its life cycle. They set up field studies where lucky graduate students get to camp out and take field notes to capture everything the animal does, what it eats, when it sleeps, etc. As much as possible they attempt to quantify their observations. Once the "data" is back in the lab, the analysis will attempt to find patterns in the behavior that will help answer some of their questions.

This perspective on behavior is for the whole system of interest. It treats the system as a "black box," that is, with no knowledge about what is going on inside the system, only about its external behavior, and, in particular, the functional relations between inputs and outputs. For example, with the cheetah, biologists are able to estimate the animal's caloric input from the kinds of foods it ingests. They can then estimate how many calories would be demanded to move the mass of the cat for the distances run in chasing down a prey, the speed achieved, and the time over which that speed is sustained. The main kind of intrusion into the cheetah's life would be capturing it to weigh it (to find the mass). Everything else can be based on observations at a distance, giving measurement estimates of distance, and stop watch readings of time. Observing what the animal eats, the biologists can gather samples of tissues from carrion and subject them to laboratory analysis of caloric content (e.g., using a calorimeter). In short, the scientists can say a great deal about the whole system behavior under given scenarios for inputs.

Any whole system can be treated in this fashion. Essentially we set up our measuring devices just outside the boundary of the system and keep track of everything that goes in and comes out. See Fig. 6.5 below for a representation of an "instrumented" system. This is the term for a system to which we have attached sensors for all of the inputs and outputs in order to collect periodic measurements of the flows. The data collected over a long time will be analyzed to show us the dynamics, function(s), and overall behavior.

6.3.2 Subsystem Behaviors: White Box Analysis

Whole system behavior can tell what a whole system does over time given the inputs. But it cannot tell us exactly how the system does what it does. Unless the system of interest is very similar to one that we have already "dissected" and appears to behave in a similar fashion, we cannot make too many assumptions about what is going on "inside."

In order to deepen our understanding of systems and how they work, it is necessary to open up the black box and look at the components—the subsystems and their sub-subsystems. By "take a look" we mean apply the same kind of behavioral analysis to the components that was applied to the whole system above. For each identifiable component we measure the inputs and outputs over time to develop an understanding of their functions. In addition, of course, we need to know the sources of inputs, which are often other subsystems, and where the outputs go—also often to other subsystems.

In other words, we do what is called a decomposition of the system in a manner that provides us information about how the whole system performs its function(s) by knowing how the subsystems work. The methods by which this decomposition is accomplished will be covered in the last section of the book, especially Chap. 13, *Systems Science Methodology*. For now we will only note that the kind of system we are dealing with determines the details of decomposition and measurement methods. For example, decomposing the subsystems in a living animal like the cheetah is problematic if you want to preserve whole system behavior while seeing how each subsystem contributes to the externally observed behavior. Suppose you want to know the contribution of the lungs to supporting these high-speed chases. You can dissect the animal and measure the volume of the lungs, but that is a pretty information-poor way to find out how they function under conditions of high-speed running, since the animal would be dead and not able either breathe or to run!

For such reasons, living systems, including meta-living systems like societies, present some of the hardest problems for decomposition analysis. However, modern measurement technology is allowing us to obtain much more detailed information about what is going on inside complex dynamic systems without disrupting the normal function of subsystems and components. For example, the functional magnetic resonance imaging (fMRI) technology used in medicine allows neuroscientist to observe the working brain from the outside without disrupting how it works. Similar nonintrusive methods are being used to get details from the inside workings of whole systems as they continue to behave.

The ideal of understanding systems and being able to explain behavior requires both the holistic black box and the subsystem decomposition of white box analyses. We will discuss these two sorts of analysis further below.

> **Question Box 6.5**
> Understanding of system behavior may be more or less theoretical. How does this relate to black box and white box analysis of system behavior? Do you think it is possible to have no black box residue in our understanding of system behavior?

6.4 Systems as Dynamic Processes

In Chap. 3 we introduced a perspective on whole systems that treats them as *processes*. By this we mean that a whole system may be viewed in terms of input and output flows of material, energy, and messages where the inputs are *processed* to produce the outputs. Figures 3.1 and 3.7 gave an introduction to this perspective. Now it is time to put some conceptual flesh on the bones of that idea. The context of a system as process grounds the various kinds of changes discussed above.

6.4.1 Energy and Work

In physics these terms, energy and work,[5] have very precise meanings which are fundamental in understanding how things change in the real world. The concepts can be extended to our more common notions of work (e.g., human labor) and energy if we are careful in how we apply them. For example, the work that you accomplish (as in doing a job) can be analyzed in physical and chemical terms that reduce to the technical definition. You expend energy as you accomplish tasks or even when you

---

[5] These terms are defined circularly as follows: energy is that which can accomplish work, and work is a measure, in energy units, of how much transformation some material object has undergone due to being acted upon by some force. Material objects can be accelerated, or atoms can be bound together, or existing bonds can be broken.

are just thinking. The brain is actually one of the main consumers of calories in the human body. In short, no energy means no work, which means no change. So dynamics, the measure of change over time, necessarily pivots on energy and work.

So when you are contemplating work and energy in everyday life, it is essential that these concepts be grounded, ultimately, in the scientific definitions from physics and physical chemistry lest we get sidetracked with fuzzy concepts like "mental energy" that many people use in a purely metaphorical sense, as in, "… that person has a lot of brain power." There is *real* energy flow occurring in brains, and *real* electrochemical work is done as neurons fire action potentials, but there is nothing in the mental domain that is a distinct form of energy (as in the difference between heat and electricity). And further, we have to be careful in how we think about energy in the everyday world so as to not get caught believing in, for example, *perpetual motion machines*—cost-free energy. It is amazing how many people, even those schooled in some physics, think about, for example, solar energy this way. Just because sunshine is seemingly freely available does not mean it is actually "free" in either the economic or physical sense. We'll provide examples of this later.

Energy flows and work (mechanical, chemical, electrical) cause systems to undergo various kinds of changes. These flows play out in systems changes, systems maintenance, and systems degradation over time. Dynamics, which studies these processes of change over time, is a fundamental area of systems science. We may be able to study the morphology or anatomy (arrangement of pieces in a system) of a system, but unless we can know what, how, and why the pieces are changing over time, we can never appreciate that system. And making *progressive* changes in any part of a system requires the flow of energy and work to be done. *Regressive* changes (see Thermodynamics below) result from the irreversible loss of energy already in the system.

Question Box 6.6

People talk about the "emotional dynamics" of various situations. Do you think there is such a thing? Why or why not? What kinds of flows would be important? Who might be interested in tracking such processes of change over time?

6.4.2 Thermodynamics

The study of how energy flows and changes forms (e.g., from potential to kinetic or mechanical to electrical) is thermodynamics. The First Law of Thermodynamics is the conservation law; it states that in any transformation energy is neither created nor destroyed. It simply changes from a higher potential form to a lower potential form. But energy is fundamentally different from matter in that each time energy flows undergo a transformation, some percentage of the energy is lost as waste heat or energy that cannot be used to accomplish work. This is the Second Law of Thermodynamics, entropy, which we take up below.

6.4.2.1 Energy Gradients

Energy can be found in different concentrations in space and time. For example, a fully charged battery contains a concentration of energy in the form of chemical reactions that could generate electric current through a wire if the two terminals of the battery were connected by that wire. The energy in the battery is potential, and the level of that potential can be measured as voltage (pressure that could produce current under the right circumstances). Energy will flow from a high potential concentration to a low potential if a suitable pathway exists, and the difference between the two is the gradient.

Question Box 6.7
How is water behind a dam like energy concentrated and stored in a battery? How would you calculate the energy gradient?

Another example is the Sun-Earth-Space energy gradient. The Sun is a source of high potential energies in many forms but mostly high energy photons (radiation). Space is cold and almost empty of energy, so that the energy produced by the Sun's nuclear furnace will flow (stream) out from the sun to dissipate in deep space. A very small proportion of that streaming energy passes through the Earth's surface, temporarily captured by air, water, rock, and living systems, but eventually being reradiated to deep space. The "pressure" that drives this process can be quantified in the temperature difference between space (very cold) and the Sun (very hot) and secondarily between the Earth (pretty warm) and space.

The origin of thermodynamics was in the study of heat engines (like a steam engine) in which work could be accomplished by putting a mechanical device (like a piston and crankshaft arrangement) between a high-temperature source (steam in a boiler) and a low-temperature sink (usually the atmosphere through a radiator device). As far as most machines are concerned, our civilization is still mostly based on heat engines such as like the internal combustion engine or jet engines. But as we have gained knowledge and experience in chemistry, especially biochemistry, and in electronics, we see that the fundamental laws of thermodynamics apply to any energy gradient regardless of type.

6.4.2.2 Entropy

At its most basic, entropy is a consequence of the Second Law of Thermodynamics that dictates the decline in energy density over time, a process tending toward the equalization of energy levels between source and sink such that no additional work can be done. Every time energy is converted from one form to another, say mechanical kinetic energy is converted to electrical energy via a generator, some portion of the energy is degraded to low potential thermal energy, from which it is not possible to obtain useful work. Heat is radiated away from systems and simply ends up dissipating to the cold sink.

All systems, on some time scale, are also processes. What entropy means, in this context, is that any organization will tend to degrade in the absence of energy input for work on maintenance. This phenomenon can be seen in many different forms. In materials it results in breakdown or degradation into wastes (high entropy materials). In machines and living organisms, it is seen in the gradual breakdown in system parts (wear and tear as well as degradation). For example, protein molecules that are active in living cells have specific shapes critical to their function, but they sit in an aqueous solution, and at body temperatures the motion of the water molecules is such that most proteins are susceptible to the bombardment and will eventually become degraded to a non-active form The cell then must use energy for work to, on the one hand, digest the degraded protein to its constituent amino acids for recycling and on the other hand construct a new protein molecule to replace the one lost.[6]

Living processes require continuous influx of high-quality energy to do the work of fighting the entropic tendency. Machines routinely require work to be done in maintenance and repair. Organizations require continuous efforts (work) to keep running efficiently.

But, because every transformation of energy in doing work loses some to waste heat, the flow of energy is one way. Energy can never be recycled with 100 % efficiency; some is always lost, so entropy is always increasing in the universe.

6.4.2.3   Efficiency

The relative efficiency of a work process/energy transformation will determine what ratio of energy input will result in useful work. Various work processes have inherent efficiencies. Some processes are wasteful and do not do a good job of getting all the work they might out of the same input of energy. For example, the first steam engines had very loose tolerances in fit of the pistons in the cylinders. As a result some of the steam leaked past the piston without contributing to the work of moving the piston. As the manufacturing of steam engines (and later internal combustion engines) got better and produced much tighter fits, more of the heat energy input that converted water to steam would end up contributing to the work process. The efficiencies of these engines, say as measured by the amount of coal needed to extract the same power, got better.

There is still the catch created by the Second Law. No machine or work process can ever get close to 100 % efficiency. One of the founders of thermodynamics, Nicolas Léonard Sadi Carnot (1796–1832) developed the concept of what became known as a Carnot heat engine and proved that such an engine working in a temperature gradient

---

[6]Cellular metabolism includes the two processes of anabolism (building up) and catabolism (breaking down), both of which require energy to do the work involved. See http://en.wikipedia. org/wiki/Anabolism and http://en.wikipedia.org/wiki/Catabolism. Also for general biology of cells and their metabolism, see Harold (2001).

could never extract 100 % of the energy available in the potential of the source. For every known work process, there is an upper limit to efficiency that may be close to the Carnot limit but is generally below it. In other words, even the most perfect steam or automotive engine could never beat the Carnot engine, and real machines won't generally come that close. This is why perpetual motion is a physical impossibility. It is a rigorous law of nature—no exceptions. So any inventor who submits patent applications for any invention which purports to be 100 % efficient (even 95 % efficient!) should summarily be shown the exit door to the patent office.

At the end of this chapter, we will provide an extended example of a system that will help in understanding these principles of dynamics. Before we can do that, however, we need to expand on the concept of a system as a process and provide a few concepts for thinking about them.

Quant Box 6.2 System Dynamics of Fossil Fuel Depletion

When drawing down a resource from a fixed finite reservoir (e.g., any mineral), the work of extraction gets harder as depletion proceeds. That means the amount of energy needed to extract the same number of units of the resource goes up over time. In the case of energy extraction from fossil fuels, this means less, and less net energy is available to the economy as the fuels are depleted. Figure QB 6.2.1 shows the situations at an early time (A) and a later time (B). In the later time you can see that more energy is required to pump the same amount of, say, oil because the resource is coming from deeper in the crust as well as getting harder to find and exploit.

Here we provide a model of this process based on a simplified system dynamics (stocks and flows) method. This model can be implemented in a spreadsheet, and different control parameters can be tested for their effects.

Models like this are important for developing understanding of the dynamics of something as crucial as the depletion of our fossil fuel resources. The question we ask is what is the shape of the production curve for fossil fuel energy (and in particular the net energy available to the economy) over time?

The Model Mathematics

Consider a fixed reservoir holding, say, 1,000 units of fuel. That is the maximum amount of fuel that could, in theory, be extracted. Let's call the reservoir amount $R(t)$. That is the amount of fuel available in the reservoir at a point in time, t. At $t = 0$ $R(t) = 1,000$. For our purposes let's imagine that the demand for energy is virtually unlimited, and so people will attempt to extract the resource at the maximum possible rate. This means the extraction volume will grow for some time. Let the direct extraction units be called $E(t)$, some fraction of $R(t)$. The growth in $E(t)$ can be modeled by an exponential such as:

$$E_{(t+1)} = E_{(t)} + \alpha E_{(t)} \qquad \text{(QB 6.2.1)}$$

where α is a rate constant between 0 and 1, exclusive.

(continued)

Quant Box 6.2 (continued)

a

initial energy investment

energy reinvestment required

extraction

net energy delivered

extraction work

extraction work backpressure

b

energy required

extraction

net energy delivered

extraction work

extraction work backpressure

Fig. QB 6.2.1 This shows a simple version of finite reservoir depletion. Here we model the depletion of fossil fuels (in aggregate) over time. (**a**) An initial investment of energy is required to "bootstrap" the extraction work process. Once started the energy needed to continue operations comes from the reinvestment of the energy output stream. Extraction creates a kind of backpressure that makes the extraction work harder as the depth (quality and location) of the fuel depletes. (**b**) After a time when the fuel is further depleted, the backpressure has increased meaning that the energy required to extract grows as a percentage of the energy output flow. The result is less net energy output per unit of raw energy extracted

However, it turns out that as extraction proceeds the work gets harder so that it is not as easy to continue the growth in extraction units. We have to modify Eq. (QB 6.2.1) to reflect that as the reservoir is depleted, it reduces the rate of growth of extraction. Or, modifying the equation,

$$E_{(t+1)} = E_{(t)} + \alpha E_{(t)} - \delta \left(R_{max} - R_{(t)} \right)$$ (QB 6.2.2)

where R_{max} is the starting reservoir value (in our example 1,000) and δ is a rate constant between 0 and 1, exclusive.

(continued)

Quant Box 6.2 (continued)

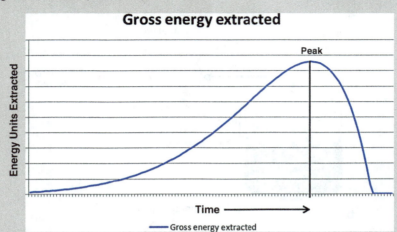

Graph QB 6.2.1 Under the assumption that the energy resource (e.g., oil) will be extracted at the maximum rate possible, the growth in extracted units per unit of time is exponential. However when the negative feedback loop of rising energy required to extract each unit comes to dominate, the rate of growth slows and extraction units eventually peak

Graph QB 6.2.1, below, shows the dynamics of such a system. Because the resource is finite and it is depleted over time, the number of units of extraction per unit of time eventually reaches a peak (also known as a peak in production rates) and then declines rapidly thereafter.

This dynamic is unsettling for its implications. Fossil fuels are the main sources of energy for our industrial economies. This graph tells us, in no uncertain terms, that the amount of fuel available to do economic work will peak and then decline rather drastically. But there is more to the story than this. What the economy runs on is the "net energy" as shown in the above figure. To see what the dynamics of net energy production is, we need to extend the model.

Energy costs to extract are a function of some basic technology factor and the work required to do the extraction. We calculate the energy cost in each time unit as

$$C_{(t+1)} = \tau \left(R_{\max} - R_{(t)} \right) \qquad \text{(QB 6.2.3)}$$

where τ is a technology factor (basically efficiency) between 0 and 1 exclusive and generally much less than 0.5. In reality τ is not a constant but increases over time as technologies for extraction improve. For our purposes, however, we will treat it as a constant (τ also is scaled so that energy units are equivalent to those of gross energy).

(continued)

Quant Box 6.2 (continued)

The graph below shows the same gross extraction rate along with a curve representing the energy cost increases as the work gets harder (the same factor that causes the rapid decline in gross energy extraction). The net energy, then, is just the gross minus the costs in each time unit. The resulting curve has a similar shape to the gross curve but with one even more disturbing difference. The peak production of net energy comes before that of gross production meaning that the effect on economic activity would be felt some time before that of peak gross production (Graph QB 6.2.2).

This model is probably too simplistic for purposes of making predictions directly. What it does, however, is expose the overall or worst-case dynamics of finite resource extraction. All resource extractions are driven by profit motives and so will tend to proceed at the fastest rate allowed by capital investment. So the assumptions underlying the extractive rates might not be too far off base. What the model does tell us is that unless we are willing to let go of our industrial economies, we had better put some seed capital to work in researching and replacing fossil fuels as our major source of energy.

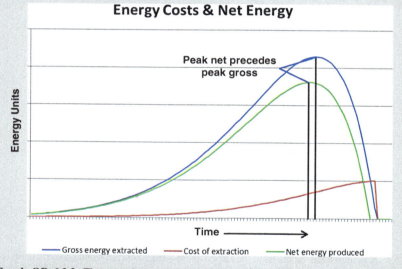

Graph QB 6.2.2 The cost, in energy units, of extraction goes up in reflection of the increasingly harder work done in order to extract. Net energy, which is what our industrial economies run on, is gross minus costs. According to this model, net energy peaks sometime before gross production peaks

(continued)

6.4.3   Process Description

In Chap. 3 we were introduced to a system description in which the system, boundary, and inputs and outputs were made explicit. Figure 6.1 replicates that description as a starting point for understanding the process perspective. Here we see the system from the outside or as a black box.[7] As we saw in Chap. 3 there are subsystems within the system that take the inputs and convert them to outputs. The system is *processing* the inputs.

The system in Fig. 6.1 is a representative of what we call a "real" system, meaning that it is the general kind of system that exists in reality as we know it. As we saw in Chap. 3, even concepts held in the mind/brain are real systems since they involve physical networks of neurons.

There is another "theoretical" kind of system. This is a completely closed system, one that has no inputs or outputs. Such a system would be isolated completely from its environment. It would be inert. As far as we know, save possibly the universe as a whole, and no such system exists. It is, at best, a semi-useful idealization that can be used in thought experiments.

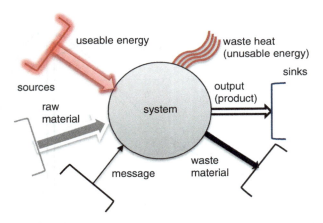

Fig. 6.1 A system takes inputs from the environment—from sources—and converts them to outputs. Inside, work is accomplished. The "useable" energy is at a high potential relative to the sink that takes waste heat ("unusable energy"). This system processes materials using input messages to output a material product along with some unavoidable waste material. At this stage of understanding, the system is treated as a "black box"

---

[7] The term "black box" originated in an engineering framework. Engineers are often faced with needing to "reverse-engineer" a device to figure out what it does and how it does it. Since most devices come in "boxes," the term black box applied to its status when first encountered. The process of taking it apart to see what was inside resulted in what became known as a "white box."

Somewhat more realistic are systems such as represented in Fig. 6.2 below. A system such as the whole Earth is essentially semi-closed with respect to material flows but open to energy flows. Actually the Earth receives a light shower of dust and the occasional meteorite from space continuously, leftovers from when the solar system formed. So, technically, even it is not really closed. Energy flow, of course, comes primarily from the Sun and gravitational effects (tidal forces) from the Sun and other planetary bodies.

Not all systems (processes) result necessarily in a material output (product) as shown in Fig. 6.1. Another result of a process can be the exertion of forces or the emitting of energies. For example, your muscles are biochemical processes that produce contraction forces along their long axis. These contractions in combination with hinges (joints in the bones) result in mechanical work being accomplished (e.g., lifting weights); a general format for this kind of system as process is shown in Fig. 6.3.

Any relationship involves transfers of energy, matter, or messages, and the input involves some kind of output. So in the relational whole of the universe, there can be no completely isolated (i.e., unrelated) system, with the possible exception of the universe itself as the ultimate whole system. Even black holes, which take in

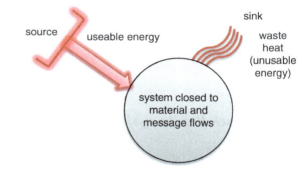

Fig. 6.2 Some systems may be found that are semi-closed, meaning they are closed to the flows of matter or messages

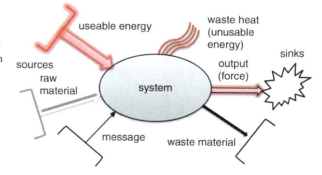

Fig. 6.3 A process may produce mechanical work by generating a force rather than producing a material output, as in Fig. 6.2. Regardless, there is a useful output that constitutes the system's processing purpose. Such a system could be an active agent if it is sufficiently complex and has goals

material but are thought not to produce material products, are now thought to return energy to the universe as a kind of energy evaporation.[8]

Any minimally complex system must always be subject to the flow of energy; otherwise the processes of natural decay and disorganization will ensue. Some energy flow is a must, but matter and information flows (which depend upon energy) may in some cases be very limited, resulting in what may be considered a semi-closed system. Though while the solar system was forming, the Earth received regular large inputs from asteroids and planetesimal bodies, but at its current age, it gets an extremely minor influx of matter from space. So for practical purposes it could be considered a semi-closed system in comparison with its earlier state. However the Earth is also constantly bathed in sunlight (energy) which is very high potential radiation. That sunlight is able to do substantial work on the planet's surface, in its oceans, and in its atmosphere. That work results in the degradation of energy to lower potential forms (heat) that is not able to perform work other than exciting gas molecules, and it is readily radiated to deep space. Thus energy flows through the Earth system, arriving as high potential photons and leaving as infrared radiation after work is accomplished on the planet's surface. That work, of course, is the driving of the climate, ocean currents, material cycles, and, most importantly, chemical processes that gave rise to life and continues to fuel the evolution of biological complexity.

Question Box 6.8
Coal and oil are sometimes referred to as "fossil sunlight." What does this mean?

6.4.4   Black Box Analysis: Revisited

Earlier we mentioned that we determine whole system behavior by setting up measuring devices on the boundary to sense the flux of inputs and outputs. Now that we see this process perspective, we can make the concept a bit more definite. Figure 6.4 shows the same system of interest shown in Fig. 6.2 but now "instrumented" to collect data.

This kind of instrumented black box analysis is useful for all sorts of physical systemic processes. But studies of human behavior typically take this form as well. Inputs are carefully measured and outputs likewise quantified. The endless controversy surrounding test scores (output) and the performance of teachers (input)

---

[8] This is theoretical, but Stephen Hawking has described the way in which a black hole can evaporate by emitting radiation (energy) transformed from the matter that fell into or was originally part of the black hole. See http://en.wikipedia.org/wiki/Hawking_radiation

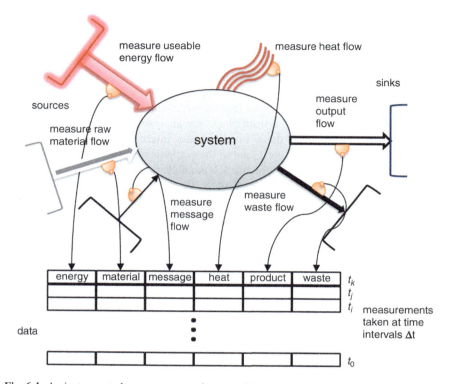

Fig. 6.4   An instrumented system: we attach appropriate sensors to the various flows, inputs, and outputs and take measurements of the flow at discrete time intervals. These measurements, contained in the slots, will be used to analyze the dynamics of the system as a black box. Note that t_0 is the initial sample, and each row represents a subsequent sample of measurements

vividly illustrates the strengths and weaknesses of the black box. Being a black box, the connections between inputs and outputs are not clear, so in complex situations such as those involving human behavior, the uncertainty regarding what inputs are relevant to what outcomes easily creates room for ongoing debate and leads to continual attempts to refine and correlate measurements. Indeed, in the absence of being able to directly observe connections, seeking statistically relevant correlations between variable inputs and outputs becomes a major feature of the social science application of this kind of analysis.

6.4.5   White Box Analysis Revisited

Again, it is important to understand that what you have with black box data and analysis is just the behavior of the whole system with respect to what is flowing in and out of it. You can understand what the system does in response to changes in its environment, but you cannot say how the system does what it does, only that it does it. The essence of understanding is to be able to say something about *how* a system accomplishes what it does. That is where white box analysis comes into play.

Let's assume we are able to see into the system's insides without disrupting its external behavior. This is no easy task, but for the moment assume that we have found various ways to discover the inner components (subsystems) of a system and are able to map out the internal flows from those coming into the system from the outside (which we already know from black box analysis) to the various receiving components, and then from them to all of the internal sub-processes, stocks (reserves, inventories, buffers, etc.), and to the components that export the flows to the outputs (already known from the black box analysis).

In essence our mapping is an accounting for all of the internal processes that ultimately convert the inputs to outputs for the whole system. Once we have the information about these components, we can find a way to treat each in the same fashion we did the whole system, that is, perform a black box analysis on each component! Figure 6.5 shows our system of interest from Fig. 6.4, now as a white box with the internal components exposed. We've also found a way to instrument the various

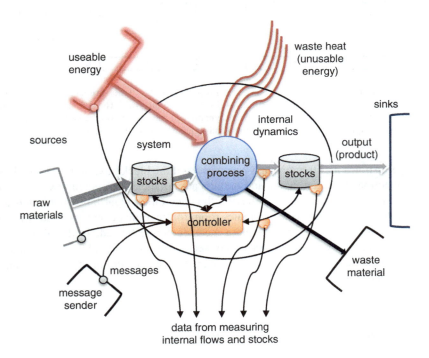

Fig. 6.5 Decomposing a system means looking inside, finding the internal components, and finding ways to measure internal flows and stocks in the same way the external flows were measured

internal flows (and stocks) with sensors and data collectors just as we did in Fig. 6.4. With this sort of arrangement, we can begin collecting data from the system and its subsystems (or at least some of them) and correlate the internal dynamics with the overall behavior. This is a major step toward saying we *understand* the system.

As we will show in Chap. 12, Systems Analysis, this method can be reapplied recursively to the more complex subsystems, for example, the "combining process" or the controller components in Fig. 6.5, to find out how they work in greater detail. The details of the formal methods we use in all sciences to determine how things work will have to wait for that chapter, but you should now have a pretty good general idea of how the sciences investigate the multiple and intertwined systems of the world. And this decomposition of layer after layer of components from black to white boxes is just why we are so tempted to think of what science does as "reductionism"—trying to reduce a system to its most fundamental components. But in reality the objective of the sciences is to understand the whole systems they study, which generally means not just dissecting the systems, but returning to understand the integration of the components in the function of the whole.

With modern digital computing we have one more way to "explore" the internals of particularly complex and sensitive systems for which we have no way to obtain internal dynamics data. In Chap. 13, Modeling, we will see how it is possible to take diagrams such as in Figs. 6.4 and 6.5 and generate a computer model of the system. We can make educated guesses about the internal processes and model them as well. As long as we have the black box analyses of the whole system, we can test various hypotheses regarding how the internals must work and then test our model against the external behavior data. This is a new way for sciences to proceed when they cannot get access to the internals of a system. For example, astrophysicists can develop models of how galaxies form and evolve. They compare the outputs of their models against the substantial data that has been gathered by astronomers to see if their models (of internal mechanism) can explain the observed behavior. Their models seem to be getting better and better!

6.4.6 Process Transformations

Processes transform inputs into outputs. That is, a process is a system that takes in various forms and amounts of inputs of matter, energy, and/or messages, and internally work is accomplished using those inputs such that output forms of matter, energy, and/or messages are produced. The outputs are different from the inputs in form.

In this section we will look at the general rules governing the transformations of systems in terms of three basic conditions: systems in equilibrium, systems in structural transition, and systems in steady state. We will then briefly examine a fourth condition as prelude to our chapter on evolution, systems disturbed by new inputs or changes in their environment.

6.4.6.1 Equilibrium

Equilibrium is the condition in which energy and matter are uniformly distributed throughout the interior. The most basic form of an equilibrium system is a completely closed system that has been sitting around for a very long time. Think of a pool of stagnant sterile water. This condition is the least interesting from the standpoint of dynamics, for such systems have no real functional organization and produce no messages. No work is accomplished internally. If the system is open, it simply means that whatever flows in, the same things flow out immediately without having any effect on the internals. The only real interest in such systems is that if we know what the composition of the system is, and its internal pressure/temperature, we can calculate the property of entropy, the degree of maximum disorganization. This is an important aspect for physicists, to be sure, but from the perspective of systems science, it holds little interest. For the most part we are more interested in the dynamic systems described as "far from equilibrium."

6.4.6.2 Systems in Transition

Nonequilibrium systems are subject to flows that cause the internal states of the system to change. Called "dissipative systems" because they take in, transform, and put out flows of energy, such systems can actually ratchet themselves up to higher levels of organization. When we take up evolution, we will see in greater detail how, given the proper conditions, they may start out in conditions that are close to equilibrium, or certainly in less organized conditions, and with an abundance of energy flow through, will tend to become more complex and more organized over time. Here we will take a quick introductory look at the basics of how a system changes internally as a function of energy flow. In Chap. 10 we will go into much greater detail and also consider what those changes mean for the system in terms of its behavior and connections with other systems in its environment.

Energy is the ability to do work, and work changes something. So a system open to a flow of energy is a system in which work is going on, and in the right circumstances, that work will be constructive and give rise to new and more complex structures. For example, in a mixture of random chemicals exposed to an energy flow, component chemicals will be brought together in ways allowing the formation of new connections based on those components' personalities (look back at Fig. 4.2). Some of the new connections will be more stable, better at self-maintaining under the circumstances (say being bumped and jostled by other components) of their immediate environment. We say such connections are *energetically favored*, more likely to endure than other more weakly formed *energetically disadvantaged* connections (Sect. 3.3.1.1) which decay more rapidly returning the original components to the mix. Over some period of time, as the energy flows through the system, it will tend to form stable structures that will accumulate at the expense of weak structures. And these more stable structures then become the building blocks for yet further and more complex sorts of combinations. Over a long enough time, at a steady flow through of energy, the system may settle into a final internal configuration that represents the most stable complex set of structures.

Question Box 6.9
Energy flow and connectivity are critical in chemical self-organizing processes. But these features function in virtually any system. Some small businesses, for example, transform into large-scale enterprises, while others barely maintain themselves or die off. What sort of flows and connectivity apply to such cases? Would describing the losers as "energetically disadvantaged" be a metaphorical or a literal description?

Structures arising under energy flows may end with new components just slightly more complex than the original ones. For example, the mineral crystals in rock are relatively simple, but they are more complex than the atoms from which they were formed. But the cumulative transformation achieved by this process can also entail extraordinary complexity. The flows of heat at volcanic vents in the ocean floors worked on a rich soup of increasingly complex inter-combining chemicals, and over millennia these vents became sites for the emergence of living cells. And on another scale, human civilization itself can be described as a very complex system arising from the components of the Earth system bathed in the flow of energy from the Sun.

6.4.6.3 Systems in Steady State

A system that has settled into a state where energy is flowing through but no new structure is being created is called a steady-state system. Being in steady state doesn't mean that at the micro level things are not happening. Even stable structures will tend to fall apart from time to time and must be repaired or eventually replaced. Things continue to happen, but on average, the steady-state system will look the same as far as its internals are concerned whenever you take a look at it.

The steady state is not the same as equilibrium. Energy needs to continue to flow through the system in order for continuing work to be done to maintain the structures. Indeed, this is the system far from equilibrium that we mentioned above. It has stable organization based on complex structures and networks of interconnections.

The steadiness in a steady-state system refers to the absence of change in the array of properties relevant to the system of interest. Thus chemical, electronic, mechanical, and economic systems all include steady-state conditions which are carefully (and differently) calculated reference points. Our own bodies exemplify one of the most dynamic and complex steady-state systems as our metabolisms continually make numerous adjustments and readjustments to maintain the overall constant internal conditions known as "homeostasis," which is Greek for "staying the same."

6.4.6.4   Systems Response to Disturbances

An easy way to visualize this subject is to start with a dynamic system in a steady-state condition when something from the outside happens that has an impact on the system's stability. Systems respond to disturbances in any number of ways. If the magnitude of the disturbance is not too great, the system may simply adjust its internal workings so as to accommodate it. If the disturbance is long lasting, i.e., represents a new normal condition, then the system may have to adapt in complex ways. Contrariwise, if the disturbance is short-lived, like a pulse, the system may respond by going out of its normal bounds of behavior for a time before returning to its prior condition.

Of course, the disturbance can be detrimental to the system's continuance. Some disturbances may be either of such great magnitude or occur within a critical process within the system causing permanent damage from which no recovery is possible.

6.4.6.4.1   Disturbances

Systems and systemic function are necessarily organized in terms of the conditions within which they arise and exist. Thus they structurally "expect" certain conditions, including inputs from sources and outputs into sinks of various sorts. A disturbance is a fluctuation or change beyond parameters expected by the system. It may come in the form of a radical shift or change in the quantity of one of the inputs over a very short period of time. It could be a similar change in the acceptance capacity of a sink causing an output to be disrupted. It could be due to what might appear to be a small change, but to an input that is so critical to the rest of the system that it has an amplified effect on it.

Ordinary disturbances are those that may be rare in occurrence (hence outside the expected), but still fall within a range of magnitudes to which the system can respond with some kind of adjustment that allows it to "weather the storm" so to speak. For example, a temporary disruption in the flow of a part to a manufacturer could be handled if the manufacturing company maintains an inventory of those parts for just such occasions. Companies that have experienced episodic disruptions in their supply chains often do keep small inventories, even in a just-in-time delivery system, for just such occurrences. An inventory is a buffer against these kinds of disturbances.

A range of variables come into play in determining when a disturbance moves from ordinary to critical. When disturbances are a matter of a change in a system's environment, what is critical is often a question of magnitude: by how much does the change or the speed of change depart from systemic expectation? It can be that the magnitude of the disturbance was much greater than the system was prepared to handle. For example, in the above case of a manufacturing company not receiving a part, if the disruption in deliveries lasted long enough to deplete the local inventory, then production might be brought to a halt. The company might even go under. A Midwestern prairie will be stressed but survive a 3-year drought relatively intact. But such an event is so beyond the expectations structured into a rain forest, it would be devastating. There are many degrees of even critical disturbances, and complex systems typically have a range of ways to compensate, but clearly there are limits.

Changes internal to a system can be critical even when they are of a seemingly small magnitude. We have already seen this in our discussion of brittle systems. What if the part being delayed is absolutely critical? What if it is used in every unit the manufacturer produces? And what if it only comes from one supplier? Clearly in this case the system is in deep trouble, brittle, and vulnerable to fluctuations in this area of inputs. Such critical disturbances can lead to terminal disruption, so brittle systems tend to emerge and survive in environments that are themselves markedly stable. There is a reason that incredibly life-intense but fragile rainforest ecosystems such as the Amazon arise only in the context of the constancy of the equatorial climate.

Disturbances also have temporal dynamic profiles that can have varying impacts on systems. A system may deal with slow change, while abrupt change in a factor can overwhelm a system that has a limited response capability. For example, in the case of the parts inventory, if the manufacturer only kept a few day's supply of the parts on hand to cover normal levels of production, but the disturbance was a complete shutdown of deliveries and it lasted for more than a few days, obviously the company would be in trouble.

Disturbances can grow stronger over time, passing from mere annoyances at normal, but still disruptive levels, to become critical over a longer time. Such disturbances can be said to "stress" the system, putting some kind of functional strain on its processes. For example, suppose the supplier of the part (above) is having difficulty getting one of the raw materials it needs to produce adequate numbers of parts per unit time (as needed by the manufacturing company). This might not stop shipments, but it could slow them down as far as volumes were concerned. If the parts supplier experienced even further delays or reduced shipment schedules, this would add problems for the manufacturer, and at the same time the stressed system would become increasingly vulnerable to other disturbances. The trouble would be building up over time and could reach a critical level.

There are a few good things about such stresses. For one, they can be a learning experience for all parties, at least if they are CASs. The one occurrence might be stressful, but if both the supplier and manufacturer design and implement mitigating capabilities to handle such stresses in the future, then so much the better. Another good thing is that if a stress builds up gradually over time, there is an opportunity for the stressed system to respond with some kind of immediate stress mitigation. For example, perhaps the manufacturing company can find a similar part from another manufacturer (maybe at a higher cost over the short run) that could be used as a substitute. Most programs of athletic fitness training are actually regimes of incremental stress to which the body adapts by muscle development and the like.

Complex systems often have redundancies of sub-processes or a certain capacity to adapt to disturbances such that they can survive many of them. Less complex systems, however, are more susceptible to disturbances and may succumb to critical ones. Living systems are, of course, the most elegant examples of complex adaptive systems, with a nested hierarchy of levels of adaptive strategies at scales from individual organisms to species to ecosystems or societies and their manifold institutions. The Earth as a whole, taking in the whole web of life in a constant interwoven flow of energy, material, and information, is probably the ultimate unit in this CAS hierarchy of life.

Question Box 6.10
Labor unions once had the power to create critical disturbances in many enterprises. How have systems changed to mitigate that power?

6.4.6.4.2   Stability

One of the more important properties of a dynamic system with respect to disturbances is their stability or ability to resist those disturbances. In passive terms, a system is stable if after a small disturbance it settles back into the state it was in prior to the disturbance.

Figure 6.6 typifies the stability of a passive system, that is, one that simply responds to excitation by returning to its lowest energy condition. But there are several different kinds of system stability depending on the kind of system. Below we cover one of the most important kinds, *active stability* or resilience in the face of a disturbance.

6.4.6.4.3   Resilience

Resilience is the capacity for an active system to rebound to normal function after a disturbance or, if need be, to adapt to a modified function should the disturbance prove to be long-lived. Simple systems may be stable, but they are not terribly resilient in the sense of flexible accommodation to disturbances. Capacity for such resiliency and complexity go hand in hand.

The examples given above of a manufacturing system adjusting or rebounding from a disruption is a good example of resilience. Basically, any system that can continue to function after a disturbance, even if in a reduced form, is resilient. Clearly, there are degrees of resilience just as there are magnitudes of disturbances

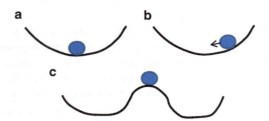

Fig. 6.6 Different kinds of stability as represented by a gravitational analog: (**a**) A passive system is inclined to stay in a particular state represented by the valley; for example, this could be a minimum energy state. (**b**) If moved by a pulsed force out of the minimum energy, the system will be pulled in the direction of the arrow and come to rest, again, in the minimum. (**c**). A system is bistable if it is disturbed slightly and ends up moving to a new minimum

and their dynamic profiles. There are also physical limits beyond which a system will not be able to adjust or adapt. The comet that crashed into the Yucatan peninsula 65 million years ago (mya) proved to be a disturbance greater than the whole dinosaur clade (excluding the bird descendants) could handle. Yet at a higher (and more complex) level of the system hierarchy, the system of life on Earth, was sufficiently complex to prove resilient even after such a disturbance: many other components (species) of the system did survive and eventually repopulated the planet in a new configuration with newer species filling new niches.

The general resilience of a living organism depends on its ability to respond to changes in its environment. Specifically an organism must be able to react to stressors and counteract their effects. For example, if a warm-blooded animal finds itself in a cold environment, it will respond by shivering to generate internal heat. Homeostasis is the term for the process of this kind of adaptive maintaining of an overall dynamic metabolic stability.

Homeostasis (Greek. *homoios*, "of the same kind" + Greek. *stasis* "standing still" or also "same staying") is the general mechanism for an organism responding to a stress in order to maintain an internal state critical to its survival. The concept was developed by Claude Bernard in the mid 1800's and later given a more refined definition (and coinage) by Walter Cannon in the early 1900's. It has come to define any biological system (though the concept turns out to be useful in prebiotic and even some non-biotic chemical systems) that responds to a stressor in a manner that maintains the internal milieu (Fig. 6.7). These include chemical responses, physiological adaptations, and behavioral responses (e.g., withdrawing from the presence of the stress). A homeostatic mechanism is the quintessence of a more general principle in cybernetics called feedback control (Chap. 9).

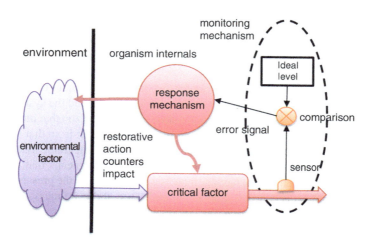

Fig. 6.7 Living systems achieve resilience with the mechanism of homeostasis. A change in or presence of an environmental factor can adversely impact an internal critical factor that is an input to some physiological process. The system includes ways to monitor the critical factor and signal a response mechanism, which is capable of restoring the factor. Responses may directly affect the internal factor or work by affecting the environmental factor. See text for explanation

Not shown explicitly in Fig. 6.7 is that the response mechanism will consume matter and energy doing work to counter the environmental factor. Thus the mechanism should only be activated on an as-needed basis.

Evolution is the long-term process by which systems attain a capacity to adjust to these contingencies when they arise. In the case of many insects and microorganisms, it is also a mechanism for the resilient response to critical disturbances, as when they evolve around our most potent pesticides or antibiotics. Human resilience is closely tied in with the way we can share conceptual information about problems and solutions. Because of their short life spans coupled with massive reproduction, insects and microorganisms perform a similar feat by sharing genetic information. The resilience in this case is achieved by the ability of a few individuals that happen to have a genetic variation that resists the new onslaught to launch a new population wave equipped with the new recipe, a different, but very effective information sharing adaptive response.

Question Box 6.11
Compare and contrast homeostasis and evolution as dynamics that provide resilience in conditions of change. In what ways do their systemic consequences diverge?

6.4.6.5   Messages, Information, and Change (One More Preview)

A critical aspect of understanding adaptive dynamic systems is understanding the propagation of change in the environment. The fact is that the environment of any real system is constantly subject to changes simply because it is embedded in an even larger environment in which changes occur. Change and response (a new change!) propagates in space and time across contiguous environmental boundaries. This introduces a temporal dimension that shapes the dynamics of adaptive systems in important, but often neglected ways.

Changes take time to propagate from far out in the environment to the system of interest. Think of the universe as a set of concentric rings around the system of interest.[9] As shown in Fig. 6.8, the inner ring constitutes a time horizon that includes all of the events that can affect directly the system of interest. But out from that inner circle is another ring constituting events that evolve over a longer time frame, but, nevertheless, affect the events closer to the system of interest. This means that events that happened in the distant past are having impacts on the events of the present insofar as the system of interest is concerned. In a very real sense, these are messages from the past that are affecting the system today. There is no escaping this essential reality.

---

[9] An incredibly interesting perspective is that described by Primack and Abrams (2006). The authors literally explain how each and every individual really is the center of the universe, by established physical principles! The view we are suggesting here really has a basis in physical science.

Messages convey information as long as they tell the receiver "news of difference." That means that messages, which are comprised of modulated energy and/or material inputs, are unexpected by the receiving system. We will explore this more deeply in Chap. 7, but here we want to call attention to the role of the layered temporal dimension of message flow in shaping adaptive systems. A dynamic system can be said to be in a state where it receives messages from sources and fulfills processes based on the content of those messages. But content is layered by time. For example, suppose a parts supplier, via something like a requisition response message, signals that the parts needed are not currently available. The manufacturer has built up an expectation, based on past performance, that parts would be available at the appropriate times (previous experience). Thus, the receipt of a message saying that parts were in short supply would be unexpected by the manufacturer, and this message is *informational* (news of difference) in the technical sense.

The manufacturer is informed of a situation that is beyond the boundaries of its normal experience. It cannot see why, necessarily, the parts are not arriving in a timely manner. It is directly impacted by the shortage, however, and so, if it can, it must adapt. But adapt how? The temporal layers leading to the immediate message become critical. The situation is very different if a flood a few days ago closed down transportation or if a revolution in some country 6 months ago is causing a global shortage of a material it supplied. The events or situations that are further out in the chain of causality that led to the supplier signaling that it could not deliver go back in time[10] (see Fig. 6.8), and this is critical for interpreting and responding to the immediate message.

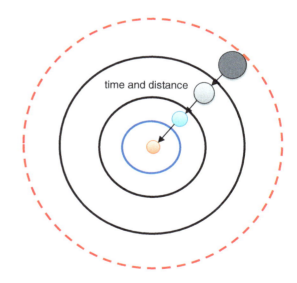

Fig. 6.8 The system of interest (the *center ring*) is impacted by events in its environment. But those events are impacted by events from an even larger environment that take time to propagate inward. In the end, the system is impacted by events which are far back in time and far away in space. Information, thus, travels inward, but always "surprises" the system of interest

time and distance

---

[10] This is another confirmation of the notion of space and time as being equivalent dimensions in Einstein's Special Theory. The further away an event is in space, the further back in time one must look for the event to have an impact on the present. You have, no doubt, heard that the further out into space we look (like at quasars at the edge of the observable universe), the further back in time we are looking.

6.4.6.6 Process in Conceptual Systems

The systems we've been talking about are physical and active (dynamic); hence we described them as processes. But what about conceptual systems? There are systems that people describe as conceptual only. Until very recently we could make a distinction between a physical system and a conceptual one on the grounds that the latter had an ethereal quality to it. When it came to concepts, you could not point to a physical boundary or components that had any kind of physical existence, and so these "mental systems" seemed to fall outside the parameters we have suggested for the analysis of systems dynamics. The rules of logic, association, and the like that govern the world of mental process seem to belong to a dimension entirely unlike the physical processes than can be assessed in terms of space and time constraints.

This picture has been greatly changed, however, by recent technological advances that allow a spatial-temporal tracking of mental processes as they manifest as processes of physical transformation and production in the brain. Take, for example, language, the epitome of a conceptual process. It turns out that neurons that participate in the activation of thoughts and words/sentences are not ethereal at all; they are spatially located and activated in complex clusters and sequences as we respond with speech or feelings, and they further modify the networked array in a feedback loop with ongoing experience. Their work (consuming energy) is very real, and it is in continuity with the overall process dynamic considerations we have been discussing in this chapter.

Of course it is not news that mental activity of whatever sort requires energy and is sensitive to physical (chemical) inputs. The mental effects of fatigue or drugs have made that more than obvious. But what is new is the extent to which we can now track the conceptual and feeling processes as complex, patterned, and transformative systemic physical processes as well. Researchers at MIT, for example, have been able to track patterns in the brains of rats indicating they were dreaming about the maze running exercises they had been doing during the day![11]

In other words, languages and all associated mental processes are very much physical systems. This might sound like the physical reductionism that gives scientific approaches to the world of mind a disagreeable feel to some. But we do not mean to go the simplistic route of asserting that your activity of reading and thinking about this chapter is *nothing but* electricity tripping synapses as it zips around a neural net. Recall from Chap. 3 that we described conceptual boundaries, and how by simply expanding a boundary of analysis, we include more systems as subsystems of a larger system of interest. In this case we can expand from patterned and measurable processes in brain tissue to see those activities as components of conceptual systems, linking function in these components with mental phenomena with new clarity. And in this chapter we have looked at the reverse

---

[11] MIT News, January 24, 2001, http://web.mit.edu/newsoffice/2001/dreaming.html. Accessed 13 June 2013.

process, transforming black box processes into white boxes by a decomposition analysis to lay open the relational dynamics and function of component subsystems. In this sense the system is conceptual, but it is composed of real subsystems so it can still be described in the above framework of physical systems with inflows, transformations, and outflows.

But in these kinds of analysis, it is important to keep the difference between performances (processes) of the whole and performances of components clear. Correlations across the levels advance our understanding, but they are correlations, not identities. One can map and analyze electrical activity in the brain and still never imagine the experience of considering a chess move or planning a birthday party. The latter are emergent functions on the level of whole persons, the product of an interactive systemic whole of which the brain is only one component. As we look at the incredibly complex systemic processes of living organisms, we find that each networked component takes in inputs from others and outputs transformed energy/matter/information to the others. And these coalesce in the emergence of new abilities at the level of the relational whole as it functions in its environment, be it a cheetah crouched in ambush or a mathematician confronting a hard problem. We will be looking at the complex process dynamic of systemic emergence more closely in Part IV. We will see that emergence counters the "nothing but" of simple reductionism with an insistence on newness, but without departing from the realm of energy-driven processes introduced here.

6.4.6.7 Predictable Unpredictability: Stochastic Processes

A purely random process would be one whose behavior would be all over the place, what is called a "random walk" (Fig. 6.9). Such processes are quite unpredictable because there is no constraint of any of the variables affecting the relevant behavior. In nature and society what we find much more often are stochastic processes. Stochastic processes generally appear random (unpredictable) at some levels, but

a   **b**

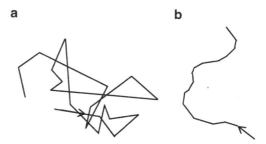

Fig. 6.9 The stochastic paths taken by two robots. (**a**) is a random walk. (**b**) is a drunken sailor walk. It is a form of "pink" noise as compared with (**a**), which is "white" noise

yield predictable patterns at others. Weather, for example, is relatively predictable for the next few days, but quite unpredictable for any given days and weeks in the future. And on quite a different scale, months or whole years may vary widely, but over decades general and more predictable patterns and trends may be traced. The term "stochastic" comes from the Greek for "to aim." If you see one arrow in a hillside, it's hard to tell what was aimed at, if anything. But a clustered pattern will reveal the likely target. Stochastic processes harbor a kind of predictability that is often teased out by statistical probability techniques. Without close personal acquaintance, for example, it's hard to say how any individual voter will vote, yet statistical polling techniques have become aggravatingly accurate in predicting the results of voting before it occurs.

A process is stochastic, then, when its behavior over time appears governed by probabilistic factors. The unpredictability at some scales does not mean that the process is random. Rather it means that there are random aspects deeper in the hierarchy of organization that create some "noisy" fluctuations in the observed (measured) parameters. Or, in a reversed perspective, in an apparently random situation there may be deeply buried constraints at some level which create unexpected pattern. That, as we shall discuss below, is what has given rise to chaos theory and the discovery of surprising pattern hidden in the apparent unpredictability of (some) nonlinear math formulas.

Since the world, at least at the scale of ordinary human perception, is such an interwoven web of mutual constraints, true random walk processes are rare in nature. Even noise in a communications channel caused by thermal effects on the atoms in the channel, say copper wire, appears to be random but still occurs with a characteristic probability distribution. We can, however, construct a system in which we can play with the variables and see how degrees of constraint move us from random to stochastic process.

Figure 6.9 shows the two different kinds of behavior. These are the pathways taken by mobile robots using two different kinds of search programs. Robot A is programmed to perform the random walk mentioned above. It goes forward for a random amount of time at a randomly selected speed. It then randomly selects a rotation to give it a new direction and starts the process over again. So all the relevant variables for a path are unconstrained, giving us a random path. Notice how the path tends to cross over itself and meander aimlessly. If this robot were looking for a place to plug in to recharge its batteries, it would be out of luck.

Robot B's path appears somewhat more "directed" (Mobus and Fisher 1999). It too chooses a random length of time and speed, but both parameters are constrained to be within limits and to be conditioned by the prior time and speed values. Similarly, it chooses a semi-random new direction, again conditioned to some degree by the previous direction. This robot will follow a novel path each time it starts from the same location, but the paths it takes are guaranteed to move it away from its starting position. It is beyond the scope of our current explanation, but it can be shown that the search program used by Robot B is more likely to find a plug-in station placed randomly in the robot's environment than will that of robot A. In fact, many foraging animals follow a search path that looks suspiciously like robot B's. The underlying program does use random selection, but it does so with constraints that prevent it from doing things like doubling back on itself. We call the robot B behavior a "drunken sailor walk," as it was discovered by one of us to resemble personal experiences from days in the Navy.

The kind of randomness displayed by robot B is called "pink" noise to distinguish it from "white" noise. These terms come from the technical analysis of the "signal." White noise is the introduction of a signal value that has an equal likelihood of occurring across a range of values. This is the sound you hear when tuning the radio between stations. Pink noise is not uniformly distributed. Instead the probability of any value in the range is inversely distributed with the size of the value by some power law. This noise is sometimes referred to as $1/f^p$ noise, where f is the frequency (size of the value) and p is the power value, a real number greater than zero. It is one of the interesting surprises in nature that so many stochastic processes should exhibit pink noise behavior.[12] In the case of the drunken sailor walk, the sizes of the distance and time chosen or the rotation selected all have $1/f$ characteristics. Most of the choices are small deviations from the prior choice, and some of the choices will be larger deviations from the prior choice, and only a few rare ones will represent large deviations from the prior choice.

Other processes that exhibit this kind of fractal behavior (see below) include the distribution of avalanche sizes, flooding events, severe weather events, and earthquakes (Bak 1966). Such behavior has been called self-organizing criticality.

6.4.6.8 Chaos

Deeply related to pink noise and self-organized critical behaviors is what has come to be known as deterministic chaos. "Deterministic" refers to the predictable link between cause and effect, so the term was ordinarily associated with predictable systems. However many systems can involve fully determined processes, but with such sensitivity to initial conditions that slight differences lead to unpredictably large divergences down the road—the famed "butterfly effect." The term "chaos" originally denoted something akin to complete, pure randomness (like the random walk). The term "chaos" originally denoted something akin to complete, pure randomness (like the random walk). It was adopted because the subject at hand had the earmarks of disorder, randomness, and unpredictability. But the real interest in this was the deceptiveness of those appearances, for it was found that, despite the chaos, there was clearly some kind of organization to all of the disorder!

Many real systems behave as semi-unpredictable ways and yet still show patterns of behavior that repeat from time to time. The weather is just such a system. It demonstrates repeating patterns of seasons affecting temperature, rainfall, etc. Yet at the same time, no one can accurately predict the exact weather that will obtain in say 10 days' time. Indeed the further out in time one wishes to make predictions, the less accurate those predictions become. Such stochastic behavior within broad patterns that repeat over long time scales is one of the principal characteristics of a chaotic system.

---

[12] Pink noise is related to self-similar or fractal behavior as described in Section 5.5.2. The drunken sailor walk shows similar winding back-and-forth behavior at multiple size scales.

Chaos theory is the mathematical study of chaotic systems.[13] It emerged as a major field of study in the latter half of the twentieth century. Computers brought a revelation in the relatively neglected area of nonlinear mathematics, i.e., the kind of equations that when repeated over and over using the results of the last iteration leads to results not in any predictable alignment with the starting point. Plugging slightly different numbers into the same equation leads to unpredictably large differences. Yet nonlinear math is a deterministic process, such that starting an equation with exactly the same numbers will produce the same chain of numbers. Computers can iterate a math equation as no human had done before, and they also can transform the results into visual correlations. These abilities early on made it an ideal tool for modeling weather. One of the seminal breakthroughs for chaos theory was made in 1961 by a researcher in weather prediction, Edward Lorenz, when he tried to rerun a weather model starting from the middle of the program. The rerun produced wildly different weather from the former prediction. It turned out that tiny rounding errors (his computer rounded things off at six digits) were enough to change the weather of the world, an amazing sensitivity to original conditions known as the "butterfly effect"—as if the fluttering of a butterfly's wings could change the weather!

Iterative nonlinear mathematical process turned out to be a wonderful tool for investigating the non-repeating but patterned world of stochastic process. Where formerly differences were the enemy—artificially smoothed out and made to look like linear processes—now the generation of endless yet pattern-yielding differences became an inviting area of exploration. The main findings of this branch of mathematics (and its impact on physics, chemistry, biology, indeed all of the sciences) are that chaos—endless unpredictable and non-repeating variations with predictable pattern at another scale—appears in systems that contain nonlinear properties (component behaviors). The iteration of a nonlinear equation is itself a mathematical feedback loop, and we've seen nonlinearity emerge from feedback loops earlier in this chapter. In fact most real-world systems contain nonlinear components to one degree or another. Is it any wonder, then, that so many systems and subsystems that we choose to study show some forms of chaotic behavior?

The stochastic processes mentioned in the previous section can all be studied under this nonlinear rubric of chaos. The most interesting systems harbor some chaos! Indeed, remember from above that information is the measure of a message that tells you something you didn't expect? It turns out that this is exactly the quality of behaviors that we deem "interesting." When a mechanical device does the same thing over and over again, we don't tend to find that very interesting. But when a living system does something a little bit unexpected under certain conditions, that we find interesting. That is because a system operating with a little chaos provides us with information that can, in the end, lead us to better understanding of the whole system. Predictable regularity has long been equated with scientific understanding. But chaos theory has opened a new window on unpredictability—and a new satisfaction in being able to predict how systems will be unpredictable! The irony of chaos is that we will never be able to predict with accuracy what a chaotic system will do in the future, but as it operates in the real world, it prompts

---

[13] See Gleick (1987) for the classic introduction to the field.

us to learn the patterns of what is possible so that we construct models (mental, mathematical, and computational) that help us consider likely scenarios as preparation for the future. This interesting twist of how unpredictability generates expectations will be covered in Chap. 9, Cybernetics.

Question Box 6.12

We often think systems are unpredictable because we do not understand the causality involved. But how is it that too much causality (sometimes referred to as "sensitivity to initial conditions") may also render system behavior unpredictable?

Think Box. The Dynamics of Thought

If you spend any quiet moments observing your own mental activities, you probably noticed how thoughts sort of pop into your consciousness unbidden, seemingly out of nowhere. They could be thoughts of some events that happened to you in the past, or thoughts of people you know, or thoughts about what you and some people you know might do in the future. When you are not busy actively and purposely thinking about something, these seemingly random thoughts emerge from the subconscious, get your attention for a brief time, and then fade away as a rule.

During waking hours your brain is always active, the cerebral cortex in particular, generating thoughts. You can do so on purpose, or you can simply let the process happen on its own. Either way you will have thoughts.

As shown in the last Think Box, the cortex is laid out in such a way that concepts (the units of thought) become increasingly complex as you go from the low levels of sensory and perceptual zones toward higher levels in a complexity hierarchy. Not all of these concepts are active at one time, thankfully, or your mind would be buzzing with the cacophony of millions of thoughts all at once. Most of the time percepts and concepts are quiescently waiting for activation to cause the clusters of neurons to become excited. They can be activated from "below" by sensory inputs or from above by more complex concepts in which they participate providing feed-down signals.

Figure TB 6.1 shows a neural hierarchy from low-level features to a higher-level concept.

You think of your friend's face. A high-level concept cluster representing the face fires signals down to the sub-concepts and percepts of faceness (nosed, eyes, mouths, etc.). You visualize, even if faintly, their particular features because those feature-encoding neural clusters are activated in concert with the higher-level percepts. All of the neural clusters that participate in this action, from the front of the brain going back toward the rear (primary sensory), fire together, essentially synchronously, and the entire ensemble of concepts and percepts and feature firing represent your memory of your friend. Figure TB 6.2, below, shows this idea.

(continued)

Think Box. (continued)

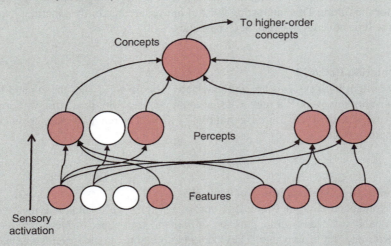

Fig. TB 6.1 Sensory inputs activate feature detectors in the primary sensory cortex. Those in turn activate percepts. Lastly, the percepts jointly activate a concept cluster that then sends its output signal further up the hierarchy. The whole ensemble of clusters acting together represents the perception of the concept (and higher)

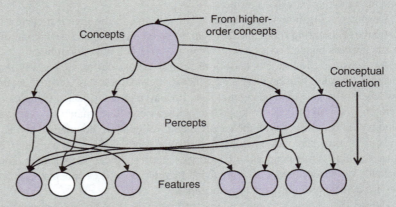

Fig. TB 6.2 When the thought of some higher-level concept is activated internally, the signals propagate downward toward lower levels. The same set of neural clusters that were activated by sensory activation, from below, are activated now by higher-level clusters sending reentrant feed-down signals back to the clusters at lower level that are part of the memory. Note that only the clusters that participated in learning the concept are also activated in remembering the concept

Thoughts as Progressive Activations

The concepts in memory at higher levels don't just send signals to lower-level clusters. They have associations with one another. For example, suppose

(continued)

Think Box. (continued)

your friend has a brother. When you see or think about your friend's face and the entire friend's face ensemble is activated, the highest level concept will also tend to activate her brother's concept cluster. However, at first her face cluster has captured the stage, as it were, of working memory and conscious attention. This latter is thought to work by boosting the ensemble that is active with helper signals to keep it strong for a time. But after a while, the thought of your friend's face may fade but leaving the echo of signal traces to her brother's concept cluster. The brief thought of your friend's face may then trigger the activation of her brother's cluster and that will become more active, leading to a cascade downward to excite those feature combinations that constitute his face. The thought of your friend ends up bringing to mind her brother. If there was some special emotional content associated with him (say you wished he would ask you out on a date), this would strengthen the new thought of him, and his ensemble would dominate your attention for a while (while also stimulating other associated thoughts, like what restaurant you wish he would take you to!).

The surface of the cerebral cortex is in constant flux as ensembles from back to front are activated and generate activations of others. Large areas of your association cortices hold extraordinarily complex wiring that forms multiple many associations between clusters at all levels of the hierarchy. Looking at visualizations of the firing patterns across large patches of the cortex reminds one of water sloshing around in flat bowl, first a wave goes one way, then a counter wave goes back the other way, and so on. Always in motion. Figure TB 6.3 shows a progression of activation of concepts that form chains and side chains across a region of association cortex.

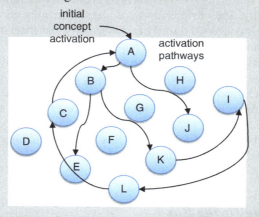

Fig. TB 6.3 Concept *A* gets activated, perhaps by sensory inputs from below or by a previous associated concept. Because it has associative links to concepts *B* and *J*, these are also activated. *B*, in turn, is shown activating both *E* and *K*. *K*, in turn activates *I*, and that activates *L*, which activates *C*, which, as happens, reactivates *A*. Loops like that probably contribute to keeping the particular pathway highly active, strengthening the connections, and triggering the learning of a new higher-level concept that all clusters in the loop contribute to

(continued)

Think Box. (continued)

Most of the time, most of this motion of activation is not strong enough to capture the attention. It is happening without our awareness. This is what we call the subconscious mind. In the above figure, for example, something else might be co-activating B, which would have the effect of strengthening B over J, and thus perpetuate the chain from B to K, etc. Much, perhaps even most, of these associations and concepts are not accessible as discrete concepts like nouns and verbs. They are rather the sort of relations we saw in concept maps—abstract—as we will see in Think Box 13.

6.5 An Energy System Example

Let us now look at an example of a complex adaptive system that will bring out most of the principles of dynamics (behavior) that we have covered in this chapter. The system we will explore is a large-scale solar energy collection system based on the photovoltaic (PV) effect. As the name implies, the PV effect transforms light energy from the Sun to electrical currents. The latter, electricity, can then be used to do work with electrical actuators like motors or electronic devices like computers.

6.5.1 An Initial Black Box Perspective

The figure below shows the PV solar energy system with its various inputs and the output of electricity. The obvious input for a real-time solar energy system is the solar energy that is converted directly into electricity, but there are other inputs to the system that must also be accounted for (Fig. 6.10).

The system itself is composed of large arrays of solar cells that collect and transform solar energy (light) into electricity, and these require work (energy) to be fabricated, put in place, and maintained. The additional work that is required to build and maintain the system is, at least initially, provided by other sources, primarily fossil fuels.

In the next section we perform a "white box" analysis on this system to see what is inside and how the parts relate to one another.

6.5.2 Opening Up the Box

In Fig. 6.11 we see five processes (four blue and one orange oval) that do the work required to build and maintain the system. The primary stock of solar collectors had to be built and installed. We show two stocks here, the collectors that are

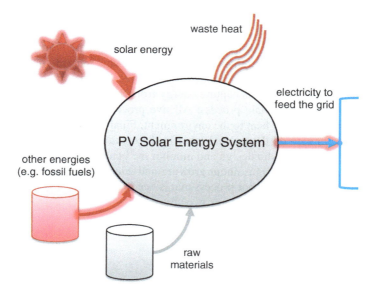

Fig. 6.10 This shows a somewhat simplified black box perspective of a PV solar energy collection system. The additional inputs of other energies and raw materials need to be taken into account as the system is constructed, maintained, and modified (or adapted) over time

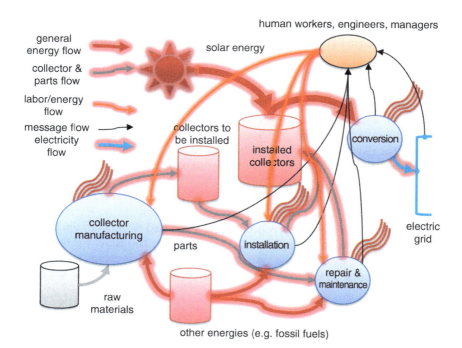

Fig. 6.11 Decomposition of the black box reveals some of the inner workings of the solar energy collection system. Note that the system boundaries were chosen so as to include the manufacture, installation, and maintenance activities that are required for the system to grow. Also, we include the human workers, engineers, and managers that work to grow and maintain the system as well as to adapt it to real environmental conditions and evolve the designs as new technologies become available. Flows of messages are shown, but controls have been left out of the figure for simplicity's sake

in inventory and those that have already been installed and are working. Installed collectors (and associated equipment that we are somewhat ignoring in this analysis for simplicity) also need to be maintained and repaired over the life of the installed base. The installed solar panels capture energy from the Sun, and it is converted to electricity by the photovoltaic process. All five processes use energy to do their work and so radiate waste heat to the environment. Finally, there is the aggregate of humans, workers who do the manual labor in the manufacturing, installation, and maintenance; engineers who design and monitor the behavior of the collectors; and managers who make decisions about growing and adapting the system over time: these constitute the human labor process that is needed to keep the system in operation well into the future.

6.5.3 How the System Works

We have not started this system from its very beginning, with no installed collectors, because we are mostly interested in its current behavior and dynamics and how that may be sustained or modified in the future. Here we can only cover the rough outline, but it should give the reader a sense of how to think about the whole system and how it works over time.

Start with the manufacturing of solar panels. This is a factory that requires input materials, labeled in the figure as "raw materials." Here we are actually representing a group of material extraction and processing operations as well as the final production of collectors. All of these activities have to be counted as part of the system so that we can account for all real costs. If we did a complete decomposition of the "collector manufacturing," you would see all of these various subsystems and how they relate in what we call a "supply chain."

In addition the factory and associated processes need other sources of energies, electricity but also natural gas, and the transportation of the panels to the warehouse ("collectors to be installed") requires diesel or gasoline. As things stand today, these other energy inputs are largely supplied by fossil fuels (with some electricity produced by hydroelectric dams), so they must be counted as energy inputs to the whole system. We will see in a bit why this is so important.

The installation process takes collectors and associated parts out of the inventory and transports them to the site and installs the equipment. This activity is responsible for "growing" the capacity of the system to obtain more solar energy by increasing the "aperture" upon which sunlight falls. Human managers[14] monitor the perfor-

---

[14] Note that we are treating all humans as if they were inside the original system. In a more refined analysis of this situation, which we will present shortly, we humans would also be an input to the whole system in Fig. 6.11.

mance of the system and the demands for electricity in the "grid" (black arrow from grid sink) and make decisions about real-time operation as well as long-term growth. Engineers are employed to monitor the performance of the existing installation and to order repairs or maintenance. They can also specify modifications, say to new manufacturing processes, to gain greater efficiencies in newer panels. In this way the system adapts to new technologies. If a truly innovative technology were to be developed, they might even specify that existing, working, panels be replaced because the cost of doing so would outweigh the sunk costs of removing existing panels. This is an example of how such a system might evolve over a longer time scale.

6.5.4 So What?

The behavior of the system from the outside, black box perspective (Fig. 6.10), involves the inputs of material and energy resources, as well as taking in the solar energy daily. It involves the output of electricity to the grid. The efficiency of the system would be measured by the power of the electricity coming out divided by the sum of all of the power of all of the energy inputs. Of course the sum coming out is always smaller than what goes in. So a system that outputs 80 units for every 100 units of input would be 80/100 or 80 % efficient.

Surprisingly, the actual total efficiency for solar systems is not very high over the long run. If we do the efficiency calculation considering only the input and conversion of sunlight, some of the best solar panels can now achieve approximately 18 % average efficiency. This means that the panels only put out 18 units for every 100 units of energy input from the Sun. This may sound pretty good, especially since we think of sunlight as "free." But in reality we must add to the energy input side of the ledger all the energies needed for manufacture, installation, and maintenance, including labor. If this is added to the calculation, the efficiency could well drop down below 10 %, and some think it is much lower than that.

There is another useful way to look at system efficiency. If we open our black box and do a careful white box analysis of humanly controlled energy inputs, we can achieve a ratio with immediate policy implications called Energy Return on Energy Invested (EROI or EROEI). This approach has been used comparatively to gauge the value of the net energy return on various energy systems such as fossil fuels, wind, and solar PV. EROI is similar to the financial return on investment (ROI) concept from economics, except the units are energy rather than dollars. It attempts to measure the ratio of the net usable energy out of the system to the total amount of energy that went into constructing and operating the system. This is especially critical in a situation when the overall availability of energy flows expected by our global system is in question. What counts to society and the economy is the net energy supplied over the life cycle of the system. With respect to our PV system example, researchers, using a careful white box analysis, have attempted to aggregate all of the energy costs going into solar PV systems relative to their useful energy production output. Note that since what is calculated is the energy *we* put into the system, the solar input itself is not included. Thus, as in the financial world, we expect to get more out than we put in. Several researchers report that the average EROI for solar PV is around 6.8:1. That is, for every unit of energy invested in construction and maintenance, 6.8 units of energy in the form of electricity are produced (Hall and Klitgaard 2012, p. 313).

This measure tells us, in effect, how much energy an energy system will give us to do anything other than just produce energy. Hall and Klitgaard (2012, p. 319) have calculated that society needs a minimum EROI of 3.3:1 just to maintain its transportation infrastructure, but that does not provide sufficient energy to grow or build new infrastructure and let alone clothe, feed, and house the population. The researchers estimate that it would take a stable ratio of 10:1 to maintain our current civilization.

We are extremely interested in the behavior—the energy dynamics—of our own energy systems, and we need to develop more comprehensive models of how they behave over extended time. At very least such models could help policy makers confronted with choices about alternative energy systems investments.

Question Box 6.13
Why would it not solve the energy crisis if we just made enough solar collectors to output the same amount of energy that we now get by consuming fossil fuels?

6.6   Summary of Behavior

All systems have some kind of behavior if observed for an appropriate time scale. We come to understand the patterns in behaviors through the study of system dynamics. That is, we choose an appropriate set of inputs and outputs from the system, put sensors on the flows, and collect data over an extended time. Using the data we can describe the dynamics mathematically. As often as not this description can be plotted on a graph that lets us "see" the behavior or at least the important features of that behavior.

Systems behave, but so do their component subsystems. We saw how you can decompose the behavior of a system to expose the internals and then apply the same kind of analysis to those in turn in order to understand not only what a system does (its behavior) but how it actually does it. If we are able to explain how a system works from a deep knowledge of how its parts work, especially working together as a whole, we can move a long way in the direction of predicting future behavior given current states of the system and its environment and preparing ourselves for the predictably unpredictable turn of events.

In this chapter we have tried to touch on some of the more interesting aspects of dynamics or kinds of dynamics that we find in complex systems. This has only been a short survey of the sub-topics that one can pursue in trying to understand system behaviors. As systems become more complex, their dynamics likewise become more layered, interwoven, and difficult to calculate. The complex adaptive systems that emerge with life have the most interesting behaviors, with a rich potential for surprise as they continually innovate and transform from the measured behaviors we think we understand. Learning and evolution are themselves critical systems dynamics, and we will address them at length in Part IV.

Bibliography and Further Reading

Bak P (1966) How nature works: the science of self-organized criticality. Copernicus, New York, NY

Ford A (2010) Modeling the environment. Island Press, Washington DC

Gleick J (1987) Chaos: making a new science. Penguin, New York, NY

Hall CAS, Klitgaard K (2012) Energy and the wealth of nations. Springer, New York, NY

Harold FM (2001) The way of the cell: molecules, organisms, and the order of life. Oxford University Press, New York, NY

Mobus GE, Fisher P (1999) Foraging search at the edge of chaos. In: Levine D et al (eds) Oscillations in neural networks. Lawerence Erlbaum & Associates, Mahwah, NJ

Nowak MA (2006) Evolutionary dynamics: exploring the equations of life. Harvard University Press, Cambridge, MA

Primack JR, Abrams NE (2006) The view from the center of the universe. Riverhead Books, New York, NY

Part III
The Intangible Aspects of Organization: Maintaining and Adapting

The subject of Part II of this book, "Structural and Functional Aspects," described how systems are structured, how they work to produce outputs/behavior, and how they interact with other systems physically. In Part III we examine how systems, especially complex adaptive systems, maintain their organization over time. Systems are continuously faced with the ravages of entropy (the second law of thermodynamics at work). In order to maintain organization, they have to continually import free energy to do internal work for maintenance sake let alone producing products.

The three chapters in this part will examine the way in which systems accomplish their work by controlling the flows of material and energy. We call this the "ephemeral" aspects because they deal with things that are real, but not observable in the same way that matter and energy are observable. Chapter 7 starts by explaining information, knowledge, and communication theories as a basis. Information is at the base of many kinds of interactions between systems. Most often information is communicated in messages that have a very low energy density and yet can have significant impact on the behavior of receiving systems. It is a real phenomenon, but cannot be directly observed. One might observe the message (e.g., eavesdrop on a telephone line), but this is not the same thing as observing the informational content of the message. The reason is a technical point which often surprises people. As we demonstrate in Chap. 7, information is dependent on the knowledge held by the observer only. You could hear someone say something as you eavesdropped on a conversation, but without the contextual knowledge of the intended target listener, you would not really have received the same information that they did. Thus, information has this ephemeral quality to it.

Chapter 8 explores a necessary kind of process that is needed in order for information to derive from messages, computation. In every case where systems receive messages (data) and process it for its information content, we have some form of computation. And as we show in that chapter, there are many forms of computation. We also show how they are fundamentally the same process, just in different kinds of "hardware."

Information, knowledge, and computation all come together in Chap. 9 where we show how these phenomena are actually employed in complex systems to coordinate, regulate, and maintain the long-term functional interrelations between component subsystems and to allow the whole system to regulate its external behavior so as to coordinate with the other systems with which it interacts in the environment. Cybernetics is the science of control. It demonstrates how information and knowledge work to keep systems performing their functions and fulfilling their purpose even in uncertain and changing environments. Here we will see the basis for sustainability arguments as well as understand how complex systems are able to keep on doing what they do in spite of the relentless push of entropic decay and the disturbances of environmental changes.

Chapter 7
Information, Meaning, Knowledge, and Communications

"In fact, what we mean by information - the elementary unit of information - is a difference which makes a difference."

Gregory Bateson, 1972

*"In classical thermodynamics we assert that entropy is a property of the microstate of the system...whereas...we are asserting that entropy is a measure of our **ignorance** of the exact microstate that system is in... We may then raise the point: is entropy a property of the system or of the observer or the relationship between them?"* [emphasis added]

Harold J. Morowitz, 1968

Abstract The physical world is understood to be comprised of matter and energy, but in the early twentieth century, science began to recognize the significance of something seemingly less physical and yet at the heart of the organization and functioning of systems. Information was defined and characterized scientifically and is now recognized as a fundamental aspect of the universe that is neither matter nor energy per se. We provide an overview of that scientific viewpoint and relate the nature of information to other nearby concepts such as how it is communicated, how is it related to meaning, and most importantly, how is it related to knowledge. Information and knowledge are, in a sense, inverses of one another, alluded to by Morowitz's quote above. These ephemeral elements are critical in the coming chapters where we see how they contribute to what makes complex systems work.

7.1 Introduction: What Is in a Word?

In the last chapter we noted that the word "complexity" has proved problematic in terms of having a single, generally agreed upon definition. Our approach was to operationalize the word by enumerating the components of complexity and providing some clues as to how they might be quantified. That is, we provided a *functional* definition which we could use in applying the concepts to all systems.

Now, in this chapter, we encounter a similar semantics problem. Here the problem arises because the term in question, "information," is used so commonly for a variety

© Springer Science+Business Media New York 2015
G.E. Mobus, M.C. Kalton, *Principles of Systems Science*, Understanding Complex Systems, DOI 10.1007/978-1-4939-1920-8_7

of purposes in ordinary vernacular conversation. There have been a number of approaches to technical definitions of the term, but even the experts in fields where information is a topic of study can revert to vernacular usages from time to time. For example, in computer science we sometimes use the word *information* when in actuality we are talking about *data*. In psychology we can sometimes refer to *information* when we actually mean *knowledge*. And in communications engineering we can use the term when we really mean *message*. Experts in these fields can get away with some sloppiness in term usage because they have a common disciplinary understanding of what their descriptions are referring to, so there is little loss of meaning among them. The problem is when experts write descriptions for lay people who do not share that deep understanding and, on top of that, are used to using the term in its vernacular senses, it can tend to muddy the waters of discourse.

Therefore, in the present chapter, we will introduce a definition of information that is consistent across many (but probably not all) disciplinary lines. Our purpose is to explicate the role of information in systems. It turns out that it is systemness, as described in Chaps. 1 and 2, that really gives functional meaning to the word in a way that both meets the criteria of technical fields and fits nicely into some of our common vernacular uses. In other words, we will focus on the *role* of information in systems as a way to unify the numerous uses of the word.

A major consideration in systems that have achieved a degree of organization (Chap. 3) is the maintenance of structures and functions over an extended length of time. Systems remain systems when they have a sustainable structure and function relative to their environments. And environments aren't always "friendly." All systems must have some degree of resilience, that is, their organization coheres over some time amidst changing environs. Some of those changes may have little or no effect. Other changes can be absorbed by suitable internal systemic modifications that leave the system changed in some ways, but still intact functionally. And other changes, either more extreme or impacting critical structures, may bring about systemic dissolution. Hence for any system, even before we get to the realm of life and its distinctive concern with *survival*, the capacity to respond to the environment is a critical issue. And the principal role of information flow is precisely to provide the means for fitting regulation of the system's structures and functions in an ever-changing world.

Information flow occurs in two related contexts. Internal to a system the flow of information is critical to maintaining structure and function. But the system is subject to the vagaries of the environment and so depends on receiving information from that environment as well. For the most part we will examine these two contexts as they pertain to complex adaptive systems. But it should be noted that the principles are applicable to all systems.

In this chapter we will develop the concept of information as it is used in several different contexts in systems science. In particular we want to develop a consistent way to measure information in terms of its effect on systems. We also want to understand how systems internalize or adapt those effects in such a way as to accommodate the changes that occurred in the environment that gave rise to the information. Therefore we will look at information, how it is communicated, and how it changes the system receiving it in a way that will not only give us a scientific definition of information but one for *knowledge* as well.

In the following chapter we will take a look at the nature of computation, both natural and man-made forms, as it plays a particularly important role in the information-knowledge process. Then in Chap. 9 we will be concerned with the ultimate role of information and knowledge in systems in general. That chapter is devoted to *cybernetics* or the theory of control in complex systems. Information and knowledge are part of the overall process of sustaining complex adaptive systems (in particular), but also they play a role in the long-term adaptation of systems, which we call evolution. These three chapters, therefore, lead into a better understanding the mechanisms of evolution that will be covered in the two chapters following these.

7.2 What Is Information?

When someone tells you something that you did not previously know, you are informed. What they tell you cannot be something totally unexpected, in the sense that the content of what they say must be in the realm of plausibility. They can't talk gibberish. They cannot say to you that they saw a rock levitate on its own, for example. Your a priori[1] knowledge of rocks, mass, and gravity precludes this phenomenon from possibility (at least on this planet). It has, effectively, a zero probability of being true. If, on the other hand, they told you that they saw a particular man throw a rock into the air (for some purpose), that would tell you something you didn't previously know, but that phenomenon, to you, would have had a nonzero probability since it fits the laws of nature. It is just something you would not have known prior to being told.

Suppose your friend tells you that a mutual acquaintance whom you know to be a mild-mannered individual threw a rock at someone deliberately. In this case your a priori expectation of such an occurrence is quite low (by virtue of what you know about this person). Assuming your friend is not a liar, you are surprised to learn of this event. Moreover, you start to reshape your beliefs about the acquaintance being all that mild in manner!

Receiving a message from your environment (and that comes in many forms), the contents of which were unexpected to some degree, informs you of the situation in your environment. Information, that characteristic of messages that inform (knowledge), is measured by the degree of surprise that a receiving system experiences upon receipt. Put another way, the less a receiver expects a particular message, the more information is conveyed in that message within the aforementioned bounds of plausibility. Information is a measure of surprise or, more precisely, a measure of a priori uncertainty reduced to certainty. But the message can only be useful, say in forming new knowledge, if it is also a priori possible.

---

[1] This word (Latin) means before an event or state of affairs. A priori knowledge is the knowledge one has prior to the receipt of an informing message.

At this point we need to differentiate between two very tightly related concepts that are often confused and interchanged indiscriminately. The first is that information is just a quantity of a unit-less measure, a probability change. The second is that messages are *about* something preestablished as having meaning vis-à-vis the sender and receiver. And the meaning has import to the receiver. These two concepts are very different, but also very intertwined such that it is nearly impossible to talk about one without the other. So it is extremely important that we differentiate between them and explain these differences carefully.

Similarly, information and knowledge are often interchanged in vernacular language. Yet these two concepts are very different. *Knowledge*, as we will show later in this chapter, *is a result* of information having a *meaningful effect* on the receiving system. Gregory Bateson (1972) coined the phrase that information is "...news of difference that makes a difference." In this one phrase he captured all four concepts that we will consider in this chapter, information, meaning, knowledge, and communication. "News of difference" connotes the receipt of a message (communications) that tells us something we didn't already know and thereby modifies our knowledge. "That makes a difference" connotes that the news has an impact on the receiver; it has meaning that will somehow change the receiver.

In Claude Shannon's landmark work on information,[2] we get our first clue as to how to handle these four concepts more rigorously. Shannon provided a very powerful definition of information[3] that makes it clear how the four interrelated concepts work in systems. Warren Weaver[4] later teamed with Shannon to explicate the idea of information as a measure of surprise and to differentiate that from the ordinary concept of *meaning*.[5] Shannon's work started a revolution in many of the sciences but is most spectacularly seen in modern communications systems. His original intent was to provide a useful measure of the amount of information carried by communications channels in light of those channels being subjected to various sources of noise. Even though his original idea was meant to be used in the context of communications engineering, it has proven to provide exquisite insights into the deeper aspects of how systems work and manage their affairs.

Shannon's functional definition provides a useful *measure*[6] of information. His insight was to approach and quantify information not as adding something not known, but as *removing* uncertainty. Information is the amount of uncertainty, held by a recipient (observer) *prior* to a receipt of a message (observation), removed *after the receipt of the message*. For example, prior to the flip of a fair coin, there is

---

[2] See Wikipedia, Claude Shannon: http://en.wikipedia.org/wiki/Claude_Shannon

[3] Wikipedia, Information Theory, http://en.wikipedia.org/wiki/Information_theory. Links to additional readings

[4] Wikipedia, Warren Weaver: http://en.wikipedia.org/wiki/Warren_Weaver

[5] Shannon, Claude E. & Weaver, Warren (1949). *The Mathematical Theory of Communication*. Urbana: The University of Illinois Press. Based on http://cm.bell-labs.com/cm/ms/what/shannon-day/paper.html

[6] See Quant Box 7.2 for the mathematical definition.

a 50/50 chance that the coin will come up heads. There is a 50 % probability that it will end up with the head side showing. Once the flip occurs the result is observed (message received) and the uncertainty falls from 50 % (for heads but also for tails) to 0 % since the outcome is known. The two possible outcomes have been reduced to a single actual outcome known with certainty.

This might seem a trivial example, but it is actually the basis for information processing in computers where the logic circuits, built from on-off transistors, have the same either-or conditions as the flipping of a coin. This is the simplest possible information process where the change of state from a 0 (or no voltage) to 1 (maximum voltage) or 1–0 represents a single "bit" of information (BIT is derived from Binary digIT). The reduction in uncertainty by the opening or closing of a transistor switch represents 1 bit of information, by Shannon's formulation.[7]

What does it mean to say that a system possesses an a priori uncertainty of an event? It isn't difficult to understand what we mean when we are talking of human systems, but what about much simpler, inanimate systems? How can we talk about a rock, for example, as having some kind of a priori uncertainty? The answer will take some work to settle, and there is a nonzero probability that you will not be completely satisfied with it. A lot has to do with what is generally meant by a probability and, thus, what is meant by uncertainty. Quant Box 7.1 provides a very quick guide to probability and uncertainty for those who have not had a course in statistics (and even for some who have had courses in statistics that failed to explicate the meaning of probability).

Quant Box 7.1 Probability and Uncertainty

In the example given in the text of a fair coin toss, we see the simplest form of probability. On any one toss of the coin, there is exactly an equal chance of heads as of tails. We say that heads, for example, has a 50 % chance of showing after the toss. Furthermore, in a series of tosses of that same coin, we expect that heads will come up about as often as tails.

We have to say "about" because a lot depends on the number of tosses in the sequence. For example, is it possible that in a series of four tosses that heads could come up all four times? Many will understand, intuitively, that when the number of tosses is small, the answer is yes. In all likelihood you have experienced such a sequence. But what about a longer sequence, say of 100 tosses? Here intuition gets a little fuzzy. How many times should we expect to see heads out of this sequence? Our immediate response is 50 % of 100=50 times. What surprises many people is that if you actually run an

(continued)

---

[7] Shannon is also the developer of Switching Theory (http://en.wikipedia.org/wiki/Switching_theory) where Boolean logic is used via binary gate circuits to produce relational logic and arithmetic (see Chap. 8).

Quant Box 7.1 (continued)

experiment tossing a coin 100 times, your result might vary from 45 to 55 heads (or sometimes even further from 50). The reason is that each toss in a sequence like this is independent of all other tosses and it is perfectly possible to get a longish subsequence in which only heads come up! Many people are willing to bet that if heads came up five times in a row, the next toss would have to be tails. They are sort of right but for the wrong reason. Many people think that the probability of next toss works kind of like pressure building up in a container. When the pressure gets large enough, the container has to burst. But it doesn't work that way at all. Believe it or not there is a nonzero probability that you could get 100 heads in 100 tosses!

What about a really long sequence, say 1,000 tosses? Would we expect to see some number closer to 500. Yes, but why? If it isn't pressure building up, what is it?

As an alternative we could do ten experiments of tossing 100 times in each experiment. If we record the actual number of heads in each experiment, we will note something interesting. A number close to or exactly 50 will come up more often than, say, a number like 45. In fact if we did this experiment 1,000 times (the equivalent of tossing the coin 100,000 times in one experiment), we would see that the most frequent heads count would be 50. Many, however, would be 49 or 51. Some would be 48 or 52, and after we account for all of the experiments, we would find that very few produced numbers like 45 or 55 (less or more).

Probabilities of events like coin tosses or die throws are governed by the "Law of Large Numbers," which essentially states that the more events you observe, the more outcomes will tend toward the "true" statistical properties of the event types. In our example, the mean (and the median) or expected number tended toward the "true" probability of 0.5 (50 %) and the variance (a statistical measure of variations in the outcome) tended toward a number representing a measure of expected spread, say 47–53.

In information theory we use set theory to determine probability and its dual, a priori uncertainty. For example, in the coin tossing the set, C consists of two elements, head, h, and tail, t, $C = \{h, t\}$. An experiment is defined as the number of tosses and counting the frequency of appearance of elements of the set. The a priori expectation can be defined simply by the number of elements in the set, its *cardinality*. We say that the probability, P, of event h (or event t) is $P = 1/n$, where n is the cardinality of the set. At least this holds when the coin or the die is fair. The physics of a fair coin or die dictate that this estimate of probability of any given event is as shown. We say each event type is equiprobable. Later we will look at some situations (the interesting ones) where the probabilities of different event types are not equiprobable. One of the primary constraints on probabilities defined on sets is that the sum has to equal exactly 1, $\sum_{i=1}^{n} P_i = 1.0$. Here n is the cardinality of the set as above. This

(continued)

Quant Box 7.1 (continued)
means that if some element has a higher probability of occurrence, then another member of the set must have a correspondingly lower probability so that this constraint is not violated.

We will be looking at examples of adaptive systems in which their probabilities of events tied to elements of their observable sets of signal states can actually change over time and experience. It may surprise you to learn that your brain actually works in a somewhat fuzzy way on this principle.

Question Box 7.1
There are 26 letters in the alphabet. If there were an equal probability of the occurrence of each letter (not the case in reality), how much uncertainty would be removed/information conveyed by the receipt of each letter? Does this measure have anything to do with what the message is about? Does the method work as one moves through the various levels of uncertainty involved in receiving messages?

7.2.1 Definitions

Before we can do an adequate job of explaining information and how it works in communication and control systems, we need to introduce a number of terms (some of which were given above in italics) and short descriptions for each.

7.2.1.1 Communication

A communication is the act of a sender inserting a message into a communications channel (the flow link) and a receiver accepting the message and then being motivated to act on the information content of the message. A rock reflecting patterns of light to your eye fulfills this description as well as a TV station sending advertisements through cables to your eyes watching the TV set.

7.2.1.2 Message

A message is a sequence of states (of the communications channel) that may or may not contain information, depending on the state of the receiver. Messages can be intentional structures, that is, the sender puts the message sequence into the channel medium (see below) for the purpose of sending information to the receiver.

Other times, a passive system may be an unintentional source of messages, as in the above example of light reflecting off a rock.

Even in the case of intentional message sending, the information received may or may not correspond with the intended amount. For example, if a friend told you something that you already knew (but they did not know that you knew), you would have received a message, but not one that was informational.

7.2.1.3 Sender

A sender is any system that can routinely modulate a message state (encode the message) for insertion into a channel for conveyance. The sender need not be "causing" the modulation with some kind of purpose (as mentioned above). Passive objects are able to be senders by virtue of their energetic interactions with the larger environment. For example, a falling tree sends an auditory signal to a human by virtue of the sound waves impacting the ear of the human. The answer to the ancient philosophical query, "Does a tree make a sound falling if no one is there to hear it," actually is quite simple. No. That is, it is no if the meaning of the word "sound" refers to the effect on the receiver. Of course squirrels and owls might be receivers to consider.

> **Question Box 7.2**
> A rock modulates light waves to create the pattern registered by the eyes, transmitted to the brain, and recognized as "a rock." We just say, "I saw a rock." How would you describe this experience of a rock (really our whole experience of the world!) more accurately? What ties the world of experience to the "real world" so that all this information is useful?

7.2.1.4 Receiver

A receiver is any system that can accept a message through a channel and for which the message state potentially conveys some amount of information. As noted above, the receiver is the one that determines the amount of information conveyed by virtue of its a priori held expectation of what message state it should receive. A receiver needs to be able to make use of the message by altering its internal structure (to some degree) when a message conveys information. This will be developed in much greater detail below.

7.2.1.5 Observer

Observers are generally *purposeful* receivers. That is, they are seeking information and receive it by making observations on various kinds of senders. Observers have an ability to interpret the messages they receive and use the information to modify their own internal organization. An observer can be either passive or proactive. A passive observer simply collects readily available data streams, such as the sunlight rays reflection off of the rock. A proactive observer probes the observed in order to elicit messages, as when someone shines a flashlight on the ground to illuminate any rocks on the path.

There is a commonly held misconception about the notion that "the act of observing something affects that something.[8]" In the case of passive observation, the impact on the object would have happened whether or not there was an observer. The light reflecting from the Sun off of the rock does not depend on an observer. In the case of the proactive observer, however, there is a nonzero impact of the act of observing on that which is observed. A flashlight beam probably has very close to zero impact on the rock. But shining a light beam on an atomic particle has a very measurable impact.

Question Box 7.3
Senders must modulate some communication medium for a message to occur, and that requires some energy change in the sender. What problems might this introduce in trying to know the precise state of atomic particles? Any analogous problems at different scales? What might this have to do with people who "let off steam" by writing angry letters they do not send?

7.2.1.6   Channel

A channel is any physical medium through which a flow can be sent. The physical characteristics of the channel determine what form a message can take and how it propagates through the medium. For example, the compression and rarification of air makes it suitable for messages to be conveyed as sounds, i.e., modulated frequencies of compression waves. However the energy inserted into the air medium propagates concentrically outward and attenuates rapidly, making this an effective channel only at relatively short distances. Metal wires can conduct current (volume) flows of electrons and pressure waves (voltage changes) in an enclosed (tunnel-like) channel with much less attenuation, making it highly versatile for electronic messaging.

Similarly, electromagnetic waves (photons) traveling through empty space can convey messages over quite far distances at very low power. For example, the Voyager spacecrafts (1 and 2) are still sending back data from beyond the planetary

---

[8] This is a common phrase used to explain Heisenberg's Uncertainty Principle. The act of measuring the position or momentum of a particle changes either of those quantities. Some imaginative physicists have gone so far as to suggest that this is tantamount to claiming that consciousness is the causal factor in things existing. But we should recognize that it is proactive observation that is the culprit here.

boundaries of the solar system with radios that operate in the tens of watts range![9] Similar to sound waves, light waves do attenuate with distance, however.

7.2.1.7 Signal

In any channel there will be disturbances that occur naturally but are not part of the message (see Noise, below). The part of the message that is genuine, as received by the receiver, is the signal. We associate signal strength with veracity. That is, the signal tells us the truth about the intended or natural message states.

7.2.1.8 Noise

Every physical medium is subject to a variety of disturbances that could mask the genuine message state or corrupt it in some way. This is noise. Noise can be inherent in the channel, such as thermal agitation of the atoms in a wire, or it may be injected by external inducing events, like voltage spikes in a wire due to nearby electromagnetic discharges. Good communications systems include selecting noise-resistant channels, various means for receivers to filter noise out of messages so as to recover a genuine signal, and methods of *encoding* that ensure the receiver recognizes the signal clearly. These are methods of modulating signals by the sender in a fashion that ensures high reliability of the proper receipt even in the face of noise. Of course the receiver must have a demodulating/decoding capability that matches the rules of encoding/modulation in order for this to work (see Protocol below).

Some kinds of noise are generated by malicious sources trying to confuse the receiver. For example, injected radio frequency noise can be used to *jam* radio communications.

7.2.1.9 Codes

Codes are methods that are used to frame a message so as to increase the likelihood of proper transmission and receipt. This was one of the main contributions that Claude Shannon made to communications theory based on his formulation of information theory. Written alphabets, for example, are codes for the visual transmission of language. Various electronic codes are based on choosing a suitable *alphabet* of message states, which are an agreed-upon limited array of specific states (i.e., a set of characters or signs) that can be unambiguously interpreted by the receiver. Codes include *redundancy* in the alphabets and/or a higher-level

---

[9] See http://en.wikipedia.org/wiki/Voyager_program. Accessed Feb 8 2013.

protocol so as to thwart attempts by malicious parties or nature to inject noise and destroy the message.

For example, the Morse code is an early example of a method for modulating current in an electric wire by simply turning circuits on and off but with shorter and longer times being in the ON state, the dits and dahs of the code. The actual transmission of a message involves the sequencing of dits and dahs, by convention, to form letters of the English alphabet. The rapid sequencing of dits and dahs followed by short pauses at the boundaries of letters allows the sending and receiving of content containing messages in words. In Morse code, an A is dit dah (* --), and B is dah dit dit dit (-- * * *). The various combinations of dits and dahs chosen for letters were based on how often a letter is used in regular English. Those letters being used most often in words are given the shortest sequences. For example, the letter E is the most common letter in English. It is encoded as a single dit (*). Why do you suppose Samuel Morse arranged his code this way? What was he trying to achieve?

Another example of a code comes from the world of biology and genetic inheritance. Genes are comprised of sequences of nucleotide molecules lined up along a backbone of sugars (deoxyribose) and phosphates. There are four *nucleotide* "letters" in the DNA alphabet (A for adenine, G for guanine, C for cytosine, and T for thymine). It takes three of these letters to form a "word" in the code. There are 4^3, or 64 possible words, called *codons*, that could be formed from these. Most of these words translate to 1 of 20 standard amino acids, which, when they are linked according to the sequence specified in DNA, form proteins and other polypeptides. Some of the codons specify message formatting symbols, like start and stop signals used in the eventual translation of DNA into proteins. Since there are more than three times as many codons as there are amino acids, each amino acid may be represented by two or more codons. Thus, the code contains redundancy to help reduce the effects of noise.

The "channel" for DNA to protein messaging is the molecule RNA which has many properties in common with DNA but is more stable as it diffuses through the cytoplasm outside the nucleus and conveys the messages to the receivers. The latter are molecular machines, organelles, called ribosomes (composed of both special RNA and proteins) that translate the message carried on the strands of messenger RNA (mRNA) into proteins. They both receive and act upon the information, serving as the manufacturing plant inside the cell for making the proteins needed to do other chemical work in the cell's metabolism.

Question Box 7.4
DNA, RNA, and ribosomes have no conscious intentions. Are we really talking about information conveying meaningful messages here, or is such language just a useful metaphor?

7.2.1.10   Protocols and Meaning

Communications can only succeed if both the sender and the receiver "understand" the nature of the messages and the meaning of the modulations of the signals. The message has to be encoded properly at the sending end and decoded at the receiving end. Both senders and receivers must have the necessary subsystems for doing these operations. The sender has to prepare the message content (sequence of symbols) and then inject them into the physical medium of the channel. The receiver has to accept the modulated energy from the channel and then disassemble the message in such a way that the energy conveyed can affect the internal workings of the receiver.

A protocol is an a priori agreed-upon matched process for achieving this. The protocol is a model of what can be sent and how it is to be sent as well as what can be received and how it will be decoded effectively. In human engineered systems this is a result of designing the sending and receiving systems. For example, the Internet is actually achieved by computers, routers, and switching devices that are programmed with what is known as the TCP/IP stack (Transmission Control Protocol/Internet Protocol). This is a set of rules for packaging chunks of a data file such that they carry destination and source addresses and other needed housekeeping data. The packets are injected into the medium (say an Ethernet cable or a wireless channel), and when received they are unpacked with the contained data being then turned over to the program that is waiting for it.[10]

One of the "duties" of a protocol is to make sure a message ends up in the "right" place within the receiving system. There the informational value of the message can have its resulting effect on the receiver, i.e., it will cause the receiving system to change in proportion with the information value.

In naturally evolved communications systems such as human speech, the same basic principles apply. There has to be machinery at both ends to interpret and route the messages appropriately. In human speech, for example, Broca's area,[11] bridging the motor area of the frontal lobe and the temporal lobe of the brain, generates the muscular signals that result in verbal sounds being injected into the air. Wernicke's area[12] (roughly bridging the temporal and parietal lobes in the brain) is responsible for interpreting the sounds received by the auditory sensory system and presenting that to the areas of the prefrontal cortex responsible for consciousness and interpretation of the meaning of the speech. These two regions of the brain (usually found on the left cerebral cortical hemisphere) function as implementations of the protocol for speech. Usually every human has both sending and receiving capabilities.

---

[10] The many services that work through the Internet actually have layered protocols on top of the TCP/IP stack. For example, the World Wide Web uses the HTTP (Hypertext Transfer Protocol) to further package data to be used by browsers and servers in working with WWW pages. There are several protocols for handling e-mail traffic.

[11] See http://en.wikipedia.org/wiki/Brocca_area

[12] See http://en.wikipedia.org/wiki/Wernicke%27s_area

7.2.1.11   Data

A datum (singular) is generally a value that represents a measure on some suitable scale.[13] Usually it is also taken at a specific point in space and time. Thus, data points have a built-in context that serves in interpreting. Messages are actually comprised of data elements encoded by means of the code and protocol. For example, the alphabet letters represent an ordinal scale (A being the lowest and Z the highest). We've already seen how the Morse code was used to encode these data, and the English language (or any pre-agreed-upon language using those characters) constitutes the protocol for interpreting messages sent using this format. In the digital world a similar code based on the ordinality of the alphabet directly, called ASCII (American Standard Code for Information Interchange), is encoded in strings of bits called bytes (8 bits constitute a byte). A binary number can thus represent an alphabetic character.[14]

A datum representing a number (on the ratio scale) may be a measurement of a useful quantity, such as sales of a given product on a given day in a given month, etc. A set of data might include all of the sales numbers for the month. In context (sales value for Sept. 12, 2012) the number is meaningful and may also convey information to an observer. Suppose the observer is the sales manager who was expecting somewhere in the neighborhood of $24,000 of sales on that date (apparently selling expensive products!). Now if the actual measured and recorded sales figure was more like $12,000, then the sales manager would have received news of difference. And that news would make a difference. She would be on the phone to the sales staff instantaneously to find out why! This is called *management by exception* and demonstrates the role of information that we will explore in Chap. 9.

Data are often called the "raw material" of information. Once data are collected and recorded, they can be processed in various ways (these days done primarily by electronic computers) to produce summary messages. In turn these messages are sent and received by interpreters who derive whatever information obtains from their a priori expectations.

Question Box 7.5
What is the difference between data and information? Is an unread computer file data or information or both?

---

[13] There are several different kinds of scales of measurement. The most commonly used are *nominal*, *ordinal*, *interval*, and *ratio*. Nominal data is categorical assignment. Ratio data is the typical measurement of a physical quantity such as temperature. See http://en.wikipedia.org/wiki/Scales_of_measurement

[14] ASCII is getting a bit out of date. Newer computer languages use 16-bit codes called Unicode to handle a much larger set of characters. Eight bits can only encode 256 distinct ordinal values, not enough to include international character sets.

7.3   Information Dynamics

Information is a dynamic concept. It constantly changes as the system receiving messages changes. Let's now put together all of the concepts just discussed to see how communications and information processes (along with knowledge construction) work. Figure 7.1 shows a model of a communication process. We will use this basic model to further describe how information is communicated and results in changes in the receiver, in the form of knowledge formation and changed behavior. The latter, as we will see, is the basis for generating additional information in other receiving systems. We will see that information begets knowledge which begets changes in behavior which begets more information. In other words, information and knowledge, unlike matter and energy, are not subject to a conservation law. It would seem that the more knowledge there is in the universe, the more information is generated. But we have to put this idea into the proper perspective—from that of systems.

In the figure below a sender is a physical system that is affected by some physical influence, a "force" causing some kind of change in the state of that system. The change results in an encoding process that encapsulates the changed state into a coded datum (or stream of data) that is (are) then injected into the channel medium as a message. It takes time for physical messages to travel the distance between sender and receiver proportional to the distance and the nature of the channel. Though not shown in the figure, we can expect that some kind of noise process is also injected into the channel, but let's assume that the signal to noise ratio is such that this will not cause a problem. The message then arrives at the decoder acceptor subsystem of the receiver and is disassembled (decoded) with the relevant content passed on to the internals of the receiver, possibly needing amplification of force to cause change in the receiver's state (see below).

Two things can result from this act of communication. The receiver's state change involves a change in its expectation of receiving that same message again in the future (this is going to be developed more fully later in this chapter). The second

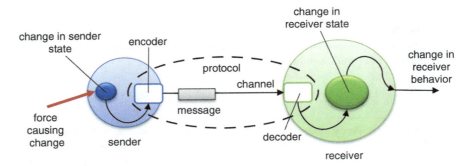

Fig. 7.1 The general model of transmitting information from a sender to a receiver. The parts of the model were covered in the previous section. A full explanation of the dynamic process of communications is in the text below

thing that can happen is that if the receiver is at all active, its behavior is likely to be modified from what it would have been if it had not gotten the message. If there were another system observing the receiver, it might have had an a priori expectation that the receiver would have behaved in a certain way. But the receiver changes its behavior from the observer's expectation, and thus information is conveyed to the third party in this transaction (not shown)!

There is much more to the dynamical aspects of information. Figure 7.2, below, shows a simplified diagram of a receiving process that gets an input from a source labeled as the sender.

In this example the input to the receiver process is a flow of some material (actually this could be electrons into an electronic device or widgets received at the receiving dock of a manufacturer) that the receiving process uses to do useful work. The relevant factor is the fluctuation the flow can have over time. Work quality, or

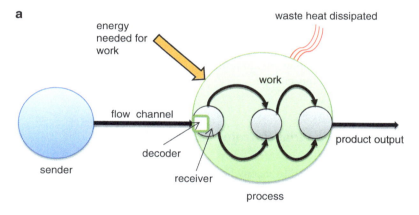

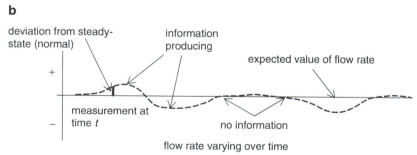

Fig. 7.2 The basic dynamics of information involve deviations of a flow from what a receiver process is expecting. The simplest version of this is a flow link between two processes (**a**). The one on the left is a sender. The process on the right has been decomposed to see that there is an internal sub-process that receives the flow and monitors its levels (decoder). The receiver has a flow rate level expectation (the horizontal line in **b**). Information results when the actual flow is greater or less than expected over time

the impact on the output of the product, can be affected by the fluctuation of the input flow. For argument's sake let's suppose the work process has been "optimized" for a long-term expected flow rate of input, and so deviations from that rate could impact the work processes' ability to produce its own optimal output.

The receiver sub-process contains a "decoder" or sensor of the flow rate. Let us suppose this measurement is made at discrete time intervals (see Fig. 7.2b, the dashed line). At each point in time, t_i, the value of the flow is compared with the expected, optimal value, and the difference constitutes the information used by the work process to possibly adjust its actions in order to compensate.

In the figure, note that the flow rates can fluctuate above and below the expected value (or a small interval around that value representing an acceptable range). When that happens the receiving process is being informed that something is happening in the sender. When the flow rate is at the expected rate, no information is being conveyed.

In this example the instantaneous measures of flow are the data that are used to extract information from the situation. The meaning of the message is determined by the role played by the particular material flow in the process.

Finally, if the work process is a complex adaptive system (CAS) capable of making internal changes in response to the information content of the message (flow), then it will change its expectations, perhaps using the long-term average information in a series of measurements, to reflect a new state of the sender. We will consider this case later in this chapter.

7.3.1   Information and Entropy

Information is a problematic concept due to it having both a colloquial and technical definition. The colloquial definition often overlaps semantically with the technical definition making it easy to slip between a purely colloquial and technical usage. Our use of surprise in discussing the dimensionality of information is useful in part precisely because it is meaningful on both sides of that divide.

On the technical side of things, the quantification of information by Shannon was deeply influenced by Ludwig Boltzmann's statistical mechanics definition of entropy in the 1870s (see Quant Box 7.2 below). Boltzmann approached entropy probabilistically, the maximum entropic condition of random distribution of gas particles, for example, being simply the most probable result of every particle following unconstrained the most likely paths available. Organization, in this framework, would be a matter of constraining particles to some less probable path or condition, that is, a reduction of the maximal probability represented by the random state or entropy. Shannon defined information as the removal of uncertainty, and the parallel of the equal distribution of probabilities in maximal uncertainty and the equal distribution of probabilities in entropic gas particles was inviting. The implications of information and entropy sharing the same mathematical framework were not lost on information theorists, many of whom now think of information as a form of entropy reduction as framed in Boltzmann's statistical mechanics interpretation.

This similarity should not be confused as an identity, but the parallelism invites interesting questions. If the ensemble of symbols (characters or message states) is similar to an ensemble of gas molecules where the energy is distributed uniformly (maximum entropy or most probable distribution), then how about other distributions, as when some molecules are "hotter" (faster moving) than others as when heat is being added to the system at a single point causing the molecules closest to that point to absorb the energy before transferring some to other particles. Such a system is far from equilibrium, and, in fact, it is theoretically possible to obtain work from such a system. What might be the analogous characters of change as information moves a system further from the state of uncertainty?

Quant Box 7.2   The Quantitative Definition of Information

We need to find a rigorous way to measure information, which means we need to formalize our notion of surprise or unexpectedness. This is done in the realm of probability theory. We first define the probability of a specific message state (member of the alphabet set) being received in the next sample period. For example, we might consider what the next letter to be read in an English sentence is likely to be. Before reading a specific sentence, we sample large quantities of English sentences and formulate a table of the frequencies of appearance of each letter in the alphabet considered independently of their occurrence in words.

Expectation of a message state is based on the a priori probability held by the receiver.

A measure of information

Suppose the probability of the receipt of a specific message state, x_i, the ith element of the alphabet set, is Px_i. The amount of information conveyed by this message state would be inversely proportional to the probability, or $1/Px_i$. The higher the probability of x_i, the lower the amount of information conveyed by its receipt. In other words:

$$I_t = f\left(\frac{1}{Px_i}\right) \tag{QB 7.2.1}$$

Shannon decided that a reasonable function for $f()$ is the logarithmic function. Equation (QB 7.2.1)

$$I_t = -\log_2\left(p_{x_i}\right) \tag{QB 7.2.2}$$

The information value received at time t is assigned the negative log (base 2) of the probability of that message state i.

(continued)

Quant Box 7.2 (continued)

For example, the a priori probability of a flip of a fair coin coming up heads is 0.5 since there are only two possibilities (heads, tails). The actual event of heads can be said to have an a posteriori probability of 1 if it happens. How much information was conveyed by the event heads? The answer is $-\log_2(0.5) = 1$ bit. The value is actually dimensionless, but Shannon gave it a dimensional quality to name an amount of information—the bit or **binary digit**.

Suppose a message has four possible states or an alphabet of four characters (like DNA), A, C, G, and T. A naïve observer would assume that a priori to seeing the next character in a message stream, it will have an equally probable chance of occurrence of 0.25. How much information is conveyed once the next character is actually received (observed)? (Graph QB 7.2.1)

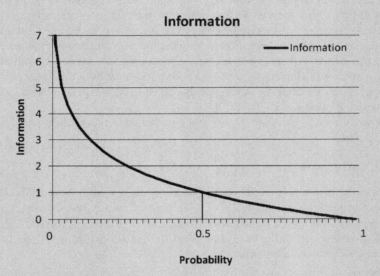

Graph QB 7.2.1 A graph of Eq. (QB 7.2.2) shows how the amount of information falls off as the probability of an event approaches one. Technically at a probability of zero Eq. (QB 7.2.2) would be at infinity

In the next Quant Box, we will look at how information defined thus is used to construct knowledge.

The average information in a message is a quantity that is useful in communications theory. One definition of average information depends on the notion that there exist a finite set of symbols or message states that each has

(continued)

> **Quant Box 7.2** (continued)
> an a priori probability of being in the message. The average information contained by a message is
>
> $$I_{ave} = -k \sum_x p_x \log_2 p_x \qquad \text{(QB 7.2.3)}$$
>
> where k is a constant of proportionality appropriate to the uses and $x \in X$, the set of symbols or message states.
>
> Given a set of four symbols, a, b, c, and d, what probability distribution would provide the largest value of average information for this set? What would the average information of this set be if the probability of a is 0.02 and the probabilities of the others are equally distributed $(1 - 0.02)/4$?
>
> The base 2 logarithm was chosen by Shannon for the simplest decision problem, a binary decision with a priori probabilities of 0.5. This scheme works very well with computer circuits based on switches that are either on or off. But natural systems need not have strictly binary choice states or even discrete ones. In such cases the natural logarithm, base e (ln), can be used without changing the basic curve form. The negative natural log of a probability of 0.5 is 0.6931 as opposed to 1.0.

7.3.2   Transduction, Amplification, and Information Processes

Now let us consider a somewhat different kind of situation in which information is being obtained by one input signal that, in and of itself, has little direct impact on the work process receiving system, but which nevertheless produces information effects (i.e., changes in the work process). Here we will introduce the notion of information per unit of power required in transmission of messages. What we will see is that, like the Voyager example above, some kinds of communications and subsequent information receipts can be accomplished very efficiently. The import of this phenomenon will become highly relevant in Chap. 9, Cybernetics, where we will see the complete role of information and knowledge in dynamic systems. Here we are just going to introduce the idea of how a very small (low-power) signal can end up having a major impact on a receiving system. It turns out that this aspect of systems is actually what allows for complexity to evolve (Chap. 11).

Our understanding of the nature of information processes starts with a look at how energy can be transferred from one scale to another in physical processes. Let's start with an example.

A physical sensor is a device that can essentially siphon off some small energy flow from a larger flow between an energy source and an energy sink (Fig. 7.3b). The second law of thermodynamics dictates that energy must dissipate to fill the space available. If there is a region of higher concentration of energy (source) and a region of lower concentration (sink), then the energy will be under "pressure" to flow (e.g., an electric voltage is such a pressure). Assuming there is a suitable channel (like an electric wire) that can convey the form of energy and connects these two regions of space, the energy will flow from the concentrated (high potential) source to the sparse (low potential) sink at a rate that exponentially decays as the energy potentials equilibrate. Take the case of a common D cell (battery). There is a concentration of electrons at the negative pole of the cell and a deficit of electrons at the positive (remember, by convention an electron carries a negative charge). If you connect a wire directly between the two poles, the electrons will flow rapidly from the negative pole to the positive pole until there are an equal number of electrons disbursed throughout the cell's interior. The cell will have been discharged.

If the wire from one pole were first connected to one lead of an electric motor of a suitable kind (direct current or DC) and then another wire were connected from the other pole to the other lead of the motor, the latter would spin and be able to do some amount of physical work as the electrons flowed through it. It would continue until the electron concentrations had equilibrated and then no more work could be done. This fundamental principle applies to all energy flows through physical devices. If the energy type (e.g., electrons moving) is coupled with some aspect of the physical device, then work is done in the process of the whole system coming to equilibrium.

As shown in Fig. 7.3, the sensor siphons off a small energy flow from the larger one. In the simple case of a monotonic declining flow rate (as happens in the D-cell example), the second channel of energy can't really be said to convey information as such.

But in Fig. 7.4 we have a slightly different situation. Suppose some external force is somehow affecting the energy source in such a way that the flow rates from source to sink are being varied in time (as in the above situation in Fig. 7.2). The flow is being modulated over time (faster-slower, denser-sparser).

We've inserted a hypothetical second process that, like our example in Fig. 7.2, expects a steady flow of energy but responds to a modulated flow. In this case the modulations convey information. However, at this point we have left out some important details to be filled in shortly. The point here is that the sensor can detect the energy flow through a channel and send messages in that energy form to another process. The transmission of energy flows, like this, allows systems at larger scales (of size, time, and power) to have influence over systems at much smaller scales.

This example uses a single kind of energy flow to produce a signal. A much more interesting case occurs when the device can use an entirely different kind of energy flow to create a signal based on the first kind (Fig. 7.5). This is called *transduction*. In other words, we are interested in devices that sense one kind of energy flow (generally at a higher power) and generate signals in another kind of energy flow (generally at a lower power). For example, we might want to sense the water flow through a pipe but have the available data converted to an electrical signal for use by a valve to change the water flow.

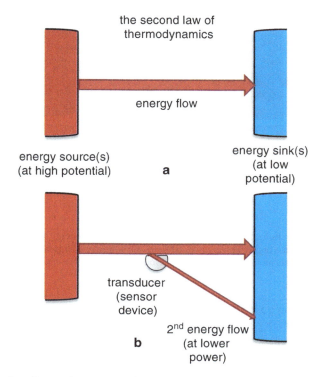

Fig. 7.3 Sensing allows a large energy flow (higher power) to result in a smaller energy flow through a different route. The large energy flow might be barely affected (attenuated) by the effect of the sensor. With the right kind of sensor attached to the large energy flow, a very tiny energy flow can be produced. (**a**) shows the conditions set up by the second law of thermodynamics and the geometry of the system that creates the energy flow. (**b**) shows the insertion of a physical device that siphons off a very small amount of energy that will then still flow to the sink, but at a much lower power (units of energy per unit of time). The second flow path can be used as a signal channel, that is, it conveys message states based on fluctuations in the large energy flow

Fig. 7.4 If another force modulates the source energy, it will propagate through the energy flow channel and through the sensor channel. If there is another modifiable process in the path, the sensor signal may now be said to convey information to this second process

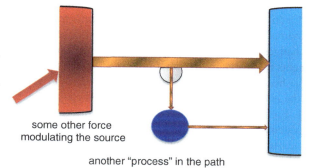

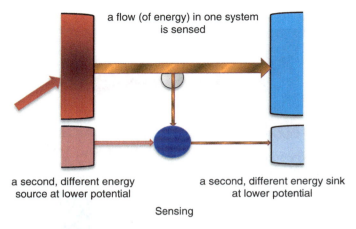

a flow (of energy) in one system
is sensed

a second, different energy
source at lower potential

a second, different energy sink
at lower potential

Sensing

Fig. 7.5 A sensing transducer is a physical device that detects an energy flow (or level as in the case of temperature) at a higher power and modulates a low-power energy flow of a different kind. The top system thus has an influence on the behavior of the bottom system

Transduction can go in the other direction as well, from a low-power source signal to modulate a high power flow.

The device that makes this possible is called an actuator. In Chap. 5 when we presented a lexicon of process semantics, it included two items that we didn't fully discuss at that time, but now need more explanation. The first is an actuator, which is any device that can modulate a higher-power energy flow using a lower energy control signal. For example, a pump is a mechanical device which pushes a fluid using a source of high power, but the controls that actuate it and regulate its output are powered by a much smaller input. Another example is a transistor used as an *amplifier*. A small modulated voltage can be used to modulate a larger voltage output or also a larger current output. The transistor doesn't create the higher-power output; it is tied to a high-power input that provides that. The small, regulating voltage acts as a kind of gate or valve. The difference between the output voltage and the input controlling voltage (also called the signal) is the *gain* that the amplifier provides.

Actuators do work under the control of a small input signal. They are a crucial part of active systems. They are the components that cause changes in the internal configuration (structure) of the system in response to the input signal. They respond to changes in that signal and so do varying amounts of work. These are the key components in understanding what is happening in a receiver. A communication is accomplished when a small (efficient) change in input through the channel is interpreted (the interpretation may be hardwired) and causes the actuator to respond (Fig. 7.6).

Figure 7.7, below, shows a communication setup based on two independent systems and a low-power communications channel. The sender uses a sensor that is measuring a flow at some nominal power. The fluctuations in that flow are transmitted to the actuator (amplifier) in the receiver system where they are used to change

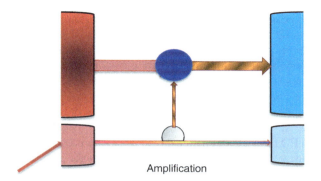

Amplification

Fig. 7.6 When a low-power signal is used to control or modulate a high power flow, then the signal is said to be amplified. Transduction is in the opposite direction from that in this figure. Devices that perform this amplification "trick" are actuators in the sense that they cause larger-scale actions to occur in the larger-scale system

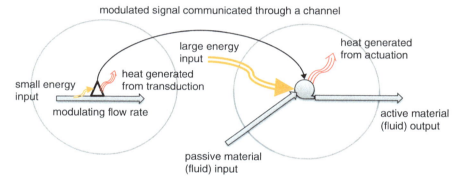

Fig. 7.7 A transducer or sensor uses energy to modulate a signal into an appropriate message flow (communications channel). An actuator modulates a larger energy flow into an appropriately modulated output flow. In this figure, if the flow rate in the system on the *left* goes up, the transducer sends a signal (up) to the actuator in the system on the *right*. The latter then pumps the material faster in response. The system on the *left* is the sender (transmitter), and the system on the *right* is the receiver

the flow rate of some other kind of material. The flow in the sender need be neither the same kind of stuff, as in the receiver, nor at the same power level. What is happening is that changes in one system are transmitted to produce changes in the other system (for whatever purpose).

All communications in active systems involve these two kinds of devices. Sensor transducers are able to detect physical changes in one system and then send a much reduced modulated signal to a receiver system amplifier transducer to affect change in that system. All of our sense organs transduce physical changes taking place in their specific sensory modality (e.g., vision) into the modulated electrical flows that are the message bearers of our nervous system. We create myriad mechanisms that function in a similar way. A *thermistor,* for example, translates heat (sensed as temperature) into proportioned modulations in an electrical current, which can then signal an actuator, as when a thermistor is coupled with a voltmeter. The thermistor is essentially a resistor, but one that changes its resistance value as a function of the temperature of its material. It transforms heat changes into voltage changes which are sent to the voltmeter. The voltmeter is sensitive to changes in voltage. It amplifies the small voltage changes it receives from the thermistor into electrical signals that can drive LED readouts or change the position of a pointer on a dial.

Sensory transducers of all kinds are inherently engaged in *measurements,* turning one sort of quantity into another. We do not ordinarily think of our senses as measuring the world about us, but that is exactly what is going on! Thus, an important subfield of information theory involving transduction is measurement theory. All readers will have taken a measurement of one kind or another of a physical property such as height or weight. Measurements assign numerical values to a transduced signal that places the signal at a point within a range of values that are scaled. The scale is given a unit designation, such as feet or meters (for length), which, in turn, can be further broken down into subunits (e.g., decimeters, millimeters, etc.). Devices that transduce the signal are never precise, nor are they necessarily accurate. Precision in measurement involves the fineness of the scaling, and the appropriate calibration of various transducers is of great practical importance. But limitation is inherent in any sort of measurement, for no scale can be continuous. For example, with a yardstick marked off in 1/8th inches, it is not possible to measure a length to more than two decimal points since there are no marks between the 1/8th (0.125 in.). Unless a length measurement ends up really close to the mark, it is safest to assume uncertainty about the last decimal place (third) and leave it at two. A measurement of 5.125 or 5.250 might be the best you can do if the end falls directly under a 1/8th inch mark. You certainly can't say that the measurement is 5.1280 just because the length appears to be just a tad more than 5.125 (the nearest mark). Furthermore, there is no guarantee that the marks are right where they should be! Even if you can be precise to within two decimal places, you could still be off the true length if your measuring device is not accurate.

The problem of noise we discussed above can also be considered a problem with measurement, for detecting (transducing!) the signal is a measurement. Noise is whatever obscures the measurement, so for different sorts of transducers, there are different sorts of noise, and some kinds of measurements are more susceptible to noise than others. Electronic transduction, such as in the case of the thermistor, is notoriously noisy due to thermal effects in the various devices and wires that carry the signals. Bright light or reflections become noise for our eyes. And an example of noise interfering with a good measurement using a yardstick might be a small scratch on the yardstick that happens to look like a mark to a casual observer.

As mentioned above, even passive objects can act as sources of messages when there are ambient energy sources acting upon them. The passive object has physical properties, e.g., the surface texture of a rock that alters the way in which the energy is reflected. In this case, geometrical properties serve to modulate an energy flow in a characteristic manner, so this is not the same as transduction. Passive objects do not make good receivers, however, since they have no actuator capabilities. For example, if you want to send a message to a rock (e.g., get it to change in some way), you are more likely going to hit it with a hammer or throw it to a new location. Rocks don't respond well to verbal commands.

Question Box 7.6

We think of properties (weight, height, density, etc.) as qualities possessed by systems. But in another perspective, properties seem to emerge when we devise processes of measurement. What does it mean to weigh 100 lb? To have a 15 % risk factor for prostate cancer?

7.3.3 Surprise!

We mentioned previously that information is news of difference (from expectations) and that it is measured in terms of probabilities for the various message states that could be received. A receiver can be said to have an inherent expectation for each state. If measured in terms of probabilities, and noting that all messages states are mutually exclusive of one another at a particular instant, then the sum of the probabilities must equal one, as stipulated in Quant Box 7.1. The amount of information received with each message state actually received is given by Eq. (QB 7.2.2) in Quant Box 7.1. Plug in the a priori probability of receipt, and that gives the value of information.

Let's use the example of a written message. All of the letters of the alphabet would be potentials, but the probability of any given letter being received next might be higher or lower. That is, they are not equiprobable, or $1/26 = 0.03846$. Suppose a receiver has an a priori probability assigned to each of the possible characters in a message alphabet (we'll just use five characters in this example; equiprobable distribution, $1/5 = 0.2$). Once again, the sum of the probabilities has to equal one by the rules of probability theory. This table lists the state of the receiver's expectations prior to the receipt of the next character (first row).

a	b	c	d	e
0.4	0.18	0.02	0.25	0.15

Which character would you say the system is most expecting? Clearly the most likely next character is "a" with an a priori probability of 0.4. Suppose the actual character received is an "a." How much information did the system receive? If we worked through the math in Quant Box 7.2, we would know that the information conveyed by the receipt of "a" is 1.32 bits ($-\log_2(0.4)$). But suppose the character received is actually "c." Then the information conveyed would be 5.64 bits (you should verify this), a little more than four times as much information.

By our convention we would say that the receiver had a greater expectation of receiving "*a*" but actually received "*c*." The receiver is "surprised"! But just how much of a surprise depends on some deeper aspects of how that table of expectancies came about in the first place. If the table is fixed and will always reflect the a priori expectancies of each character, then we can say that the surprise is just equal to the information conveyed by the arrival of any character.[15] On the other hand, if some letters are more likely than others to follow a given letter, then the probability topography is modified with the receipt of each letter.

The quantitative value of surprise is especially important for active systems because this is used to drive processes within the receiver through the effects of an amplifier subsystem such as in the prior section. Work is accomplished internally, assuming energy is available, and the receiver modifies its behavior as a result. We will also see that this work can include modifying expectations as a result of ongoing experience. This amounts to altering the table of expectancies to reflect that experience.

7.3.3.1 Modifying Expectations: An Introduction to Adaptation and Learning

A system that can change its internal structure to some degree as a result of messages it receives from its environment is said to be adaptive if that change allows it to continue, essentially continuing to perform its basic functions. Conversely, if a change results ultimately in the breakdown (breakup) of the system, then the change is clearly not adaptive. Brittle networks, those that have rigid connections between the components, are not, generally speaking, adaptable. If you hit a rock hard with a hammer, its internal connections will be fractured, and it comes apart, no longer being the same rock. On the other hand it is possible to traumatize living tissues, damaging them, but not necessarily killing them. They have the ability to heal if the trauma does not go beyond a certain limit. Living systems are adaptable, maintaining their primary functions in spite of many kinds of surprising messages they receive.

Adaptive systems use information to modify their internal structures slightly, or within limits, so as to sustain their basic overall structure and functions. Such systems can be said to be preparing for the future. In this section we turn our attention to complex adaptive systems (CASs) rather exclusively to focus on the way in which the receipt of information prepares a system for its future through adaptation. We will also introduce the theory of learning systems, that is, systems that perform ongoing adaptations that *improve* their likelihood of dealing successfully with whatever that future might hold. For a living system, this learning ability is a vital part of its ability to maintain fitness in an ever changing environment.

---

[15] An alternative formula for surprise value is to subtract the amount of information provided by receipt of the most expected character, in this case "*a*" = 1.32 bits from the information value of the actual character received, "*c*" = 5.64 bits, giving 4 bits as the surprise.

It may seem odd to claim that forgetting is a vital part of learning. However the environment in which causal relations are established may change. If it does, then some learning might actually become obsolete at a different time. If a learned relation is maintained in spite of such changes, then the CAS would end up making incorrect decisions based on the old relation. It is therefore adaptive to not maintain what has been learned indefinitely.

7.3.3.2 Adaptation as a Modification in Expectancies

We have seen that every modification to a system also involves information in the form of a change in the potentials and probabilities, the expectations, associated with the system as it moves into the future. The theory of information quantification depends on the notion of a probability and how much a message entails a modification to what was expected, that is, it assesses the **unexpectedness of the message**. For passive systems, probabilities simply are what they are for the system in a given state in the context of its entire range of environmental relations. But, systems that have evolved the ability to adapt and learn move into a probabilistic future in a way more complex than, for example, a rock. Living organisms engage the future actively, with needs that must be fulfilled for survival, and they are primed to act adaptively upon information received. They not only move into a future, they act into a future with action that anticipates what conditions will be encountered.

In order to understand systems, we have been at pains to push familiar words like expectation, uncertainty, and surprise back to application at presentient and even pre-living levels. Based upon the fact that every system always has potentials and probabilities that constitute the topography of an expected future, there is a next step, the emergent capacity to actively use this expectation in a way that amounts to proactively moving into the future. This comes to fullness with the evolution of creatures that have the ability to cognitively anticipate the future. This cognitive expectation is so basic to our own way of being in the world that it is the first thing we think of when we hear the word "expectation." Now we return to this more familiar ground of moving with sentience into a future about which we hold expectations.

It is well known that human expectations track rather loosely with the world of calculable probabilities. That is, the world of cognitively processed expectation cannot be totally detached from the nonsubjectively processed world of expectation, but there is a significant difference insofar as anticipation involves an imagined world based upon our cumulative interpretation of our life experience to date. We change our expectations as a result of experience with real-world occurrences, but those experiences are interpreted through lenses of hopes and fears in a landscape of personal experience sufficiently common that communication is possible, but sufficiently unique that misunderstandings and even mutual incomprehension are part of that common human experience.

We frame our activity in terms of a richly textured fabric of expectations, including many degrees of anticipated probability which are continually modified in a feedback loop with our ongoing actual experience. As varied as our interpretive frameworks, degrees of stubbornness, etc. may be, the ongoing modification of expectation by experience is a common denominator that can ground a more general and useful formulation of this process. An eighteenth-century British theologian and mathematician, Thomas Bayes, developed a mathematical formulation for how expectancies can change as a result of information.[16] He started with the proposition that our expectations change as a function of experience. Prior to a specific event (out of a distribution of possible events), we hold some expectancy that that event's occurrence would be just as shown in the above table. Upon the actual event, we adjust our expectancy of that event in the future to reflect its actual occurrence. Intuitively, the occurrence of an event suggests that it is likely to occur again in the future (as compared with other possible events). Also, if an event that is expected does not occur, and continues to not occur in each opportunity period, then we adjust our expectations downward.

Effectively we are saying that more frequent encounters with some event should cause us to hold greater expectations for that event in the future. Similarly, but in the contrary direction, the more rare the occurrence of a specific event over time, the lower our expectations for its occurrence in the future should go.

Bayes provided us with a probabilistic formulation that uses a priori probabilities, such as given above, and the actual observation of an event, or nonevent (of those that are possible), to generate an a posteriori probability that will be used as the a priori probability the next time an opportunity for observation presents itself (see Quant Box 7.3).

Real systems are processes whose internal structures and responses to input messages reflect their expectations. Adaptive systems, sentient or not, will modify their behavior on the basis of information flows changing their expectations. A human being will react to surprise (information) by learning and changing behavior. This is accomplished by biophysical work taking place in the brain. A tree will also modify its growth or seasonal behavior (e.g., when it drops leaves) based on environmental experiences.

Let a system make a sequence of observations or receive events (message states) along a single channel conveying messages that have the five states represented by the five characters shown above and repeated here. At each observation the system expects to see, in order of decreasing probabilities, "a," "d," "b," "e," and "c." Let us see how the Bayesian intuition can be implemented. To generalize, let x be an observation event and p_{xi} be the probability (second row of the table) that x will be the ith element in the set of event states (as characters in our table). Then the *a posteriori* probability is some function of the information obtained by the actual observation event.

---

[16] Bayes' formula applies to the probability of one event given the observation of another "associated" event (see Quant Box 7.2). We are using the probability of an event given the observation or non-observation of that event at a prior time.

$$p_{x_i}(t+1) = f\left(p_{x_i}(t), I(t)\right) \tag{7.1}$$

The function $f()$ is "computed" (see the next chapter for an explanation of what this means) by a process in a real system. And because different systems will have different kinds of processes, the function could be just about anything. However, we should suppose that if the information value is low, then the change in the a posteriori probability (which will become the a priori probability in the next iteration) should be to increase it, but by a small amount. If the information value is high, then we would expect to see the a posteriori probability increase by a greater amount. In other words, the new expectation should be proportional to the information value obtained from the last observation.

Since the p_{xi} is a value in a set, the sum of probabilities of all elements in the set must equal one. Thus, the change in the probability of the actually observed event means that the probabilities of the other elements must be adjusted so as to compensate for the increase in the one observed. Since the other elements were not observed, their probabilities will go down in some manner proportional to their weighting in the set.

As an example, consider the system with expectancy vector as given above. For a definite function of the form in Eq. (7.1), let us use

$$f_{t+1}(o) = \left(p_o + \frac{I_o}{\max(I)} * \frac{1}{n}\right)_t \tag{7.2}$$

where:
o is the observed event (e.g., a, b, c, d, or e).
p_o is the a priori probability of observed event.
I_o is the information computed by Eq. (QB 7.2.2).
$\max(I)$ is the highest information value of any item in the set (again by Eq. (QB 7.2.2)).
n is the number of items in the set (in this case 5, so $1/n = 0.2$).
$t + 1$ is the time step of the next observation.

Hence $f_{t+1}()$ provides the a posteriori probability after the current event is observed. But we must now normalize the set so that the sum is still one, so we need a function that will push the other probabilities downward in proportion to their weights within the set. We know that the total non-observed probability has to be $1 - f_{t+1}(o)$, so as a reasonable approximation we can multiply each a priori probability in the set, other than that of the actually observed event, by $1/p_t(i)$, where i is the index ($i \neq o$) of the other event elements. This formula does not quite normalize the set, however, in that $1 - \sum p_i > 0$. This remainder can be distributed evenly among the other elements without too much distortion.[17]

---

[17] This formulation is being kept simple for illustration purposes. More realistic formulations for normalization tend to be a little too complicated.

At time t_0, the start:

a	b	c	d	e
0.4	0.18	0.02	0.25	0.15

Suppose the system receives a "c" in the current observation. This event was somewhat unexpected. We have already seen that the information conveyed by its receipt is 5.64 bits. Using this information we have guidance in how to modify the a priori probability for "c" in the next observation opportunity. Plugging our a priori probabilities into the above formulas, we get

$p(c)=0.22$, $\sum p_i = 0.764$, and remainder, $r=0.016$. A very simple adjustment to the table would be to add $r/(n-1)$ or 0.004 to each of the other elements.

At time t_1 with observation "c":

a	b	c	d	e
0.316	0.144	0.22	0.199	0.121

As we can see the observation of a "c" character at time t_0 causes a large increase in the a posteriori probability assigned to "c" which will be the a priori probability used in the next observation opportunity, $t+1$. At the same time the other probabilities are downgraded to reflect that their nonoccurrence at t_0 can mean that they are somewhat less likely to be observed at the next observation.

The actual functions for adjusting the probabilities with each observation can be much more sophisticated than given here. Indeed, Bayes' formula can be used directly if the processor being used is a digital computer. However, whatever functions are employed, it should be clear that given much larger alphabets or event possibilities, the time needed to recompute a table like this becomes large as well.

This example gives us our first and simplest view of how a system will adapt its expectations as a result of information received. Next we need to develop a sense of the ongoing dynamic of adapting expectations based on an iterated process of message receipts.

Suppose the system receives an "a" next. By how much would the probabilities change? The reader is invited to use the above equations to recompute the table at t_2.

Quant Box 7.3 Bayesian Inference
Thomas Bayes' theorem is generally given in the form of a conditional probability, $P(A|B)$, where the probability that event A will occur is conditioned on whether or not event B has occurred. The formula is

$$P(A|B) = \frac{P(B|A)P(A)}{P(B)}$$

(continued)

Quant Box 7.3 (continued)

The inverse probability (of B given A) times the prior probability of A and divided by the probability of B. The theorem provides a rule by which an inference can be made if the prior probabilities are known.

Bayes rule can be applied to computing the probability modifications that are the result of actual message state receipts at a particular time, t. One way to do this is to substitute the prior probabilities of A and B for x_i at $t+1$ and x_i at $t-1$. That is, we want to compute what a future probability of x_i will be given that that same symbol is received at time t.

$$P\left(x_i\left(t+1\right)\mid x_i\left(t-1\right)\right) = \frac{P\left(x_i\left(t-1\right)\mid x_i\left(t\right)\right)P\left(x_i\left(t\right)\right)}{P\left(x_i\left(t-1\right)\right)}$$

Unfortunately this computation also means that you would have to compute the posterior probability for all x_j, $j \neq i$ symbols as well in order to normalize the set. This puts an extraordinary computational load on the system. This is why Bayes' theorem, while conceptually useful (and mathematically rigorous), is not very useful in many practical applications, such as a real-time learning algorithm. Rather, a Bayesian-inspired method such as in the above text section can be used to approximate a Bayesian formulation.

We should conclude this subsection by noting that real systems do not possess numbers representing values of a priori probabilities! Your brain, for example, does not have special neurons that are the locations of actual real numbers. Rather it is the activity rates along certain pathways between neurons and neuron clusters that "represent" numerical relations to expectations. The above example only illustrates the effective mathematical model of changing expectations. Below we will dig into what it means from a mechanical perspective as the numerical value is represented in the structure of the system.

7.3.3.3 Internal Work in the Receiver

As we have seen, active systems can change their internal structure, and hence their functions or behaviors, as a result of receiving messages that are informational. The internal change comes from doing work on those structures. How much work should be done? Assuming a parallel between structure (including cognitive structure) and systemic expectations, how much difference in the structure should be made as a result of information receipt? We can use the above mathematical model as a starting place.

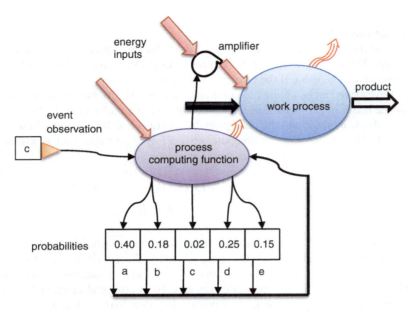

Fig. 7.8 A system that uses a set of a priori probabilities to generate work and adjust that set (compute a posteriori probabilities) in preparation for the next observation cycle is a model of an adaptive system. The information generated from the observation is used to modify the probabilities and to actuate a work process. Waste heat is dissipated from both processes

The Figure 7.8 can actually represent many different receivers. If the probabilities table is fixed, then the diagram could represent a decoding machine (including a computer). But in the more general case where the computing function does modify the probabilities, we have the beginnings of a complex adaptive system (CAS).

Probabilities have a foundation in the systemic structures and relations of the real world, but insofar as they belong to the future, they inhabit the cognitive world as mathematical concepts, intuitive calculations that, given the requisite process, can be given numerical expression. Our brains, for example, don't actually compute probabilities in a strict sense. There is no set of memory slots in our brains that store numbers as shown above. But all living systems rely on approximation computations (discussed below). In some ways our neurons approximate fuzzy logic, where the sum of probability sets does not have to strictly equal one. Nevertheless, brains do compute expectancies that approximate probabilities, which is why we humans can even think about the concept of information in terms of removing uncertainty.

7.4   What Is Knowledge?

The system that we explored in the previous section is able to modify itself as it receives ongoing inputs. It doesn't take much to see that if the frequencies of the various input characters are stable over a long time frame, the system will have a distribution of probabilities that reasonably represent those frequencies. This is what is meant by a subjective/frequentist interpretation of probability. With each new observation the probabilities will change, especially when a rare probability is observed, but over the long haul, the probabilities will be in the neighborhood of their frequency-based values. Some systems might even be capable of invoking an aged-based formula that diminishes the amount of modification that any message state can generate, emulating the increased work that must be done to dislodge more deeply established expectations. It would be easy to introduce another time-based factor that would reduce the information value that any message would generate, imitating the diminished surprise factor that goes with a greater accumulation of experience. The system could mature! Suppose a system is currently adapted to conditions in its environment. What this means is that the system has had time to use prior information to change itself and the current flows of materials and energy are what it has come to expect. It did this by modifying its own internal structures when the flows first came into the levels they are at the moment of our supposition. What we are describing is that the flows were changing at one point in time which demanded the system to accommodate them. If the flows now remain constant at the new levels, the system is in steady-state equilibrium with its environment, and the flows themselves no longer convey any information. That is, the system is advancing into a future for which it is perfectly prepared.

We might consider knowledge as the cumulative expectations with which a system moves into the future. In this sense we say that the system *knows* what it needs to know in order to exist comfortably in the flows of its current environment. Knowledge, then, is that internal structure in a system that matches its capacity to dissipate flows in a steady-state equilibrium. We use the steady-state example here to make clear that knowledge is the fit between structure-grounded expectation and the actual situation as it unfolds. This will prepare us to see the relation between knowledge and the information that comes with less expected modifications of the flows.

This is a highly abstract, narrowly functional approach to knowledge. Most people think of knowledge as something that one possesses about the nature of the world that helps them navigate successfully in that world. We are looking with special focus on the navigation component of expectation, its functionality in enabling a system to move along handling the future with systemic adequacy. Nonconscious metabolic components have this kind of knowledge, and failures of knowledge as well, as in the cases where overactive immune system causes life-threatening allergies. And even our conscious forms of knowledge are grounded, like metabolisms, in physical structures and flows, for they are actually embodied in the intricacies of the connections of neurons in your brain. When you learn something new, by receiving information that is about something which affects you, the neurons in your brain literally undergo some rewiring or at least a strengthening of some existing connections. When you receive information in the strict sense that we have defined it above, the actuator biochemistry in your brain cells goes to work to generate new axonal connections where needed to support the representation of what you now know.

This restructuring is generally very tentative. The new connections are made weakly at first. If you continue to receive messages of the same sort, however, these will reinforce the learning and strengthen the new connections so that the new knowledge goes into longer-term memory. Such complex structures as the habituated patterning of neuron firing enables organisms to have varying degrees and kinds of memory capacities. This adds layers of complexity and flexible capacity to what we have so far discussed as a rather simple structure of systemic expectation. After establishing new knowledge as the new structuring of our brains, we are able to use that knowledge to form or reform our more immediate expectations about the world (no knowledge exists in isolation, knowledge is always associative). Having such knowledge, then, in the future, if we receive similar messages, we are not particularly informed by them.

Question Box 7.7

In simple theory, the more unexpected (but still in the range of possibilities) an occurrence is, the greater its information value. But in practice, outliers, rare events, are often discounted because of their very unexpectedness. How might you modify the pure surprise metric to include the weight one might give information? How does the measure change as applied to different situations?

We can actually suggest a formal definition of the knowledge dimension based on the formal definition of information (esp. in Quant Box 7.1), namely, that the measure of knowledge is inverse to the surprise value, or information value, of a message. The more you know, the less surprised you are by the receipt of that message.

Inspired by the work of Harold Morowitz, quoted at the beginning of the chapter, a formal approach to knowledge is to simply quantify it as $1/I$, the simple inverse of information! In essence this formulation says that the more you know, the less you will be surprised. Not only does this conform to our intuitive sense of knowledge as opposed to ignorance or uncertainty, it also provides us with a mathematical approach to describing systems in terms of their capacity to either adapt to a changing environment or deal with an environment that they have knowledge of. Graph 7.2 captures this notion of knowledge.

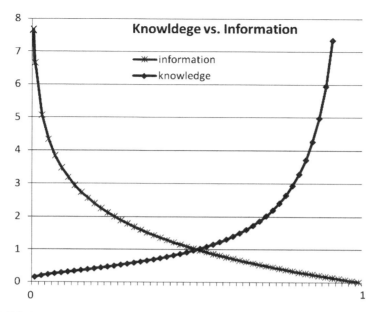

Graph 7.2 Knowledge is plotted as a function of information (as in Graph QB 7.2.1 (in Quant Box 7.2))

The theory suggests that it is impossible to have absolute knowledge of any event. The knowledge curve approaches unity (1 on the x-axis) asymptotically, while, at the same time, the information approaches zero.

7.4.1 Context

Thus far, except for our cursory discussion of memory, we have presented the information/knowledge relation for the simplest case of a first order message process. That is, each message state (character or symbol) received is considered independent from the other states that are received at different times. The only effect of any given message state being received is to change the probabilities of the set of states for adaptive systems.

In a second order message process, such as characterizes systems with memory, each message state received depends on message states already received. In other words, context is constructed during the receipt of a message stream that affects the probabilities of future message states.

A simple example of this is the use of letters to construct written words (the same holds for sounds or phonemes used to vocalize words). Words convey more meaning than just letters taken singly. Suppose a receiver (reader) first gets a "c" in a new message stream. Of all of the other 25 characters in the alphabet, what letter is the most likely next one to be received? That is somewhat of a trick question, because it depends on both the context of the new message stream and the conventions of English spelling. If the discussion was about cats, "*a*" would have a higher probability, while if it was about children, "*h*" would be more probable. Also there is a subset of letters that are allowed to follow "*c*" in the English language. They would all have nonzero probabilities associated with them, while the disallowed letters would have effectively zero probabilities (or vanishingly small probabilities, more likely). As the first letter in a word "*c*," for example, is more likely to be followed by an *h* than a *z*, and any vowel would be more likely than a consonant.

Suppose the next letter received is an "*a*," so the message stream so far is "*c*," "*a*." The receipt of that letter now has an impact on the probabilities of all of the possible following letters. In English, the likelihood of the next letter being a "*t*" or an "*r*" will be elevated, say relative to a "*q*." Thus, the amount of information (not the meaning) of the word being "cat" or "car" would be lower than if the next letter were "*m*" ("cam" is a word or could be the prefix set for "camera"). The information in the message by receiving an "*e*" would be higher, especially if that was the end of the word (there are no words in English spelled "cae").

As you can see from this example, the information content of a message based on a language protocol with the semantics of combinations (words) dictating the rules for computing probabilities is much more complicated. It can depend on the preceding letters but also depends on the length of the message stream itself.

Because of these complications our brains adopt a somewhat simpler approach to learning and using language. As we learn language, we learn the frequency distributions of two-, three-, and possibly four-letter combinations that we encounter in spoken (first) and written language. So our mental model of language resembles the sets above but for combinations of characters rather than individual characters themselves. We can use this shortcut because languages have conventions built into them that restrict legal combinations to a much smaller subset. All other combinations are assumed to have near zero probabilities, so if one encounters such an illegal combination (like "cae"), being so contrary to expectation (except a residual expectation for typos!), it generates considerable information causing the reader to do some work to figure out what the spelling was supposed to be. For example, consider the sentence, "the cae in the hat." A reader would have a near zero expectation for seeing "cae" as a word, but could do a little mental work to guess that it was supposed to be "cat." In this case the reader is using the sentence itself, plus some higher-level knowledge of Dr. Seuss's works. Or if the word "cat" appears in nearby sentences, the reader can infer that this was the originally intended communication. This example is another case of noisy messages. The "*e*" character was what we would call a "glitch," a mistake, or simply noise in the message.

Complex, adaptive systems, like human brains, use the information (surprise) content of messages to spur information processing work (thinking), and they use context to both form expectations and resolve mistakes as needed. Both of these are aspects of the more general activity of interpretation, the process by which message content is transformed from data to meaningful information. The more complex the information processing system is, the higher the orders of message processes you will find in it. And, as a result, the more complex the interpretive net becomes. Verbal context and lexical and semantic factors such as order of letters are but the more calculable tip of a contextual iceberg. Consider the way one's emotional state, relationship with the source of the message, life experience with persons of this sort, and a broad slate of other expectations for a context for understanding or interpreting the constant stream of sensory and verbal messages within which we live our daily lives.

7.4.2 Decision Processes

One way to frame the work that processes do as a result of receiving information (as in Fig. 7.3) is that they make decisions. That is, the information received along a specific channel is used to guide which of several possible work actions the system will take. But for truly complex adaptive systems, decisions are based on the system's model (expectation!) of how the world works and what effect an action has on that world. Elements in a computer (switches) make the simplest decisions of all, turn on or turn off (see Sect. 7.2 for more on computation). At a much higher level of complexity, components of our bodies such as neurons in our brains make far more complex decisions, and whole assemblies of neurons are responsible for making incredibly complex decisions.

Decision making assumes that there is a goal state that the system should attain after doing the work. That is, there is a result of one or a series of decisions that the system seeks to achieve.

In this section we will only examine three aspects of decision processing that have been investigated rigorously: decision trees, game theoretic decisions, and judgment. All are examples of how information plus a model are used to guide the work processes.

7.4.2.1 Decision Trees

Some kinds of decision domains can be organized using a particular graph structure called a tree. When the work actions are well known and the information can be readily ascertained, it is possible to construct, a priori, a tree of decision nodes, each with a set of rules that are activated by the information received by observing the state of the environment at a given instant in time. For example, a game of checkers can be characterized as a tree of state nodes linking to player moves that change the state of the game (Fig. 7.9).

Decision trees are frequently used in business and finance to guide which option should be taken in the pursuit of profits and earnings. These are generally weighted with risk factors which are essentially the probabilities (of success or failure) that have been estimated or learned from history.

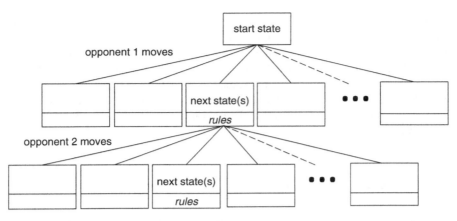

Fig. 7.9 A game tree. The start state is the setup of the board (of checkers, say) at the beginning. Opponent 1 makes a move (possibly based on a rule, but also possibly random), which takes the game state to the next level down. There are any number of possible moves indicated by the number of nodes at the first level. Similarly opponent 2 takes a decision based on some rules (and an educated guess about what opponent 1 will do in response). Opponent 2 received information from the move actually made by opponent 1. Prior to the move, almost any new state could obtain, but as the game proceeds, each player develops some likelihood estimates about which rules (moves) the opponent might take, thus changing their expectations. Surprising moves convey more information

7.4.2.2   Game Theory

A more sophisticated approach to decision making is based on game theory, which is used extensively in battle and business competition planning activities. Game theory goes beyond simple board games to look at constrained options and decisions taken under uncertainty. One of the more interesting results of using game theory in research of human decision making has been to show that humans are not really "rational" decision makers as would be suggested by, for example, decision trees. Unlike Commander Spock on the television series, Star Trek, humans are prone to consistent decision mistakes and biases owing to their reliance on heuristics such as stereotypes, rules of thumb, or common sense as opposed to algorithmic processing.

7.4.2.3   Judgment

Real life is far too complicated for an individual or an organization to "know" all of the relevant rules they should follow in making decisions. Rational thinking, moreover, is really hard to do; it takes a lot of energy. The heuristics mentioned above often are employed, by individual humans, in the form of intuitions and judgments that guide or help making choices at decision junctures (the nodes in a decision tree). Judgment comes from having very elaborate models of how the world works (including other people) that are held in what is known as tacit memory. That is, these models operate in sub- or preconscious thinking, and one experiences their "nudge" in one direction or another in the decision process as a vague sense of rightness or wrongness. While personal experience plays a significant role in shaping these models, so does the network of one's significant interpersonal relationships, which often share in and reinforce major features of the model. Thus, while judgment seems the most personal conscious activity, many sorts of judgments typify groups, subcultures, or whole cultures.

Legal systems are more explicit collections of judgments that have been made historically in a society. Judges expect lawyers to bring arguments to the bench based on what has worked and what hasn't in the past.

Question Box 7.8
The role of models in judgment is instanced fundamentally in the plausibility of a piece of information: is it believable or not? Sociologists of knowledge speak of a "plausibility structure," essentially the model comprised of everything we think/expect of the world as it is daily reinforced by the activity and interactions of our lives. This model makes some information inherently more believable or less believable. Social psychologists speak of "cognitive dissonance," the condition of tension one undergoes when facing the question of what to do with information not in keeping with one's overall model. Some people believe information about global warming, and others dismiss it. What other items are likely to be features of the models of the world carried by believers and disbelievers? Is how strongly we hold on to a model itself a feature of the model or just a personality characteristic?

7.4.3 Anticipatory Systems

Anticipation is not the same as prediction. The job of prediction is to specify a future event or state with a probability that implies high confidence. The point of prediction is to "know" what the future is likely to bring. Anticipation shares some common attributes with prediction in that it also attempts to know something about the future, but only in terms of what is possible. Clearly many things can happen in the future (many different states) in a complex system. There are too many variables to consider in order to do a good job of predicting. So the objective of anticipation is to assess likelihoods or a range of probabilities for members of a set of possible outcomes (in the discrete version). But then anticipation serves another purpose as well.

Technically speaking, a prediction works only if the event predicted comes to pass. With anticipation the point is to have a different outcome! In a way it is the antithesis of prediction, for anticipation is a way of feeling one's way into the future, both alert for opportunities and perhaps even more alert for dangers. Anticipation of future events serves both purposes for complex adaptive systems. Anticipation allows such a system to alter its behavior so as to exploit opportunities or to avoid threats. A cheetah can anticipate that a running antelope will veer to the right while being chased. The cheetah's brain and muscles are geared to make the turn at the slightest hint that the antelope is doing as anticipated. The cheetah was primed by anticipation to rapidly turn to head off the antelope and by doing so to capture it more quickly. The anticipation is similar to a prediction that the antelope will arrive at a more distant point on the map at a future time. The cheetah's purpose is to make sure the antelope *doesn't* reach that point! Of course another antelope may have anticipated that a cheetah might be crouching in the tall grass near a watering hole waiting to pounce. This antelope would be alert to the slightest hint of a trap and bound away to prevent an otherwise predicted sudden end. Thus, anticipation is meant to allow a CAS to change what might otherwise happen. Most of us don't want to predict our demise. We want to avoid it.

We have mentioned the feedback loop by which information mediates the continual modification of expectations as systems move into the future. Gene pools use this loop as populations are selected in terms of fit; a selected fitness of what has worked so far projected forward a likely fit for an expected (not too changed) future. There is rudimentary anticipation in this insofar as an anticipatory system is one that uses information to build a model (knowledge) of how its world seems to work, for the selected fitness is indeed a model of a world such that the given genetic recipe worked well enough for reproduction, which will produce another iteration of the cycling probe into the future. The same dynamic, in which a selective past becomes an informed base for probing an uncertain future, is repeated at another level with the emergence of individual organisms equipped with memory and consciousness. These abilities yield the kind of anticipation with which we are most familiar. That is, they enable the construction of what amounts to a model of causal relations that can be used to conjure up possible and/or likely future states of the world given current states and dynamic trends. In animal brains the time horizon for how far into the cumulative past the memory goes and how far into the future the anticipation goes depends on the amount of brain power available. Humans seem to have the most brain power of any animal, and with the emergence of symbolic language with its ability for self-reference, memory, anticipation, and causal thinking/modeling all move to a yet higher level. Thus, we humans not only frequently think about what might happen in a more distant future, but we also reflect on the future as such and so raise anticipation, strategizing, and planning to a new and distinctive level. There's all the difference in the world between looking for lunch—which most mobile organisms do—and wondering how to ensure a steady supply of lunches for an indefinite future, which is what humans do. Of course the reliability of these modes of anticipation depends on the efficacy of the model constructed. We will return to this subject later in the book.

Creatures that anticipate consciously can do so because, coupled with memory, they have evolved the ability we sometimes call imagination. Based on its tacit knowledge from the past, the brain has the ability to project memory-like images that represent the future. In essence it is running a model in fast forward in order to get a glimpse of what the future might be. Imagination can have great causal efficacy in strategizing for the future and especially so when it is linked with the power of rational thought to explicitly consider cause and effect in seeking to accomplish an end. The more efficacious one's knowledge about particular circumstances and how that part of the world works, the more confident one can be that their models are truthful. In other words, one can have a feeling of near certainty that what they anticipate in imagination will come to pass in reality.

Anticipation is in fact so important to human well-being that the construction, refinement, and testing of models have become one of the most sought-after applications of computer technology. The military models war strategy, FEMA models the major disasters we can imagine, and businesses have sales projections that are based on different condition variables, like the state of the economy, a competitor's offerings, etc. They run these models (in fast forward compared with real time) to see what will happen under different conditions. They want to enhance success and avoid disaster, whether that means winning wars and avoiding defeat, minimizing the effects of storms and floods, increasing sales and profits, and avoiding threats from competition or market conditions. These models are often thought of and argued about as predictions, but the motive is exactly the same as the anticipatory efforts of predator or prey to find lunch or avoid becoming lunch.

Think Box The Brain: Receiving Information and Constructing Knowledge

The view of information presented in this chapter is at once overlapping with most vernacular and scientific concepts, but at the same time differs in subtle aspects. We insist that knowledge and information are not one and the same and argue this from our systems viewpoint. It turns out that the contrast between what is actually information and what is knowledge can be best appreciated in terms of what transpires in animal brains. As argued in the chapter, information is a value of a message that is unexpected (or of low expectation relative to other messages that could have been received along the same channel). The higher the information value of a message, the more impact it has on the receiver; in this case we are talking about the brain. Specifically, the information content of a message drives the formation of new knowledge in such a way that the future receipt of a similar message will be better anticipated and thus contain a lower value of information and in turn have a lesser impact on the receiver, i.e., it generates less new knowledge.

Concepts, as we have seen in prior chapters/think boxes, are knowledge structures formed in neuronal networks, particularly in the neocortex. Such structures are constructed by the strengthening of connections (synapses) between neuron clusters that co-fire when excited by sensory inputs (percepts). Concepts are malleable, that is, they are plastic in that they can be modified by the receipt of messages (through the senses but also from other concepts) that are informational.

The brain stores concepts in these neuronal networks. Suppose a sensory input derives from an encounter with an instance of a concept that is not in accord with the current concept. For example, suppose yourself to still be a child

(continued)

Think Box (continued)

who has encountered many instances of dogs and all of them had furry coats. Your concept of dog-ness includes furry-ness, and generally, when you see a new instance of a dog, your concept is confirmed. That is, seeing another dog with fur provides no information because your brain expects that a dog is furry.

Then one day you come across a doglike creature, smallish compared with other dogs with which you are familiar, but nevertheless, dog shaped. There is just one startling fact about this dog; it has no fur. Its body has bare skin! This is something new and you may actually be surprised by it. Your brain tries to fit this new thing in with what it already knows. But nothing quite fits except that its other features (aside from size perhaps) are all doglike. Then someone tells you it is a dog called a hairless Chihuahua, a breed of dog that is an exception to the rule that dogs have fur.

Your brain now takes this information (i.e., the difference between your expectation that all dogs have fur and the current instance) and modifies the neural networks involved in the memory of dog-ness. It has to be structurally modified so as to include the notion that there are smallish dogs that have no fur, and they are called Chihuahuas (a new name that needs to be instantiated in a network in the language processing part of your brain). Figure TB 7.1 is a rough sketch of how we think this works in the cerebral cortex (association areas).

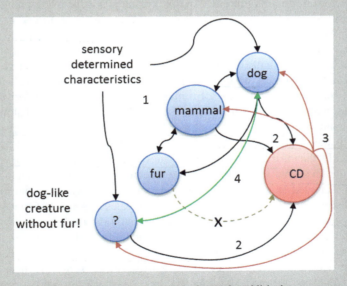

Fig. TB 7.1 Step 1 shows the mutual activations of several established concepts as a result of seeing a doglike creature (?). The activation of the dog-ness cluster activates mammal and fur, all part of being a dog or so thought. Concurrently some other cluster is being excited by the same set of features that excited the dog cluster, except it has no fur. A special neuronal assembly is thought to reside in a part of the brain called the hippocampus that detects coincidences between clusters that are firing but are not "wired" together (step 2). This assembly issues a signal back to all of the cells involved (step 3) that essentially tells them to form new links (step 4). Not shown is that a link to a new "noun" cluster associated with the name "Chihuahua" is also formed to the new "doglike without fur" cluster to help differentiate it in the future

> **Think Box** (continued)
>
> The next time you encounter a Chihuahua, it might be a bit bigger or smaller. Its skin might be more or less wrinkled. But your knowledge of Chihuahua-ness has been formed, and you can begin to learn variations on that theme to expand your understanding of dog-ness. Each time you encounter a new Chihuahua, you are much less surprised by the hairlessness and eventually take it to be quite normal. But the first time you saw one after seeing and incorporating so many instances of dogs into your knowledge of the concept, you were surprised.

7.5   Summary of Information, Learning, and Knowledge: Along with a Surprising Result

This chapter has attempted to cover a tremendous amount of ground. What we want the reader to take away is that the amount of information in a message is actually a property of the receiver (observer) and not of the sender (or the observed). This is quite restricted in comparison to broader usage, where it is common for people to use the term information in expressions such as "… encoding 'information' into a message…." The sender cannot know in advance what will be informational to the receiver. It can only send a message based on its own state. Whether the message informs the receiver depends entirely on the receiver's state, meaning the receiver's expectations of that specific message. What is happening is that the receiver is being informed of the sender's state by virtue of the message. But if the receiver already knows that state, and expects that message, then it is not particularly informed, only reinforced.

Such a restriction on the definition allows us to tackle the important issues of what is meaning and what is knowledge in ways that apply inclusively to all systems. Far from diminishing the significance of these terms, this approach allows us to see the evolving role of information as new modes of processing emerge, the emergence of life and the human use of symbolic language marking especially important thresholds. Meaning is established by virtue of the actual message channel linkage from the sender to the receiver's internal organization, which constitutes its interpretive framework. Each message can cause the receiver to respond in particular ways, the magnitude of response based on the amount of information in the message. Some of that response can include, in adaptive receivers, changing the future expectation or prior probability of the message to reflect the impact of current information on future information. This ability to change the set of probabilities based on actual experience is what we call learning, the process by which an adaptive system can modify, amplify, and confirm the expectations which comprise its knowledge.

We have seen the critical role of probability as the encompassing topography of possibilities in the expected world. It is possible to relax the strict constraints of probability theory (in computing) using fuzzy set theory. The restriction of probability theory, namely, that the sum of all probabilities in the set of message states must equal one, puts a difficult burden on computations as shown in Quant Box 7.2. In natural systems we find a relaxation of the rules of probability so that, for example, the sum of all of the representations of likelihood does not have to equal one; it can be a little more or a little less. The math emulates the more flexible relative ordering of probabilities in conscious processes, but that sum, over time, will tend toward one if the set is fixed in size (things really get interesting when set sizes are themselves adaptable as happens in the human brain). In other words, on average the set will be essentially normalized. So, natural systems like neurons in brains, that are adaptable, can approximate probability calculations. They can appear to behave as Bayesian machines even though they are not doing computation in the strict sense (as below). When we simulate natural systems by building artificial systems that use Bayesian math to learn, for example, patterns for pattern recognition, it works because natural systems do approximate the math. They just don't actually DO the math.

There is a surprising result of the relationship between information and knowledge as discussed above. Adaptive systems that receive information (regardless of meaning) change their external behaviors as a result. But changed behavior means that they are now sending out messages that are different from previous ones. That in turn means that other observing systems are now receiving information in turn. They modify their behavior which starts the cycle over again!

In other words, in complex worlds of adaptive systems, information begets knowledge, but knowledge, in turn, begets information. Thus, information (and knowledge) unlike matter and energy seems to obey a "nonconservation" law. Information (and knowledge) increases over time rather than decreases. As we shall see in Chap. 10, this is an important element in the dynamic evolution of systems to greater and greater complexity. There is, however, an important caveat that relates matter/energy to information/knowledge. Information (and knowledge) can increase so long as the embedding system is experiencing a flow of energy in which there is more usable energy available. Once a system is in steady-state equilibrium, knowledge can be maintained, but no new knowledge will be generated because nothing really "new" can happen. In other words, information will not increase. Furthermore, if the system is experiencing a reduction in the availability of usable energy, then knowledge will decay (remember knowledge is structured in the organization of material and energy in the system, so the second law of thermodynamics takes over and the system decays). Paradoxically, when knowledge decays, then any messages that are generated between subsystems may become unexpected or, in other words, provide information. However, since there is no energy available to do the work triggered by that information (to reform the knowledge structure), it is transitory information.

Bibliography and Further Reading

Ash RB (1965, 1990) Information Theory. New York, NY: Dover Publications, Inc.

Avery J (2003) Information theory and evolution. World Scientific, Hackensack NJ

Barrs BJ, Gage NM (eds) (2007) Cognition, brain, and consciousness: introduction to cognitive neuroscience. Elsevier, Amsterdam

Bateson G (1972) Steps to an ecology of mind: collected essays in anthropology, psychiatry, evolution, and epistemology. University of Chicago Press, Chicago, IL

Hartley RVL (1928) Transmission of information. Bell System Technical Journal, July 1928

MacKay DJC (2003) Information theory, inference, and learning algorithms. Cambridge University Press, Cambridge

Morowitz HJ (1968) Energy flow in biology. Acdemic Press, New York, NY

Rosen R (1985) Anticipatory systems: philosophical, mathematical, and methodological foundations. Pergamon Press, Oxford

Shannon CE, Weaver W (1949) The mathematical theory of communication. Univ of Illinois Press, Champaign, IL

von Neumann J, Morgenstern O (1944) Theory of games and economic behavior. Princeton University Press, Princeton, NJ

Wiener N (1950) The human use of human beings: cybernetics and society. Avon Books, New York, NY

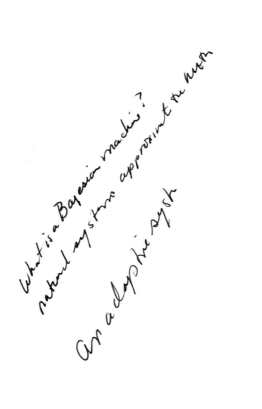

Chapter 8
Computational Systems

"Calculating machines comprise various pieces of mechanism for assisting the human mind in executing the operations of arithmetic. Some few of these perform the whole operation without any mental attention when once the given numbers have been put into the machine."

Charles Babbage, 1864 (From Hyman 1989)

"The most important feature that distinguishes biological computations from the work carried out by our personal computers – biological computations care. ... valuing some computations more than others."

Read Montague, 2006

Abstract Computation is to the processing of data, information, and knowledge, what physical work processes are to material transformations. Indeed computation *is* a work process of a very special kind in which energy is consumed to transform messages received into usable forms. Here, we will consider several kinds of computational processes and see how they all provide this transformation. Everyone is familiar with the digital computer, which has become one of the most ubiquitous forms of man-made computation. But we will also investigate biological forms of computation, especially, for example, how brains perform computations such as transforming sensory data streams into actions. One especially important use of computation (both in machines and brains) is the construction and running of simulation models.

8.1 Computational Process

The word "computer" has taken on a rather specific meaning these days. Most people think of the modern digital computer that sits on or under their desk. Many people have computers that they carry around with them to do work on the fly. And anyone who has a cell phone has a computer in their pocket. But the digital computer that is so commonplace today is not the only system in the universe that does computation.

Digital computing machines perform a very specific kind of computation based on algorithmic processing. Algorithms are a lot like recipes, telling the machine

© Springer Science+Business Media New York 2015
G.E. Mobus, M.C. Kalton, *Principles of Systems Science*, Understanding
Complex Systems, DOI 10.1007/978-1-4939-1920-8_8

what to do and when to do it. There are strict rules for what the operations are, so the steps in an algorithm are unambiguous as to what needs to be accomplished. It turns out that this type of computation, while wonderfully powerful for solving many kinds of problems, even controlling electrical and mechanical machines, has limitations. Not all problems that need solving can be solved algorithmically.

We need to pause for a moment and consider what we mean by the word "problem." The full import of how we are using this word will not come to light until the next chapter where we show how the subjects of the last chapter (information, etc.) and the subject of this chapter operate in systems. For now, we propose a fairly general description of a problem as being any situation in which a decision maker needs information in order to take action. That decision maker can be anything from a simple switch (on or off) to a world leader deciding on a course of action in world diplomacy. The "problem" can be framed as a "what should I do?" question. And a computational process is needed to provide the information needed to answer it.

The word "compute" combines the Latin *com* "with" or "together" and *putare* "to think." So the basic idea involved in computation has to do with combining, putting things together so they can be used for producing some kind of outcome—just the kind of thing we (or any process) needs to do in order to answer the "what should I do?" question.

So brains are computational processors too, but they are not digital computers.[1] They solve wholly different kinds of problems, putting together a wide variety of data, though the human kind of brains can, to a limited extent, emulate a digital computer doing arithmetic and logic! When Montague says that "…biological computations care," he means that there are biological consequences to wrong decisions. He also infers that the computation is an integral part of the system that does the caring. Living and nonliving computations thus have an interesting interface: while it is true that a digital computer doesn't care what computation it is doing or what the consequences of its reports will be, the human organization that is employing the computations certainly does care!

Question Box 8.1
Caring introduces differential weighting to outcomes: success and failure become consequential when we cross the threshold from chemical to living processes. We often hear the advice to try to approach problems "objectively," removing the torque that caring introduces into calculation. The Star Trek series had great fun playing off Captain Kirk and Spock as different sorts of information processors. But what would it mean to say nonconscious processes, such as our metabolisms, "care?"

---

[1] Throughout the history of neuroscience, the workings of the brain have been compared to, if not explained as, the workings of the best-known technology of the time. They have been treated as complex plumbing systems, telephony switching exchanges, and, most recently, digital computers. Thankfully, we now know that it is none of the above (although in some sense, it is also all of the above!).

In this chapter, we will consider multiple kinds of computational processes that are geared to provide the information needed to the decision maker to answer the question.[2] It turns out that there are numerous kinds of computational processes, both mechanical and biological, that are employed in systems for this purpose.

Mechanical forms of computation come in many forms, of which we will restrict this chapter to basically the digital form. The field of computer science provides a comprehensive framework for understanding all forms of mechanical (these days that means electronic) computation from an abstract mathematical perspective.[3]

A computational process is any process where the primary inputs and outputs are messages encoding data and where the process performs specific kinds of transformations on the inputs to produce the outputs. The process will also have an energy input since all transformations, even on data, are a form of work. The process will produce waste heat. In some cases, like the living brain, the computational process involves machinery that requires ongoing upkeep, meaning that a certain amount of material inflow occurs and a certain amount of waste material outflow occurs as well. In this chapter, our main consideration will be with the data flows and transformations.

In Fig. 8.1, we see a generalized computation processor. The data streams could be any sorts of message conveyance mechanisms, as discussed in the last chapter.

8.1.1 A Definition of Computation

Computation can be accomplished by any number of mechanisms such as gears and levers, electronic switching, or chemical reactions. What is important about computation is that the output of the process is similar in form to that of the inputs and the "value" of the output is determined by the values of the inputs, as measured in the same units. Value is generally interpreted to mean a numeric measure associated with the signal (i.e., the data) of the message. As per the last chapter, we normally associate computation with low-energy message processing; however, computational devices can be built out of any suitable dynamic process.

---

[2] Computing processes are generally divided into digital and analog, but these generally refer to "machines." Analog computers were extensively used prior to the development of the integrated circuit (chips) which allowed more computation power to be developed in small packages. Human-built analog computers were used for control and simulation purposes, which is why digital computers could replace them. However, in nature, there are a large number of evolved analog computing processes meeting the definition given here that cannot be easily emulated with algorithms. These are discussed in the chapter.

[3] According to David Harel (1987), the field is inappropriately named. He likens the name of computer science to toaster science. The science is not about the machine per se, it is about algorithmic computational processes, which just happen to be implemented in mechanical/electronic machines!

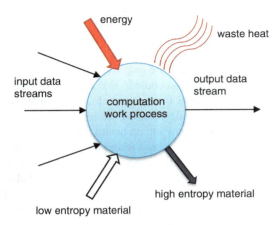

Fig. 8.1 A computational process is like any ordinary work process except that the work being done is a transformation of input data streams into output data stream(s). The work requires energy as with all processes. It radiates waste heat to the environment. It may also require occasional or small low-entropy material inputs to maintain the internal structures against the ravages of entropic decay. In that case, the process will produce some outflow of high-entropy material waste. The main focus, however, is on the transformation of data through the work process

For example, a mechanical computer composed of gears and levers may involve the transfer of relatively large amounts of energy required to turn the gears or push the levers. The input messages could involve the measurement of the angle of rotation on an input gear. The output would similarly be the angular position of a mark on the readout gear. We will discuss several different examples of computing process types as well as types of computations they can perform in the next section.

Computation is the combining of data inputs to produce data outputs. The combining is governed by a set of rules that are enforced by the internal structure of the processor (the "wiring diagram" of the internal network of sub-processing elements). Figure 8.2 shows the white box view of the processor in Fig. 8.1. The three input signals, here labeled A, B, and C, are combined according to the interaction network shown to produce output E. This figure is extremely generic and could represent any type of computational process.

The interaction network, along with the types of "atomic" sub-processors, constitutes the rules by which the input data is transformed into the output data. For example, in the below figure, the computation taking place might be a simple arithmetic formula, say $E = (A + B) \times C$. The a, b, and c sub-processes are atomic buffers; they simply hold the data as needed to ensure the correct timing of combinations. Sub-process d is an additive combiner (the addition operator of arithmetic). Its job is to take the two values of A and B and output their addition value. Sub-process **e** is a multiplicative combiner. It multiplies the value of the output of d with the value of C to produce E. Most people understand that this kind of arithmetic operation is computation. But what few people might realize is that the inputs and outputs are

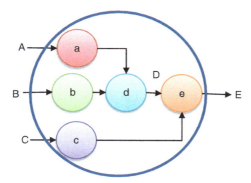

Fig. 8.2 The internal workings of a computational process involves message processing atomic processors and an internal communications network that allows them to interact in a "rule-based" way. In this diagram, signals A, B, and C are received by processors a, b, and c, which buffer the inputs. Process d combines signals A and B according to its internal function to produce an intermediate signal, D. That signal is combined with signal C from processor c to produce signal E. All signals are of the exact same form, in general, so that the value of E is determined by the actual values of A, B, and C at the time of receipt. This process network should not be interpreted as only mechanistic (e.g., logic gates as discussed later) but can include stochastic or nondeterministic processes as well

conveyed as data (messages) represented in an encoded form and not just the numeric values themselves. In other words, A, B, and C could all be encoded in pressure levels, modulated flow rates, or any physical dynamic phenomena for which a suitable transduction process exists (and as we saw in the last chapter, that is just about anything). There are more subtle ways to encode data into input signals as we will shortly see.

What makes this an effective computation is the nature of the interconnection network and choice of atomic operations. In the above figure, if sub-process d had been the multiplier and sub-process e had been the adder, we would have gotten a very different result from E, namely, $E = A \times B + C$ (which, parenthesized, would be $[A \times B] + C$). The choices of atomic operations and how they are networked together determines the *function* of the computation.

Computations are not restricted to "hard" rules such as the arithmetic operators and deterministic communications, as in the above example. Combinational rules (processes) can be probabilistic or fuzzy (as we will discuss below). The internal communications network could be subject to noise injections just as discussed in the prior chapter. In other words, the computation can be nondeterministic as much as deterministic. Much depends on the physical devices employed in constructing the computation process. However, even if a computation is nondeterministic, it is not completely random—such a computation process would be worthless to the systems that employ computational processes. Specific outputs from these processes will always fall inside a range of possible outputs given specific inputs so as to approximate a "proper" result. We'll see this in more detail below.

Question Box 8.2
Typical mystery stories often conclude with a scene in which the detective sits in a room full of worried people (mostly suspects), reviews all the bits and pieces of facts in the case, and concludes with the identity of the perpetrator. In terms of the above description, is this a computation? Why or why not?

8.2 Types of Computing Processes

In this section, we will demonstrate the universality of computation by showing a number of examples of computing processes from different physical domains. As just mentioned, we will see that computations can be nondeterministic and still provide useful data outputs, that is, useful for the larger embedding system that depends on them (as shown in the next chapter).

8.2.1 Digital Computation Based on Binary Elements

Computation boils down to a process where a state change in one elemental component of a system can conditionally cause a state change in another elemental component. In its simplest form, these components can have just two states, usually called "ON" and "OFF." This means that the element is either sending a message state of one value (say '1') or an alternative value (say '0') and no other values are possible. Figure 8.3 shows several forms of this simple two-state computation process, known as Boolean logic[4] rules after George Boole (1815–1864), an English mathematician who developed this form of logic.

The simplest Boolean rules, or functions, are AND, OR, XOR, and NOT. These rules can be combined by an internal communications network (deterministic as a rule) to produce various combinational functions, such as in Fig. 8.3c. It is somewhat remarkable that Boolean logic and various sequences of Boolean processes (generally called logic "gates" in computer talk) can be used to accomplish arithmetic (see below). We call these combinations "logic circuits" (they are implemented using electronic devices called transistors). And circuits can be designed for many computational purposes, including (and importantly) storing a state for a duration, something called a memory cell.

The rules for computation of each of the Boolean logic gate types are shown in Table 8.1 (or several tables). For the three binary input gates, the A and B columns designate the input values by ascending combinations. There are only four possible combinations of two values in two inputs, 00, 01, 10, and 11. The output

---

[4] See Wikipedia, Boolean Algebra: http://en.wikipedia.org/wiki/Boolean_algebra.

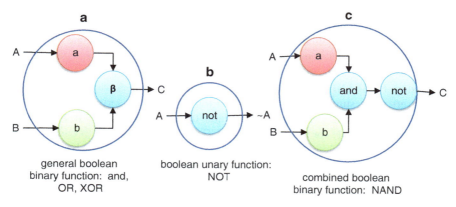

a general boolean binary function: and, OR, XOR

b boolean unary function: NOT

c combined boolean binary function: NAND

Fig. 8.3 A binary computation is based on the sub-processors having only two possible states, generally represented as "ON" and "OFF" or "1" and "0". (**a**) The **β** sub-processor is the combiner that produces an output, C, based on one of the Boolean functions (rules), AND, OR, or XOR. (**b**) A "unary" operation (having only one input) is the NOT function. It inverts whatever the input at A is. (**c**) Combinations of the three binary Boolean functions and the NOT function produce many more Boolean logic computations. This one inverts the output from an AND function to produce a NAND (Not AND) function. Boolean logic processes are the basis of the modern digital computer

Table 8.1 These "truth" tables show the computational rules for Boolean logic gates A and B are inputs, either 1 or 0 (ON or OFF). C is the output. AND will only produce a 1 if both inputs are 1. OR will produce a 1 if either or both inputs are 1. And XOR will produce a 1 only if just A or B are 1 but not if both are. NOT inverts the value of the single (unary) input A

AND			OR		
A	B	C	A	B	C
0	0	0	0	0	0
0	1	0	0	1	1
1	0	0	1	0	1
1	1	1	1	1	1

XOR			NOT	
A	B	C	A	~A
0	0	0	0	1
0	1	1	1	0
1	0	1		
1	1	0		

corresponding to these inputs is given in column C. The NOT gate merely inverts the value of the input, A. The output value is designated as ~A (the tilde is used to indicate a NOT'ed value).

Another intriguing property of logic gates and their implementation is that only two kinds of gates are absolutely necessary and all other gate types can be derived from them. You need either an AND or an OR gate and a NOT gate. With, for example, just AND gates and a NOT gate, you can implement an OR gate.

First, we combine the AND and NOT, expressed in Boolean logic format, in an expression, NOT(AND(A,B)). This means take the AND of A and B as above and then invert the result. This corresponds to the logic of all sorts of disjunctions, including mechanical switches or gates and true/false questions. If AND(A,B) would produce an ON (or also YES, or TRUE, or 1), then invert it to OFF (or NO, or FALSE, or 0). This logical element is called a NAND rule (NOT (AND)), and we can now construct an OR rule by using just NAND rules! It goes like this: C = NAND(NAND(A,A), NAND(B,B)). The reader should try this combination to see that the results of this logic replicates the truth table of OR as above.

8.2.2 Electronic Digital Computers

Since we have introduced the Boolean logic gates as examples of computational processors implemented in electronic circuits, we will start our look at large-scale computational processes by examining the modern digital computer, which uses bits as the unit of data (information).[5] The word digital refers to the representation of numbers (hence "number crunching") in binary code as discussed previously. All computations involve numbers even when the numbers are, in turn, used to represent something else like letters and punctuation in word processing[6] or colors put into "pixels" on a monitor.

In an electronic computer, the inputs and outputs are conveyed by wires with a voltage level representing a 1 and a lack of voltage representing a 0 (these days, the usual voltage range is 0–3.3 V DC). A transistor is actually an amplifier as described in the last chapter, but the voltage levels for all inputs and the output are kept in the same range and the transistor acts more like a switch than an amplifier per se. The output wire is either ON (1) or OFF (0) based on the state of the output transistor. Logic gates are built from a small number of transistors wired slightly differently to produce the outputs shown in Table 8.1. Each wire carries one bit of information since it can be in only one of two possible states. Next, we will show how logic gates are used to construct a more complex processor circuit that will give us arithmetic using binary numbers. A binary number is similar to a decimal number in that the position of a particular digit determines its summative value. The big difference

---

[5] For interested readers who would like to learn more about computers from the bits to the programming level, a very accessible (text) book is by Patt and Patel (2004).

[6] For example, natural numbers (integers 0, 1, 2, etc.) are used to represent letters and many different written symbols in various languages. See American Standard Code for Information Interchange (ASCII), http://en.wikipedia.org/wiki/ASCII. This code has been supplanted in newer computers by the Unicode, which allows the encoding of 65,536 different characters allowing it to represent most all characters from most written human languages!

between decimal and binary is that in the former, there are ten digits (0–9) to work with whereas in the latter, there are only two (0–1). For example, in decimal, a "10" represents the integer value of ten. In binary, a "10" represents the integer value of two. In decimal, the value comes from the sum of the position value of each digit. "10" $= (1 \times 10^1) + (0 \times 10^0) = 10 + 0 = 10$. In binary, the value is derived in exactly the same way except the base value used is 2 instead of 10. Hence, "10" $= (1 \times 2^1) + (0 \times 2^0) = 2 + 0 = 2$. The value 10 in binary would be "1010" $= (1 \times 2^3) + (0 \times 2^2) + (1 \times 2^1) + (0 \times 2^0) = 8 + 0 + 2 + 0 = 10$. One of the authors has a sweatshirt with the following inscription on the front: *There are only* **10** *kinds of people in this world, those who understand binary and those who don't.*

In computers, number values are represented in devices that hold a string of binary digits in a row. These devices are called registers or <u>memory lines.</u> They are usually arranged in units of eight cells, called a *byte*. As with the example of the integer value 10 above, a byte representation of the number 10 would look like this: 00001010. There would be a wire coming from each digit (cell), and those wires would be routed to various other devices for use in computations.

The basic function of a computer is to compute numbers! That is, a computer manipulates numeric representations according to the list of steps and rules programmed into it to produce a final numeric result representation. Now it is true that we have learned to use computers to do more things that on the surface appear to be more than just manipulating numbers. Anyone who has seen the latest adventure movie using computer-generated imagery (CGI) animations may readily wonder how that very realistic looking truck, crashing through the Brooklyn Bridge, could just be a bunch of numbers in a computer memory. But that is exactly what it is. And here is the really surprising thing about computers. Fundamentally, all they can do is add! But, technically speaking, that is all you have to do if your representation method (binary numbers as above) includes a way to represent a negative number. Fortunately, such is the case for a computer; it is very easy to literally flip a binary number to make it represent its own negative value (see Quant Box for details). Adding a negative number to a positive number is just subtraction. Once you can do addition and subtraction, assuming you have a way to do them repetitively, you can do multiplication and division. You can do all of the operations of arithmetic with only two actual operator circuits, an adder and what is called a negator (change to negative representation). Everything else is just repetitive uses of these two operators.

Figure 8.4 shows one of the most important functional components of a computer's central processing unit (CPU), the heart of a computer. It is the circuit which adds together two binary digits (bits) and produces a binary result and a carry bit. Also in the figure is the truth table for this circuit.

Adding together two single-bit binary numbers (1 or 0), A and B along with a carry-in bit are accomplished by this logic circuit.

The lines labeled A and B are the inputs coming from two different memory devices or registers. For example, let's say you have two 8-bit registers labeled RA and RB. Each has eight wires coming out of each cell, respectively. These can be represented, from the right end to the left, as RA_0, RA_1, RA_2, ..., RA_7, and RB_0–RB_7. Each of the two paired lines from the two registers goes to the A and B inputs to one of the eight full adders, ADD_0 through ADD_7 (see Fig. 8.5).

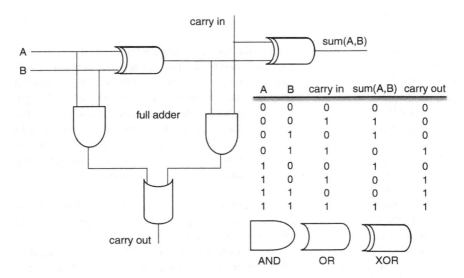

Fig. 8.4 A full adder is a circuit built from logic gates that can take two binary numbers, A and B, along with a carry bit (e.g., the value carried out of a previous column addition just as in decimal addition) and add them together to produce a binary sum and a carry-out bit. See text for explanation

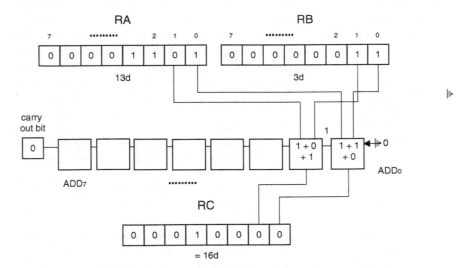

Fig. 8.5 Adding two binary numbers (8-bit values) using eight full adder circuits (ADD$_0$ through ADD$_7$, as in Fig. 8.4) produces an 8-bit answer in the C register. Register RA contains the binary representation of decimal integer value 13(d) and RB contains decimal integer value 3(d). The adders work from *right* (ADD$_0$) to *left* (ADD$_7$) in sequence. The carry-in bit to ADD$_0$ is hardwired to value 0 (ground). The carry out of ADD$_0$ is wired to the carry in of ADD$_1$ and so on. This is called a *ripple carry adder* since each subsequent full adder has to wait until its right-hand neighbor has finished its computation. 13d + 3d = 16d. Only a few wires are shown, but there would be a similar pairing for each memory cell of both registers. It is possible that the addition of two binary numbers of eight bits each could lead to a result that is nine bits in length with a carry out from ADD$_7$, so most machines have a one-bit register to capture this. It is normally set to 0 and changes only if there is a carry out of 1

Each adder, in sequence starting from bit position 0, takes the two corresponding bits from registers A and B and produces a sum value. The rule is very simple: $0+0=0$; $1+0=1$; $0+1=1$; $1+1=0$, carry out a 1 (as in the truth table in Fig. 8.4)! This is exactly the same thing that happens when you add, for example, two 5 s together in decimal (or any combination that generates a carry out). The lowest digit will be zero, and there will be a value of 1 carried into the next higher column. For example, if we add 1 to 1, we should get 2 (in decimal). And that is exactly what the rule produces. Similarly, the carry-in bit will either be a 0 or a 1 depending on the carry out of the prior pair addition. For the lowest bit pair, there is no carry in, but for every pair higher in order, there will be a carry-in bit of either 1 or 0.

The Arithmetic-Logic Unit (ALU) of a CPU contains a set of these adders where the carry out of each full adder is wired to the carry in of the next higher-order full adder. Addition of two binary numbers is carried out by starting with the lowest order pair and then computing the next higher order pair once the machine has resolved the carry-out/carry-in state.

It is rather amazing that from such a limited set of components and interconnections that so many useful circuits can be built. Logic gates can be wired together in many different ways. Circuits that operate like the full adder are composed of layers of gates and are called combinational circuits. The potential for developing circuits that can perform all kinds of computations (like the adder) is essentially infinite. It is because with each layer you can expand the circuit complexity combinatorially. Unfortunately, you also then slow the system down because the signals propagate from the starting layer through each subsequent layer and that propagation takes a finite amount of time. It is also the case that it is physically impossible to get more than a certain number of gates etched on a chip of silicon. Thus, there are practical limits to how much work can be expected from the hardware. To go beyond that point, you need software. Later in the book we'll explain how software comes into the picture to add new kinds of components to computing and new kinds of interconnections as well.

There is another important circuit that is needed in order to build a computational process, and that is a *conditional decision processor*. In this circuit, the objective is to test a condition, actually a relation between two (or more) input states, and to output one value if the condition is met and a different value if it is not. Figure 8.6 shows two versions of this circuit, one a very generalized form and the other a specific Boolean form using logic gates.

The important point here, however, is that computation, a way of combining states comparatively, is a universal principle. Recall Bateson's description of information as a difference that makes a difference. The notion of "difference" is inherently comparative, meaning it involves somehow bringing elements, states, or whatever together; rules, such as discussed above, construct the possible ways of doing this combining. The most fundamental of such comparison is binary, an either/or, on/off sort of difference, and we have seen above how Boolean logic allows us to build derivative sorts of differences from this foundation. So computational circuits handle binary differences and are the basis for information processing. From this, you might be able to understand why computer scientists don't always make the distinction between a message (or data) and information when

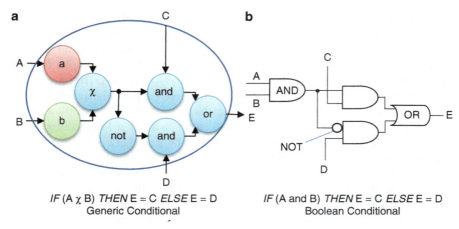

a

IF (A χ B) *THEN* E = C *ELSE* E = D
Generic Conditional

b

IF (A and B) *THEN* E = C *ELSE* E = D
Boolean Conditional

Fig. 8.6 A conditional decision processor is used to make a choice between two alternative outputs based on a condition being true or false (in the simplest case). (**a**) This is a generalized decision processor, which could be receiving any kinds of messages at inputs A, B, C, and D. It performs a conditional test, χ, on the two inputs, A and B. If the test is met, then the processor ANDing C and χ will send the value of C to the OR gate and thus output, E, can be the same value as C. Otherwise, E will be the value of D. (**b**) This is a Boolean logic gate implementation of the conditional decision processor using an AND gate for the χ operation. The circle placed on one of the inputs to the lower AND gate inverts the input value (NOT)

talking about what goes on in a computer (recall in the last chapter we noted how the two words are really not the same thing).

British mathematician and computer scientist (before there were practical computers!), Alan Turing, demonstrated the universality of computation with a thought "device" that has come to be known as the Turing Machine.[7] This machine incorporated all of the various elements that we have just described, registers for storing data, circuits for changing data, and conditional decision processing to produce an effective computer. Turing provided a somewhat more elaborate version of this basic set of these parts and introduced the notion of a "program" that would allow a machine to compute any computable function.[8]

The Universal Turing Machine (UTM) consists of a mechanism that slides along an infinite tape (the infinity solving the question of having to state any definite limit as determining computability). The mechanism includes a means of interacting with the tape and an internal set of rules for interpreting the markings and producing a result (just as the simple computation of outputs based on inputs described in the devices above). The tape is comprised of a series of squares (registers), each of

---

[7] See http://en.wikipedia.org/wiki/Turing.

[8] Though it will be beyond the scope of the book, it should be pointed out that not all functions that might be expressed are actually computable algorithmically. Below, we explore other kinds of computations that are less rigid than machine types (deterministic) that can approximate non-computable functions.

which can be marked with a one (1), a zero (0), or blank.[9] The tape can be read from by the mechanism, and it can be erased and written to as well. Some of the tape has markings representing the "program" and "data" and other parts that are blank. The machine works by the mechanism sliding along the tape, reading the mark in the square just below, following the interpretation rule(s) and depending on the rule(s) either erasing the mark, overwriting it, or simply moving to a blank portion and writing a result.

This seems like a very simple and limited process, but as it turns out, Turing was able to prove that this machine, with sufficient iterations of these steps, could execute what we call an effective procedure, which means the machine is capable, in principle, of running any algorithm for solving computable functions. Modern computers don't exactly work like this (they don't have infinite tapes for one thing), but they embody the principles of the UTM sufficiently to do practical computation.

With Turing's introduction of an effective procedure, we get the notion of a program where the operations to be performed can be arranged and sequenced through based on conditions of the data inputs. During the time when the first actual digital computers were being built, it was realized that a few very general hardware circuits could be used to perform any more complex operation by using the conditional circuits to sequence through which processes should execute and in which order. In other words, it was possible to use data inputs to tell the sequencer circuits which operations to perform based on the conditions in other data. One set of data, written beforehand in a memory, could be used to decide how to operate on another set of data, also in memory. This recognition led to what we call the "stored program" computer generally attributed to the work of John von Neumann.[10]

The invention of "software" was a major advance in digital computing. The term refers to the fact that the stored program can be changed as needed. This means programs could be fixed if they had an error (a process affectionately referred to as "debugging"). It also meant that many different kinds of programs could be run on the same computer by simply changing the stored program. A remaining problem, however, was that using numbers to represent instructions was a difficult cognitive task. For example, the instruction to add the two values that had been "loaded" into the A and B registers in Fig. 8.5 might have looked like this in binary: 10010011. That is 147 in decimal. Programmers had to remember all the codes for all the various instructions, including the instructions for getting the values into those registers in the first place. The data numbers had to be stored somewhere in the main memory, which is essentially a large array of registers. Each element in the array has an index or address. The data could be stored at a specific address, and an instruction would cause the wires coming out of that address to be connected to the input wires of, say, the A register of the adder. This is called a load operation, and a great deal of computer programming involves moving data to and from memory addresses to

---

[9] This third "state" introduces the notion of trinary (ternary) logic as opposed to binary logic. But all of the process rules still hold.

[10] See http://en.wikipedia.org/wiki/John_von_Neumann.

the various computational units (like the adder). A very simple piece of a program might be:

- Load register A from address 100.
- Load register B from address 101.
- Add.
- Store register C to address 102.

But in binary numbers, it looks like:

00001000 0000 01100100

[The first group of binary digits is the load instruction, the second group is the numeric address of the A register, and the last group is the binary representation of the decimal 100 address of the data.]

00001000 0001 01100101

[Another load instruction, this time designating the B register, and the last group is the binary representation of the decimal 101 address of the data.]

10010011

[Add the two registers together and put the result in the C register.]

00001001 0010 01100110

[Store the contents of the C register into memory location 102.]

These values, in binary as shown, are what is stored in the computer memory. The explanation in square brackets helps you keep track of what they mean. As you can see, remembering what these numbers mean, even being able to work with them in binary, is a daunting task.

Computer programmers were nothing if not persistent. As the sizes of memory spaces grew and the number of kinds of instructions did too, programmers figured out how to write programs that could take text representations of what the programmer wanted to be done and translate it into the correct binary codes for loading into the computer's memory. The first such programming languages were called "assemblers" since they assembled the machine codes from human readable mnemonic codes. For example, the above program could be written in *assembly* language as

```
LD      A, 100
LD      B, 101
ADD
STR     C, 102
```

The assembler program, the original version having been written in binary (!), would convert this human readable, if cryptic, code and convert it to the above binary (machine) code. Once programmers had a much easier time specifying program instruction sequences, it wasn't long before they wrote even more elaborate and sophisticated translators called compilers that would take a much less cryptic text and translate it down to something like the above assembly code and then let the assembler program convert that to machine code. Languages like COBOL (Common Business-Oriented Language), FORTRAN (Formula Translation), and C (C) started popping up and making life for programmers so much easier.

Today, software is produced in languages that implement the systems approach, called object-oriented languages (OOL). Languages like C++ and Java allow programmers to specify the structure and function of objects (subsystems) that can be integrated to form whole software systems. In this day and age, both hardware (such as above) and software follow the systems-oriented approach to design and construction. We will return to some of these concepts in Chap. 14.

Computer programs are developed from instructions by virtue of procedures called algorithms.

An algorithm is a sequence of unambiguous instructions, including instructions to repeat key steps, that is guaranteed to (1) terminate with a solution and (2) produce the correct solution on any instance of a problem. The first condition means that given a suitable (computable) problem, the algorithm will always terminate with a solution—given enough time (recall Turing's infinite tape!). The second condition says that regardless of how big the specific instance of the problem, the computation will produce the right answer every time. As an example, consider the problem of sorting a list of names into alphabetic order. The first condition means that a properly designed algorithm (program for our purposes) will always complete the mission. The second condition means that the algorithm, say a name sorting algorithm, will produce a list with all the names in exactly the right order, no mistakes, independently of how many names are involved.

If you give the algorithm (running as a program in a computer) a list of 100 names, it will get the job done correctly in the blink of an eye. If you give it a list, that is, one million names long, it will take a bit longer, but the result will still be correct, though you may not be inclined to check one million names to see.

Computer-based computation aka deterministic logic is based on coming up with an effective procedure as expressed in a suitable language. The language must include instructions for reading data from and writing data to various sources such as disk files, communications ports, or keyboards/video monitors. It has to have instructions for doing basic arithmetic and logic, e.g., adding, subtracting, performing AND and OR operations, etc. And it needs instructions for conditional processing (the "if" rules). All computer languages support these "constructs" as well as specifications for types of data such as integers, characters (letters of the alphabet), real numbers (or approximations of real numbers called floating point numbers).

The rules are embodied in the instructions for conditional processing. For example (keywords are in all capital letters),

IF x is true THEN $y \leftarrow$ true ELSE $y \leftarrow$ false is a rule that assigns either true (1) or false (0) to a Boolean variable y based on the truth or falsity of the variable x (remember Fig. 8.6). Variables, as used here, are similar to the components with binary states that we mentioned above. However, variables can be defined for all data types we might be interested in. Another example might be

IF $a > b$ THEN GOTO line 2000,

which means that if the numeric value of b is greater than the numeric value of a, then jump the computation to line 2000 (meaning a specific program address in the computer memory) where a different instruction will be found; otherwise, continue to the next instruction after this example.

The IF-THEN-ELSE rule allows for the kind of conditional computation that we saw embodied in the simplest rule above. All computer programs consist of these kinds of statements along with assignments (LET $a = 32$, sets the numeric value of a to 32), input/output (PRINT a), arithmetic ($c = a + b \times d$), and several other types that allow programmers to express any algorithm desired.

These very simple rules turn out to have extraordinarily powerful application. Early developers of computers thought of them largely in terms of number crunching abilities, as the term computation suggests. The notion that we would be using forms of these devices to watch movies, listen to our favorite music, or record endless pictures and videos of every aspect of life was nowhere on the horizon. But recall that information comes through differences and that computation is a way of processing differences, and the explosive development of a digital age becomes intelligible. As discussed above, our sense processes are making measurements, and the way they process data is translatable into and out of digital codes and computation processes. So in a few decades, the world of math and digitized computation has encompassed and unified the diverse informational flows which shape our daily lives and activities.

General-purpose electronic digital computers have shrunk considerably in size since the early days (say before the mid-1950s). Today, extraordinarily powerful computers are etched onto tiny silicon chips. The prices of computers continue to drop thanks to the effects of Moore's Law.[11] One way in which computers (and thus computational processors) are affecting everyone's lives in the developed nations is by being *embedded* into so many common machines. In a high-end luxury car today, you might find over 20 computer chips, each doing a particular task with respect to controlling the effectiveness of the internal combustion engine or the comfort of passengers inside the car. All of your modern electronic devices are controlled by these embedded processors.

Cheap computing is also responsible for the capabilities to communicate through the Internet. All storage, switching, routing, and other communications services are performed by computing devices embedded within the fabric of the Net. There is not one aspect of your life today that is not impacted to one degree or another by computational processes.

---

[11] See http://en.wikipedia.org/wiki/Moore.

> **Question Box 8.3**
> We use the distinct information channels of our senses to move and behave responsively in a space-time environment. Our manifold responsiveness to the world about us has always been a computed process. The digital revolution translates any sort of information into the single digital medium, capable of being stored, manipulated, and variously read out through increasingly miniaturized electronic devices. What are some of the ways in which this new computation augments, infringes on, or modifies the (always computed) way we move and behave and live in the world?

Algorithmic computation has become a vital part of our lives over the past five decades. Modern society could not function without it. Yet, as we marvel at the digital age, one of the surprising, and little appreciated, results of the study of computation has been to recognize that not all expressible problems can be solved by appropriate algorithms. It turns out that problems come in classes of solvability! The "easy" problems, like the sorting of names, can be solved readily and rapidly making their exploitation feasible and in many cases "profitable." Then there are problems that can be solved for small instances but not practically (i.e., in reasonable time) for large instances. For example, the traveling salesman problem asks whether there is a sequence of cities that a salesman could visit without having to go through a city she/he has already passed through while being guaranteed to visit every city on their itinerary. Think of an air flight itinerary that would cost more if one had to make a connection in a city that they had already visited. While this is a practical problem, it turns out that it is "problematic." There exists an algorithm that meets our criteria above that can solve the problem, but as the number of cities grows linearly (from say 30 to 40), the time it takes to solve the problem grows exponentially! A solution for 30 cities can be computed on a modern high-speed computer, but the time needed to compute a 100 city itinerary would take more time than the universe has been in existence!

Then there are problems that can be expressed in a human language but for which it is possible to prove (mathematically) that there is no effective procedure (algorithm) that can compute the solution. We can't go into the details in the scope of this book, but this issue is related to several other surprising results from physics and mathematics. In physics, the Heisenberg's Uncertainty Principle[12] tells us that it is impossible to know both the momentum and position in space of a quantum particle. In mathematics, Gödel's Incompleteness Theorem[13] tells us that a mathematical system, like arithmetic, can either be complete (i.e., provides truthful statement for all theorems) or consistent (you cannot prove that a false statement is true) but not

---

[12] See http://en.wikipedia.org/wiki/Uncertainty_principle.
[13] See http://en.wikipedia.org/wiki/Incompleteness_theorem.

both! Alan Turing (again!) provided another example of a non-computable problem known as the Halting Problem.[14] This problem asks if a particular kind of algorithm, given a particular input, will meet criterion 1 from above—will it terminate with a solution (never mind if the solution is correct!). It turns out that there are problems for which there is no answer to this question (which Turing proved).

All in all, these kinds of results demonstrate that there are real limitations to computing processes that depend on deterministic rules, as are found in axiomatic systems. One of the burning questions (philosophical as well as practical) in computation is: are these kinds of questions insurmountable?

From the standpoint of deterministic computation, the answer, so far, appears to be *yes*, they are! But are we constrained, in systems science, to think only of computation as a deterministic (which is to say axiomatic/algorithmic) process? Might there be other approaches in nature that get around these limitations? Happily the answer is yes, BUT. There is a substantial trade-off or price to be paid for pursuing other routes. The beauty of algorithmic problem solving or deterministic computation is that when there is a solution, it is *guaranteed* to be right (and hopefully quick). With the approaches we are about to discuss, this is not the case. However, it turns out that most systems in nature work more like the descriptions which follow below than like the above deterministic processes. Humans have learned to conquer the above with computer science (applied mathematics) and mechanical machines to do our bidding, and that provides a substantial benefit to our pursuit of solving problems. But systems science must also include a recognition of naturalistic problem solving. We will see this especially in Chap. 11 when we explicate the nature of evolution, what we might consider the ultimate problem solver regardless of problem type.

8.2.3 Probabilistic Heuristic Computation

Computation as a process requires time; it advances into a future to arrive at the solution. Algorithms composed of deterministic rules are good for advancing into determined futures. Determined futures are those that can be constructed in advance by rules, much like games. We do not know who will win the world series, but we do know how the game will begin, progress, and end, with every step including determining who has won covered nicely by a determined IF THEN kind of algorithmic procedure. If we give up the requirement that an algorithm have a guaranteed solution, if we are willing to accept a solution approach that generally works even if sometimes it makes a mistake, then we are suddenly cast into a realm of possibilities that corresponds with most natural systems. Such systems move into futures heuristically (from the Greek verb *heurein*, meaning "to explore"), for as we

---

[14] See http://en.wikipedia.org/wiki/Halting_problem.

all know, the future does not necessarily obey the familiar rules of the past. We will anticipate what will be covered more deeply in the next chapter. But for completeness in understanding the nature of computation and problem solving for information processing, we need to cover *heuristic computation*, which seems to be the way in which brains go about solving the kinds of problems living creatures encounter as they go about making their living.

We can actually introduce this idea from the perspective of computer-based problem solving as already covered above. Heuristics, in general, are what we could describe as "rules of thumb" or rules that work most of the time but are not guaranteed in the same way that algorithms are guaranteed. Because they are not guaranteed, good heuristics involve coming up with alternatives when what usually works does not. Our cats and dogs do this every day, but it is a real challenge for preprogrammed machines!

Classical logic differentiated two types of reasoning procedure, deduction and induction. Deduction links propositions with necessity. In its simplest three-step form, an example would be, "All birds have wings. X is a bird. Therefore X has wings." Induction takes the form, "The first bird has wings. The second through 100th bird had wings. Therefore bird 100+x has wings." Alternatively inductive reasoning might say "The next bird I see will have wings because all 100 birds previously encountered had wings." The necessity and power of mathematical reasoning comes from its deductive nature. But most of life is more inductive, where experience teaches us what is usually the case, so we have definite expectations, yet are not totally overwhelmed when it turns out some birds do not have wings (or maybe they all do—who knows?!). Probabilistic heuristic computing involves using rules that are approximate and nondeterministic. These are called probabilistic because they use probabilities to quantify the approximation aspects. But there are other ways of doing this as well. Below, we will see a heuristic approach that is more like what is going on in living brains.

Whereas algorithms deal with deductive rules, heuristics are useful in the realm of inductive and abductive inference. Induction involves logic that builds generalizations from multiple specific instances that seem to point in a general direction. For example, if every time we see an animal classified as a mammal, we note that it has hair, and then we might conclude (and use as a predictive rule) that all mammals have hair. This rule will fail for naked mole rats, but it will work most of the time.

The second form, abduction, is derived from induction. When we have constructed a causal chain based on inductive rules, we can also abduct, working backward, to infer a cause. For example, if we have developed (through induction) a rule that most drug abusers commit crimes of robbery to pay for their addictions, we might conclude that a known drug abuser has also committed robberies (remember Bayes formula?) or, even more loosely, that a suspect of a robbery case is a drug abuser. In the case of deduction, the latter proposition would be simply ruled out as bad logic, the equivalent of saying, "All birds have wings, X has wings, therefore X is a bird," which is possible but not *necessarily* true. Abduction works with probabilities and possibilities suggested by common (but not deterministic/necessary) causal linkages and so includes less probable as well as more probable inferences for consideration. Neither inductive nor abductive logics are guaranteed to produce true inferences, in the same way that deductive logic does. However, these forms of inference are incredibly powerful in the real world where causal chains often are involved in our observations of effects.

Heuristic rules tend to look a lot like algorithmic rules. They use the same IF-THEN-ELSE structure but with an important caveat. When we say

IF x THEN $y \leftarrow n$ (0.85) ELSE $y \leftarrow m$

If x is true then approximately 85 % of the time y should be set equal to n and 15 % of the time it should be set equal to m. A random number generator can be used to turn the "roulette wheel," so to speak. Another interpretation of the rule is that x is true about 85 % of the time it is tested. There are actually several different approaches to introducing probabilities into heuristics. One very popular approach in the field of artificial intelligence (AI) is a rule-based expert system, in which series of probabilities derived on the base on expert experience can be used to diagnose diseases or mechanical system failures. Rules along the lines of "IF the heart has stopped there is a 90 % chance that the patient is dead; test the breathing" (though not so blatant) are used to guide nonexperts through a diagnostic procedure. You can see how this is related to the inductive and abductive reasoning mentioned above.

These kinds of approximation heuristics are simulated on deterministic computation machines. However, even these simulations suffer from a certain amount of "brittleness" since ultimately the computation, a math procedure, has to be based on deterministic algorithmic rules.

In nature (see below), nondeterminism is built into the nature of the computation. All of the elements are nondeterministic in the true sense of that word. Computers can *simulate* such nondeterministic computations, but they cannot truly *emulate* them (meaning to work the same but in a different representational medium). To do so would require the embedding of truly random processes into the algorithmic process, which is feasible, but often not practical.[15]

In nature there is evidence that evolution has favored brains that have embodied hardwired versions of this kind of heuristic computing (with caveats to be explained below). Instincts, found in all "lower" animal life forms, provide a good example. Instinctive behaviors are governed by genetic controls over the development of neural circuits that cause an animal to respond to environmental cues (inputs) with programmed responses. These responses are not guaranteed to produce a favorable outcome, but over the evolutionary history of a given animal species, they have been found to generally provide a good outcome. Instincts are heuristic approaches to survival. They are tested by natural selection and if found wanting will eventually go extinct. Nevertheless, while proving fit for the animal's survival, such instincts represent a "quick-and-dirty" way to solve problems for which no algorithmic solution

---

[15] Computers use pseudorandom number generators to approximate stochastic processes. These allow us to simulate stochastic processes but not to actually emulate them. There are add-on devices that use, for example, line noise, to produce true random variable values, but these generally have a uniform distribution output and so are still of limited value in trying to emulate real-life randomness.

might exist. Because of the long period of time over which such instincts have been hardwired and the huge sample space used, these heuristics have a probabilistic appearance; that is, they seem to meet the laws of probability when executed by animals. However, we know that brains cannot represent real numbers or intervals from zero to one (inclusive) and do not do rigorous calculations. So they cannot, in any real sense, be doing the kind of math required, for example, to compute Bayesian posterior probabilities. It just looks like they are doing something like that in terms of instinctive behaviors. Real brains work on a much more general principle of computation that can, under the right circumstances (such as in the case of instinctive behavior), appear to be probabilistic or even non-heuristic or algorithmic (as when we do math problems).

Question Box 8.4
Gene pools appear at least partially to play a mathematically probabilistic game as they manage a species' advance into the future. That is, the genetic recipes that succeeded well enough to reach reproduction are represented in strict proportion to their relative rate of success. But what does rolling forward of random variations or genetic mixing by sexual reproduction do to this heuristic process?

Here, we circle back again to the difference between prediction and anticipation discussed above as ways of advancing into expected futures. We imaginatively program our machines with ranges of flexibility and alternative responses, but all of these are in fact only more complex versions of determined present thinking, a *prediction* of the future. The well-programmed machine may appear to make the flexible moves of an exploratory heuristic similar to the way an anticipatory system moves into the future, but the behavior is constrained to a predictable, rule-determined game, as it were. It is hard to see that this has to be the case in principle, but so far, it is the line that has not been crossed.

8.2.4 Adaptive, "Fuzzy" Heuristic Computation

Finally, we will look at a form of computation which is applicable to most of the higher forms of intelligent life on Earth, mammals and birds in particular. Extending the ideas we just visited in the above section, there are approaches to computation (information processing) that not only embody heuristic rules but also adaptive heuristic rules. That is, as adaptation takes place, to some extent, the rules themselves change, and this makes all the difference. This form of computation is effective, but not at all guaranteed, and effectiveness in the broad sense is a moving target insofar as circumstances of the future are always shifting—in part by one's very adaptation to them! Thus, instincts that generally serve their possessors well eventually need to be stretched and modified, or they become maladaptive, a loss of fitness.

Short-lived and massively reproducing organisms such as microbes and insects heuristically probe the future in forms hardwired by gene pools. But most longer-living and more moderately reproducing animals extend the notion of heuristics into a realm that allows modification to the rules based on actual experience in life. A rule, such as "look for good tasting food," depends on what exactly is meant by "good," an important but fuzzy guideline. The brain has a genetically encoded version of "good" tastes (or smells) that evolutionarily have served the species (and the species from which it came) well over a long history. But there can be variations on the exact nature of "good" if the species has a capacity to explore other possibilities (this means that the species can digest less than desirable foods that still supply nutrition). In this case, the individual animals need to be *able to learn* what variations do supply food that is nutritious. The reason this is needed by many kinds of animals is that it makes them more able to adapt to variations in the environment that attend, for example, climate changes. Animals that are locked into a single (or few) foods can suffer extinction if the food supply is exterminated by environmental changes.

It has been shown that it is possible to approximate emulation of this kind of computation with what is known as "fuzzy" logic. Formally, a fuzzy system is one in which some of the rules of probability (as in the previous section), such as the "excluded middle" (e.g., something has to be either true or false), are relaxed, allowing that elements can be partial members of different sets. That is, an element, a, can be 20 % in set A and 80 % in set B. This works even if set B is the complement set of A! For example, consider two characterizations of how tall people are: SHORT could be defined as everyone under a certain height and TALL would be everyone else. While this might sound good to someone who believes that everything should fit into hard defined categories, it doesn't always work in practice. Suppose you are 1 in. taller than the cutoff point but you happen to be in a team of basketball players that are all nearly a foot over that point. Are you tall or short? Your membership in the basketball team (presumably because you are really good at the game) suggests that you are considered generally taller than the average person and you are one inch taller than the cutoff. But compared with the other players, you might be given the nickname of "Shorty."

So, often, the category (set) that an element (like you) is placed in is not entirely clear or what we would call "crisp." Depending on the situation, you could be both tall and short at the same time.

Fuzzy computation allows a kind of approximation that is not strictly probabilistic. A probability of 85 % is still a masked either/or proposition, not a fuzzy both-and or more-or-less proposition. Fuzzy heuristics actually provide ways to smoothly transition from one set of conditions to another, and these systems have found wide usage in complex control systems. It appears that our brains use some kind of fuzzy heuristic approach to problem solving even though, oftentimes, we would like to believe we are rational decision makers, in the crisp logic sense.[16]

---

[16] In the field of psychology, a considerable amount of work has been done showing that human beings do not usually make rational (logical) decisions or judgments but rather use a discernible set of heuristic rules for most judgments. These heuristics can and do lead to consistent and detectable biases and result in judgmental errors. An excellent compendium of the state of research on this can be found in Gilovich et al. (2002).

In the prior chapter, we described how a system that represents probabilities for message states can use a Bayesian-like computation to alter its expectations, which is a form of learning. With fuzzy set theory and fuzzy logic, we can extend these ideas to include non-probabilistic learning. It now seems that neurons are able to represent fuzzy approximations by virtue of their varying rates of action potential firing and their ability to link up with neurons belonging to different categories or concepts (Chap. 2). This involves a linkage of immediate processing with the deep accumulated stored patterning (expectation) we have described as the knowledge base, the context for interpretation. Neuronal networks (i.e., the kind in real living brains) do not compute in either the deterministic (algorithmic) or the probabilistic heuristic methods, yet brains obviously produce computational solutions. Indeed, brains can solve problems that we do not presently have a clue how to solve with either deterministic or probabilistic heuristic computation. We can now incorporate fuzzy logic into programs that adapt the sharp on/off true/false dichotomies of mathematical computation to the gradual and indistinct transitions that characterize most real-world processes. But brains (and the neurons from which they are built) still do even better than that. They have the capacity to construct new fuzzy heuristics.

8.2.5 Biological Brain Computation

We will finish this discussion of computation by looking at how real biological neurons and networks appear to work in computing information necessary for life to be successful. For this, we will first take a look at a fairly simple neuronal network that allows an animal to learn (encode in a memory trace of) a conditional relation between a stimulus that is always meaningful (say the presence of food) and one that has no necessary significance with respect to the meaningful stimulus but comes to be associated with it, playing the role of a cue stimulus that will prime the animal for action. In the next chapter, we develop the theory of anticipatory response in the context of control theory. We use Pavlovian (also called classical)[17] conditioning as an example and show how that form of anticipation greatly improves the fitness of the animal by lowering energy and material costs associated with responding to stimuli.

After considering how neurons compute, we will turn to more complex networks that comprise the cortical structures of vertebrate animal brains to see how, especially in humans, perceptions are identified and concepts are constructed (learned). We will finish this section by considering some possibilities for how brains build networks that act as models of systems in the world (Principle 9) and models of one's own self (Principle 10).

---

[17] We will save the details for the next chapter but curious readers might want to take a look at the Wikipedia article: http://en.wikipedia.org/wiki/Classical_conditioning.

8.2.5.1   Neural Computation

Here, we will take a look at the neurobiological basis for learning and computing a causal relation such as in classical conditioning. We need to start with the way in which neurons record a memory.

8.2.5.1.1   Synaptic Potentiation[18]

Neurons communicate in a one-way channel whereby an output from the sending neuron or sensory organ, a pulse of excitation called an action potential, travels along a long extension called an axon (see Fig. 8.7). It arrives at a terminal junction with the receiving cell (a synapse) called a synaptic bouton. The arrival of an action potential causes the release of a chemical called a neurotransmitter from the presynaptic membrane of the bouton. The neurotransmitter diffuses (rapidly) across the very narrow gap and the molecules attach to special receptor sites on the postsynaptic membrane of the receiving neuron. Those receptors activate channels through the membrane and allow ions in the surrounding liquid space to enter the postsynaptic compartment. This raises the excitation of the postsynaptic membrane and that excitation then spreads out from the synapse along the dendrite and cell body.

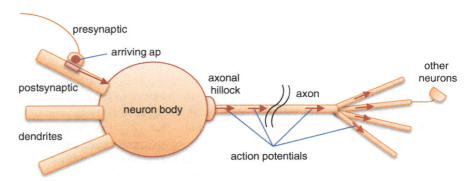

Fig. 8.7 Neurons process signals called action potentials, which are pulses of excitation that travel along the cell membrane. The signal arrives from a sending neuron at a synaptic junction, here shown on a structure called a dendrite. If the receiving postsynaptic membrane is sufficiently stimulated, the signal transmits along the cell body membrane and reaches the root of the axon (the outgoing "cable"), called the axonal hillock. This piece of membrane acts as a threshold trigger. If the sum of the excitation reaching it is high enough, then the cell will produce another action potential that will travel out along the axon to other neurons. This figure shows a history of action potentials that were generated by previously received incoming action potentials

---

[18] Readers interested in neural processing are directed to a very comprehensive reference: BarrsGage, NM (2007). Chapter 3 gives a detailed description of synaptic processing.

Synapses can act like signal filters. They do not necessarily reach sufficient excitation to send the signal on to the cell body with just a single or short burst of pulses. Their "willingness" to do so is called their efficacy, which is a variable that changes value depending on the history of recent excitations. Synapses of this type are said to be "plastic," (adaptive to change) in that they can become more efficacious with a higher rate of pulse bursts received. This is called activity-dependent potentiation. A synapse that has been recently, strongly excited for a period of time, even after a period of non-excitation, will remain potentiated, but with an exponential decay curve. The level of potentiation is the result of the frequency of incoming action potentials and the concentration of certain ions (especially calcium) that have entered the postsynaptic compartment. The synapse behaves like a leaky integrator over time.[19] The higher the frequency of the input pulses, the larger the "charge" of ions that accumulate. The concentration of those ions, in turn, drives a cascade of biochemical reactions, each step operating over a longer time domain that makes the postsynaptic membrane increasingly sensitive. However, since the ions are actively pumped out of the compartment (the leaky part), the drive for those reactions is reduced over time so that the amount of sensitivity—the potentiation level—is limited within a range.

If a new signal (say a short burst of pulses) arrives after a period of time has elapsed since the last burst, but before the potentiation can decay back to a ground level, then the synapse is more likely to reach a level of excitation that will travel all the way to the hillock. And, if this happens, sufficiently many times that level of potentiated excitation may exceed the hillock threshold and produce an outgoing action potential or even a short burst of them (as in Fig. 8.7).

The potentiation of a synapse is the effective recording of the memory trace of recent activity. Depending on the exact type of synapse (and cells), it may retain a trace of potentiation for some time (say seconds) after a short burst over several milliseconds. In some cases, some kinds of synapses will retain their higher potentiation for longer periods if the incoming signal is repeated over a more extended period, say short bursts of action potentials every few seconds.

Consider the axonal output signal as a response to a synaptic stimulus input signal. The general rule is the stronger the stimulus, the stronger the response. For example, suppose the stimulus is coming from a skin pain sensor in a finger. Suppose the response activates a neural circuit that reflexively pulls the hand back. If you put your finger on a hot surface, a strong pain signal is sent to the mediating neuron. It in turn sends a strong signal to the motor response circuit to cause the movement. If the surface is only warm, then the signal will be weak and the mediating neuron will not pass the signal on to the motor circuit for action. In this particular example, the synapse will perhaps record a slight trace of potentiation just in case you put your finger back on the hot surface (you will respond much quicker in that case).

---

[19] A capacitor, an element in electronic circuits that stores a charge, is an example of a leaky integrator. See http://en.wikipedia.org/wiki/Capacitor.

Actual memory trace recording in brain cells involves a bit more than simple activity-dependent potentiation. Memories, as we normally think of them, are associations between neural clusters that represent things in our world.[20] Therefore, there needs to be some kind of associative encoding scheme that is a bit more long lasting.

8.2.5.1.2 Associative Potentiation with Temporal Ordering: Encoding Causal Relations

We will now see what Montague meant in the opening quote when he claimed that biological (neural) computations care.

In classical conditioning, the idea is to associate a previously meaningless stimulus (like the ringing of a bell) with a very meaningful stimulus (like the serving of food) such that the prior becomes a cue event that primes the animal to take more proactive action. In the language of classical conditioning, an unconditioned stimulus (UCS) causes an unconditioned response (UR). In the case of Pavlov's dogs, the presence of meat would cause the dogs to salivate and Pavlov could measure the amount and timing of that instinctive response. The UR is automatic with the onset of the US. If the experimenter rings a bell just prior to serving a hungry dog some food, and does this over many trials, then the dog comes to associate the bell with the presence of food and will begin to salivate just after hearing the bell. The bell is the conditioned stimulus (CS) and the salivating with the bell (and before food is presented) is the conditioned response (CR). Before this regular pairing, bells have no meaning vis-à-vis the presence of food. Yet after a number of trials where they are paired in the proper temporal order and duration, the bell comes to represent a cue that food is on its way. Figure 8.8 shows a neuron at three points in time as CS and UCS signals are processed to produce a long-term potentiation (LTP) of the CS synapse.

---

[20] Here is where real brains work quite differently from classical artificial neural networks (ANNs) based on what is called distributed representation work. In the latter, representations are encoded in a distributed fashion throughout all of the neural connections. Each synapse has a weight value corresponding to the notion of synaptic efficacy. But every synapse participates in every memory trace. In these ANNs, the whole network encodes every pattern and requires extensive training to get all of the weights just right. In real brains, we now know that neurons and clusters of neurons represent patterns and objects.

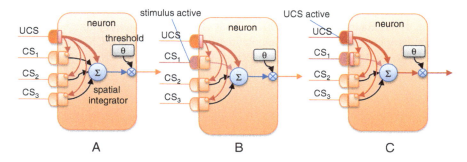

Fig. 8.8 This sequence of events shows how a CS synapse can be long-term potentiated by a UCS input following the excitation of the CS (CS₁). See text for details

The figure above shows a schematic view of a neuron and its computational functions. In Fig. 8.8a, we see the situation just before signals start coming in. The UCS synapse is a special, nonplastic synapse that if excited will generate an action potential output from the cell. Such synapses and their relation to plastic synapses (CS₁–CS₃ in the figure) have been studied in both invertebrate and vertebrate models.[21] The reddish postsynaptic patch of the UCS synapse represents the fact that it is already and always efficacious in generating an action potential. In the figure, the thick (dull) red arrow from this patch represents this fact. Even a small burst of action potentials can immediately generate a strong enough signal to activate an output of action potentials.

The blue circle with the Σ symbol represents the spatial integration accomplished by the cell membrane, as discussed before. Input signals from all of the synapses are continually summed to determine whether or not an output action potential is to be generated (the θ symbol represents the threshold above which the integrated excitation of the membrane must reach to generate an output AP). The thin red arrows from the UCS to each of the CS synapses represent a special effect that the activation of UCS can have on the postsynaptic compartments of each of the CSs. In panel a, there are no current inputs. The black arrows from each postsynaptic compartment indicate no contribution will be made from them—they are not sufficiently potentiated even if a burst of action potentials arrived at any one of them.

In panel b, CS₁ input has been activated (dull red arrow). This is still not sufficient to contribute to a spatial summation that would exceed the threshold. So no action potential is generated from just this input alone.

In panel c, a short time after the onset of the CS₁ input, the UCS signal arrives and activates the cell to produce an action potential output (brighter red arrows). It also sends signals to the postsynaptic compartments of each of the other CSs, representing a broader web of potential associations. Since CS₁ had already been activated just prior to the arrival of the UCS signal, the effect of the UCS is to gate the activity-dependent potentiation at CS₁ into a longer-term storage in the compartment (reddish patch in CS₁). The details of how this is accomplished are

---

[21] See Alkon (1987), ch 16 for a thorough description of the model.

beyond the scope of this treatment but can be found in the bibliographic references (Alkon, 1987; Baars and Gage 2007). What is important to recognize is that the biochemistry involved in gating the potentiation only works if the CS signal arrives just prior to the UCS signal. If the reverse is the case, then the CS compartment is prevented from achieving a longer-term potentiation. It is this feature that produces the encoding of a causal relation.[22] *Post hoc propter hoc* (after this therefore because of this) has long been recognized as a logical fallacy, but it seems to have sufficient heuristic value to have gotten itself wired into our brain processes!

In this interpretation, the CS is assumed to happen a short time before the UCS and so must somehow be associated with the "cause" of the UCS. Of course this isn't necessarily true in any kind of objective sense. What it really means is that there is some chance that whatever triggered the CS signal (recognition of a bell ring) is causally associated with the onset of the UCS (presence of food). Moreover, a single occurrence does not make a pattern. The longer-term storage of the potentiation at the CS will, itself, decay with time unless the pairing, in the right temporal order, is repeated many times. Each time the CS-UCS pairing occurs, the potentiation at the CS compartment strengthens and is gated further into even longer-term storage. With sufficiently many such pairings, the CS will become efficacious to the point that it alone is capable of generating output action potentials—the CR (salivation activation). The CR will occur even without a UCS signal following.

The plastic increase in CS potentiation, however, is not permanent. The encoding mechanism does not assume that because two things were associated for some longer period of time (e.g., say the dogs heard the bell prior to being fed for 2 weeks) that the association will continue indefinitely into the future. Obviously, in the case of the bell-food pairing, this was totally arbitrary insofar as the normal events surrounding getting fed and the experimenter can stop the experiment at any time—the dog has no way of knowing this. So the CS synapse potentiation will decay if the pairing is not continued. The rate of decay will be very slow as compared, for example, with the rate of decay mentioned above in the context of the leaky integrator. The dog will continue to salivate upon hearing a bell for several days or even weeks after the experiment stops. If the bell is rung at random times and never paired with following of feeding in the time window required for association encoding, then the association will slowly fade away until the ringing of the bell no longer produces salivation. What can be learned can be forgotten!

That isn't, however, the end of the story. It turns out that a non-detectable remnant of a memory trace will remain for a very long time. The potentiation of the synapse has decayed down below the threshold level, but there is still a very long-term trace remaining. This is shown to be the case because if the experiment of careful pairing of bell and food is reinitiated after the memory trace has "extinguished," the dog will start showing a CR much more quickly than it took to get that response in the first set of experiments. In other words, the synapse retains a small expectation that if this association was valid at one point in time, it might become valid again so don't completely forget it!

---

[22] Mobus' PhD thesis (unpublished) provides a complete analysis of this model. A book chapter in Levine and Aparicio (1994) was taken from that thesis. See Mobus (1994).

> **Question Box 8.6**
> Associative before-after causative thinking carried to excess is sometimes called "magical thinking." What sorts of threshold might be advisable in how seriously we take associations that, for reasons just discussed, keep popping into our heads?

Encoding association traces in neurons is the fundamental basis of all memory phenomena. Even in extremely complex brain circuits, such traces are what link various sub-circuits together to create associations between percepts and concepts. We will now take a look at these more complex neural computations as they apply to the kinds of cognitive processes that we experience, from recognizing objects in our environment to building conceptual models of how all of those various objects behave and interact in the world.

8.2.5.2 Neuronal Network Computation

In Chap. 3, we briefly discussed the nature of pattern recognition in the context of how we recognize objects as organized wholes. Here, we will describe the current understanding of how animal brains (vertebrates generally and primates, including humans, specifically) perform this computation using the above neuronal trace encoding to establish circuits and causal relations in larger networks.

8.2.5.2.1 Cortical Subsystems

The cortices of higher mammals (e.g., the cerebral cortex) are extremely complex structures. They are essentially sheets of neural tissues organized into relatively discrete modules called cortical columns. These modules are called columns because the sheets are multilayered with the modules running vertically across these layers. The details are way beyond our scope here; they can be found in Baars and Gage (2007): 126. We will treat these modules as subsystems that are, themselves, CASs (of course you should by now realize that neurons are CASs and, indeed, synapses are CASs, so we have CASs made out of component CASs!). This will be an abstract, simplified version of what you will find in the cited references.

8.2.5.2.1.1 Feature Detection in Sensory Cortex

As discussed in Chap. 3, features are elemental constructs that can be used to build a sensory representation. For example, in that discussion, we showed that a larger visual pattern is composed of an array of relatively small features (line segments). The array structure sets up the relations between the features.

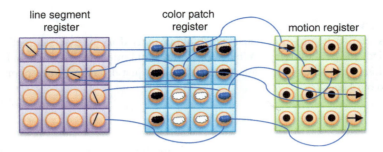

Fig. 8.9 Arrays of cortical columns are used to compute spatial relations between several visual sensory components. Signals coming originally from the retina(s) are broken up into these component modes and sent to aligned registers in the occipital lobe. This view is as if we are looking down onto the cortex from above; the *orange circles* represent the columns and the *arrays* represent the topologically registered "addresses" of the columns. The *blue lines* show the correspondence mapping between columns. The visual system is processing a moving *blue-colored curved line* against a *black* and *white*, stationary background (the *dots* in the motion register represent no detectable motion, whereas the *arrows* represent motion in the pointed direction for that column)

More broadly, the sensory cortices in the brain form such arrays of columns. For example, in the visual cortex (the occipital lobes at the back of the mammalian brain), signals arrive from structures deeper in the brain sorted into feature types such as lines, color patches, motion, etc. (Fig. 8.9).

Figure 8.9 shows three kinds of feature detection arrays. These registers are also called feature fields. There are registers like these associated with every location in the visual field (or other sensory modalities as well). They are laid out on the two-dimensional map of the visual cortex in a particular region. Other regions of the primary visual cortex are responsible for analyzing the detected features for relational features, such as shown above where the line segments are found to be contiguous (suggesting a long, single-line curve) and all moving in the same direction. The job of the feature detection area of the cortex is to identify which features predominate in various areas corresponding with the visual field of the retinas.[23] Those predominant features are then passed to a secondary area of visual cortex where the computation attempts to integrate the various feature types in preparation for identifying the composite features as a recognizable object.

---

[23] The mapping from retinal locations, i.e., the two-dimensional array of photosensitive cells in the retinas, is retinotopically preserved. See Wikipedia article on retinotopy for more details: http://en.wikipedia.org/wiki/Retinotopy. Other sensory modalities have similar topologically preserving layouts, e.g., touch sensors in the skin. Auditory sensing requires a transformation of complex audio signals that breaks the signal into its frequency components. This is accomplished by the hair cells arrayed along the cochlea of the inner ear, where each group of hairs along the linear array is responsive to a specific frequency (range). Primary auditory cortex then uses a very similar mapping scheme to preserve what is now called tonotopic registration. Auditory features also include amplitude of the various frequencies composing a tone at a specific instance in time.

8.2.5.2.1.2 *Perception: Feature Integration in Association Cortex*

A secondary area in sensory cortex is responsible for integrating these features and passing the information on to yet other areas that do a higher level of pattern recognition (as discussed in Chap. 3). The cortex processes data from many micro-feature detectors combining those that seem to be causally linked (using the same associative causal inference method discussed above). The resulting macro-features (e.g., object outline, colors and textures, etc.) are then combined.

Before patterns can be recognized, however, they need to be learned. In Fig. 3.21, we showed a two-level neural network in which the feature fields (as in Fig. 8.9) map to specific neural clusters. This mapping results from repeated exposure to the patterns of association that are encountered in the naïve observer (e.g., in an infant). The mapping settles into long-term associations based on the associative encoding discussed above. In Fig. 8.10, we provide a three-level association mapping where micro-features taken from various feature type fields excite specific macro-feature representing cortical columns (more likely a cluster of columns). In turn, the macro-features map to objects based on the repeated exposures associating those macro-features. When a naïve observer starts life, there is an exuberance of axonal connections from low levels to levels above. Many, even most of the micro-features, may be connected to many of the macro-features. But as actual observation experience proceeds and reinforcement of the type discussed above strengthens specific synapses, the nonparticipating connections are lost (reabsorbed by the sending neurons) leaving just those connections that code for the higher-level feature or object.

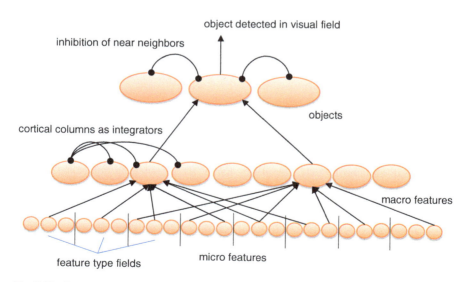

Fig. 8.10 Objects are identified in terms of the aggregate of micro- and macro-features currently active and exciting the level above. This mapping has been learned by the causal association mechanism discussed previously. The *black arrows* represent very long-term potentiated synapses

It is important to note that while we show these columns as levels in a hierarchy from micro- to object-level detectors, these different levels are actually different patches laid out on a two-dimensional cortical sheet. The microlevel represents the primary sensory cortex, the macrolevel represents feature integration in secondary cortex areas, and the object level represents the object detection tertiary cortex. In general, this layout of patches of primary to higher levels of sensory cortex proceeds from the posterior areas of the brain lobes responsible for that modality toward the anterior areas. For example, the primary visual processing columns are found at the very back of the occipital lobes (back-most lobes). Integration of simple features to complex objects and their relations in the sensory fields increases as we move toward the anterior-most portions of the lobes.

In the figure, we show another aspect of neural computation that is very important with respect to preventing misidentification of nearly alike stimuli. Clusters of columns tend to learn somewhat similar associations, but where there are differences in contributing features, there has to be some way to let the cluster that comes closest to identifying the compositions (called a best-fit excitation) to inhibit the neighboring clusters that would compete for activation. So each cluster of columns that code for unique objects (or features) has output processes that send inhibitory signals to its neighbors. Then when excited from below, the various clusters can compete, and the winner, the cluster best matching the fields of features that are most active, will have a stronger output, thus inhibiting the others so that it becomes the one most active in its output.

Question Box 8.7
Levels of experientially learned associations characterize this process from synapses on up. This ability to make associations must have been selected for and honed in evolutionary time because it made organisms more fit. What is the fitness utility in being able to make associations?

8.2.5.2.1.3   Object Conception: Percept Integration in Association Cortex

Figure 8.10 includes a higher level that integrates many macro-features into sensory objects. In reality, there are probably many levels just getting to objects that we would be able to consciously identify by name. Every object in our world that we can interact with consciously is a hugely complex composite of micro-features and their various spatiotemporal relations. Human perception is remarkably capable of distinguishing objects that are in a similar category from one another on the basis of tiny differences in feature relations. All ordinary human faces have pretty much the same set of macro-feature types, e.g., nostrils, eyelashes, etc. But what makes every face different is the large number of combinations of micro-features that can go to make up those macro-features.

As we showed in Fig. 8.10, close neighbors in clusters of columns coding for kinds of macro-features and kinds of objects actually have a great deal of similarity in terms of which clusters in the next lower level contribute to their excitation. In other words, super clusters code for categories of things. This is the basis for the human brain's capacity to categorize objects and relations. There is, for example, a category called "face" that is a cluster that is activated whenever a sufficient number of face features are present in the visual field and sending excitatory signals upward.[24] The category includes various kinds of subcategories and sub-subcategories along the lines we described in Chap. 3 regarding the nature of objects. These nested subcategorizations lead to the identification of a specific person's face without the need to have the codes for "faceness" and old faceness, wrinkled, etc. repeated with every old person's face we recognize. Thus, the brain achieves an incredible efficiency by "reusing" codes at multiple levels. This is also the reason we can all recognize a happy face ☺ as a face. It has the minimal number of features that cause our face neurons to fire. We can recognize faces of all kinds of animals and they all cause this single cluster to activate.

Interestingly, some of our computer encoding algorithms, say for image files, are starting to employ somewhat similar approaches to save space and time.

Figure 8.11 shows a bit more detail regarding the nature of the level-to-level processing that occurs in the sensory cortices. We show only the features that directly activate the object cluster. As with the previous figure, the features are activated from actual sensory detectors (not shown) and, in turn, activate the object cluster that is the best fit. That cluster then sends its signal upward to yet higher levels (to be discussed below) where more complex concepts are encoded. But in addition, this figure shows an interesting feedback loop or, rather several loops, one between the object detector and the feature level and one from the higher concept level downward.

Very often some features that contribute to a given clear identification might be occluded or distorted by environmental factors (e.g., looking at a face through a dense fog). Even so, if there are enough features activated, we can still recognize the object. This is accomplished by the object detector (after subduing its competitors) achieving a weak output. However, that output goes not only up to the higher levels, it also activates a very sensitive local helper neuron (or cluster) that sends excitatory signals back downward to all of the feature detectors that contribute to the object's recognition. That is, this helper makes sure that all of the features that should be active are so as to improve recognition. The success of this scheme depends on there being enough features activated to get the loop activated, even if weakly, as long as

---

[24] "Face neurons" have actually been detected in human brains! Using a procedure called unit recording (listening to individual neurons or small clusters for higher than background activity), neuroscientists have determined that there really are neurons that get excited with either the presentation of specific celebrity faces (like Bill Clinton's) or when the individual was asked to think about the celebrity. This along with many similar findings from animal recordings have revived the theory of neuron encoding (presented here) of features, objects, and concepts, as opposed to the theory called distributed representation (Rumelhart and McClelland 1986, ch 3). For a quick description of how individual neurons can encode specific objects, like grandma's face, see http://en.wikipedia.org/wiki/Grandmother_cell

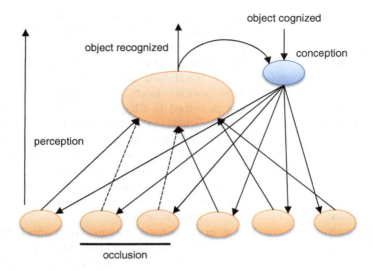

Fig. 8.11 Once an object identification from its component features has been learned, it is possible for the object to be identified even if some of the features are occluded and are not contributing to the excitation of the object detector (*dotted line arrows*). The helper, which has learned to fire when the object is recognized (*arched arrow*), sends signals to all of the feature detectors, activating the ones not originally participating. Since the others are already participating, there is no further effect on them

there is the slightest non-ambiguity with other competing objects (through the local inhibitory network in Fig. 8.10). Once the object detector is excited, the helper will activate the missing features thus completing the features. This has an interesting benefit that has been a great mystery in neuroscience until quite recently.

When you are asked to think about someone you know (like Bill Clinton in the above discussion!), you invariably can picture that person in your mind. If you close your eyes and think about them, you can sometimes picture them clearly. When we dream, we often see seemingly clear images of people we know. How does the brain do that? This is not easy to answer. Computers are able to store images in files, and when those images are needed, they are literally copied into main memory for processing. Surely the brain does not store static copies of images which are then copied into working areas when called to mind. This would present an enormous storage size problem for biology!

The answer is hinted at in Fig. 8.11. As fairly recently determined in neuroimaging experiments on cognition of images, it turns out that the same neural clusters that are used in sensory perception are used in conceptual imagining. That is, when you imagine someone's face, the same neurons that are activated when you are actually looking at them are activated in a top-down fashion (the downward arrow in Fig. 8.11). The brain reuses the feature/object detection neural networks to create images in our minds. This would be like somehow having direct processing access to the static image file on the hard drive of a computer without any copying.

It gets even better. The helper neuron could be activated from any number of higher-level concepts, especially those that have established some kind of associa- tion between the specific object and many other concepts. You have experienced thinking about something and then find yourself thinking about something else entirely and realize you got to the second thought because it is, to some degree, related to the first thought. For example, you might be thinking of your favorite song briefly only to find a second song by the same artist comes to mind.

The top-down activation circuits are part of the brain's ability to recall memories stored in the cortical structures. This includes everything from episodic memories (specific events) to implicit or tacit memories, those that do not involve conscious recall per se. Everything that contributes to a memory of events, people, things, rela- tions, etc. is stored distributed throughout the cortices all the way down to the specific sensory contributions. The efficacy of this recall capability is a contribution to what we call general intelligence (as in IQ tests), but it might also be the case that if we have not adequately constructed the needed level-to-level upward connections that we will not be very good at recall. Now you know why repetition is an essential part of memorization. It takes time and repeated exposures to the same patterns of activity to establish both the upward (perception) connections and the downward ones (recall).

Question Box 8.8
Knowing the brain's associative architecture of memory, how would you explain two person's different memories of events both witnessed? What about strategies for memories we would like to forget?

8.2.5.2.1.4   Relation Conception: Object Integration and Causal Direction

By now, you should be getting the idea that the vast complexity of the brain is actu- ally based on a few simple mechanisms. What makes the brain complex is that there are numerous variations on these mechanisms that permit huge differences in how the neurons process various kinds of data. The basic multi-time scale potentiation of synapses has the same kinds of dynamics for all neurons. But variations in a large number of parameters such as specific ion channel proteins in different neuron types (there are hundreds of cell types!) can make tremendous differences in the details of timing in those dynamics.

The wiring diagram of cortical columns is basically similar throughout the vari- ous cortices (e.g., cerebral, cerebellar, cingulate, hippocampal, and several other "minor" structures). We see very similar cell types, such as pyramidal cells doing the basic integration work, with numerous kinds of helper types, long-range com- munications substation cells, and many more. But the basic theme is preserved across all of these structures. Cortices process patterns in various hierarchical levels of complexity.

We have been mostly talking about spatial patterns so far. But temporal patterns are also important. Objects, taken by themselves, relate to nouns in language. They are things. They have names, both generic ("dog") and proper ("Fido"). But things move and affect one another. They have positional and affective relations with other objects that must also be encoded. As it turns out, the brain (the cortices) uses the same basic tricks for encoding relations and changes in relations (verbs) as they use for encoding objects. That is, there is a two-dimensional mapping that allows neural clusters to organize relations in the same way it organizes things, i.e., by categories. The details would end up repeating some of what we have covered above, plus there are a few more complications that require knowledge of some advanced mathematics. So we will leave those details to your advanced study of neurobiology!

8.2.5.2.1.5 *Mental Models*

The payoff is that big brains like ours have evolved the ability to encode complex models of things in the world and how they work. In other words, we have the ability to build models of systems in our minds. This ability is crucial, as we will see in the next chapter, in that models can be used to anticipate the future. Running a model of the world in fast forward means thinking through what might happen to us under different starting conditions and scenario evolutions. The more knowledge we have gained about the world and its workings, the more likely our models are efficacious in providing realistic scenarios

Brains exist to manage the body and the body's interactions with the world it finds itself in. It does this by learning and storing knowledge of all kinds for use in moving successfully into the future. The human brain has reached a significant level of sophistication in being able to not only learn the world but to communicate that learning to others and to ask "what-if" questions wherein we try out new scenarios and new configurations of matter and energy—we invent stuff. Recent research suggests that (in most people) the left hemisphere of the brain builds and stores our basic models (concepts of all degree of complexity) as we learn through experiencing the world. The right hemisphere acts as a novelty detector and novelty generator, basically trying new, never-before tried combinations of concepts which can be tested, a process accomplished mainly in the prefrontal cortex—the judgement and executive control centers. What if there was a horselike creature that had a single pointy horn sticking out of its forehead? You can put together your memory of a horn and of a horse and see if it resonates! If motivated by the need to tell a particular kind of story,[25] then it certainly might.

---

[25] Storytelling is really what the brain does. We experience the world as a sequence of events that flow, generally, from one to the next over time. When we try to communicate with others and with ourselves, we actually construct stories. One important aspect of stories is that they take on meaning based on the affective quality of the events and their content. A unicorn might elicit a feeling of mysticism since no one has ever seen one, but the idea that it could exist in a different place or time helps generate a sense of wonder.

Question Box 8.9
We ordinarily know the difference between imagining and remembering, but sometimes the distinction is blurred. What is the difference, and how do you suppose we generally are aware of it, but sometimes not?

8.2.5.3 Other Biological Computations

Neural processing of data to produce actionable information is by no means the only form of biological computation. The hormone (endocrine) system, the immune system, and indeed even the digestive system all process input data (chemical concentrations, antigens, and food composition, respectively) and produce information that causes something important to happen. Such computations are purely analog in nature (neural processing is a hybrid between discrete and analog!) but computations nevertheless.

8.3 Purposes of Computation

In the above, we discussed various kinds of computational processes, arranged in terms of certainty of outcomes. That is, algorithmic computation, when it is applicable, produces certain results. Heuristic (probabilistic) computation generally produces good results but with much less certainty. Finally, adaptive, fuzzy, heuristic computation is a process that adjusts itself (its rules) in accordance with experience. It is the basis of learning behavioral modification in animals.

In this section, we will consider the uses to which computation has been put in general, mostly in terms of localized processes within larger systems. In the next chapter, we will provide a much deeper understanding of the role of computation in systems where the objective is to allow that system to perpetuate into the future and find a stable, sustainable function that keeps it viable into that future.

8.3.1 Problem Solving

The use of computation that most readily comes to mind for most people is that of solving problems. It turns out there are many kinds of *problems* that are solved with computation. Here, we survey just a few and discuss how various kinds of computation (as described above) work for solving them. Figure 8.12 shows a generic diagram of problem solving. A problem is represented as an abstract and unorganized set of subproblems. The job of computation is to sort things out into a

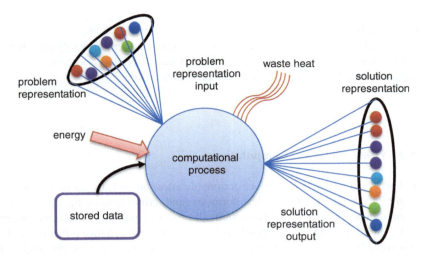

Fig. 8.12 A computational process takes, as input, a representation of a problem, for example, as messages from its environment; uses energy and, perhaps, stored data relevant to the problem domain; and generates a solution representation as output. The solution is presumed to be meaningful to a subsequent process (not shown)

more ordered set so that a subsequent process can make use of it. When we say we need to "figure something out" or "see how it all fits together," it points to the computation dimension of problem solving, even when the matter at hand seems far removed from the numerical domain we associate with computation.

8.3.1.1 Mathematical Problems

Clearly, the most familiar notion of a problem is the kind we encounter in mathematics. For these we rely today on digital computers (like the one embedded in your calculator). As we described earlier in this chapter, computers are supreme at doing arithmetic. This is because the system of arithmetic starts with counting, addition, and negation. The first is just a form of the second. And subtraction is made possible because we can operationalize the last one and then apply addition. Multiplication and division are procedural applications of addition and negation (do you remember how to do longhand division?)

But mathematics involves a good deal more than just arithmetic. It turns out that with high-level computer languages, we can easily describe complex arithmetic expressions, for example, $a = c \times (b + d) - 32$. Using variables to hold different values (a, c, b, and d) allows the solution of this expression for any value combinations we care to consider. In computer languages, this is accomplished via *functions* or *procedures*, subprograms that compute specific values based on being given specific values held in those variables. Even though digital computers do only integer arithmetic, their range is expanded by using functions that implement well-known mathematical approximation techniques, such as by using a specific Taylor series[26] (polynomial) to approximate transcendental functions. For example, the function sine (x), where x is in radians, can be approximately computed by the Taylor polynomial, $x - x^3/3! + x^5/5! - x^7/7!$. Exponentiation and computing the factorial of a number are straightforward (in this case, the factorial values of constants are simply stored as constants rather than repeatedly computed). So the problem of finding a sine value for a number becomes a problem in arithmetic. This "trick" opens up the whole area of trigonometry problems to computing!

Indeed similar kinds of tricks, called numerical methods, allow computers to be used in solving many more sophisticated math problems such as those presented in calculus. There is a cost in this, in that these tricks cause us to give up some accuracy and precision in the computations. But from a practical standpoint, such costs are generally minimal and solutions are sufficiently accurate and precise so as to allow us, for example, to fly a space probe near enough to a moon of Saturn to take pictures.

In essence, digital computers have made it possible to explore a much wider world of mathematics than was ever possible in the pencil and paper era. Computers allow physicists, for example, to go deep into theories in hours or days that would have required years, even centuries, of hand calculating. The realm of applied mathematics has been completely transformed by the advent of practical digital computing.

8.3.1.2 Path Finding

In a sense, all other problems that can be solved with computing are just various versions of mathematics. However, some of these are best described as abstract mathematics, or approximate mathematics. The latter comes from recognizing that animal brains do not DO mathematics when they are solving problems relevant to living. What they are doing may be simulated by mathematics or at least represented by some mathematical formula. But the computational processes in biophysical systems like brains are not at all like the deterministic rule we find in digital computers. Let's look at a few examples of problem types that can be solved by animals and by computers even if in different ways.

One such problem is that of finding a way through a complex environment. Animals are often faced with navigating and negotiating their way through unknown terrain while trying to find food or find their way home. Brains have evolved elaborate modules that handle various aspects of the way-finding problem and working collectively provide the computations needed to succeed (from an evolutionary perspective, if these computations did not generally work, then the animal would not survive to procreate!). These brain modules, such as the visual system's ability to recognize and classify objects in the environment, tag environmental cues that the animal has learned (or instinctively knows) to follow going from one loca-

---

[26] See http://en.wikipedia.org/wiki/Taylor_series.

tion to the next. For example, a foraging scout ant will follow the scent of food along a gradient of increasing odor until it contacts the food item. Even bacteria use this kind of analog computation to find food and avoid danger. One of the authors (Mobus) has done research on how to get a computer-controlled robot to behave in this fashion.[27] Another version of path finding is demonstrated by the autonomous vehicles that DARPA (Defense Advanced Research Projects Agency) has funded through their "Grand Challenges" program.[28]

Digital computations are at the heart of how messages are passed around the Internet, finding their way through an incredibly complex network of router nodes and switches, to arrive at the destination efficiently and effectively. Routers are constantly solving a dynamic problem of which route to send a message packet given the traffic load, distance to travel, and other factors.

8.3.1.3 Translation

The general problem of translation is to find and use a mapping from one set of symbols and their organization in a grouping to another set of symbols (and their organization) that preserves the semantics of the original set/organization. In other words, the problem is to take, say a sentence, in one language and construct the same meaning in a sentence in another language. Translators at the United Nations and other international deliberation bodies have to do this on the fly. Their brains have learned the syntax and semantics of two or more natural languages, and they can construct the mapping such that what they hear in one language can be translated into the other language in real time.

Above, we saw that the solution to writing more sophisticated computer programs was to let the computer itself translate a program written in a "high-level" language to a program written in machine language through a mapping process called compiling. Today, computers can do a reasonable job of translating spoken or written human languages into other languages reasonably well. You can even invoke a service from a major search engine that will translate web pages written in one language into text written in your preferred language. Here is a sample sentence in English translated to French (in real time) by an online translation service:

The rain in Spain stays mainly in the plain

translated to

La pluie en Espagne reste principalement dans la plaine

---

[27] A full accounting of the approach can be found in Mobus, G. Foraging Search: Prototypical Intelligence   (http://faculty.washington.edu/gmobus/ForagingSearch/Foraging.html.   Accessed September 24, 2013).

[28] See http://en.wikipedia.org/wiki/DARPA_Grand_Challenge.

Admittedly, that is probably not a difficult challenge today, but 20 years ago, it would have been nearly impossible, especially in real time. Our computer scientists' understanding of languages and translation has made it possible to write very robust algorithms to accomplish this feat.

It turns out that there are a huge number of problems in nature that amount to translations of all sorts. Any time a pattern in one physical system must be interpreted and acted upon in another physical system, we are looking at a problem in translation. The interpretation (constructing a mapping) may be guided by instincts or hardcoded rules, or it may be learned, as shown above. For example, an animal must recognize patterns that represent food or danger and compute a valid interpretation translating that into action (eat or run). Lower species on the phylogenetic tree have this mapping built into their brains genetically, whereas higher level animals have an ability to learn and flexibly interpret the many variations that might exist in patterns in a more complex environment.

8.3.1.4   Pattern Matching (Identification)

Before a computing process can perform a translation, it needs to identify the pattern of symbols (configuration of the physical system it observes) as belonging to a valid arrangement. For example, a monkey needs to recognize a specific fruit shape and color as one it knows to be safe to eat. The recognition phase may operate through what is called pattern matching. Even though this process must precede translation, for example, it is not necessarily a lot easier computing wise.

To match an input pattern with a previously stored variation, it is necessary to test correspondence of points or nodes in the two patterns to verify alignment or nonalignment. A simple example is finding a sub-string of letters in a longer string. For example, in doing a search for a term or word that might exist in a page of text, it is necessary to check the sequencing of letters, starting at the beginning of the page, with the sequence of letters in the target term. This is actually compute-intensive regardless of the kind of computer being used. In the brain the work is done by massively parallel circuits of neurons that simultaneously test subsets of the pattern against subsets of the text. When you read, for example, you do not scan every letter, space, and punctuation mark one after the other in a linear manner. Your brain blocks out chunks of the text, using spaces and punctuation marks as delimiters, and does a holistic recognition on the "word" or "phrase" at one time. Your brain then connects these into larger chunks and clauses, for example, as it starts the translation process to extract meaning from the text.

Digital computers cannot do this trick yet. They are constrained to examine each letter sequentially while trying to build up a likelihood of the letter sequence so far being a specific word. For example, after examining C A and T in sequence, the computer might conclude there is a 90 % likelihood that the word is "cat." This would go to 100 % if the next character were a space! But if the next character were an A, then the possibilities widen out a bit. The word could be "catapult" or "catastrophic" or something else. Thus, the poor computer is forced to continue examining the next letter to try and narrow down the possibilities. As soon as it determines that there is a 100 % probability of the word being a specific one, it can skip examinations of the next letters and just proceed to the next space.

The language translation trick done above starts with identification of the specific words and checking them in a dictionary. Patterns need not be just words in a sentence, of course. Any pattern that is a persistent feature in nature (like fruit shape and color) can be identified with a suitable method of matching the observed instance with a stored "archetype." Either the stored pattern can exist in the form of an isotropic map (point for point similarity explicitly represented) as is often done in computers or as a set of reconstruction rules that allow, for example, neural networks to reconstruct images from a set of features represented in synaptic weights. The latter is the way the brain does it and accounts for the parallel and extremely fast way it recognizes previously encoded patterns. Artificial neural networks simulated on a computer have achieved a limited ability in this direction but can have issues when it comes to learning such pattern reconstruction rules.

> **Question Box 8.10**
> Humans tend to match visual patterns not only with categories but with names: "Oh, it's a leaf!" Symbolic or verbal patterns seem to at least double the pattern recognition task, but somehow it seems having the names speeds the recognition. In terms of network memory processing, could this indeed be the case?

8.3.2 Data Capture and Storage

Digital computers capture and store data explicitly as binary-encoded patterns having very specific addresses in multidimensional arrays of "memory" (see above). Animal brains, on the other hand, do not store explicit data but only relational excitations of neurons through varying excitability at different synapses in the networks of neurons. This leads to very different methods of learning and recall in machines and animals.

In biological neural networks, data is stored in the form of synaptic excitability being strengthened or weakened. The excitability of a synapse changes with frequent usage and with time-correlated excitation at neighboring synapses, usually of a different type. They are said to be potentiated, meaning that even after a period of quiescence, they can be more readily excitable by subsequent action potentials. Potentiation covers multiple time scales. The synapse may be short-term excitable only, which means it will contribute only weakly to the overall excitability of the neuron cell (thus, not particularly contributing to that neuron's output signal). It can be longer-term potentiated, if excited more frequently in the past and having had some correlated input nearby. In this case, it will contribute more excitation to the neuron's excitation and may contribute more strongly to that neuron firing.

Synapses that have been long-term potentiated are then targets for further strengthening of their excitability through morphological changes in the postsynaptic membrane. These changes are very long term, essentially permanent. The prior kinds of potentiation will fade with time (short-term potentiation will fade rapidly, a matter of minutes; long-term potentiation will fade over hours or days). They are associated with short-term and intermediate-term memory traces, respectively. Morphological changes, however, seem to represent very long-term memory traces that persist for years (even these may fade but are more easily restored with re-excitation).

Memory (capture and storage along with retrieval) is the basis for adaptive systems. We will go into much greater detail of these mechanisms in the next chapter.

> **Question Box 8.11**
> What kind of process do you go through when you try to remember something? How does that relate to the way the brain stores data?

8.3.3 Modeling

A model is any encoded representation of a real-world system that can be dynamically computed at a rate faster than the real-time dynamics of the physical system. The medium in which the model is constructed and run will determine the rate at which it can be moved forward in time. But the important point is that the system that employs the use of a model of another system is in a position to look ahead, so to speak, and see what lies in store for that system in the future. Models allow the possessor the ability to predict or at least consider plausible scenarios.

Modeling is the epitome of computational processing. Modeling, assuming the models are veridical (truthful), provide the possessor system with the ability to anticipate situations in the future such that the possessor can become proactive rather than merely reactive. Furthermore, if the possessor has the ability to construct models from scratch or modify existing models (an advanced form of learning), then that possessor has achieved the ultimate in adaptability. This is precisely the power of the human brain that has made humans so successful in colonizing nearly the entire Earth. There are, however, some important caveats.

> **Question Box 8.12**
> What is the relationship between models and adaptability?

Models are always incomplete and less detailed as compared with the systems they attempt to be models of. Put another way, models will never be completely veridical. Their behaviors will diverge with the real systems under conditions not accounted for in the model. Models must start from some minimal representation. They may be improved over time (our knowledge of the world may tend toward greater veracity). But they will always make mistakes due to there being some knowledge not incorporated in them. This means the model possessor/user will always be ignorant to some greater or lesser degree. Couple that fact with the fact that models take time to process and in general the more complex (more veridical) a model is, the more time it takes to generate solutions. This gives us another potential problem with models, namely, we might not have enough time to run the model before we need a solution. In human decision making, this is called bounded rationality.[29] The bounds are established by the limits of time and space. The demands of the real world are in real time and must be responded to in that time frame. Knowledge is stored in space (computer memory or brain networks or other structures) and only so much can be accommodated and accessed in a given space and time. On top of that, the adequacy of the model depends on what knowledge has been learned and incorporated into the model. If the system has not been acquiring knowledge rapidly enough, the model will be very weak.

Even so models are essential to complex adaptive systems. In the next chapter, we will establish the way in which models are used and learned, especially in autonomous CASs.

Question Box 8. 13

Models are inherently utilitarian: they exist to guide activity. So what does it mean to say a model is veridical or truthful or that one model is more true than another?

Think Box. Why Can Brains Do Math?

As seen in this chapter, brains are a kind of *computational device*, but not really the kind we think about when we use the word "computer." This is kind of ironic since the term was originally applied to people whose job was to calculate formulas, for example, for constructing tables of logarithms (see this Wikipedia article: http://en.wikipedia.org/wiki/Human_computer). However, the people in this scheme were simply following computational algorithms to calculate results and record them in the table. In Frank Herbert's *Dune* books, human *mentats* were capable of performing computations in their brains (see this Wikipedia article: http://en.wikipedia.org/wiki/Mentat). And then there is another science fiction character from the *Star Trek* franchise, Mr. Spock, who is essentially a biological computer (completely logical most of the time).

But it turns out that brains don't actually compute using numbers the way a calculator or a computer do. Some people have the remarkable capability of

(continued)

---

[29] See http://en.wikipedia.org/wiki/Bounded_rationality. Take note in this article how the word information is substituted for knowledge! We warned you that even the experts make this error.

Think Box. (continued)

doing complex calculations in their heads, but it is actually a trick of memory rather than actual computation. Computers perform math operations algorithmically; they actually add or multiply numbers, for example. But humans perform math-like operations by using what we call a massive lookup table in what is known as associative memory. Given a neural representation of the two operands (numbers that are represented in the same way a name is represented in language cortex) and one for the operation, say multiplication, these people are able to use these to generate an activation trace to a mental representation of the "answer" that is pre-stored in the network.

You, yourself, have experienced constructing such a lookup table-like knowledge structure when as a grammar school student you were cajoled into memorizing your "math facts." Multiplication facts (from 0 to 10 perhaps) were committed to memory for fast retrieval when doing the mathematical operations of multiplication, similarly for addition and subtraction.

Only a very small portion of arithmetic can be so committed to memory by most people. There are some people who have the ability to commit much more to memory and they are superior at doing arithmetic in their heads. But what about people who can do truly amazing kinds of computations, seemingly in their minds, people who are called (usually unfairly) *idiot savants*? There are numerous cases of people who seem to do much more elaborate kinds of computations, for example, multiplying 10 digit integers. The jury is still out on this, the psychologists who study the phenomenon are pretty sure they are not performing machinelike computations. But there is a clue to a possible explanation in the nature of mistakes that they sometimes make in arriving at the right answer. Occasionally, those mistakes suggest that rather than computing the results, they have developed a keen ability to "estimate" an answer and use that to then home in on a more precise result. That is, they come close to an answer and then, perhaps, use a lookup process to zero in on the right answer.

But what about mathematicians who are able to "think" mathematically about abstract relations? Once again it may be that we are looking at a phenomenon of estimation, but one bolstered by an incredible ability to "visualize" the relations being worked on. Most people have heard the story about Albert Einstein's claim that he could "see" the relations in the phenomena he pondered. He is said to have mentally viewed a person in an elevator that was falling or being escalated rapidly to work out the nature of gravity in terms of inertial frames of reference. He did not do math in his head. He "saw" it.

There is a facility in the brains of mammals and birds that allows them to discriminate small differences in numbers of things and another that allows them to have a sense of accumulation. Rhesus monkeys, for example, have been shown to get "upset" if they are given just two food treat morsels while a neighbor gets three or four. Dogs get curious when they see four toys disappear in sequence behind a screen and only three come out the other side. It seems the brain has some very primitive counting ability but the math is still done with paper and pencil.

8.4   Summary: The Ultimate Context of Computational Processes

Why are there special informational processes that involve computation? Where did they come from? What are they for?

These are ultimately the kinds of questions that require an evolutionary answer. We will examine these questions both in the next chapter on Cybernetics and in the following section on evolutionary processes. To introduce the next chapter's subject, however, let us review what we have considered so far in this and the previous chapter. Then we can start to establish the context in which information, knowledge, and computation operate in systems. To anticipate a bit, in the next chapter, we will use computation and information processes to establish one of the most important set of properties of complex adaptive systems, and that is resilience and stability in the face of changing environments. Systems, to have any meaning at all, must persist over some relevant time scale. If systems broke at the smallest instances of changes in their environments, then systemness would be meaningless. From a different point of view, we see that in fact systems do persist under change. What we will see in the next chapter is how this comes about using information, knowledge, and computation.

Recall that information is news of difference conveyed in messages sent from one system and received by another. The news of difference quality is the result of the ignorance of the receiver not a property of the sender. And the amount of information (in the form of what we called surprise) causes material differences to manifest in the receiver. The receiver reacts to the difference (say in the form of an *a priori* held model of the state of the sender) by doing internal work on its own structure, usually using amplifiers (actuators) to convert low-power message signals into higher-power energy flows to accomplish the work needed. The nature of the work process already exists within the receiver and is part of what we mean by the *meaning* of a message; the message receipt is predesignated to accomplish that specific work in proportion to the information content of the message.

The work in question changes the system's knowledge in that it makes the future receipt of the same message more likely and hence less informational. We can say the system comes to expect that message (representing the state of the sender), which is another way of saying the receiver system knows better what to expect. We saw, though, that due to entropic decay, systems do tend to forget what they learned.

At this level of description, almost any physical transfer of energy (conveyed even by material transport) constitutes an information process. But we also saw that information messages can be transmitted using very low-power channels because real systems include transducers (in the sender) and amplifiers (in the receiver). High-powered transformations that take place in the sender are converted to low-power signals that are transmitted through appropriate channels (such as via electromagnetic waves) and when received transformed back into high-power, effective, energy flows.

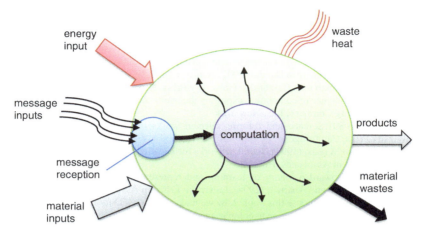

Fig. 8.13 A summary of information, knowledge, and computation in the context of systems. Computation provides a low-cost (in energy and material) way to process informational inputs (messages) and then drive effective changes in the system. What those changes are, and how they affect the system in its continuance of existence will be the subject of Chap. 9

But another kind of process can capture and use the input signals before they are converted into high-powered form. And that is the computational process which can use the low-power form of signal in an efficient mechanism that can derive additional information from input signals. For example, computational processes can combine separate streams of input signals (data) to derive relational information that is more *informative* than the raw data itself. Two pieces of information are more valuable than either piece taken separately.

Computation allows a system to be more effective in using information from the environment. The power is synergistic and comes at a relatively low cost. In one sense, computation makes complex adaptive systems feasible. Without the ability to manipulate and combine data flows to derive information, the power requirements for complex processes would preclude their existence!

Figure 8.13 summarizes the situation that we have been describing so far. Systems that can employ computation based on information input are capable of *adaptivity* in ways not achievable by other kinds of system. Living systems and their evolved social systems are the prime examples of complex adaptive systems, though there are interesting examples of such systems simulated in computers.

We now turn our attentions to the actual effective results of information flows and computational processes in systems. The benefit of these sub-processes is, first, to obtain stability in a changing world; second, to provide a mechanism for resilience in that world when things change a great deal; and third, to provide a way to learn so as to be preadapted to future changes.

Bibliography and Further Reading

Alkon DL (1987) Memory traces in the brain. Cambridge University Press, Cambridge

Baars BJ, Gage NM (2007) Cognition, brain, and consciousness. Elsvier AP, New York

Gilovich T, Griffin D, Kahneman D (eds) (2002) Heuristics and Biases: the psychology of intuitive judgment, Paperbackth edn. Cambridge University Press, New York

Harel D (1987) The science of computing: exploring the nature and power of algorithms. Addison-Wesley Publishing Company, Inc., New York

Hyman A (ed) (1989) Science and reform: selected works of Charles Babbage. Cambridge University Press, Cambridge

Levine DS, Aparicio M (eds) (1994) Neural networks for knowledge representation and inference. Lawrence Erlbaum Associates, Hillsdale

Mobus GE (1994) Toward a theory of learning and representing causal inferences in neural networks.In: Levine and Aparicio (1994), Lawrence Erlbaum Associates, Hillsdale, ch 13

Montague R (2006) Why choose this book: how we make decisions. Dutton, New York

Patt YN, Patel SJ (2004) Introduction to computing systems: from bits to gates to C & beyond, 2nd edn. McGraw-Hill, New York

Rumelhart J, McClelland D (1986) Parallel distributed processing: explorations in the microstructure of cognition. MIT, Cambridge

Chapter 9
Cybernetics: The Role of Information and Computation in Systems

> *Information is a name for the content of what is exchanged with the outer world as we adjust to it, and make our adjustment felt upon it. The process of receiving and using information is the process of our adjusting to the contingencies of the outer environment, and of our living effectively within that environment.*
>
> Norbert Wiener, 1950

Abstract Information, as defined in Chap. 7, and computation, as described in Chap. 8, will now be put to work in systems. Cybernetics is the science of control and regulation as it applies to maintaining the functions of systems. In this chapter, we investigate basic cybernetics and control theory. But we also investigate how complex systems have complex regulatory subsystems that, unsurprisingly, form a hierarchy of specialized control or management functions. These subsystems process information for the purpose of managing the material processes within the system and to coordinate the system's behaviors with respect to its environment. This chapter covers what might be argued to be the most crucial aspect of the science of complex systems such as biological and social systems. It is the longest!

9.1 Introduction: Complex Adaptive Systems and Internal Control

In Part II we examined the structural aspects of systems, how matter and energy are organized in whole systems so that the systems, as processes, perform work as in producing products and behaviors. We provided a "process semantics" for describing complex systems as dynamic systems but said little about how the structures of the systems provide the necessary internal regulations to maintain those processes in functioning order.

In this chapter, our focus will stay on complex adaptive systems (CASs) and examine the informational processes (sub-processes) that provide the internal regulation as well as giving the systems the capacity to interact fruitfully with their environment. Environments invariably have non-deterministic dynamics. Environments in nature

© Springer Science+Business Media New York 2015
G.E. Mobus, M.C. Kalton, *Principles of Systems Science*, Understanding Complex Systems, DOI 10.1007/978-1-4939-1920-8_9

also have a tendency to change their characteristics over longer time scales, also in a non-deterministic way. In the face of these kinds of challenges, CASs still persist. They continue to exist and perform their functions in spite of non-normal interactions with their environments. They adapt and show resilience to some more extreme changes when necessary. If they did not do so, then the story of systemness would end rather abruptly with very uninteresting systems taking up the whole landscape.

We humans are complex adaptive systems. We are also parts of larger CASs. We are autonomous to a large degree as are some of those larger systems. But we are also embedded in the world system, the environment that includes all living and nonliving components of our Earth. And even though we enjoy a certain amount of autonomy in our lives, we are not free to interact with that environment in any way we feel. As we are sadly discovering from past unwise interactions (e.g., pollution), there are definite constraints on our behaviors imposed by the larger environment. As a result of our own actions, we are changing the environment radically, and it remains to be seen if our capacity to adapt will allow our species to persist. How resilient are we?

The basis of regulating processes, resilience, the capacity to adapt and persist, is found in the subjects we have been discussing in the prior two chapters. Namely, information, knowledge, and computation (especially involving models) will now be seen as playing their roles in making CASs stable, resilient, and adaptive to changes. In this chapter, we will explicate the nature of cybernetic systems, which are the control and management subsystems that CASs employ to keep themselves stably functioning and able to deal with their environments as the latter produce new challenges and opportunities.

CASs have two fundamental problems to solve. They have to have internal regulation of the actions of many multiples of subsystems that keep the whole system operating to produce its optimum output. This is a particularly difficult problem when multiple subsystems use the same resources and those resources are in short supply for some reason. Some competition might be useful between such subsystems, but it has to be of benefit to the whole system, otherwise it can lead to malfunction and loss of stability. The most successful CASs have internal coordination mechanisms to keep otherwise competing subsystems from becoming greedy and disrupting the whole system.[1]

The second problem involves the interactions a CAS has with its environment. By definition the system is a whole unitary entity that does not have control over the sources of its resources nor the sinks for its output products and wastes. And in almost all natural environments, including, incidentally, those of human societies, change in conditions and dynamics is the rule not the exception. Sources and sinks can change in unpredictable ways that create problems for a system. This is why systems have to be adaptable in the first place. They need to have built-in mechanisms that allow them to respond to changes by changing their

---

[1] As an example of what happens when internal controls over resource deployment breakdown, consider cancer. Cancerous cells have broken internal regulating mechanism and as a result begin to grow and multiply uncontrollably. Unless the cancer is removed or treated, the animal dies from this breakdown.

own internal processes to accommodate these changes while maintaining their basic functioning and purpose.

Question Box 9.1

The notion of control is layered and subtle—as are systems themselves. We certainly often think of ourselves as controlling our environments (including societies, organizations etc.), but often "fail" to do so. What differentiates the "inside" control of an organic system from its manipulation of its "outside" environment?

Since the system does not have control of the environment, and in the general case anything can happen, it needs to have an assortment of responses that it can use to mollify the otherwise detrimental effects of change. As we will see, simply reacting to changes is not enough either. The systems that are most adaptive actually anticipate changes based on more extensive models of how the environment works in general, a deeper grasp of causal relations that will allow them to preemptively modify their responses to avoid the higher costs of repairing damaged subsystems that can occur as a result of the changes. We will see that anticipatory systems are, in fact, the most adaptive of all CASs and, in general, the most successful in sustaining themselves over long time spans.

The starting point for understanding how information, knowledge, and computation play a role in keeping CASs "alive" and well is to understand the dynamics of cooperation between two or more systems embedded in a larger environment.

9.2   Inter-system Communications

CASs have to find ways to communicate with one another in order to cooperate. In larger meta-systems, one CAS may produce a product that is needed or used by another CAS. This may work reciprocally as well. If the two or more systems become dependent on one another, then continuance of the meta-system depends on their abilities to signal one another and respond to those signals such that the overall set of processes will achieve stability and persistence. In this section, we explore the mechanics of this to show the purpose of communications and information/knowledge.

9.2.1   Communications and Cooperation

Consider the simple case of two systems that are coupled by the flow of a product from one to the resource input of another (see Fig. 9.1 below). Recall that in our lexicon, system and process are identical and interchangeable terms, so we will

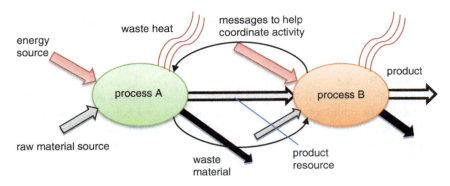

Fig. 9.1 Two processes that interact with one another so as to cooperate with each other. Process B uses process A's lower entropy (product) material output as one of its inputs. Source and sink symbols have been left out to avoid clutter. Both processes receive inputs in the form of higher entropy materials and low entropy energy. Work is accomplished as per above, and both low entropy material outputs and high entropy material and energy (heat) outputs are generated. In addition to process B's ability to interpret the material flow from A and derive information from that, the two have established a high-efficiency communications system in which both inform the other of changes that might impact A's output (e.g., B cannot accept A's product for a while) or B's output (e.g., it is too low and needs more material to process)

more often refer to these systems as processes to emphasize the fact that they are processing material inputs using energy inputs to produce products that can be used by other processes. In this first pass, we will assume that the two processes, A and B, have established a cooperative flow where B needs the output from A as a resource.

Communications subsystems allow intra-process coordination or cooperation. What is required is that two or more processes have established a channel link. As we saw before, a communications channel is of value when it most efficiently provides a mechanism for sending and receiving messages, as covered in Chap.7. Cooperation is possible when the communications are two way, meaning that one process can send messages to the other and vice versa, creating a feedback loop that enables them to work in terms of one another.

Processes can communicate, after a fashion, in a one-way message sent through whatever substance is being sent by one and received by the other (e.g., product from A to B). What is necessary is that the receiver must have some means to interpret the message and act on whatever information is contained. For example, process B in Fig. 9.1, above, can monitor the actual flow of a low entropy material that it receives from process A. Fluctuations in the flow rate may encode information that will activate reactions by process B. Indeed most complex systems have subsystems that receive and monitor their input flows. A raw material inventory system in a manufacturer, for example, has to monitor the flows of received material in order to provide information on things like late deliveries.

But one-way communications can only suffice in purely reactive systems, those that may take action but are not, technically speaking, cooperating with their supplier.

Two-way communications evolved as a way to allow complex processes to coordinate their activities. This figure shows a basic set up for two-way communications.

Internally, both processes need to have subsystems as covered in Chap. 7 that allow them to encode and receive/decode the messages. Each needs a transduction capability that includes a way to encode energy flows through the communications channel to the other (partner) process. Each also needs a way to interpret the incoming signal and actuate responses based on the information content.

> **Question Box 9.2**
> In what ways is this communication also a computation process, as discussed in Chap. 8?

In this chapter, we will not attempt to explain how these communications systems come into existence. That will be covered in Chaps. 10 and 11, having to do with system auto-organization and evolution.

9.2.2 *Informational Transactions*

The point of communications is to provide processes/systems with information that they can use to respond and do something differently in order to maintain themselves and their functional competence. As long as process A is producing its product at a rate commensurate with that needed by process B to produce its product in the most efficacious manner, then all is well and no information need be exchanged between processes. Indeed, to drive home, again, the distinction between messages (in the form of encoded data) and information, process A could be continually sending process B a code that means "flow steady as requested." Process B, having an expectation that the flow will be steady at the level desired, is receiving the message as sent, but there will be no change in its behavior. It does not need to do anything different from what it was already doing. The only information in such a message comes from the difference inherent in different moments of a temporal process, so a reconfirmation of a state of affairs is meaningful, just not terribly informational. But if process A sends a new code that means "flow going down, sorry!", then the difference is in the content of the message itself and process B has to take some action to accommodate that change.

One of the actions it would probably take is to send a different message from the one it had been sending while the flow was steady, namely, "thanks....thanks...." Now it might send, in response to the flow going down message, a message that pleads for more flow, "more please."

The two processes have established a protocol for sending and receiving messages, an interpretation of the meaning of those messages, and they respond when the messages received are different from what their expectations had been.

For example, process A might send another message saying "can't do that right now," to which B might respond, "now see here A, we have a deal," or something to that effect.

As you may recall from Chap. 7, protocols are mutually expected conventions that enable the transmission, receipt, and interpretation of information. Many layers of protocol may be required, for signals to get translated from medium to medium in the process of transmission and receipt. Before words and grammar (high-level protocols) can be conveyed, there must be some alphabet of signs and some agreed-upon way of encoding them for transmission and receipt. Protocols are a set of rules for transacting messages that, along with low-level encoding, help ensure the timely sending and receiving of messages and especially the receipt of information. Time is critical. The receiving process needs to respond as quickly as it possibly can in order to minimize delays (as we will see below, delays can be very bad for business). Protocols in nature evolve (Chaps. 10 and 11), or in human communications systems are designed (and then later evolve!). A great example of a human communications protocol is the hypertext transport protocol or HTTP which is used to transfer many different kinds of documents (but mostly World Wide Web pages) between computers (called clients, the requestors, and servers, the senders). Actually in complex communications channels like the Internet, multiple levels of protocols are involved in transactions and transfers, from the very low-level electrical, radio frequency (WiFi) or photonic specifications of the "pipes," to the Internet message packet protocols, to something like the HTTP protocol.

Protocols handle the meaning of message traffic, establishing the intent of sender and the needs of receivers. The expectations of the receiver of what is IN the messages that are sent determine how informational they are. Take, for example, a typical WWW request for a page from an online newspaper. The page is sent and displayed by the client program, your browser. Other than the formatting information contained within the document's hypertext markup language protocol (HTTP), the browser doesn't have any clue (or expectation!) about what kind of news is in the page. And if the page contains stories that you had just heard all about on the radio, then you would not be getting anything different from what you already knew, ergo, no real information. Clarity regarding information as *difference from expectation of the receiver* is important so we do not become confused by the fact that the same information may come from different messages (e.g., different words or languages), or the same message may convey different information to different receivers. We realize we are being very redundant at this stage; we've harped on this point many times before. But it is very important to understand, and the sloppiness with which the concepts are typically handled compels us to continue to harp on the difference between mere messages (data) and information at every turn.

Question Box 9.3

Explain, in terms of the above, how what is *information* to a computer is *data* to the computer user, and the kind of *information* the user gets may again become *data* for a researcher studying the use of the web, and the researcher's *information* may again become *data* for businesses seeking advertising opportunities, and on and on.

9.2.3 *Markets as Protocols for Cooperation*

Markets are typically the subject of economics, and we think of them as essentially a human construct. But the human version of markets picks up on a form of systemic organization widely represented in the nonhuman-built world as well.[2] In Chap. 2 we identified a class of systems called "ecosystems" in which cooperation mechanisms predominate with little in the way of hierarchically imposed coordination. The mechanisms that develop for cooperation can be described as a market.

The key characteristic of a market is that a group of processes cooperate by exchanging materials and energy based on inter-process communications. Distribution of various forms of utility is the function of the market, as evident in barter systems, which trade one good or service for another good or service. Some, like the energy distribution system inside living cells, have a complexity that brings to mind our contemporary economic markets based on money. Indeed, the endocrine system in the body is an intricate communications system that regulates flows of matter and energy in essentially all tissues by using more subtle messages in the form of hormones, chemical signals that have responsive impacts on target tissues.

In order for a market to work, there has to be a general format or protocol for how communications are managed, meaning that receivers are organized to handle messages and senders are organized to encode messages appropriately. The market is a set of protocols for establishing senders and receivers (of different kinds of messages) and the encoding/decoding mechanics needed to generate effective behavior in each process. Figure 9.2 below shows a simple market structure.

Messages flow from processes accepting flows of, say, materials, back to processes that provide those flows. The figure does not show the forward messaging system to avoid clutter. But every supplying process would be sending messages to their "customers," In an advanced market structure such as the commercial markets in advanced economies, the feedback messages would be money (and its informational equivalents). Feed-forward messages would include purchase orders and bills of lading, etc.

Question Box 9.4
Market systems, human and natural, usually involve competitive dynamics and have rules, man-made and natural, against forms of cooperation that would mess up the competition. Think of cooperative price fixing or prey just offering themselves up to predators; what happens to the system? How then, can markets be described in terms of "mechanisms that develop for cooperation?"

---

[2] Howard T. Odum (1924–2002), one of the founders of systems ecology, developed elaborate energy and material flow (exchange) models for natural systems. He discovered that these systems resembled human economic systems or rather discovered that human economic systems resembled those natural exchange systems. Human-developed markets are an extension of natural markets. See Odum (2007), chapter 9.

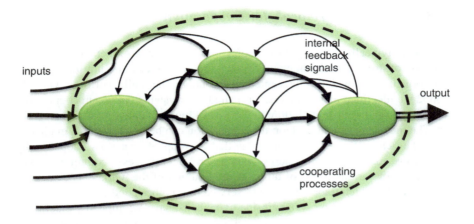

Fig. 9.2 A market exists when a group of cooperating processes communicate in a way that helps regulate internal flows. These processes cooperate by virtue of communications channels that provide feedback from the output end back through to the input processes. The feedback signals help regulate the various internal flows. Money is such a signal device in human economies

9.3 Formal Coordination Through Hierarchical Control Systems: Cybernetics

As complex systems become even more complex, (next chapters) their ability to rely on simple cooperation through market-like mechanisms starts to diminish as complexity increases. At some point in the hierarchy of complexity, cooperation is no longer viable, and new mechanisms for coordination need to be in place. Throughout nature we see this pattern emerging again and again. It is not atypical for us (humans) to deride the notion of hierarchical control, especially in Western nations where the concepts of individual freedoms and rugged individualism prevail. The thought of controls over our lives from some structure "above" is onerous. But it turns out that nature's solution to the coordination of a complex set of processes is always through the emergence of hierarchical control subsystems. Like it or not, hierarchical management is a fact of nature. We will see several examples where it comes to play in nature, and, perhaps, if we understand better why it works, we will be a little less suspect of its role in human societies.

Hierarchical control systems are ubiquitous in nature. They even permeate our human systems of governance. We should attempt to understand their purpose and role in the concept of providing sustainability to complex systems. Perhaps if we did, we would do a better job of designing governance for our human societies.

Hierarchical control systems are based on the fact that there are different layers of control or management in systems based on the kinds of management decisions that are required to keep the underlying processes (production systems) operating in coordinated fashion and at optimum performance. These systems address the two basic problems for CASs posed above. They address the low-level operational control of individual sub-processes, the coordination of those processes with one another, and coordination with the environment in which the whole system is embedded. The rest of this chapter is devoted to explaining how these systems work, individually and collectively, to produce highly adaptive and sustainable systems in the world.

The subfield in systems science that studies control systems is called cybernetics. Cybernetics gets its name from the Greek term for the helmsman or steersman of a ship, the one who controls where it is going. It is the science of control processes, originating in WWII projects such as automating the control of antiaircraft guns and spreading to become the computerized automation that enmeshes every aspect of contemporary life.[3] The term "control" carries some sociological baggage, unfortunately, when it is conflated with terms like "command" or "dictator!" Many control structures are imagined as top-down dictatorial processes. The objectionable connotation easily attaches to the whole notion of hierarchical structures. One can picture the "boss" at the top of an organization giving commands to mid-level sub-bosses, who in turn bark commands to their workers who have to blindly follow those commands, asking no questions.

Of course there are organizations that have very strong top-down, hierarchical command and control structures. The military could hardly function in war time if it didn't have a fairly rigid authoritarian structure. Many for-profit companies may operate in this way, though their employees may not be particularly happy. Even these top-down processes depend on feedback. One might suggest it is less the necessarily hierarchical structure, a feature inherent in layered systems, than it is a selective and exploitative use of feedback that gives rise to the objectionable character of some hierarchical systems.

The reality is that control is an absolutely necessary aspect of all stable complex, dynamic systems, particularly if they are adaptive. The balancing side of the picture emerges if we consider what is associated with the notion of things being "out of control." Systems offer a variety of ways to achieve the ordered mutual coordination and functionality signified by being in control. This may or may not involve a hierarchical structure leading to an ultimate "in charge" function. In this section, we will develop the general principles of control in the cybernetic sense. Then we will start to describe the principles of what is sometimes called distributed control, principally the mechanisms of competition and cooperation that allow multiple interactive processes to form markets as above.

Question Box 9.5
Paternalism and dictatorships are both top-down hierarchical governmental systems. How do they differ in their use of feedback information? What makes it such a slippery slope from paternalism to dictatorship?

---

[3] A collection of definitions for cybernetics can be found at "Definitions of Cybernetics." A Larry Richards Reader 1997–2007. 2008, http://polyproject.wikispaces.com/file/view/Larry+Richards+Reader+6+08.pdf, accessed 9-5-12. These come from the classic writers on the topic, showing some slightly different perspectives, but all agreeing on the fundamental nature of the subject.

9.3.1 Hierarchical Control Model Preview

We will first provide a preview of where we will be going in this section. Hierarchical control refers to the structure of a control system where decision processes of particular kinds are distributed in a hierarchical fashion. At the lowest level in the hierarchy is the operational control of all of the main work subsystems in a large complex system. The level above this consists of coordination control, which consists of two subsystems, logistical and tactical. The top level is dedicated to long-term planning and oversight of the whole system. It executes strategic control.

Figure 9.3 provides a basic layout of this control system.[4] The lowest level is where the actual subsystem processes are working away. Resources enter at this level and products and wastes exit. Process control involves subsystem self-regulation in real time. The green oval in this level represents just one process. The arrow exiting the process and re-entering represents information feedback. All of the arrows represent information flows.

The next level up from operations is the coordination level. Coordination requires the observation of processes over an intermediate time scale and making decisions that will result in coordinating the interactions between processes. Internal coordination (between processes within the CAS) is called logistical control. It involves monitoring the behaviors of internal processes and issuing control signals to adjust those behaviors as needed to realize an overall optimal performance of the system. Logistics includes finding optimal distributions of resources and the prevention of internal competitions that might produce suboptimal results.

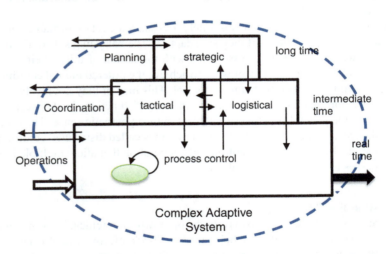

Fig. 9.3 This is the basic layout of a hierarchical control system for a complex adaptive system

---

[4] This model is an extension of the work done by Findeisen et al. (1980). That work synthesized results from earlier models of control in complex systems, distributed across levels and time domains. See, especially, pp. 7–14.

Systems also need to coordinate their overall behavior with external processes that constitute the sources of their resources and sinks for their products and wastes. Tactical control involves monitoring the environment and issuing signals to key operational processes that have direct interfaces with those external processes. The tactical controller must also communicate with the logistical controller in order insure coordination between the two coordinators.

At the highest level, planning refers to the fact that CASs are faced with changing environments that require adaptive responses in the long run. Typically, strategic control involves planning and executing change control over much longer time scales.

Hierarchical control systems have developed as CASs have evolved to be more complex. Think of the small, one-person bakery where the owner is also the worker (process). Suppose she is so successful that she soon finds it necessary to hire an additional worker as her operations expand to serve more customers. If the operation continues to grow, it will involve more workers and the owner will need to start being more of a manager. She will need to do the bookkeeping as well as ordering supplies and handling shipping. As growth continues, the operations become more complex, perhaps the bakery will need to be segregated into departments to handle the volume of sales. At some point, the owner decides to hire a sales staff and a marketing manager to find new outlets for the products. Before long, she is spending most of her time thinking about the future for her company. She hires department managers and a general manager to put together operations manuals and oversee all of the functions.

Hierarchical control systems also emerged in the evolution of more complex life on this planet. We will be using examples from the human social world as well as from biology, especially the human brain, to show how cybernetic systems achieve the management of CASs such that they are able to sustain over time.

We will explicate the hierarchical control model from the bottom up, following a developmental-like trajectory. That is, the sections will follow the same kind of path that our baker business owner above went through.

9.4 Basic Theory of Control

We will start our examination of control systems by looking at a single process model. Every process (as a subsystem) should be considered as a semiautonomous entity that finds itself embedded in an environment of other processes. Even so, that fact alone does not mean the process of interest is going to have an easy time of doing its job. The world, even one in which one has a purpose to serve, can be a dangerous place! The fact is that every process faces the same kind of problem in that the environment is always finding ways to disrupt the optimal operation of that process. Systems are always in danger of being thrown off in performing their functions. Entropy is fed by probability, for there are far more paths to disorganization than to organized functionality. Yet, most systems somehow seem to persist and perform their functions reasonably well most of the time. They achieve this because of a very simple, but non-obvious aspect of complex systems. The statistical edge of disorganization triumphs inevitably only when there is no flow of energy to put into maintaining or even increasing order. They can use energy to control their own behavior even in the face of potentially disruptive interference from their environments.

In this section, we will explore the principle of control or the general principle of cybernetics, namely, system self-regulation through information feedback. We will expand the small green oval and information arrow in Fig. 9.3 above to see the details.

9.4.1 Open Loop Control

Rarely it occurs that a process can be controlled deterministically simply by issuing a command and assuming that it will be carried out as desired. This is called "open-loop" control because there actually is no information "loop," as will be clearer below. In open-loop control, the controller simply issues an order, and the process is assumed to carry it out as given. As stated, examples of this are quite rare in real life, because a guiding or controlling function must ordinarily be continuously informed of the conditions over which it is presiding. If you restrict the definition of the system of interest to just the fuel injection system in an automobile, then the position of the gas pedal providing a control signal to the injector is sufficient to force the fuel flow rate to a specific value. The injector blindly follows its orders mechanically.

But the greater reality is that the pedal/fuel injector system is part of a larger system that includes a human being, making decisions about what "speed" they want to travel. The human is getting information about what speed they are currently moving, and if they want to go faster, they have to order the injector (through the pedal) to pump more fuel into the cylinders. In fact, absent this element, one might better describe the injector-fuel-piston process as simply causality. All causal processes, in a wide sense, control their effect, i.e., make it happen, but this is not control in the ordinary sense of the term, which has to do with achieving some particular function.

Thus, while there are examples of simple open-loop control systems, they are extremely limited in scope and number in real life. The more general principle of control is that given by the human in the loop described above. This is a closed-loop system.

9.4.2 Closed-Loop Control: The Control Problem

In order for any system to produce a particular result (output) in spite of the influences of environmental disturbances, it is necessary to continually monitor and adjust flows and related process performance as in the example of the human pushing the gas pedal above.

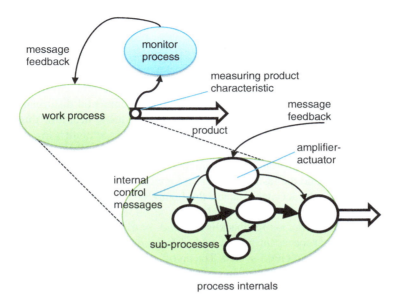

Fig. 9.4 Feedback or closed-loop control consists of an output monitor process whose job is to measure the value of the output and compare it with the desired value. It then sends a message back to the work process that causes the latter to adjust its internal sub-processes to obtain the desired result. Within the work process (exploded view), there needs to be sub-processes that are able to respond to control signals from the "amplifier-actuator" such that the desired response is given. The objective is to get the output back to its desired value

In this case, the work process must include affecters that can change the operations according to the kind of command being given. What is required is some kind of monitoring of the work process' output that determines if the output is what is required/desired in order to fulfill its function. In the automobile example above, the human monitors the speedometer and makes decisions regarding the "right" speed. The monitor then adjusts the command signal that causes the work process to change its behavior in order to fulfill the requirement/desire. The human presses the pedal more to increase the speed if that is what is desired (or ease off of the pedal if the car should slow down). This is called feedback control. Figure 9.4 shows this concept.

The work process must be capable of altering its behavior in response to a control message, just as the monitor must be able to modify its output signal in response to incoming measurement data. This will involve some form of actuators that can modify its action as a result of the information content of the control messages. While we think of control as a one-way process with a focus on a given outcome, on close inspection we find this information feedback loop involves a mutual process. Does a thermostat control the furnace, or does the furnace control the thermostat? As a mechanical cybernetic system, both are true, though we favor the thermostat because that is the point where *we* intervene to exercise our control. But then, is the furnace controlling us by making us feel too hot or too cold? You can see that notions of one-way or open-loop control are generally far too simplistic to capture the systemic dynamics of ongoing process adjustment and resilience!

One of the simplest examples of a closed-loop controller is in fact the thermostat that monitors the internal temperature and turns the furnace or air conditioner on or off to keep the temperature within comfortable limits. The problem for most systems is maintaining a stable state in the face of environmental disturbances. The temperature of a living space is subject to fluctuations due to heat flow through the walls, windows, and ceilings of the space. If the outside temperature is low compared with the internal (desired) temperature, then heat will conduct through the barriers in proportion to the temperature differential. The colder it is outside, the faster the heat will transport and the faster the internal temperature will drop.

An open-loop controller cannot handle this highly variable situation because a single command cannot adequately reflect the changing situation. So a closed-loop controller working through information feedback is required. In the past, the monitor/controller would be the human occupant who would stoke or damp the fire accordingly—the equivalent of flipping an on/off switch. The on/off switch, uninfluenced by the consequences of its signal, is in essence an open-loop controller, necessitating the constant human intervention to make this a functional closed-loop system. The dawn of cybernetic control was when we figured out these systemic information feedback dynamics and began rigging mechanical devices that would monitor and respond to feedback by an output signal to a suitably equipped receptor. We have thermostats (and myriad other automated devices!) so we need not constantly ourselves be the monitoring feedback component in the process.

The thermostat detects changes in the internal temperature. It has a built-in thermometer (transducer/sensor) that provides the monitor signal. It also has a value called a "set point" that provides the ideal temperature (as determined and set by the occupants). When the internal temperature falls below the ideal, by some margin (remember measuring processes must be calibrated!), the thermostat sends a control signal to the furnace. It sends a low voltage to a relay (an actuator) that closes a high voltage circuit to the fan motor and to a valve that opens to allow gas into the burners. The gas burns and produces heat which the fan blows into the heating ducts. When the temperature reaches a slightly higher value than the ideal, the thermostat turns the circuit off so that the space does not overheat.

Figure 9.5 shows the internals of a generalized monitor process. The sensor and comparator components come directly from our lexicon given in Chap. 5. The set point is a settable "memory" that provides a value for the comparator.

The sensor measures the product value of interest (in this case the temperature of the space). The comparator is a simple computational device that subtracts the sensor value from the set point memory to generate an error signal. The set point is the ideal and, depending on calibration, any variance recognized as a deviation from that value constitutes information that needs to be acted on. This is used by the control transmitter to send the actuator inside the process the command to act. The transmitter works from a model of which control action matches specific error values. In the case of the furnace, this is simply a switch that is activated when the error

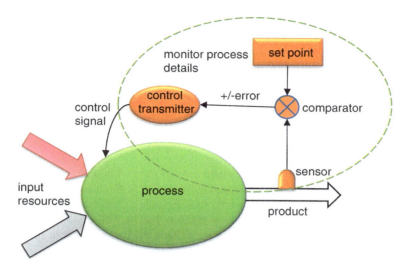

Fig. 9.5 The monitor is a computational process that provides error feedback to generate a control signal back to the work process. The latter then takes action to decrease the error

value shows the temperature too low. The model is extremely simple. In more complex process, the control model used to determine the control signal transmitted will be correspondingly complex. We will see examples below.

The closed-loop control model is the core of the field of cybernetics.[5] The term was created by Norbert Wiener from the Greek *kybernētēs*, which translated means "steersman," to describe the form of counteracting feedback that corrects errors in operations (Wiener 1948). A steersman detects an error in the heading of a boat and turns the tiller in the opposite direction to correct the heading. Another example of the early use of this principle comes from James Watt's use of a "governor" mechanism to control the speed of a steam engine. This device worked by closing down the steam valve if the revolutions per unit time of the engine exceeded the desired value.

Closed-loop control is the basic form of operational control. It is found in every complex dynamic system. We incorporate this structure in the automation of all sorts of mechanical processes, but it is equally pervasive and even more complex in living systems. In organizations, the responsibility of managers is to monitor their operations on particular quality and quantity parameters and to take action when the values of those parameters get out of their desired range. In biological systems, the mechanism of *homeostasis* (mentioned in Chap. 5) is precisely a form of closed-loop control.

---

[5] This term now covers a very large field of study, often called control theory. The reader is encouraged to visit the Wikipedia page on Cybernetics at http://en.wikipedia.org/wiki/Cybernetics.

Homeostasis is found at all levels of organization in living systems, from inside cells to physiological subsystems in the bodies of multicellular organisms, such as the pH (acidity level) regulation mentioned in Chap. 5 or the blood sugar maintenance subsystem in animals that trigger hunger-driven feeding behavior to satiation that turns it off. The term is derived from Greek for "staying the same." Any system is conditioned by parameters beyond which it cannot be maintained, so every system is in a sense vested in things staying the same. But living organisms present such complexity coupled with such tightly conditioned parameters that must be maintained for life to continue, that the term "homeostasis" was minted to describe this striking feat of self-regulation.

Living systems could not maintain their organization without the work of homeostatic closed-loop control. The environment is forever in fluctuation with respect to the effects on critical parameters in the maintenance of healthy cells and tissues. Thus, the earliest problem that living systems had to solve was how to react appropriately to these fluctuations in order to keep the internal system functioning properly to sustain life.

Yet another example comes from manufacturing where a quality control function is used to monitor the "quality" of given parameters for a product. There may be many features of a product that must be inspected to catch defects or out-of-specification measurements. Some defects may be ordinary and are simply corrected in the product before it is packaged and shipped. But QC also keeps a record of the kinds of defects found and the frequency (number of defects per product count). If the kinds of defects found or the frequency indicates, then the QC manager signals the production manager that there is a problem. The latter manager uses that information to investigate the production process and fix anything that is not working properly. It could be a machine used in the manufacture has gotten out of order or it could be a new employee who did not get proper training (or any number of other problems). The manager is acting like a homeostatic mechanism or a general feedback controller where the QC inspection provides the necessary error information.

Question Box 9.6
Why does control demand closed-loop feedback processes? What happens when government becomes "out of touch"?

9.5   Factors in Control

Before moving on to the higher levels in the hierarchy, there are some details that apply to all control mechanisms. These factors need to be taken into account especially when designing a control system for, say, a complex machine. In living systems, these factors were "designed" through natural selection. They constitute some of the most important characteristics that evolution could work on to produce fit species over time.

The factors fall into three broad categories. There are temporal factors or issues of timing that are extremely important. The second category concerns the power to affect changes in the controlled process or how capable is the actuator in causing the needed change and at what cost. It turns out that more powerful actuation is also more costly. We will look at the various costs associated with control and see how they can be minimized in certain circumstances. The third category considers the kind of computation that is needed to obtain the best control response. As with the power of the actuator, the complexity of the computation also has a cost. However, as we saw in the last chapter, since computation is done at much lower energy levels than actuation, the costs associated with computation may have more to do with time (remember bounded rationality?).

9.5.1   Temporal Considerations

As every comedian will tell you, timing is everything! So it is with control, to a degree. Timing may not actually be *everything*, but it is supremely important for success. Below we will consider a number of temporal issues that must be taken into consideration either when analyzing a control system or designing one. Some of the more difficult things that go wrong in controlling a process have to do with timing issues. This is not a book on control theory per se, so we will only take a look at the more qualitative issues involved in temporal considerations. Many very thick books have been written about these considerations in control engineering, a sampling of which appears in the chapter bibliography. So we won't try to cover all of the details. Our intent is to introduce the reader to what the issues are and give a sense as to why they are important.

9.5.1.1   Sampling Rates and Time Scales

All control systems face the same kinds of issues when it comes to dynamics. How rapidly do the parameters being monitored change? How rapidly does the control system need to respond to changes?

A quick diversion regarding measurement theory, however, is in order. Control depends on measurement of physical parameters and measurements can be accomplished in one of two ways, depending on the physical nature of the measuring device and the computation device being used. Processes can be measured either continuously or in discrete time. An example of a continuous measurement would be the old style thermostats that used a coiled bimetal spring to measure temperature. Such a spring coils tighter or uncoils in response to changes in the temperature. Attached to the spring is an elongated bulb of mercury (what we didn't know in the old days about the toxicity of mercury didn't hurt us—much). This bulb would change its angle with the coiling/uncoiling and it contained, in one end, a pair of electrical contacts. When the mercury flowed to the point of bridging the gap between the contacts, it closed the electrical circuit causing the furnace control relay to turn on the gas and the fan. This is an example of an analog control working in continuous time.

Modern thermostats work on a different principle entirely, though they accomplish the same task. A thermistor or thermocouple sensor is used to measure the temperature. Technically these devises are analog, meaning they smoothly track the temperature change in time. But their electronic value (resistance or current flow, respectively) is only sampled periodically and that value is converted to a numerical value that can be used by a digital computer chip to track the changes in temperature numerically. The measurement is taken at discrete intervals in time, but is taken so rapidly that the computer can approximate the analog value and respond as well as (or maybe better than) the older analog thermostat.

In some systems, the control loop can operate continuously, such as in the case of the steam engine governor or the simple thermostat. Mechanical systems are generally of this type. Of course, at molecular scales we discover that such macro-continuity is only apparent; molecules react in bursts when you look really closely. It is still in many cases useful to speak of continuous control loops, if only to distinguish them from the many cases where control loops operate with intermittent gaps evident even at the macro-level. In such cases, the control loop operates in discrete time, meaning that it takes measurements and actions at discrete points in time, and understanding the consequences of these gaps is essential to comprehending the nature of the process. Even though the examples from biological homeostasis might appear to be continuous when looked at from the scale of human real-time, in fact, on the molecular scale, the events are discrete. What lends to the perception of continuous monitoring and acting is that these operations are carried out by numerous asynchronous parallel mechanisms. Each channel is activated in discrete time, but there are many, many such channels operating over the surface of the cell and their spatially averaged activity is effectively continuous in time, a structural mitigation that fills in the unavoidable gaps.

However, there are many examples in the human-built world of clearly discrete time control loop mechanisms. Today so many of our machines are controlled by digital computers. Even our humble mechanical thermostats and engine speed regulators have been replaced with digital equivalents. Digital systems have to necessarily operate in discrete time, but we can contrive ways they can do it that approximates continuous time.

There are many timing issues involved in closed-loop control. The ideal situation would have the controller track or follow the fluctuations in the critical parameter instantaneously and respond with exactly the right counter forces so as to minimize any deviations from the ideal set point. However, in real systems, there will be time delays in reactivity that cannot be avoided. For example, in the case of cells responding to changes in critical parameters, it takes time to open the channels and for the cascade of reactions to occur even before the cell can react. It then takes more time for the level of reaction to build. And finally, it takes time for the response to start taking effect by seeing a measurable counter effect.

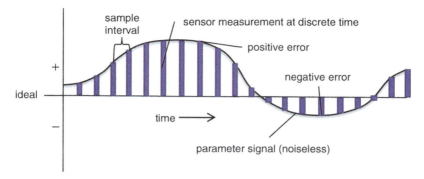

Fig. 9.6 An output parameter can be measured at discrete time intervals to approximate the continuous values that will then be compared with the ideal in order to compute a control signal (see below). The *purple bars* represent measurements taken at sample intervals. The values can be converted to digital integers via a device called an analog-to-digital converter. In this graph, the parameter signal (produced by the sensor) is smooth. Below we will see that real signals are generally noisy and that can cause measurement errors. As the signal goes above or below the ideal value (the *horizontal line*), the comparator will compute a positive or negative error that will be used to determine the control signal value

Digital systems typify these problems: they have to take these time delays into account in terms of how often they measure the critical parameter, how much time it takes to compute the best control signal, and how quickly the system responds to that signal. Today there are exceedingly fast digital converters (called analog-to-digital converters, ADC) that can take a simple sensor reading, like a voltage, and turn it into a digital number for use in a computation (the comparator and control model). The figures below show some of the attributes of the control problem as handled by a discrete time computation. Perfect tracking, computation, and response are not possible, but if handled well, the unavoidable discrepancies of these three interdependent parameters will still track with the process closely enough to keep it in control.

Figure 9.6 shows an idealized picture of a parameter (say temperature) measured at discrete time intervals. The parameter value oscillates around the ideal value, and the measurement is used to compute the error signal that will result in a countering control signal to the process actuator. If the system is properly designed, the control signal will cause the error to drop quickly but not so quickly as to cause it to go in the opposite direction.

In Fig. 9.7 we see a situation in which timing issues cause a system to enter a deteriorating control condition. If the control signal is delayed too much in countering the parameter error, then the controller may overshoot. In a case where the control actuation is very effective, this could then cause the parameter to overshoot the ideal in the opposite direction. If the actuation is not as quick and the parameter response lags, then the two signals will end up chasing each other with some danger that the whole system may experience increasing amplitudes that drive it out of control completely.

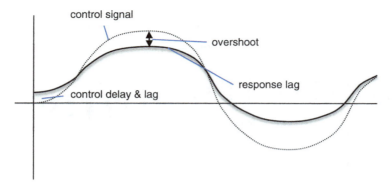

Fig. 9.7 Timing issues and actuator ineffectiveness can cause systems to lose control and oscillate wildly about the ideal set point

9.5.1.2   Sampling Frequency and Noise Issues

We are interested in controlling all sorts of parameters in systemic processes that unfold at very different time scales marked by very different periodicities. How often should the control system sample the parameter signal? How often should a grade school student be tested for progress in reading? How often should one visit the dentist? What would be a frequency curve for checking on whether a cake is baked, or wine fermented? The issue is one of reproducing the signal of the parameter we wish to control with adequate fidelity. The discrete time computation is attempting to produce an accurate, and hopefully reasonably precise, replication of the parameter signal in order to compute the best control signal. In Fig. 9.7 above the parameter is shown as naturally oscillating around the ideal value (desired value). The objective of the control is to minimize the deviations from the ideal (see below for discussion of "costs" associated with deviations). Therefore, it is best for the measurements to be made fast enough so as to capture the slightest change necessary to achieve this objective. Notice that the oscillations or variations from the ideal line shown are typical in that there is no simple frequency or amplitude for such fluctuations. Most systems undergo such deviations due to natural, stochastic disturbances, and hence the oscillations are actually a mix of many elements that condition the process we wish to control. The oscillations of a communications signal, for example, prove to be a mix of many adventitious frequencies. Harry Nyquist (Swedish engineer, immigrant to the USA, 1889–1976) devised a rule (proved as a theorem) that maximizes the fidelity of measurements meant to recapitulate such a complex signal. The rule is to measure at a rate twice the frequency of the highest component frequency in a signal. This sampling rate assures that the measured value at each point in time will be as accurate as the measuring device (sensor) to within an acceptable margin of error. Sampling at a faster rate would be uneconomical. Sampling too much slower would open up the opportunity for a sudden change in the signal to be missed. And control could be lost (as in the above figure).

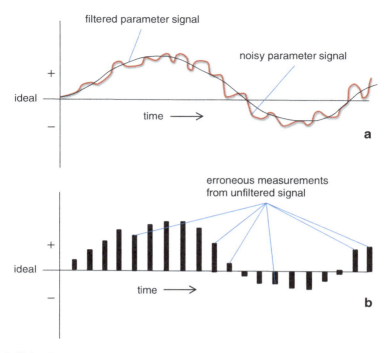

Fig. 9.8 Using the same sampling rate on a signal containing a higher-frequency component could lead to measurement errors that could affect the performance of the control system. (**a**) An unfiltered (noisy) signal is compared with its filtered version (*smooth black line*). (**b**) If measurements are made on the unfiltered signal measurement, errors are likely, which will have an impact on the entire control system

There can be a problem associated with this approach that also needs to be addressed. In many real systems, parameter signals are corrupted by noise, which produces high-frequency components that are not actually part of the signal.

Figure 9.8 gives a sense of the possible errors in measurement that might occur if the parameter signal contained a higher-frequency component from noise. The composite wave form is not smooth as was shown above. Furthermore, it might not make sense to sample the parameter any faster (which Nyquist requires). The reason is that this puts much more load on the computational process which must now run faster and deal with much more data. If there is noise in the data, one way to handle this is to filter the signal through a device that removes the high-frequency portion, thus leaving the smoother real parameter signal. In communications parlance, we are going to improve the signal-to-noise ratio before taking measurements.

Noise is ubiquitous in nature and that includes human-built systems. Filtering or smoothing the parameter signal helps keep the error and control signals true to the actual situation, thus improving the overall control of the system. Today, with modern very high-speed microprocessors, it is actually possible to not prefilter the signals but to sample at the higher rates that include the noise component and then perform digital signal filtering. This is turning out to be less costly than building analog filters that condition the signal before measurement. See Quant Box 9.1 for an example of a digital filtering algorithm that is simple, fast, and yet works surprisingly well in many situations.

Quant Box 9.1 Filtering High-Frequency Noise As seen in Fig. 9.8, noisy signals can cause difficulty when trying to compute an accurate control signal. There are a number of ways to filter out high-frequency components through what are called "low-pass" filters (the pass through the low-frequency components).

One simple approach to doing this is called smoothing to remove jitter from fairly low-frequency signals such as that from measuring room temperature with a thermistor. This process can be readily done in software obviating the need for special (and more expensive) analog filters. We use a method that computes the time average of the signal and replaces the original with the averaged value. What we are looking for is a signal like the black line in Fig. 9.8a.

Time (moving) averaging can be accomplished in several different ways, but what we seek is a fast, efficient method that gives a good approximation. The exponential weighted averaging (EWA) algorithm fits the bill very nicely. The basis for this algorithm is

$$\hat{s}_{(t+1)} = \alpha x_{(t)} + (1-\alpha)\hat{s}_{(t)} \qquad\qquad \text{(QB 9.1.1)}$$

$\hat{s}$ is the time averaged signal
x is the measured parameter value at time t
α is a constant, $0 < \alpha \le 1$

Equation (QB 9.1.1) is iterated, with each sample of the variable x taken at each time step t. The value of x is the measured sample. If α is relatively large, say greater than 0.5, then the average will change more quickly in the direction of x, so the average, $\hat{s}$, will tend to be more like x. If α is small, say in the range of 0.25, then $\hat{s}$ will tend to deviate more from values of x. This is what smoothes the trace line. Effectively Eq. (QB 9.1.1), with a smaller α, will ignore deviations from the ongoing average, treating short time span deviations as high-frequency noise.

This algorithm has the advantages of not requiring much memory, just two variables and two constants ($1-\alpha$ is usually pre-computed at startup and treated as a second constant), and very simple arithmetic. The EWA formula can be shown to come arbitrarily close to any time window moving average formula, but the selection of the value for α is sometimes hard to derive.

Sampling rates may be established with great precision for cybernetic control of many sorts of mechanical processes. But the challenge posed by determining a suitable sampling rate is as broad as there are kinds of process to be controlled. Whether it be managing a business, a garden, a relationship, or frying a fish, the closed loop of control begins with some form of monitoring, and finding the sweet spot between what we commonly recognize as sloppiness on the one hand or over-control on the other is commonly a question of the appropriate sampling rate.

Question Box 9.7
Managing social relationships of various sorts can be one of the most challenging "control" projects. How frequently should you "touch base"? How do you distinguish noise in the communications channel from the serious messages? How do people's set points differ? Are there any rule of thumb heuristics you use for this incredibly complex calculation?

What has been described in terms of man-made digital control systems is really just as true for naturally evolved systems (e.g., living systems) as well. Evolution produced measurement devices (sensors like taste buds and retinal cells but also proprioceptors[6] that measure internal conditions such as tension in muscle tissues), communications channels (neuronal axons), and many different kinds of mechanical and chemical actuating devices (muscles and hormones) all of which must operate in an appropriate time scale and deal with a good deal of noise (e.g., from thermal vibrations). This was accomplished, as we will see in the coming chapters, by the fact that any of these subsystems that didn't work well would have resulted in the kinds of control problems discussed above (and below), leading to loss of control and death. Thus, poor control systems were weeded out by natural selection, leaving only well-tuned controllers to serve the purposes of living systems.

9.5.1.3 Computation Delay

An analog controller such as a mechanical shut-off valve can react immediately to the incoming values of the parameter. Even so, a mechanical device has inherent inertia in the mechanisms doing the computation. And in digitized systems, a discrete computation must be done algorithmically, and as we saw in the last chapter, this has to be done with sequential instructions. Computing a solution can thus take more time. Fortunately modern computers used for embedded control are exceedingly fast, making it possible to have fast sampling rates and produce a result, a

---

[6] See http://en.wikipedia.org/wiki/Proprioception for information regarding self-sensing.

control output signal, quickly enough to respond to the changes in the parameter in what we call "real time." We now trust the computerized braking systems of our cars to produce a more accurate, modulated response to conditions than could ever be achieved by older mechanical methods.

In living systems, the problem is a bit more difficult to resolve. Living systems like animals have to respond to real-time changes in their environment. Their brains have to collect the data and process it to generate decisions on what actions to take, when, how, and how much. Time delays in getting data, sufficiency of data, and time taken for brains to perform their computations are limited when compared with the need to take rapid action. This is known as *bounded rationality*.[7]

A decision that needs to be made quickly but for which there may not be enough time or available data to compute a completely rational answer is bounded by time and data limits. Living systems (animals in this case) have evolved support computational systems that operate heuristically (as in Chap. 8) and produce quick and dirty solutions. Heuristics, recall, are rules of thumb that usually work but are not guaranteed to do so in every instance. Thus, animals might make the wrong decisions in some cases and wind up dead. Once again evolution will weed out heuristics (think of them as instinctive behaviors) that don't work well enough most of the time to allow a species to successfully pass genes to subsequent generations.

In the last several decades, control engineers have actually started to borrow ideas about how to get around bounded rationality in machine control systems from our growing understanding of animal brain computations. We can now find heuristic rules used to supplement algorithmic computations when time is of the essence and data is limited.

9.5.1.4   Reaction Delay

The controlled system will also have an inherent inertia or lag between the time a control signal is given and the parameter itself starts to respond. As long as the error signal shows a need for a control, the system may try to overcompensate, which, in turn, will produce too strong a control signal when the parameter does start to respond. This can either create oscillation problems, for example, if the system is inherently sluggish in one direction but not in the other, or it simply results in ineffectual control with wasted efforts (costs).

A control actuator has to have the power to respond to the control signal as quickly and effectively as possible. A weak actuator cannot get the job done and will end up costing the system more resources than it can afford. On the other hand, stronger-than-needed actuators might be more expensive to install and maintain. Later we will

---

[7] We are using this term a little bit more broadly than the context in which it was first introduced. See the article http://en.wikipedia.org/wiki/Bounded_rationality for more details on the psychological uses of the term.

discuss adaptive control systems and show how one kind of adaptation, the longer-term strengthening of the actuator (e.g., muscles), can produce cost savings over the life of the system being controlled.

9.5.1.5 Synchronization

We are inclined to think that faster is better, but proper synchronization of sampling, computation, and response is most fundamental. Sampling, computation, and response are interdependent not only analytically but as sequential steps of a control feedback process. The rate of computation and response enters into the appropriate sampling rate, for many processes are undermined when a new round of sampling-computation-reaction is launched before the results of the last round can appropriately enter the new sample. Think of what happens when you impatiently hit a command key or a combination of command keys as you impatiently "over-sample" the screen of a slow computer program!

Question Box 9.8
Human processes are full of computation and reaction delays—just look at the US court system, for example! It's no surprise that when speed and efficiency are the measure, our machines have outperformed us even before the current electronic/digital revolution. Yet our machine-enhanced lives can get pushed to tempos that are uncomfortable and counterproductive. What kind of systems have our bodies and brains evolved to keep in time with? As we speed up through technology, our agriculture machine assists crops, cattle, and pigs, etc. to keep up. What about the rest of the life community?

9.5.2 Oscillations

As the Fig. 9.7 shows, the nature of control involves reducing the amplitude of the parameter trace. That means the control is attempting to keep the parameter value as close to the ideal as possible. Along with controlling the amplitude of deviation is the control of the frequency of what is called the "zero-crossovers" (crossing the axis or zero point, also known as the average wave length of the signal). Ideally the parameter signal would be exactly at the ideal, meaning that the deviation or error value would be zero. But, unfortunately, this is never possible in practice. We have already shown that the signal will cross from positive to negative and back to positive error values in natural course. If the control system is very effective, this crossover will be minimized over time with very low amplitude.

Unfortunately there are conditions, mostly driven by the time delays we have just examined, where the application of a miss-timed control signal can cause the system to oscillate more, in terms of zero-crossover or amplitude, or both, as time goes on. The system goes *out of control* as a result of attempts to control it. We often experience these dynamics, for example, when trying to carry an overfull cup of liquid, where a perfectly timed and calibrated correction can keep it from sloshing over, but one overcorrection leads to a bigger counter overcorrection and we're looking for a sponge to clean up the spill.

9.5.3 Stability

The ultimate objective of a closed-loop feedback control system is to achieve operational stability in the face of disrupting environmental influences. In producing stabilized functionality, control mechanisms, whether evolved or created by human ingenuity, confront the challenge of the unexpected and therefore disruptive event, and often the more unexpected the event, the more disruptive. So the question that establishes the limits of closed-loop control is, how good is the control in the face of highly unexpected events? When a very rare but extreme event pushes the system far from the ideal for a short period of time, how quickly can the control system respond and restore acceptable levels of the parameter value? As we will see below, control necessarily comes at a cost, and extending the effective reach of control involves increasing costs. At some point, the logic of diminishing returns sets in. The cost of installing the next increment of control (measured linearly in terms of units of error responded to) begins to cost more than the previous increment. Or, put alternatively, each new increment of cost buys much less actual improvement in control.

Figure 9.9 shows two situations. In Fig. 9.9a, the system is stable and returns to a normal operation after being disrupted by a step event (a short-term extreme disturbance). In Fig. 9.9b, the system cannot recover from the disruption and, due in part to delay conditions or weak actuation, goes wildly out of control. In "natural" systems, such a situation results in the system failing and disappearing from the scene. We will see this situation play out in the evolution of systems (next chapter). Those systems that are "fit" are able to restore stable operations after such a disturbance.

Question Box 9.9
Governments sometimes have to respond to disruptions of civil life by demonstrations or even riots. What does Fig. 9.9 suggest about handling these situations? How does the strength and security of the government factor into the situation?

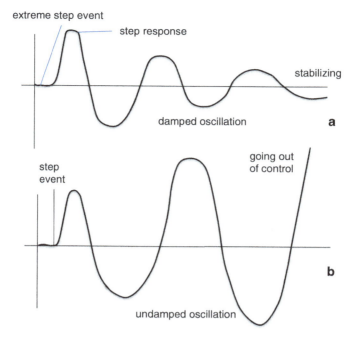

Fig. 9.9 Two different systems can respond to the same extreme event (step event) in different ways. (**a**) This system responds in a measured manner and has a strong actuation capability. The disturbance, as measured in the trace of the error signal, is damped out and the system is restored to normal operations. (**b**) The system may overrespond too quickly or too forcefully and go wildly out of control. The control feedback becomes positive rather than negative, causing the undamped oscillation to explode

9.6 Control Computations

The control signal that will be sent to the in-process actuator has to be computed from the error signal generated from measuring one or more product parameters. These computations can be quite daunting. Here we will look at a few kinds of computations that are used in the case of a single parameter feedback loop situation. And we will introduce a variation on the feedback form of control that uses additional information to get a head start on computing a control signal.

9.6.1 PID Control

PID stands for *p*roportional, *i*ntegrative, and *d*erivative control. These terms come from the mathematical nature of the treatment of error signals in deriving a control signal. The first term, proportional, is simple algebra. The control signal is proportional to the error signal, usually, though not necessarily, linearly. The second and third terms derive from the calculus, where an integral describes the way a function "accumulates" over time (at least in the context of control) and a derivative describes a rate of change (velocity). We will explain these terms below without resorting to the calculus necessarily. Readers interested in the more formal mathematics can refer to Quant Box 9.2 for details.

PID controllers are computational processes that combine these three ways of handling error. It is often possible to get by with just proportional or proportional and derivative or proportional and integrative components and still achieve adequate control. Each method has strengths and weaknesses in different sorts of situations, so in combination they can complement one another and enable controllers with a wide range of applicability. PID controllers, or subsets thereof, are therefore the most widely used form of control in industrial applications. Here we will give a brief review of how this model of control is generally applicable in most controlled systems.

We should note, however, that PID-like controllers are to be found in natural CASs as well. We will give a few examples as we go. Here we are focusing on the computational aspects and so are using automated control systems to show the algorithmic aspects of the control.

In many cases, a control signal can have a simple proportional relation to the error signal generated by the comparator. In mathematical terms, the control signal, c, is directly proportional to the error signal, or $c_{(t+1)} = k_p \, e_{(t)}$. In this equation, $c_{(t+1)}$ is the control signal issued in the next time increment, k_p is the constant of proportionality, and $e_{(t)}$ is the error signal at time t.

What this equation says is that the amount of control is directly proportional to the amount of error. If the error at time t is e, then the value of the control signal will be based directly on this error by some parameter, k. All control systems will have some computational component that is based on this equation. Proportional control is, therefore, basic to all forms of control. If the deviation increases (either plus or minus), the contravening control signal will increase in proportion. This is well suited to systems where the degree of deviation is the essential control question: if the flow is 2 gal per minute and it should be 1.6, a signal to decrease the flow by 20 % is a sufficient correction.

Unfortunately, in many control situations merely reacting with proportional control can give rise to inefficiencies that have real costs to the whole system. It is possible to estimate the rate of change in the error (its velocity) so that the control signal might be modified to take account of that rate of change (in calculus this is the first derivative). If a deviation is small, but increasing at a high rate of change, then it would be useful to increase the control signal somewhat by an amount proportional not to the absolute error but to the rate of change of that error. If the error is increasing rapidly, then the control signal should be more powerful than the simple proportional signal would have been in order to subdue the tendency to overshoot. In other words, the control signal should be slightly larger as a result and the response should start to decline faster than by proportional output alone.

Quant Box 9.2 The Mathematics of PID Control

Here we will show the basic form of the PID control formula. Figure 9.10 demonstrates the advantage of using the derivative term to produce a more effective control response. In these equations, c is the value of the control signal that will be issued in the next time step $(t + 1)$, e is the error value (the desired set point minus the actual reading) at time t, and the constants of proportionality k_P, k_D, and k_I are values determined empirically, in general.

Proportional component:

$$c_{P(t+1)} = k_P e_{(t)} \qquad \text{(QB 9.2.1)}$$

Derivative component:

$$c_{D(t+1)} = k_D \frac{d}{dt} e_{(t)} \qquad \text{(QB 9.2.2)}$$

Integrative component:

$$c_{I(t+1)} = k_I \int e_{(t)} dt \qquad \text{(QB 9.2.3)}$$

Composite:

$$c_{T(t+1)} = k_P e_{(t)} + k_D \frac{d}{dt} e_{(t)} + k_I \int e_{(t)} dt \qquad \text{(QB 9.2.4)}$$

Computers don't do calculus, exactly. The derivative of error can be estimated by taking the difference between error at time t and that at time $t-1$. This works if the error is filtered as in Quant Box 9.1. The difference represents a very short interval slope of a line tangent to that interval. Similarly the integral term can be approximated by keeping a running summation of the error.

Figure 9.10 shows two graphs of the same parameter trace under different control signals. Figure 9.10a shows a trace of both the parameter and a simple proportional control response. Note that the deviation of the parameter trace grows to a larger size than in Fig. 9.10b as the lag in the proportional signal consistently underestimates the escalating deviation. In Fig. 9.10b we see the results of including a derivative component in calculating the control signal. The trace is labeled with the points in which the derivative is positive, meaning the error is growing larger; zero, meaning that the error has ceased to increase; and negative, meaning that the error is starting to decrease. Note the dynamics of the PD control makes it much more responsive to the error change. In essence, the derivative signal boosts the proportional signal so as to more quickly have an effect on the error trace. As a result, the deviation is not nearly as pronounced as it was in Fig. 9.10a.

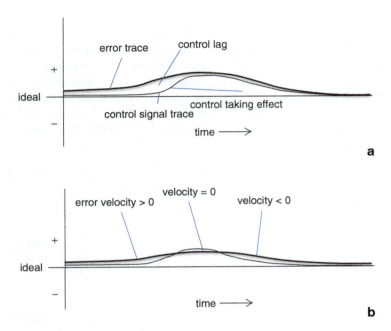

Fig. 9.10 A comparison between a P (proportional, **a**) and a PD (proportional-derivative, **b**) controller over a parameter trace (error) caused by the same level of distortion force. A PD controller helps to reduce the amplitude of the error and can greatly improve the system dynamics since it responds more quickly with a restoring force. In **b**, the derivative of error, or velocity of error change, is used. When it is greater than 0, the control signal will rise faster, providing a stronger counter force to the distortion force. As the rate of error increase slows to 0, the signal only rises in proportion to the actual error value. When the rate of error increase goes negative, the control signal goes down accordingly

PD control is widely used in many practical control systems designed by humans and can be found in many natural systems. But there are situations where the real-time proportion and the rate of change are not sufficient to determine an optimal control signal. Some systems may be subject to transient pulses that are not noise, per se, but also do not need to be responded to because the transient will be damped by extrinsic factors. For example, a short-lived gust of wind should not cause a wind turbine to feather its blades more than the average of the steady wind. The integral term in a PID controller provides some information about the history of the error so that the controller will not overreact to these transients.

PID control theory has been used broadly in many, perhaps most, control designs for human-built machines. However, it also provides a model basis for analyzing natural control systems such as those found in living systems.

9.6.1.1 PID in Social Systems

The language and conceptualization of PID control strongly reflects the background of cybernetics as an endeavor to automate the control of mechanical function. We can measure with precision far in excess of the capacity of our senses, iterate computation processes, and mechanize responses in temporal and spatial dimensions both smaller and larger than available to human bodies. But the basic dynamics and issues of control are common to systems at all levels of organization, including the levels we experience in daily life and in participating in the institutions and processes of the social world. The changing array of circumstances to which we must constantly respond is so varied and complex that they can hardly be captured in the various artificial systems we create to measure and control the many sorts of "deviance." Social engineering may or may not deserve its dubious reputation, but its field of application falls far short of the systemic complexity of real life. Rather, to see PID in action, we should look at how we use these categories as guides for the way we respond to situations.

Proportionality is an almost constant concern, reflected in notions such as working too hard, eating too much, getting too angry, or spending too much money. The word "too" expresses some degree of deviance, and the appropriate response in mitigating the "too" should be proportional. We see our departure from some sort of norm as proportional and transfer that sense of proportionality to our response to control the problem.

Rate of change is such a vital part of guiding the response to many situations that we seem almost hard-wired to react with concern and alarm as rates of change accelerate. Living organisms in general are structured systemically in ways that expect their environment, and the same systemic role of expectation functions with the systemic elaboration of sensory and conscious worlds. Complex adaptive systems learn from and adapt to change, but the rate of change is itself a critical parameter. Too much, too fast, too sudden, these are rate issues that naturally inform the responsive sensitivity by which we control and maintain conditions in which we can flourish. We become suspicious, for example, of people who profess deep friendship after a brief acquaintance.

The integral measure takes account not only of what's going on but how long it's been going on. The change-resisting character of human affairs on all sorts of levels is intertwined with the patterning effects of repetition through time. This is how personal habits, institutions, whole cultures are formed, and their ability to resist attempts to reform and correct them is well known, so this integrative calculation is an automatic element of assessing a strategy to bring about a course correction. This awareness joins forces with our anticipatory ability in a kind of feed-forward responsiveness embodied in the common notion that we must "nip it in the bud," intervene before time, and repetition allows a problematic pattern to form. This integration not only with past but also anticipated history can add force to corrective measures that might otherwise seem out of proportion to the problem.

Question Box 9.10

How does PID factor into the "get tough on crime" enthusiasms that periodically arise? How would different situations (in PID terms) suggest different degrees of response? What would be a reasonable sampling rate to monitor for adjusting expensive tactics such as heavy enforcement, long prison terms, etc.?

Fig. 9.11 More advanced forms of operational control can be achieved with information fed forward from monitoring inputs directly. The control model will necessarily have to be more complex and the computation task will likely take longer

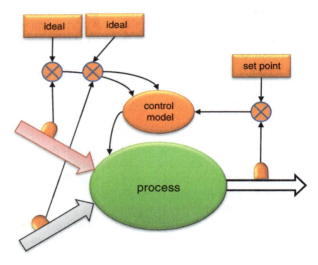

9.6.1.2   Information Feed-Forward

Humans, as we have seen, move into the future with anticipation. This is an important advantage when it comes to control processes: if one can somehow calculate and begin a response to a change *before* it occurs, the problem of lag time is greatly reduced. We can emulate this in some processes by introducing monitoring and feedback on flows even before they enter the process, a strategy described as information feed-*forward* rather than feedback. It is possible to refine the operational control of a process by using this feed-forward information regarding changes in the levels of input flows. Using a similar arrangement of sensors, comparators, and ideal set points, the control model can use this information to start making adjustments to the work process even before the output is adversely affected. This works best when there are sufficient delays in the actual work process itself such that this early information about fluctuations that *will* affect the output can be used effectively to respond and attempt to maintain the quality/quantity of the output product (Fig. 9.11 above).

Of course, the control model must be more complex than it was for simple feedback. The use of feed-forward is an early version of anticipation. As an example of a control system with this form of anticipation, consider the thermostat we've already looked at. More advanced thermostats, such as those used in more complicated heating, venting, and air conditioning (HVAC) systems in commercial buildings, include sensors outside the building that measure outside temperatures. Since the rate of heat loss or gain of a building is proportional to the temperature difference between the inside and outside, having the outside temperature, and knowing the heat transfer rate constant across the building boundary, the controller can compute the anticipated temperature at some time in the near future and turn on the heating or cooling system in advance of that time in order to do a better job of keeping the temperature within a narrow range of comfort. In an even more advanced version of this system, infrared sensors can be placed in public spaces to measure the amount of heat being generated from inside sources (like human bodies) and incorporate that information into the computation as well.

Obviously the computer model of the building is getting more complicated. But the gains in cost savings (see below) often make it worthwhile. Moreover, since these kinds of systems serve large commercial spaces, the maintenance of comfortable temperatures will keep the customers happy!

Note that feed-forward control of this sort acts similarly to the derivative term in PID control. It is trying to get out in front of the change before it happens. Note also that the PID approach is just as useful when looking at the use of feed-forward information. That is, PID can be applied to the feed-forward error just as it was for the feedback error.

9.6.1.3 Multiple Parameter Algorithms

PID control is extremely useful for many systems at a very low level of operations. The automatic velocity control on most automobiles is a case in point, where the only objective is to maintain a constant speed regardless of whether the car is climbing a hill or on a flat road. But consider the control problem of obtaining the best gas mileage under varying driving conditions. Gas mileage is affected by several parameters that can cause it to go down, including humidity, oxygen levels, temperature, and the weight of the foot on the gas pedal. The latter parameter is nearly uncontrollable as long as humans are doing the driving, so we won't worry about it just now. The control actuation for this problem is the fuel-air mixture, which is handled by the fuel injector subsystem. Changing the ratio of fuel to air mix can help keep the consumption of gasoline at a feasible minimum at a given speed and under those driving conditions. There is a computer in most cars these days that handles this. In fact in many newer models, the computer has a readout that the driver can see that shows the current results and also shows how the mileage goes down when the foot gets heavy. That kind of feedback puts the driver back in the loop—if the driver chooses to use the information!

As additional parameters are added, the control problem becomes more complex, and simple PID feedback won't really be sufficient. In the next section on logistical control, we will be examining the problem of optimization of an objective function under multiple requirements and constraints. We can see that the gas mileage problem is essentially such an optimization problem, only being in this case applied directly to an operational process.

Controlling a process is essential to maintain stability and sustainability while minimizing costs to the system. We next look at the costs that a system incurs as compared with the cost of having good controls.

9.6.2   Systemic Costs of Non-control Versus Costs of Control

Achieving a more optimal control, meaning keeping the amplitude and phase shift of error to a minimum, is needed because in physical systems there are real costs associated with too much deviation or deviations lasting too long. We start with the premise that sloppy control puts demands on system resources (costs) and those demands would be lessened if controls kept the system operating within its desired parameters. However, we also recognize that control does not come for free. There has to be a control subsystem with sensors, communications networks, computational processes, and actuators that will consume energy to do their work. What we need to consider is what these cost trade-offs are and how we might minimize overall costs. In the biological world, the "design" that produces this objective is worked out by evolution. In the human-built world, engineers and managers have to figure it out.

Fundamentally there are three basic cost components. All three can be considered in terms of energy required to "fix" the situation.

Cost of damage from loss of control: Any time the system is operating too far from its ideal, it is likely to incur some kind of damage that will take energy to repair (assuming the system isn't damaged beyond repair). This energy has to come from reserves and will not be available for other purposes (i.e., there is an opportunity cost with respect to using energy to repair damage as opposed to, e.g., building new structures). Therefore, it is a cost-saving measure to prevent damage from occurring.

Cost of control actuation (*responding*): Assuming a control system, with actuator(s) in place, that can counter the effects of damaging deviations, it still takes energy to do the work of responding. The control system has to do this work at a low enough energy cost that it will not overshadow the costs of damage. This trade-off is governed largely by risk factors. The degree of control and kind of control are determined by how frequently damaging events might take place and what the cost of that damage might be. This cost is not unlike the deductible one pays on an insurance policy. Control actuation costs each time it is used. But hopefully it won't be needed that often to combat large deviations.

Cost of maintenance of control system: The final cost has to do with maintaining the control system in ready condition. The control system, including the actuator, is subject to the second law of thermodynamics just like every other physical system. It requires work be done to maintain it in functioning order. Typically this is a low-level, but long-term cost.

Ideally the cost of repairing damage will be kept low by virtue of incurring the costs associated with response and maintenance. However, the maintenance costs for having a more complicated controller have to be considered. The greater the complexity of a control system, the more things can go wrong and therefore the greater need for energy used to maintain it. Happily, since these costs are incurred over longer time scales, their short-term value tends to be small in comparison with the cost savings enjoyed by using them.

The same set of parameters can easily be seen in the mechanisms by which we exercise control in human organizations. Accountants, bureaucrats, personnel departments, police, military, and a myriad other ways we organize for control all come at a price. We usually think of the costs in terms of dollars and cents, but monetary expenses can all ultimately be pushed back to the kind of energy expenditures we considered above. Even the well-known "time is money" dictum is just another counter for energy, since only time spent without energy expenditure, literally doing nothing (if that were possible), would indeed be free time.

PID introduces us to the basic objectives of control and the dimensions through which they are most easily addressed in mechanical systems. Living systems, as we have seen, move into the future adaptively, with potential for learning and anticipation. These abilities enhance what can be done with PID and can even achieve degrees of cost minimization, so not surprisingly we are finding ways to emulate them with our mechanized/computerized controllers.

9.6.3 More Advanced Control Methods

Now we will take a look at control architectures that while employing the controls we have looked at so far go beyond those simple feedback and feed-forward information loops to achieve even greater levels of persistence in changing environments. There are many circumstances in which those simpler control schemes will not suffice over the long haul.

One could easily argue that one of the distinguishing characteristics of living systems is their ability to employ these advanced sorts of controls to the problems associated with living in changing environments. It may be possible to learn how to emulate these controls in human-built systems. There has been a modicum of progress in this direction.[8]

---

[8] For example, the use of machine learning, especially Bayesian networks, is creating a whole new category of adaptive machines (robots). See Sutton and Barto (1998).

9.6.3.1   Adaptive Control: The "A" in CAS

Integrative control involves a kind of "memory" being used to modify a control response based on recent history. In living systems in particular, memory-based control modification allows a system to use history to be more responsive to errors.

Memory impacts the control situation insofar as it provides a means of not just mechanically advancing into a future but of preparing for the future. As an example, consider what happens to a person who gets a job that involves a lot of physical work. In people who do not routinely stress their musculature, the muscle tissues develop only to deal with everyday living and the kind of demands that are placed on them (e.g., sitting behind a desk all day). As one begins working at physical labor, the muscles do not immediately respond with greater work capacity and that person gets worn out quickly. Over some time, with continued stimulus, the muscles start to grow new additional fibers because the regular work is telling those muscles that they should expect to do more work in the future.

All living tissues respond to repeated (reinforced over time) stimulus by building up the machinery needed to respond to the stimulus. They are using the information in the stimulus to modify their expectations and preparation to reduce the cost effects incurred by greater stimulus effects. This is also the case in social systems. For example, if a city is experiencing a rise in crime rate, they might respond by hiring more police officers. If a company is experiencing higher demand for a product, they might add manufacturing capacity to keep up (and profit there from). These are all representative of adaptive response to changes in the environment that put greater demand on the system. The point of adapting the response mechanisms for future expected higher demand is to reduce the cost of repairing damage, as covered above. The system keeps the amplitude of the error trace low by having a stronger and quicker response mechanism even if doing so incurs additional response and maintenance costs.

Even in some machines, we have engineered adaptive controllers. For the most part, these are simpler kinds of adaptations than we find in natural and social systems. For example, one simple adaptation would be to just change the set point (ideal) value of the controller. We will see this at work in the next section. But a more interesting approach that somewhat emulates natural systems is to modify the constants associated with the PID control components discussed above. This is tricky, and theory on when and how to make adjustments is still weak. But you can imagine, for example, if a control signal were routinely overshooting the parameter trace, as in Fig. 9.7 above, then one possibility would be to reduce the constant, k_P, value a bit to see if that helps. Of course this would entail including computations to monitor the overshoot and keep a running history of it to use to decide when and by how much to lower the constant.

Let us turn back to living systems, however, to look at how they adapt their controls. After all, living systems are the epitome of adaptive systems. We shall introduce here the language of *stimulus* and *response* as this is the language that biologists use in describing the phenomena. Up until now, our main descriptive

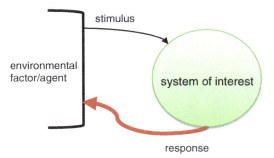

Fig. 9.12 The system of interest is stimulated by some factor or agent in the environment. It responds with an output that is meant to influence that factor or change the relation between the two. For example, the system's response might be to move away from the factor if the nature of the stimulus is harmful

language for systems as processes has been input flows and output flows. Here we will be using stimulus as a specific kind of informational input flow and response as a particular kind of influencing output flow. A stimulus causes the system of interest to respond specifically to the informational content of the stimulus. This is shown in Fig. 9.12.

The relation shown in Fig. 9.12 is very general. In living systems, the stimulus is a message that informs the entity that a situation is developing in its environment. The entity has a built-in response mechanism that is able to counter the stimulus in the sense of reducing the information measure. That is, the system acts as a cybernetic negative feedback mechanism to bring its relation with the environmental factor or agent back to the expected one. For example, say the environmental factor is a bad smell somewhere in the vicinity of the system. The smell is bad because the factor (some aromatic compound) might be associated with something particularly harmful to the system. The system responds by retreating from that area. If the smell level diminishes, then the system continues to move in the direction down the odorant gradient (away from the source of the smell). If it increases, the system can reverse direction in an attempt to remove itself from whatever is causing the odor. Sessile plants and animals have responses to threatening stimuli too. A plant, for example, can secrete an alkaloid substance that will taste bad to an insect munching on it. A barnacle can retreat into its shell and close its front door in response to a bad taste floating by in the water stream. The essence of the stimulus-response (S-R) mechanism is homeostasis, which we ran into before. Graph 9.1 shows a simulation of an S-R system and the various costs that are incurred. Notice the time delay between the onset of the stimulus and the onset of the response.

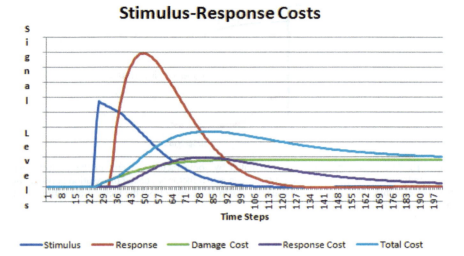

Graph 9.1 A simulation of a typical, "pure," S-R mechanism shows that costs accumulate with each occurrence or episode. The stimulus onset starts the accumulation of damage costs and builds rapidly. The cost of responding starts accumulating slightly after the response onset (which suffers an eight-step delay). The cost of maintenance is not used in this simulation since it is relatively constant at a low base level. Total cost is simply the sum of the two costs as they accumulate. Units for signal levels and cost levels have been normalized

What we will be looking at here is the way in which the S-R mechanism can actually be modified over time and regimes of stimuli such that the system increases its likelihood of success while also attempting to lower overall costs. Graph 9.2 provides traces of the same costs seen above but using an "adapted" response. This response is stronger by virtue of having been modified according to the history of stimuli episodes. The system started from that in Graph 9.1 and changed to be able to respond with greater effect.

Recall the autopoiesis discussion in Chap. 4. There we were using it as a model of complex networks of processes that were irreducibly complex and maintained a system far from equilibrium (alive). Now let us take another look at that model in a slightly different light here. Figure 9.13, below, shows a slightly different version of the autopoiesis model. We have included what we call a response constructor or

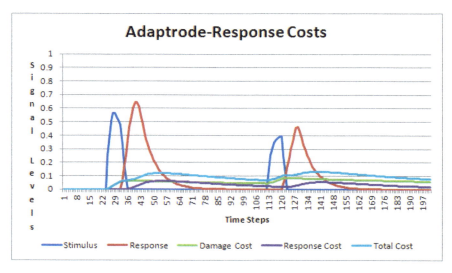

Graph 9.2   Here the system has adapted to respond more quickly to a stimulus episode after already experiencing an episode. This graph shows two episodes of the same stimulus. In the first episode the adaptrode memory is potentiated by the stimulus event (not shown). The adaptrode represents a memory trace and adds its weight to the response after a small time delay. The second episode shows the effects on both stimulus and response with the potentiated adaptrode. The response still has a time delay but has a greater power due to the added weight of the adaptrode memory trace. It therefore damps the stimulus much faster than was the case in the prior episode. The overall effect is for there to be lower cumulative costs incurred from damage and response effort. There is a marginal total cost increase from the additional resources needed by the adaptrode

maintainer process that operates to keep the homeostatic core up to its ability to respond to external disturbances in its critical factor (the stimulus). This process operates on the response mechanism itself, taking the current state of the mechanism as "input" and producing a new, more developed copy as "output," at least conceptually. The material arrow into the constructor and the curved black-outlined arrow out of it capture this idea. Like all processors, this one needs material and energy to do its work.

The constructor is guided in the work it does in building up or maintaining the response processor by the history of errors kept in some form of memory. While we haven't shown it explicitly in this diagram, that history involves not only the integral of error but also the relation between the error and the actual response given and the

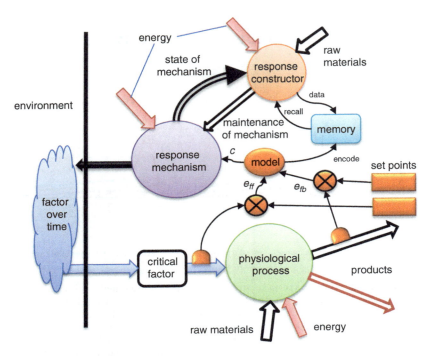

Fig. 9.13 A basic homeostatic mechanism is a model for the stimulus-response mechanisms of living systems. In this version, the response mechanism is modified by a response constructor process as a result of a built-up memory of the error. The response mechanism is being adapted to reflect the longer-term demand placed on the system. Such a system then is able to respond more quickly in future episodes and thus lower overall costs. e_{ff} is error feed-forward, e_{fb} is error feed-back, and c is control

results obtained over time. This is what guides the constructor process. If the error tends to be large over time and the response of the current mechanism is not as successful as it should be (as determined by what actual costs are incurred over that time scale), then the constructor is activated to do its job. And it will ramp up doing its job in proportion to the cost history. The latter can be gotten from the history of usage of material and energy by the constructor.

It turns out that the constructor is controlled in exactly the same manner as the operation (physiological process + response mechanism) with feedback and a control model. The only difference is that the constructor's time scale of operation is much greater than that of the response and it uses time-averaged information instead of real-time information.

Over longer time scales, if the physiological process is being stressed, say in episodes like regular workouts, then the memory of the integral of error is used to activate the constructor to build more equipment, so to speak. Or in the muscle example, more muscle fibers per bundle to increase strength. The response mechanism is adapted to the increased demand and in the future will be able to respond with greater force to whatever external force is stressing the process. Adapting responses based on the running history of stimulus experiences is called demand-driven plasticity. That is, the tissue or cellular mechanisms strengthen if called upon frequently and can diminish in readiness if the demand drops. This is generally hysteretic, that is, the diminishing is slower than the increase in strength, and does not necessarily diminish all the way back down to the starting condition. There will usually be a residual improvement left over, a kind of long-term memory trace.

Question Box 9.11

We often speak of becoming "sensitized." This covers a range from chemical substances to types of sensation to irritating behaviors. Figure 9.13 depicts how this happens. How about correcting oversensitive responses?

We have introduced a fourth category of cost as a result of this adaptive response capability. The new cost is that of constructing more or better response mechanisms. At the same time, we will have an incremental increase in the maintenance cost of the response mechanism, since there is more of it to maintain. But the tradeoff is the same as before. If this increase in costs to build and maintain are substantially less than the costs of damage repair as a result of inadequate response, then it is worth it.

In truth, we have also introduced what might be called a second-order set of costs in terms of now needing to maintain the response constructor itself (not shown in the above figure) and the memory structure as well, along with the computational processes needed to encode and recall data stored there. The response constructor is just another specialized kind of response mechanism, one that responds to the needs of maintaining or building more response mechanism. In real systems, particularly in living systems and economic systems (where we will see these same phenomena), response constructors are not called into action for only specific response mechanisms, and so single constructors can be shared among many mechanisms, thus reducing the costs of maintaining the constructors or amortizing it over many single-response mechanisms. The other kind of second-order cost is that of keeping track of costs! This is all overhead in a business operation, but the same principles apply to living systems. As we will discuss later in the chapter, these kinds of computational processes are involved in what we call logistical management, so we will save further discussions on the subject till we get to that section.

9.6.3.2 Anticipatory Control

As complex adaptive systems evolve (see Chaps. 10 and 11) the need for more mechanisms to maintain and control, the primary mechanisms increase the total cost of operations over time. But also, as we have seen above with the PID controller, the addition of history can help shorten the duration of the potentially destructive deviation and thus reduce the costs of repair. What if we could use history in a manner so as to project into the future, such that we could anticipate a deviation even before it occurred? In some respects, this is what is done by the adaptive control described above, but that control works in a different time dimension, constructing better adapted response abilities. Anticipatory control would be addressed to the loop of ongoing activity involving response to error signals. Done properly we could initiate a control signal even before the error became apparent with the intent of minimizing the error to a much greater extent. We saw this to some degree with the anticipator thermostat up above.

An anticipatory controller is one that can use a forward signal (e.g., feed-forward from the inputs) that gives a warning that deviations in the primary output are about to be felt. The anticipator can then use this warning to initiate responses even before the trouble begins and, in essence, nip the problem in the bud.

Figure 9.13, above, also shows the potential for a form of anticipatory control that uses changes in the input values to provide this forward signal. Systems with this kind of arrangement can experience yet further cost reductions because the cost of maintaining such a system is far less than the costs incurred if the system had to do major repair work.

> **Question Box 9.12**
> Muscle building constructors equip us incrementally for heavier work once they get the message that such work is being called for. What strategies might be available to anticipatory control as possible alternatives to muscle building (we're pretty good at avoiding strenuous work!). What sort of costs go with various alternative strategies?

This is a great example of the principle of investment and return on that investment. Here the system invests in a more complex control system but enjoys considerable cost savings as a result. Hence there is a net saving representing a return on investment. This works in the competitive economic arena, but it also works, as we shall see, in the competitive evolutionary arena of natural selection: there is a good reason the living world is well populated with organisms that have hit upon heritable capacities for various sorts of anticipatory control.

In Fig. 9.14 a sensor is placed on the flow of the critical parameter (influenced by something in the environment). The amount of warning that this parameter can provide is better than nothing, but it is so specific that it is limited in terms of early warning. Complex living systems, particularly animals, have developed even more complex capabilities wherein the sensor is placed on the boundary and senses broad changes in the environmental parameter itself. All of the normal external senses such as taste, smell, touch, vision, and hearing provide means for receiving information from the outside world and using it to anticipate things that will affect the physiological process after some time lag. The challenge is to link this sense-mediated information to the meaningful world of systems impacts even before the

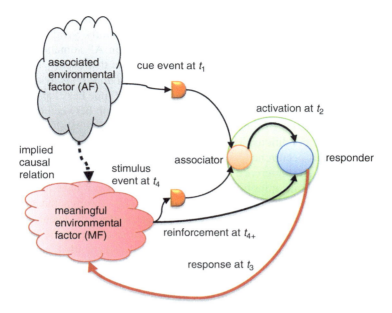

Fig. 9.14 An associative anticipatory system is one that can exploit implied causal relations (time-lagged correlated events) to respond to stimuli from non-impacting "cue" events rather than waiting for the impactful stimulus event. The system uses a new sensory system to detect changes in the associated environmental parameter (a cue event at time t_1) to trigger, via an associator subsystem, an activation of the responder (at time t_2). The response (at time t_3) starts before the stimulus event (at time t_4). That event along with the actual deforming force from the meaningful environmental parameter provides a reinforcement signal (a time t_{4+}) that strengthens or maintains the associator's association link between the associated and meaningful parameters. The requisite time lag between them must be maintained over the experience of the associator in order to continue to imply the causal relation

intrinsically meaningful impact takes place. How do the differences registered by senses become "differences that make a difference," that is, *information* that guides activity? The data gathered by sensory input has to be processed interpretively in order to become meaningful information, and that is where nervous systems come into the picture.

We saw in Chap. 8 that neurons are adaptive systems that can encode a time-lagged correlation between a cue event from a "non-meaningful" environmental parameter and a meaningful (impactful) event from the meaningful environmental parameter. "Time-lagged" means the cue event occurs first and then the meaningful event follows. A time-lagged correlation can imply a causal relation between these two parameters, namely, that the associated parameter has some kind of causal influence over the meaningful parameter. In fact, this temporal association is one of the main ways we recognize causality. Meaningfulness attaches to parameters that have impact on a responding system, and the temporal-causal linkage encoded by the neurons extends the meaning of the impact to the (prior) cause (cue event).

Suppose that the associated environmental parameter (AF for short) has a causal link to the meaningful environmental parameter (MF for short). Symbolically, even-$t_{AF} \rightarrow \Delta_t$ event$_{MF}$ means that an occurrence of an event in AF somehow causes an event in MF. Cause here has a very specific restriction in that event$_{AF}$ must always precede event$_{MF}$ in time, by some nominal Δt, and it must *almost* always do this, and it must never follow event$_{MF}$ unless a sufficient amount of time has elapsed since the event$_{MF}$. In other words, there will exist a strong correlation between events if and only if there is a time lag of time Δt. We say that event$_{MF}$ probabilistically causes event$_{AF}$ under these conditions. The duration of Δt and that between the offset of event$_{MF}$ and the next occurrence of event$_{AF}$ are system specific, but the form of the rule for making a causal inference is quite strong.

The event$_{AF}$ is non-meaningful in the sense that it does not have any impact on the critical parameter in the responding system, except through the meaningful environmental parameter. But if it occurs sufficiently before the event$_{MF}$, and reliably so, then it can act as a cue event. The associator must recognize this cue and trigger the response, which would now precede the actual eventMF and its impact on the system, thus not only anticipating it but taking preemptive action against it so as to really minimize the impact.[9]

One can see in this ability to form time-lagged linkages an essential foundation for the emergence of imagination. The stimulus of the cue fires the neuron-embedded association with the effect, which becomes present in anticipation. In their rudimentary form, such associative preemptive response can be seen in the famous conditioning experiments done by the Russian psychologist Ivan Pavlov in 1927, as discussed in Chap. 8. In his protocol, he measured the feeding response in dogs by measuring a surrogate response, that of salivation. Pavlov built an association by sounding a bell several seconds prior to offering hungry dogs some food. After so many trials of doing this, the dogs started salivating at the ringing of the bell. This physiological associative response was easily measurable, but any dog owner has observed similar imaginative associations when they pick up a dog's leash.

Graph 9.3 shows the final result of an associative anticipatory adaptive response. Note that the amplitudes of all signals are substantially below those above. Also note that the use of associative information has tremendously reduced overall costs to the system. A very small investment (marginal cost factor) produces a huge cost-saving and tremendous reduction in risk to the entity.

---

[9] One of us (Mobus) has shown how a learning (adaptive) associator, called an adaptrode, is able to code these causal relations between cue events and meaningful events and reduce the overall energy costs for a responding system. See Foraging Search: Prototypical Intelligence, The Third International Conference on Computing Anticipatory Systems, Liege, Belgium, August, 1999. Available at http://faculty.washington.edu/gmobus/ForagingSearch/Foraging.html. The graphs in this section were replicated from that work.

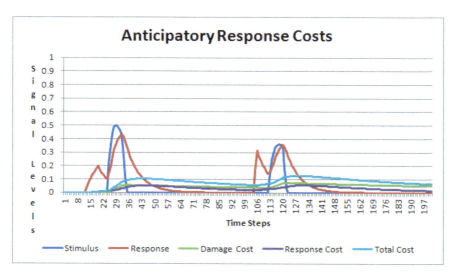

Graph 9.3 Using an associative predictor (like the ringing bell) causes a system to start responding even before the stimulus gets started, thus greatly minimizing the cost of damage. Even the cost of responding is less since the response need not last as long. Note that the response here starts shortly after the onset of a cue event stimulus

9.6.4 Summary of Operational Control

In this section, we have explored the many aspects of process operational control or what keeps a process performing its function. We've seen a number of parameters that must be considered and many variations on the fundamental cybernetic principle of using information to activate a change in behavior to compensate or counteract a disturbance in nominal function. All of these variations are like units or versions of control architectures that can be used to control any process.

Since systems are actually processes that are made up of subsystems, or subprocesses, it stands to reason that we will find these cybernetic principles at work at both the level of the whole system and at the level of the sub-processes. We now proceed to examine this architecture as some new requirements are introduced at the level of sub-processes. Namely, when two or more processes need to cooperate within the boundary of a whole system, then it is necessary to introduce control mechanisms that facilitate this cooperation. Naturally, as the number of subprocesses proliferates in very complex (and especially adaptive) systems, then the problem of facilitating cooperation becomes more complex. It turns into a need for coordination.

9.7   Coordination Among Processes

Large complex systems are comprised of many smaller component subsystems. Our concern here is with subsystems that are, themselves, sufficiently complex that they all have determined functions within the larger (whole) system and that these functions need to be maintained if the whole system is to itself function properly. The system must do so to survive in its functional form.

Until now we have considered individual processes and their operational control, mostly through feedback, but also using feed-forward signals when appropriate. We have hinted, as in Fig. 9.5, that each subsystem process has its own operational control apparatus. That is, operational control is actually distributed among the various processes rather than having some master control trying to handle all of the operational control decisions for the whole system. Distributed control is more efficient and more responsive in general. But it also introduces a new problem with respect to the whole system, for these many separately controlled processes need coordination. There is, however, a problem with this when it comes to large systems with many internal processes that must interact with one another in a coordinated fashion. Figure 9.15 provides a hint of this problem.

In a complex larger system, every subsystem has to perform its function in a coordinated way simply because the outputs of some of these subsystems will be inputs to other subsystems, so needed input must be matched with appropriate output. And then breakdown and maintenance present another challenge. As we learned in the chapters on dynamics and complexity, subsystems are subject to all sorts of entropic decay issues that can cause disruptions in normal functions. Since subsystems that work together may be subject to very different rates and degrees of wear and tear, not only their functioning but the systems for keeping them functioning must be coordinated. Among such systems are the subsystems that act as interface

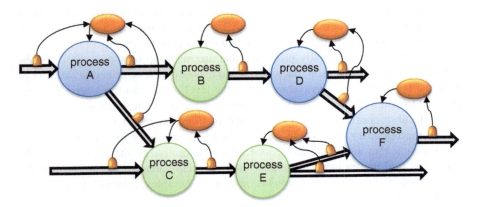

Fig. 9.15 When many processes are interconnected, where the outputs of some are inputs to others, there is a basic problem with ensuring that all processes are coordinated. Note that each process has its own feedback and feed-forward control, but there is no communications between them as in Fig. 9.1

with the surrounding environment. Some of these act as the receivers of environmental inputs to the whole system, which they then need to "distribute" appropriately to other subsystems, while other interface subsystems mediate output back into the environment. Every organism, for example, takes in food, processes it as specifically diversified flows of matter and energy for an array of subsystems, and then expels waste material back into the environment.

Complex systems therefore need to have multiple forms of coordination control. Here we have identified three of the main categories: one form provides internal coordination in operations; another provides internal coordination in repair and maintenance; the third form provides coordination between the "interface" subsystems, the ones receiving inputs or the others expelling outputs for the whole system into the environment. The first two kinds of coordination have been lumped together under a single title called "logistical control" as it involves maintaining the overall internal functioning of the subsystems and the distribution of resources to ensure it. The third kind of coordination is called "tactical control" as it entails processes of interaction with the environment such as getting the system the resources it needs, keeping the system out of trouble, and controlling the deposits of wastes or exporting of products. This tactical control is a form of coordination that maintains a sustainable fit between internal subsystems and the external systems that are sources and sinks.

When we considered process operational control, which is the basic level of control infrastructure in a hierarchical management architecture, we saw that it was efficient and effective to distribute operational control to the various subsystems. Now as we look at the need to coordinate this distributed control, we move to the next higher level in this hierarchal architecture (take a look back at Fig. 9.3 to recall the architecture).

Systems that have significantly many operational subsystems, such as animal bodies and large commercial operations, need extensive coordination internally. But we also find that as systems grow in this kind of complexity, they also tend to have operational clusters, or sets of operations that are more tightly interacting or otherwise sufficiently similar in their control models so that they can be coordinated somewhat independently of other clusters. In such a situation, we invariably witness the emergence of not only many coordination controllers, one for each cluster, but coordination controllers for the operations coordinators! This is how structural hierarchies grow upward. Layers of middle managers are needed to bring together the coordinators.[10]

---

[10] This principle seems to only apply to large entities that retain a unified set of operations. When companies diversify into multiple lines of business and into distributed locations, they lose this basic entity-hood at the operational level. In a sense, they are simplifying and thus do not require deep hierarchies. As multicellular organisms evolved in complexity, on the other hand, we see deep control hierarchies retained since each individual is a single entity. Interestingly, however, the case of eusocial insects might provide an example of diversification, at least into castes, where there is no longer a need for coordination of coordinators. In ant colonies, for example, the control of work is highly distributed with no central coordination controller. Ants work it out with cooperation mediated through chemical scents (pheromones).

Just as distributed operational control is efficient and effective on the level of various operational subsystems, we will find that distributing the coordination of clusters of similar subsystems to the various layers in the hierarchical architecture is also efficient and effective (see footnote 7). In systems where the distribution of coordination responsibilities is done well, by evolution or by design, this form of hierarchy is optimal, more effective than the alternative of massive centralization across layers. Unfortunately, we can all think of examples of organizational hierarchies where the distribution of management functions has not been done well, so, as they say, it seems the right hand does not know what the left is doing. This is what gives bureaucracy a bad name!

Question Box 9.13
One part of control is responsiveness to changing internal and external conditions. As layers of coordinating control deepen in an organization, what tends to happen to responsiveness? How about instituting specialized response constructor units, as in Fig. 9.13?

9.7.1   From Cooperation to Coordination

In Chaps. 10 and 11, we will provide a more detailed explanation for the emergence of coordination among processes, but here we must provide a little context for the move from cooperation to coordination in systems. We have seen that two or more subsystem processes can cooperate when the output(s) of one is (are) a necessary input(s) to another. With simple distributed operational control, each process will have its own feedback control loops, including a control model that is error driven as we saw above. But the control models can include message interfaces with one another so that they can send messages back and forth as needed to inform one another as to the current control actions, and each controller can include a model of what the other process will do under various circumstances of disturbances to their flows. A parts supplier, for example, has a pretty good idea of the needs and processes of the businesses it supplies, and they in turn have a pretty good sense of the constraints and capacities of the supplier. The two each take account of this knowledge in controlling their own processes since they have mutual benefits from their processes functioning properly (Fig. 9.16).

Such a relationship, though cooperative, is still informal, that is, it is not yet fully structured as collaboration. In a collaborative system, the looseness of the freely cross-referenced controllers would be replaced by a more formal and therefore predictable cross-referencing, as when, for example, both operations report to a single boss. Thus, automotive manufacturers, for example, have sought to ensure the flow of parts by buying up their suppliers and making them subsidiaries. As we will show in Chap. 10, cooperation can transition to coordination when one sub-process takes

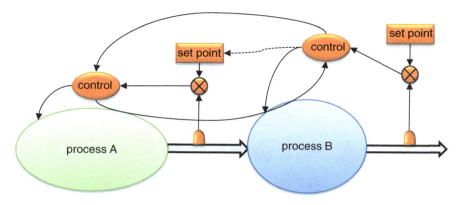

Fig. 9.16 Cooperation between processes may be achieved by the two controllers sharing information directly. Alternatively, process B's controller might provide "suggestions" to the set point setting of process A to directly "nudge" A's output as input to B (*dashed arrow*)

on the role of coordinator in systems having more than just two sub-processes. The coordinating function may have such importance that in some sense the process that takes on the coordinator function may emerge from being a material/energy processor to become a message/information processor. On a mega-scale, we even see whole economies transformed by this dynamic from manufacturers of material products to information and service economies! There are many pathways to accomplish this. At this point, we only need to see that cooperation can evolve into coordination, which produces the beginnings of the control hierarchy.

Question Box 9.14
As systems become more integrated, they move from looser cooperation to more formal modes of coordination. The UN seems to be a compromise, a cooperative coordinator. What kind of systemic factors leads to the organization of that kind of control?

9.7.2 Coordination Between Processes: Logistical Control

Logistical control coordinates both the actual functioning of various subsystems and also the maintenance of that functioning. We will start with a very simple version of logistical control. The coordinator monitors the information flow from all controllers at the operational level. It has a basic model of all of the processes it is attempting to coordinate and a memory of historical performances. It uses this model and time-averaged information about the actual performance of each process

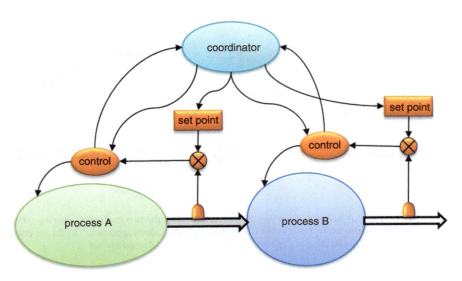

Fig. 9.17 A simple version of coordination between two cooperating processes mediated by a higher-level controller (coordinator) is the adjustment to set point values as a result of variations in the outputs of both processes. The output (product) of process A is the main input to process B (all other inputs/outputs have been left out so as to not clutter the diagram). If there are problems with A's product flow, relative to B's process requirement, at B's given set point, then the coordinator may reset either or both set points as well as give directions to either process to modify their control programs

to obtain an optimal solution regarding maintaining the final output by adjusting the internal flows as needed. So the automobile manufacturer, who has bought up the supplier of brake parts, knows the average flow history of both the making of brakes and their incorporation into new vehicles on the assembly line. He can use this to maximize the coordinated efficiency of both processes, perhaps by seeing excess capacity on the brake, making side and moving resources to speed assembly accordingly. Figure 9.17 shows a very simple diagram of this kind of coordination. Messages from both process controllers are monitored and performance values over time are averaged. The coordinator has a model of what needs to happen in each process in order to maintain some level of optimal output from the final process, and it uses the data to determine what the best set point value for each process should be in order to keep the final output within some specified range. It should be obvious that there are no guarantees about the final output simply because there is no control over the environmentally sourced inputs. A coordination scheme such as this depends on general stability of the overall system/environment. It works best when fluctuations are small and generally from internal disruptions. Internal coordination of brakes and assembly line does not much mitigate earthquakes, floods, or riots!

Some degree of coordination with the environment, however, is both necessary and possible. The controller already has feed-forward information from the controllers of the earlier processes in the stream, so, for example, planned maintenance in the

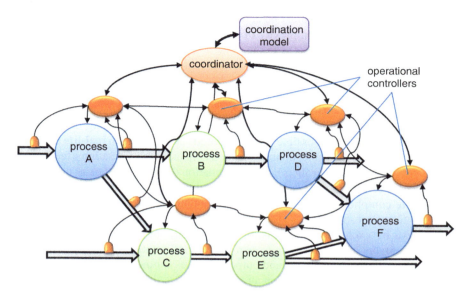

Fig. 9.18 The same system as seen in Fig. 9.15 is seen here with the addition of a coordinator. Represented here is a combination of operational control (feedback from output and input sensors to controllers), cooperation (two-way messages between sub-process controllers), and coordination (two-way messages between sub-process controllers and the coordinator). The cooperation control augments the operational controls in the very short-term (near real-time). Coordination control takes into account the somewhat longer-term time horizon and the relative behaviors of all of the processes. The coordinator has an explicit coordination model (possibly adaptive) for the whole system under a wide range of fluctuations in any of the flow variables

brake division can be factored into activities on the assembly line. And the coordinator may also use feed-forward information from the environmental inputs as shown in Fig. 9.11 above, with the advantage that it can use information from the changes in inputs to the whole system (not shown in Fig. 9.18) for a more adequate response.

The situation depicted in Fig. 9.17 is a linear supply chain (just part of a system such as in Fig. 9.15) without material feedback or parallel flows. Process B simply works with the output of process A. More realistic systems will have multiple flow paths internally and multiple functions need to be coordinated. As we will see shortly, this introduces additional complications in that flows from splitting processes need to be regulated somewhat differently from simple flow-through processes as shown in the above figure. Once we have introduced parallel flows and combining processes (e.g., manufacturing semifinished components going into a final product), the situation is much more complex. Nevertheless, the basic role of coordination is the same: make sure all of the processes receive their necessary input flows in order to supply necessary output flows.

When we come to coordinating many parallel as well as serial processes, as shown in Fig. 9.18, we enter an extremely interesting realm called optimization. A given process may have its own optimal flow rate, differing from the optimum of another process. If the two must be coordinated, what will be the optimum flow? It is easy to see how the complexity of finding an optimal solution ramifies with the addition of each new process to the system. In mathematics this is modeled by the formulation of a set of functional equations representing the set of requirements and constraints for a system to produce its output (called an "objective function"). See Quant Box 9.3 for an example of a very useful approach for solving a logistics problem using a mathematical system. Figure 9.18 shows a simplified (believe it or not) diagram of a system comprised of six sub-processes, including dividers and combiners, with the three possible control methods as are found in real-life complex systems. The three possible control methods include the operation (feedback), cooperation (between controllers), and coordination (from controllers to the coordinator). These control models, some of which include feed-forward information, are much more complicated than simple feedback controllers as depicted in previous sections. Every sub-process requires its particular control, but its control can also be cooperative by exchanging information with other controllers, enabling each to include the status of the others in its own guiding information. But each process is still involved with its own inputs and outputs. It takes another level, the coordinator, to take into account overall function and output. It is at this level that the optimization of the entire process can be addressed.

What makes this a simplified diagram is the fact that there are no material or energy feedback loops shown. Every process produces products that are fed forward to the next processes in two parallel streams. Such forward flows are called a supply chain. Process F is the recipient of two such streams (a combiner), which is actually the more typical situation.

Question Box 9.15 If in Fig. 9.18 the workers in process E figured out a way to make the process more efficient and faster, what would be the information/approval route they would have to go through before the improvement could be implemented? What difference would their proposal potentially make for process A?

Also in this figure, we have depicted the explicit coordination model used by the coordinator to optimize the behaviors of all of the processes in order to produce final products that are within range of their "ideals." Quant Box 9.3 provides an example of such a model and a basic algorithm (the coordinator's computation) for computing the optimal solutions given the constraints on the system.

9.7.2.1 A Basic Logistic Controller: Distribution of Resources via Budgets

On top of the general coordination problem, or as part of it, we must consider a very particular logistical control problem with which almost everyone is familiar. The problem is to distribute (usually precious) resources among the various

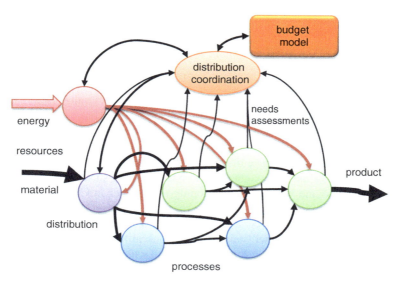

Fig. 9.19 Distribution coordination uses a budget model and needs assessments from processes (controllers now assumed within each process, other message channels not shown) to regulate the distribution of resources, primarily material and energy. Resources are received by special splitter processes that then distribute their particular resource to the processes that need them. The *red process* distributes energy to all other processes, including the material distributing process (*purple*)

sub-processes that need them. Part of what a coordinator has to do is prevent, or at least manage, competition for those resources. The general method that is used, especially in complex adaptive systems, is a *budget*. Almost everyone has been faced with the problem of how best to allocate resources (e.g., household income) among many demands. Most people develop personal or family budgets for how money is going to be spent on necessaries and discretionary expenses. Companies routinely develop formal and often elaborate budgets in order to control expenditures.

Over time the coordinator, in the role of distribution controller, develops, or at least operates according to a budget that details the proportions of resources that need to be available to each and every sub-process for most effective overall operations. Process controllers can supply information regarding their real-time and/or near real-time needs, which allows an adaptive coordinator to build a budget model from experiences gained over time. This model details how distributions of resources are to be made under nominal operating conditions. The resource receiving processes (in Fig. 9.19 below) provide the coordinator with information about the long-term behavior of resource inflows.

If the environment is more stochastic, the distribution coordination problem can also involve distributions under conditions of stress, e.g., when inflows are out of the normal ranges and the system must make internal decisions on restricting some less critical functions to supply critical ones. Families often model this behavior when faced with significant resource fluctuation (droughts, famines, recessions, etc.).

The controller has to decide which functions will not get their normal share. So, for example, the budgeted amount for new clothing purchases may be reduced in a given month in order to meet the rent or mortgage payment.

9.7.2.2   Modeling Process Matching and Coordinated Dynamics

In order for a coordinator in any kind of complex system to successfully perform its duties, it needs a model of the entire set of sub-processes over which it has control.[11] All of the methods that we have discussed in Chap. 5 on developing a lexicon and syntax for modeling can be brought to bear on constructing such a model for human-designed systems, and we will be developing this further in Chap. 12, Modeling. But more generally, such models evolve with experience over time and are open to continual modification. This is certainly the case for natural complex adaptive systems such as organisms, and we will take a closer look at this evolutionary process in Chap. 10. But it may come as a surprise to learn that the evolution of models is also the main method for human-designed systems such as organizations. For example, since the eighteenth century, modern armies have been at the cutting edge of large-scale organization, with coordinated logistical control one of the major challenges. The models of how to manage and move supplies for armies in support of conflict have had a long history of development. Today, military logistics is a finely honed part of keeping armies (and navies, etc.) successful in the field, and it must continually keep pace with (i.e., coordinate with) a myriad development in tactics, materials, technologies of transportation and preservation, maintenance needs, etc. as they affect numerous sub-processes. Typical of this sort of large-scale organization, many models and budgets are maintained in computer systems to provide very fast response times, to project future needs, and to enable constant review and updating.

In all model building, it is first necessary to understand the micro-flows between processes (see Chap. 13 for systems analysis). Since the success of every process is dependent on its inputs, a model can be developed in which the required set points (ideals) for outputs of the supplying process can be determined by working back from where that output becomes an input. That is, we can calculate set points by moving backward to the direction of flows, from the end product and the output processes back to the initial suppliers (distributors) and input processes. This process allows the outputs of forward processes to be matched with the input requirements of rear-end processes and a general balance formulation to be developed. A model of the whole system is thus built by knowing all of the internal requirements for all of the processes and computing the dynamic distributions under varying conditions of inputs, etc. In formal human organizations such as enterprises and militaries, this kind of model building and refining is explicit, allocated to a formal department for logistics that builds and maintains such models.

---

[11] Some authors have adopted the term "second-order cybernetics" to acknowledge the role of model building and using in making complex decisions for control purposes. See Heylighen and Joslyn (2001, p. 3).

In living systems such as cells and in whole multicellular organisms, there is a very similar framework for balancing flows according to distribution requirements. For example, there is a center in the brain stem that monitors oxygen levels in the blood and adjusts breathing to keep the level just right. This is a basic homeostatic mechanism. But when the body starts to exert effort, the need for oxygen increases to support muscles and brain, while blood flow to other tissues may be restricted to just what they need. There is coordination between breathing control (operations) and blood flow directions (distribution) in order to service the short-term changes in needs. This coordination is mainly accomplished in the brain areas that are monitoring the activities and needs of processes, a budget model established by evolution in the trial-and-error product process of selecting what works for the success of the species and encoding it in the genetics thereof.

> **Question Box 9.16**
> Budgets of various sorts reflect the interdependent functional structure of systems. In our complex contemporary lives, time is often budgeted as a necessary resource in short supply. What does your time budget reveal about the way your life is organized? How do you deal with optimization issues?

9.7.2.3   Regulating Buffers

Coordination of flows involves not just quantities but also timing, and this is where buffers play a critical role in coordinated control. Timing is essential in matching inputs to processes with the outputs of other processes. Almost as a rule, one can find that the timing of outputs from one process does not always coincide with the input timing needed by the process that receives the outputs. As a result, it is common to find processes that have internal stores (like stocks in the stock and flow models) or buffers to hold excess amounts while the work process "catches up" and can use the input resource at its own rate. Sometimes the input has to be "pumped" into the buffer or storage at the times it is available. Pumping gas into an automobile's tank for later burning in the engine is a simple example. Figure 9.20 shows this kind of internal buffering that is so common in human-built as well as natural complex systems.

A very important version of this buffering is something every human (animal and plant) does every single day, namely, eat and store energy. Of course the process details are far more complex than suggested in the figure, but the general concept is the same. We eat at meals and the energy in the food is converted into a form that can be stored (in the liver, say) for later release as we go about our activities. As simple as this may seem, it is a crucial part of controlling the quality of the work process.

Coordination models will often include details of these buffers and their internal regulation. For example, manufacturing companies maintain a detailed inventory list providing information regarding the kind and number of every single item

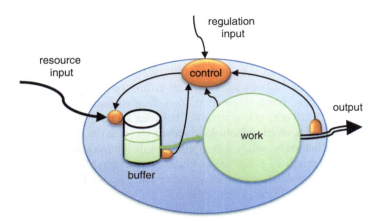

Fig. 9.20 In most processes, there is a need to match the timing of the availability of resources with the timing of the need for use of that resource. A buffer is used to hold the resource when it is obtained. In this diagram, the resource is gained actively (e.g., by a pump symbol). In other situations where the resource flow is under pressure (i.e., being pushed into the buffer), the pump symbol would be replaced by a valve symbol, meaning that the input flow can be shut off or reduced when the buffer is full

needed for production. When parts inventories get below a certain point (part of the model), then a signal is sent to the purchasing system (actually part of the tactical coordinator, see below) to obtain more parts. But semifinished parts are also counted and inventoried as part of the logistical coordination. The principles of regulating the buffers (inventory counts) is the same but completely internal to the manufacturing system.

In living systems, say in cellular metabolism, examples of buffering to coordinate internal biochemical processes abound. It is safe to say that organizations like manufacturers and living cells could not begin to function properly for long without proper buffer regulation.

9.7.2.4 Regulating Set Points

When someone feels too hot or too cold, they go to the thermostat and change the desired temperature setting. It is not always obvious why they feel uncomfortable, especially if they had been sitting in the room for a while or were comfortable at that setting yesterday. It could be that the humidity has changed and that will have an effect on the sensible heat experienced by the skin, thus changing the subjective feeling of comfort. In other words, some environmental change other than the temperature, which is monitored by the thermostat, may cause a need to change the set point of the thermostat.

This is one way that coordinators have of adjusting the performance of subprocesses so as to balance the needs and supplies of all. Starting with the needs to produce an optimal output given the changes in flows resulting, say, from an external disturbance, one possible solution might be found by resetting some or all of the set points of the various process controllers. For example, an inventory manager, seeing that production has increased its rate, might anticipate a need for more parts and reset the desired level of those parts in inventory, thus setting off a requisition process.

Question Box 9.17

Humans have long coordinated set points on interdependent and complex processes. Now we find more and more ways to mechanize the coordination. So, for example, it is now cheaper to get a point-and-shoot camera than one that gives you the option to control the various set points yourself. We may even automate traffic systems with vehicles that drive themselves. This transfer of where the coordination function is exercised is a matter of advances in mechanizing flexible adaptive response. Are there any limits inherent in this process, or is it just a matter of what we're comfortable with?

9.7.2.5 Coordinating Maintenance

Remember from above, we saw that controllers require maintenance or even build up. This required another processor that was responsible for keeping a primary process in shape and functioning. Processes that maintain processes need to be coordinated just as primary processors do. In most complex systems, a subset of maintenance processors actually operate over several primary processes. That is, their activity is shared among numerous primary processes which require similar maintenance operations and only require them sporadically or episodically. Thus, well-designed (or evolved) maintenance processors have a low cost of operation and can be, themselves, maintained at low cost.[12]

Nevertheless, applying maintenance, which requires resources, has to be budgeted and scheduled in light of the overall system operation. Thus, logistical coordination is required for these second-order processes.

---

[12] Some maintenance sub-processes are so general that they can maintain one another! This is what prevents maintenance from becoming an infinite regress. Who maintains the maintainer?! As an example consider the manufacturing plant mechanic. She/he has a general set of skills and tools that can be used in any kind of manufacturing plant since all machines operate on the same basic principles. A lathe, in the hands of a good mechanic, can produce parts for another lathe! In biology there are sets of general purpose enzymes that are used and reused in cell metabolism (e.g., phosphorylation). At some basic level, the same tools can be used regardless of the mechanics of the process being maintained or built up.

9.7.2.6   Time Scales for Coordination

As we have hinted, logistical coordination operates over a longer time scale than operational control or cooperation. Essentially the coordinator has to keep a history of operational performance over time and solve a time-averaged version of balanced flows and set points. In part this is to prevent wild changes in operations. It would not be a good idea to change set points, for example, every real-time sample period. The system would never be able to find a stable operation. This is very similar to the use of integrative terms in real-time PID control, but set at the level of multiprocessor coordination.

Coordination models are, therefore, based on longer time scales than the operational or cooperation controls. Just how long those time scales are will be very dependent on the detailed dynamics of the specific system. In general, however, it is safe to say that the time scales will be at least an order of magnitude greater than those of the operational level. In many cases, it will be more.

But to complicate the picture even more, for truly complex systems (like living systems), there are going to be multiple levels of coordination control at the multiple levels of organization. So, for example, in a multicellular organism, we will find coordination time scales at the level of metabolic events. But we will also find coordination time scales at the level of tissues (e.g., masses of cells of the same type) that need to integrate with the metabolic level for the larger system to work properly!

This leads us to an even more complicated picture of the principle of coordination, for as the timing of processes become coordinated with each other at a given level, that coordination will also need to be incorporated into the coordinated routines at both higher and lower levels. Time scales typically vary from level to level, and it is critical that these variations mesh. This brings us to the problem of *coordinating coordination*, an unavoidable complexity that will be generally applicable to all that follows.

9.7.2.7   Process Control of the Coordination Process and the Coordination of Coordination!

Do you remember the principle of recursion? Or do you remember the concept of self-similarity? We are about to make life more complicated by revisiting how CAS control hierarchies can be recursively complicated!

An example might help here. Accounting is a subfunction of any enterprise that is part of the logistics management level. Accounting gathers data from all of the other operation units (and from other management units) and processes the data to produce balance sheets and income statements (along with many other informational instruments used by management). In a large organization, however, accounting is itself an operation. That is it is composed of many subfunctions that have to be coordinated in order for the whole accounting process to perform its function. Thus, there will be an accounting manager who acts as the overall coordinator for accounting. Indeed all

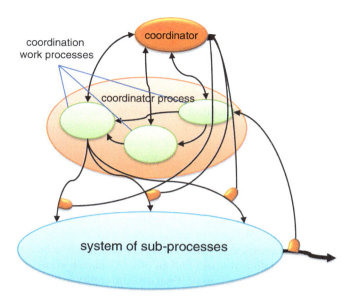

Fig. 9.21 A complex coordinator process contains sub-processes that must, themselves, be coordinated. The coordinator process's output comes from its product sub-process (far left sub-process) after other information sub-processes do their jobs (explicit coordination model not shown, but these sub-processes obtain their inputs from that model). Something has to coordinate those processes. A "master" coordinator coordinates the sub-processes within the operational processes coordinator. Moreover, each coordinator sub-process has a feedback control loop!

of the subfunctions in a large organization will have many people working in that function and that will require a supervisor to coordinate everyone in the subfunction. For example, one of the subfunctions of accounting is to manage payroll. This function can require a number of accounting clerks and specialty payroll accountants. They all need to be supervised by a payroll manager.

What we have described is a very complex situation in which one logistical coordination controller (the accounting department) requires the very same kinds of operational control and internal coordination control that we have described above. Think of it as a coordinator inside a coordinator! Figure 9.21 shows a diagram of this kind of relation.

The dynamics of control and coordination we have been discussing in this chapter apply to all sorts of systemic organization and to our ways of conceiving systemic wholes. Remember how we gave an example of a coordination controller as an optimization computation? This computation is implemented by a computer program (e.g., as given in Quant Box 7.3) that takes as input the states of all the variables and computes an optimal control (balancing) solution. But the computer program is itself a fairly complex process that is running in a computer memory. When we think of the program as a whole, we realize it is comprised of many operational modules called subroutines (or functions). And each subroutine has to be called by a master control program, which coordinates the execution of each sub-process. We do not get caught in the trap of infinite regress because the coordination of the calling of each subroutine is fixed by the master program.[13]

In really complex systems, each process controller is itself a process. And, just like the coordinator we've discussed above, there is a recursive sub-process control along with some kind of coordination control. These cybernetic processes are exemplified in human as well as mechanical organization. Consider the inventory manager in a large manufacturing company (as above). This manager is not a single individual making coordination decisions or operational decisions about the inventory. Rather she/he has a staff of people who are responsible for different aspects of the operation of the whole inventory department. The manager's responsibility is to coordinate that staff. But in order to fulfill her/his complex duties, they also have an office staff to handle basic operations of the office. So, picture the organization chart: a manager to whom a number of operational managers (e.g., managers who keep track of different kinds of parts) report and an office operations staff (e.g., secretaries, assistants, etc.). These latter personnel need to be coordinated so that in a very large operation you will find an office manager whose job is to make sure the office staff are doing their jobs! Then each inventory assistant manager will have some staff to assist in making sure their functions are being performed. As you can see, administrative costs escalate as an organization grows in size!

But it actually works the same way in smaller organizations. The systems dynamics demanding control/coordination remain the same. The difference is that personnel in smaller organizations are multi-talented and cover all of the above functions themselves, or with minimal staff. Humans have a capacity to fulfill different control functions at different times. Of course problems may ensue when a manager doesn't do a good job of differentiating between when she/he is being a coordinator or a coordinator of coordinators and a direct feedback controller of operations. But all of these functions must be fulfilled. Most complaints you might come up with regarding how bad bureaucracies are can be shown to resolve to this sort of confusion of functions (when it isn't a result of sheer laziness!).

Question Box 9.18
One can see how administration seems to grow by its own dynamic! The old complaint of "Too many chiefs, not enough Indians" has its reasons. But there are also systemic factors that drive the need for increasingly complex layers of coordination. What are some of the systemic factors that drive the increasing complexity of administration?

---

[13] But, you must be warned, that isn't really the end of it! Each program consists of a sequence of machine instructions that are coordinated by a control unit in the central processing unit (CPU). And to add insult to injury, that unit is controlled by microinstructions. But, trust us. There is a bottom to this recursion. Those interested should read the summaries in Wikipedia: http://en.wikipedia.org/wiki/CPU and http://en.wikipedia.org/wiki/Microcode.

9.7.3 *Interface with the Environment: Tactical Control*

Coordination of internal processes is a major concern for management of systems. Logistics are crucial for smooth operations and hence for success or fitness of the system in its environment. But there remains the problem of how to interact with that environment in such a way as to succeed in the intermediate term. We will consider the long term below. But the fact is that a system must succeed in the short- and intermediate-term time scales if it is going to have an opportunity to succeed in the long term.

As we have seen, control and coordination take place within the boundaries of a system. The environment, therefore, is precisely what is NOT under the organizational control of the system. The environment is what it is, with its own kinds of processes and controls. It will act as it will act. As a result, a system can only survive and thrive by coordinating its activities with that environment. In distinction from the coordinated meshing of internal processes, the coordination that meshes systems with their external environment is referred to as adaptation. With this form of coordination, the system has to coordinate with its sources of resources and its sinks for products and wastes if it is to be successful over the long term, or often even in the short term. Coordination with the environment is the realm of tactical control.

We have already been introduced to the notion of feed-forward information as it can be used in operational control (and to some degree in logistical control). Now we need to develop this notion into a set of principles for interactions between complex dynamic systems and their environment.

Fundamentally, the system must monitor the sources and sinks of its environment in such a way as to be able to adjust its own overall behavior such as to compensate for fluctuations in those sources and sinks. Principally, the system must be able to adjust its own behavior so as to optimize its access and receipt of resources and make sure its products are fulfilling the needs of its "customers" while providing for the disposal of wastes in a "sustainable" fashion.

Simpler dynamic systems, like rivers or continents, react passively to changing conditions and so don't have to concern themselves with these issues. This is the realm of complex adaptive systems, those that take in information and react with flexible response. Living systems and human-built institutions are primary examples. So we will, henceforth, concern our examples with those realms.

9.7.3.1 Interface Processes

Interface processes are flows of matter, energy, or information between a system and its environment. Within a dissipative system, there are processes that are responsible for receiving the inputs from environmental sources and for outputting products or wastes to environmental sinks. Figure 9.22 shows a diagram of a typical system embedding in an environment, which, from our perspective, is a meta-system. That is, we can see the sources and sinks and the interfaces of the systems. From the

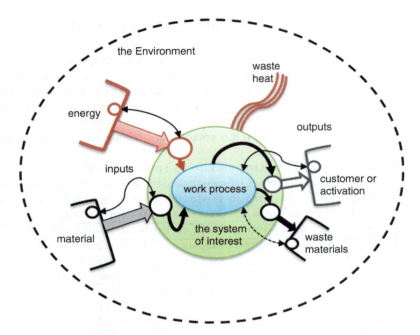

Fig. 9.22 The system of interest has to coordinate with the environment of sources and sinks. Since these sources and sinks are under control or influence from other parameters, the system has a difficult problem to solve. Here the system consists of a work process that converts inputs of energy and material into outputs of either products or the movement of actuators (e.g., muscles). The system has to have interface sub-processes with which to receive inputs and expel outputs (*small open circles* within the boundary of the system of interest). The various sources and sinks are un-modeled (*open rectangles*, same as clouds in other diagrams) except for the message receiving and sending interfaces (*small circles* inside the sources and sinks) that facilitate communications with the system of interest

perspective of the system, the interface is comprised of border conditions which constrain flows of matter, energy, and information into the system. These conditions are a critical element of a system's expectation of its environment.

9.7.3.2 Active and Passive Interfaces

A passive interface is any sub-process that receives inputs from an external source where that source provides the energy (pressure) to move the substance. Osmotic forces moving ions across the cell membrane are a good example. Most message receivers are passive, which is good because it allows nonselected messages, whatever the world has to communicate in a given medium at that time and place. The case can be a bit more complex, since organisms may actively search out information, but even though I may move my eye in some direction, the energy by which the light enters the eye comes from the light, not the activity of the eye. Such passive interfaces may have an active capacity to seek out or block inputs, but they do not "retrieve" or actively import substances from the environment.

Exports back into the environment are ordinarily part of system activity, and the system's active interfaces may also include importing from sources as well. An animal that forages for food is an example of a system with an active interface (eating). A cell may get its nutrition passively, as the nutrients flow on their own through appropriate receptors in the cell wall, but the forager uses its own energy to actively seek out its lunch. Its muscles are the tactical control elements it uses to actuate its own body in moving through the environment. It uses vision and other senses to detect the presence of food or danger.

Many systems cannot control their environments, but they can change their own situation relative to the environment, thereby coordinating themselves with that environment. Other animals do actually have some control over some aspects of their environments. Birds make nests, for example. Beavers make dams. Humans, of course, exercise the epitome in coordination with their environments by exercising control over the configuration of that environment.

9.7.3.3   The Use of Feed-Forward Information

Above we mentioned adaptive controls in the context of the operational level. But, of course, this is just the kind of control needed to adaptively coordinate with the environment.

The process of coordinating adaptively with sources and sinks in a fluctuating environment requires a constant flow of data and processing of information. In many cases, and most obviously in creatures that can move about, feed-forward information allows a system to react more quickly to real-time events taking place in the sources and sinks. Messages from senses are interpreted with cause-and-effect anticipation (remember neurons' prior-posterior associative ability), making foraging, hunting, hiding, and all sorts of purposive or actively adaptive environmental interaction possible. Other sorts of interfaces can be controlled by using feed-forward information supplied by the sources or sinks. A source may signal readiness to supply. An active interface is then activated to import the substance. A passive interface may be notified of higher pressures which would cause more flow than the system needs. That would, in turn, cause the interface to throttle down its admission mechanism (like a valve).

Interfaces can also supply feed-forward information to the coordination controllers (both logistical and tactical) on their status.

Controllers utilizing feed-forward information can minimize disruptions, especially by not waiting for the events to cause disruptions or deviations in the output products—reacting to feedback with delays that could cause larger amplitude and/or longer deviations and oscillations.

Question Box 9.19
Feed-forward information spans a temporal interval between the origin of the information and its application. What is the relationship between the utility of the information and the rate of change in the environment? In a very fast changing culture, what happens to the guiding value of the past? Could you formulate a kind of speed limit on the pace of change as it relates to experience providing guidance?

9.7.3.4   Coordination with External Entities

A more sophisticated tactical coordination approach is to use models of the sources and sinks that provide longer-term anticipation of deviations in supplies and output acceptances and communications with key sources and sinks. The models tend to be of the simpler correlation type, not necessarily attempting to model the internal workings of these entities (see below re: strategic control/management). But armed with such models and communications, where appropriate, systems can develop maneuvering capabilities in order to maintain their own access to resources and handling of outputs.

For example, a manufacturing firm may have a model of a given parts supplier in which the latter has an annual slowdown in deliveries during summer months. The manufacturing purchasing department is tasked with maintaining the flow of parts regardless, and so plans to buy some inventory from an alternate supplier who may tend to charge more per part, but is more reliable during the off months of the other supplier. When the first supplier gets back into full production, the buyer restores orders from it. The communications are handled through order forms and notifications as well as contracts and prices. This is a tactical maneuver to keep production going in spite of the practices of a particular source.

In mobile living organisms, tactical control basically revolves around coordinating movements with entities in the environment that are either food or are looking for food. Feed-forward sensory information, as we have seen, enables the tactics that animals use to forage or avoid predators. But this immediate stream of messages is critically framed by models, some hardwired, some experientially formed, regarding relevant behaviors of the environment. Wolves learn to hunt a wide variety of prey, for example, and the learning involves building up a model of what may be expected where and how it will behave. In a similar way, companies research and model consumer behavior and use it in their hunt for customers.

In addition to a sustaining nutritional flow, living organisms have a need to procreate, so tactical controls are needed for finding and attracting mates and, sometimes, in offspring rearing. Some animals display very elaborate mating rituals. These rituals involve many hardwired tactical behaviors shaped by selective pressures over evolutionary time. But in higher animals, we often see many individually learned variations on the central themes. For mating rituals, even adaptive ones, there is actually a strategic purpose, more than a logistical one. Often the rituals are done in the context of mate selection, particularly the female selection of a male, and the point is to choose those mates that do the best performance. The theory is that performance is an indicator of genetic quality: the male who can enhance his performance and present a fine physical image is likely a possessor of superior attributes. The female of a species is investing all of her energy into a few eggs, while a male is capable of producing millions of sperm. So the female needs to be choosy, whereas the male needs to be a good performer.[14]

9.7.4 Summary of Coordination and Its Relation to Operations

Before launching into the final kind of control architecture for very complex systems, let us summarize what we have seen so far.

Very many complex, dynamic systems that maintain their forms and functions over longer time scales do so by having not just internally cooperating subsystems, what we call component processes, but by developing coordinating subsystems that help keep the internal operations balanced in terms of their production and the flows of resources. To achieve more optimal performance over time, these systems not only coordinate their internal affairs but also coordinate their external behavior with respect to the sources and sinks in their environments. These various subsystems and their coordination (through communications of information) can be shown schematically as in Fig. 9.23 below.

As can be seen in this figure, which only shows the fundamental relationships, coordinated operations, both internal and external, are already very complicated. In a real system, say even something as seemingly simple as a single-celled organism, a more complete diagram of all of the sources, sinks, and processes, along with the controls, would be orders of magnitude more complex. Even so, there are advances being made in systems analysis tools that enable scientists to visualize such complex networks and dynamics. This is a very exciting time for many sciences involved in understanding dynamic complex systems, not only for understanding metabolisms or social or ecological organization in life sciences but also in other fields where complex and changing interrelationships between not-so-simple entities exist. Below we will make a brief foray into brain science to give one very important example.

---

[14] For a really eye-opening example of this, read about the bowerbirds (New Guinea and northern Australia) and the male nest-building performance: http://en.wikipedia.org/wiki/Bowerbird.

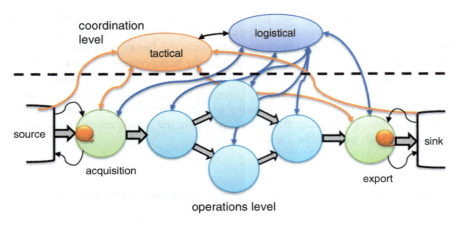

Fig. 9.23 This is a schematic representation of the (mostly) complete control problem/solution for complex, dynamic systems. The basic problems being solved are (1) acquisition of resources from a source (the pump symbol represents an active acquisition process), (2) internal distribution (budgets) of resources and flow adjustment controls, and (3) coordinated export of product (or wastes) to an external sink. The actual work of moving and changing material/energy through the system is accomplished by work processes employing feedback and cooperative feed-forward communications (not shown). The problems of coordination are solved by tactical (external interactions) and logistical (internal interactions) coordinators. Logistical control messages are shown as *blue arrows*, while tactical messages are *orange arrows*. These two will generally have a very strong intercommunication link (*dark black arrow at top*) and coordinate with each other as the dynamic, stochastic environment/system changes

9.8 Strategic Management

So far we have discussed control and coordination with the environment on the relatively simple and short-term level of tactical control. The more sophisticated tactical controllers are still only able to respond to relatively short-term actions of external entities and only to those with which the system directly interacts. Generally speaking the tactical coordination models tend to be simple, such as an adaptable anticipatory response model. For systems sustained by a few relatively constant flows, there is little cost benefit, biological or financial as the case may be, from maintaining an expensive computing capability that would address complex causal relations as opposed to straightforward correlations. Blue-green algae have pursued a tactical course photosynthesizing sustenance from water and sunlight for over three billion years, hour by hour, day after day. In relatively stable environments where sources and sinks do not significantly alter their behaviors over the lifetime of a system, correlation is as good as causation.

But this is not a sufficient situation for much more complex entities existing in more complex environments with more extensive dynamic processes and an element of non-stationarity. At the opposite end of the spectrum from the blue-green algae, an electronics firm that does not constantly reinvent itself will soon be toast. Such systems are faced with both rapidly and, at another scale, more slowly changing relations with their environments. It is no longer possible to solely rely on tactical models and communications because these cannot readily adapt to changes that would ultimately alter the entity's access to resources and avoidance of dangers while maintaining its systemic integrity. Additionally, these more complex systems, i.e., higher-order animals, humans, and human institutions, have many more degrees of freedom with respect to how they maintain their integrity. As opposed to relatively constant needs met by a limited range of behavior, they tend to have many more modes of behavior and mixes of drives, which mean they have many more ways to interact with their environments and fulfill their needs. Such systems are said to be *purposive* in that they display behavior which is goal oriented beyond just obtaining a meal.

Such systems move beyond short-term tactics to become strategic: they have developed the capacity to plan out alternative behaviors that are directed toward a given objective. Furthermore, strategic objectives tend to be more far reaching than an immediate interaction with sources and sinks. For example, a predator may use tactics of stealth and speed in the immediate hunt while developing a longer-term strategic plan for going to a location or a series of locations where it might expect to find an abundance of prey. Rather than be just concerned with the next meal only, the predator may calculate the prey density in relation to the possibility of many future meals.

Human institutions exist by meshing adaptively with what can easily be considered the most complex environments of all. In order for these institutions to survive and thrive long into the future, they need to plan their internal operations, their logistical activities, and their external tactical activities in light of the complex dynamics of their environments. Those environments include not only resource sources and product/waste sinks but also other strategizing institutions.

9.8.1 The Basic Strategic Problem

It is something of a truism that complex systems exist in complex environments. It is also a truism that for a complex system to continue to exist and function in a complex environment it had better be able to do more than just react to short-term fluctuations in that environment. The basic problem of such systems is depicted in Fig. 9.24.

A complex environment will be full of dynamic, stochastic influences from other entities. The sources and sinks with which a system may be tactically interacting are subject to unknowable changes in their behaviors due to these influences. A familiar food source, for example, may become tainted from distant sources, as the *E. coli* infection problem that crops up regularly. Even a system with a tactical coordinator that includes nice models of the sources' and sinks' past behaviors is not free from this effect. The environment is "big" and anything can happen (see Quant Box 9.3 below)!

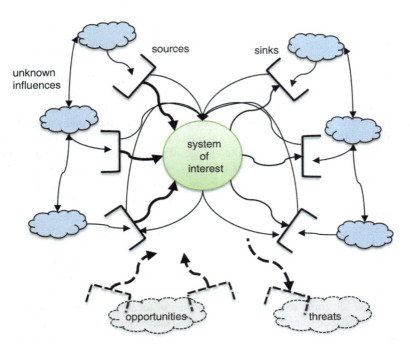

Fig. 9.24 A complex system of interest is embedded in a complex environment in which its direct contacts with sources and sinks cannot help it deal with unknown influences from other entities that can influence those sources and sinks. The *clouds* represent unknown, to the system, but real entities that have influence on known sources and sinks. Additionally, if the system is adaptive, there may be other resources (shown at *bottom*) unknown to it just based on its tactical interactions with the known sources and sinks. There may also be unknown "threats" lurking about (shown as a sink that would consume the whole system!)

Systems that only know about and interact with a limited subset of environmental entities are at risk whenever the environment changes in unpredictable ways. And in addition to the possibility of unforeseen changes in the sources and sinks, the environment might also contain unknown resources it could use or threats that could destroy it. So the challenge of strategic coordination, then, reaches beyond the known and predictable to include the question of how to deal with the unknown.

Quant Box 9.3. Stochastic Environments
Complex adaptive systems (CASs) such as most macrofauna and humans live in environments that are subject to change all the time. There are two kinds of change that directly affect such systems. The first is the kind of environmental dynamics we discussed in Chap. 5—things move about. The second kind is the longer-term change in composition of the environment—its evolution over longer time scales.

Both of these kinds of change, in the general case, are subject to probabilistic effects. That is, the motions and compositions in environments must be described using statistical representations rather than strictly deterministic ones. We refer to statistically described processes as *stochastic*. For an example of a stochastic dynamics problem, consider a baseball that has just been hit into left field. If no other influence were to affect it, the left fielder could readily calculate where the ball will land (or be caught) and move to the proper location in preparation. But other things, not taken into account, could, in theory, cause the ball to veer from its trajectory. A strong gust of wind might, for example, cause the ball to change course slightly. There was no prediction of that gust of wind, so the fielder could not have just dumbly gone to the location computed prior to the gust. Rather he must adapt his placement in response to the change in direction. The stochastic nature of real processes in the world is precisely why we CASs have to be adaptable in the very short run.

The stochastic nature of composition is a lot more complex. A simple version of it has to do with a new object entering into the environment that has an impact on existing systems. For example, a new invasive species of plant may be able to out-compete a native food source for a consumer species. Both the food plant and the consumer may be affected by this kind of event that would change the overall ecosystem, with possible ripple effects throughout the environment.

Stochastic processes can be modeled mathematically by statistical methods. Suppose we noticed that the annual flooding of a certain river delta (like the Nile in Egypt) was never exactly the same depth. Rather we note that it can vary as much as 3 ft from 1 year to the next, but it generally falls within a certain range of depths. Suppose we had 1,000 years of recordings of these depths (actually there are interesting ways in which the Nile floods were recorded by both humans and natural markings for that amount of time!). Using that data we could easily compute the average flood depth as well as the frequencies of discrete depths to get some idea of the distribution across the range. If we were to divide the data up into ten 100 year groups and compute the mean and variance as well as look at the distribution (it turns out that flooding distributions are not the typical normal form, but follow a power law distribution—there is a very low frequency of deep floods, a moderate frequency of medium floods, and a high frequency of shallower floods!), there are several things we might be able to note.

First, if the various statistical moments (e.g., mean and variance) were essentially the same over each of those time periods, then we would say the long-term behavior of the process was *stationary*. That is, the statistics of the dynamic system would not change over time. From the point of view of the Egyptian civilization, this was largely the reason that they could remain stable for so long.

But if there was a recognizable difference (one that is shown to be statistically significant) from one period to the next, then we would say that the process is *non-stationary*. That the statistical properties of a stochastic process are undergoing change over longer time scales indicates that something on a grander, longer scale is shifting and having an effect on the more local process. We might not be able to say what, however, because that would require collect correlated phenomena data over a much longer time scale, say one million years versus one thousand.

If we find that the mean, say, of the flood depth is getting larger in each subsequent period, then the non-stationarity is said to be *homogeneous*, meaning that the changes themselves have a fixed rate. This is what we call a trend. There is something big and very long-term going on that is causing the trend, perhaps climate changes in the water shed. But suppose that we find the data to be jumping up and down from period to period, in a more erratic fashion. One year the mean could be large, the next smaller, the next smaller still, and then suddenly back up to above where we started. Clearly this indicates that whatever is impacting the flood depth over those 1,000 years is, itself, highly variable. The flood process is now said to be nonhomogeneous, non-stationary.

This situation is the most general for natural environments. Climate change and continental drift are usually very long-term processes that produce both homogeneous and nonhomogeneous non-stationary changes in local processes. The central problem for CASs is to be really adaptive and resilient one has to be capable of dealing with the nonhomogeneous, non-stationary case in general. This is what is meant by the phrase, "anything can happen, and usually does!"

Fortunately such processes are interrelated and causally correlated with other processes that may show changes before the one that impacts the system of interest. The trick of anticipatory adaptation can be further extended to provide an advantage to a CAS having the capability of building complex models not just of its tactical interactions but the interactions of those other associated elements in the environment shown in Fig. 9.20. This is the strategic adaptation problem.

This strategic challenge varies with the complexity of the sources and sinks by which a system is sustained. If we take a survey of the animal kingdom with respect to what we would call phylogenetic complexity (e.g., from low-level, simple sponges to high-level, complex primates) and note the animal's relation with its environment, this emerges clearly. Simpler animals such as blue-green algae live in situations where (1) the immediate environment is simple and (2) these environments have remained relatively unchanged with respect to the sources and sinks needed by that animal species for survival (as described in Quant Box 9.3). Similarly we note that complex animals have their needs met by complex environments, and those environments are subject to many changes over multiple time scales (e.g., during the life of individuals as well as over multiple generations—same Quant Box). Note how the meaning is tied to the needs of a given organism: New York City is not a more challenging environment than a creek in the countryside if you are a bacterium! The next chapter, on evolution, will delve into this phenomenon of the availability and emergence of more and more complex lives. Our interest here is to note this proportional relationship between the complexity of systemic need, the complexity of related environmental sources and sinks, and the consequently complex strategic challenge of correlating and maintaining sustainable relational dynamics between the system and its environment. There is good systemic reason for strategic abilities to play a greater and greater role as systemic organization becomes more complex and sustainability more challenging.

9.8.2   Basic Solutions

Tactics and strategy characterize systems that have the ability to in some way anticipate the future. We have seen that feed-forward information is critical for anticipation. Very complex dynamic systems need to have a great deal more advance information about many more aspects of the environment than just the immediate, apparently capricious, behaviors of their sources and sinks. In fact they need information about what else is going on in the environment that might impact those sources and sinks. Moreover, they need models of the environment that will allow them to compute likelihoods of changes based on current information. They need knowledge of much more of the environment in terms of both space (other entities) and time (history) than is available at any single given place and time.

Such systems need to develop longer-term scenarios in order to plan actions for the logistical and tactical coordinators to take over various time scales. They need to have objectives that, if met, provide resources and avoid threats over the longer time scale of their duration as functioning systems. We shall see various solutions to the challenge of strategic anticipation at other systemic levels such as gene pools or markets, but at the level of individual organisms, anticipation seems to demand that they need to *think* about their future and how to succeed.

At this point, it might seem that we are describing humans and human organizations. This is indeed a highly developed human capacity, and we have used it to transform and elaborate an incredibly complex global environment to satisfy our needs and wants. But as we have seen, the grounds for the strategic challenge are systemic, so such future "thinking" is not limited to humans and organizations. Complex individual lives in fluctuating circumstances emerge on the condition of anticipating the future. Thus, many high-level organisms (mammals and birds mostly) have been shown to formulate what we can call strategies for achieving objectives in the future. Some kinds of birds and squirrels cache nuts or other foods in secret places that they "plan" to come back to in the future. Orca whales team up to startle seals off of safe rocks into the jaws of their companions. Wolves learn different hunting strategies in different packs. Chimpanzees have been known to wait for nearby companions to move away from a spot when the subject chimp had spotted a nearly hidden morsel of food that the others had not seen. Rather than go right over to the food, reveal it, and risk another chimp trying to take it away, the subject chimp bides its time. It can imagine a near-term future when the others are not nearby and it can retrieve the food safely. This last example may seem more tactical than strategic, and it does demonstrate how strategic control may have evolved from tactical control as a way to account for more than just the interaction with an immediate environmental situation. But for the chimpanzee, waiting even an hour is like a human planning what major to take in college! It is a long time scale by chimpanzee standards and shows the emergence of more strategic-like thinking.

Such examples illustrate the varied range of strategic thinking. It is coupled with the ability to learn, remember, and especially to apply that in picturing at least a near future possible scenario in their minds. The essence of strategy is being able to control their behavior based on that picture. In the short term, this is tactics, but as the anticipated span and relational complexity of the situation ratchets up, we have the beginnings of strategic thinking. We will discuss this more below when we talk about the human brain as a hierarchical control system.

> **Question Box 9.20**
> Evolution is able to "hard-wire" many organisms with tactical responses to their environment, and humans have mastered cybernetics to do the same with their machines. "Learn, remember, and apply" can be wired into the circuits of control mechanisms and robots work well for specific situations. What, then, is the problem/challenge in coming up with robots that can operate strategically?

9.8.3 Environmental and Self-Models

Figure 9.25 depicts a strategic manager (SM) with access to two interrelated models, one of the relevant external environment and the other of the relevant internal operational organization (the self). This combines the kinds of control we have looked at thus far.

The system has been given (or evolved or learned) a model of the environment that includes those entities which had previously been the sources of unknown influences on the sources and sinks (in Fig. 9.25). Strategic management hinges on some anticipatory abilities that can extend along this network of causal influences and, ideally, parameter in not only the less expected but even to some extent the not-yet known. Systems employ observer processes to collect data on both the resource behavior and the other entities, and the observer can develop an internal model of the

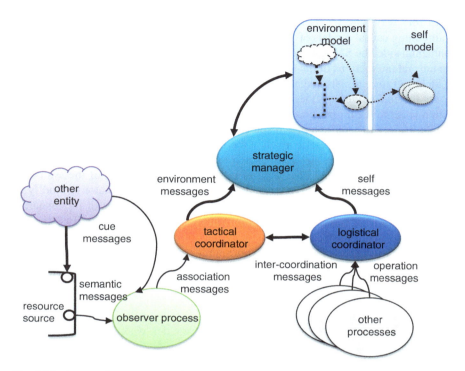

Fig. 9.25 A strategic manager requires models of both the environment and the self and how the two interact (e.g., what effect do changes in the environment have on the system or self and vice versa). See text below for explanations

causal associations as we saw in Sect. 7.4.2. The observer reports such associations to the tactical coordinator for any tactical control requirements, and the latter reports to the strategic manager which then builds the model (seen as the smaller representation in the model). The SM is not concerned with tactical information, since the tactical coordinator is already handling the appropriate tactical responses to changes. Rather it is concerned with much longer-term behaviors of entities in the environment which might affect the tactical response. The SM assumes that the associative relations between critical sources and its environment will change in both predictable and unpredictable ways over longer time scales than are useful for tactical decisions. Routine maintenance schedules exemplify predictable but long-term perspectives. And critical flows and processes must be reviewed with an eye to a range of possible fluctuations ranging from the likely but temporally unpredictable to the unlikely but still possible. For example, the SM may have analyzed alternative sources for critical resources in the event current sources fail to provide what is needed. The SM then instructs the tactical coordinator to realign resource inputs with the alternative sources when the resource flow rate drops below a certain level.

In this way, the SM is responsible for long-term planning by building and maintaining causal models of the environment as well as of the self. From the environmental side, the SM analyzes the situations that are developing that might impact the normal interactions of the system with the environment in the future. It attempts to identify future threats and also future opportunities in that environment.

On the self-modeling side, the SM identifies internal strengths and weaknesses with respect to the threats and opportunities. There has to be a match between, say, an opportunity to acquire a new resource and the capacity of the work processes to actually use that resource effectively. Of course good strategic management is also flexible: if the SM identifies a mismatch but deems that taking advantage of the opportunity is still a good idea (e.g., it will increase the strength of the system in some other area of operations), it can direct the logistical coordinator to modify or build anew the work process capabilities that can take advantage of the resource. This is part of what is meant by adaptability in systems. They can modify their own structures if need be. Even some bacteria have a capacity to alter their internal metabolism to exploit an alternate energy source, especially if their primary source is not available.

Question Box 9.21
Why does strategic management require a self-model, when it seems tactical management can get by without one?

9.8.4 Exploration Versus Exploitation

A key aspect of strategic management is deciding on the trade-off between exploiting resources and exploring for new ones. To the exploring dimension, we can now add the elements of vigilance and caution, additional ways in which strategy accounts for the unknown. Like the costs of time and energy for exploring, there are likewise costs and trade-offs for cautiously slowing down and for not doing something as well as for doing it. For most living systems, this trade-off is basically determined by genetic propensities. That is, the species has evolved a basic behavioral schema that is hardwired into their brains. In higher animals, such behavior can be modulated based on circumstances. In one set of situations, it might be more adaptive to continue exploiting a resource, while in other situations, it might be a good idea to go exploring more, or build defenses, or stay close to home. In humans and human-built organizations, the trade-off computation is highly dynamic, and there are few, if any, fixed rules for how it is computed.

Since it is an adaptable response to future fluctuations and possible fluctuations, strategy itself is widely variable. Psychologically it does seem that different personality types have different baseline or default trade-offs and tipping points between them. Some are much more exploratory than others. Similarly some companies invest more heavily in research and development (R&D) than others. In these systems, the determination of the tradeoff, and when to go one way versus the other, is made strategically based on the analysis of threats/opportunities-strengths/weaknesses.

> **Question Box 9.22**
> How do different models of the self affect the strategic analysis of threats and opportunities?

9.8.5 Plans (or Actually, Scenarios and Responses)

The output of the SM's analysis and decisions is a long-term plan that sets goals to be achieved by the whole system and then translates those goals into a series of sub-goals that can be handed off to the coordination level controllers. The tactical coordinator can take an intermediate-term goal and further translate it into a series of operational goals that can be encoded into the directions passed down to interface processes. The latter then executes the steps of that operational goal (such as find acceptable nesting site, build nest, lay eggs, etc.). Similarly, the logistical coordinator can be given intermediate-term goals in the form of budget adjustments or instructions for modifying operational units as needed.

These plans are not hardwired or deterministic since the environment may not actually do what such a plan might assume. Rather they are an adaptive set of scenarios with likelihoods attached to the stages. For example, the SM planner could develop a decision-tree-like scenario with modifiable payoff weights attached to the nodes. This can be incorporated in mechanized cybernetic process. The challenge is what happens when the intermediate choices bring a system to a particular state, but the environment suddenly does something improbable according to the plan; then the plan has to be modified on the fly and a modified scenario is generated. This is what more complex forms of life do well (it is perhaps a condition for the sustainability of some levels of complexity), but proves very difficult to emulate in nonliving processes. Once again we see the value of information and associated interpretive strategies for adapting responses. More information with greater latitude of interpretation results in greater modifications and reassignments of probabilities in the decision tree.

9.8.6 Summary of Coordination and Strategic Management

As we have seen, complex adaptive systems are comprised of many semi-independent processes that need to cooperate in terms of their behaviors in order for the whole system to maintain itself over its normal lifetime. Simple cooperation, say through a market protocol, is generally not sufficient to maintain balance between the outputs of processes and the inputs to other processes as materials and energy flow through the whole system. It then becomes necessary to have a new kind of control system operating over a longer time scale that can help coordinate the behaviors of all of the sub-processes.

Internal work processes require some form of logistical coordination so as to match behavioral set points and allocate resources in an optimal fashion. The system's processes for interacting with the environment, with the sources of resources and the sinks for products and wastes, need to coordinate their behaviors with respect to those external entities. This requires tactical coordination of the interface processes to maintain optimal interactions. And the logistical coordination has to cooperate with the tactical coordination in order to balance inputs and outputs, from and to, the external world, with internal work processes.

As we have seen, coordination is achieved by computational processes using information from the operational processes and producing information (directives) for those processes. But as processes in their own rights, these computational processes require their own internal coordinators. In other words, coordinator processes require internal operational sub-processes that are, themselves, tactically and logistically coordinated. We avoid an infinite regression by virtue of the fact that computational processors ultimately are built from atomic processes (as covered in Chap. 5). For example, computers are built from transistor-based logic gates, and those are coordinated by virtue of how they are wired together.

Coordination alone is not sufficient for complex adaptive systems existing in highly dynamic, non-stationary environments. Such systems need to monitor the whole of their environments, not just the immediate entities with which they directly interact, in order to project future states of the world and how it might affect them. Strategic management is the most sophisticated and highest level of control. It produces longer-term scenarios, probabilities, and response plans that translate into tactical and logistical action plans. Good strategic management is highly adaptable since the non-stationarity of real-world environments almost guarantees that any plan must be modified in light of actual ongoing experience.

Life really gets interesting at the more collective level such as colonies, packs, families, ecosystems, and societies, where many (relatively) whole organisms or whole organizations of whole organisms operate with strategic competitive/cooperative dynamics. This yields a yet higher-order level of systemic coordination brought about by the dynamic interplay of innumerable feedback loops among the members of the system. Such higher systemic levels may or may not coordinate their many subsystem processes through strategic management, depending upon the nature and availability of anticipatory goal-oriented capacities at that level. Human organizations and societies easily coordinate (and compete) at these more collective systemic levels, as evident in markets or in international politics. We also like to think of "Mother Nature" as operating nonconsciously but with her own strategic objectives, though the border between auto-organizing coordination and strategy remains difficult to assess. Traditional science rejected such notions out of hand as smacking of "teleology" (no strategy with no goals!), but current understandings of auto-organization offer alternative perspectives.

We have seen that cooperation can coordinate processes, but that to make such coordination more stable, coordinators emerge, and such coordinators operate at a next level up in order to be inclusive of the processes they coordinate. This is the dynamic that gives rise to hierarchical structuring, one of the most common solutions to control and coordination issues.

9.9   The Control Hierarchy

Complex adaptive systems exist in complex environments and it takes information and knowledge about how to also get energy and material resources in order to keep those systems maintained, functional, and long-lived. That is, in order for CASs to be sustainable (at least for their expected life spans), they must have the resilience to adapt to many kinds of unforeseen changes in their environments. In our considerations above, we described the emergence of layers of systemic control in pace with increasing complexity of processes and extended time scales. We now conclude with a more explicit consideration of the hierarchical management architecture that is key to how complex adaptive systems succeed.

In order to sustain the integrity and operations of a whole complex adaptive system, nature has invented a hierarchical management (control) subsystem that distributes various control tasks according to their decision requirements. Figure 9.26 summarizes this.

This depiction shows that control, that is, the insurance that operations (functions) will be performed appropriately for very long times compared with the "life" of the system, is distributed among levels of a hierarchy based on functional requirements. Every successful system (see Chap. 11 for a definition of what this means) can be found to follow this architecture.

Hierarchy emerges from the inherent dynamics of coordination, for coordinated units become merged in a higher-level unity. Hierarchical control should not be denigrated just because there are functions that are distributed by levels. Every element in these structures is as vital as every other element. It is only the scope (in time and space) and the computational power needed that makes any difference from level to level.

Hierarchy becomes a more complex phenomenon as it functions in the organization of corporate systems composed of members that are whole systems (organisms) in their own right. In social insects, this may simply parallel the kind of inborn hierarchy of metabolic organization such as the function of a central nervous system. But it is open to competitive sorting out as well, so chickens have their pecking order and social mammals often organize in terms of alpha males or females. In human organizations, the structural rationale for rewarding upper management more generously is that it takes more capacity to manage coordination and strategic levels, even though all levels are critical for the whole process. But this easily gets mixed up with typical primate competitive dynamics and with the inevitable tensions between goals of the individual and goals of the group: we do not necessarily welcome having our personal strategic objectives subsumed by some higher authority into a more collective goal (which too often is really just the individual goal of the higher authority!).

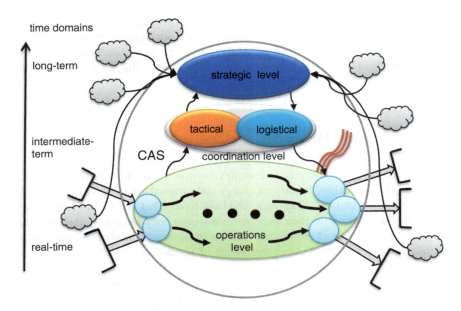

Fig. 9.26 Complex adaptive systems (CAS) are organized to have internal control processes in a layered architecture. Operational controls perform feedback and feed-forward (cooperative) control over the system's working processes in real time through various market mechanisms. The coordination level includes tactical and logistical coordination processes that operate over a range of intermediate time scales. The strategic level uses its model of the environment and of the system itself in making strategic decisions that translate into tactical and logistical execution. The strategic level operates over a range of long-term time scales

So it is easy to see that while hierarchy may be crucial to complex systemic function, at the social level, it often gets a bad name. But it is not the nature of hierarchical control per se that leads to privileged class structures. "Bosses" aren't bosses because they are inherently more important as human participants in society, but because they, in theory, bring more vision to the process of managing an organization. But the role of competitive dynamics in bringing about a hierarchical social structure and its attendant privileging of controllers inevitably motivated by personal as well as collective goals are also systemic phenomena. The inherently cooperative functions of the organs and members of an individual organism should not be mistaken as a sufficient model of the dynamics of the more complex level of organization represented by systems (societies) composed of those individuals!

Question Box 9.23
In the abstract discussion of control, coordination and management are functions carried out by components of a system. What complication enters the picture when these are also whole organic systems (such as persons) in their own right?

9.9.1   Hierarchical Management

We can see the rudiments of strategic management in even simple living systems such as single-celled organisms. In those simpler cases, however, the strategic elements are not actually adaptive as individual strategies. They have been determined by evolution of the species choosing those strategies that best suit the animal or plant (especially discernible in plants since they are non-mobile). And we can see how that same strategic force of evolutionary selection begins to endow higher animals (mammals and birds) with primitive adaptive strategies, which gives them greater latitude in adapting to fluctuating situations. In primates we see the impact of adaptive models that can significantly affect strategic choices. In humans as individuals, we see the penultimate form of adaptive strategy, and in human organizations, we see the ultimate form (that we know of). Our organizations appear to be indefinitely adaptable as long as there are resources to utilize. Corporations, for example, are forever reforming themselves to take advantage of emerging markets and technologies.

Individual humans both shape and, by their participation, constitute the management structures within organizations. As mentioned above, human individuals need to think strategically for their own futures as well as for the future of their organizations, an unavoidable systemic tension. How successful these individuals are in accomplishing this dual role depends on many parameters, such as education, connections, intelligence, and wisdom. Recent events in corporate entities and in governments around the world illustrate the difficulties that go with this level of corporate strategic management and call into question just how successfully human individuals are accomplishing this dual role. As companies like Enron[15] and various national governments that have been unable to meet the economic and political challenges of recent years have shown, the capacity for individuals (leaders) to rise to the requirements of strategic thinking for extraordinarily complex adaptive systems appears to be limited. Knowledge of hierarchical management as adaptive handling of internal and environmental flows and conditions is an aspect of systems science that may become extremely important in coming up with more workable governance mechanisms in the near future!

9.9.1.1   Examples of Hierarchical Management in Nature and Human-Built Organizations

The world is full of examples of both successful hierarchical control or management systems and those that are just emerging. Living organisms provide the most cogent examples of hierarchical control that succeeds in meeting the overall objective of keeping the organisms alive and functioning over a life span that allows for propagation of the species. The human brain is an example of such a control system that exists at the intersection between internal biological organization and

---

[15] Interested readers should take a look at the Enron story here http://en.wikipedia.org/wiki/Enron!

external social organization. The human brain is clearly successful as a hierarchical biological control system, but, especially in the unprecedented complexity of a globalizing systemic environment, its management of the supra-biological level of social organizations turns out to be an ever-mounting challenge.

Living Cells

We have often appealed to living cells as a fundamental example of biological organization. More complex or nucleated eukaryotic cells evolved as coordinated assemblages of smaller and less complex prokaryotic or nonnucleated cells. Contained within the protective bubble of a cell membrane, these once stand-alone units came to take on distributed and coordinated roles as subunits, homeostatic mechanisms for performing basic metabolic work such as capturing sunlight (photosynthesis) to drive the synthesis of high weight molecules that are needed in maintaining the structure and functions of the eukaryotic cell. Eukaryotic cells in turn may function as individual organisms or play a role as specialized subunits in multi-celled creatures, again calling for another level of coordination under a much tighter controlling protocol. In terms of control, coordination, and strategic responsiveness to an environment, we see here a pattern of growing complexity, as system nests within system which nests within a yet larger system. With the emergence of each new more inclusive level, much of the basic control and coordination apparatus of the former units remain, but the whole is subjected to a new set of requirements as its functions are more tightly coordinated into a larger functional whole. This exemplifies the typical layering of a hierarchy of control that accompanies the emergence of a more complex functioning unit.

It is of interest to note how strategic control shifts as the type of integration into the environment changes. As we shall see more in detail in Chap. 10, environments exercise their own kind of selective control over what can fit. For a free-living nonnucleated cell, control by selective fit is simply in terms of survival and reproduction; then as an organelle in a nucleated cell, fit becomes a matter of fulfilling a function other than just surviving and reproducing, and survival becomes rather the control over the nucleated cell. And the same scenario is repeated as the nucleated cell becomes a subunit of a multi-celled organism. At the level of cells, genes are the strategic management mapping a system functionally responsive to environmental conditions, and those conditions select for the fitting genetic strategies. While we are far from the kind of conscious strategic decision making with which we are most familiar, its systemic structure is already visible here as we see these micro-units genetically shape and reshape themselves to fit with changing environments.

The Brain

One strategy for controlling and coordinating the escalating complex functionality of evolving life has been the emergence of central nervous systems, with brains taking on major control functions. The brains of all primates (indeed all higher mammals) have been described as "triune." That is, the brain is composed of three distinct layers derived from essential functions demanded as evolutionary history progressed. The brain stem involves all of the low-level functions that control basic body processes like breathing and heart pumping rates. This is the most primitive part of the brain, evolutionarily, shared with fishes and reptiles. The midbrain of reptiles, and of higher levels in the phylogenetic tree, processes sensory data to extract basic patterns and issues basic orders to the body for responses. This is the coordination level as discussed above. The primary function of this part of the brain is to coordinate the activities of all of the low-level functions as well as deal with tactical interactions with the environment, such as identifying prey or enemies, so as to allow the animal to survive. The third level, seen in mammals and birds, is the neocortex of the cerebrum, which provides a medium in which more complex models can be learned and used for strategic purposes. In humans the prefrontal cortex (PFC) has expanded considerably compared with our nearest simian relatives or with other mammals, and it is now thought that this enlarged PFC is the seat of our exceptional strategic thinking abilities.

Organizations

Human beings are inherently social, and much of the coordination required in living together is achieved by informal cooperative strategies. But as communities and their enterprises become more complex, they grow beyond the strong familiar face-to-face bonds that make cooperative coordination sufficiently predictable and functional. This is where the dynamic of coordination by formal hierarchy takes root. Both military and enterprise organizations have developed hierarchical management structures to deal with the complex environments in which they work. Armies demand complex coordination on a large scale, so the military was probably the earliest human organization to formalize the differentiation between operational, logistical, tactical, and strategic levels of management or control. The payoff for coordination and the costs of failure in the military are such that militaries tend to be highly stratified, rigid organizations with well-defined distribution of control processes. Corporations have tended to follow suit as they evolve from simpler forms, where one founding manager has to fulfill many roles, to take on the distribution of controls such as we see in large organizations. This process is not always smooth, and especially in newer corporations the division heads and CEOs may not fully grasp their specific roles in the same delineated way that a general or corporal might.

Government

Social governance is subject to the same dynamics as corporations and militaries. Public goods, such as food supplies, transportation, education, sanitation, etc., each reach a complexity that demands its own control and coordination and then must be coordinated at a higher level with the others and with budgets and revenue streams. It is easy to see how governments, regardless of political theory (e.g., capitalism versus socialism), are inherently layered into hierarchical structures. The relationship between, for example, operational control and the government's interactions with the economy are not at all well understood, and there are fierce debates about where and how to regulate. Nevertheless, it is clear that insofar as societies need reliable coordination in the flow of goods and services, some kind of hierarchical management will be required. Whether it is a tribal chief or a prime minister or president, the emergence of some kind of hierarchy attempting to emulate nature's solution to system integrity and longevity is being attempted.

The question of the fitting type of governance is conditioned, like any question of functional fitness, by concrete circumstances. But perhaps with some insights from systems science regarding the inherent necessities and dynamics of systemic coordination and control, the designs of future political/governance systems might benefit.

9.10   Problems in Hierarchical Management

Complex adaptive systems with refined hierarchical control mechanisms achieve stability, resilience, and sustainability. But things can go wrong. No system is completely immune to various forms of dysfunction. In this section, we will examine several problems that arise in CASs when, for whatever reason, control structures fail. These can be categorized in three basic ways. The first involves situations in which the environment of the system simply changes too much or too quickly or in completely unpredictable ways that overcome the built-in resilience. Examples include the extinction of the dinosaurs 65 million years ago. The second involves the breakdown of critical internal mechanisms in the control hierarchy. The breakdown can come from simple decay (entropic) processes or from targeted disruptions. One example is cancer, wherein the internal regulations that keep cell growth in check breaks down and the cells divide profligately robbing healthy tissues of resources, eventually killing the organism. The third category is best exemplified by the management of human organizations and governance. Here the control mechanisms involve human agents who are not mechanical decision makers. Essentially everything we like to complain about in bureaucracy involves the inability of human agents to implement rational decision-making and especially sticking to the level of decisions apropos to the level within the hierarchy in which those agents work. We'll take a look at the phenomenon of "micromanagement" as an example of how hierarchical management breaks down due to "imperfect" agents making decisions.

9.10.1   Environmental Overload

Sustainability of any system ultimately depends on a certain level of stability in the fluctuations of forces and conditions in the environment. If some critical factor in the environment goes out of the range that was stable over the time when the system evolved to become fit for that environment (see Chaps. 10 and 11), then the test of resilience on the system's operation will be strained and may go beyond the system's ability to respond. Subsequently the system will incur internal damage that it may not be able to repair even if the factor comes back into range after a time. We will examine three kinds of these environmental stresses that can cause even a hierarchical control system to fail. Technically, all three can be characterized as information overload (the first one we will examine) since they all come under the heading of changes that are unexpected by the system. However, we will treat information overload as a separate category to focus on how purely message-based (communications) inputs to a system can lead to a more subtle form of breakdown. The other two, force overload and loss of resources, are related more directly to the clearly physical cause of breakdowns.

9.10.1.1   Information Overload

The human brain is capable of dealing with an incredible amount of novelty in terms of what is happening in a person's environment. Messages that convey that novelty are, by definition, low probability and hence informational (Chap. 7). Our brains have evolved to handle the processing of informational messages, which is why we can adapt so readily to novel technologies like computers and televisions, things that were certainly not present in the environment of the Pleistocene epoch.[16] It can be argued that the human brain has become capable of modeling any kind of environmental situation that might come into existence. Indeed we are often characterized as *informavores* or seekers of information (meaning novelty) because we seem to actively seek out what is new. Think of how we eagerly absorb the nightly newscasts or can't wait to get our hands on the latest technology gizmo.

While we seem able to deal with a nearly infinite amount of novelty over our lifetimes, the fact is that, as individuals, we are not able to deal with novelty or information that comes at us too quickly. And there is a limit to how much information per unit time (or per message state) an individual can absorb and process. The rate of information flow can exceed an individual's capacity to deal with it in the sense that they are able to convert the information to knowledge as discussed in Chap. 7. It is a simple matter of computational load. The brain is a physical process after all, and its rate of processing capacity is limited to the underlying biochemical rates imposed on neurons. If too much new is happening at too high a rate, our brains cannot keep up with it and we suffer what is known as *information overload*.

The normal flow of information a person receives during a day is stored in multiple levels within many parts of the brain. Memory traces of events in the environment as well as a person's feelings and emotional states are temporarily recorded in

---

[16] The Pleistocene epoch preceded the current Holocene, which roughly started about the time of the invention of agriculture. Humans evolved to their modern form during the latter part of the Pleistocene. See http://en.wikipedia.org/wiki/Pleistocene for more details.

neural circuits that can retain those images, sometimes for several hours or even days, depending on how intense the events were and how emotionally impactful they were (e.g., everyone of a certain age at the time remembers where they were when the news of John F. Kennedy, 35th president of the Unites States, was assassinated). It is now thought that one of the major reasons we sleep at night is so that our brains can process those memory traces, probably during REM sleep.[17] The details of likely processing are beyond the scope of this book, but if we take the systems approach to thinking about this, then it is clear that as a physical process, memory processing has a natural rate limit. Along with the limits on the amount of time one sleeps, this imposes a constraint on just how much information a person can incorporate into knowledge over time. It appears that we are able to store, in temporary form, a good deal more information than we can process in any one night. Presumably this has an evolutionary advantage under the conditions that short-term high information periods are rare or infrequent so that the average person could "catch up" with processing over several nights.

But what happens when the din of information is incessant and every day? Alvin Toffler coined the term "Future Shock" in a book by that name in the 1970s when he recognized a social phenomenon that he characterized as information overload.[18] He posited that people in the modern world had passed a threshold of natural ability to deal with the rapid pace of technological change. He claimed that people were suffering various forms of mental illness resulting from this overload. What made it insidious is that the load was just enough to harm our decision-making capabilities but not enough, except in extreme cases, to debilitate the majority of people. No one was particularly immune to this phenomenon, though some could function better than others. The decision-making capabilities of managers and government officials were starting to be compromised little by little.

Since decision making is part of the computational function within a hierarchical control system, it is easy to see that control itself could be compromised if that function is not working very well. Some technology writers have speculated that the advent of using computers and now the Internet as part of the computational processes in decision making have relieved our human brains from some of the stress caused by information overload by taking over some of the heavy lifting of processing data. However, there are counter voices in social critics who say that this is only hiding the problem, that the information produced by computation is still an issue, but that many decision makers are abrogating their responsibilities to computers in a faux semblance of making decisions.[19]

---

[17] Rapid eye movement sleep, which occupies between 90 % (babies) and 25 % (adults) of sleep time has been associated with memory consolidation processes. See http://en.wikipedia.org/wiki/Rapid_eye_movement_sleep for more information.

[18] See http://en.wikipedia.org/wiki/Alvin_Toffler with links to describe Future Shock.

[19] See http://en.wikipedia.org/wiki/Information_overload for more information.

Question Box 9.24
As we become deluged with available information, we avoid overload by becoming more and more selective, with the result that we live more and more in our own information worlds. What are the systemic consequences as the social and political world of shared information becomes more and more tailored to individual filters?

9.10.1.2   Force Overload

By definition, selection forces are those elements of the environment that stress (physically) individuals in a population in a given setting. There are many kinds of stressors in this domain, such as temperature or even salinity (for non-aerial biological systems). We have already mentioned homeostasis above as a mechanism for a living system to respond to unfavorable changes in some critical factor or force so as to counter its effects. If the force or factor gets too far out of the optimum level for the organism, then the homeostatic mechanism can no longer effectively counter it and the organism will succumb.

Our machines are also subject to this kind of overload. Every time we crash a car, we are subjecting a machine to forces it was not designed to resist or compensate. A more cogent example with respect to control systems would be the sudden overwhelming gust of wind that overpowers the autopilot of an aircraft. An autopilot is designed to deal with many wind-shear contingencies, to be sure, but there is always a possibility of one really fierce force that the autopilot cannot compensate for. Hopefully the pilot would be able to take over and land the plane safely.

Global warming provides us with a possibly unhappy example of a stress force, temperature, causing increasing variance in the frequency and severity of weather events. We are already seeing some of this impact now. Communities are suffering physical damage as a result, and it is not really clear yet that government agencies are able to either respond appropriately (think about the repairs that have not been accomplished in New Orleans after Hurricane Katrina) or take actions to prevent or lessen the damage. Even the market mechanisms of human society may be adversely affected by the damages caused by severe weather. It is too early to make any predictions, but from a systems perspective, we can definitely say that the current control mechanisms are being highly stressed by these forces. Below we take up the failure of decision processes when the internal components are themselves flawed or incompetent with respect to the needs, a factor which also plays into societal responses to climate change.

Question Box 9.25
Force overload challenges systems at all levels or scales, calling for controls appropriate to the operative scale of space and time. Our abilities to remember and to anticipate and act proactively to meet such challenges have evolved to function at what scale? How elastic is the scale? Our intellects can reach far into the future; how about the emotions which function to motivate us?

9.10.1.3   Resource Loss

A final example of breakdown in systems and particularly their control structures comes from the phenomenon known as resource depletion. If a system is growing and developing, it will increase its rate of extraction of low entropy resources from its environment. If those resources are renewable and are being replenished at a rate greater than the system can extract and sequester them, then there is generally no problem. In the best cases, the system will eventually slow its growth to match the replenishment rate and the system, and its environment will come into a dynamic equilibrium or steady state (Chap. 6).

On the other hand, if the resource is nonrenewable, then a difficult problem ensues. If the system develops a control mechanism that allows it to recycle the material parts of the system and it is the material that is finite and limiting, then, again there may be no problem if the system ceases growth and establishes a dynamic equilibrium. But if it attempts to continue growing, it will find that the material resource becomes either depleted or sufficiently depleted that the cost of further extraction is prohibitive.

Consider our society which is increasingly reliant on computers to provide computational power for control decisions. Those computers, in turn, are constructed using some rare-earth elements, which, as their names imply, are rare! If we build more and more computers to support our continued economic growth, at some point, those elements are going to become prohibitive and our control structures will not be able to keep up with the needs.

There is a general law of nature that growth in nature is limited by the scarcest resource required by the system. Liebig's law of the minimum[20] is an observation that growing systems cannot grow beyond a limit imposed by this scarcity. The law not only applies to material resources but to energy as well.

As we explained in Chap. 6 on dynamics, energy is different from material as a resource. It is absolutely required to accomplish useful work, but it also gets used up, so to speak, in the process of doing that work. It is conserved in an absolute sense, but it is converted into low-grade heat with every unit of work, and no additional

---

[20] See http://en.wikipedia.org/wiki/Liebig%27s_law_of_the_minimum

work can be gotten from it. Energy always goes downhill! Materials can be recycled, in principle, given enough energy to pump out entropy. But no such luck with energy. It can only be used once for any level of work and then it dissipates into space, forever gone. The problem arises when the source of energy is a nonrenewable, fixed, finite source such as fossil fuels. Sunlight is finite in quantity per unit time (power), but we are pretty safe in saying we will get a new batch of it every single day (discounting cloud cover). It is effectively renewable. But coal, oil, and natural gas will be depleted (are already showing signs of such) eventually and then the energy flow from those sources falls to zero. Our current economic system is highly dependent on fossil fuels (over 80 % of our consumed energy, worldwide, comes from fossil fuels). Liebig's law is going to assert itself one day if we don't find some kind of viable alternative energy source(s).

Perhaps a simpler example of this is what happens when your car runs out of gas? It stops. No amount of sophisticated control can make it go without energy.

Question Box 9.26
The flexibility of human adaptive and anticipatory capacities has made resource substitution a useful strategy for overcoming the limits imposed by depletion. Economists have even theorized unlimited resource substitutability. How do factors of time and scale figure into control by substitution?

9.10.2   Internal Breakdown

The second major category of problems involves the breakdown of control elements within the system. As long as the world operates at a temperature somewhat above absolute zero, −273.15° on the Celsius scale, things can go wrong. Crystals of ice can maintain their shape so long as the ice stays frozen. But then ice doesn't actually DO anything interesting. The world is in motion because energy flows, but those motions can sometimes cause trouble.

9.10.2.1   Entropic Decay

All real systems age. This means that their internal structures are subject to entropic decay. For nonliving systems, like our machines, this means parts wear out. If we provide newly built replacement parts and do the work of repairing the machines, we can usually get a bit more "life" out of them. But over a sufficiently long period of time, other parts wear out, and then others. Anyone who has owned an automobile for more than 10 years knows that maintenance is an ongoing and mounting problem. Eventually the cost of maintenance is greater than the perceived utility of the vehicle, and it is time to buy a new one.

So too the parts of living organisms wear out and need repair. A living system has a certain, but limited, capacity to do self-repair. You see this when an injury occurs and the body heals itself over time. But in addition to repairing overt injuries, the cells of our bodies must continually "fix" small breaks in internal components. Just as with a new machine, when we are young, our capacity to make these repairs is much greater than the rate at which they take place. But as we age past a certain point, the rate of decay starts to outpace the rate of repair and we start to show and feel our age. This is, unfortunately we might think, the way the universe works. Entropy is always increasing in the universe even while it seems to decrease in local systems like the surface of the Earth (evolution of life). A famous saying is: "Rust never sleeps."

Even if your body was never attacked by disease, or you never had an injury in your life, you would still wear out. This goes for the brain as much as the body. The brain is an obvious hierarchical control system but so is the metabolism of every cell. Once any of these mechanisms starts accumulating what are effectively micro-injuries, then there is no known way to reverse the process. And control starts to break down. Fortunately life consists of many redundant control mechanisms so there is a certain amount of backup available to keep things moving for a while. But ultimately, there is no bargaining with the Grim Reaper it seems.

9.10.2.2 Point Mutations

Ionizing radiation and chemical pollutants as well as microbial agents (viruses) are able to alter the DNA at the point of a single nucleotide, thus changing the message encoded in that DNA. If the DNA is part of a protein-coding gene, then the function that the protein performs in the cell can be altered and may be rendered inoperable or, worse, harmful. This kind of situation occurs in both somatic (body) cells and germ (reproductive) cells. In the latter case, the "mutation" can be passed on to a conceptus. In most cases, the mutation is either harmful or inert, and the resulting embryo will either die or there will be no perceptible effect on the phenotype (the individual's form). Occasionally the mutation renders the protein more effective and confers greater fitness onto the individual with the result that the mutation may eventually propagate into the population.

Somatic mutations, however, lead to different results. Most cells in most tissues in adult organisms are programmed to reproduce by mitosis only when the tissues need to be repaired. Most of the time, the cells are held in check from reproducing by an elaborate system of signaling so that the body as a whole retains its integrity of function and form. This system includes a number of redundant pathways so that failure of one, or a few, signal receiving mechanisms will not result in loss of checks on growth. But every so often, the right combination of DNA point mutations can cause such a loss, and the cell fails to respond appropriately to the signals. It begins uncontrolled reproduction that leads to cancerous growths called tumors. Cancer is now seen as the result of loss of control due to damage to key elements in an otherwise finely honed control system.

Ionizing radiation can also be responsible for the damage to a single-memory cell in a computer system. Depending on where in the system the memory cell sits (e.g., if it is part of the memory of the operating system), this can cause the entire computer to go down! More modern computer designs, following the lessons of living systems, include some amount of redundancy in elements in the computer that allow this kind of damage without bringing the whole system down. Performance may be degraded, but the system will not fail catastrophically, thus allowing the user to save valuable data.

9.10.3   Imperfect Components

As noted above, working systems are usually found at some temperature much above absolute zero. As such their components are in a constant state of agitation (e.g., Brownian motion in a liquid is the random jostling of molecules); the higher the temperature of a system, the more agitation. Thus, components, especially at the microscale level, have a stochastic behavior rather than a perfectly deterministic one. This phenomenon can lead to occasional misbehaviors, which, if they occur at a critical moment, can lead to control failure.

Likewise extremely complex adaptive systems (such as cats or humans) become increasingly difficult to function as part of a control system where predictable performance is requisite.

Control failures in these systems are like noise effects in a communications channel (indeed could be categorized as special cases of noise!). They are the inherent in the components themselves and not the result of decay or interference from outside.

9.10.3.1   Stochastic Components

Stochastic means probabilistic behavior. That is, the actions of a component can best be described statistically rather than precisely deterministically. Mechanical and electrical machines are designed to have the absolute minimum of stochasticity in their component parts so that their performance is highly reliable (until entropy takes its due). Biological systems, however, are much more subject to the problems of working with stochastic components since they are based on biochemical reactions and operate in an extreme complexity. The control systems in living cells and organisms have evolved to keep stochasticity from disrupting normal operations, but the mechanisms of control are themselves also subject to stochastic process and so can, on occasion, fail. Fortunately this is rare in that biochemical processes operate within relatively narrow boundaries of deviation from the norm. But in probabilistic terms, a large deviation, no matter how improbable, is still possible. And the result could be loss of control of the rest of the process.

This phenomenon is advanced more on theoretical grounds than observation. By definition such events would be difficult to capture because of their rarity.

9.10.3.2 Heuristic Components

Somewhat related to the idea of stochastic components (perhaps, in fact, because in biology so many system components are stochastic) is that of heuristic functions. A heuristic is sometimes characterized as a "rule of thumb" or non-algorithmic rule (covered in Chap. 8). They show up in animal and human behavior and judgment. Our brains do not compute solutions to problems algorithmically. Rather some of our more primitive neuronal networks are programmed to respond to external stimuli based on evolutionary success in following such a rule. For example, when we see a lion, we run. When we see something that we take to be a lion, we run. Evolutionarily this proved to be a successful rule to follow, so it is wired into our limbic system. In fact the neural activation of the motor response is faster than the activation of conscious response, so consciously we humans feel fear as a signal that comes after we are already taking action to avoid the lion. Occasionally, however, that heuristic rule might not serve us well. If what we saw (or actually our ancient ancestors saw) was really a deer in the brush, but we mistake it for a lion, we would run and miss out on a meal.

Heuristic components of control systems (computation) work most of the time on average. They only fail occasionally so that decisions based on heuristics are acceptable. The alternative, being absolutely right all of the time, would require more computing horsepower than our brains could ever muster and would be restricted to problem domains that are solvable algorithmically. Since most domains are not so solvable, living systems have little choice but to rely on heuristics.[21]

9.10.3.3 Internally Motivated Agents

That leads us to the purely human situation, that is, when humans form the main components of a control system as the major decision makers. As noted, humans are just as subject to heuristic thinking and biased decisions as our animal predecessors. Moreover, they are operating in a substantially higher-dimensioned space of decision options in almost all problem domains. And if that were not complex enough, consider the wide range of motivations each individual human has with regard to their own interests in all that they do.

In Fig. 9.2 we depicted a market mechanism in which a number of subsystems cooperated for the good of all. We noted that as long as the number of subsystems (components) was kept small, it was possible for such a market, regulated only by inter-subsystem communications, to function reasonably well. In that situation, we assumed that all of the subsystems were "motivated" to cooperate. We argued that the need for coordination arose mostly because the number of subsystems grew,

---

[21] An excellent work on how heuristic thinking plays a role in ordinary, everyday decision making can be found in Daniel Kahneman's (2011) book, *Thinking Fast and Slow*, Farrar, Straus, and Giroux, New York.

and the complexity of the overall market became such that information would be unavailable to some, or most, subsystems. The latter would then not set their own operations to optimally coordinate with the other subsystems and the whole system would suffer. It turns out there is another possibility when dealing with internally motivated agents as subsystems. That is, these agents (human beings) may be primarily interested in their own well-being and possibly maximizing some value for themselves. We now know that in spite of Adam Smith's dictum regarding the "as-if" invisible hand[22] turning these self-interests into global optimization, real internally motivated agents will, from time to time, do harm to the global system if they think they can profit and get by with it.

In a social organization of humans (like a tribe, or village, or company), most of the time, most of the individuals will have an earnest sentiment of cooperation with their fellows that helps market systems work well enough. Unfortunately every once in a while, a "cheater" emerges who will attempt to maximize his own value (say net worth) at the expense of others.[23] Actually it seems that most people will, from time to time, do a little cheating here and there (as when you drive 5 mph over the speed limit!). We are a complex species! That being the case, control systems that are reliant on human judgment (flawed by biases) and subject to noncooperative spirits are bound to be less than completely effective. The reader is directed to any textbook on human history for ample examples of problems with social governance systems.

Question Box 9.27
Considering the variability and strength of individual motivations, it is easy to see how mechanisms of social control can be broken down. Yet it is also true that we flourish in communities and organize on an unmatched scale and complexity. What characteristics account for the incredible combinatorial potential and proclivities of these human components? Are we individual first and then systems—as in social contract theory, for example—or systemic first and then individual?

9.10.4   Evolving Control Systems

Effective hierarchical control systems such as are found in living cells and whole organisms are not built from whole cloth. They must evolve, as described in Chaps. 10 and 11, which means there must be multiple versions of the systems operating to be tested for effectiveness against the selection forces operating in the

---

[22] See http://en.wikipedia.org/wiki/The_Wealth_of_Nations and Smith (1776).

[23] See, for example, the treatment of the evolution of cooperation in Bourke (2011), Section 1.4, "Inclusive fitness theory and the evolution of cooperation."

environment. As we have demonstrated in this chapter, control evolves from the bottom up, from the operations level to the strategic level, by virtue of increasing complexity in the operational level, requiring increasing amounts of coordination. This is because at some level of complexity, simple cooperation begins to break down or falter. For a system composed of many interacting subsystems (operations), the problem of maintaining resilience and stability is one of optimizing overall operations internally. The problem of sustainability is one of effective performance with respect to the environmental elements with which the whole system interacts. Those elements are obviously not under the control of the system so the system must evolve an ability to respond to changes in the environment.

Strategic control emerges from the coordination between tactical and logistical control subsystems. It is the final stage that provides the maximum possible levels of adaptability and hence resilience and sustainability. For most organisms, strategic control is not computed on board (so to speak), but the strategies employed by these organisms are pretty much hard coded into the brains (or computational subsystems) of those organisms. Animals lower on the phylogenetic hierarchy follow instinctual behaviors to sustain themselves in their environments. This is what we recognize as their ecological niches. Animals higher in the hierarchy have brains that learn models of their environments and that provides them with a wider range of behaviors from which to choose when the environment varies from some long-term norms. Humans appear to have the broadest range of possible strategies to choose from, and there are those who contend that humans have the capacity to invent new strategies under extreme circumstances. Much more research on the capacity of the human brain will be needed to explicate this claim, but it can certainly be said that the human brain has achieved a level of strategic thinking that goes far beyond any other animal on this planet.

The question of complete and completely functional hierarchical control in societal organizations remains open. As we saw above, human agents are not completely rational or sufficiently knowledgeable agents to function perfectly as components in a hierarchical management system. It is intriguing to speculate a bit on what might be the case in some distant future if humans continue to evolve better judgment capabilities as a result of selection pressures originating in the need for better functioning control systems. Criteria basic criterion b for coevolution (see the end of Chap. 11) of better judgment in human agents and better management of organizations is met in that there are so many different organizations for selection to work on. There is a very high level of variation in organizational management, from corporations, to nonprofits, to government agencies, to military units, so that selection has much to choose from. The question remains, what kind of selective pressures might emerge that drive human control systems to evolve in this direction. This possibility is, of course, purely speculation, but it is based on the observations presented in this chapter and later ones. And holding to the contention that systems obey the principles we've been writing about, we contend it is not an idle speculation.

Think Box. The Brain: The Ultimate Hierarchical Cybernetic System
Animals, for the most part, have to move and interact with their environments.
Even animals like corals go through a stage of life where they move from one
place to another to find a suitable place to settle down. During such times,
even these creatures have more elaborate brains (which later get simplified to
match their sedentary life styles).

Predators have to hunt. Prey has to escape, and many animals usually have
to forage for food sources. These behaviors require elaborate hierarchical
control systems in order to be successful. Operational level controls are, in
general, handled by innate circuits in the brain, found in brain stem and the
lower level of the brain. Logistical control, that is, the balancing of distribution of internal resources like energy, can also be mostly found in the lower
brain. The autonomic nervous system is responsible for monitoring and commanding automatic functions like breathing, heartbeat, blood chemistry, etc.
Most of these functions are homeostatic during normal operations. But on
occasion, the organism may find itself in danger, or exerting extra effort that
require temporary changes in the distribution of resources (e.g., increasing
heart rate and blood flow to the peripheral musculature at the expense of the
viscera when running from danger).

Tactical control is that which manages the organism's behavior given internally generated drives (e.g., hunger signals from the logistical system prompt
the tactical system to start looking for food). Tactics involve matching the
organism's situation in its environment to its internal needs. In more primitive
vertebrate brains such as fish, amphibians, and reptiles, the control circuits are
primarily still in the lower brain (what has been called the "reptilian" brain).
These circuits are "hardwired" by genetics and constitute what we generally
think of as instinctive. They are automatic, responding to external stimuli with
minimal adaptation capabilities. In reptiles there evolved more capacity for
learning in the form of primitive cortical structures such as the amygdala.
These allowed a limited amount of encoding (probably short-term memory)
that expanded the reptile's ability to deal with more complex and non-stationary environments.

In mammals and birds, the emergence of much newer and larger cortical
structures (e.g., the neocortex in mammals) furthered the organism's capacity
to encode contingent associative patterns, that is, the changing, stochastic
environment with which these animals were then able to cope. The number of
contingent patterns coupling perceptual inputs to motor outputs (behavior
matching the demands of the situation) was greatly increased. Tactical control
was largely turned over to the cortices but retained connections with the older
reptilian brain to provide a two-tiered system; one that could handle evolutionarily established behaviors, such as responding to an immediate threat,
and one that could process contingent conditions and intelligently choose
among possible alternative behaviors.

The prefrontal cortex where perceptual and associative cortices gave way to response planning and pre-motor cortices, further evolved, in primates most particularly, to produce a relatively new "super-tactical" form of planning, namely, for the future. Strategic control is based on having elaborate models of the environment and of the self that can produce possible future scenarios—what the environment may do, what the self may do in response. Ordinary tactical control is limited in the time horizon it deals with. One way to think about tactics is in the nature of playing a game like chess. Tactics are the near-term moves that opponents make to try to gain advantage and ultimately win the game, but they are often responses to what the other player does. Strategy, in chess, operates over the entire life of the game and involves trying to take an initial configuration that long-term experience has taught the player provides an overall advantage. That is, tactical moves are made to enforce a strategic plan.

The human brain underwent a huge expansion of the prefrontal cortex relative to the other areas of the brain, even while the whole brain increased in size relative to other apes. The result is the human capacity to formulate plans for the more distant future—what you want to be when you grow up, what kind of mate you would like to find, where you want to live, and so on.

Figure TB 9.1 shows a diagram which maps the hierarchical cybernetic system of controls for the human body and its interactions with the world to general parts of the brain. The evolution of the human brain shows the pattern of adding higher-level management structures (e.g., strategic thinking) such as the prefrontal cortex (see TB 5) as the brain expanded in the hominin lines.

All of the human brain's abilities to function as a complete hierarchical cybernetic system are reflected in many social institutions. For example, companies are generally structured with top management responsible for long-range plans such as what to produce and how to grow the business. They have tactical management such as finance and marketing where decisions about how to interact with the company's environment to further its strategic goals are made. It has several logistical management functions such as the managerial accounting system that monitors the flow of "value" through the organization. Finally it has many forms of operational controls such as shop floor supervision that makes sure work orders are carried out and products are shipped to customers.

The structure of governance also reflects the hierarchical cybernetics of the brain. The executive branch acts as an overall coordinator, and we generally expect the president or prime minister to be thinking about the long-term interest of our countries as well as taking care of tactical issues involving trade and conflict with other governments. The legislative branch is concerned with making laws meant to regulate the internal logistics of a country. The judicial branch acts as corrective when something goes wrong internally, much like homeostasis. Governance is further hierarchical in being broken down into smaller similar units on regional and local bases.

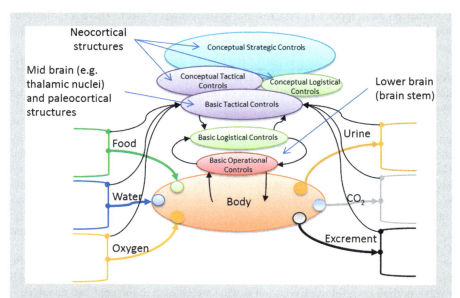

Fig. TB 9.1 The body is a basic system that gets inputs of food, water, and oxygen, processes them to construct and repair its internal structures and outputs waste products (other outputs like movement and reproduction not shown). The most primitive parts of the brain stem and older lower brain structures control body operations (*red oval*). Coordination of these basic operations is handled by basic logistical processing, also in the older parts of the brain (*lower green oval*). In reptilian brains, newer structures were added on top of the older brain to provide more behavioral or tactical control (*lower purple oval*). These structures (many cortical-like) allow the processing of more complex environmental information (*black arrows* from sources and sinks to the *purple oval*) and control of more complex behaviors related to obtaining resources and eliminating wastes. The higher-level *purple*, *green*, and *blue ovals* represent the evolution of the neocortex and its elaboration into what is called conceptual processing for control. The highest level achieved in a few mammal types and greatly expanded in humans is the strategic controls

The reader may recognize that the mapping of hierarchical cybernetic functions is not perfect in governance institutions. Indeed one might wonder if this "almost" system suffers many of the dysfunctions we so often see because it is a poor model of hierarchical cybernetics. That would probably be the subject of not one, but many other books!

9.11   Summary of Cybernetics

On the one hand, the role of information in systems science is relatively simple (news of difference that triggers response). On the other hand, the details are quite complex. In this chapter, we have provided a survey of the issues surrounding information as it applies to systems and system management. There are, unfortunately, many details and nuances when we consider how information turns into guidance for the control and coordination of all sorts of systemic processes.

Nevertheless, information is one of the key concepts underlying systems theory. Along with its companion theory of knowledge, properly understood, information and its role in cybernetic systems is one of the main foundations for understanding systems science.

Information is news of difference that makes a difference. That difference is embodied in the work that information triggers by virtue of its magnitude, that is, by the dimension of surprise as measured by its difference from the already expected (known) information of the system. The specific amount of work (which does involve the meaning of the message) changes the internal organization of a system so that its behavior changes and, as a result, it becomes a source of additional information feeding back from those changes.

Information and its role in generating knowledge is one of the most important concepts in systems science. It is every bit as important as the concepts of matter and energy. Indeed it might be, in some senses, even more important, for as the modification of systemic expectations, it tells us something of how systems will evolve in their own future. Evolution is what we take up in the next two chapters.

We have seen how computation in its universal sense is an integral part of making knowledge out of information, of making sense out of messages, and providing additional information to decision processes for purposes of controlling the long-term behavior of complex adaptive systems. The science of cybernetics brings these various threads together in terms of how systems actually *are* in the world. The world (the environment of a system) changes over time, requiring the system of interest to adapt and respond. In CASs, particularly living organisms and social systems, we see how a hierarchical structure supports different, important kinds of decisions that work collectively to keep the system functioning in spite of those environmental changes. In the best cases, we see that this organization has proven exceptionally successful in maintaining living organisms, populations, species, and higher genera of life for well over two billion years of Earth history. Social systems have been in existence only during a mere blink of an eye in geological time, so they are still evolving toward greater integration and coordination. We are at a peculiar time in the evolution of hierarchical control in human social systems. We observe some successes, but also many failures, as nature sorts out what works from what doesn't. We can only make forward inferences based on having seen the pattern of emergence and evolution of hierarchical systems many times before in lower levels of organization. We now turn to a more detailed examination of the processes of emergence and evolution to get a better handle on how this is so.

Bibliography and Further Reading

Ashby WR (1952) Design for a brain. Chapman and Hall, London
Bourke AFG (2011) Principles of social evolution. Oxford University Press, Oxford, UK

Findeisen W et al (1980) Control and coordination in hierarchical systems. Wiley, New York

Heylighen F, Joslyn C (2001) Cybernetics and second order cybernetics. In: Meyers RA (ed) Encyclopedia of physical science & technology, 3rd edn. Academic Press, New York. Accessed at: http://pespmc1.vub.ac.be/Papers/Cybernetics-EPST.pdf

Kahneman D (2011) Thinking fast and slow. Farrar, Straus, and Giroux, New York

Odum HT (2007) Environment, power, and society for the twenty-first century: the hierarchy of energy. Columbia University Press, New York

Smith A (1776) An inquiry into the nature and causes of the wealth of nations. W. Strahan and T. Cadell, London

Sutton RS, Barto AG (1998) Reinforcement learning. The MIT Press, Cambridge, MA

Wiener N (1948) Cybernetics: or control and communication in the animal and the machine. MIT Press, Cambridge MA; 2nd revised edition 1961

Wiener N (1950) The human use of human beings: cybernetics and society. Avon Books, New York

Part IV
Evolution

According to philosopher Daniel C. Dennett, the concept of evolution is "universal acid" that eats away at every area of knowledge.[1] He contends, and with good arguments to back him up, that nothing we know is immune to the impact of understanding how evolution changes our world. In this he echoes Theodosius Dobzhansky (1900–1975) who said "Nothing in biology makes sense except in the light of evolution."[2] Dennett expands that sentiment to much more than biology, and he has much company as discipline after discipline now introduces evolutionary process as providing an essential window for understanding their subject matter. Over the last several decades, the details of evolution theory have informed subfields such as evolutionary ecology, evolutionary behavior (and the combination evolutionary behavioral ecology!), and, of course, evolutionary morphology (e.g., comparative anatomy and phylogenetics). Evolutionary medicine looks at how the evolutionary processes that shaped human populations contribute to things like the different rates of diseases in various subgroups, an obvious example being sickle cell anemia in black populations owing to the selective pressures brought on by malaria in tropical Africa. Evolutionary psychology looks at how intelligence and other human mental capabilities came into being and developed to what they are today. There is even a field called evolutionary economics, and sociologists are investigating how the dynamics of the evolution of social institutions show striking similarity to biological evolutionary processes.

Of course the mainstream of biology has long been centered on evolution as first successfully outlined by Charles Darwin (1809–1882),[3] who has received the lion's share of credit, and Alfred Russel Wallace (1823–1913),[4] who both announced their theories in 1858. Even before Charles Darwin (and Wallace) provided a "mechanism" for explaining evolution, many people had come to accept that the

---

[1] Dennett (1995).

[2] Essay in: *American Biology Teacher*, volume 35, pages 125–129, 1973.

[3] See: http://en.wikipedia.org/wiki/Charles_Darwin.

[4] See: http://en.wikipedia.org/wiki/Alfred_Wallace.

Earth had been undergoing profound changes over many more years than had been assumed. And they came to accept that along with these changes, the biota of the Earth must have evolved, with different forms existing at previous times, and somehow giving rise to newer forms. What Darwin and Wallace did was help explain *how* this might be the case. That is, they provided a scientifically based explanation[5] in the form of natural selection, the idea that the fittest members of a species would produce a larger number of offspring that would reach reproductive age themselves. They posited (but had no direct evidence) that offspring inherited traits from their parents and that those traits might have some variability in how effective they made the possessor in fitting into its environment successfully. Some variants of a trait were superior to others and led to greater success of those possessors. Since competition for resources was always an issue in nature, the less competent variants would lose out, in the sense that they would produce fewer offspring. Eventually the more fit variants would come to dominate the breeding population and, in Darwin's view, the species would gradually shift toward that more fit form.

Darwin never actually explained how species would come into being.[6] He assumed that after enough time had passed, the new forms would emerge. He did allow that divergence (the splitting off of two or more species from a single root stock) was possible due to migration and isolation. The migrating population would become isolated in a new environment, and selection then would produce forms more fitted to the contingencies of that new environment even while the root stock continued on in its original environment. Thus, eventually two species derive where one had been before. But what constitutes the actual act of speciation was never made clear. Even to this day, the boundary line between any two closely related species is fuzzy at best.

Biological evolution provides us with a core model of the process of change in whole systems over time, but the evolutionary dynamics of living systems do not answer some of the more universal questions about systems that are important to understanding their change in form and behavior over time. Thus, in the second half of the twentieth century, the understanding of evolutionary processes reached back beyond life sciences to encompass physics and chemistry. In other words, the challenge of understanding systemic evolution in its fullest sense reaches back to the beginning of the universe, not just the beginning of life.

---

[5] We call this a scientifically based explanation in part because Darwin and Wallace were attempting to stick to known scientific principles while providing a logically consistent mechanism. In today's terms the "theory" of evolution was more of a hypothesis that had yet to be tested. Indeed, it has only been in the last several decades that really adequate empirical and observational research demonstrating the mechanism of selection in action has been done, confirming the theory as a true scientific theory.

[6] Even though he titled his first, and most famous work, *On the Origin of Species*, 1859. See http://en.wikipedia.org/wiki/On_the_Origin_of_Species.

The view that we are adopting is called "big history." Ordinary history is based on written records and is therefore limited to the last 4,000 years or so of human life. Evolutionary biology extends our vision of history back by reading the signs in the phylogeny of life. Big history takes the giant leap to extend our interpretation of what happened before humans ever existed, and indeed, before the Earth existed, back to the origin of the universe itself. This approach attempts to reconstruct the origin and evolution of the Universe, as we understand it, in order to place our own origin and evolution into a larger context. Such a reading depends on the work of astronomy, astrophysics, and especially cosmology. That is, needless to say, an ambitious program. But it may be within reach, especially with a systems perspective.

Note that we are not talking about just dynamics, as in Chap. 5. We are talking about wholesale modifications that transform systems as they age. The range of questions we should consider include:

- Origins—How do systems come into existence in the first place?
- Complexification—How do things (as well as the inverse) get more complex over time?
- Emergence—How do groups of complex things interact to form new levels of organization with new properties?

The *origins* question deals with how systems come into existence and achieve stability sufficient for evolutionary processes to begin operating on them. Once those forces do come into play, we are curious as to how systems transform over time, most notably how they evolve to become *more complex* in form and function as time goes on. This is what Darwinian, and so-called neo-Darwinian evolution, is about for living systems, where it takes the form of speciation. But we will see that this question is also appropriate for prebiological entities: how could systems of matter, energy, and information arrive at such complexity that they could become alive in the first place? Finally, *emergence* is concerned with what we recognize as levels of organization and complexity: we are interested in how evolving systems of systems, aggregates of complexifying entities, interact so as to create new, even more complex wholes, with new properties, at a higher level of organization. So our three questions mean we will be studying the overall question of how systems organize over time with these three closely related but shifting focal points.

The general model of universal evolution can be said to start with the process of auto-organization, in which disparate components in an otherwise unorganized system with low realized complexity (cf. Chap. 6) are naturally attracted or repelled in ways that lead to the formation of assemblies. At some point, these assemblies create new aggregate personalities so that new properties and interactions between them emerge. Chapter 10 will take up these two aspects of evolution, with special attention to the process of organization and the emergence of new properties and interactive capacities as organization mounts. All the while components/assemblies are tested by the environment in which they exist, including the competition and cooperation between and among the components themselves.

Chapter 11 contains the story of the unfolding of the entire process of evolution. We will look at the process leading to the emergence of life and then its branching into divergent trajectories, including the emergence of human level intelligence, with its concomitant creativity and inventiveness, and then consider the incredible articulation and coordination of the networked interdependent relation of all these lives in ecosystems and societies down to the present day. It undertakes to describe the grand sweep of evolution as the spiraling cycle of increasing complexity that seems to somehow go against the second law of thermodynamics in producing ever greater levels of organization.

Of course the second law still rules; as we will see, the seeming contradiction is no contradiction at all. In fact, the operations of the second law contribute to the whole process! Without entropic processes breaking little pieces of the system down from time to time, there could be no progressive buildup of new organization. The key, as we will see, is that energy must be available and flowing through the system (dissipative systems again!). Once life emerged from organized chemical processes, it was subject to a rigorous process of selection, and this "natural selection" became the keystone concept at the core of the nineteenth century discovery of evolution. But we will see that life here simply exemplifies with particular clarity, a selective dynamic that is critical to all evolutionary process.

Chapter 10
Auto-Organization and Emergence

"...nature yields at every level novel structures and behaviors selected from the huge domain of the possible by pruning, which extracts the actual from the possible."

Harold J. Morowitz, from *The Emergence of Everything*, page 14.

"...the flow of energy through a system acts to organize that system."

Harold J. Morowitz, from *Energy Flow in Biology*, page 2. italics in the original.

"The hallmark of emergence is this sense of much coming from little. This feature also makes emergence a mysterious, almost paradoxical, phenomenon...."

John H. Holland, from *Emergence*, page 2.

Abstract This chapter introduces and explores the fundamental organizing principles of systems. Up until now, we have been describing how systems are, how they are organized, how they work, etc. Starting in this chapter, we will be examining the general processes that account for how systems, particularly complex ones, actually come into existence, how they get to be how they are. Underlying the development of complex systems are the twin processes of auto-organization and emergence. These, in turn, are part of an overarching process, that of evolution, which will be covered in the subsequent chapter. Auto-organization explains how components of systems first start to organize and interoperate. Emergence explains how functions come into existence at a level of organization built upon what new structures have auto-organized and have survived within their environments. Perhaps the paradigm example of these processes at work is the origin of life.

10.1 Introduction: Toward Increasing Complexity

Complex adaptive systems (CAS), including all living systems, are the poster child of evolution, for, as the name implies, they exhibit the dynamics of probing for an ever-changing adaptive fit. But these systems are not formed from whole cloth in

© Springer Science+Business Media New York 2015
G.E. Mobus, M.C. Kalton, *Principles of Systems Science*, Understanding
Complex Systems, DOI 10.1007/978-1-4939-1920-8_10

461

their final forms; they themselves have developed on the face of the Earth through a long and arduous evolutionary process that started with the self-assembly of marginally complex molecules. The interactions between these molecules gave rise to the emergence of new properties for material systems, and we move from physics to chemistry. Those new properties were tested within the contexts of their environments and were found acceptable or wanting. The latter disappeared since the testing destroyed them. The former went on to explore new possibilities built on a mounting base of increasing complexity until some systems closed the loop on reproduction, and we move from chemistry to biology. And as we have seen, each step in this process of nested systemic emergence is accompanied by new mechanisms of control and coordination. This chapter will explore the dynamics of auto-organization,[1] whereby simpler systems ramp up into more complex systems. We will then consider how it is that new levels of organization with new properties and capacities for interaction emerge from this process.

At this point, we need to resolve a seeming discrepancy. In Chap. 5, we defined complexity in terms of levels of organization. If you recall, a whole system can be said to be more complex when it is composed of nearly decomposable components. Such components are subsystems and thus may recursively be composed of nearly decomposable components. As will be demonstrated in Chap. 12 (Systems Analysis), this property is exploited in functional decomposition when conducting a reductionist analysis of whole systems. What comes out of this is a hierarchy composed of levels. The number of levels is a rough index of the complexity of the system. In this chapter, we will work in the opposite direction. That is, we will proceed from an essentially unorganized system of atomic components to show how this kind of organizational hierarchy obtains.

In essence when we refer to increasing complexity, we are referring to this kind of index applied to the whole system. At the same time, at any designated level, the components (subsystems) will have their own index of complexity. For example, in the origin of life problem (see Sect. 10.4.5), we recognize that Earth started out with very simple molecular structures in the various "spheres" (litho-, hydro-, etc.). Life eventually emerged from the processes we will be covering below. There came a new level of organization containing living cells, which are extremely complex. The new "biosphere" now came under the laws of evolution through genetic variation in copying and natural selection. Eventually multicellular life emerged, but unicellular life forms persisted as well. Moreover, simple molecules continued to exist so the system (Earth) had reached a new level of organization, but only small segments of the internal system achieved higher levels of organization for themselves. They became the complex adaptive systems that will be the primary focus of our attention. When we refer to increases in complexity over time, we are actually referring

---

[1] Many authors have used the term 'self-organization' to describe a process of a potentially complex system evolving toward an organized (realized complexity) system. The difficulty with the term "self" is that it can convey the meaning of intention which is inappropriate. In many scientific contexts, the term "auto" implies that the system itself contains the necessary explanation for its own dynamics. We prefer to use this term, so readers need to be aware that when they run into the terminology "self-..." in the literature, it means essentially "auto-...".

to the combined effect of increasing levels of organization for the whole system and increasing complexity in some of the subsystems within that whole system. Note, however, that an increase in the complexity of some components contributes to the organization at that level and hence the complexity of the whole system is also increased.

Our interest, now, is in how did the world get to be complex? How did we get from a world full of molecules to a world of life, species, tribes, organizations, economies, and cultures? How did the complex emerge from the simple?

10.2   The Basic and General Features of Increasing Organization Over Time

Complex systems not only come into being, they remain in existence and may even become more complex as time passes. In spite of the presence of continual collapse and decay, there seems to be some kind of ratchet gear on complexity. In culture our experience is certainly one of constantly escalating complexity, and at virtually every scale of space and time, a similar process of increasingly complex forms of organization can be identified. How do we explain this?

In the earliest days of standard "history," people who thought about the nature of the world, the earliest philosophers, accepted the fact of complexity without really wondering how it got to be so. They might see complex phenomena as needing some explanation, but they did not identify complexity itself is a dynamic and ramifying process to be understood. Thus, early religious explanations tended to simply observe that a God or Gods created all that is, as it is, and that seemed a good enough explanation. Indeed, a number of traditions also identified great cycles of temporal transformation, but these generally served as just large frameworks to understand where we are located, where we are headed, and often, why life seems so messed up and difficult (the golden age is always in the past!).

In the eighteenth and nineteenth centuries, as people started to think more scientifically, that is, looking to empirical inquiry, the notion that the world had always been the way it was started to unravel. Geological investigation began to yield evidence of a new scale of change, showing that rocks in various formations had changed in form, position, and constitution over long ages. If the Earth was much older, and had been quite different in a distant past, what else might have changed? And while examining some of these rocks and other formations, students of geology discovered fossils of ancient animals (and plants) that no longer existed to anyone's knowledge. At about the same time, many naturalists were noting how species of plants and animals tended to be clustered in groups having similar anatomical features and behaviors.

Process was emerging as a new way of thinking. Geological organization might not be a simple fact but the results of a process of vast change over a long period of time. Maybe the same was true of biological phenomena—a notion that brought out new and controversial implications in the observation of groupings and similarities.

One of the most famous examples was the similarities in anatomical and some behavioral features between humans and great apes. The suggestion was that these species must be related in some way, but the process or mechanism that would explain the relationship was not obvious. Naturalists started noting how there were representatives in these groups that appeared to be more "primitive" or simpler in form, suggesting that they had come into being much earlier and then, somehow, gave rise to more complex or "advanced" forms. Thinking of process started to be framed as a directional movement from the more simple to the more complex. Charles Darwin's famous grandfather, Erasmus (1731–1802), had actually worked on this concept, which came to have the general name of "evolution."

10.2.1   Definitions

Before delving into the process of increasing organization, we should make a few clarifications regarding some potentially problematic words that will be used here in a technical sense. Just as we saw with the terms "complexity" (Chap. 6) and "information" (Chap. 7), in addition to the loose or fuzzy common uses of the terms, there are variations on even the technical definitions of some that could cause trouble when we are trying to be precise and specific in our models.

10.2.1.1   Order and Organization (or Order Versus Organization!)

We will start with a pair of related terms that often can cause confusion but are used interchangeably because they have close associations. The terms are "order" and "organization" (which was the subject of Chap. 3). Quite often the word "order" is used to describe complex structures such as "a living system is highly ordered." In such usage, "order" becomes virtually synonymous with "organization."

The problem is that the term "order" is used very precisely in thermodynamics to describe a state of a system in which entropy is minimized. Under the usual interpretation of the third law of thermodynamics, order refers to the condition of a system in which the components find themselves in a minimal energy state. A minimal energy state would mean being near or at a freezing point! Ice is a more ordered state of water molecules than is the liquid. In general crystalline structures are highly ordered, whereas amorphous, dynamic structures are less so in this thermodynamic sense. The actual situation involves what we can think of as the "rigidity" of the interconnections between the component parts, the tightness of the coupling (Chap. 3). The tighter the coupling and the more regular the structure (such as a lattice), the higher the measure of order. In *this* sense, order is the antithesis of entropy.

Highly organized living systems are likewise far from entropic disorder, but their dynamic order is nothing like the kind of static minimal energy of a frozen crystal. This is where the confusion starts because both our intuition about living systems being low entropy, far from equilibrium, and that such a condition implies order leads to a possible misapprehension of what is actually going on. Living systems are clearly not frozen; their molecules are in constant and generally high rates of motion and combining/breaking bonds.

Yet this kind of organization likewise is subject to entropic decay. If you place a living system inside a thoroughly sealed container, keeping the temperature the same for a very long time, that system will die and eventually all of the complex molecules will decay into much simpler molecules. Given a long enough waiting time in this condition, the system of molecules will be in thermal equilibrium, and there will be no real organization, which tallies with the classic description of entropy.

A critical systemic and thermodynamic difference distinguishes the organization of life from the minimal energy order of ice crystals. The former is a far-from-equilibrium dissipative system, an open system through which energy flows. The latter is a closed system. Illya Prigogene's work showed how under conditions of energy flow the open systems can ratchet up to higher levels of organization—just the phenomenon we will study in this chapter as "auto-organization."[2] Such organization or order is maintained only by a continual input of energy, which is quite different from the order of minimal energy relationships, which on the contrary are closed to further energy input.

A resolution to this conundrum was provided by Harold Morowitz. He devised a broader definition of thermodynamic order to include systems far from equilibrium at a given temperature.[3] Morowitz's solution allows us to find an equivalence between the organization of the molecules in a living system that implies complex structures that are comprised of components at minimal energy *for that temperature*. In other words, a living system has achieved a structure that allows its components to operate in what we could think of as a "comfortable" level of activity at physiological temperatures. But the key, as Morowitz pointed out, is that energy must flow *through* the system and not contribute to an increase in internal energy levels which would move it away from the measure of order. So it turns out, if we are careful in our formulation, the two notions of an organized structure with ongoing stable function and an ordered system can overlap.

10.2.1.2   Levels of Organization

The phrase "level of organization" will be used frequently in this chapter and others. We need to be very explicit in what this phrase means. We already have a fair idea of what is meant by organization, having spent a whole chapter devoted to it, but the notion of levels has not been given enough attention. It becomes important here because we will be examining the principle of emergence later, and a more rigorous definition of a level is needed for that.

---

[2] Prigogine and Stengers (1984).

[3] Morowitz (1968). The technical details are well beyond the scope of this book, but the arguments here should give the reader a better sense of why organization and order are related concepts.

As a first approximation, we can consider the structure of the physical world. To the best of our current knowledge, the smallest components of the universe are "fundamental" particles, such as the quarks and electrons, and force mediating particles, such as the photon and the gluon. There are a relatively limited number of kinds of these particles, but they can combine in different ways to produce an extraordinary "zoo" of second-level particles. These second-level particles have their own unique "personalities" which are open to many more kinds of interactions. Thinking in terms of organization as ways in which components may relate to one another, we then have here two levels of organization: the fundamental particles constitute the first (known) level of organization in physical reality, and the secondary particles, such as protons, neutrons, and various mesons, along with their antiparticles,[4] constitute a second level of organization with a distinctive mode of relationship. These particles in turn participate in the interactions that produce atoms. And atoms in turn interact through the electromagnetic force to form molecules and crystalline structures and also exchange photons as means for communication of states. Each of these levels arises by relational combinations of components at the next lower level, hence the notion of "levels" rather than just different systems. The difference between levels here is that level by level the resultant combinations comprise components with new personalities in regard to their potential for relation and interaction.

The levels of organization, then, are largely defined by the scale of components and their interactions. Our capacity to perceive these scale differences with our direct senses starts primarily at the molecular level. However, the notion of "level" starts to get a little less easy to define once we get down to the molecular, since the molecular scale itself encompasses a huge range of different sizes, shapes, and combinatorial potentials. At the next lower level, the numbers and kinds of interactions between atoms are already vast (knowable but vast), and so the numbers and kinds of molecules and other material substances they combine to constitute become nearly infinite. Organic chemistry alone can produce what would appear to be a limitless variety of molecules, given the binding nature of carbon atoms. Nevertheless, we generally consider the "molecular" level of organization as a single level with a number of sublevels depending on the kind of chemistries possible.

The chemistries of nonorganic molecules are important in the physical processes that constitute geological, hydrological, and atmospheric systems. Everything from the shaping of mountains, to the volcanic vents in deepwater trenches in the mid-Atlantic ocean, to the formation of clouds and weather may be considered another level of organization, the geophysical. And embedded within that framework is the organic chemistry that we call life! Living systems share in the geophysical scale, but many authors put them on a level just above what we would call the geophysical due to the complexity of biological processes and of the structures organisms build. It can be useful to think of life as a distinct scale, but it has the drawback of obscuring the continuities of interaction and relation at this geophysical scale. Life acts as

---

[4] For example, an electron-like particle but having a positive electric charge is called a positron.

a geophysical force in the processes of the Earth, and the other geophysical factors such as climate and continental drift shape the community of life. Again we prefer here to take recourse in a definition that is based on scale, but it is also clear that at the scale we are talking about the ranges in space and time are quite extensive.

It becomes clear that levels are a functional concept: while based on real factors of organization, the sort of levels identified may vary with the sort of factors selected as criteria to suit a given purpose. For other purposes, for example, levels of organization might be identified on the basis of which of the fundamental forces (strong, weak, electromagnetic, or gravity) rules the overall behavior. At the molecular level, the electromagnetic force dominates. At the level of primary particles, the strong and weak forces do so. At the geophysical level, clearly gravity is the most noticeable force in play. Electromagnetism keeps the components together and still mediates the personalities of those components, but gravity is what keeps the components bound in the same system.

More nuanced versions of levels of organization might be based on an assessment of the modes of interaction between components at each level. We know that the kinds of interactions depend on the fundamental forces. But it turns out that we can discern more elaborated interactions that emerge only at a given level of organization and complexity. For example, at the molecular level, a new factor, the shape of the molecules, emerges as critical in catalyzing otherwise unlikely sorts of chemical interactions. And at another level, the interactive organization of living organisms has such distinctive complexity that many consider it a distinct level. Extending this approach, there are those who would distinguish humans and their cultures as another distinct level.

The human social level of organization becomes a virtual cornucopia of new kinds of interactions and the emergence of interactions not predicted from basic instincts. The creation of an organization like a corporation cannot be predicted by simply knowing that all humans need to eat! Yet at the same time the drive to obtain resources to keep one's self alive is at the root of acquisition and the desire for profits, and corporations emerged over a long history of trying new ways to organize to make profits. So we need to keep an eye on the purpose for which we identify distinct levels in order to avoid the pitfall of losing sight of systemic roots that carry across what we for our purposes identify as distinct levels. New forms of interaction and organization indeed occur, but they do not simply leave behind the base from which they arise. See the discussion on emergence for further elaboration.

Question Box 10.1
Joe's parents organized an elaborate birthday party for his sixth birthday, inviting eight of his best friends from school. How many systemic levels can you identify in this scenario? Clue: consider how it might look from different academic disciplines.

10.2.1.3   Adaptation

Another term that can cause problems is adaptation. In the most general sense, adaptation means *changing a form or function in response to some force from the environment acting on the system.* But it turns out there are many different flavors of adaptation, the definitions of which are highly context sensitive.

The problem arises in respect to being clear about the mechanism of the adaptation the level of organization we are considering. For example, a person can physiologically adapt to being cold by shivering to increase body heat. This is a short-term process: It is the body's immediate response to an external condition, but it cannot last forever since it is draining the body of energy. However, the same human can adapt, behaviorally, to the same cold by putting on more clothing or seeking shelter. Going even further, that person can "learn" to associate cold weather with a need to dress warmly. The brain then has adapted, changing in some way reflecting a learned response so that in the future, when the weather is cold, the person will automatically dress accordingly.

And then there is a form of really long-term adaptation which we think of when we consider evolution. It is conceivable that there are genetic variants that make people hardier and better able to function in the cold. We speak of these individuals as being "adapted" to the cold weather even though they did not actually change their constitutions.[5] And if cold weather were to become more prevalent for the population, then the species might slowly "adapt" to that fact, with those hardy types surviving better and producing more children with similar constitutions until eventually the majority are hardy.

So adaptivity means different things, or at least is implemented through very different strategies as the scale and level of organization change. As you can see, there are multiple subtle meanings attached to this term which makes it necessary to ensure that the context of its use is well understood. Learning and evolution are certainly forms of adaptation. But for systems analysis, a less open-ended form of inherent adaptability is an important consideration.

In a systems analysis context, we prefer to use adaptation to refer to the capacity of an individual system to respond to an environmental shift by reallocating internal resources to the responding mechanism, as in the example of shivering above. This implies two things. First the shift has to be of a temporary nature so that the system will be able to relax back to a "normal" state eventually. Second, the range of the shift has to be within boundaries dictated by the system's inherent range of response, meaning that it possesses the ability to reallocate those internal resources without doing permanent harm to the system's other mechanisms.

---

[5] Some authors prefer to use the term "preadapted" to distinguish it; however, the problem of interpreting what causes the change remains.

10.2.1.4   Fit and Fitness

As with adaptation, the term "fit" can apply to a number of levels of organization and time scales. For example, the term can be applied to an individual, a population in a specific ecosystem, or a species over some duration of time. A biological system is said to be fit if its traits ("personalities" from Chap. 3), compositions, behaviors, etc. match well with the attributes of a particular environment. It fits in, in the same sense that a hand fits into a glove or a person fits into a social group. In evolution theory, those systems that prove more fit than any other competing system will tend to increase in prevalence and possibly come to dominate in numbers, or even push out competitors from the "niche."

The key to understanding the competitive evolutionary advantage of the more fit is the association of fitness with superior success in reproduction, so the more fit genetic recipe spreads exponentially over the generations. Strategies that outreproduce others thus have superior fitness, even if they seem to exact a higher price on individuals. In some spider species, for example, females eat the male immediately after mating; a case in which a well-fed mother-to-be must, in that environment, have some kind of reproductive edge that outweighs a shorter life span for males.

To get a better sense of what "fit" means for systems in general, consider Figs. 10.1 and 10.2. These depict systems of interest and their environments as networks of components and interactions. The depictions are not meant to represent actual geometrically proportioned relations, only functional ones.

Figure 10.1 depicts a system of interest (SOI) that has some degree of fitness with respect to its environment. Its boundary components have direct, though generally loosely coupled, interactions with the environmental entities that directly impact it. In one area of the boundary, however, there appears to be no interaction with the environment, and there could even be a negative interaction happening at that point. This SOI might enjoy a moderate amount of success in its overall interactions with its environment.

Now we need to consider the situation over time. Imagine that the SOI and its environment are dynamic; they both tend to vacillate, sometimes stretching the coupling interactions in various ways, straining them. Interactions have a range of tolerance for variation, and in many cases, the interactions are somewhat elastic. That is, they can be stretched, but in general they hold and bounce back. What this actually describes is the effect of fluctuations around a set of mean interactions due to energy flows variously exciting some components or environmental entities. They jostle with activity, but stay generally within the vicinity of what is shown in the figure. Nutritional flows, for example, often vary in type and/or quantity around a functional mean, with sustained departure from the mean proportionally taxing the system (e.g., starvation or obesity).

But what this flexible dynamic means is that it is possible for some interactions to occasionally be displaced, opening up opportunities for new things to happen. Suppose, for example, that some SOI that is essentially the same as the one depicted in Fig. 10.1, but has one fortuitous component right where the SOI in Fig. 10.1 has a poor fit. For instance, the SOI in Fig. 10.2 might be a type of plant that manages

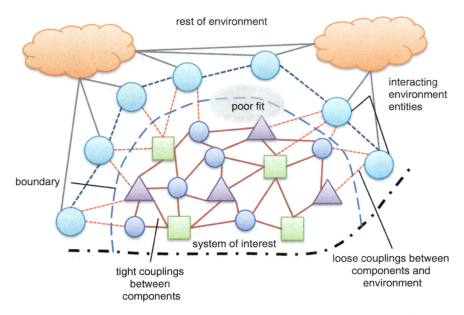

Fig. 10.1 A system of interest (SOI) interacts with its environment through its boundary, here considered as the specific interactions between components lying in functional proximity to the environmental entities with which they interact. The *dark black links* indicate tight couplings between system components that provide system organization and integrity. The *dashed lighter links* represent the loosely coupled interactions between SOI components and the direct environmental entities. The *clouds* represent the rest of the environment that affects the coupled entities but not the SOI directly. The *grayed oval* indicates a place where there is little or no interaction that could indicate a less-than-ideal fit (see next figure). The *bottom dotted/dashed line* just represents a visual cutoff; the SOI extends further down with a similar environmental configuration

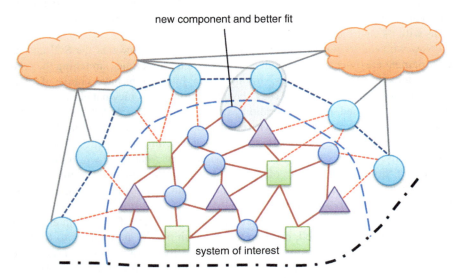

Fig. 10.2 Another SOI that has acquired a component that interacts better with the environmental entities forms a better fit

to survive but struggles with salt in the environment, and SOI in Fig. 10.2 is a variation that happens to be more salt tolerant. With the jostling process of time and reproduction, it is easy to imagine the new salt-tolerant variety displacing the old less salt-adapted variety since it has an extra component that forms a much better interaction with the environmental entity. It would look like the fit in Fig. 10.2.

The new SOI is essentially the same as the old one, but with one little change, that can make a big difference. The new SOI is better adapted to the environment because it has a better fit as defined by stronger interactions with the entities in that environment. In our example, in a salty environment, salt exercises a selective pressure that makes salt tolerance a superior fit. A further variation that found a positive benefit in the salt might be an even better fit—at least until salt became a scarce resource in the environment. You can see that the notion of "fit" or "fittest" is an inherently relational term and has no meaning without considering a system's environment.

The fitness competition among comparable systems typically hinges on this resource relationship, so it is common to speak of competition for resources. The critical factor, however, is overall fit with resources, not just grabbing more—which may or may not be a better fit. Adding to the complexity is the fact that systems typically fit in multiple ways with their environment, with multiple degrees of tolerance and criticality that change relative to the situation. In a competition between species in Figs. 10.1 and 10.2, the superior fit of the latter would tend to enable it to make use of a slightly wider variety of resources and be in a less-strained or critical condition. Both varieties might do well enough for a period, so neither would appear more fit. But then if something should happen to other resources, say the white circle (environmental entity) in the lower left corner of the figure were to be taken out, the species in Fig. 10.1, being already under more strain, might have a harder time absorbing the change and would be at a disadvantage compared with the species in Fig. 10.2.

For living organisms, fit is another way of talking about sustainability, a notoriously difficult question (Appendix A will provide a systems science perspective on sustainability). Insofar as fit is a relational matter, it is necessarily a two-way street. In more complex organisms which learn from experience, individual adaptation is a marked feature. In effect, the system in Fig. 10.1 is able to transform itself into an 10.2 system when that challenging ill-fit circle looms into view. Humans have so perfected their moves in the dance of fitting themselves to circumstances and fitting circumstances to themselves that one often hears that challenges, those looming circles of potential ill-fit are really opportunities for creative advance. Moving from ill-fit to a better fit is undoubtedly an advance, but this returns us to the way the question of fit hinges on scales of organization and time.

Fitness is a matter of flows enabling and sustaining various sorts of complex organization. On the level of physics and chemistry, this has given us the phenomenon of selective auto-organization to the point where life emerged. With life comes the especially creative selective process leveraged on reproduction, which selects for whatever fits with the circumstances well enough to survive long enough to make copies of itself. At a much larger systemic scale of fit, we would be considering the fit of the Earth within the cosmos as shaped by the fundamental laws of physics. But that kind of fit is more or less automatic, while fit in the ever-changing dance of the living community is constantly negotiated and renegotiated.

In the community of life, relatively short-term and long-term fits are not necessarily congruent, and we are accustomed to using the longer-term to critique short-term strategies. The reason is that without this framing, the short term tends to leave out the larger scale of interdependent relationships for a narrow version of fit that cannot last. Predators that become so clever they wipe out their prey are not sustainable, for they have destroyed the fitness of the prey. Businesses that consume all competitors destroy the market system in which they arose and excelled. The question of fit is constantly in a state of flux, but also systemically involved in a dynamic feedback loop in which long-term consequences, generated in a multi-causal system, circle to become conditions that in turn determine the imperatives and possibilities of immediate fit. Species that, like ours, modify flows mightily best have in mind the fit of those flows with the rest of the life community within which they must fit.

Question Box 10.2
Issues of policy often amount to arguments about what policy is the best fit. When a goal is shared, the fitness question probably has an answer. For example, will technique X or Y enable us to catch the most fish next year? How would enlarging the time scale possibly change the answer to which technique is most fit? If one party was arguing fitness assuming a few years, and another was thinking about his children becoming fishermen, both could be right but still disagree. Is there any end to such arguments? What guidance might come from a systems perspective?

10.2.2   Evolution as a Kind of Algorithm

Processes get the direction to head somewhere by selection; in the absence of some kind of organizing selective factor, one has just a wander, not an evolutionary trajectory. We will see that there can be a number of factors that can constitute a selective dynamic, but what is selected for is always a matter of fit with some sort of criterion. In a sense, one might even argue that the notion of fit is already present in the concept of selection. In the case of systemic evolution, whether of life or of prelife, the criterion is not resident in some mind but in the conditions imposed by the nature, shape, and functioning of the surrounding system.

Selection, then, is the key to the process of evolution. Daniel Dennett suggests that Darwinian natural selection can be understood as an algorithm. Recall from Chap. 7 that an algorithm is a set of unambiguous steps—actions—that solve a problem. Dennett may have taken just a bit of license with the word algorithm, but if we think of natural selection more as a recipe for fit, we don't have to face any nit-picking mathematical constraints. For one thing, there isn't necessarily an a priori "right" solution to be gotten in evolution, just degrees of better fitness. And whatever fitness may be attained is never final, for the environment can always be counted on to change, thus changing the fitness relations. Rather than an algorithm for finding a solution, natural selection is an effective procedure for chasing better fitness endlessly!

The basis for the "algorithm" is what Richard Dawkins calls a *replicator* (Dawkins 1976). This is a system that makes copies of itself with as much fidelity as is physically possible. Both asexual and sexual reproduction require that the genetic codes for the construction and maintenance of a living organism are replicated and separated into the two resulting cells. The algorithm starts with the replication of many individuals in a population of similar replicators. Since every physical act of making copies will be subject to some inherent noise, mistakes will be made occasionally. These mistakes are the basis for what we call genetic mutations, and they account for a degree of variety available in the gene pool. Such mutations have to be very minimal, of course, otherwise the accumulation of many mistakes in repeated replication cycles would lead to unrecognizable entities after a short time rather than the relatively stable species we observe. DNA replication in living systems has remarkably low error rates. Indeed living cells have evolved protein-based machinery that detects and tries to fix errors! Even so, the occasional error will get into the population of the next generation of entities.

An error may affect the functionality of whatever protein the DNA codes for. With such random change in a highly organized system, there are far more ways to mess things up than to be neutral or to even improve things in some respect. So odds are random mutations that will be deleterious and the recipient organism is very likely to expire early in its development. On the other hand, on some rare occasions, a mutation may lead to an improvement in function. As diagrammed in Fig. 10.2, the mutation may lead to having a "better" fit with its environment.

This process could be expressed as the following algorithm:

1. Every entity in the population replicates, with there being some number of errors in copying in that new generation (generally this is taken as the gene unit of replication).
2. Each new entity is exposed to the factors of its environment to test its fit.
3. Entities compete for resources within that environment (this is where *selection* comes to play).
4. Less fit entities lose the competition more often than the more fit.
5. Entities get the opportunity to replicate themselves roughly in proportion to their overall fitness.
6. The next generation will tend to have proportionally many more of the better fit copies of entities that proved more fit.
7. If a copying error increases the fitness of the possessor, then that entity will be even more successful in getting copies of itself into the next generation.
8. Over many generations of replication, the varieties having the greatest fitness will tend to dominate the population.

Computers excel at running algorithms, so not surprisingly there are computer simulations of this process called, variously, genetic[6] or evolutionary[7] algorithms. These are sometimes distinguished in terms of what they are seeking to accomplish, but they are essentially attempts to have a computer emulate the evolutionary process on the theory that evolution pursues optimal designs. A varied population of solutions is put through a selective "fitness function" which will sift for superior performance of whatever target task is set up, then the sifted generation produces another generation with a few random modifications and that is resifted, and the superior performers reproduce with random variation again until some optimum emerges or a preset time limit is reached.

As it turns out, this theory is not exactly correct unless one is careful to qualify the notion of "optimal." Natural selection can only work on what is there to start with and its potentials to mutate in various ways. What remains after selection is only the best of the possible choices. The original designs upon which mutation and selection work may have been very nonoptimal as a solution to the then-current environmental fitness problem, but it was the only raw material that evolution had to work with. As compared with an engineered design, evolved designs are frequently found to be lacking, say in efficiency of energy use or even functionality. The human eye, for example, is an evolved marvel, but an optical engineer would have no difficulty finding design "flaws" that could be improved. The vertebrate eye involves an inverted retina, in which the photo-receiving neurons (e.g., rods and cones) are in an inner layer and the axons from those cells point inward and toward a single convergence point, producing a real blind spot where they punch through the retina and then form the optic nerve. No one yet understands why this should be the case in terms of an evolutionary adaptive explanation, unless it involves some sort of limitations inherent in the starting point for the evolution of the vertebrate eye. In the parallel evolution of the cephalopod eye—a "convergent evolution" arriving at a similar solution (an eye) from a different starting point—the design is closer to what our engineer would recommend: in the cephalopods (squids and octopuses), the design seems more rational in that the axons exit the retina directly to the back of the eye and so produce no blind spot.

Evolutionary algorithms run on computers tend to find optimal solutions (as analyzed after the fact) because the dimensionality of the fitness problem is very low. Real evolution is operating in a much higher-dimensional space on many different and interrelated traits simultaneously. And insofar as changes in the fitness of one organism also modify the fitness function for all kinds of related organisms, which in turn evolve to match the new fit and so change its fitness etc., evolutionary change in the real world is a coevolutionary dance in which the selective fitness function is continuously renegotiated.

Many authors argue that ultimately natural evolution does compute a global optimal solution, but for a vastly more complex problem than our engineers are used to

---

[6] See http://en.wikipedia.org/wiki/Genetic_algorithm

[7] See http://en.wikipedia.org/wiki/Evolutionary_algorithm

tackling. They point out that the mammalian eye design may have problems, but it works sufficiently well.[8] Mammalian evolution was simultaneously working on many more design solutions at the same time, and the whole enterprise could not wait around for a completely new eye model to progress. Working well enough to fit and survive and reproduce is the bottom line functionality selected for in this kind of evolution, whether or not that produces the kind of optimum engineers might look for. While there may be merit to the argument for some sort of global optimization, it may also ultimately be impossible to test since "working well enough" is so vague and flexible, and conditioned by multiple changing relationships, it is hard to conceive what "optimal" might even mean. What we do know is that evolution does a reasonably good job of working out a complex systemic organization of mutual but continually changing fit. Individual species may eventually lose their fit and go extinct, but life continues to survive and complexify.

10.2.3   Increasing Complexity Through Time

If we take a sample of organisms living in specific time periods across the big history of life, we find the following fact: organisms that appear on the scene first are in general simpler (as per Chap. 6), and later organisms display a range of complexity from the same kind of simple to far more complex. Some of the original simple recipes have continued to work down the ages: blue-green algae have been around for over 3 billion years. But they are now surrounded by very complex multicellular organisms such as us mammals. Starting from the origin of the first cells, the range of complexity of organisms has expanded, apparently exponentially. At the beginning, only simple life forms existed. At the present, a wide range of complexity in life forms exists.

Moreover, it isn't just the complexity of species that has increased. It is also the overall complexity of the biosphere—the globe-encompassing system of life—that has increased as relational interdependencies become richer and more complex. For example, life strategies for simple organisms are no longer necessarily so simple. Interrelations between the simple forms and the complex ones abound. Bacteria have learned how to become parasites on complex organisms. Some Achaeans may have learned to give up free-living capacity to actually simplify and become viruses, mere DNA or RNA strands packaged in proteins that could burrow into cells and take over the genetic machinery for replication—a personal simplification that is in fact a more complex strategy that shifts the costs of reproductive complexity to a host population. And even though simple worms (annelids) may have remained

---

[8] Herbert Simon, in 1956, coined the term "satisfice" to describe a suboptimal but satisfactory solution that could be found in reasonable time. He suggested that satisficing was a more appropriate approach to solving very complex optimization problems. See Simon (1983) for a more recent description of this and related concepts.

simple worms, their systemic role takes on new dimensions as they can now help organic farmers improve rich soils.

Not so long ago, most evolutionists were loath to consider the idea of "progress" in evolution, since the idea was usually associated with the teleological or goal-oriented idea of purposive movement to some fixed end. This sort of process, as we have seen, is typical of conscious systems that act with anticipation, and evolutionists were concerned to keep it clear the process they were describing does not depend on some mind giving it a planned direction. Creationism (a spiritual version of engineering) and intelligent design (another, possibly more sophisticated, angle on spiritual engineering) have been proposed as alternatives to Darwinian evolution and preferred by some religious communities as explanations, while others stress the compatibility of Darwinian evolution and a creator God without claiming that evolution as such demands a mind. Natural selection was introduced as a new, nonconscious process to explain changing and complex design, and evolution thinkers have been anxious to defend the sufficiency of this process to explain the design of life.

While the question of the origin of design was the initial focus, the notion of an evolving fit led to the deeper understanding of interdependent and changing mutual fit in the community of life. Such a process, as reflected both in the fossil record and in contemporary understanding of systems dynamics, moves on an inherent trajectory toward increasing complexity. Initially many evolutionists were resistant to any notion but change, since a "trajectory" seems to be headed somewhere, which implies a goal and thus reintroduces teleology. The question is much different today, since advances in the last half of the twentieth century have disclosed how systems can self-organize (i.e., become more complex) in the far-from-equilibrium context of an energy flow. Now most evolutionists have come to accept the obvious evidence of the increase in the complexity range, and evolution is understood as a systemically produced trajectory of increasing complexity that need not be teleologically headed anywhere in order to nonetheless spiral upward.

The process of evolution may now be seen as driven toward more complexity as a natural consequence of the geometry of space, the galaxies and stellar/planetary systems, and the composition of matter itself. On the scale of living organisms, increasing complexity can be explained by the chemical and mechanical work that is accomplished on the surface of (some) planets by virtue of the flow of high potential energy flowing through the planetary surface system from their Suns and out again to deep space as waste heat. No other force need be implied. What Darwinian natural selection provides us with is a clear explanation for how the results of these work processes are sorted out, the ones that are doing something useful in terms of systemic fit being preserved, while the not so useful are broken up into pieces that will fit elsewhere in the process. His insight into the role of selectivity in this process has, as we shall see, broad applicability to all sorts of evolving systems, be they physical, chemical, biological, or social.

In Fig. 10.3 we provide a schematic summary of this universal process.

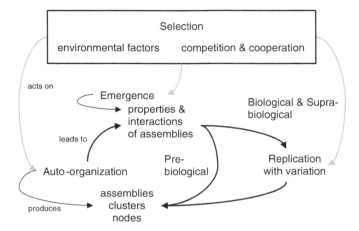

Fig. 10.3 Universal evolution is founded on auto-organization, emergence, and selection. Biological (neo-Darwinian) and supra-biological (cultural and ecological) evolution includes replication with variation

10.2.4   No Free Lunch!

Darwinian and neo-Darwinian evolution applies to biological systems that have already reached some level of complexity and organization. It depends on a biological entity's capacity to reproduce and to self-maintain. It does explain the exploration of fitness space and the continuing improvement in form/function as well as increasing complexity, but only once some primitive kind of life is established. Biological evolution as an algorithm can even be applied to non-biological (strictly speaking) systems such as human culture, with the appropriate caveats for details of how such systems operate. Cultures and their component subsystems do not have quite the same mechanisms as genes and genetic mutations, but they do have patterns that, like genes, can be copied and spread, and to which alterations can be made, either by accident or by human purpose, with subsequent new constructions incorporating the changed plan. The term *metabiological* has been used to describe systems composed of both biological and non-biological components, such as people and cultural artifacts (society), and ecosystems compounded of microbes, botanical and zoological species, and geophysical objects. Such systems show many characteristics of evolution. We will cover these in the next chapter.

But Darwinian evolution does not have anything to say about how the most primitive biological entities, those first displaying autopoiesis[9] and interaction with environ-

---

[9] Autopoiesis or self-regeneration/construction is a property (or rather a function) of living organisms and so must have emerged at the origin of life.

ments, came into existence in the first place. But it is difficult to expect random chance could lead to the complex organization found in even the simplest forms of life. Consider the problem of establishing the first primitive bacterial-like cell strictly as a matter of chance. Suppose for a moment we assumed that the origin of life could be explained as a highly improbable statistical fluke, but one that once having occurred gave rise to Darwinian evolution. We can analyze the probability by the following thought experiment. Say we take the simplest bacterial cell that we know about.[10] As a live system, it is composed of many complex molecules interacting in a complex web to produce the typical properties of a living system. Now put that bacterium in a tiny vessel and heat it at a high enough temperature that you destroy all possible complex structure. All of the complex molecules will break down to their constituents. Now cool down the vessel to normal room temperature and let it stand for a very long time. Open it. Do you find a living bacterium? Why not? All of the necessary ingredients are there. The temperature has returned to one that supports life. Why didn't the system reassemble as a bacterium? If we had waited a bit longer, might it have done so?

Airplanes are a popular example with those who critique notions of life as the product of random chance. Take all of the parts of a Boeing 747 aircraft as a jumbled pile on the ground. Using a giant crane, throw all the parts into the air at once. What is the likelihood that they will fall to the ground as a fully assembled and functioning airplane? How many times would you need to repeat the experiment so that in one of those times the aircraft would appear fully assembled?

You see the problem. From the standpoint of plain old-fashioned probability theory, there is a nonzero chance that order will spontaneously appear. Yet our intuition tells us that neither the cell nor the airplane would likely assemble into functional structures even if you had started doing experiments at the beginning of time up to the current supposed age of the universe (about 14 billion years, give or take several million) and did one experiment every second! In very plain words, the structures of complex systems like cells and airplanes (the former being many times more complex on the molecular scale) is so improbable that we now conclude that the origin of life could not have been a chance occurrence.

So when it comes to complex organization, there is no free lunch. How then should we think about the origin of a living system such that biological evolution (which also does involve some elements of chance) could lead to the vast array of biodiversity and complexity that we witness today?

We must ramp up, with some sort of organizing process (evolution), to explain the development of the ramp. Greater organization (complexity) depends on prior organization and the availability of energy to create further relational structure. We do not get order for free in an absolute sense. There has to be an initial seed of organization and the right conditions for things to develop. That is what we take up next as the process of "auto-organization."

---

[10] Actually there are cells that are simpler than bacteria in the domain Archaea. But for our purposes, any bacterium will do.

10.3 Auto-Organization

The term, *self*-organization, which is very often used in this context, can be somewhat misleading to the uninformed reader. Yet, applied to the not-yet living stages of evolutionary process, it has become popular as a rubric for a process that gives us the original investment in organization (buys us our lunch). It might seem to imply that a system can somehow, almost magically, assemble itself without something outside of itself providing any support. The insight into auto-organization ("self-organization") has indeed revealed that an outside designer is not required; design can emerge from within. But as we have seen, it can do so only in the context of flows of energy, matter, and information into and out of the system. That is, auto-organization as we now understand it is a process that takes place in the context of what we have described as far-from-equilibrium dissipative systems. Living organisms are such systems, but so too are many physical and chemical systems. Organization requires work, which requires an energy input that ultimately will come from beyond the system. Otherwise systems, even if they have all the right components in the right proportions, cannot, on their own, just assemble as a fully functioning complex structure.[11]

In this section, we will apply our systems process approach to show how a system will auto-organize or assemble based on both its component interaction potentials and the kinds of inputs it gets from its environment.

10.3.1 The Organizing Process

Physics tells us that from the moment of the Big Bang the matter-energy content of the universe is a constant; the universe itself is therefore not a dissipative system. But within the universe, the distribution of matter-energy is far from the randomized equilibrium of entropy. This means it is available for a tremendous process of inner transformation, with the emergence of new forms (e.g., quarks, protons, neutrons, electrons, etc.), new combinations, and new sorts of relationships and processes among these. This is the emergent organization of which we speak. The universe evolves not from the disorganization of total entropy but from the not-yet organized condition of vast energy flows among emergent subunits or components. The path from the Big Bang to galaxies, solar systems, and planets such as Earth with complex chemistry and eventually life is a process of moving not from disorganization to organization but from an original simpler organization to the more complex and highly differentiated forms and relationships of the contemporary cosmos. We understand many parts of that process, such as how we get complex heavy atoms by the nuclear fusion burning

---

[11] There is a potential contradiction to this in the form of self-assembly in which the personalities of the components are so structured that as connections are made, no other kind of organization is possible. Self-assembly still requires the flux of appropriate energy in order to proceed. In some cases, however, it may involve first agitating the components by heating, followed by cooling to allow them to settle into their "preferred" structures.

of hydrogen and helium in stars, and how supernovas disperse these heavy elements to eventually reassemble under the force of gravity into planets with interesting chemistry (due to the complex interactive potentials of heavy elements) such as the Earth. The cosmic systemic question, the search for a Grand Unified Theory (GUT), now focuses on the first few millionths of a second of cosmic emergence.

But within this overall orderly cosmos, we have many sub-processes where new sorts of relationship emerge, where sectors move from less to more complex interrelationship. As we said above, organization ramps up; it does not simply appear by chance from disorganization. This form of emergent organization is exemplified by planetary scale processes, that interesting chemistry where heavy atoms relate and form molecules which in turn relate in new ways until, at least in the case of the Earth, the organization reaches the complexity that crosses the threshold to autopoiesis and life. In this discussion, we will focus mainly on this scale of evolutionary organization.

Calling a system unorganized refers to internal structures only, since some minimal boundary organization must exist or we have none to refer to as unorganized. The Earth, for example, is bounded by conditions of mass and gravity, but its components have ample room to find new forms of relationship. Let us assume that a system exists by virtue of having a boundary inside of which there are just a collection of disassociated components (Fig. 10.4). You will recall from Chap. 4 that such components have personalities (Sect. 4.2.2.2). Also from that chapter (Sect. 4.2.2.1), we introduced the concept of potential complexity (amplified in Chap. 6). The collection of components, having linking potentials with one another in various ways, is primed for increasing complexity of structures (networks of linked components) by dynamically interacting and becoming interconnected.

There is a minimum starting condition for all systems that requires a boundary to keep the components in proximity to one another. This boundary will not be absolutely impermeable (energy flow is required!), but it does need to generally keep components in over a long enough time scale. The boundary should be energetically insulated in most areas except for some kind of energy input window for high-power energy inflow and a radiative/convective outflow window in contact with a sufficient heat sink to allow waste heat to be dissipated. The geometrical arrangement is important. It should be such that there is an energy gradient set up across the system such that energy flows from the input to the output through the system. Finally, there needs to be a power source that emits energy in a form that can be absorbed by at least some of the components (Fig. 10.4).

A second condition is that the components assembled within the boundaries and exposed to this energy flow have personalities that can indeed be interactive. Separation in space and time are what we think of as the ordinary barriers to interaction, and that problem is solved by invoking boundaries to bring the assemblage together. But an interactive potential between component A and component B already points to some systemic commonality of origin. To think of the universe as a single evolved/evolving system, we begin with a single origin from which emerges a multiple ramifying expansion with mounting diversity of components that interact with increasing complexity over multiple scales of space and time. Even having emerged from a single systemic source, all elements may not be directly interactive, but some will be primed for further connectivity and able to enter into increasingly complex forms of organization.

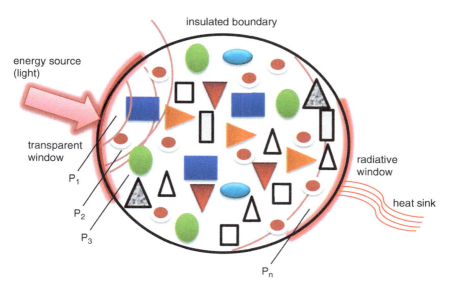

Fig. 10.4 The minimum requirements for a system to evolve from an unorganized state involve a containing boundary that is essentially closed to material inflows and outflows (though not absolutely so), a boundary that is insulated everywhere except for a transparent window that can admit high-power energy (say light radiation at relatively high frequencies—components of white light), and a radiation window that can emit low-power waste heat (low-frequency light such as infrared). The geometry of this arrangement matters a great deal. An energy source provides input to the input window, and an essentially infinite heat sink accepts heat radiation from the radiative window. Across the system, an energy gradient from high power source to low-grade heat sets into motion the actions that cause the system to organize. The potential for energy to do organizing work on the components diminishes as energy transports through the system. Potential $P_1 > P_2 > P_3 >>> P_n$. Note that the system contains many components with many different "personalities"

Question Box 10.3
Draw a circle. Then draw many and differently shaped lines within the circle. Do all the pieces still fit together? Now draw two circles, and draw lines within each. If you draw without very careful design and calculation, will any component section of one circle fit in the other circle? Which of these experiments fits with auto-organization theory? Which fits with the thesis that fit requires a designing mind? Why is the idea that the universe issues from a single origin (e.g., the Big Bang) important for auto-organization?

The Big Bang theory presents this kind of model: it presents the universe as a single systemic process starting at a unity and exploding into diversity, thus encompassing and containing every other system that emerges within it. So a subsystem emerging from new relationships among components already involves some kind of organization which serves as its initial conditions—the down payment on lunch so to speak. The explanation of how this minimum amount of external structure came into being takes us deeper into the unfolding systemic properties of the cosmos.

In the case of the Earth and other planets, that explanation is basically gravity and its systemic proportionality with forces of attraction and repulsion in the nucleus of atoms. The force of gravity is what rules the shaping and dynamics of the universe at least on the scale of galaxies, stars, and planets, for the gravitational attraction of mass is what assembles particles for the most basic stages of interactive connectivity and organization. It is responsible for aggregating the original hydrogen and helium clouds that eventually condense into stars, and its proportionality with nuclear forces dictates the mass that must be assembled before the internal pressures and temperatures (due to gravitational energy) overcome the force of repulsion between protons and ignite fusion reactions. The stars, like our Sun, burn by the collapse (fusion) of simple light elements like helium and hydrogen into the more complex nuclei of the heavier elements. The size of stars necessary to ignite this way is dictated by the proportions between the forces of attraction and repulsion: Change the proportional force of gravity much in either direction, and we would have a universe so scattered and simple nothing very interesting would happen, or so small and fast burning there wouldn't be enough time for complex processes such as the emergence of life on Earth.

Some of the early stars were so massive that they generated heat and pressure too rapidly to remain stable. They exploded (supernovas), creating in the process even heavier elements and lots of interstellar dust, now a virtual stew of complex atomic elements. Some of this dust condensed under the force of gravity to flattened clouds around newly forming stars. Within these rings, gravity operated on the dust to produce solid bodies and some of these coalesced into planets.

In essence, then, the interrelated forces that emerged from the Big Bang, as we currently understand it, have been responsible for setting the initial conditions of systemness, supplying the geometry (planets circling stars), the components (the elements in great quantities), the space-time dimensionality, and the energy flows from stars that radiates across other solar subsystems and into the virtually infinite heat sink of deep space. The interaction of these forces is as close as you come to having a free lunch because they are constitutive of the universe itself. Their differentiation and proportional interrelation provided the seed conditions of systemness and of all subsequent organization. We see these conditions operative in critical features of the Earth, which has mass and gravity such that it has a temperature gradient from a very hot molten core[12] to a surface cool enough that water can exist in its liquid (even frozen) form, with ongoing energy flows in a range allowing or even enhancing interesting chemistry based on the new interactive potentials of the heavier elements.

---

[12]Actually the very central core is solid, probably iron-nickel composition, because the pressures due to gravity at that depth overcome the liquefaction of those metals. The solid inner core rotates and is thought to be the source of the electromagnetic field around the Earth. That field is important in shielding the surface of the planet from some forms of harmful radiation.

The surface of the Earth, and probably many Earth-like planets in the universe, is the system of interest for exploring the more immediate ramping up of complexity. The boundary for this planetary process is supplied by gravity, keeping an interactive mass of gases and liquids tightly bound to the surface, while within the surface, gravitational forces create the thermal pressures and flows that cycle elements and move the great plates which constitute the dynamic crust of the planet. Gravity also supplies tidal energy flows (from Sun and Moon) that help create further dynamic cycles organizing surface conditions. And fusion-produced solar energy supplies the bulk of a continual energy flow in the form of white light. The atmosphere is the revolving input window and exit gate for this solar energy, and the atmosphere has systemically transformed through feedback loops with oxygen respired by early photosynthesizers to create an oxygen-rich atmosphere with a shielding ozone layer. This development was critical insofar as the ozone layer allows just the right wavelengths to reach the surface while filtering out more harmful wavelengths that would prohibit life developing on dry land.

Life now flourishes on land, but it undoubtedly originated in the chemistry made possible by a watery environment. We have mentioned the complex chemistry made possible by the expanded relational potentials of heavier elements. Yet the hunt for such chemistry in the cosmos is almost identical with the search for the presence of liquid water. Chemistry involves elements getting together in a process of combining and recombining, and water is almost uniquely suited to this. For elements to get around gases are too mobile, solids too static. Water remains liquid at an unusually large temperature range, including the temperatures that facilitate chemical reactions. Finally, water is a powerful solvent, breaking down molecules and setting the scene for new combinations—just the right dynamic environment for interesting new things to happen.

How the Earth evolved geologically from its early formation to the point when liquid water could accumulate in vast quantities is thus an important part of understanding biological evolution later. However, the story is long and is probably best obtained from a geology textbook. What we can say is that a fair amount of physical evolution (of rocks, mountains, oceans, etc.) preceded the origin of life, which current estimates place about 3.8 billion years ago. The energy flows from the Sun evaporated water, while gravity pulled condensed water back to the surface as rain and governed its distribution and flow in streams and rivers back to the oceans and lakes. Heat from the core created by gravitational pressure rises in convection loops, driving continental drift and continual recycling of the mineral crust, while gravitational forces of the Sun and Moon wash the planet with continual shifting tides. All of this ongoing jostling of the physical components of the Earth system has organized and reorganized the Earth in an ongoing evolutionary process that shaped the planet in such a way that life could come into existence and evolve.

Above we said that the boundary cannot be impermeable, but it helps if it is nearly so. The surface of the Earth has not been completely isolated materially. It receives a fairly steady rain of space dust and small bodies called meteorites. Generally this influx does not add anything new, though meteorites have varying proportions of elements depending on where in the dust cloud they condensed or where they were created when a larger body was broken up. But we need to mention the import of some special molecules that it now appears are formed in space and have rained down on Earth just like the dust and meteorites. These are very basic organic molecules. Normally, "organic" means coming from living systems. But it turns out that some basic molecules that are instrumental in the origin of life were produced through ordinary chemical processes in space. Their formation, however, required the light radiation from stars (our Sun) so their organization was still the result of the minimal system requirements we have just outlined.

10.3.2   The Principles of Auto-Organizing Processes

We now turn to consideration of the basic principles involved in the dynamics of internal organization and the increase in complexity over time. As the system ages, energy from a high potential source drives linking processes internally to produce structures of varying degrees. Figure 10.4 depicts the starting of this process. Here we will consider the physical principles whereby this energy flow eventuates in increasing structural complexity (Fig. 10.5).

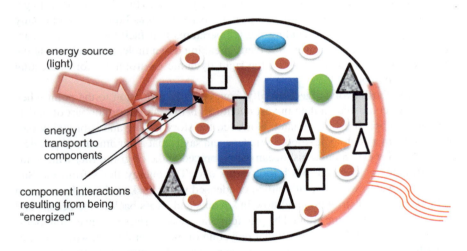

Fig. 10.5 The first step involves energy at high work potential (near the transparent window) which is captured by some of the nearby components. Their energy levels are elevated; they become mechanically activated but also tend to give up some amount of their absorbed energy to their neighbors that are at lower energy potentials. The high potential energy propagates across the system from source to sink. With each energy transaction, however, some high potential energy is degraded to low potential heat. The latter can still perform some mechanical work on the system in the form of moving components as long as there is a temperature differential along the gradient as well. At least some of the high potential energy is used to form component interactions like bonds (*black two-way arrows*)

10.3.2.1   Energy Partitioning

Energy is never just energy, but some *form* of energy, and different forms of energy drive different sorts of work processes. Energetic processes involve not only energy flow but energy partition, that is, transformation into alternative forms. This partitioning is a critical feature in the organizing of systemic process. Thermal energy, for example, can boil water, transforming to kinetic energy as steam pressure drives turbine blades moving a coil through a magnetic field, thus transforming the energy of movement into electrical energy, which can transform into heat in the element of an incandescent bulb, with a final radiant transformation into light energy.

Once energy of the right type, such as high temperature heat or light photons of the right frequencies, enters the system and interacts with components, it ends up in one or more modes with respect to the kind of component. One mode is that of raising the vibrational activity of the component, as when it becomes the heat of the water or the higher energy state of an atom which absorbs a photon causing an electron to jump to a higher energy state. Alternatively the energy can change the translational (momentum in a given direction) activity of the component, as when it causes the turbine blade to move faster. These energy transformations are dynamic, causing components that are free to move in the system to bump into or otherwise interact with one another. Depending on their personalities, this interactive jostling may result in two or more components forming links. The electron jumping to a higher energy state might be followed by the formation of a covalent bond with another atom, with the energy then being stored in the new bond.

The kind of partitioning that takes place depends on the nature of the components involved, and the components may belong to any systemic level. As we just saw, energetic interactions with atoms and molecules can lead to movement and bonds. Energy in the form of electricity can run a machine's electric motor, or a computer, or the lighting. Or at a very different level, we can perform a very similar analysis following energy inputs to a corporation as they are partitioned in different ways and follow relational pathways in forming or maintaining various sorts of organizational bonds.

> **Question Box 10.4**
> Virtually everything said about energy flows and their consequences in physical systems has a close parallel when it comes to money in a social system. Cash flows play a critical role in organization formation and maintenance, and too much cash, like too much energy, breaks down organization. In what ways could money be regarded as another of the many forms of energy?

Note also that according to the second law of thermodynamics, every time energy transitions to a different mode, some of the energy is lost to waste heat and is no longer contributing to any mode that might contribute to structure building. There is no perfect energy cycle that would allow, for example, the electricity produced by a turbine to transform into the same quantity of water-heating thermal energy that produced it originally.

10.3.2.2   Energy Transfer

As mentioned, components that receive energy can interact with other components and transfer some of that energy to those components, either in a transformed or similar form. In turn those components may interact with still more components and the transfer will proceed.

In general (and on average) the transfers go in the direction of the flow from source to sink. Imagine that components near the source are excited to a high level and end up transferring some of that excitement to other components nearby in the gradient (Fig. 10.5). Again depending on the link potentials between the components, some of that excitation may end up in the form of coupling (like chemical bonds) rather than merely transferring momentum.

And as with partitioning, every time energy is transferred from one component to another, some is lost as waste heat.

10.3.2.3   Cycles

Repetition is fundamental to pattern formation and thus to organization. A particularly powerful form of repetition is cycling, where a dynamic process bends back to its source and thereby initiates an iteration. Cycling dynamics include a vast range of scales and complexity, from intermittent ice ages, to the atmospheric processes we call "weather," to the behavior of economies, to the complex metabolisms of living bodies and individual cells.

The essential dynamics of cycling are evident in common thermodynamic processes. In an energy flow, if the components, or some portion of them, are mobile, then energy partitioning and transfer can cause them to move, on average, in the direction of the gradient decline, i.e., components will move away from the source and toward the sink. This is easily seen in the case of thermal gradients causing, for example, a liquid to enter currents that move away from the hot input and toward the cool output. Water in a pot on a stove is a simple example. If the heating (or energy input) continues and heat is removed to the sink, and the geometry is favorable, then you will see these currents turn into convective cycles, creating a pattern of cells of rising and sinking water. Components give up their excited states as heat at the sink end and will tend to be displaced away from the direct path of the gradient, say around the sides of the pot. There they sink (in this case due to gravity as much as displacement) back toward the energy source. Once there, they will become re-excited and start the journey toward the cool end again. If you watch a pot of water being heated from below, just before the bubbles start to form, you can make out the developing currents because warmer water diffracts the light slightly differently from cooler. Once the water starts to boil, there is too much kinetic energy to maintain structured patterns and chaos reigns.

Physical cycles where components move as described are common when thermal energy dominates. Another common type of cycle occurs when linkages or series of linkages form, dissolve, and form again. Linkage cycles are common in chemistry and biology. In these cases, energy is stored in linkages between components. But linking is never a one-way process. Links can spontaneously break or be broken by other factors, such as a nearby component that attracts more excitation, displacing the previous linkage. Then if more energy is added, the former linkage may reform and start the cycle over again. In living cells, the combination of physical and chemical bond cycles constitutes the phenomena we observe as metabolism. Cycles may also drive larger cycles, as when the cyclic pumping of the heart drives blood cycling throughout a multicellular organism. And, as we will see, really large geographical scale and long time scale cycles are found in ecological systems. The hydrological cycle driven by the Sun and gravity, for example, cycles water through its solid, liquid, and gas forms to give us the conditions of atmosphere, ocean, and ice that pattern the climate and with it the life conditions of the Earth.

A special case of the cycling phenomenon is the temporal or pulsed cycling found in many systems. One example is the fact that since the Earth spins on its axis, it has a day-night or diurnal cycle that drives increases and decreases in activities on the surface. For example, green plants photosynthesize during the day, and more so during the summer depending on latitude, and respire at night, and more so during the winter. Another very important example of pulsed energy flows is the lunar tidal pulse that raises and lowers the local oceanic levels with two ups and two downs every (roughly) 24 h. These cycles are superposed over the solar tidal forces and the precession of the Earth, all of which produces a complex tidal schedule. All intertidal zone organisms have adapted to these tidal effects, and the rich and complex life of the intertidal zones has far-reaching consequences for the entire oceanic ecosystem.

Pulsed energy cycles along with the physical and linkage cycles produce a pattern of complexity in the general environment that becomes a part of the natural selection process.

10.3.2.4   Chance and Circumstances

An internally unorganized system is in a state of high entropy. With the flow of energy through such an unorganized system, chance encounters must therefore constitute the major mode of dynamics. So randomness, a correlate of the absence of organization, plays a role, and especially so in the initial stages of ramping up organization. Energy transfers and link formations occur through those chance encounters and alter the future dynamics, so the probabilities of further linkages moves away from the statistical base line of random chance. For example, when components link, their combined inertia will alter how fast they may be moved, and if the link is particularly strong, these linked components may be out of linkage circulation, affecting the probability of other linkages taking place and also opening a possibility of new levels of linkage occurring among more the more complex already linked components. With a flow of energy available, the expectation of the system for even further organization increases as linkages are added.[13]

Chance will always play some kind of role in what interactions take place, and when and where in the system they take place. Systems can organize not only to tolerate but even to expect and utilize this element of randomness. For example, even in the extremely organized cytoplasm of living cells, at the molecular level, random motion and chance encounters of low weight molecules (like the distribution of ATP molecules) are an essential part of the chemical dynamics of the system. And in biological evolution, chance gene mutations make available a range of modified traits critical for the fundamental process of natural selection.

10.3.2.5 Concentrations and Diffusion

One of the laws of nature called the law of mass action, actually a variant on the second law of thermodynamics, is that components move from a point of higher concentration toward lower concentrations. If either by chance or by some energy flow process a cluster of the same component type happens to develop, and those components do not form strong linkages, then there will be a "pressure" from within that cluster for the components to spread out in the space, including into the interstices between other components. In an open space, this is seen as a diffusion process, as when a drop of ink spreads to evenly tint a glass of water. The rate of spread is a function of the concentration remaining (as diffusion proceeds) and the distance from the center of the concentration.

While diffusion is a process of randomizing disorganization, concentration is a matter of organization, for the components are in some way constrained to be more tightly concentrated. It therefore takes physical work to force components into a concentration and to keep them there. When you see a concentration that persists over time, then that is evidence that some work process is actively maintaining it. A container (boundary) can be used to prevent diffusion, but it takes work to construct the container, and it takes work to push components into it. No free lunches here.

10.3.2.6 Dissociation

Linkages not only form, they also come apart, a breakdown of existing organization that may also be a necessary part of maintenance or growth and reorganization. Various forces within a system can work at dissociating linkages or disrupting

---

[13] For a study of the changing probability dynamics as linkage emerges, see Kauffman (1995), ch. 3.

cycles. For example heat, at normal physiological temperatures, can denature proteins (make them dysfunctional) which are then broken up into their constituent amino acids for recycling. Water excels as a medium for many sorts of chemical interactions because it dissolves bonds and frees up components of compounds for new interactions. Businesses and institutions must be able to "reinvent" themselves, adapting by dissolving old organization to reorganize to fit new circumstances. A new kid in school can have a disruptive impact on existing cliques.

Dissociation, the process of breaking apart linkages, thus occupies a deeply ambivalent place in the world of process. Systems do break down and decay. They also recycle, reorganize, and grow. Especially in life processes, death—the dissolution of vital organization—at one scale is commonly an expected life-maintaining event at another. In biology there is even a term, "apoptosis," for programmed cell death in multicellular organisms. And where the untrained eye may see waste in a fallen and decaying tree, a forester sees an important moment in a complex nutrient cycle that enables the forest to flourish.

10.3.2.7   Higher-Order Principles

By "higher order" we mean principles (applied specifically to emergence and evolution) that provide appropriate operations on the substrate to shape and organize it.

10.3.2.7.1   Cooperation and Competition

Some kinds of structures, emerging from the disorganized state (see below), are organized in such a way that they mutually enhance the probability of further component formation. By cooperation they can form what amounts to an amplifying structure (as we saw in Chap. 7). One very important case of this, seen at the chemical level, is that of catalysis. A catalyst is a cooperating molecule (or even an atom) that has properties that affect other components in a way that makes them more likely to interact. Thus, their presence means that the specific reactions they promote tend to be favored over other possible reactions and will possibly succeed at the expense of the other possibilities. The random emergence of substances that happen to enhance the likelihood of other reactions massively tilts the world of chemistry from the probabilities of pure chance linkages to the probability of specific reactions, which may in turn serve as catalysts in their own right making other reactions more likely, in a cooperative chain producing otherwise highly improbable forms of organization.

Interactive cooperative dynamics that promote further organization can occur at all sorts of levels. Symbiotic strategies are common among organisms: the coral reefs that promote complex communities of life in the sea are themselves the product of the symbiotic cooperation of photosynthesizing algae lending their energy to the coral polyps they inhabit. In human organizations of all sorts, certain personality types interact to get people together and "make things happen." When we refer to them as "catalysts," we think it is a metaphor, but the cooperative facilitation of otherwise unlikely organization is a common systemic dynamic.

Cooperative dynamics tilt a competitive playing field, and competition is itself an important dynamic. When energy is available, work can be done, but who gets it, for what work? In chemical catalysis, we see that certain combinations, acting cooperatively, can be more potent at competing with other combinations that are attempting to obtain more components from the same pool as the others. Competition is an active work process that pits two or more structures, able to do work, against one another in an attempt to grab resources. The competition is a matter of resource usage when there is not enough resource available for everything to happen (all work to occur) that could possibly take place. Some takes place (the winner) at the expense of something else not being able to take place (the loser). In metabolism, for example, some molecular machines, like ribosomes, compete with other machines for ATP molecules which serve as energy batteries circulating around in the protoplasm. When there are enough ATPs to satisfy every machine's needs, all is well. But when the energy pump that produces them in the mitochondria runs short (e.g., when sugar inputs diminish), then the machines will compete.

Complex organizations structure priorities, so competition for resources is rarely simply a level playing field. If you recall from Chap. 7, this is where the need for an arbitrator, a logistics controller, comes in to decide who has priority. In the metabolism world, this is handled by specific feedbacks and pathways that favor certain critical processes over others. At another level, in human organizations, budget constraints may give rise to destructive competitive dynamics, especially when prioritization is not clearly established. Markets are premised on competition among organizations, and endless disputes about "fair" and "unfair" practices make evident not only the power of competitive dynamics but also their sensitivity to numerous relational factors that may themselves be competitively or organizationally manipulated. Successful competitor organizations win out or dominate the market, and their dominance tends to modify the system in ways that favor their continued competitive success.

Question Box 10.5
There is a familiar saying, "The rich get richer and the poor get poorer." Explain this in terms of auto-organization dynamics relating to cooperation and competition. The saying is often, but not always, true. What kind of dynamics might work to organize the poor so they too might get to be better off?

10.3.2.7.2   Forced Moves

Daniel Dennett notes another higher-level principle, the forced move (Dennett 1995, p. 128). Evolution is a stochastic search approach for solutions to fitness problems. Often there may be myriad ways that are all nearly equally satisfactory in terms of providing a solution. In such cases, because of the random element of the search process, there is some likelihood that many, most, even possibly all of these solutions are found and tried out. Indeed, this accounts for the huge variety in some traits we see in related species and within a species.

The forced move is when there really are only a few, maybe even just one, ways that combinations can be made. At the auto-organization level, this means that the components' personalities provide only a limited number of ways for combinations to be made. But it also implies that once made, those combinations are potent as competitors. Dennett provides an example from human cognition regarding the nature of arithmetic. Given the nature of numbers and basic operations on them (e.g., two oranges added to a pile of twelve oranges can only ever make a pile of fourteen oranges, no more, no less), it turns out that there really is only one way to do arithmetic. We should expect that if there are intelligent alien life forms out there in the universe, they will have discovered the very same arithmetic. On the other hand, he points out, the exact kinds of symbols chosen to represent numbers and operations are probably close to infinite. If we did contact aliens and found that their numbers looked like ours, 1, 2, 3..., we should be utterly astounded because there are no deep reasons why these symbols are the only ones that work.

Forced moves reduce options. In the world of process, one can think of moving toward relatively more forced moves as a narrowing of the field of systemic expectations. Organization as it progresses may constrain systemic expectation, introducing greater predictability. Thus, in seeking more predictable mechanical behavior, we constrain pistons and other moving parts to finer and finer tolerances, in effect rendering their field of motion closer and closer to an ideally forced move. Starting from a more disorganized state, organization shrinks the field of systemic expectation, moving successively in a literally less and less expected direction, the organized predictability of the forced move being the least expected of all. This is the logic that led nineteenth-century thinker to regard the predictable forced movements of clockwork mechanisms as intuitive evidence that random process could never lead to designed organization.

Except that hypothetical state of random disorganization never existed. Instead, the nature and proportions of the constitutive forces of the universe already are pregnant with organization insofar as the resultant laws of physics constitute the deepest source of forced moves. The organizing role of gravity we mentioned above exemplifies an evolution rich in forced moves. And in a similar way, chemistry is rich in illustrations of the range of relative forcing and the organizational consequences it entails. For example, oxygen is notorious for combining preferentially and spontaneously with many other atoms like hydrogen (to form water), carbon (carbon dioxide), and iron (rust). Moreover, these bonds are highly stable. It takes considerable energy to break them apart. So knowing the relative plentitude of these elements and the forcing rules of their combining, we can predict a lot about the kinds of molecules we encounter in nature. Or alternatively, knowing the kind of molecules common in nature, we can work back to understanding what the forced moves are at the auto-organizing level of the world of atoms.

10.3.2.7.3   Path Dependency

Forcing is not the only way in which systems move from wider (less organized) to more constrained expectations. The history of a system's development at each step alters the range of expectation for its future development. This factor, known as path dependence, commonly works in combination with degrees of forcing to explain the combination of chance and necessity we find in the organization of the world of nature and of culture. We trace path dependency when we review how our choices over time open or close the field of options that constitute our present situation. But the dynamic of path dependency is present in all sorts of evolutionary processes. We have seen that every change to a system is new information, a modification to its field of potentials or expectations. When the initial number of choices of combinations is large, and the probabilities of each kind are effectively distributed uniformly (i.e., unforced), then mere chance decides which ones obtain. But then the chance choice affects the subsequent range of choices of combination, and the system is on a historical (path dependent) trajectory. In a chemical solution, after chance initial combinations are made, the likelihood of future combinations is altered both by the changed proportions of the concentration of reactive elements and by the reactive, potentially catalytic, effects of the new molecules. As systems organize, the establishment of specific combinations dictates the kinds of combinations that can be made in the future (see progression below). Thus, when we say that a system is path dependent, we mean that its unique history plays a role in the way it is organized: starting from the same basic mix of components and rerunning the auto-organizing experiment again, we might end up with a different mixture of combinations.

Path dependence plays a very important role in biological evolution, for frequently the establishment of a given trait, early in the phylogenetic tree, fixes the form that all possible future variation on that trait might have. For example, the early fixing of five spines in the lateral fins of proto-fishes seems to have no deep forcing principle—any other reasonable number might do. But once in place, the five spines led to the evolution of five digits in amphibians and in later phyla. So we have five fingers and five toes simply because that was a successful combination when fishes were first evolving! If we met aliens from another world that were upright, bipedal with two arms, there is no a priori reason to expect them to have only five fingers. They could, of course. But they could just as easily have four or six.[14]

---

[14] That is to say, we don't think there is any necessary reason why early proto-fish had to have five spines! If we found other animal life forms in the galaxy that all had exactly five digits, then we would have to consider a necessary forcing condition that must have existed in early evolution.

10.3.3   Organizing, Reorganizing, and Stable Physical/Linkage Cycles

We are now ready to consider auto-organization, not as a magical process but as the result of components interacting and forming structures, at first seemingly randomly, but later in a more organized fashion. It turns out that this process is exactly a matter of chance generation of variations followed by selection of the most fit. Darwin described how this process works to bring about biological evolution. Now we see the universal scope of these evolutionary dynamics.

10.3.3.1   Order from Chaos

A dissipative system is one that has exactly the starting organization covered above in our discussion of the organizing process: it has a boundary, a source of energy, components that can interact with that energy and each other, and a sink into which the waste heat produced from all of those interactions can dissipate. Under these conditions, the system will transition from a chaotic, random inner organization to an ordered organization. That is, it will go from a system with little internal structure and function, to one with much greater structure. And, more importantly, that internal structure will mature and persist over time, as long as the energy flows through.

Recall that in systems of any complexity, it takes work, i.e., energy, both to create and to maintain structure. Should the source of energy diminish, the system will give up organization as the second law (entropy) begins to dominate. As we experience in virtually every area of daily life, without maintenance, things fall apart. Alternatively, even though there may be a continued energy flow, the sink may become unable to absorb the produced heat, so the system will begin to heat up, temperature will rise, and the second law will rule again, this time because structures that were stable at the previous temperatures may be unstable at higher ones. This is evident, for example, when a car radiator runs out of the water it needs to absorb engine heat. But recalling the partitioning of energy into many forms, we can see the principle applies to many forms of flows and sinks. As we have discovered, when flows cannot be absorbed by sinks appropriate to a system, they become forms of pollution that begin to break down the system. This can even apply at the symbolic level, where energy takes the informational form of money and can become a flow that overwhelms a social, political, or economic system. And when economists speak of "overheating" an economy, perhaps their language is not as metaphorical as we think!

On the other hand, some authors argue that the rules of selection are so tight that the fact that vertebrate life forms have five digits means that there was some difference in fitness between four, five, and six. In explicating organization, the mix of necessity (forcing) and path dependence (chance) is a good question to contemplate.

10.3.3.2   Selection of Minimum Energy Configurations

As energy flows through a system, there is an important difference between routine maintenance and ramping up organization. Any system can be considered as settling in to a relative getting-by-as-it-is mode, a minimal energy state where the work done maintains the status quo. If it is more energized, some other sort of work, perhaps some further linking and structuring, may take place. Or maybe the extra energy will just be dissipated and nothing much will happen. In any case, you can see the importance of energy accounts: energy in all its forms is real, so where it goes and what it makes happen is a fundamental question as we consider how systems can become increasingly complex.

It is easiest to follow the energy accounts at the level of basic chemical reactions, where the structure of excitation, incremental organization, and/or dissipation, and settling toward a minimal energy level—the default tendency—is basic for any reactive process. Since all molecular assemblies will tend to settle into a minimum energy configuration consistent with their composition and inter-component interactions, if a given assembly is excited, say by receiving a jolt of energy (e.g., an electron on one of the atoms in a molecule absorbing a photon that raises its energy level), something must happen. If it is otherwise a stable assembly, then the energy will be dissipated and the assembly will, once again, settle to its minimum energy level. Or if the opportunity is present, the energy may be used to create new connections, thus forming a more complex molecule. Energy not directly used to create connections is always dissipated. Dissipation is actually yet another face of the second law of thermodynamics in a slightly different guise. If an assembly is excited, this means that energy is at a higher potential than the surrounding environment; if it is not locked up as new organization, it will, therefore, radiate to the nearest low potential sink.

We tend to think such dynamics are literal in the world of chemistry, but fail to recognize that they apply at all levels of organization. The energy processes of human interactions, for example, are so complex that we cannot begin to do the exact accounting we can do with interacting molecules. But terminologies such as "excited," "energized," "depleted," "settle down," etc., are not as metaphorical as we may think when applied to social situations. People can get energized in a variety of ways. Our food intake unleashes processes that can be described as complex chemistry. We may get hyper charged up, for example, by drinking four shots of espresso in the morning. The actual energy is coming from our store of sugar (actually glycogen, a starchy substance that the liver converts into glucose for use by the body), which is activated by the brain being stimulated by the caffeine in coffee. A person can get very active as a result and do more work, more physical activity, as a way of dissipating the excess energy.

More complex and incalculable still are the intertwined energy/information flows that create and maintain all sorts of social organization. The realm of energy-dependent activity includes sensation, emotion, thought, and all sorts of information processing, so we find the energy-governed dynamic of organization manifest in the pattern of interpersonal bonding and organization as well. The high energy level that

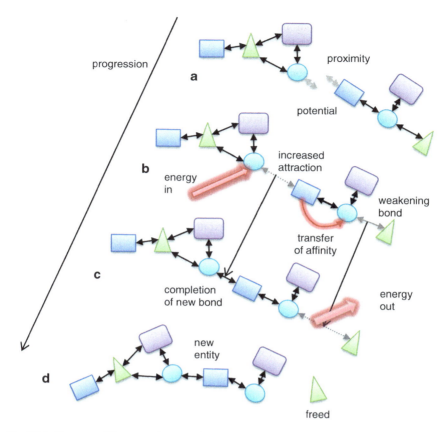

Fig. 10.6 Two assemblies come into proximity (*top to the right*) with each having some exposed personality that provides a combination potential. If energy is available (next down the progression), then that energy can excite both assemblies and set up an attractive force. In the general case, these assemblies may either form a temporary attraction that would dissipate leaving each assembly as it had been before (not shown) or possibly through a transference of affinity weaken the connection of some other component. If the latter occurs, the disengagement of that component would dissipate energy (at a lower potential). A new entity is created in this process, and the freed component can now participate in other combinations

goes into creating new relationships and more extended organization subsequently settles down to maintenance mode. Then individuals and organizations may find they are overextended and unable to come up with the input to support the new structure, and it begins to disintegrate—or, as we will see below, the new structure may catalyze yet more extended structures of relationship with an even greater energy demand.

Figure 10.6 illustrates this process as it might occur when more complex entities (assemblies) encounter one another in the presence of an exciting energy.

Such assemblies can recombine to form new assemblies and by-products. The notion of by-products (with their own new organizational potential) is important because at more complex levels assemblages reconfigure in the process of combining and the new configuration may have no place or energy for a formerly related element. By-products, be they atoms, molecules, workers, or whole organizational units, are spun off into new fields of systemic expectation, which may in some cases include the possibility of important feedback loops with the new organizational unit.

The attraction shown in the figure (step B of the progression) is transitory. It is generated from the excitement of energy at a high potential. The initial energized attraction can transition to loss of attraction by virtue of the energy dissipating away immediately. Or it can, as illustrated here, form a connection and transfer affinity (via the excitation energy) to another, less attractive connection point (the circle, step C). Either way, the energy of excitation is dissipated back to the environment, to a sink of lower potential than the original energy input. In the case illustrated here, the energy dissipates in weakening the subsidiary connection between the new affinity component and its affine (the triangle). The relationship is broken and that component becomes a by-product, now freed to enter into other combinations.

As mentioned above, this process can be traced quite easily in chemical reactions. It is less obvious, but true, that this exact pattern of activity takes place on *all* levels of organization. For example, take interpersonal relationships. Two people may have formed an interpersonal relationship that is reasonably stable, marry, and have children. But if another person of greater attraction to one of the partners comes along, and that person also is married, Fig. 10.6 could well illustrate the highly charged dynamics of remarriage and merging two families. And the same pattern prevails when businesses or corporations merge and spin off duplicate units or sacrifice some weaker links to balance the new budget. Regardless of the level of organization, from atomic to society, these same patterns prevail in complex, dynamic, and stochastic systems as they undergo growing organization.

Question Box 10.6
Space-time used to be an important constraint on social relationships, since it required a considerable energy/time input to span distances separating people who otherwise might connect well. The Internet, especially with the development of social media, has thus transformed this relational world, since the globe can be traversed with minimal energy or time input. What kinds of energy flows continue to constrain and shape the new world of web networked relationality? What are stability/instability factors? Are there by-products spun off, similar to what is described in Fig. 10.6?

10.3.3.3   Hyper-Cycles and Autocatalysis

We have seen how the availability of a flow of energy alters the probability terrain of a system. With available energy, work can be performed, and in the right circumstances, work will increase the complex structure of the system. In this situation, the improbable (further order) actually becomes the more probable. Now we turn to a further dynamic that tilts the playing field of probabilities even further in the direction of auto-organization. The dynamics of a networked structure of relations can give rise to cyclical behavior. With available energy, a system may wander through a chain of transformations. But within the range of possibility is the chance that such a chain will happen to loop back and produce the components with which it began, as illustrated in Fig. 10.7, below. Such cyclic behavior may seem relatively less likely initially, but it has a particular systemic resilience in that, once it emerges, unlike linear chains, a cyclic process ensures conditions for its recurrence, a form of stability that serves as a wedge for incorporation into yet further forms of complex organization. Cycles of combination and recombination are commonly found in nature and in the human-built world, and they become especially common in pulsed energy systems, where the pulsing provides the pattern of initiating stimulus and eventual dissipation common to such cycles. As long as energy flows through the system, there will be opportunities for these kinds of cycles to obtain.

Cyclical combinations and recombinations are a dominant feature of dynamic systems. Figure 10.7 shows a schematic representation of such a cycle.

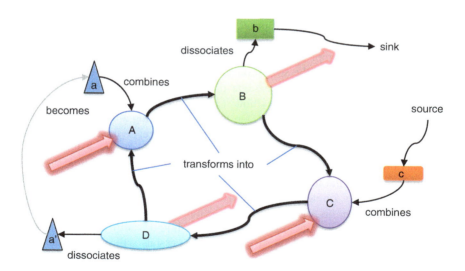

Fig. 10.7 Entities can recombine with components to form new entities (as seen in Fig. 10.5), but those can then give rise to the precursors of the same entities. This figure shows a cycle of entity A combining with component a to form entity B. After a while, B dissipates energy (heat) and gives off component b decaying to entity C. In the presence of component c and input energy, C produces entity D. Again after some time passes, D dissipates energy and gives off a', which decays to a, and A. If energy is still available, the whole cycle starts over

Resilience and predictability are important building blocks of organization. The cyclic process described above is a significant advance in that direction, but it still has a weakness: for C to produce D, it needs (with an energy input) to combine with component c. But what if C's expectations at this point include the equal possibility of combining with d, e, or f, (not represented in diagram), which would lead in a non-D direction and break the cycle.

This is where the phenomenon of catalysis plays a critical role in the emergence of complex auto-organized systems. A catalyst is a chemical molecule that, because of its shape or some other feature of its personality, relates to several other elements in a way that encourages their combination. So with the right catalyst, the reaction of C with c becomes much more likely than alternatives (d,e,f), so the catalyst drives the process more predictably toward D and the closure of the cycle. Catalysis, then, critically transforms organizational probability by making some combinations more favorable than others, so that in a given energetic flow, those combinations are favored and drive a cycle as in Fig. 10.7 in a given direction.

Such catalyzed cycles play a very important role in all sorts of metabolic processes. But while they make function repeatable, they do not of themselves increase and multiply, filling the world with their organization. This aspect of emergent auto-organization requires a slightly different twist of the catalyzed cycle. What if a catalyst happened to catalyze a process that led to the formation of more of the catalyst? This catalytic cycle, or "autocatalysis" as it is called, would act as a positive feedback loop, with more leading to yet more until finally the supply of reagents was exhausted.

One of the simplest examples of such autocatalysis is the formation of a crystalline structure in a supersaturated aqueous solution. You can do an experiment at home for this. Take a container of water. Warm it to near boiling, say 180 °F. Now start adding sugar (regular table sugar—sucrose) and stir. Keep adding sugar as it dissolves into the solution. You can add a lot more sugar because the warmer water will dissolve more of it. If the sugar collects on the bottom, stop adding and warm the water just a tad more until the sugar disappears. Now, very carefully, let the container cool for a while, until you think it is at about room temperature. You must not disturb this solution in any way. Don't even risk putting a thermometer in it to test the temperature! After it is cooled to room temperature, do the following simple action. Drop a single crystal of sugar into the solution. What happened?

If you did this carefully (and you may need to try it several times to get it right), then you should see the following: dropping the crystal of sugar into the container should cause a very immediate crystallization of sugar around the seed crystal! Another variation on this, one that is actually employed by candy manufacturers, is to attach some sugar crystals to a string while the liquid is still warm and lower it into the solution. What will happen is that sugar crystals will grow up the string forming a solid tube of candy.

What is happening is that the seed crystal works as an autocatalyst, providing a correctly shaped surface onto which the molecules of sucrose can attach in a preferred pattern to form yet further crystals. As the crystalline structure grows, it encourages more crystals to form. Letting the solution cool down and the water to evaporate keeps the solution supersaturated (this is the source of energy to cause the formation to occur).

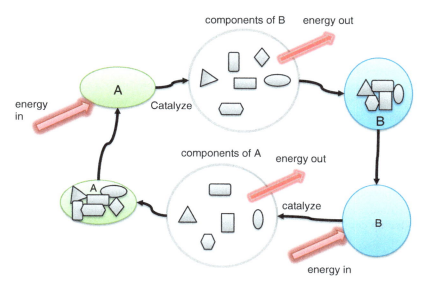

Fig. 10.8 Mutual catalysis occurs when entity *A* catalyzes the formation of entity *B* out of the available components AND entity *B* then catalyzes the formation of entity *A* out of a similar pool of components. This is a positive feedback loop that needs to be balanced (moderated) by some other process damping it down. Processes that control the availability of components or energy are found to perform this function in living cells

A more complex form of autocatalysis is a cross-referenced process in which components of a system mutually catalyze each other's production in a cycle that maintains or even reproduces a system. The emergence in increasing complexity of such autocatalytic systems is one of the precursor processes for life. Figure 10.8 shows a conceptual diagram of a mutual catalysis in which entity A catalyzes the formation of entity B which, in turn, catalyzes the formation of entity A. This is yet another example of mutual causation, but it only works because of energy constantly supplied to drive the reactions.

One of the most important autocatalytic processes discovered that has implications for the origin of life is that of ribozymes (ribonucleic acid (RNA)+"zyme" is the suffix used to name biologically active catalysts called enzymes, usually protein molecules). A certain form of RNA molecule called an "RNA polymerase ribozyme" can autocatalyze its own synthesis. One theory about the origin of genetic encoding is based on this process. RNA is thought to have been the first molecule to play the role of genes (now mostly handled by DNA). That RNA may have been able to also play the role of a catalyst for itself suggests that this kind of process was the first establishment of a cyclical process that later evolved into the DNA-RNA-protein hyper-cycle that makes up life today.[15]

---

[15]For a discussion of such hyper-cycles in the emergence of life, see Kauffman (2000), pp. 120–125.

The emergence of hyper-cycles takes us to the level of complex autocatalytic organization required as a stepping stone to the complexity of the simplest living organism. Autocatalytic cycles composed of mutually catalyzing components can be far more complex than the kind of simple autocatalysis of sugar crystals with which we began. But while interdependent mutual production extends organization to a new level, unless reinforced by some type of redundancy, such cycles are as vulnerable as their weakest link. One form of such redundancy, a kind of insurance policy for components, would be an autocatalytic cycle composed of components which not only mutually catalyzed each other but also were their own autocatalysts. This is a hyper-cycle, an autocatalytic cycle composed of components which can also self-replicate. Different strands of RNA sequences, for example, might both have the ability to replicate themselves and become catalysts for other strands, eventually forming a self-reproducing autocatalytic loop of self-reproducing components!

10.3.3.4   Self-Assembly

One of the more remarkable forms of auto-organization involves entities assembling on their own, as in automatically forming structures with little or no random variation or trial and error. We saw an example of such self-assembly in the auto-catalyzed supersaturated sugar crystallization discussed above. On closer examination, the "trick" is revealed to be just a special case of auto-organization, a systemic condition that is so highly constrained by the nature of the components that no other kind of assembly can occur; it is assembled by a series of forced moves, as discussed above. In the case of the sugar crystals, the catalyst seed crystal provided a surface shape that triggered a reaction tightly constrained both by its own shape and the reaction potential of the sucrose molecules.

The reaction in the supersaturated solution proceeded on potential energy already present due to the supersaturated state of the system, with the crystallizing reaction releasing the potential energy and moving the system toward a minimum energy configuration. Self-assembly is often differentiated from auto-organization on the basis that in self-assembly, as in the sugar crystallization example, the system is actually undergoing a dissipation that would lead to a system closer to equilibrium. In contrast, auto-organization is more often thought of as something that occurs because the system is being driven further from equilibrium due to an energy flow. We have chosen to relate these two because both are dissipative. Moreover, there is nothing that prevents a self-assembly process from operating within an energy flow.

Self-assembly, then, occurs when energy is available and component personalities permit only a single configuration. An extremely good and very important (for life) example of this is the cell-membrane assembly shown in Fig. 3.12 in Chap. 3. In that diagram, we see a regular and highly ordered bilayered assembly of molecules consisting of phospholipids and cholesterol chains. Cholesterol, a pure lipid, is water averse, and so, when these components are in a water environment, they will tend to seek an association that gets them away from the H_2O molecules. The

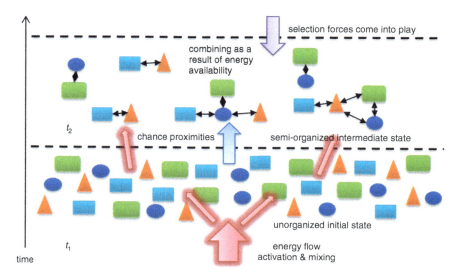

Fig. 10.9 The beginning of organization results from an energy input to a system of components with the potential to combine through linkages. Various structures appear as a result to chance encounters between components. These structures will be subjected to selection forces imposed by the nature of the internal environment

phospholipids are part lipid, so that part can form a bond with the cholesterol and together seek a position far from water molecules. The phosphoric end, however, loves water! So the combination of a hydrophobic lipid component and a hydrophilic phosphoric component causes these molecules to self-assemble into the bilayer membrane as shown in that figure. All that is required is that they are in a water environment and they will automatically find that configuration without any outside energy flow needed. This is an example of a system that is seeking a minimal energy state.

Materials scientists are quite excited about the self-assembly process as applied to a number of advanced material designs. They are investigating nanoscale[16] assemblies of carbon-based substances that have extraordinary properties compared with ordinary molecules. Some of these assemblies can actually do mechanical or electrical work. They have been given the name of nanobots (nanoscale robots).

10.3.3.5 Auto-Organization and Selective Pressure

Auto-organization and self-assembly can be represented more generally as in Fig. 10.9. Even in the case of self-assembly, energy has to be supplied at some early stage in order for the basic components to be in a form that will combine

---

[16]Nanoscale refers to objects measured on the order of a nanometer or one billionth of a meter.

dissipatively. In this diagram, we show an auto-organizing process in which basic components first form primitive assemblies by chance, with some being more complicated than others. The specific assemblies depend entirely on the personalities of the components (e.g., the bonding potentials, say, in atoms and molecules) and the availability of energies needed to form the connections. Selective pressures begin to act to create an intermediate stage of organization.

In every environment where combinations are taking place, there are local conditions that tend to favor or disfavor certain combinations. These can be viewed as selective forces. For example, typical for aqueous solutions of organic molecules, the temperature, acidity, or alkalinity (pH), the presence of various salts such as sodium chloride or other chemically active agents can disrupt some combinations of the organic molecules while leaving others untouched. Such a process will tend to concentrate the favored molecules and make components from the disrupted combinations available for incorporation into the favored assemblies. Figure 10.10 shows the continuation of this kind of chemical selection process.

This kind of process can occur in simple aqueous solutions, but it is precisely captured within the most important cycle in living cells, the citric acid cycle, which produces the energy packets (ATP) from the controlled oxidation of sugar molecules.[17]

Notice the emergence of competition and cooperation as selective/organizing factors in Fig. 10.10. In dynamic environments where assemblies are undergoing energy flows and selection forces, the dynamics of competition and cooperation develop most clearly. Cooperation can be thought of as the tendency for components to form strong, non-disruptable connections between themselves. Competition is when two or more assemblies try to obtain another component, either one that is freely in circulation or that is weakly bound to another assembly. A strongly connected assembly may very well be able to win such a competition because of the strength of its connections.

10.3.4   Auto-Organization Exemplified in Social Dynamics

We have exemplified these organizing processes mainly at the level of chemistry in order to illustrate systemic properties that tilt physical flows far from random disorganization toward seemingly improbable complex structure. But the principles apply to other levels of organization such as organisms in various sorts of communities or other sorts of social organizations just as well.

Lest you think that prelife and organic systems are the only examples of everything we've been discussing about auto-organization and emergence, consider the Puget Creek Restoration Society.[18] Its history started with several separate groups of

---

[17] See Wikipedia's page on the citric acid cycle: http://en.wikipedia.org/wiki/Citric_acid_cycle

[18] This story of how the society came into being was provided by the PCRS Treasurer Scott M. Hansen. For more information about the PCRS and Puget Creek, see http://www.pugetcreek.org/

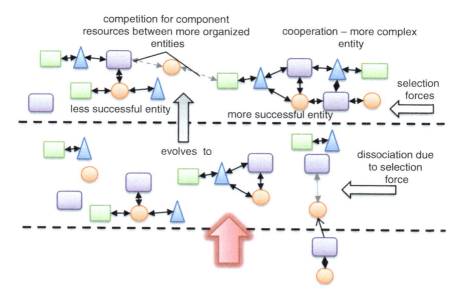

competition for component
resources between more organized
entities

cooperation – more complex
entity

selection
forces

less successful entity   more successful entity

evolves to

dissociation due
to selection
force

Fig. 10.10 Organizing principles continue to drive a system toward higher complexity as energy flows and higher-order principles come into play. Selection forces can cause dissociation of weakly bound components. Cooperation between some structures can increase their mutual capacity to compete with yet other structures for component resources

individuals, each with its own interests in a natural creek that empties into Puget Sound near a largish city in Washington State. The creek runs through what is now a heavily residential area except for a narrow margin of wild plants and trees that, the local's had hoped, would protect the creek from harm. Unfortunately things like sawmill, timber harvesting, sewer/storm line projects and development have affected the creek and nearly destroyed it as a habitat for fish species, especially salmon.

Three groups emerged, each self-organizing in terms of shared interests, one of the most powerful human connective factors. One group had formed from a mutual interest in what it would take to restore the salmon run, while a second had become concerned that human hikers were trampling the vegetation. Yet another group saw the effect of the change in the creek on the spawning small fish and shellfish populations at the creeks outfall. The groups formed spontaneously through individuals meeting one another and discovering their mutual interests. Each group began holding meetings and discussions regarding how they could improve the fate of the creek. Some of the members would make routine trips out to the creek to investigate the conditions and explore ways to achieve their goals. That is, they began to independently work on different aspects of the creek's restoration. This was auto-organization of social groups around different aspects of a common issue.

As luck would have it, two members of two of the groups encountered one another during their independent forays to the creek. As they chatted about what they were up to, they discovered that they both belonged to these informal community groups that had, at base, a common goal—the restoration of Puget Creek. The two groups ended up holding a joint meeting at which they discovered how they could both cooperate and pursue their independent goals but with mutual support. One group set out to get permits to build constrained pathways that would encourage hikers to not trample the vegetation. The other group started chemically analyzing the outfall water for contaminants. The third group had, independently, started to talk to the state fisheries experts about the feasibility of restoring salmon runs to the creek. One of the experts had also been talking to the outfall water quality group and gave the salmon restoration group the contact information.

From there all three groups began to communicate and cooperate in their various ways. Before long they were joined by an environmental biologist who suggested they form a nonprofit and go for grant money to fund restoration efforts. Soon the society was formed and succeeded in getting a state grant (a form of extracting energy!) with which they commenced planning a full restoration program. Today the organization is functioning well, and the creek is on its way to supporting the wildlife that had been there before the residences.

The forming of the independent groups was not without its difficulties. The forces of selection worked on these groups, just as the forces of thermal agitation works on molecules in the origin of life. Family demands, work demands, etc., were pulling the attention of members even while they sought to work together. If even one key actor had been removed from one of the groups due to some accidental situation, the formation of the greater organization might not have transpired at all. Indeed several members did have to leave for various reasons they were replaced with others who were better able to develop the means to help the group succeed. Still the cores of the groups retained their connections. The environment for these groups, daily life, tested the bonds that kept them together working on a common interest. Fortunately for the creek those bonds prevailed.

All of these independent entities formed spontaneously through auto-organization. Their bindings were social and informational. Nevertheless, they formed nuclei of interests that happened to overlap with those of the others. All it took was for some chance meetings and contacts to allow the next phase in the development of a full-blown process—the emergence of a system. Emergence is the phase when new organizations and functions arise from the interactions of smaller, less complicated entities. We turn there next.

10.4 Emergence

Emergence has to do with something new occurring—in this case a new form or level of organization likely marked by new properties and new functionality. In considering auto-organization to levels of higher or more complex organization, we were in fact looking at the emergence not just of additive complexity (the whole is just the sum of its parts) but at new sorts of systems with characteristics that are other than simply a summation of what was there in the unorganized components (the whole is more than the sum of its parts). Depending upon the numbers, personalities, and potential connectivity of components, in systems that age with appropriate energy flows, the components may assemble into larger-scale patterns due to auto-organization. These patterned assemblies are tested for stability and durability by various forces inherent in the internal environment so that over time certain patterns persist and resource components tend to be absorbed, so the system settles at the new level. In this way, auto-organization, as we have described it above, produces sets of assemblies that are, in essence, at a new level of organization.

These new assemblies can participate in their own rights with one another at this new level as entities or processes. And the kinds of interactions between these new entities may be completely new and unexpected by an outside observer. Given a starting system with sufficient potential for complexity, these new entities, with their new interactions, may start the process of auto-organization over again at this new level and move to a yet higher level of complexity as the system matures further.

This is the essence of the phenomenon of *emergence*. New organizations and levels of organization are made possible because the emergent entities have new properties not seen in the assembly of components prior to the initial auto-organization and selection process.

10.4.1   Emergent Properties

When components at one level of organization interact with one another and form assemblies, these new entities can display an aggregate personality that is not entirely predictable from simply knowing the personalities of the components taken independently. Salt (sodium chloride) is a common example: both sodium and chlorine are deadly poisons taken separately, but when compounded as salt, it becomes a required part of our diet. This nonadditive change of properties that emerge as components enter new relational assemblages gives rise to the common dictum that "the whole is greater than the sum of the parts."

Systems are constituted by relationships, so it should be no surprise that altering relationships changes the properties or that creating a more complex network of relationships may result in the emergence of a new sort of system with new properties. Even as simple a system as a salt crystal illustrates the transformative power of relationship. As assemblages become more complex, combinatorial potential and the possibility of new properties escalates. At the high end of complex combining, the interpersonal relating of humans in systemic units of families, friendships, interest groups, businesses, etc., the novel consequences of interpersonal "chemistry" are striking. Two organizations with similar structure and purpose may behave quite differently due to different personnel and the way they function together, or a single person may change markedly depending on the sort of group that forms their relational context of the moment.

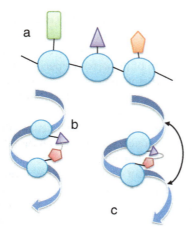

Fig. 10.11 A protein molecule is formed by linking a number of amino acid molecules together via peptide bonds. In (**a**), amino acids are the *circles* with various shaped side chains representing the different kinds of amino acids (there are about 20 different amino acids that are important in living systems, all differences due to different submolecules attached to side chains). Those bonds are represented by the *angled lines* running between the circles. In (**b**), the peptide bonds cause the chain to coil such that some amino acid side chains are brought close together and may, under the right conditions, interact weakly through electromagnetic attractions (*gray arc* between the *triangle* and the *pentagon*). These kinds of interactions can cause the primary (alpha helix) coil to bend further, as in (**c**), forcing the molecule into a more complex shape

10.4.1.1 The Molecular Example

The example of protein molecules illustrates the complexity of emergence at an intermediate level characteristic of the constituents of life and even prelife. Twenty kinds of amino acids are available to form an array of proteins by combining in different combinations and orders in long chains joined by peptide bonds. Proteins, once assembled, will coil in a predictable fashion, called an alpha helix, due just to the conformational energies in amino acids of all kinds. The peptide bonds link the core of an amino acid molecule to the core of another amino acid, but surrounding the core each amino acid has a characteristic side chain of submolecules. As the alpha helix coiling brings the molecules of the side chain into closer contact, these side chain submolecules can further interact weakly with one another enough to cause the protein to contort into more elaborate shapes. Often the exact form of the shape cannot be predicted just by knowing the sequence of amino acids in the chain alone.[19] Indeed the very same sequence can assume different shapes depending on many other factors in the aqueous solution in which it is formed. So the shape that a particular protein assumes is an emergent property in this sense (Fig. 10.11).

---

[19] This is known as the protein folding problem. It has proven nearly intractable except for smaller, more regular polymers of amino acids. Recently, however, some progress may have been made using game programming-style problem solving!

Different proteins formed in this way have different and newly emergent properties. Most strikingly, at this scale of organization, shape takes on a functional significance such that the distinctive complex folding of each protein in its given environment becomes a critical and unique property. We have already discussed catalysts, and how they make less likely interactions or combinations more likely. It turns out that shape is one of the most important properties in catalysis. Proteins, called enzymes, catalyze the combinatory chemistry of cell metabolism by which organisms subsist. In simple molecules, the primary operative property is the location of valence electrons and "holes" (where an electron could occupy an orbital location but doesn't keep the atom electronically neutral). Getting together by random collision means not only happening to collide but having to collide in just the right way (and with sufficient energy). An enzyme molecule may be contorted in such a way that it has one or more "pockets" shaped to fit and bring together the molecular components or "substrates" upon which it acts. It is this fit that sets up the enzymatic chemical reaction that will then take place. The shape of the enzyme literally makes the reaction more favorable (this includes both splitting molecules and synthesizing them). The shapes of proteins thus turn out to be a major property for determining their abilities to interact with other molecules.

10.4.2 Emergent Functions

New, unexpected properties emerge at a higher level of organization, and the most notable properties are those that, like the shape of enzymes, are coupled with new functionality. But functionality can also take a lead in endowing the material components of a complex system with new properties. This is especially evident in the symbolic world of human culture. Money is a wonderful example of an emergent function in human societies. A dollar bill, or a credit card, has properties/functions that emerge on the level of symbols and information rather than material structure; though a material substrate is a component. The story of money is complex and covers much ground in the world of economics. Here we will look at the early story of money as it emerged in ancient history as an emergent function of growing social complexity.

10.4.2.1 An Example from Society: Money

Based on archeological findings, early human settlements traded with one another by direct barter. After the advent of agriculture, about 10,000 years ago, settlements started to grow in size and complexity as individuals became more specialized in the type of work they did. Some became farmers while others became agricultural supporters, such as specialists in producing the tools needed in larger agricultural operations. As the variety of kinds of exchanges and their scope increased, barter became increasingly difficult in terms of making fair trades. How many chickens equal one goat? And what if you didn't need all of those chickens at the same time?

In ancient Mesopotamia, during the early Bronze Age (about 5,000 years ago), farmers started using clay tablets into which marks were made to indicate the contents of storehouses and the amount of various products like grains. These tablets were the first known accounting system in that they gave an account of inventory of these products. Barter practices still involved trading volumes (or weights) of products directly, as in a market. But at some point, traders started accepting the tablets as representing the value of the traded products rather than the products themselves. No one knows when there was a shift from trading merchandise to trading clay tablets symbolizing the merchandise, but a few such tablets have been found that seem to be broken carefully between markings indicating that someone traded a fraction of the inventory, possibly the first kind of monetary representation of true wealth!

Markers provided a convenient means of keeping track of real physical wealth—grains, animals, etc. Those markers could be traded in place of the actual goods and redeemed at a convenient time to the holder. Thus, the notion of money was born as an emergent function in an increasingly complex economic world.

Looking at a simple clay tablet (or portion of one), seeing some innocent-looking marks on it, how might one have predicted that this emergent object, fulfilling what seems a simple function, would one day become coins, then paper, then bits in a computer memory, all serving elaborations on that simple function?

And function can also take on new directions, emergent applications that feedback to transform systems. In this case, the early accounting markings appear to be the earliest form of "written" representation, and they eventually evolved into cuneiform, a kind of script used to record not just accounts but also ideas. One of the earliest written clay tablets found seems to be marking the number of clay urns of a precious liquid that resulted from the fermentation of hops and barely. It seems incredible that our modern ability to write numbers and words started as a way to keep track of how much beer one owned!

10.4.3   Cooperation and Competition as Emergent Organizing Principles

As systems auto-organize to more complex levels, the dynamics of inter-system relationships take on new potentials. Let us revisit the issues involved in competition and cooperation. We saw above, in auto-organization, that when some components interact, they form strong linkages that provide structural stability. They persist. In network parlance, these components form a clique. Other assemblies or cliques form from other components and their linkages. Between there are still potential interactions in the form of competition for unattached or less strongly attached components (Fig. 10.10). Those assemblies that have the most cooperative linkages can be "stronger" or more "fit" in the internal environment of the system and thus be more successful at whatever competition takes place.

We see this dynamic at every level of organization, but there is an important emergence in the role played by information in cooperation and competition as we proceed through different levels of organization. At one end of the spectrum, in sports, businesses, and all sorts of areas of human life, people share information based on experience and plan cooperative strategies to enhance future performance, often enough in an environment of competition with other groups. Here, in a way typical of and perhaps exclusive to the world of life, information is used to actively engage and shape the field of system expectations.

At the other end of the spectrum might be atomic particles engaged in a cooperative interactive dance with other particles in the environment orchestrated by the shared information we describe as the laws of physics. Here the field of expectation seems simply the future that fits with the information describing the nature, momentum, coordinates, etc. of the particles at any given moment. Information here, as far as we know, simply guides into the future rather than being a platform for some kind of forward-reaching intervention. So if we speak of carbon or oxygen atoms "competing" to bond with a limited supply of free hydrogen, for example, we know we are measuring relative rates of "success," but the language of competition and success seems almost totally metaphorical. Or to put it another way, we use the metaphor because it does describe some pattern common to the dynamics of these two very different situations, even though the future-oriented essence of what we experience as "competition" is absent in the case of the atoms.

Question Box 10.7
What are shared or common elements in the dynamics of competition at these two different levels that might justify using a common term, "competition," for both?

Somehow, from the self-less bond-acquiring competition of the world of atoms, the goal-oriented teleological world of life and self-maintenance has emerged. The process of this emergent transformation can be traced at least in part in terms of emergent differences in the function of cooperation and competition as these fundamental organizing dynamics operate on more and more complex systemic levels. Much is now understood about how cooperation and competition organize and transform living species, ecosystems, and, at the most complex end of the spectrum, human cultures. And in physics and chemistry, there have been great advances in understanding the role of these dynamics in basic bond creation and the recruitment of components into more complex assemblages. We have also considered the critical emergence of catalysis, autocatalysis, and autocatalytic cycles and hyper-cycles at the most complex levels of chemistry. If there is some intermediate emergence bridging atoms proceeding without a care into a lawful future, and organisms that dither and strategize ongoing survival as they enter a treacherous and uncertain future, this is perhaps the place to look for it.

We might begin by asking how cooperation and competition function in a catalytic environment. We noted above that catalysts tilt the playing field, changing the expectations of a system to favor certain kinds of interactions. Whatever competition for resources is going on, they weigh its outcomes drastically by their participation, though the catalysts themselves are not winners or losers but mediators. But although the function of catalysts is a new factor, in terms of information and systemic expectations, initially catalyzed systems seem little different from more simple systems—the presence of catalysts is just a new factor to be considered in the information regarding the total system environment, which is what determines the system's expectations.

Autocatalysis, however, introduces a new twist, in that the catalyst itself becomes a "winner," creating by its activity not just a future but a future for itself. Note that there is an "it" that has any particular stake in the future—or more precisely, we might be just on the threshold where such incipient selfhood emerges. There is an emergent new dimension in that a system's activity is such that it tweaks the field of expectations in favor of its own extension. In autocatalysis a systemic formation has the effect of causing or maintaining its own existence, a new kind of self-referential feedback. Up to this point, systemic stability has related to the enduring power of strong bonds: it's still that way because it sticks together really strongly. But in autocatalysis, where A catalyzes B which catalyzes C which catalyzes A, we have a new kind of dynamic in which components endure by structures that link them in a process of mutual production: it's still here because it keeps a complex process going which reproduces itself.

The traditional critique of teleology has been a hardheaded resistance to the notion that a nonexistent future can influence a present reality. But here we have a complex process which endures as a form of organization precisely because of its future consequences: it produces itself. Living organisms in their behaviors and their metabolisms use information in a distinctive anticipatory way to engage and manipulate the future. Evolution selects for and refines among the living those that succeed in this competitive dynamic engagement by living long enough to reproduce. With the emergence of autocatalysis, we are not yet at the full active anticipatory utilization of information to engage the future that becomes the hallmark of life, but the basic selective structure has emerged that marks evolution at that level. That is, the characteristic of this cross autocatalytic structure is that it is stable and endures *because of the way it works*, and there is *a difference between working and not working*, a difference that can be selected for and that transforms evolution from the level of auto-organization to the survival of what fits, that is, what works.

Cooperation and competition are systemically interwoven dynamics. Cooperation between linked components has an impact on how successful the assembly, which thereby becomes a component at a higher level of organization, will be in competing against other assemblies at that level. The details of the natures of the cooperation and competitions will vary across levels, as we have seen, but we see this dynamic playing an important role in system evolution at every level of organization.[20]

10.4.4 Emergent Complexity

The emergence of complexity has attracted considerable attention.[21] In dynamic systems that are aging under the influx of energy, complexity, in the form of increasing connectivity among components at different levels of organization, will develop.[22] As long as energy flow provides more available energy to do work in the system, then more complexity will emerge. Probabilities reflect the easiest thing to happen in a system. With no energy flow, the easiest thing to happen is paths that increase disorder: things fall apart all by themselves. But, somewhat counterintuitively, where energy is available in a system of components with potential connectivity, the easiest thing to happen is more connection, meaning an increase in complexity.

The primordial Earth, prelife, provides an exquisite example of a system driving toward increasing physical and chemical complexity as a result of multiple energy fluxes. The Sun provided a continuous flux of radiation energy to the surface of the planet as it cooled and formed oceans and atmosphere. Tidal energy from the Moon, and to a lesser extent from the Sun, provided a pulsing energy flow that particularly affected the interface between oceans and land masses, the intertidal zones. Finally, gravitational and nuclear decay energies, heating the interior of the planet, provided long-term fluxes that affected the position of land masses and produced volcanic events across the face of the surface for billions of years. And all these energy flows had plenty of molecules of potentially high connectivity to work on.

On a geologic time scale, simple molecules such as methane (CH_4) and ammonia (NH_3) along with water, nitrogen, and others were compounded into more and more

---

[20] Group selection as a mechanism in evolution has been hotly debated. Darwin actually considered it as a valid type of selection that might help explain cooperation in social animals. But since the primacy of the gene theory rose, many evolutionists rejected the idea. Today more evidence has emerged that group selection is a very important selection mechanism underlying the evolution of altruism and cooperation. From a systems science perspective, since the cooperation/competition mechanism is found universally, we can't imagine it not being operative in human evolution.

[21] For example, see Melanie Mitchell's description of emergent complexity in the work of Stephen Wolfram (Mitchell 2009, pp. 151–159). Then see Wolfram's own description in Wolfram (2002). We'd recommend starting with Professor Mitchell's version!

[22] In the just-mentioned work of Wolfram and in numerous versions of cellular automata and artificial life, there is often a failure to mention that energy is implicitly being supplied by the computer circuits! Indeed, most of the simulations of these kinds of systems do not include the fact that the computer is being supplied with energy flow continuously, which, along with the energy flows supporting the human designers, is the motive force behind the emergence process. So read these descriptions with some caution. Physical reality must be paid its due.

complex molecules and molecular assemblages in chemical reactions driven by these energy fluxes, including also the electrical discharges of lightning that frequented the early atmospheric environment. In 1953, Stanley Miller and Harold Urey at the University of Chicago conducted experiments[23] using glass containers holding what were then thought to be the major molecular components of the primordial Earth. They used electric sparks in these flasks to simulate an energy flux that seemed likely to see what sorts of chemical compounds would emerge. Dramatically, they found prebiotic compounds such as amino acids and sugar-like molecules.

10.4.5   The Emergence of Life

The kinds of experiments that Miller and Urey conducted (and many similar ones have been conducted since, using a different set of assumptions regarding the chemical makeup of the primordial "soup") showed conclusively that prebiotic chemicals could be generated by energy fluxes under the right conditions. But since that time, it has been determined that those conditions may actually be common in dust and gas clouds in orbits around stars! Current theory favors the view that prebiotic chemicals may actually have developed in space and later rained down on planets like Earth during early formation. This is a similar process of auto-organization, just occurring in another venue with a longer timeline, which helps account for the relatively rapid emergence of life when the planet was a barely cooled 700-million-year-old newborn. Either way, the conditions on the primordial planet were right for the development of increasingly complex molecules, including nucleic acids, basic polymers of amino acids, fatty acids, and all of the basic molecules found in living systems today.

Although there is no precise agreement on the defining features of life, Fig. 10.12 presents a minimal version in terms of contemporary life components.

Of course one cannot begin life with the complexities of ATP, mitochondria, and ribosomes, but the interlocked systemic functions provided by these highly evolved components somehow had to come together in a functioning, evolvable unit.

There are many theories being floated today about how the first systems of chemical reactions that would pass as living came about. And there are many open questions about how the process could have been "bootstrapped." A systems approach might well begin by considering the functions involved in being alive and then consider possible temporal priorities in the emergence of those functions and what kinds of processes could lead to that emergence. An advantage of this somewhat abstract functional approach is that there can be multiple ways of realizing a given function. 3.8 billion years have allowed evolutionary process to arrive at some very complex and effective solutions to functions that may well have initially had a far

---

[23] See Wikipedia: Miller, Urey Experiments: http://en.wikipedia.org/wiki/Miller%E2%80%93Urey_experiment

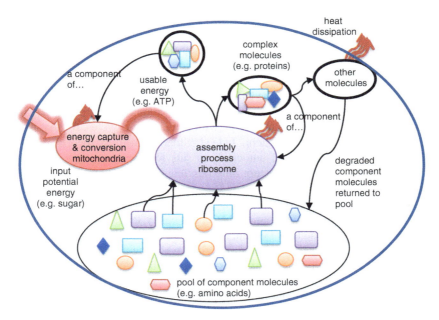

Fig. 10.12 A minimal system that meets the criteria of a living system is capable of capturing and converting an energy flux (like sunlight) to a usable form for other chemical processes. A principle process would act to assemble components such that the assembly process would be maintained (autopoiesis) as would the energy capture process. Other molecular processes (the origins of metabolism) would use these complex molecules but, in doing so, degrade the more complex molecules so that they could be returned to the component pool. Living systems are masters at recycling components in this manner

different form. For example, reproduction is necessary, but do we need to come up with some identifiable ancestor of the complex ribosome-enabled template reproduction found in all present life to account for the origin of life, or is that an emergent variation later on in the evolution process? Functionalism opens the door to considering far more simple origins.

While there is no agreed-upon strict definition of life that determines a clear threshold between the living and the not yet alive, there are a number of traits generally associated with living. Being alive is a process that includes some kind of self-maintenance, self-repair, and the ability to reproduce. As such, it belongs to the realm of what we have described as far-from-equilibrium dissipative processes, those that continually take in energy, use it, and return it in some form to the environment. Whirlpools maintain and even repair their form by taking in and expelling water, but they do not reproduce themselves. As we saw above, this reproductive moment is a unique systemic threshold, for it closes the causal loop by seeding the future with more systems of such abilities. The whirlpool endures and pops up anew any time conditions are appropriate; the living organism not only endures; it has to endure in a manner that, if successful, will plant more of itself in the future.

Whirlpools just are, while living organisms are constantly selected for and honed for a constantly changing fitness for this reproductive success. Living is in this way a new kind of self-referential systemic process, one that enters the future actively with a strategy shaped by success and failure, not just as the passive manifestation of the laws of physics. Perhaps here we see the roots of that "self" reference, without which it is difficult to describe living processes.

Self-maintenance and self-repair would seem to require, minimally, the closed loop dynamics that we have seen emerge with autocatalytic cycles, for that is what introduces stability into the system. Indeed, Stuart Kauffman theorizes that we should look to cross-catalyzed autocatalytic sets for the emergence of life and claims that such an emergence is not only possible but almost inevitable given the rich catalytic potentials of virtually any large array of diverse and complex molecules (Kauffman 1995, ch. 3). Terrence Deacon takes this suggestion a step further, considering environmental dynamics that could easily lead to the disruption and dispersion of the autocatalytic system. This might require the protection of some kind of containment vessel, a simple form of the enclosing cell-membrane structure now common to all forms of life. Contemporary cells form all sorts of structural elements by producing various sorts of macromolecules that, due to properties of shape, minimal energy configurations, etc., self-assemble into the required cell wall structures. The intertwining of this double dynamic—autocatalysis and self-assembly—seems to characterize all life. Deacon suggests then that the critical emergence would be not just closing the autocatalytic cycle, but such a cycle that also happened to produce, as a by-product, molecules that self-assemble to encapsulate the autocatalyzing cycle that produces them (Deacon 2012, ch. 10).

Deacon's "autogen," as he calls it, is able not only to thus maintain and repair itself, it can even reproduce. Enclosed in an impermeable membrane, the autogen will soon exhaust its substrate, but being protected by the membrane, it will not break apart. But when environmental wear and tear eventually do break down the membrane, the autogen may also break and then reassemble itself catalytically from substrate available in the environment. Pieces may not simply reassemble, but separately reconstitute the cross-catalyzing autocatalytic unit from environmental resources, and with the production of molecules for new self-assembling capsules, it will have managed the fundamental reproductive feat of increasing and multiplying. And, critically, the nature of such a breakdown-reassembly process would be open to the presence of a few adventitious molecules in the mix, just the kind of random variation to supply mutation for the selective process of evolution to work upon (Deacon 2012, pp. 305–309).

The autogen is not yet alive; it uses energy from breaking down bonds in available substrate molecules and then essentially goes dormant. It maintains coherence, but this is not yet the coherence of a metabolism maintaining itself by recruiting energy and resources in a far-from-equilibrium dissipative process. Fundamental to evolving to this fuller version of life would be transforming this relatively self-enclosed system to more continuously capture and convert sources of energy flux into usable forms. Recall that auto-organization requires an energy flow through a bounded assembly of components with sufficient potential for complex connectivity.

Evolving an appropriately semipermeable form of membrane would open the autogen to this kind of environmental flow and make possible the emergence of the evolved complexity of structure and process that marks the way contemporary organisms maintain their lives in constant interaction with their environment.

Growing complexity of structure and process tends to demand dependable coordination of energy flows. Today, in metabolic processes, the predominant form of energy distribution is through the adenosine triphosphate molecule (ATP) which acts as the portable batteries needed to power all of the other metabolic processes. Most of the key processes so powered are syntheses of complex molecules used in other metabolic sub-processes, including the kind of structural self-assembly mentioned above (see Fig. 10.12).

As we will discuss further below, evolution, as the process of weaving a transforming and increasingly complex network of systemic organization and suborganization, reaches to the beginning of the universe. But as the components are thus woven into increasingly complex networks and hierarchies of organization, the process of the weaving itself takes on new, emergent properties. Physics is not chemistry, chemistry is not biology, and biology is not ecology or sociology. Here we have been attending in particular to the critical threshold between chemistry and biology. At each of these levels, systemic expectations, or, in terms of information, the differences that make a difference in those expectations take on new dimensions. In physics, momentum, mass, charge, etc. are the differences that make a difference. In chemistry it becomes all of these plus matters of relative shape and related properties that enter into the phenomenon of catalysis that is a new difference that makes a difference in systemic expectation. In biology autocatalytic closure creates cycling organizations whose dynamic product is more of themselves, i.e., their systemic expectation is self-referential.

With this looping, information becomes a difference that makes a difference to an organizational unit that may or may not "succeed" in continuing to produce its own existence. For the first time, information becomes not just a descriptor of determination but a potential guide of what works and what works better, a selectively guided rather than simply determined way of entering the future. Information, as a difference that makes a difference in how well something works for an organism, sets up evolution not only as an organizing process but as a *selectively* organizing process that will weave ecological and social systems of mutual living fitness.

The details of the process of crossing the threshold of life are complex and a matter of perhaps unending speculative hypothesis. We have seen that complex systems must be ramped up, not simply emerge full-blown, and the overall process of evolution provides the necessary framework. Within that framework, there are clear systemic questions that provide focus even though they may be addressed by multiple paths. What are the essential functions of life? Why those and not some others? Are they related? What sort of organization can support such functionality, and how could it arise from available components and system dynamics?

Looking at life as an ongoing process of systemic evolution distinguished by selection for workable fit grounds the minimal functionality that must be accounted for. Without units that maintain, repair, and reproduce themselves and copies of themselves with some degree of random variation that can be passed on, the kind of evolution that has shaped the world of life about us could not take place. The authors we have presented here exemplify cogent, even if necessarily speculative, attempts to address these questions. Whatever the final answer to how the threshold to life was crossed, it will bear a systemic family resemblance to this.

10.4.6   Supervenience and the Emergence of Culture

Supervenience refers to the way systemic layers of greater complexity, often with distinctive properties of their own, emerge in dependence on a prior level of less complexity. Living organisms relate to one another and to their physical environment, a supervenient web of relationship that gives us an ecosystem. In a similar manner, human beings in their manifold social relationships weave a supervenient level referred to as culture (or cultures). In many respects, cultures are the human ecosystem, though for good reason we commonly distinguish the world of culture from the world of nature. The reason is that abilities developed to a unique degree in humans give rise to a systemic organizational dynamic and complex structure unlike any found in the "natural" world. Above we looked at the emergence of money as a distinctive human-level phenomenon. But money is just one of the many unique cultural products that emerge from our combined abilities to converse with one another (and with ourselves!) and to turn this to strategically weaving all kinds of technologies into tools to realize individual and collective goals. Here we will discuss the emergence and effect of these two critical human abilities, language and tool making. Note that both are inherently relational and information centered, and these are the features that constitute the fast-evolving and distinctive world of human culture.

10.4.6.1   Language

One of the more important evolutionary advancements for humans was the emergence of the language facility. A true language is based on arbitrary but shared symbolic representations (words) of meaningful objects and actions in the world, that is, nouns and verbs that can be produced from one's mental map and can produce a corresponding mental map in the mind of another. Not only do words carry lexical meanings, but the way they are strung together in sentences conveys higher-order meanings or semantics: "the man caught the shark," is quite different from, "the shark caught the man," even though the very same words are employed. The way they are strung together has to follow some rules of grammar, or syntax, in order to consistently convey the semantics. We think of language as referring to something, but these rules, the skeletal structure of every language, refer to nothing; instead they embody the way every word modifies every other word in the process of constructing a meaning embodied in no single word. So words not only refer to objects, they also refer to and modify one another.

Both the arbitrariness—and hence immense flexibility—of associating a given sound symbolically with some content and the self-referential loop of words enmeshed in syntax make language far more complex than the basic animal communications that we observe in a wide variety of species. Bird songs are now understood as communications that allow male birds to proclaim their territories and attract mates. Chimpanzees have relatively elaborate sets of grunts, hoots, and screams that they use effectively to communicate emotional states or important aspects of the environment at the moment. A first communicative strategy is to associate things that go together: just as smoke signals fire, so a scream of alarm signals danger and can be used as well to signal approaching danger. But once associating sounds with objects or situations is in place, a degree of latitude can enter in, as in some species of songbirds that learn different versions of their hallmark songs in different places.

In recent years, we have discovered that some animals, especially primates, can be trained to communicate using symbols and sign language. Building associations between sounds and objects is relatively easy, but the syntactical interweaving of sound meaning has proved a much more formidable barrier. So far we have succeeded in teaching language restricted to very simple syntax and fairly simple semantics, usually related to something very immediate such as their emotional or motivational state. For example, "Washoe wants fruit" could get transposed to "wants Washoe fruit" without loss of semantics (a possible model for Yoda's syntax in Star Wars!).

That chimpanzees are able to communicate some semantic values via these means suggests strongly that the primate brain possesses an underlying neural basis for speech. It is suggested that our brains actually coevolved with language (see Deacon 1997). Speech and language appear to have emerged during human evolution, perhaps 250,000 years ago or more. Early humans probably had the same basic communications capabilities as chimpanzees, but human evolution took a somewhat different route with respect to the kinds of environments humans could adapt to and foods they could eat. Humans became much more generalists in their adaptive capacities and as a result were exposed to more complex environmental and social situations. As a very social species of generalists in a complex environment, it appears that more nuanced human utterances emerged from the necessity of cooperating with fellows and sharing more complicated resources. At the same time that early humans were working at inventing and combining new words (components), their brains and vocal apparatus were evolving by selection for the capacities to form those words and to string them together meaningfully.

And as the lexicon/syntax capacity emerged, it was tested by the context of the social milieu of the tribe. Both talking and listening developed jointly, but were handled by different parts of the brain. Language complexity coevolved with a whole new level of brain organization that successfully encoded messages that could be efficiently transmitted, received, and interpreted. Once that basic brain-language coevolution became more stable, a further evolution took place. Languages themselves evolved in a manner very similar to how biological evolution took off after the initial emergence of life. This is not surprising, insofar as the powerful evolutionary dynamic that emerged with life, an ongoing selection for fitness, selected for linguistic fit in widely differing environments. As human migrated around the world, their languages changed as they needed new words and new ways of constructing mental models. Even the syntactic structures could be modified to some extent, creating a new kind of feedback loop in which world views became enshrined in distinctive languages and language became a vehicle for instilling a world view.[24]

It is not a stretch to claim that language is the keystone feature of human culture. The key to biological evolution is the emergence of traits or variations that can be passed on by genetic inheritance and sifted for fitness. Linguistic communication does the same for the realm of individual human experience and discovery, which becomes the shared possession of the society, not by shared genes but by being rapidly passed around by language. Early humans could teach children not just by example but by explaining processes such as how to make an arrow point. Later in our prehistory, humans began to represent noun objects with pictures and pictograms. Then, as we discussed above, the representations evolved to become abstract markings on various media such as clay tablets or papyrus. Language evolved to be not just spoken and heard, but written and read.

The emergence of writing released language, and with it the potentials of information, from the constraints of immediate space and time. A carried written message could have effects hundreds of miles from its point of origin. Preserved writing could carry information across generations. Many creatures have evolved abilities to learn from experience, enabling them to both adapt to changing conditions and engage the oncoming future with proactive strategy. But with language and writing, humans transform personal experience and learning into a collective reality, and they accumulate this learning in a trajectory of exponential growth over centuries and thousands of years. This language ability and the steady and increasing introduction of technologies that expand its power and reach have produced a wave of interactive, inventive, and self-augmenting information that in the eyeblink of 10,000 years has carried our socially created cultures from small hunting and gathering groups to globe-encompassing civilizations. Now six and a half billion of us use shared information to engage and adapt the global system to our own purposes.

---

[24] For the classic study of language and world view, see Whorf (1956). Whorf was the pupil of Edward Sapir, and the so-called Sapir-Whorf hypothesis about how language shapes thought became widely influential in anthropology and many related fields.

10.4.6.2 Tool Making

A tool is any artifact that allows a human to apply leverage to a work process so as to accomplish that process more quickly or more finely. Early humans, like chimpanzees and many other kinds of animals, learned to use naturally occurring objects, like stones, to assist in some particular work, such as breaking open nut shells. All primates and many other mammals and birds have a capacity for "affordance," which is the ability to mentally draw a relationship between an object and a possible use for an intended purpose. For example, seeing a log on the trail as a "seat" upon which to sit and rest is an example of an affordance. The object "fits" the need, in the hiker's imagination, and so can be used.

Chimpanzees (along with most great apes) and some birds are also able to modify certain naturally occurring objects to fit the role of a tool. Chimps can strip leaves off of a branch to use it as a "fishing" device to capture termites from inside their nests. Crows, similarly, have been observed breaking a branch using their beaks and feet and then using the branch to pry rocks so as to get the grubs underneath. Once again we see that a behavioral capability is already built into the animal brain. And again we see it emerge in humans as a more advanced capacity to modify objects for specialized tasks, that is, to make tools.

Some of the first tools humans developed are also the best preserved, namely, stones shaped as axes, knives, and arrow points. At about the same time that humans were developing their language capacity and perhaps either because they were using more linguistic communications, or through some evolutionary pressure that caused both to develop simultaneously, they started elaborating tools and tool designs. These included clothing, shelters, baskets, ovens, and many other artifacts that enabled them to adapt to ever more challenging environments.

Initially humans developed a standard set of tools that provided them with a more successful capacity to fit in many environments. They could more readily compete with other predators to become the "top dogs" on the food chain. Most of these "cultures," as anthropologists call them, persisted for much of human prehistory, changing very little over long stretches of time. Then between 50 and 100,000 years ago, the capacity to be much more inventive, to explore alternative designs for tools, and to create new tools and new uses seems to have taken a giant leap. Over this time period, humans learned to master fire for many purposes beyond cooking, such as firing clay pots and later smelting bronze. They discovered alternative energy sources (Sun and fire being the first) such as water flow and animal power, beginning a long synergistic positive feedback in which societies and cultures became increasingly shaped both by and for their energy resources and technology.

Tools have always allowed humans to work faster and better, to do some things that they could never do with just hands and muscle power. Much of education—communicating ideas to children—has been about teaching the next generation how to make, use, and invent tools. Today the concept of tools includes procedures, abstracted instructions on how to do things of importance. Just as with language, the basic components of tool making emerged and interacted in early man, but once under way, natural selection within and among cultures for the efficacy of their tools elevated the further process of change to one of cultural evolution.

Cultural evolution, which today involves tremendous complexities of artifacts, understanding, and interactions between humans, and between humans and their environment, is running at an incredible and seemingly ever-increasing pace. We will discuss this subject further in Chap. 11. It is important to recognize how such evolution got going by the emergence of what would become the basic components of culture, two of the most important being the language, which gives us the ability to organize and modify organization with incredible flexibility and scope, and tool-making, which has finally made the technological mindset of seeking a better way of doing virtually anything the shared character of the aggressively changing and expansive global culture.

Quant Box 10.1 Increasing Complexity Over Time

In Quant Box 5.1, we provided an approach to a "rough" index of complexity by recognizing some properties of systems as defined in Quant Box 3.1. To recall we defined a system as a 6-tuple of sets:

$$S_l = \{C, N, I, B, K, H\}_l, \quad l = 0, 1, 2 \ldots m$$

where l is the level index and m is the number of levels in the hierarchy of organization. Please refer back to Quant Box 3.1 for the definitions of these sets. In this Quant Box, we show how this formal model can be used to observe the auto-organization, emergence, and evolution of more complex systems from simple ones, i.e., from potential complexity to realized complexity.

To formalize the concepts of auto-organization and emergence, we show the starting conditions of a whole system, S_0, and introduce the role of time and energy flows that will drive the system toward complex organization.

In order to somewhat simplify this approach, we assume that the energy flow from an infinite reservoir at a high potential to another infinite reservoir at a low potential through the system is constant for all time. We then introduce a new index, t, to the system and start at $t = 0$. Thus, we have $S_{0,t=0}$.

The multiset $C_{0,0}$ contains all of the component types and the number of each type as in Quant Box 3.1. The composition of the multiset may actually change over time as a result of fluxes across the boundary (see B below). In Chap. 3, Sect. 3.3.2.1, we introduced the concept of component "personalities." Each type of component has a set of interfaces with which it can potentially interact with other components. These are represented by the different shapes of components in the various figures in this chapter and are formally encoded in the multiset K. This set may also change with the influx of new components over time if that occurs. Since these personalities obtain from

(continued)

Quant Box 10.1 (continued)

what is internal to the component, i.e., its own organization, we can forego further consideration of what this set of interfaces looks like—we are assuming the initial state of our system is composed of non-decomposable atoms. At $t=0$, let all components be independent of one another. What does this mean? It means that the set $N_{0,0}$ is empty! Similarly, the multiset $H_{0,0}$ must be empty by definition since no history has accumulated!

Referring back to Fig. 10.4, we get some idea of the nature of the boundary conditions that are encoded into the set $B_{0,0}$. This includes provisions for the insulating, nonporous sections as well as the two "energy windows" that allow the passage of energy through the system. Once more to keep things simple (maybe too late), we assume that B is fixed for all time.

Up until this point, we've said nothing about the geometrical aspects of the system components except that they must be contained within a volume defined by the boundary. Now, to model auto-organization, we need to rectify that omission. Every component of every type must have associated with it a Cartesian coordinate that uniquely identifies it by position (thus C could then be treated again as a regular set). In other words, we label each $\{c_i,n_i\}$ element with a triple, $\{x,y,z\}$ which can be a position in space relative to a convenient origin within the system boundary. Now we are ready to describe what happens in terms of a model of auto-organization.

Using good-ole graph theory with its extensions in evolving graphs, we construct a graph of all components (with their attendant personalities and positions) in which both $N_{0,0}$ and $I_{0,0}$ are empty sets (see Fig. 10.4). As time progresses, energy is available to the components to actualize their potential interactions with neighbor components according to the rules in K (see Fig. 10.5). As interactions are realized (or broken), a time-stamped recording of such can be added to H so that the history of the auto-organization process can be traced.

This formal approach may seem extremely abstract (which it is) and difficult to imagine. One example is the model of the brain under development by one of the authors (Mobus). The components are the neurons of various types. Of particular interest is the way in which a neocortical-like model can auto-organize as a function of time and historical experiences. Sensory inputs are modeled in the I set, K contains the rate constants for each type of synapse and neuron, and N contains the growing set of tuples representing the connections between source neurons and receiver neurons as the network evolves. At specified intervals, the state of the system is recorded in a file, H, for later analysis.

This abstract, formal approach can be used to capture any system of any kind. As with any model, certain details are omitted for tractability sake. But the S structure captures the essence of any system and allows us to watch the system evolve over time.

(continued)

Quant Box 10.1 (continued)

In order to capture the essence of levels of organization, the next step is to treat each dense sub-network in N as a subsystem (see Chap. 4, Sect. 4.2.3.4) and construct an S structure for it as well. That is, we subsume all of the components in the network under the new S, construct an N for all the internal connections and an I for the sparse connections to other sub-nets. In this manner, the organizational hierarchy is built from the bottom up. We end up with the tree structure as that seen in Chap. 12, Fig. 12.3.

Auto-organization produces new structures that represent greater complexity per object. Those new objects can have more complex personalities and behaviors as a result. As well these objects can then interact in new ways not inherent in the original system. This is a new level of organization.

Think Box. Brain Development as Auto-Organization

Embryonic development takes an organism from a single cell—a fertilized egg—to a neonate with perhaps billions of cells and hundreds or thousands of differentiated kinds of cells. The brain also develops, but in addition to the emergence of various neuron and glial cells (support cells), it develops a basic wiring structure that is consistent within a species. The wiring scheme adds a level of complexity to brain development many orders of magnitude above, say heart or lung development. Some have estimated that the specification of such intricate structures as in the brain would require 10–100 times as many genes as are in the human genome if that structure were determined by genetics.

What can account then for the elaborate structure of the brain that develops in the embryo, the fetus, and the child if not genetic programming? The answer may lay in the same way in which neurons wire themselves together when learning is taking place. In a sense the neurons in one area *learn* through a process of auto-organization which neurons in another area to send their axons to and make synaptic connections with.

The lower brain areas (the more primitive part of the brain) are very much organized and developed under genetic programming. This part of the brain is composed of specialized modules, each with its own set of tasks to perform relative to the operational control of the organism. Neurons in these modules are predisposed to send their axons to specific other modules. For example, there are clusters of neurons in a structure that is one of the thalamic nuclei that receives nerves from the eyes. The neurons in the eyes grow their axons

(continued)

Think Box. (continued)

along the optical tract directly to the thalamus. The thalamus, in turn, grows axons from its neurons to various other primitive structures in the lower brain and also to the primary optic processing neocortex, a specific region in the occipital lobes called V1. This is where visual information first enters the neocortex. But from there, the complexity of processing explodes.

Cells in V1 send projections to an area called V2 of the early optic sensory neocortex, and you might think that implies that the same kind of genetic predisposition is at work. Instead, the projections from V1 to V2 develop due to a generalized stimulation of both regions. The stimulation isn't from the eyes, which are still, themselves developing. Rather the brain spontaneously generates waves of activity that "simulate" stimulation. Neuronal clusters in V1 are mapped topographically to the same layout as the retinas, that is, retinotopically. But from V1 up to V2 (and for all of the rest of the non-primary sensory and motor regions, e.g., auditory, neocortex), the projections seem to go everywhere in the latter. That is, V1 sends many more projections than are needed for discrimination in V2.

After birth, when the eyes open, then actual images with low-level feature detection in V1 start activating V1 and from there to V2. Along with the images, the neonate is experiencing other primitive limbic senses, such as hunger, satiation, and the warmth of mother's caress. These act as the unconditionable stimuli that are used to reinforce the patterns of stimulation that are occurring in V2 (see Sect. 8.2.5.1.2). The baby learns to recognize basic features in its environment that are important to its survival. The learning, however, isn't due to the creation of new synapses from V1 to V2. Rather it is the pruning of unused synapses that were formed in that earlier development stage. The rule of organization is: "wire everything up in a sequence from the back of the brain (primary optical sensory) toward the front and then if you don't use it, you lose it!"

The original wiring seems to be almost random. It is certainly nonspecific in that retinotopic regions in V1 do not have corresponding topographical regions in V2 (at least they are not as localized if they exist). The process is reminiscent of that depicted in Fig. 10.9. Energy drives trial combinations, and then actual experience (environmental experience) determines which combinations will last. Organization of structures and memory trace encoding are emergent phenomena that have an element of chance and of selection (for one interesting version of this phenomenon, see Gerald Edelman's theory of Neural Darwinism 1987).

10.5   Summary of Emergence

Auto-organization and self-assembly processes are enabled by energy flow driving systems, internally to greater complexity. Under whatever conditions the system finds itself, some of the assemblies will be favored (e.g., are stable). Some will provide structural support for further stabilizing the system. Some will provide new functions. That is, they will be sub-processes that convert some inputs into outputs. If those outputs serve a purpose for other assemblies, then they will be favored and persist.

Some sub-processes will become involved in cycles that are mutually reinforcing. An extremely important type of cycle involves autocatalysis or mutual catalysis, a positive feedback that causes some of these processes to grow in number.

As the tendency toward greater complexity proceeds, these new assemblies become functional and interactively relate to form a new level of organization. New properties and functionalities can then emerge. These emergent properties still depend on the lower-level processes, but they take on a reality of their own. The objects that comprise the new properties become the components for interactions in the new level of organization.

Bibliography and Further Reading

Bourke AFG (2011) Principles of social evolution. Oxford University Press, New York, NY

Casti JL (1994) Complexification: explaining a paradoxical world through the science of surprise. HarperCollins Publishers, New York, NY

Dawkins R (1976) The selfish gene. Oxford University Press, New York, NY

Deacon TW (1997) The symbolic species: the co-evolution of language and the brain. W. W. Norton and Company, New York, NY

Deacon TW (2012) Incomplete nature: how mind emerged from matter. W. W. Norton and Company, New York, NY

Dennett DC (1995) Darwin's dangerous idea: evolution and the meaning of life. Simon & Schuster, New York, NY

Edelman G (1987) Neural darwinism: the theory of neuronal group selection. Basic Books, New York, NY

Gleick J (1987) Chaos: making a new science. Penguin Press, New York, NY

Gribbin J (2004) Deep simplicity: bringing order to chaos and complexity. Random House, New York, NY

Holland JH (1998) Emergence: from chaos to order. Addison-Wesley Publishing Company, Inc., Reading, MA

Kauffman S (1995) At home in the universe: the search for the laws of self-organization and complexity. Oxford University Press, New York, NY

Kauffman S (2000) Investigations. Oxford University Press, New York, NY

Küppers B (1990) Information and the origin of life. The MIT Press, Cambridge, MA

Miller JH, Page SE (2007) Complex adaptive systems: an introduction to computational models of social life. Princeton University Press, Princeton, NJ

Mitchell M (2009) Complexity: a guided tour. Oxford University Press, New York, NY

Morowitz HJ (1968) Energy flow in biology: biological organization as a problem in thermal physics. Academic, New York, NY

Morowitz HJ (1992) Beginnings of cellular life: metabolism recapitulates biogenesis. Yale University Press, New Haven, CT

Morowitz HJ (2002) The emergence of everything: how the world became complex. Oxford University Press, New York, NY

Nicolis G, Prigogine I (1989) Exploring complexity: an introduction. W. H. Freeman & Company, New York, NY

Prigogine I, Stengers I (1984) Order out of chaos: man's new dialogue with nature. Bantam Books, New York, NY

Sawyer RK (2005) Social emergence: societies as complex systems. Cambridge University Press, New York, NY

Simon HA (1983) Reason in human affairs. Stanford University Press, Stanford, CT

Wolfram S (2002) A new kind of science. Wolfram Media, Inc., Champaign, IL

Whorf BL (1956) In: Carroll JB (ed) Language thought and reality: selected writings of Benjamin Lee Whorf. MIT Press, Boston, MA

Chapter 11
Evolution

"One general law, leading to the advancement of all organic beings, namely, multiply, vary, let the strongest live and the weakest die."

Charles Darwin, *On the Origin of Species*, Chapter VII.

"Nothing in Biology Makes Sense Except in the Light of Evolution"

Theodosius Dobzhansky, 1973 essay

Abstract Biological evolution turns out to be a special case, albeit an extremely important one, of a more general or universal process of increasing levels of organization and complexity of select systems over time. So long as energy flows there will be a tendency toward greater organization and increases in complexity of various subsystems. We examine several examples of evolution at work in several contexts beyond just the biological one. Even our technologies are subject to the rules of evolution as are culture and societies in general. Of particular interest is the nature of coevolution, that is, the way in which mutually interacting systems affect the further developments of one another.

11.1 Beyond Adaptation

We would humbly submit an amendment to Dobzhansky's proclamation: "Nothing in *Systems* Makes Sense Except in the Light of Evolution." In this chapter we will provide evidence and arguments for this case.

In the previous chapter we provided a preview of evolution as a form of change in capabilities of a system. Thus far in the book we have identified three basic kinds of changes in systems over time. The first was dynamical change or behavior (Chap. 6). The second kind of change was adaptability (Chaps. 9 and 10) wherein a bounded system adjusts its internal workings in response to environmental changes. The third kind of change involves the construction of new structures and functions first introduced in Chap. 10 where auto-organization and emergence give rise to new capabilities. We now expand the notions of this kind of change in systems by rounding out the explanations we started in the previous chapter.

© Springer Science+Business Media New York 2015
G.E. Mobus, M.C. Kalton, *Principles of Systems Science*, Understanding
Complex Systems, DOI 10.1007/978-1-4939-1920-8_11

As noted before, many authors do use the term adaptation when applied to species as describing evolution. We have no argument with that usage so long as it is clear that the reference is to species and not to individuals. In this book we have restricted our descriptions so that the term adaptation primarily applies to essentially reversible changes in structures or behaviors that individual systems undergo in response to their immediate environments. In truth the distinction is one of scope in scale and time. If one chooses a species as a system, then, of course, the long-term adaptation of the system is evolution. If one chooses an individual member of a species as the system of interest, however, then adaptation is not the same as evolution. We will now take up the case of species or populations of similar individuals in which permanent modifications to structure and/or behavior arise over longer time scales.

Below we will provide a more formal distinction between adaptation and evolution and introduce a system characteristic called *evolvability*. We find that as one changes the choices of boundaries—what is the system of interest—one discovers variability in how evolvable the selected system is. But first we consider why evolution deserves a revered seat at the table of systems science principles. We have long felt it deserves the head seat at the table for the simple reason that it is at the heart of all identifiable phenomena in the universe. That is a strong claim. We will try to provide convincing arguments in this chapter.

11.2   Evolution as a Universal Principle

The arguments we present in this book are based on the idea that there is a universal principle of evolution that applies equally well to the universe as a whole as to the subatomic world of particles and quanta.[1] The whole of the core concepts of organization (structure), network relations, dynamics and process, complexity, and cybernetics/information are all subject to the overarching notion that systems of all kinds and at all scales evolve as reflected in each of the these conceptual frameworks. Networks change, complexity increases, dynamics modify, information is communicated, and organizations develop and then, in turn, interact at a new, higher level, leading perhaps to yet further emergent organization. Some systems are highly evolvable, as we shall shortly discuss. Others have very low levels of evolvability but can still be shown to be subject to the same rules of evolution in general. In this chapter, and this section in particular, we want to bring all of this together as a holistic framework for understanding how systems not only change but get on trajectories of development in space and time.

---

[1] In this case we are claiming that the evolution of particles took place in the Big Bang and the evolution of atoms, as another example, took place in stellar furnaces and supernovae events. In this book, however, we are more concerned with the evolution of living and supra-living systems, organisms, ecosystems, and social systems.

Except for the critical first few millionths of a second, the evolutionary trajectory of a cooling and expanding (locally condensing) universe can be sketched as follows: the universe is itself evolving from an initially formless mass which expanded at phenomenal speeds while at the same time being organized by the interplay between the known four forces of nature, gravity, electromagnetic, weak, and strong nuclear. The universe has undergone the coalescence of nuclei of galactic and stellar masses, gravity providing the basic boundaries. The other three forces, working within these boundary conditions, have worked to produce a plethora of particles (components) with remarkable personalities (e.g., electron shells) and essentially unlimited potential complexity. And working together, the matter in the universe began to auto-organize into planets and complex chemical molecules.

Gravity supplies the gradient for the energy flow that drives the whole process. Gravity compresses the particles in sufficiently large masses (the proto-stars) such that, in cooperation with the strong nuclear force, nuclear fusion converts large amounts of mass into extraordinary amounts of high-powered energy ($E = mc^2$!). And as long as energy flows from these point sources, stars, to fill space according to the dictates of the Second Law of Thermodynamics, passing through planetary systems and clouds as it dissipates, the whole universe generates pockets of increasing organization and complexity, even while entropy increases for the whole.

One of the lucky recipients of the flow of energy was a smallish multi-compositional rock, large enough, happily, to retain an atmosphere and hydrosphere. Biochemistry emerged, at least here on Earth, as a new level of organization that would give rise to the emergence of a new form of selectivity in the evolutionary process. All the while the extant physical and chemical conditions operated to sort out the viable assemblies to become components at that new higher level.

11.2.1   The Environment Always Changes

Since the universe as a whole is undergoing dynamic change as just described, every subsystem of the universe is exposed to change over time. From the perspective of any system within the universe, the environment is constantly, though not necessarily rapidly, changing. Until our Sun burns itself out and no more energy flows through the Earth system as it does now, the Earth will continue to be dynamic. Sometime in the very distant future even the heat coming from the Earth's core will dissipate, and the convective processes through the mantle will no longer drive continents to drift.

In the meantime the environment for living systems is constantly undergoing change at all scales of time. The environment is non-stationary for all evolving systems within it. This is a technical condition that means the statistical properties of fluctuating processes are actually changing over time. Consider the climate, which is a major factor in the evolution of life on Earth. The climate has changed and will change as its composition modifies. We are currently quite concerned about the changes that are going on in our climate as a result of burning fossil fuels and

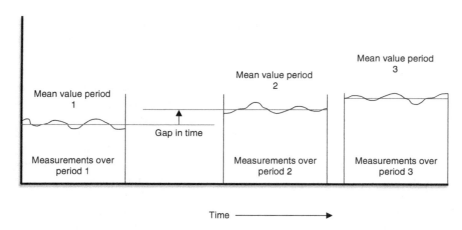

Fig. 11.1 A non-stationary process (e.g., average global temperature) is one in which the measured statistical value is changing over long time scales. In period one measurements are made frequently enough to obtain the time series curve of the property. The average value (*blue line*) appears stable over that period. After a gap of time a similar period of measurements produce a higher average value. Similarly, after another gap the average value in period three is higher still. This system is trending upward over the whole span of time of measurement

increasing the proportion of carbon dioxide in the atmosphere. CO_2 is known to be a "greenhouse" gas, that is, it is able to trap heat from the Sun in the atmosphere. Global warming is the effect of accumulating slightly more heat than is radiated back into space, leading to higher average temperatures. When we say higher average temperatures (e.g., on an annual basis), we are talking about a non-stationary process. The average temperature is trending upward rather than remaining stable over long time periods. Figure 11.1 shows this graphically.

Finding a trend, a trajectory of change, as opposed to ordinary fluctuations around a stable mean, can be a tricky project. The graphic in Fig. 11.1 represents the situation where measurements of a value, say temperature, are taken frequently, say daily, for a long period of time. The wavy lines in the graph show that the value actually fluctuates, but the time average can be computed and appears to be stable over that period. However, suppose the measurement takers wait for a very long time before resuming measurements. This is shown as a gap (not to scale necessarily). The measurement procedure is repeated and a new average is computed. But this time the average is higher than the previous one. It would be tempting to simply average the two averages and assume that the really long-term average would have been found to be that. There is, however, a problem with this. We do not know what happened between the two periods that would give the difference (black vertical arrow). It is possible that we are observing simply the normal variation of the process, but for some reason the within-period variation (amplitude of the waves) just happened to be low. If this is the case, then we would expect the average at a yet latter period to be somewhere between the two blue lines.

So, after another gap of time (again not scaled) we take a third period measurement and discover it is still higher. At this point we can only conclude that either the first sample period produced an abnormally low average (over that period) or, now more likely, that the value is trending upward over the very long haul.

In fact many natural processes are *non-stationary*, which means that the environment in which they operate is going to be non-stationary itself. As systemic relationships change, there is new information, a change in the field of expectations or probabilities, a difference that makes a difference. There is a loop here that is critical for the system-constructing nature of evolution: as a difference makes a difference (Bateson's definition of information), the difference made becomes a new difference making a difference. The world, the universe in process, is not just a series of unrelated events, but a building of intersecting and interactive differences which shape differences. And the selective feedback loop of differences that make a difference build the trajectories of change which we describe as evolution. Things are always changing, and sometimes they change with dynamics that articulate into the emergence of new levels of complex development: they evolve.

And this is why there is evolution that seems to produce more complex forms over a very long period of time. But just how long is "very long"? The answer depends on the system in question. In the above example of average global temperatures, over geologically pertinent time scales, the temperature has been much higher and much lower than it is today. So over that length of time the change we are presently undergoing would be absorbed in an average, and the process might be considered stationary. But what matters is the length of time relevant to living systems which must respond to changing conditions. That is months and years for organisms and hundreds of millennia for species. As long as over those kinds of periods the relevant factors (to the systems) remain reasonably stable, the individual or the species can remain basically stable in its phenotypic form. But since we know the environment is non-stationary relative to these shorter periods, the living systems are under pressure to evolve in order to meet new conditions.

11.2.2   Progress: As Increase in Complexity

The term progress generally is interpreted as a system being better off in some ways as time goes on. For example, economic progress implies that the standard of living for people increases over time. The proxy for the level of the standard of living is usually a person's income. This is a materialistic interpretation that generally translates to "more income means more material goods." Of late more people are becoming interested in more humanistic measures of progress.[2] Biologists generally do not like to talk about progress, since the notion of betterment implies a value judgment from an anthropocentric view. But regardless of the metric, the notion that something is increasing over time is at the base of progress.

---

[2] For example, see Gross National Happiness, http://en.wikipedia.org/wiki/Gross_national_happiness

In the case of biological evolution, there has been an obvious increase in the complexity of multicellular life forms over evolutionary history. The first recognizable life forms in the fossil record appeared in the Archean Eon (2.5–4 billion years before the present); the fossils represent what we would describe as simple single-celled organisms grouped under the kingdom Prokaryota, cells with no nucleus. During the next eon, the Proterozoic (0.5–2.5 billion ybp), nucleated eukaryotic single-celled forms made their appearance. Eukaryoteic cells are more complex than prokaryote cells such as bacteria. Their name means "true nucleated"; they contain membrane-bounded nuclei and also other organelle that are bound in membranes of their own. These organelles perform specialized functions critical to the maintenance and reproduction of the much larger eukaryotic cell. They were originally themselves independent prokaryotic cells which evolved, it is believed, by a form of cooperative coevolution to become components of the much more complex eukaryotes. We will discuss this further later in the chapter.

Along with complexity of morphology another metric starts to show itself as important in evolution, biodiversity. The meaning of the term "species" is somewhat ambiguous in the realm of prokaryotes.[3] But in the eukaryotes we start to see a wide array of definitive species and higher-level classifications. Recall from Chap. 10 that this is what is meant by emergence of new forms and functions at a higher level of organization. And eukaryotes included in their new functionality the potential to join in larger cooperative systems, multicellular organisms. In the Cambrian period the fossil record explodes with multicellular life forms of extraordinary diversity. In terms of complexity these fungi, plants, and animals are at a much higher level of organization than the single-celled forms. Multicellular life involves an organization of differentiated tissues and cell types within a single body. These different tissues perform more specialized functions and have to cooperate with the other tissue types in order for the whole organism to function appropriately in the environment. That is, they need to operate in a complex, coordinated choreography of multiple behaviors in order to fit into their environment.

The simplest version of tissue differentiation is that between germ cell lines and somatic cell lines. Germ cells, and the tissues that generate and protect them, are the sperm and egg cells that are part of sexual reproduction. Lower life forms reproduce for the most part by asexual mitosis. The earliest forms of multicellular life were largely just connected colonies of somatic (meaning body) cells responsible for obtaining nourishment, movement, etc. But a small contingent of aggregated cells took on the responsibility for producing the sperm or egg cells through a different form of cell division called meiosis. The resulting cells, as a rule, contain half of the genetic complement each. But when sperm and egg fuse, the genetic complement is restored in the resulting single cell that is capable of dividing and developing into a complete new individual.

---

[3] The concept of species is used in microbiology to describe recognizable characteristics in groups. However as we learn more about genetic transfers between cells in this kingdom we realize that the boundary for species is a bit more fuzzy than originally thought.

As evolution progressed simple multicellular forms developed new, more specialized, tissues/cells in response to the demands of the environment. Thus, complexity grows and so does biodiversity. Sometime during the Paleozoic animal forms developed forms of mobility to obtain food. In order to control their motion for seeking food and for avoiding dangers, the first primitive "central" nervous systems emerged, with light and chemical sensing organs at the "head" end and excitatory nerves innervating muscles along the body to coordinate undulatory "swimming" motion. This basic bilateral body plan (one of several to evolve) proved to be especially full of potential for further elaboration: it became the basis for further evolution in animals, giving us the familiar pattern of a head housing a "brain" and a body performing the actions.

For what we now will refer to as higher animal life, the evolution of behavior became the focus of innovations. Brain complexity evolved in response to the increasing complexity of the environment (see below). But the complexity of the environment was due to the increasing biodiversity and the impact of other life forms on selection. So it was no longer just a physical environment that drove evolution toward higher levels of organization, complexity, and biodiversity. It was now the evolving biodiversity itself along with the dynamics (behavior from Chap. 6) of animal life that became major factors driving further evolution.

In this sense, then, evolution is progressive, a developmental trajectory that builds upon itself.

Question Box 11.1

Is it necessarily "better" to become more complex? Why or why not? What criteria are you using to judge whether something is better? Does "progress" or "progressive" as used to describe evolution imply things get better and better?

11.2.3 The Mechanisms of Progressivity

In biological and supra-biological evolution, there is a peculiar fact that is sometimes forgotten. Changes tend to be cumulative. Nature doesn't start from scratch with every new species: it works by simply reshaping spare copies of old equipment to fit new requirements. We humans are no different as we go about changing (evolving) our organizations or our technology. Every technological product is based on the ones that came before. Usually the new design is a modification of and improvement on the previous one. Then occasionally a major discovery will be incorporated that leads to new ways of working or new functions. For example, when the transistor was invented, the nature of electronics equipment changed forever. Radios were still radios but they no longer used vacuum tubes to accomplish their amplification of very low power radio waves into sound waves that people could easily hear. But everything new has its predecessors; it has a history of changes.

Biological evolution doesn't often invent new genes so much as mostly tinker with the ones that are already working. Genes specify the building of proteins, the workhorse chemicals in the cell. Proteins are very complex polymers of amino acids (20 in all as far as life is concerned). Their complexity is multi-dimensional, but in all cases each protein has a critical region of activity where it gets its work done. Very small changes in the amino acid sequences in those regions from DNA mutations can lead to altered functionality. We will cover mutations as sources of variation later. For now we just want to point out that quite a lot of tinkering can get carried out over the history of a species and population. Every once in a while the tinkering pays off.

Yet another peculiar phenomenon is found in biological and supra-biological evolution that contributes to increases in complexity in new species. Occasionally, during a replication operation where the two daughter structures are supposed to separate and go their own ways, they don't. They stick together and continue to function as before. Sometimes when genes are being replicated (see below), an extra copy of a gene gets made and is inserted into the chromosome. Most of the time this is disruptive, but occasionally the cell can tolerate the extra gene and go on functioning. What happens next however is really interesting. The two copies are essentially independent of one another, and through the accumulation of entirely different mutations, one of them can diverge from its original function. Of course one of them has to maintain its original function, but the other initially is just redundant and is free to explore design space (see discussion of evolution as a search through design space below). And that is where the potential for innovation comes from. Unneeded redundancy plus time can lead to a new gene with a new function that confers new abilities onto the possessor.

This mechanism appears to be an important contributor to progress in evolution. The new capability may lie dormant for many generations of the species, but because the environment is non-stationary in the long run, eventually it may emerge as a critical factor that makes the possessor more fit than the rest of its clan.

Take the evolution of more complex brains in animals as an example. During embryogenesis tissues that make up specific regions of the brain are actually derived from previous existing tissues. The derived tissue then differentiates to become the region it was supposed to become. The brain develops from the progressive differentiation of new layers from the back forward (Striedter 2005). One theory of brain evolution involves a redundant production of an extra layer of tissue forward of the needed one. Thus, a new layer that is not needed for normal functioning of the animal is available for further tinkering. New levels of intelligence can be developed as a result. Figure 11.2 is a simplified representation of this process.

The accretion of new tissues by accidental replication provided the raw material for selection to work on in terms of increasing the information processing competence of the brain. Information in the environment was increasing due to the increasing complexity of the environment. There was more information to be received if a competent receiver came into existence that could do so. Bigger, more complicated brains could deal with more novelty in environments and hence the exploitation of more niche resources.

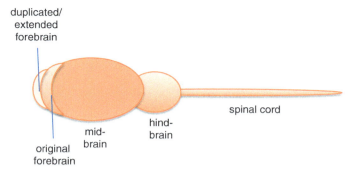

Fig. 11.2 Brain evolution seems to have proceeded from the back forward as newer "modules" emerged from older ones. The hindbrain contains the basic decision processing for physiology and motion control. The midbrain processes primitive sensory and association signals and generated motor commands through the hindbrain. The original forebrain developed to do more sophisticated processing of sensory inputs and coordinate more complex processing in the midbrain. It would eventually evolve into the cortical structures of higher vertebrate brains such as the neocortex, hippocampus, and olfactory cortex. Shown is an extended layer that resulted from an accidental duplication of the original forebrain. This would evolve into the most anterior portion of the brains of mammals and birds, the prefrontal cortex

This is an admittedly, perhaps overly, simplified description of what is thought to have occurred in the evolution of progressively more intelligent species over the history of animal life. But it provides another sense of how evolution has progressed cumulatively.

We actually see the very same phenomenon occurring in human organizations. Businesses, for example, can diversify their products to address new (emerging) markets. They don't necessarily throw out the old products, especially if they can be "cash cows," still selling but without incurring additional costs. New divisions bud off from old ones. Innovations are incorporated in both products and processes.

The generation of novel assemblies followed by the sorting (selecting) of viability and stability in order to allow the emergence of the next higher level is the essence of universal evolution. Biological[4] evolution is a special case of this universal principle applied to very special kinds of assemblies, cells, organisms, populations, etc., operating within a larger system that provides the necessary volume of material and flow of energy. Universal evolution is a process of looking for ever more complex organizational structures that are dynamic enough to adapt to changes in their surroundings and survive long enough to serve as patterns for more of the same. If those structures are sufficiently complex, as long as energy flows they will convey an ever expanding repertoire of personalities that will then let the process step up to a new level of search.

---

[4] Shortly we will include what we call "supra-biological" evolution as being the emergent layer built upon biological or neo-Darwinian evolution. There are new mechanisms that emerge in biological evolution that work at that level, but also at, say for example, the cultural level as well. Later we will examine even newer distinctive mechanisms that emerged at the supra-biological level.

Question Box 11.2
Duplicate tissues become free to explore new types of functionality. Compare that dynamic with the effects on cultural evolution of spare time created by labor-saving devices.

11.2.4   Evolvability

Individual organisms may change and adapt, but they do not, strictly speaking, evolve. The selection process for the fittest adaptations takes place only *among* individuals. And not every species of organisms or organizations or other potential units of evolution has the same potential for evolution. *Evolvability*, the potential for evolution to take place, depends upon many factors both internal and external to the unit. Adaptive capacities vary widely, as do the situations that call for some sort of adaptive response. Brittle systems, for example, are not good candidates for evolution, nor are conditions of such changeability that they offer scant target for adaptation. A number of such characteristics enter into considering a system's potential for evolution.

The following factors contribute to evolvability:

- Many functionally redundant components that are being continually generated from a base pattern. The components age by the second law so that there is a continuous replacement. Reproduction in biological systems is the paradigm example, but so is employee replacement in organizations, and so is the generation and development of galaxies in cosmogenesis. The redundancy creates room for probing new variations of the base patterns, possibly leading to new functionality and greater complexity.
- The generation process should be of fairly high fidelity, but occasional heritable ("copy") random errors should occur to generate variation in functionality. Genetic mutations fulfill this requirement in biological reproduction. In organizations, new employees filling an existing job are quite variable in terms of personalities, so even when they have the requisite (same) job skills, they can tend to alter how the job is actually done. The huge variation in the kinds of galaxies that have been cataloged and how they fair in terms of their life cycles is another example. Varity enhances the chance for successful adaptation to new selective pressures arising from changing environments.
- The environment of the components (in essence the rest of the embedding system) should be effectively stationary over the time scale of component life cycles, preferably over several generations. But it must be non-stationary on longer time scales. In other words, it needs to change. The geophysical/climatological environments on Earth fulfill this requirement nicely (usually). If the environment is insufficiently stable, the adapting system never has a chance to settle into an integral pattern, while too stable conditions simply select for more and more of what has already been the successful pattern.

- When changes do occur they push the critical factors toward one or the other extreme or can push the range of stressors on the factor. Thus, over the life cycle time of an individual component, they live in stressful conditions relative to what the previously normal environment had been. As an example, when a company adopts some new technology to process its work, some employees may find the change uncomfortable, possibly to the point of quitting their job. Stress on *critical* factors means changes in what does and does not work of a magnitude that eliminates some less adapted units while it multiplies the better adapted.
- Among the variations in the response capabilities of individual components, there will be some that are pre-adapted to the new normal stressor, i.e., the environmental factor while changed is less stressful for some individuals. Those then are able to function better in the new environment relative to those components that are being stressed.
- There needs to be a mechanism that ensures that the prototypes (blueprints) of the more successful components are used differentially in the generation process. In Darwinian evolution this is embedded in the idea of the more successful reproduction of the fittest. Those members of a species better able to compete for limited resources will generate more offspring over time, resulting in the domination of the favorable variation (mutated allele) in the system. In a company, management may note that as the nature of the work changed a certain set of skills possessed by some employees led to better productivity. They will then make sure human resources screens new replacement employees for those skills.
- Over a longer time scale the new capacity becomes the norm.

Question Box 11.3
There are over 6,800 species of ants that have evolved to fit a wide range of environments. There is only one species of humans who likewise adapt to fit a wide range of environments. Compare the evolvability of these two species.

11.2.5 Evolution as a Random Search Through Design Space

There are a number of astronomers and exobiologists engaged in a very serious search for extraterrestrial intelligence (called the Search for Extraterrestrial Intelligence or SETI). They filter through radio signals received from different sectors of the sky (deep space) looking for patterns that cannot be mere noise. With modern radio telescopes and computerized methods of analyzing the data gathered, these researchers are able to sift through terabytes of data, but only by a very clever method.

The method they use is called a *massively parallel search*. It relies on the fact that batches of data can be independently filtered for patterns. What the folks from SETI did was to invite millions of personal computer owners to "loan" some of their computer's cycles to run a program that would do the filtering. They then parceled out the batches of data and sent them out to the volunteers. Millions of computers were working in their own time frames to process the data and then report their findings back to the main computer at SETI.

Each computer, by itself, has to treat data in a serial manner. It is limited by what we call the Von Neumann architecture to just execute a single instruction at a time.[5] But if a problem, like the search for patterns in huge data bases, can be broken up into pieces, then many computers can carry on the analysis at essentially the same time. This is called a massively parallel search.

Evolution has been characterized as a massively parallel search through what Daniel Dennett called "design space." This is the space of all possible biological designs for morphology (body form) and behavior. Evolution is a search through this space for designs that "work." That is, they are fit given a particular environment.[6]

The use of the term "design" requires some comment. We typically think of design as an intentional process of constructing something based on human desires. Using the word design in the context of evolution often raises hackles among biologists who worry about the juxtaposition of the ordinary semantics of the word with what is a totally unguided (in the mental sense) process.[7]

The computers, in the case of life, are the massive numbers of individuals that are produced with each generation.[8] The computation is determining the fitness of each individual. The winners, the ones who find viable patterns, are those who have a greater success in producing the next generation of individuals.

This selective search is the essence of evolution. Darwin and Wallace were the first to give scientific accounts of how this search is conducted in nature. Today we have extended the concepts to provide a universal principle that helps explain all transformations of systems over time. In Chap. 10 we considered how it applies as

---

[5] This should not be confused with the fact that most modern computers have multiple "cores" or independent microprocessors that allow it to do a limited kind of parallel processing. All personal computers are still relatively sequential when it comes to performing computations.

[6] See also Kauffman (2000), pp. 142 ff, where he describes what he calls the "adjacent possible" to describe the nature of the universe of possible configurations that have not yet been tried, but are near current configurations, essentially describing the way in which multiple variations on, for example, the genetics of a species that after selection has acted to favor a variant giving rise to new species. See our discussion of speciation below.

[7] Richard Dawkins (1987) coined the phrase "Blind Watchmaker" to highlight the notion that nature produces systems that could pass as designed by a designer, but, in fact, are the result of blind processes. This is closely related to auto-organization.

[8] Some species, particularly invertebrates, rely on spawning many, many embryos, whereas many vertebrates, adopting parental care for the offspring, produce fewer offspring per generation. Even so, from a population standpoint, there are many different individuals "testing" themselves in the extant environments.

auto-organization in ramping up complexity in physics and chemistry. In this section we will focus on biological evolution, of course, as it is the paradigmatic example. But we will also look at cultural and institutional evolution as other clear examples of the universal principles in action.

11.2.6 Biological and Supra-biological Evolution: The Paradigmatic Case

We will first focus on biological evolution since it is the field which first brought intense scientific interest to the nature of changing assemblies (species) and their capacity to complexify over time by generating new levels of organization.

This section is not meant to provide a deep education in biology or even biological evolution itself. That is simply too huge a topic and there are literally thousands of books and perhaps millions of scientific papers devoted to the topic. The bibliography will contain numerous references to very good introductory and intermediate books that the interested reader may find helpful. Our assumption is that most readers of this work will have a basic (high school biology) understanding of the main points of biological evolution. What we will be doing is to build on that understanding and point back from examples in biology to the principles operating in nature in general, i.e., the universality of evolution as represented in the particulars of biological evolution.

After quickly reviewing the roles of auto-organization and emergence in the biological framework, we will look at the details of the evolutionary process as it relates to the algorithm-like definition given the steps of the process. These are, basically, replication of entities, fitness (against the environment and in competition), relative successes in terms of replication (or extension of kind into the future). We will go deeper into the nature of replication.

11.2.7 How Auto-Organization and Emergence Fit into the Models of Biological and Supra-biological Evolution

As we have seen in the previous chapter, auto-organization and emergence are the mechanisms responsible for setting the stage for levels of organization (recall Fig. 10.1). Auto-organization shows us how assemblies of more complex subsystems are formed. Emergence helps us understand how these new assemblies interact against the background of a particular environment to form wholly new processes. Of especial importance, we have seen, are mutual catalysis, autocatalytic cycles, and hypercycles. Taken together, these processes constitute the internal dynamics of a larger system. The larger system of greatest importance to us was seen to be the origin of life.

Once the original living systems of the type known to us, primitive cells using
DNA or RNA to encode the knowledge of metabolism in genomes, came into being,
the processes of auto-organization and emergence gave rise to what we now recognize as biological evolution. Auto-organization and emergence still play a role in
the ongoing process of universal evolution in which new levels of organization
obtain from lower levels. We have seen how atoms auto-organize into molecules
and molecules into yet more complex assemblies with emergent catalytic and autocatalytic dynamics. New levels of organization emerge as simpler units interact and
form new relational units. The same processes are at work in biological evolution.
For example, the symbiogenesis[9] theory of eukaryotic cell evolution (true cells having nuclei as opposed to bacterial cells with free-floating chromosomes) describes
the symbiotic relations that developed between primitive cell types to form more
complex cells, including, ultimately, the nucleated cells.

Biological evolution, as first described by Darwin and Wallace, dominated the
further development of life until organisms evolved that interacted in more social
ways. In other words, some living systems evolved to have particular interactions
that would lead to the emergence of something new in the Earth system—societies.
Sociality involves organisms of the same species forming colonies, or tribes, or
clans, based on behavior, as opposed to simply being physically connected, as, for
example, is the case with coral colonies. Organisms as diverse as insects (e.g., bees
and ants) to mammals (especially primates like chimpanzees and humans) form
social structures in which the whole group acts in ways that suggest coordinated
behaviors evolved to ensure the survival of a larger number of the group.

At the same time that sociality emerged, and demonstrating the auto-organization
principle at another level of organization, the brains of animals were evolving to
deal with the informational complexities of social organization. Evolutionary selection in some species favored those brains that provided greater communications
capacities between members of the social organization. And that led, in turn, to the
emergence of culture, a web of relational/behavioral agreements based on communication and learning rather than biology. In the human animal, this achieved its
epitome.

Auto-organization in biological evolution involved the same dialectic between
competition and cooperation, competition between conspecific individuals, for eco-
niche resources, and between species for ecological resources in general. Social
coordination emerged as a mechanism to increase the efficacy of inter-specific
competition.[10]

As we saw in the last chapter it is the flow of energy available to do the work of
moving components and binding them together that is a necessary condition for

---

[9] Symbiogenesis was a theory first advanced by Lynn Margulis to explain why certain cellular
organelles, like mitochondria, contain their own working DNA (extra-nuclear genes). See Margulis
(1993). "Origins of species: acquired genomes and individuality". *BioSystems* **31** (2–3): 121–5.
Also see http://en.wikipedia.org/wiki/Symbiogenesis

[10] There is a hierarchy of sociality that emerged over biological evolution. Cf Eusociality: http://
en.wikipedia.org/wiki/Eusociality

auto-organization and emergence. The same is the case for evolution. So long as there are unexploited energy resources available to an evolvable system, then that system will continue to evolve toward progressively higher levels of organization, as is the case for social and cultural emergence. This is called macro-evolution and it is what gives rise, in the biosphere, to new genera rather than just new species (e.g., mammals as a class rather than species of mammals as particular subclasses). On the other hand, if all of the available energy is being exploited already, then macro-evolution cannot proceed. However, micro-evolution, in which new species can arise through simple replacement of older species (of the same genus) due to minor changes in the environment, can still take place. The situation is quite different in the case where available energy is actually declining, as would be the case for the onset of a glacial period—an ice age. Initially the increasing cold temperatures will cause a decline in biodiversity. Since the ice ages have been cyclical, however, there is an eventual resumption of energy flows that then leads to an increased rate of speciation.

In the previous chapter we provided an example of auto-organization and emergence in a cultural institution, the Somewhere Creek Restoration Society. This same kind of story can be told for small businesses, churches, and all other socialized organizations that form from individual people finding mutual interests with others. The cooperative behaviors they exhibit lead to a synergistic accomplishment of work that no one individual could accomplish on their own. These organizations, as they form, replicate a template model of similar organizations. Businesses have specific models that involve production, sales, profits, etc. In this sense, social organizations are the equivalent of replicants in biological systems.

Question Box 11.4
Reproduction is a critical step in biological evolution, since it is the means by which fit adaptations are rolled forward. Organizations do not reproduce in this sense, but they are nonetheless evolvable. What is their analog for biological reproduction? How do different organizational approaches to selectivity impact their evolvability?

11.3 Replication

Replication is an active process of making copies of systems that have proven successful in the past. As noted above, auto-organization generates systems of some complexity, and the most fit within an environment enjoy stability and longevity. Not until mutual catalytic cycles emerge do we find the necessary precondition for biological and supra-biological evolution to be fulfilled, namely, the making of copies. Replication is a necessary condition for biological evolution (we will generally use this phrase instead of including supra-biological too, but we will generally mean to include both levels). Replication is necessary because it involves making high fidelity copies of systems that have proven fit in the past, but with minor variations from time to time. Selection among those variations provides the exploratory opportunity underlying evolution's massively parallel search.

Replication is different from auto-organization in that the latter involves a greater degree of random chance encounter of components with personalities that happen to be complementary, followed by the environment testing their combined stability (see Chap. 10 to review). Replication is a much less stochastic process. It is a guided process in which the components are brought together through specific mechanisms that are "preordained" to form a new copy of the original.

Copying of a complex system could proceed in several different ways. The simplest, conceptually, but also the most difficult to operationalize, is for a copying processor, or also called a "constructor" process, to get information regarding all of the absolutely necessary details of construction, e.g., the component parts, the boundary mechanism, and the minimal connections between part (i.e., the minimal network structure of the system needed to function). With those details, the constructor then picks up components from a pool of free components and does the necessary work of producing a copy of the original system. The new system is in an initial, minimally required starting state and ready to "go out into the world on its own." Figure 11.3 shows this kind of operation. The idea is called the "Xerox" approach to replication.

There are several important points to make about this "simple" (straightforward) approach. First it becomes incredibly complex when the system to be copied is complex. It would take more time to complete the more complex the existing system.

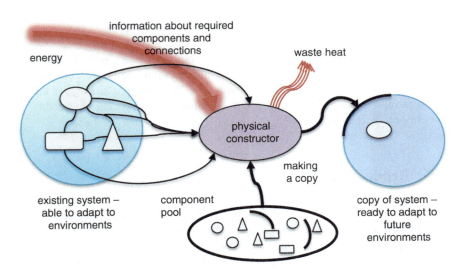

Fig. 11.3 Replication involves copying an existing system's structure (minimally required for function). This is accomplished by an independent "constructor" process assembling a new system using information about the structure of the existing system. Energy flow is required to support the work of getting components and linking them according to the information supplied. We call this the "Xerox" process because the new system is constructed by simply copying the existing system

Second it requires a dedicated constructor. This latter is not a serious problem since we have already seen that catalysis (including auto- and mutual forms) has this logical structure. With autocatalysis, it is not hard to imagine a constructor that is capable of constructing itself. We will expand this idea below.

The biggest problem with this scheme is the first one. It takes too much time and too many resources to have a constructor actively obtaining messages from the system regarding each and every necessary component and connection every time a copy is to be made. This problem can be avoided by the constructor not working from actual systems but by using a stored representation of the basic system that allows for fast and efficient access to the information needed to construct the new copy. The stored representation would thus be some form of knowledge representation of the system to be reproduced.

11.3.1 Knowledge Representations of Systems

We have said that the structure of an adaptive system at a given time constitutes a kind of general knowledge held by that system about its environment. That is, if the system receives messages from its environment that are informative, then internal work is done to modify the very structure of the system to reflect this. Thus, the knowledge of the system regarding its environment changes in response to information such that in the future the same message will not convey information, or at least convey less since the system is "prepared."

What we have not covered is the concept of knowledge of the construction of the system itself. This is a necessary consideration for several of the key elements of the evolution algorithm.

This separate structure contains a "representation" of the system to be constructed in the form of instructions for combining components. An interpretive process, a decoder, reads out the knowledge and transduces it to a message that is sent to a constructor process (see Fig. 11.4 below). The constructor uses the information in these messages to activate the mechanical systems required to construct the system just as before. The difference now is that the system constructor works from a model or interpreted representation rather than the original system. This is much more efficient. We call this the "blueprint" version of replication. The new system is constructed from a "plan" rather than directly from the original system. The approach is much more efficient since the original system needs to be "read out" only once to produce the blueprint but many copies can be made from that one read out.

For this approach to work it is necessary to have another kind of constructor that builds the knowledge structure from a template (the original system). It has to have the capacity to encode the information available from the original system into messages that are recorded in the "medium" of the "model." The latter is just another system and so its internal structure represents the information it received from the model constructor, just as the existing system's instantaneous structure represents

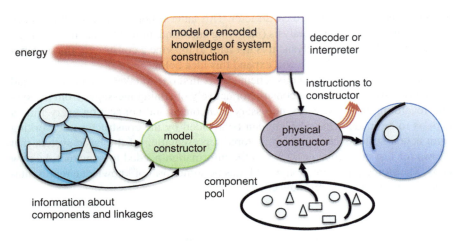

Fig. 11.4 A more efficient method for making copies of systems would be to work from a model or knowledge representation of the structure of the original system. The construction process for the new copy is the same as before, but the source of information is now the model (through its decoder/interpreter). A new kind of constructor is needed to encode the information from the original system, acting as a template, into the structure of the model. We call the model an encoded knowledge structure

its environment! What we see here is nature's way of reusing simple principles to do different jobs.

There are additional requirements regarding the nature of the model as a medium. One reason we call it a medium rather than refer to it as another kind of process (though from Chap. 2 you know that that is a completely legitimate term to use!) is that it is relatively passive; it is a structure that must be organized (written to by the constructor) and translated into suitable messages (decoded or interpreted) for usage.

As we will see below, the nature of the medium for recording and reading out is extremely important to make the whole scheme work. It must be highly stable over long time scales relative to the average lifetime of the active processes shown above. As we will show below, the model constructor, the decoder, and even the basic constructor can all be reconstructed as needed. But each of these requires its own knowledge encoding structure, and if the latter were as unstable as the work processes, then there would be a danger that vital knowledge (and hence construction information) would be lost. In point of fact, this does happen in real systems, as we shall soon see. It is both a curse and a boon! It keeps the evolving systems from being perfect, and it produces some opportunities for finding new kinds of perfection (i.e., fitness).

Examples of media that are highly stable and easily written to and read from include nucleic acids (DNA and RNA), paper, and computer disks/tapes. We'll provide more details below.

11.3.2 Autonomous Replication

Let's take this a step further. Remember autopoiesis from Chap. 10? We can now be a little more explicit about what this "self-creation" is all about. Figure 11.5 extends the basic construction process from above but now includes other necessary features.

In this version the constructor is not only responsible for creating a copy of a system from the knowledge code (as above) but is also capable of constructing other necessary subsystems that are processes, do work, and have specific functions. In the figure we show an extractor and a destructor as additional processes. The extractor is needed to get resource materials into the local pool of components. It is built and maintained by the "general" constructor from knowledge for extractor systems in the knowledge code system. Similarly, for the destructor the knowledge code includes instructions for copying (or repairing) that processor. The latter system is responsible for destroying nonfunctional systems and recycling the usable components back to the pool and removing (expelling) any unusable. Decayed and un-recyclable materials have to be disposed of, and this is why an extractor is needed, to replenish the component pool by capturing usable (fresh) components from the environment.

In addition to constructing the extractor and destructor, the constructor is shown able to construct the knowledge representations (knowledge codes), decoders, and even itself. That is some kind of constructor! Living cells are of this sort. They contain everything needed to replicate themselves when the conditions are right. The "exported" system shown in the figure is really a complete copy of the whole system that did the construction, so it also includes the ability to replicate itself. This is what is needed for self-reproduction.

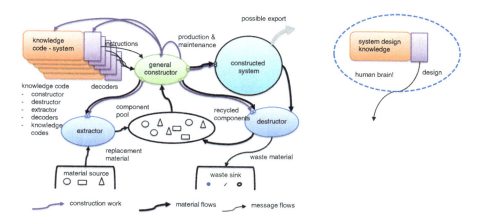

Fig. 11.5 An autonomous construction system that can build the desired system (which might be exported) as well as build component machines needed to keep the whole system working. This is the basic concept of an autopoietic system as first encountered in Chap. 10. Note that the "constructed system" for "possible export" could in fact be the same as the system doing the construction. This would be representative of a self-reproducing system

The idea of a universal constructor is obviously tricky. We don't actually have too many machines in the human-built world that fit this concept. But we have a basic set of machine tools that can be used by human operators to build every other kind of tool, including themselves! In the above figure we have compressed a more complex set of component machines into a single "general" constructor to simplify the diagram. In more realistic systems there are cooperating constructors practicing an expanded version of mutual catalysis. One constructor might "specialize" in building destructors or some other subsystem. Another might specialize in building the "destructor-constructor"! And another might build the destructor-constructor-constructor. Somewhere in the mix one constructor has to build another constructor that builds the first constructor or operates through a chain of these relations. We will mention a few examples below.

Assuming that there exists an initial set of knowledge structures and minimally necessary constructors, one can create an autonomous replicator, indeed a self-replicator.[11] The original problem for living systems had been how to get to this stage. How do you bootstrap an incredibly complex process of self-replication? There are a number of working hypotheses for how life began on Earth. They are all centered on the concepts covered in the prior chapter.[12]

11.3.2.1 The Knowledge Medium in Biological and Supra-biological Systems

Key characteristics of knowledge encoding involve compactness (e.g., code efficiencies), accessibility for readout, structural stability over time, and construction and maintenance efficiencies. Ideal knowledge encoding takes little space and energy resources from the whole system, yet provides sufficient fidelity that a new copy of the system, including copies of the knowledge structure itself, will allow a reasonable facsimile to be generated.

In biological systems this function is provided, for the most part, by nucleic acids (DNA and RNA), but also to some degree by structural proteins that act as message translators. The knowledge conveyed from one generation of organism to the next is encoded within the genetic material comprised of DNA and packaged in various proteins that help store and manage (including repairing occasional damage) the DNA.

---

[11] The polymath John Von Neumann (famous for many systems science concepts) showed how to build such a self-replicating machine. See http://en.wikipedia.org/wiki/Self-replicating_machine. Several others have done so as well. See also von Neumann (1966).

[12] We wish to recall your attention to the work of Morowitz (1992) covered in that chapter, giving a good account of how proto-cells may have likely emerged and then evolved into primitive bacteria and archaea.

In the living cell DNA is the principle medium (structure) that has all of these characteristics of an ideal medium.[13] It is typically wrapped in a sheath of proteins and other molecules that protect it from the cellular milieu while regulating its interactions with specific decoding (readout) enzymes. The readout process involves constructing a complimentary molecule of RNA (called, appropriately, messenger RNA, or mRNA) that is an effective copy of the DNA code. That code specifies a sequence of amino acids that constitute a particular protein or polypeptide molecule to be made in the cell's cytoplasm. The mRNA molecules travel (communications) to a special organelle called a ribosome[14] which is a general protein constructor. There the mRNA code is used to put together the amino acids in the proper sequence as at least a starting point for the production of functional proteins. Some of the proteins that are assembled are the very enzymes that operate to regulate the readout process mentioned above! In other words, the DNA contains instructions for building the proteins needed to maintain and replicate the DNA itself. The DNA in the genome stores what we might call self-knowledge, or the instructions needed to fully replicate the cell and its genome as well.

From the origin of life an extraordinary mutual catalytic cycle emerged in which proteins and associated RNA molecules auto-organized and through the process of chemical selection formed stable structures. Later a three-way mutual catalytic cycle, incorporating the more physiologically stable DNA molecules and probably coupled with the original adenosine triphosphate production cycle (the original energy capture and conversion process), were able to enjoy long stability in the aqueous solutions of the primitive world. Along with the protection of bi-lipid membranes (auto-organized), what emerged were the first primitive cells. They used RNA or DNA as the basic knowledge template, or blueprint for producing all of the components, especially the protein enzymes, that would be needed to produce an exact duplicate. The first genes were organized. DNA has the critical ability to be split down its spiral form and to attract free-floating deoxyribose nucleotide molecules to link up according to a consistent protocol: adenosine links with thymine (A-T), and cytosine links with guanine (C-G).This preserves the encoding of the amino acid chains for the proteins, providing the most incredible mechanism for copying knowledge and preserving patterns. The lengths of DNA, encased in protective proteins, and under the right conditions splitting and replicating, became the ancient genes, very many of which have been preserved in essentially their original forms and are present in living systems today (Carroll 2006, ch. 3).

Genes are the basis for biological evolution, for the most part. What we mean by this is that genes are the biochemical mechanisms for encoding the accumulated knowledge of protein chemistry that is most fit given the environments of the organisms

---

[13] Stable means the knowledge encoding does not degrade easily, say due to entropic decay. As an example of this consider DNA strands. Scientists have found still viable DNA from extinct animals, including *Homo neanderthalis*! We say that DNA is the "principle" encoding medium because in some Archaea RNA plays this role.

[14] Indeed the ribosome is the best model of a general constructor we can imagine, at least for protein construction. See http://en.wikipedia.org/wiki/Ribosome

in which those genes operate, and they inform the reproductive process by which that fitness shapes the future.

The concept of a gene emerged slowly in our understanding of evolution. Darwin understood the need for a pattern of biological inheritance, which he acknowledged as needed to make his theory of natural selection work. In the neo-Darwinian synthesis, which incorporates the gene theory first advanced by Gregor Mendel, genes are the encoded knowledge in DNA molecules.[15] This theory gained prominence from the work of James Watson and Francis Crick who first illuminated the structure of DNA molecules and the way in which their structure could carry genetic knowledge.[16]

A gene is a segment of DNA within a chromosome that basically encodes a polypeptide or protein molecule (a long polymer of amino acids) that is assembled by a subsystem in the cellular protoplasm called a ribosome. The gene is transcribed in the nucleus, in eukaryotic cells, by a process involving copying of RNA molecules (messenger RNA or mRNA) that are able to travel from the nucleus to the cytoplasm and link up with the ribosomes as templates for the construction of proteins (or polypeptides). Essentially, the DNA-RNA-protein transcription process tells the cell what proteins are needed to fulfill the various biological functions to be performed by the cell.

In multicellular organisms this process is complicated by the fact that different cell lines perform different functions and so different proteins are transcribed at different times during development.[17]

Genes encode the knowledge for maintaining and replicating biological organisms. But how is such knowledge encoded into structures in supra-biological systems? Consider human organizations, such as a relatively large corporation. In large-scale enterprises all processes are typically encoded in procedure manuals and operating instructions. Manufacturing companies will have libraries of product plans (like blueprints), parts inventories, etc. as records of the state of the system. Essentially all of the records of such an organization, including its accounting records, collectively, encode the current state of that organization. Periodically, those records are used to report on the average state over some time period, such as an operating quarter or a year—as in the quarterly financial statements, for example.

That knowledge is built up over time through the accumulated experiences of the organizational personnel who "learn" from their experiences and develop solutions to problems that they then record for the betterment of the organization. In days past the medium of recording was basically paper. Today, of course, just about everything

---

[15] See Wikipedia: Gene, http://en.wikipedia.org/wiki/Gene

[16] We continue to differentiate between information and knowledge as developed in Chap. 7. Genes encode knowledge rather than information. That genetic knowledge, however, can be treated as information when looking at the way it is interpreted by cellular mechanisms that have to interpret the messages conveyed from the nucleus via transfer RNA (tRNA) molecules.

[17] Would that there were enough pages in this book to explain this fantastic phenomenon. Anyone truly interested in the nature of biological knowledge encoding is directed to the nature of development. Cf biological development: http://en.wikipedia.org/wiki/Developmental_biology for a glimpse of how this works.

is recorded digitally, and paper is only used as a communications medium between computers and humans. Gone are the large heavy filing cabinets of yesteryear. These have been replaced by computer files and heavy bulk disk/tape reading/writing machines! One day even these will likely disappear to be replaced by solid-state file memory not unlike the device known as the memory stick, which behaves similarly to a disk drive but with no moving parts.

Unfortunately for too many organizations the found solutions may not be committed to paper or computer records and are "stored" only in a human memory. It is safe to say that, not counting malicious behavior, much of the dysfunction experienced in organizations comes as a result of the weaknesses or loss of human memories. On the other hand, some organizations have worked reasonably well as long as there is someone who does have the organizational knowledge in memory and has good recall! Small organizations can run for as long as the "boss" is alive.

Question Box 11.5
Fads replicate quickly and spread through a culture or sub-culture, but do not last long. Other changes replicate, spread, and last. What factor or factors account for the difference?

11.3.2.2 Copying Knowledge Structures: The Biological Example

Replication of a system depends on making a copy of the knowledge structure to be used in constructing the new system. In externally stored systems, as in Fig. 11.3, this could be as simple as making a new blueprint from the master plan (the Xerox solution) or sending a copy of the machine control program to the milling machine computer on the shop floor. But biological systems store their knowledge structures internally, in the DNA (sometimes RNA) itself.

The mechanism for replicating a double helix strand of DNA is the epitome of elegance. Figure 11.6 shows a cartoon version of this mechanism. Put as simply as possible, life evolved a set of enzymes (protein molecules that act as catalysts) that split the double strand into two single strands. This is triggered just as a cell is about to replicate itself. First it needs to replicate its genetic complement. Almost as soon as the split occurs another set of enzymes goes to work. Recall that enzymes are catalysts, molecules instrumental in promoting combinations of other molecules. In this case, the enzymes are capable of grabbing free-floating nucleotides from the aqueous medium surrounding the chromosome and joining them up with a newly forming double strand according to the protocol discussed above (A-T, G-C).[18]

---

[18] A similar mechanism is responsible for the transcription of the genetic code into a strand of mRNA. The DNA double strand is opened when a transcription is needed. The enzymes responsible for transcription extract ribonucleotides from the nuclear medium and attach them to the "sense" half of the DNA strand. Transcription is controlled by an elaborate set of control mechanisms coded in the DNA itself. These are short segments of DNA that signal such things as start and stop positions for the genes.

original double strand of dna

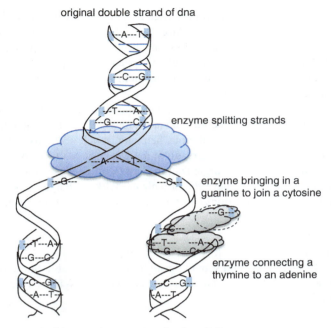

enzyme splitting strands

enzyme bringing in a
guanine to join a cytosine

enzyme connecting a
thymine to an adenine

two new double strands – exact copies (mostly!)

Fig. 11.6 A DNA molecule in a chromosome is being split (during reproduction) by one enzyme. At the same time another enzyme is grabbing complementary nucleotides (A) from the general pool of nucleotides in circulation and attaching it to the free nucleotide (T) while attaching the phosphate backbone molecule (*small blue parallelogram*) to the forming new backbone. A second enzyme is bringing in a G to link up across from a C in the one half of the original strand. The exact same process is working on the other forming double strand. When these processes are complete, there will be two DNA molecules of exactly the same sequences (almost) where before there had been one

Because of this complementary coding scheme the two resulting strands are exact copies. There are now two double helices where before there had been only one. The whole process is under the control of an elaborate set of control signals that orchestrate the entire cell division process.

Now here is the really wonderful part of this scheme: the DNA contains the code for the enzymes that do this work, so it includes the instructions for implementing its own reproduction. This is a truly autonomous replication system. All that it needs is to be present in a pool of materials, deoxyribonucleotides, and the right conditions.

DNA replication is a fairly high fidelity copying operation. As we said, high fidelity in copying is essential to long-term stability and maintaining functions. But occasionally a copy error can occur. When that happens, there are other enzymes that can detect and correct errors. So the vast majority of the time the copying goes correctly. But as with all stochastic processes, every so often a mistake is missed and something new is introduced into the gene where it occurred. If this type of error occurs in a gametocyte (a cell that produces sperm in males or ova in females), the resulting division during meiosis results in one mutation. The mistake will potentially end up in a fertilized egg. Later we will discuss mutations, their sources and effects, and most importantly how they contribute to the evolution of systems.

11.3.2.3   Copying Knowledge Structures: The Supra-biological Example

We've been using organizations such as businesses as examples of supra-biological systems. Continuing that theme, we can examine knowledge structure replication in such organizations. Here we see some major differences between the biological example, DNA, and human organizations as examples of systems. In the biological example copying is of high fidelity and is somewhat like the "blueprint" version in Fig. 11.4 above. In organizations we see a much wider variety of media for knowledge storage and differences in the application or function of the knowledge as well.

There are so many examples of "bylaws," "contracts," "accounting methods," and other template documents that can be used to establish a new organization by tweaking the details to fit the new situation. This is similar to the blueprint model of DNA function, and in this way organizations share a large body of stored knowledge that can be decoded and used to establish new instantiations fairly fluidly.

Human memory is perhaps the most basic knowledge storage mechanism, and it plays a large role in replicating and modifying organizations. For example, consider the employees who decide they are going to leave a company to form their own. They take with them memories of the policies and procedures that were established in their employer's company. When they establish their own business, they can try to replicate those that are applicable to the new organization. But people who rely on their memories frequently make slight mistakes in replicating knowledge. Some of those mistakes may prove serendipitous, leading to some kind of improvement in a process, say. Most often they will probably be neutral because humans have an extra ability to recognize a mistake and compensate for it. Either way, novelty can come into the replicant leading to some new capability.

But human organization also has another path to variation. Unlike the DNA blueprint model, humans actively imagine (model) a future and strategically modify the knowledge they apply accordingly. So in addition to the inevitable modifications introduced by faulty memory or misinterpretations of various materials, human organizations are also open to a dimension of deliberate newness or creativity. Every business that produces the same kinds of products or services has slightly different ways of doing the same task internally and tweaking the product to bear their special stamp. These "proprietary" processes help differentiate organizations and contribute to their fitness in the marketplace.

Question Box 11.6
Cultures exist as systems of shared memory about all sorts of social relations
and conventions. But memory is a slippery medium, circumstances change,
and in addition humans tend to be innovative. "Write it down!" is the common
cure, which produces contracts, laws, constitutions, scriptures, and many
other documents. Unlike living memory, however, writing is a much less
adaptable medium. What are the advantages and disadvantages of this? How
do different attitudes to various sorts of written documents relate to cultural
change or evolution?

11.4 Descent with Modification

The replication that we have been discussing is necessary for the generation of a
population of similar entities. It must take place regularly to assure that a population
is either growing or is at a stable size relative to the exploitable energy sources.
A population is characterized not only by the number of individual entities, but,
more importantly, by the variation that may exist due to those occasional copy errors
discussed above. Darwin did not have direct knowledge of a mechanism for this, but
he realized that it was a necessary aspect in order for there to exist various degrees
of characteristics upon which natural selection could work (see Selection below).
He called this descent with modification. In our modern terms this means replication of generations with *mutations* that produce variations in the characteristics.

Figure 11.7 shows the process for asexual reproduction in bacteria or mitosis in
the cells of multicellular organisms. One cell divides to give rise to two daughter
cells, which in turn grow and divide. This sort of reproduction leads to exponential
growth of a population unless there are environmental factors that will trim the
excess, so to speak, or shut down the mechanisms for copying. The former, such as
predators, keeps a population in check. The latter is seen in various internal signaling channels that turn off the growth impetus. We will look at that in the next section
on Selection.

Darwin realized that species must produce many more numbers in their population than could possibly survive given the amount of available energy (as well as the
dangers of predation). Living systems reproduce often enough to generate those
large numbers. Darwin's real genius however came in recognizing that all living
species generate variants in each generation. Each variant is then subject to the
selection pressures of the environment and will be tested for fitness ranking (recall
fitness from the last chapter, Figs. 10.2 and 10.3). Variations that give rise to more
fit individuals will tend to out-compete those less fit, and so over more generations
will tend to supplant or dominate the population. In the figure above we show two
variants arising. The green variant is a small fraction of the nth generation size. The
light blue variant is just getting started at the $n-1$ generation. What happens next

Fig. 11.7 Asexual reproduction (or mitosis) leads to an exponential increase in population as a function of time. Each generation the population doubles. Shown here is a representation of descent or giving rise to subsequent generations. The lower cells are descended from the higher level. Also shown is the effect of the occasional mutation that gives rise to variation in some trait or characteristic (both structural and behavioral). In this depiction the green variants are about as fit as the *dark blue* "native" types. The *lighter blue* variant has just arisen by mutation and its fate is yet to be determined

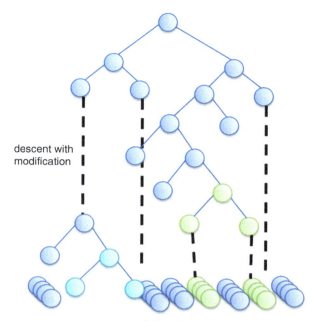

descent with modification

population after *n* generations

depends on the environment and how well these variants fit into it. If the green variant is as fit as the dark blue, then it has no advantage, and its future distribution in the population is subject to purely random events. If it has even a slight advantage over the dark blue, then over some number of generations it should come to dominate the population distribution. Similarly, if the light blue variant turns out to be less fit, then it will tend to be eliminated from representation in the population.

However, should the environment change in any way that gives either of the new variants an advantage, then they will be more successful in reproducing. After some number of generations they will come to dominate in the distribution. Thus, both genetics and environment interplay to determine what traits will be found in the population.

Question Box 11.7
Microbes and insects evolve around antibiotics and pesticides, and the more they are used the faster resistant strains appear. Mustard gas was introduced to kill humans in WWI and is still quite effective. What factors enable microbes and insects to evolve around critical challenges that would simply wipe out many large-scale organisms?

11.4.1   Mutations: One Source of Variation

In all systems that make copies of knowledge structures as a prelude to replication there is a nonzero chance that an error will occur somewhere in the copying process. In biological systems these inevitable errors are called mutations. A mutation occurs when, during replication, a nucleotide substitution takes place in one of the codon triplets, leading, in the case of a protein coding gene, to a possibly different amino acid insertion in the final protein sequence. Most of the time a substitution of this nature will be neutral. Occasionally it will be detrimental if it occurs in a section of the protein that is essential to its function. In the latter case the malfunctioning protein (say an enzyme) can cause a premature death in the offspring that inherit it. In biological systems there are numerous backup systems or redundancies that can sometimes compensate for such malfunctions. But sometimes not, and the offspring death assures that that mutation will not be propagated into the next generation. This phenomenon is true for both mitosis (ordinary cell division) and meiosis (sexual reproduction). In the case of cellular mitosis (within tissues in the body) the mutation is most often of no consequence. However, occasionally mutations in a number of specific genes can lead to cancers. Mutations in meiosis, however, provide the raw material for evolution by selection to work on.[19]

In human-built and supra-biological systems, knowledge structures must be similarly copied in order to hand the knowledge down to the next generation, so to speak. One of the most used pieces of equipment in any organization is the copy machine (today supplanted functionally by simply printing out new editions of computer files). Knowledge is reproduced just about every time a worker takes on a task. For example, plans are reproduced and sent to the shop floor for execution of the construction of a product. Copying by machines is not generally the source of errors, however. More often the copying that results in errors is the brain-to-brain kind that happens through the facility of language. And natural language is notoriously ambiguous at times, leading the hearer to grasp something that the speaker did not say! When one worker, say, a supervisor, tells another worker to do job A with a set of verbal instructions, there is a nonzero chance that the worker will get something wrong. And the more complex the instructions, the more likely something will get lost in translation. As with biological systems, whatever the source of the error, most of the time the results will be more or less neutral (non-harmful). Occasionally the result could be harmful, but just as with deleterious mutations in genes, there are redundant subsystems to pick up the slack or an error-correcting method to fix things before the error causes serious trouble. Occasionally though the worker will do something different and discover that it improves the job performance. With any luck the worker will record the changed procedure and notify the supervisor of the good news. Too often, however, such things go unnoticed. The picking up and

---

[19] We have been describing what is known as a "point mutation" in the gene code. There are actually several additional mechanisms whereby changes to the DNA can occur and they operate on varying scales. These mechanisms are beyond the scope of the book. Their ultimate effect is still similar to what we describe here.

transmission of a favorable mutation is a critical element of the process of evolution, which is why it is always a good idea to include useful changes in written procedures in organizational processes.

Human organizations, as supra-biological, have an additional source of knowledge structure alteration that biological systems don't have. Humans can consciously alter a procedure or design if they can see how an improvement would result. As mentioned above, they anticipate the future and move strategically to shape it to their desires. What causes a human to consider such changes is beyond the scope of this book. It involves subtle psychological factors that, while fascinating, are better read from a book of psychology. We will discuss some aspects of this in the subject on the methodologies of analysis and design in Chap. 12. For now we will have to leave it at the fact that small changes in knowledge structures, whether random accidents or intentional designs, can lead to greater or lesser fitness for the organism or organization. And that fitness is tested by the selection imposed from the environment of individual biological or supra-biological systems.

Question Box 11.8
Too little mutation can lead to brittle uniformity, while too much endangers reproductive stability. Biological evolution finds the sweet spot by selecting out what does not work as well. What mechanisms might function to control the rate of change in human organizations and cultures? Humans have a wide variety of attitudes toward change, and different attitudes have a way of pervading various institutional and organizational cultures. Is there an ideal "sweet spot" attitude, or does the variety itself serve a purpose?

11.4.2 Mixing

In populations of essentially similar systems, e.g., species in biology or retail merchant stores in a town, there is generally a fair amount of variability in the individual knowledge structures, i.e., gene alleles in biological systems and retail employees' experiences (which they gain by working in different retail stores) and personalities. We know that knowledge structures cluster in correlated units and interact with one another operationally. Genes are clustered in chromosomes and the genes on the same chromosome often interoperate for various purposes (see Epigenetics below). Retail employees are obviously working in the same environment and must interact with one another routinely.

A valuable source of variability, or novelty, is when knowledge structure units are intermixed and wind up in different members of a population. Gene alleles that end up in offspring and that had not been previously paired might work together differently than the previous arrangements in their parents. During meiosis[20] in sexual reproduction, there is a point at which the chromosome pairs can interact through what is called crossover.[21] Essentially pairs of chromosomes exchange segments of DNA at one or more crossover sites. This doesn't necessarily always happen, but when it does genes that are on one segment in one chromosome are separated from partner genes (of particular alleles) on the other segment and are brought into association with the same gene, but a potentially different variation, on the other segment of the sister chromosome. These exchanges create a mixing of gene alleles so that there is an opportunity for alleles that may work better together to get the opportunity from time to time.

Employee turnover in the retail store is not terribly different in the sense that it mixes up the operative elements and allows different experience levels and personalities opportunities to work together. This is generally useful for these stores since there are opportunities for new hires to bring in talents and knowledge that can help other employees do a better job. Of course it could also produce personality clashes that might disrupt the business. But just like mixing in the biological knowledge structures, mixing in the organizational knowledge structures (and not just for retail) also can lead to novel variations that can then be subjected to selection for the whole system.

11.4.3   Epigenetics

It often comes as a surprise to many people who learned their evolutionary biology in high school more than, say, a decade ago to learn that the genes themselves are not the only heritable knowledge structure units. They are also surprised to learn that the effect of gene transcription and protein construction is not as was previously thought, a one-way flow of genetic information: DNA → mRNA → proteins (called the Central Dogma of molecular biology[22]). Biologists, over the last several decades (and just now starting to enter the textbooks), have discovered an incredible and elaborate network of control mechanisms that is comprised of numerous kinds of molecules, including those that convey messages from sources external to the cell itself. These mechanisms are involved in all aspects of embryogenesis and development, switching specific genes on or off based on the location of the cell and the stage of development. They are also involved in what we now understand as genome adaptivity, the ability to turn on or off specific genes in mature cells based on environmental conditions. We've already seen one form of this phenomenon in the long-term engram encoding in neural synapses. The mechanism appears to be that in

---

[20] This is the form of cell division in which the sperm and egg cells are produced. Each gamete cell has just one of the two pairs of chromosomes in ordinary diploid species, or half of the genetic complement. See http://en.wikipedia.org/wiki/Meiosis

[21] See http://en.wikipedia.org/wiki/Chromosomal_crossover for a description of this phenomenon.

[22] See http://en.wikipedia.org/wiki/Central_dogma_of_molecular_biology

long-term potentiated synapses messages sent to the cell nucleus cause specific genes to be activated and produce protein factors that will wind up reinforcing the potentiation of the synapses.

The mechanisms involved are far beyond the scope of this book, but it is important to recognize them and how they fit into the general schemes of systems science and evolution. Basically the turning on or off of protein-encoding genes can be handled by a process in which a special kind of molecule from outside the chromosome attaches to a specific DNA group in a gene, effectively making the gene closed to transcription, i.e., turning it off. The gene that got turned off may, in turn, be responsible for repressing another gene, in which case the latter gets turned on! It is becoming clear that these "epigenetic," meaning "on top of genes," mechanisms are involved in every aspect of gene regulation and expression of specific genes under variable conditions. Readers who might be interested in this important field should consult Carroll (2005) and Jablonka and Lamb (2006).

From the perspective of systems science the regulation networks (Chap. 4) of genes fits nicely into the information/knowledge issues covered in Chap. 7 and the cybernetic models we looked at in Chap. 9. The reader is left to contemplate the complexity issues!

11.5   Selection

The structure and dynamics of the environment of a system act to either inhibit the success of a system or enhance its success. This is the key to universal evolution and is the very same process we saw in our discussion of how auto-organization and emergence lead to further organization (Fig. 10.1). Selection is the environment gradually, perhaps subtly, passing judgment[23] on the fitness of a system. It results in the system either promulgating more of its kind into the future or struggling for any survival and losing out to competing systems.

At this point we need to take a small excursion to explain something that is very important for understanding the full scope of evolution. Fitness is *NOT* a simple function. Every gene in a biological system stands for a factor (like a protein or a regulator) that contributes to what is called the "phenotype" or form and behavior of the individual. The actual phenotype results from the interactions of thousands of genes being turned on and off at different times. This could be in response to environmental factors, such as temperature, or the chemical milieu surrounding the cell during development, so phenotypes are further differentiated by such factors. We normally think of the phenotype as being based on the specific collection of genes that an individual possess ("nature") in combination with those factors in the environment ("nurture"). We can represent this relation as in Fig. 11.8.

---

[23] Of course this is meant metaphorically, not literally. Selection is blind, no judgment per se is involved.

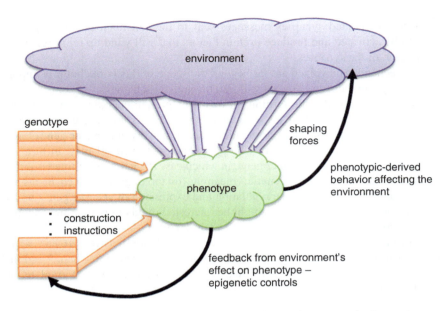

Fig. 11.8 The phenotypic form of an organism is dependent on the construction instructions provided by the genotypic set (active genes in the chromosomes) and the environmental forces that help shape the form during development. Throughout life the effects of the environment will continue to affect the phenotype through various epigenetic feedback signals. It is even more complicated because the phenotypic-derived behaviors of the system will affect the environment and may lead to additional shaping forces and additional epigenetic feedback signals, which could, in turn, affect behavior. And it goes on in an endless loop

As seen in the above figure the genotype-derived construction of the phenotype leads to behavior in high a dimensional space. In turn that behavior operates on the environment to change it and thus leads to changes in the shaping forces acting on the phenotype through both positive and negative feedback loops. The environment is selecting in every dimension simultaneously. Some forces will be stronger at times than others.[24] Fitness is the optimal match between environmental selection forces and phenotypic form in this high dimensional space.

Until relatively recently this very important message channel was unknown. The old "Central Dogma" of genetic biology held that the genes built phenotypes and that was a one-way instruction channel. But now we know that genes can be affected, turned on and off, or interfered with by what are called "epigenetic" factors that are

---

[24] In fact, there are situations where selection forces are relaxed to a point that the underlying gene for the phenotypic response is no longer useful. In those cases mutations can accumulate in the gene and render it inactive. Carroll (2006) calls these genes "fossils." A good example is the inactivation of genes associated with eye formation in embryos and function in adults of certain cave-dwelling fishes and amphibians. Sight is no longer required in these species so the selection forces (light availability) have relaxed and those genes have "fossilized."

triggered by factors outside the nucleus including environmental factors such as psychological or chemical stresses. The altered functioning of the genome can then produce different construction instructions to the phenotype leading to altered behavior. This can be either a vicious or virtuous cycle depending on mostly environmental factors. Brought up in a favorable environment most juveniles of a species will develop normally and behave according to the species norms. But juveniles raised in suboptimal or stressful environments can have permanent rewiring of their genetic control circuits through epigenetic mechanisms, some of which can actually be passed on to offspring![25] This phenomenon might, at first, look like Lamarckian evolution, but the epigenetic tag does not change the underlying gene; it simply marks it for different activation.

Question Box 11.9

Eugenics was the notion that you could engineer a genetic recipe that would produce superior human beings. Could you modify this notion to take into account epigenetic factors, or does the whole notion just break down?

Generally, selection is the process of retarding the reproduction of less fit individuals (or organizations) and allowing more fit ones to be more successful in their reproduction. This is the essence of Darwin's insight. Reproduction produces more individuals than can survive given the nature of the environment. The offspring are then faced with living in a difficult world where they may have to compete with their brothers and sisters, cousins, and so on just to keep living and to reproduce themselves. Many generations are produced in time scales much smaller than those of the changes in the environment, so species may remain relatively stable over many, many generations. Mutations, which are infrequent within a single individual, become far more likely in extensive populations, given the turnover in generations. Sheer numbers ensure that mutations will arise that are then subject to selection.

It turns out that while competition is the main operational mechanism to generate "improvement," in the sense of greater fitness, there are other mechanisms, related to auto-organization and emergence, which are at work in biological and supra-biological evolution. Two of the most important are cooperation and coordination. Both have been important in the evolution of stable unitary biological and supra-biological systems such as multicellular organisms, tribes, enterprises, and even nations. We will examine these three mechanisms and provide some examples of their working in both biological and supra-biological systems.

---

[25] See http://en.wikipedia.org/wiki/Epigenetics. Also, for a thorough treatment of epigenetics and the role of regulatory genes, see Jablonka and Lamb (2006) and Carroll (2006).

11.5.1   Competition

We first ran into explicit competition in our discussion of auto-organization in the last chapter. Components of different types could compete with one another for the ability to link up with other components. Given a stable environment and energy flow, the components can jostle around trying to find the most stable set of configurations. In this case it was a blind sorting process, passive in the sense that the components were not actively vying for resources. When we now look at the larger evolution process, we see competition again, but with what looks like a more active mechanism. Organisms actively compete.

Competition occurs when two or more individuated systems require the same resource and must adopt tactical behaviors to acquire what they each need. If the resources are abundant, and there are not too many entities taking the resources, then competition is not a strong selection factor. Contrary-wise, if the resources are sparse or there are too many entities competing for what is there, then competition becomes a very strong selection factor. This is the general case in all biological and economic systems, but can also be found in emergent systems such as the pre-biological Earth.

Between Dissimilar Systems: A good example of competition between dissimilar systems is the predator–prey or grazer-vegetation relationship. The resource here is biomass, specifically of the prey or vegetation. The predator needs the resources of the prey body as food. The prey needs its body in order to successfully procreate. Plants need their leaves to produce biomass. There is a balance point at which the number of eaters and the number of those to be eaten is sustainable over time. But this point is rarely maintained exactly. Evolution itself can move the balance point. For example, if the genetics controlling running capabilities in the prey animal undergo a mutation that makes the possessor run faster, then that capability will spread through the population over some number of generations. The reason is that the possessors will be eaten less often and reproduce more often than their slower conspecifics (see next paragraph). But as the trait gains traction, the predators will catch fewer prey and thus face resource decline. The solution for the predator is to either increase its running speed through evolution or, more likely, to increase its cunningness. This actually seems to be what happened in vertebrate evolution. The predators tended to get smarter, while the prey got better at hiding or running![26]

Between Similar Systems: Biological systems of the same or related species often compete for the same resources. Here too, any improvement in a system's capacity to out-compete its rivals will be selected for under the conditions of limited resources. Though it is a little fuzzier in the world of commerce, we still see the effects of competition between firms that are attempting to sell a similar product or

---

[26] This phenomenon of evolution producing an oscillation in holding, temporarily, the upper hand in fitness is known as the Red Queen hypothesis. The name is based on Lewis Carol's "Through the Looking Glass" where the Red Queen complains, "It takes all the running you can do, to keep in the same place." See http://en.wikipedia.org/wiki/Red_Queen%27s_Hypothesis

service in the market. Winners in the competition are often found to have developed greater efficiencies or put more features into their products/services for competitive prices. They become more fit in the marketplace by selling more and making greater profits. Companies that do so tend to grow in size and, through the positive feedback loop of "economies of scale," become more competitive. The economic version of competition is similar to the biological model in that it promotes innovation: new inventions, like new mutations, that provide a benefit are adopted and change the balance between participants in the marketplace.

This is the general benefit of having systems compete: they tend to improve their fitness. Of course fitness itself is a moving target since every improvement of one system's fitness is thereby a change in the fitness landscape for all competitors. Those that don't innovate eventually fail in the environment, while those that adopt innovations advance in the sense we discussed above. Such advances can be the basis of the next round of emergence, e.g., speciation and eventually whole new genera or whole new markets.

11.5.2 Cooperation

In recent years the role of cooperation, long neglected, has been more deeply investigated. The original version of evolution theory, Darwinism, especially when applied to the commercial world, tended to emphasize competition as the main or even the only real factor in progress. Survival of the fittest was generally interpreted in the frame exampled by poet alfred, Lord Tennyson's coining, "Nature, red in tooth and claw." The prevailing version of strong selection had been viewed as fierce competition with lots of killing going on among competitors. Social Darwinism,[27] while not advocating killing the competition, except in metaphorical terms, nevertheless adhered to the notion of struggle.

Biologists, sociologists, and economists have more recently begun documenting the power of cooperation as a strong selection factor. In fact some might now argue that cooperation has been a far more potent organizing principle in the evolution of life and societies.[28] Quant Box 11.1 provides some background in the mathematical tools being used to explore models of how cooperation can arise in an evolving system.

Fig. 11.9 below illustrates the basic nature of cooperation.

Cooperation exists when both entities are sending and receiving messages to one another in order to facilitate the exchange of material or energy. We introduced this idea in the chapter on cybernetics to demonstrate how complex adaptive systems use this form of self-coordination to achieve an objective that neither could achieve alone. The processes in the figure have to have internal decision agents that can

---

[27] See http://en.wikipedia.org/wiki/Social_Darwinism

[28] For example, cooperation in human societies as asserted in Sober and Wilson (1998).

participate in the communications and make decisions about how to regulate their own process in order to achieve the desired outcome. As a result both entities are better off than either would have been alone. This is synergy.

Quant Box 11.1 The Evolution of Cooperation

Altruism is the sacrifice of potential reproductive fitness by one individual in order to promote it in another individual. Altruistic behavior can be observed in a number of species including humans. Evolutionists had long been puzzled by how altruism could emerge in the evolutionary context until William D. Hamilton (1936–2000) proposed that the emergence of altruism could be explained by his mathematical theory of *inclusive fitness*. The theory is based on what two other evolutionists, Ronald Fisher and John B. S. Haldane, in 1932, called *kin selection*[29] or the tendency for social animals to help those most closely related to them (see Question Box 11.10 below) and explains the phenomenon completely in terms of genetics. The theory is based on the idea that closer relatives share more genes than do distant relatives or strangers (this is based on the way chromosomes are distributed during sexual reproduction). Thus, close relatives are more likely to assist one another than are strangers. Siblings, for example, share half of their genes and cousins only 12.5 %. The theory supposes that an individual would be willing to sacrifice itself by helping a sibling or even a cousin because that would assure that that percentage of its shared genes would make it into the next generation.

Hamilton provided a simple inequality that would hold for this situation. He defined something called *reproductive cost*, C, meaning the loss of reproductive fitness associated with an altruistic act. For example, if an older sister puts off having children while helping her parents raise her siblings, then there is some minimal cost associated with that act. Worker ants, on the other hand, sacrifice all in the sense that they give up ever reproducing for the good of the colony and so that the queen can do so. Darwin had some misgivings about this fact. The general idea of natural selection is that competition is what weeds out the lesser beings. Altruism doesn't seem to fit into the picture.

Hamilton's inequality is given as $rB > C$ in which r is the probability that a gene selected at random (in a chromosome) is shared with another individual and B a benefit gained by being altruistic. The ant worker, it turns out, enjoys a substantial benefit from not reproducing itself because it can be shown that the success of the colony, and thus the queen, is related to the fact that all workers share the same genes and more baby ants (!) will be produced, say as compared with other insect species that only have a few young succeed in staying alive and reproducing out of generally large broods.

---

[29] See http://en.wikipedia.org/wiki/Kin_selection

(continued)

Quant Box 11.1 (continued)

While the notions of kin selection and inclusive fitness theory do seem to explain altruism's emergence, there are some problems. Alternatives to that single mechanism have emerged lately that show the more general phenomenon of cooperation can evolve in several different ways.

Using game theoretical constructs, Martin Nowak, for example, has been exploring models of populations of learning agents (see Chap. 13 for a discussion of agent-based modeling) that interact in what is known as the prisoner's dilemma (PD) (Nowak 2012). The PD game is set up with two players (the suspects or accomplices) who are accused of a crime. Both suspects are going to be interrogated individually, and they have two choices. They can *defect* against the other suspect (rat on him) or they can *cooperate* and admit nothing (clam up). The catch is that there is a penalty schedule that determines the penalty that each will suffer depending on which choice they make. Here is what is called a payoff table (though these payoffs are "least worst") of jail time.

| | | Suspect 1 | |
		Clam up	Rat
Suspect 2	Clam up	2 years for S1	1 year for S1
		2 years for S2	4 years for S2
	Rat	4 years for S1	3 years for S1
		1 year for S2	3 years for S2

If you work through this table, you will see that it would be in either suspect's best interest to rat on the other. To each one the idea of ratting on the other and only spending 1 year in jail looks pretty good. But if they both rat they will both get 3 years in jail! That is just 1 year less than if they clam up but the other one rats. What to do? Clearly they should both rat. Since neither one knows what the other will do, each has a chance of getting a 1-year sentence, but the worst that could happen is a 3-year sentence.

So for this single-shot instance, the best thing to do is rat or defect. In evolutionary terms, this strategy will be the "fittest" approach. But what happens if these same two players are faced with this decision over and over again (as appallingly bad thieves who always get caught)? A version of this game called the "iterated prisoner's dilemma" (IPD)[30] allows the suspects to learn from prior experiences and adjust their strategies in hopes of minimizing their prison time. It turns out that in simulations of this game, a strategy called *tit for tat* emerges as the optimal way to avoid excessive jail time. If a suspect rats on his buddy in a round, then in the next round the buddy will do so (to get even?).

---

[30] See: http://en.wikipedia.org/wiki/Iterated_prisoner%27s_dilemma#The_iterated_prisoners.27_dilemma

(continued)

Quant Box 11.1 (continued)

Over many iterations this strategy shows that the upper left hand corner is an attractor, that is, the system will tend to favor both suspects cooperating.

A further variation on this game involves the evolution of agents in the population over very many generations. The agents encode propensities to cheat (e.g., rat or defect) or cooperate. In randomized simulations it is always possible for both kinds of agents to dominate, but in versions of the game where agents can cluster into groups of cooperators and cheaters, and then these groups compete with each other, the cooperating groups invariably persist in the long run. That is, they are more successful (fit) in producing more agents of their ilk in the next generation.

Though there is still some disagreement among evolutionists regarding this "group selection" for cooperation phenomenon, there is growing evidence that it was a key factor in human evolution and a reason that we have become a truly social or "eusocial" creature (to be discussed later). See Wilson (2013).

Between Dissimilar Systems: Sometime around 1 billion years ago some bacteria started forming associations in which several smaller, specialized bacteria lived inside a larger bacterium. The bacteria cooperated with one another and made the emergent eukaryotic (true nucleated) cell more fit than any of the individual bacteria had been before. This form of cooperation is called endosymbiosis (*endo* within, *symbiosis* cooperation for mutual benefit).[31] Symbiosis in the biological world is well understood as imparting a unique and powerful fitness on the participants, but we are just starting to appreciate its scale and scope. And we are finding increasing examples of cooperation among different kinds of organization in the human world. Indeed the nature of a contract, say between suppliers and users in a supply chain relation, is to serve as a formal method for establishing an ongoing form of corporate symbiosis. And a corporate merger looks a lot like endosymbiosis in many respects!

Between Similar Systems: Sober and Wilson (1998) have argued that cooperation among early humans within a tribal group led to a strong selection factor between groups. This is called group selection and is related to but not quite the same as sexual selection. Group selection theory, while put forth by Darwin himself, did not fare well in neo-Darwinian thinking. Most evolutionists bought into the primacy of the gene as the only unit of selection, especially as espoused by Richard Dawkins' famous "selfish gene" metaphor (Dawkins 1976). Sexual selection (usually where the male of the species sports some advertisement that is meant to attract females

---

[31] The endosymbiosis hypothesis was proposed by Lynn Margulis and she writes about it in *What Is Life?* with Dorion Sagan (2000). See Chap. 5, "Permanent Mergers," especially page 119, "Twists in the Tree of Life." Also, for a broader analysis of symbiotic relations that have emerged throughout biological evolution, see *Principles of Social Evolution* by Bourke (2011). Social here means how entities of both like and unlike kinds can form persistent interactions, i.e., be social.

for mating) and generic natural selection work on gene frequency and reproductive dynamics in a rather obvious way. Opponents of group selection could not see how groups could be selected for in a comparable way. However Wilson and Sober and now Edward O. Wilson (no relation, 2013) have made a compelling case for group selection promoting altruistic behavior among tribe members who are then better able to compete against other tribes, when needed, and against all of the rest of the natural world. The way in which group selection does affect gene frequency over generations is beyond the scope of this work. The resolution has been called multi-level selection.[32] Cooperation between similar companies has not been seen much in the commercial world until recently, and then only weakly. It shows up in the media businesses such as movies and television production studios forming cooperatives to produce a product that both could do independently if they had the resources, but find it easier to finance when they work in cooperation.

Question Box 11.10
Cooperation is facilitated biologically by shared genes, since helping others who share your genes, even at the price of failing to reproduce oneself, none-theless advances those genes in the evolving gene pool. This is the "selfish gene" explanation for eusocial insects such as ants and bees. Humans may exhibit similar self-sacrifice even beyond their familial groups of shared genes on the basis of altruism, a mental rather than genetic sharing of identity. Are there necessary scale limits to this kind of identity (Tribes? Nation states? Human kind? All living beings?), or is it open to cultural evolution? If the latter, what could be the selective pressure—since there is no evolutionary trajectory without a selective pressure guiding it?

11.5.3 Coordination

Cooperation between adjacent entities as shown below in Fig. 11.9 is a reasonable mechanism for coordinating the mutual activities of the entities. But as we saw in Chap. 9, as the number of entities participating in an attempt to cooperate as a larger whole increases, we need a new kind of decision processing entity acting as a coordinator. We need a hierarchical cybernetic system to process much more information than the individual entities could manage.

Evolution has repeatedly solved the problem of complexity that comes from having a large number of smaller, highly differentiated entities trying to cooperate with one another. The hierarchical control structures of Chap. 9 solve this problem. They also reflect the definition of complexity as hierarchical depth that we studied in Chap. 5.

---

[32] See http://en.wikipedia.org/wiki/Group_selection

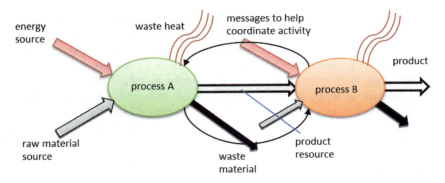

Fig. 11.9 Cooperation between entities requires communications that facilitate the exchanges of material and energy

Animal brains evolving as described above are exactly the kind of hierarchical cybernetic system needed. As environments got more complex (or had more information potentially available) animal brains evolved to process more and more information and encode more and more knowledge. Bodies, of course, evolved more and more variations in behavior capacities in order to exploit the newly accessible resources.

In the case of social evolution as described above (eusociality), eusocial species groups show various amounts of hierarchical coordination and organization. For example, leafcutter ants from South and Central America have four "castes" of workers with very different morphologies and an elaborate top-down hierarchy of pheromone signaling that orchestrates the colony work. Ants have brains but function more like mindless components of a social organism than the more individualistic functioning of higher animals.

In humans there was an early form of coordination hierarchy based on age and life experience. Coordination among members of a tribe, in terms of sharing the global work load, was facilitated by wise elders who could give verbal advice and counsel. As human populations expanded and competition between tribes increased (and especially after the invention of agriculture and a settled lifestyle), there was a shift to a stricter top-down command and control hierarchy. Strong men became leaders but also tended to give orders. They provided protection for territories and land. The evolution of human culture shows how these hierarchies grew in depth and the complexity of societies increased, especially with the inventions of machines and the exploitations of new energy sources.

As we discovered in Chap. 9, hierarchical cybernetic systems reach their ultimate in strategic management and planning for the future. These capabilities need substantial information processing and knowledge storage capacities. The human brain with its capacity for conscious decision making and imagining the possible future represents the emergence of strategic thinking as a mechanism for achieving yet greater fitness even in the context of a very dynamic, very non-stationary, and oftentimes chaotic environment. Individual humans show some capacity for strategic thinking, but it shows itself best in larger organizations such as militaries and corporations. At our current level of understanding of the strategic decision-making process, it seems that our institutions of governance have still not quite evolved the ability to "think strategically." Perhaps that is the next stage of social evolution, for government to be designed along the lines of a complete hierarchical management system. Nature discovered the secret for managing complexity in biological systems (animals specifically) and in smaller social groups. It is recapitulated in organizations of reasonable size (not too big) and with very focused missions. It may yet emerge in the larger societal governance mechanisms.

Question Box 11.11
Strategy limits the field of systemic possibilities to certain objectives—there is no strategy that simply does everything. Strategy thus involves sacrifice: some possibilities fulfilled while others are foregone. Hierarchical coordination works wonderfully in biological whole organisms, eusocial insects, and human organizations of limited size. What limits its effectiveness, its ability to strategize for the whole in the case of larger units—as reflected in the statement that governments are as yet unable to think strategically?

11.5.4 Environmental Factors

Selection is driven in the long run by major environmental factors. Here we briefly mention two that are particularly prominent in biological evolution and one that can be seen in both biological and cultural evolution. The first two do have economic impacts which also play out in cultural evolution.

Geological Effects: Continental drift is the very long time-scale force that shapes oceans and continents. As continents have moved; crashed into one another, causing mountain building; or spread apart over billions of years, the local environments have shifted gradually. One can find similar genera on separated continents, for example, the prehensile tailed monkeys in South America compared with the non-prehensile tailed ones in Africa, showing that separation by geological barriers (the South Atlantic Ocean) can lead to many levels of speciation by ongoing selection. On an even shorter time scale speciation can be due to more rapid shifts in local geology. A severe earthquake can rift open canyons that separate two populations of a given species which then are under slightly different selection forces and diverge from the population norm of the single species population. Geology plays a role in economics from the standpoint of competitive advantage due to, say, the presence of some valuable natural resource in concentration in one region owing to geophysical and chemical forces acting over time. For example, consider the concentration of light, sweet crude oil in places like West Texas or Saudi Arabia.

Climate Effects: Sometimes throughout the Earth's history there have been strong correlations between geological events and climate changes that impact species' evolution. Very often the impact can involve extinction rather than just a drift in fitness to match the conditions of an altered climate. A lot depends on how fast the climate change comes about. A slow shift in climate in Africa due to the ice advances in northern Europe during the last several glacial periods is thought to have been a strong factor in the evolution of *Homo sapiens* from *Homo erectus* and, possibly the evolution of the genus *Homo* from whatever preceded it, most likely, given current data, *Australopithecus*. In general climate plays a strong role in selection. As you might guess, climate also has played a major role in human cultural evolution. A straightforward example is in the area of agriculture. Areas of the world that enjoyed long periods of mild and stable climates also developed greater wealth and led to higher civilizations.

Coevolution Effects: As noted above, species may evolve in relation to competition forces such as predation. The predator-prey example above is an example of coevolution, where two or more species or types of complex adaptive systems act as mutual selective agents. We will finish the chapter with an elaboration of this idea and show how it plays out in numerous evolving systems.

11.6   Coevolution: The Evolution of Communities

Natural selection winnows out metabolisms and ways of making a living that do not fit changing environments, and it reinforces inheritable adaptations that work better by rolling them forward with a statistical edge in the gene pool. This gives us a window for understanding the adaptive process that has honed the metabolism, morphology, and behavioral traits of any species. But the focus of this window on species evolution can be misleading if one does not go on to clarify that species do not just evolve, they co-evolve with other equally evolving, adaptive organisms. Coevolution is already implicit in the notion of selection for environmental fit, for a critical element of adaptation to environment is fit with the other organisms and physical features that together constitute the environmental community, the ecosystem.

Coevolution is really just evolution, but the term developed as a way to clarify and emphasize the dynamics of evolution as a mutual dance among a multitude of networked partners. In the dance of coevolution every organism is continually adapting to changes in others as they adapt in turn to their changing environment. A dynamic that produces a continually changing but mutual fit is guaranteed at least in the negative sense: extinction awaits any species that becomes too ill-adapted to its fellows in the ecosystem.

11.6.1 The Coevolution of Ecosystems

Selective pressures, the demand to fit, are inherent in relationships with the environment; even the inner mutual structuring of the components of a metabolism can be understood in terms of fit with surrounding conditions. But evolution occurs not from the pressure itself, but from adaptive response to the pressure. Thus, the adaptive unit becomes an important consideration. All sorts of adaptive moves, both within an organism and in its way of responding to environmental conditions, contribute to its life or death. This basic form of working/not working is selectively registered in terms of the unit's presence or absence in the pool of reproducers, which effectively rolls forward what works better as a statistical edge in an ever-evolving status quo. It is in the nature of things that ecosystems are often described as food webs, in which a complex relational network of eating and being eaten is the functional life of the system. Insofar as every species, from micro-organisms to elephants, maintains itself only by engaging in this process, they are effectively interwoven in a complex of interdependent interactions selected at the species level for what works.

Ecosystems, like any system, have an always adjacent field of systemic expectation. In addition to the expectations governed by the laws of physics and chemistry, these systems are especially characterized by the dynamics of species' ongoing evolutionary selective probe into what works. These interwoven selective dynamics amount to an ecosystem level expectation of how things will work, i.e., what sort of life community, with what sort of distribution, reproduction rates, predator–prey relationships, etc. will be at work. This systemic expectation is resilient, able to both demand fit in the face of change and find restored balance after disruption. But unlike non-living systems, it can also work and not work in various degrees, as measured in terms of the interdependent well-being of the life community. Thus, we hear of healthy, unhealthy, and even dying ecosystems.

Information regarding what works shapes species through natural selection. Projected into the future through reproduction, this amounts to a prediction for each species of what the future will be like. Thus, an ecosystem, like the species that make it up, is critically leveraged on a certain stability of conditions. Diversity in the ecosystem plays a role similar to diversity within a species when it comes to absorbing and adapting to change. System degradation or collapse is typically a matter of something that contravenes systemic expectations in some critical respect: invasive species, major changes in the physics and chemistry of the system, sudden alterations in flows within the food net, or conditions of habitat are typical factors in the collapse of ecosystems.

The cross-referenced interdependence woven by the coevolutionary dance amounts to a system of mutual constraint, where behaviors, abilities, and even metabolisms are limited and shaped to fit with the larger system. The coevolved web of mutual constraint plays an essential role in ongoing evolution insofar as it shapes the selection of what must or what can fit. Recall the example of the marine ecosystem of the Aleutian Islands we saw in Chap. 4, a coevolved food web of fish, seals, orcas, sea otters, sea urchins, and kelp beds. Kelp beds support fish, which support seals, which support orcas, while sea urchins eat kelp and are in turn eaten by the sea otters. One does not easily see the mutual web of constraint, i.e., the limits inherent in these interdependencies, until something unfitting transforms the dynamic. In this case, when over-fishing depleted the larger fish population, the seal population declined and orcas were pressured to adopt a new food source, the sea otters. Sea otters, however, had not evolved to either elude such predators or to reproduce rapidly enough to support such a high rate of becoming dinner, so their population went into steep decline. Sea urchins, on the other hand, had a new opportunity to multiply unchecked by the otters. Who would have thought that their evolved reproductive capacity "expected" predation by something like otters? In any case, they multiplied happily and munched on the kelp in numbers and with a collective appetite the evolved kelp reproductive capacity likewise did not expect.

Here we see how one critical surprise—the overfishing of larger fish—cascaded through the system as violated expectations, rendering even the apparent adaptive activity of some species (the orcas) an expectation-violating degrading factor on the system level. A naturally evolved ecosystem is a network of functional relationships that give expected kinds of life and well-being to its component species. The integrity of this relational network, the product of the coevolution of the community, is what commands respect when people talk about following what is "natural" as some kind of norm. Because there is an integrity that can be violated, an expected healthy condition that can be destroyed, intervening and introducing changes into natural systems produced by coevolution calls for serious forethought and consideration.

Coevolution takes place on the level of whole organisms, and it results in ecosystems which weave together the life strategies of the many members of the community of species into a web of cross-referenced expectation. Changes challenge expectation, so changes in a given species, or in the composition of the community, or the physical environment, all drive further evolution—or, when the surprise exceeds systemic adaptive capacity, devolution and breakdown.

Nowadays it seems that all too often we humans are the source of problematic surprise, as in the above case of the intersection of the strategy of making a human living by fishing with all the other strategies for making a living in the community of the Aleutian marine ecosystem. But why should this be the case? Do we not also belong as members of the coevolving community of life on Earth? We will find the answer to this when we understand some critical systemic differences between ecosystems and human cultures.

11.6.2 The Coevolution of Culture

The term "nature" is often used to mean everything as it is, untouched by humans. We are described as living in a world of human-made culture rather than in the natural world, so "culture" and "nature" can be terms of contrast. In a systems science perspective we and all our works are part and parcel of the whole evolving natural system. As we have discussed, the evolution of life is a massive parallel probe exploring the many unfolding strategies of what works and can work as conditions emerge. Whales are one such probe. So are hummingbirds. So are we.

But there is a meaningful systemic reason to functionally distinguish human society or culture from the world of nature. Above we described the coevolution of ecosystems as weaving communities of mutual constraint. In the Aleutian example, the Orcas were free to change their diet, but they are also ultimately constrained by the conditions of the system. Of course if the system collapses, they are also free to just move on. Still, they are sufficiently constrained to be responsive to their conditions, so it would be unlikely that they would ordinarily be a shocking surprise wherever they go. Or if they were, we would expect their future as a species to be rather limited.

What distinguishes the human community from the natural community, then, is a matter of degree rather than absolute difference. We are even less constrained than the orcas by the conditions of whatever life system in which we find ourselves. More than any other species of which we know, we can flex our own conduct and also actively modify the world about us to such a degree that, in the short term, we are relatively unconstrained by the systemic web that tightly selects the viability of strategies for most of the life community. The world of society and culture is a system of self-imposed constraints specific to the human community, for we too require that our lives be constrained by some form of relatively immediate systemic expectation. That is how we get predictability, and without predictability we would not be able to make a living.

We have a framework of expectations regarding the natural world, so we can anticipate, for example, the need for warmer clothing as winter approaches. But human strategies are so flexible, that how we go about meeting even such basic needs is unpredictable. What kind of warm clothing is available and acceptable? Who makes it? How and where do I get it? In our shared social world, we need common expectations regarding almost every aspect of life, for we have evolved to meet our needs with great latitude of behaviors. As a social species, we typically live and survive in close concert with other humans, but in order to do that we construct for our daily guidance (constraint) a human cultural world, an interwoven tissue of explicit and implicit shared agreements about what's what and who's who. Culture emulates the systemic predictability of the natural world in allowing us to have a framework of expectation in which we can function.

Human culture is a complex and always evolving system. Its evolutionary trajectory is marked by thresholds, transitions that mark new emergences that transform the relational system in important ways. In the geologic eyeblink of 6,000 years we have moved from hunting and gathering in tribes of hundreds, to agriculture and urban groupings in the hundred thousands, to industrialized communities of millions. We have organized our politics and economies locally, regionally, in empires, nation states, and finally globally. The logistics of this vast expansion of organizational scale have required a steady pushing back of the constraints of space and time by a series of revolutions in transportation of physical goods and in the speed of the information flows, which enable coordination and control of a now global network of production and consumption.

As a systemic organization of life, cultural evolution shares the basic dynamic structures of biological evolution, but with important differences. Like biological evolution, cultural evolution is a process of transmitting models which control the construction, maintenance, and productive activity of the system. But where genes are carried in cells and shape metabolic systems, the models of culture are carried in minds and shape social systems. Biological evolution is measured by life spans and rates of reproduction; cultural evolution is less predictable, being a matter of how conditions render minds more or less receptive to change. Cultural memory may be stubborn and slow, or slippery and habituated to expect rapid change.

Both biological and cultural forms of evolution select for fitness through a process of relative statistical frequencies in a pool of models with variations that are selectively passed on, be it the gene pool or the collective cultural memory. These similarities are grounded in the fundamental structural dynamics of any process of evolution that selects in terms of what works.

Question Box 11.12
Organisms coevolve with their sources of nutrition. They do not ordinarily have appetites for foods that decrease their lifespan; that is exactly the kind of thing evolution selects out. We have great appetites for sweets, salt, and fatty foods. In human evolutionary time these appetites must have pointed us to behaviors that worked. Now they lead us into diabetes and heart disease. What changed and why? What does this tell us about the selective process that shapes cultural evolution?

Like an ecosystem, cultural evolution is a coevolutionary dance. Instead of self-organizing competitions and cooperation among many species in a communal ecosystem, we have the competitions and cooperation of multiple individuals, institutions, and organizations of a social system. We construct our own individual lives in terms of the society around us, and that society in turn is the organization of a myriad individual lives and their interwoven and shared expectations about how to live.

11.6.3   A Coevolutionary Model of Social-Cultural Process

The coevolution of the social system is more than simply the dynamics of a community of individuals. On one level, culture exists fundamentally as shared models of a common life. As such it is basically a mental phenomenon. But as we act and live in terms of these models, we alter them continually in accord with new ideas, changed circumstances, and a restless quest for better ways of doing things. Richard Norgaard has modeled this process of social/cultural change as an interdependent web of mutual feedback relations among environment, technology, organization, values, and knowledge (Norgaard 1994, ch. 3) (Fig. 11.10).

Fig. 11.10 Norgaard's model
of interacting components in
socio-cultural systems

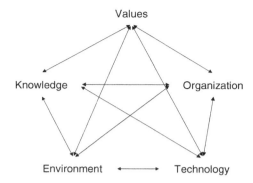

The strength of this model is its concise inclusiveness: it includes systemic flows from and into the environment, along with the human mental capacities (knowledge and values) which especially distinguish human culture from other living systems. It further includes the way we objectify these capacities through our organization and technologies, two features that especially feedback and shape the minds that originate them. This network of feedback loops means that all of these elements are both causes and effects of the shape and functioning of the others.

Recall our initial consideration of feedback loops (here represented as double-headed arrows). Does the thermostat control the furnace or the furnace control the thermostat? Do values shape technology, or does technology shape values? One can see that when dealing with coevolutionary dynamics, either/or questions regarding causality are misleading. But questions in a given case of *how* influence flows in this network of mutuality can be quite instructive. How do these systemic elements selectively shape one another?

Selection in natural living systems sifts a shifting recipe of what works by an invariable criterion, survival long enough to reproduce. This defines fitness when selection works as a kind of gatekeeper on what gets transmitted across generations. But the world of culture is not biologically transmitted but learned. Cultural memory is transmitted from mind to mind,[33] for culture is fundamentally a shared agreement about what's what, creating an expected, shared world for human activity. These worlds of tacit agreements among people include numerous partially overlapping groups and subgroups, each with its own coherence and boundaries. We understand cultural fitness intuitively because we all have the experience of fitting and not fitting in different social situations. When we are outside our familiar social groups, we feel unsure about just what's what: how to speak, dress codes, manners and demeanor, jokes, music, relative values, and much more can all have differences that, like dialects, range from subtle nuances all the way to mutual unintelligibility.

---

[33] Richard Dawkins, in *The Selfish Gene*, introduced the term "meme," a parallel with "gene," as a way to think about units of cultural memory and how they get passed around and stick in minds.

Each of these cultural worlds is a system in constant but coordinated flux. As Norgaard puts it: "Everything is interlocked, yet everything is changing in accord with the interlockedness" (Norgaard 1994, p. 26). Being interlocked provides critical system stability. Minds and their expectation of the world are inherently changeable, and the social-cultural world endures only as it stays in active memory, the category we save for "the way things are" versus the layer we label "the way things used to be." Every element already in the culturally formed mind serves both as an interpretive framework for new information and a selective braking mechanism on how information modifies the whole.

Question Box 11.13
Another way of putting this interpretive loop of knowledge and information is that we tend to see what we expect/are looking for. How does this stabilize the culturally constructed world? In what ways can it be problematic?

In the circumstances of any given moment there is a range of potential and serviceable behaviors available, but we culturally expect only a small subset of those possibilities, and this, when it works, gives us a relatively predictable and controllable social world. These selectively constructed cultural worlds are so successful that, until the relatively recent revolutions in transportation and communication, people commonly assumed their cultural world represented the only truly fit options, the *real* way things should be done.

The above diagram of coevolutionary social dynamics is, in effect, a model of the process of cultural selective dynamics. The range of expectations, i.e., possibilities and probabilities, associated with any one of these systemic areas—values, knowledge, environment, technology, and organization—is intimately affected by the others. They affect one another by the kinds of development or emergence they call forth or constrain, with every change rippling to affect the others, and change in any area being screened for fit with the status of the others. This more or less guarantees a status of short-term gridlock and long-term unpredictability, frustrating both zealous reformers and those who would keep things just the way they are.

In the early twentieth century, for example, after major advances in scientific knowledge, accompanied by new technologies such as railroads and telegraph, progress was a high value in Western society, and the expectation that changes would lead to a better world smoothed the path of social reorganization from rural to urban societies. National and international markets subsumed older, more local dynamics. Expectations regarding how to make a living changed, as did the complex flows needed to sustain all sorts of business and manufacturing, and in the new environment education systems became critical elements of social organization. Everything made sense in terms of everything else, values, knowledge, technology, and forms of organization all in concert.

In this sort of cross-section analysis of a social system at a given time, then, everything helps explain everything else. This is because, as in an ecosystem, everything is constantly selected in terms of its systemic fit with everything else. Even contradictions and tensions, resistances, and subversive movements can be analyzed this way, and over time they also play out in familiar patterns: rise and fall, power and opposition, success, and overshoot are familiar themes in the history of human affairs. Such patterns suggest the negative feedback dynamics in which systemic excess and deficit alternate in maintaining a fluctuating equilibrium. Thus the proverb, "The more things change, the more they remain the same."

Coevolving systems, however, can also have positive feedback dynamics, in which change accelerates as more leads to more. The familiar analysis of capitalist growth, for example, expects values of competition, growth, profits, and investment to subsidize competitive research, expanding knowledge and technology for more productive organizations, which will become yet more productive through reinvestment in research and technology. This is the virtuous positive feedback loop in which consuming products produces more profits to fuel yet more knowledge, technology, and productivity, for yet more consumption. The pie gets bigger the more you eat it.

The only element of the coevolutionary system left out in this scenario is the environment. The environment is not, like the other categories involved in this positive feedback dynamic, explicitly human and social; it is, nonetheless, necessarily involved as the ultimate source and sink for the flows which constitute society. The positive feedback of growth is presently the imperative dynamic of humans organized socially into a global market economy; how this will coevolve with the complex adaptive environment of the natural world is the urgent question of sustainability. We will address this further, but first we must look more closely at features of the evolved structure of contemporary society.

11.6.3.1   Social Evolution

The present global market system is only the contemporary edge of a long process of social evolution. Evolution, we have said, is more than just change: it is change with a long-term trajectory, typified by mounting systemic complexity. In the broad sweep of its history, human society has undeniably evolved. The present global network of human society is far more complex than it has ever been in its values, knowledge, technology, and organization. And this complexity is the culmination of ever accelerating and amplifying processes rooted in the exercise of our basic abilities to strategize, communicate, and manipulate.

If one were to chart the growth of social complexity over the course of *Homo sapiens sapiens'* 100,000 years on Earth, we would find the familiar curve of exponential growth: a long slow takeoff, followed by a rising curve that soon seems to head almost straight up. First we have small bands of hunters and gatherers for about 90,000 years. Then comes a gradual transition to agriculture, giving rise to settled farming villages by about 5,000 years ago, then early city states emerge about 3,000 years ago, followed closely by conquest and empires. The modern unit of society, the nation state, emerged only about 300 years ago. Now we have entered the era of globalization, in which the processes of daily life increasingly depend on a networked flow encompassing the globe in a single fabric of interdependence.

> **Question Box 11.14**
> Is any item of the clothing you are wearing made locally? How much of it is made in places far distant? How did people in those distant places get the information to make clothing that would appeal to you?

Exponential curves are produced when processes increase by increments proportioned to the growing base. Systemic complexity increases as components become richer in connectivity and thus in their interwoven functionality. For the human community, linguistic communication, our uniquely powerful connectivity, grounds this exponential growth in cultural complexity. We become uniquely flexible learners and adapters because language makes it possible for any individual experience to become the shared experience of the group. And with the invention of writing, this sharing is expanded horizontally across arbitrarily large reaches of geography and vertically through any number of generations and centuries. Successive layers of technology have amplified the speed and scope of this communication process: paper, printing, telegraph, telephone, fax, computers, and the internet. This has allowed the sharing and coordination of ways of making a living, making war, and expanding and controlling territory. The reach and consequentiality of such activities mushrooms because information and inventions have a way of triangulating to produce more information and invention. This is the dynamic of exponential growth.

In the last half of the twentieth century this process crossed a new threshold as information itself became the focus of technological innovation. Now all originally discrete forms of information have been synthesized in a digital format, so a unified cyberspace of digitalized information becomes available, in principle and increasingly in fact, to everyone everywhere through a range of devices that grow in speed and power exponentially. Measured in bytes, in 2006, a single year, we produced three million times as much information as recorded in all the books written up to that time.[34] And in the world of hi-tech especially, one sees information from a myriad sources looping and intersecting to produce yet newer devices networking yet more information: a single smartphone now may involve as many as 250,000 separate patents. That in turn supports legions of lawyers and ensures that intellectual property laws are a vigorous and growing element of international law.

---

[34] "The Expanding Digital Universe: A Forecast of Worldwide Information Growth Through 2010," an IDC white paper, http://www.emc.com/collateral/analyst-reports/expanding-digital-idc-white-paper.pdf

The space-time constraints that once made distant connections slow, uncertain, and therefore weak gradually lessened and have now virtually evaporated in a burst of communication and transportation technology that has rendered humans an interdependent global economic unit of many distinct languages and cultures. Sometimes the new accessibility of all to all produces reactionary shocks. Defensive dynamics set in motion by the new proximity and availability of formerly remote options sometimes harden and accentuate the distinctive values, customs, and expectations of historically diverse communities, so the resurgence of fundamentalism is itself a global phenomenon with deep systemic roots. At the same time, the systemic economic organization of the globe manifests itself in an advertising industry that has emerged as a unified shaper of the values that increasingly unite the human community in a shared cosmopolitan identity as consumers.

More leading to more has become the norm of a global market system structurally premised on growth. More research poured into more technology which enables more productivity by more complex organization, enabling more consumption and a value orientation focused on comfort, speed, convenience, and the wealth to make these possible. People need more and more education to keep up with the technology, especially the information technology, to be more productive in jobs that will pay enough to support the demands of increasing consumption. And the environment has thus far cooperated in this expansive burst by supplying more and more cheap fossil fuel for the work to maintain and expand this vast organization, with more research and new technology working in a continual feedback to supply the expanding need.

This is only a rough sketch, but these positive feedback dynamics are structured into the normative functioning of our evolved global system. That is, the absence of such positive feedback is the equivalent of systemic malfunction. The alternative to a positive feedback of growth in our present global system is a positive feedback of contraction, giving us the phenomena of a recession: less consumption leading to less production, loss of jobs, less support for education, poorly maintained infrastructure, less reliable technology, lowered productivity, etc. In one direction or the other, positive feedback dynamics seem to control the behavior of this system.

We both expect periodic recessions and expect the return of patterns of growth. A resilient, complex organization can recover: indeed, complexity can enhance resilience, for in the extended global network it is less likely contraction will occur everywhere, and healthy areas can be an "engine of growth" for those caught in a downturn. A global contraction presents a much more serious systemic problem, but complex connectivity also increases possibilities for further new emergence—the systemic base for our confidence we can always creatively find a way out of whatever problem we are in.

We have seen that language and communication ground a shared cumulative process in which our knowledge, technology, organizational reach and connectivity, and the motivational values proportioned to them now coevolve with an essentially positive feedback dynamic. Positive feedback dynamics usually characterize systems in transition: the system may grow until it crosses a threshold and settles into a new level of complexity with new characteristics, or it may devolve and disintegrate.

Many social commentators have understood these dynamics and drawn the appropriate (but opposite) conclusions. Some say we are on the cusp of a new stage of social evolution, while others point grimly to the edge of the precipice. Complex connectivity supports both expectations. Increased complexity may also increase systemic probabilities for moving to new levels of organization. At the same time, greater complexity also increases maintenance costs and presents more ways for malfunction to occur and so increases vulnerability and instability. The relative weighting of these factors shifts as complexity increases. Joseph Tainter has described the growth of social complexity as subject to the law of diminishing returns, and some commentators feel we are already on the downward slope of getting less and less from the further increments of complex organization (Tainter 1988). But clarity in such matters comes mainly with hindsight. Anyone carefully studying the process by which a single-celled egg becomes a human infant, for example, would intuitively conclude the process is far too complex to work, too riven with junctures where a tiny error would give rise to grave malfunction. Yet it somehow does work well enough in most cases.

Another group of social commentators—perhaps the majority—simply assume a future of more of the same, much the way we assume tomorrow will ordinarily be pretty much like today. That assumption seems to ignore the transformative effects of positive feedback, as if an acceleration of change can go on indefinitely. That would appear unlikely, especially since it strains the fitness criterion which the environment imposes on any evolving organism.

11.6.3.2 Society's Fit with the Environment

In the systemic organization of life on Earth, evolution selects for fitness in a given species, while coevolution cross-references this into the mutual fitness of all the organisms that have evolved together in an ecosystem. Assessing the mutual fit of A and B would amount to an inquiry into how B is factored into the expectations structured into A and how A is factored into the expectations structured into B. So by considering expectations structured into human society and considering how the structural expectations of the environment fare as the expectations of society are met, we can assess comparative fit. In these terms, what can be said of the fit of human society and the environment?

The structural expectations of an organism can be considered in terms of how it makes its living. What are the conditions it expects, in terms of flows from and to the environment and in terms of its own abilities and strategies for maintaining its life in that environment? For most organisms this assessment could be carried out entirely in terms of local flows as processed through the physical-sensory capabilities of individuals or their permutations as ramified by various forms of group organization. In the human case, for anything close to this kind of direct interaction and living from readily available local environmental flows, we would have to go back to the first 90 % of our history, when we were organized as hunting and gathering tribes. And even at that level of organization our communication abilities allowed an accumulation of strategic technologies that already set us on a distinctive trajectory.

But the most decisive structural shift away from relatively direct dependence on available environmental flows came with the emergence of agriculture and the domestication of animals. Here the human interface with nature became mediated by the technology of crop planting, growth, harvest, storage, reseeding, and techniques of animal care and breeding. Instead of depending on the uncertain availability of food stocks, humans now had the advantage of a more sure supply based on their own intervention. The price for this advantage was a shift to dependence on kinds and quantities of plants and animals that depend upon our labor rather than dependence upon the "free" products of the environment. Here the distinction between the natural world and the human world takes on meaning as we begin to make our living more and more out of the products shaped by our own activity. Technologies have their own expectations, and these now become an essential element in any assessment of expectations structured into how humans make a living.

Technology, then, inserts a layer of proactive human products between humans and direct adaptive response to the natural world. In addition to technology, humans have introduced one other critical mediating layer between themselves and making a living from the environment: money. For humans the expectation now is that a living is made by having a job which will produce money, and the money can transform into all the actual flows needed to survive and flourish. Money has a long and fascinating history, from direct barter to "commodity money" backed by a currency of grains or precious metals, to the present purely symbolic unit for the quantification of any sort of exchange. This evolution carries immense implications for forms of social organization. But for the question of expectations structured into making a living, the important point is how money inserts another layer of mediation between the strategy of making a living and the environment, which still remains the ultimate systemic source of the flows that support any life.

Having a job thus produces two things: first, some input into the productive flows which sustain the economy and, second, the money which we require to maintain our well-being as consumers of those flows. Since it is the expected means of access to consumption, there is a tendency to equate money with well-being. But its nature as a systemic mediation is evident if one considers that there are many ways of making a living in terms of money that have little or nothing to do with actually contributing to either personal or communal well-being.

Question Box 11.15

All organisms that exist as dissipative systems amidst flows to and from the environment have thresholds beyond which the flows that sustain them become toxic. In other words, enough is enough. Is there such a thing as enough money? Does a flow of too much money become toxic? What might provide a measure for "enough" or "too much"?

Understanding the intervening mediation of technology and money in how we are presently organized to make a living, one can understand how the expectations of a globalized economy can become so poorly aligned with the environment. The selective feedback for technological evolution has for over 300 years been industrial: better means faster, more efficient, and more productive with less cost. Costs to the environment enter this industrial feedback loop only when degradation assumes forms and proportions that affect balance sheets, as when fisheries collapse or air pollution drives up health-care costs. Money works in synergy with the industrial technological dynamics, for increased productivity means more profit. In the last decades of the twentieth century this evolved to the point that schools of business management routinely taught that the main duty of a corporation is to maximize the share price for stockholders, confident that maximizing money could stand in for maximizing corporate well-being and productivity. But since money can be maximized in many short-term ways that degrade actual industrial well-being, the financial world becomes a thing apart from the world of productive industry, much as technology and industry becomes a world apart from the environment.

Although we ultimately depend upon the environment, our technologically and financially mediated dependence no longer involves the same kind of effective shaping constraint that selectively forms ecosystems as communities of mutual fit. Rather we are organized in a way that expects massive flows of food, energy, and materials from nonlocal sources in quantities made possible only by increasingly sophisticated technology and the money to pay for it. This global market system is sensitive above all to profitability, which in turn depends upon consumption and hence upon human preferences. Far removed from the basic metabolic needs that ultimately ground them, preferences now are shaped by our market-dominated social organization and its use of advertising and media technologies to ensure a continuing feedback loop of demand. In this technological and financially driven world, more productivity demands more consumption, and more consumption supports yet more productive growth.

Dissipative systems, such as human society, exist situated between flows from environmental sources and the sinks in which those flows end up after being processed by the social system. Our ability to leverage flows and move on when sinks transform in problematic ways has allowed us to structure expectations that may not be indefinitely sustainable, adding to the feeling that this may be a transitional era. Energy is of course the most basic flow, for it makes all other technologically mediated flows possible. Richard Norgaard notes that contemporary society has essentially coevolved with the discovery of plentiful fossil fuels, and especially oil (Norgaard 1994, ch. 4). Many are concerned that oil depletion could bring our world of plastics, chemicals, mining, industrial processing, fertilizer- and pesticide-based industrial agriculture, and transportation to a crisis. And as the expected flow of oil gets higher and supplies diminish, our efforts to get at the more difficult residues require more energy and more drastic environmental impact. Foreseeably we will with mounting urgency seek to transfer this dependence to some other source, confident there will be a technological fix for virtually any sort of supply problem.

While energy gets a lot of attention on the sources side of the fitness question, sinks are equally important and often present urgent problems stemming from unanticipated effects of processes that once seemed beneficial. In the mid-twentieth century, we still thought just building higher smokestacks made things go "away." Now with the globe wired with all kinds of sensors, we realize there is no such place as "away" for the outflows from the process of humans making a living. Everything ends up somewhere—carbon, toxic chemicals, fertilizer and pesticides, plastics, heavy metals, etc. And insofar as these flows are of major proportions, they often are a major transformation, a shocking surprise to the expectations of ecosystems and social systems alike. Dead zones in oceans, algae blooms, warming atmosphere, acidifying ocean, polluted rivers and lakes, acid rain, mercury in breast milk, toxic drinking water, etc. are all sink problems, the home of unintended consequences.

As we more aggressively and efficiently go after sources and increase effluents into sinks, we create extensive and sudden alterations of the life-world expected by many organisms. The unintended consequence here is often referred to as "the sixth mass extinction."[35] Estimates vary, but we are surrounded by species vanishing at, depending on the type of species, perhaps 45 to over 250 times the average or "background" rate of extinction. If this continues, some estimate that within a century up to 30 % of the present community of life will have failed the test of fitting in the environment as modified by humans.[36]

This depressing review could be extended, but the point is clear. By major systemic measures such as a sustainable relationship to environmental sources and sinks and our fit with the coevolved community of life which supports our nutritional flows, our present fit with the environment is highly problematic. This is not news: we have for some decades been able to track global changes, and the alarm has been sounded so often groups invested in the status quo have had to develop organized strategies to dismiss it or at least postpone action.

The systemic structure underlying such an environmental misfit is the way technology and money form layers that insulate the social structure of human cultures from environmental concerns. Resistance to change takes a predictable form: "It will hurt the economy and cost jobs." The way we make a living is an expectation woven into concrete and steel and structured into businesses, civic, and government organizations. The system welcomes change if it means more money and more jobs because jobs and money are not only the organized way we make a living; they are also the metric, we think, for improving the way we make a living. Corporations, seeking profit, shape jobs to maximize efficiency, and they resist any environmental restraints that might make them less competitive and therefore less profitable. If lopping the tops off mountains and dumping them into valleys is the most efficient

---

[35] Sixth because the fossil record reveals that over the 3.8 billion years of life there have been at least five major extinction events. The resilience of life, which seems to bounce back with new bounty and variety after a few million years, is one of the encouraging aspects of this history of systemic collapse and recovery.

[36] The PBS Evolution Library, http://www.pbs.org/wgbh/evolution

way to get at coal in West Virginia, the corporation claims it must do so as a duty to stockholders, and coal miner unions accept it because they need jobs.

And globalization, in the absence of any universal regulating agency, always provides the competition argument: if we don't do it, "they" (often the same company in another place!) will, and we will lose market share and/or jobs. Effects on the environment, in comparison, are "side effects," incidental to the system until they loop back to threaten human making a living at a basic level such as toxic runoff into ground water. This structure is geared to produce a fierce criticism of offshoring jobs but a weak analysis regarding how various sorts of jobs or practices affect the environment.

We recall Norgaard's dictum about the coevolving cultural system: in the short term, everything is gridlocked. In the longer term, everything changes.[37] And change is wedging into the system: the feedback loop of environmental problems and challenges looms too large to be ignored. Values, knowledge, technology, and organization all are now more deeply shaped by environment than any time in modern history. But still the governing dynamic is the technologically and financially mediated corporate market and mass consumption structure described above. Every sector of the coevolutionary socioeconomic system harbors potential levers of change that, in the right circumstances, could tip the system over a threshold into a new condition. We are masters of expedient flexibility. But complex, coevolved systems of components which mutually produce one another also have tremendous resilience. After a close brush with global financial meltdown in 2008, for example, the financial industry emerged, after widespread calls for drastic reform, with only marginal change. Collapse, of course, would have changed everything, but if supermarket shelves emptied because frozen finance turned into frozen transportation, that would be a chaotic and dubious passage to a more sustainable human social structure.

Since we regard ourselves as thinking, motivated decision makers, many urge a change of mind and heart as the key. Values and ways of thinking are most in play, however, in times of immediate crisis and fear, as after 9/11, for example. Short of a major environmental catastrophe (protracted heat waves and droughts can be persuasive!), a more viable alternative might be locating leverage points within the systemic expectation of the existing market system, strategic modifications that would redirect the system from what are at present the most probable forms of business as usual. Thinkers such as Amory Lovins and Paul Hawken have numerous suggestions about how to move to a more environmentally fit capitalism.[38] In systems terms, the point would be to use the profit motive already fundamentally structured into the system, but connect profit/price to environmental costs in a way that would guide production and consumption in the direction of a better environmental fit. Government taxes and subsidies create the topography of a commercial playing field. So subsidizing what is a better fit, such as clean, renewable energy, makes it a more competitive way to make money, while subsidizing oil well depletion keeps

---

[37] Norgaard (1994), p. 46, paraphrased.

[38] See their jointly authored book, with Hunter Lovins, *Natural Capitalism: Creating the Next Industrial Revolution* (1999).

the system doing the wrong thing. A significant vehicle tax proportioned to miles per gallon would change the shape of the automobile industry. If foods grown with fertilizers and pesticides were more expensive than organic crops, it would reshape agriculture. Instead of needing to exercise heroic virtue, doing the environmentally fitting thing should be, as Paul Hawken says, "as easy as falling off a log" (Hawken 1993, p. 56).

Changing values and the way we think would modify organization and the way we use technology; alternatively, changing organization and market dynamics would change values and ways of thinking. Or a technology breakthrough might shift the economy versus the environment question in ways that would free up the market-dominated values and thinking to assume new forms. Or some combination of these coevolutionary dynamics might change our unstable fit with the environment, upon whose flows and sinks we depend. The historical path of our cultural and social evolution has brought us to a systemic organization that functionally insulates us in our own subsystem of technology and finance. But understanding the path-dependent nature of our present global system frees us from both the positive hype that sees this as somehow the necessary culmination of a long evolution and the debilitating negativity that sees a flawed, greedy human nature inevitably running us over a cliff.

In a systems perspective, human behaviors, the way we think and the way we prioritize the values that guide us, are less programmed by bodily form, metabolism, or genetic heritage than they are by our cultural heritage and the organization of the societies within which we make our living. By the participation of our lives we contribute to the shaping of the ongoing trajectory of that culture and social organization. But there is also a larger systemic interplay of multiple coevolving cultural factors, the environment, and contingent, unpredictable events. The dance of coevolution is given shape and form by all, but controlled by none. This is true even for the environment, which can transform and degrade to fit with our cultural misfit, in the process exerting mounting pressure for changing the dance, until some limit is reached. Our cultural and social system is in no way a fixed, determined quantity. At some point we will come up with our new moves, and the sooner we find the leverage to lessen our tense and overextended expectations of environmental sources and sinks, the richer the field of expectation of the ongoing system will be.

Think Box The Fitness of Intelligence
The evolution of animal species shows a distinct trend toward larger and more complex brains. Later evolved genera tend to have larger brains (relative to total body size) and more complex behaviors. Intelligence might be defined as the competence to operate in more complex environments where there are more "other" factors and entities in the environment to deal with. In part this competence depends on larger memory capacities to hold more knowledge about the environment. Another factor is the capability to build causal models

(continued)

Think Box (continued)
of how the world (the specific environment of the species) works. Such models allow an individual to make anticipatory guesses about what is going to happen next so as to preemptively act rather than always being reactive. Preemptive actions can be shown to reduce energy costs since the animal avoids, for example, damage that needs repairing or can be more direct in acquiring a resource like food (see Sects. 9.6.2 and 9.6.3.1).

Greater intelligence from larger more complex brains allows the possessor to function in highly dynamic and non-stationary environments. It provides the possessor with the ability to adapt through learning new behaviors to take advantage of new opportunities (e.g., new game food). As such it is not at all hard to recognize the selective advantage to increasing intelligence in evolutionary terms. Greater knowledge capacity allows an animal to deal with greater complexity and manage the information through creating that knowledge (Chaps. 7 and 8). That leads to higher survival rates and greater potential to produce offspring.

The human brain has given our species the capacity to adapt to nearly any environment and to exploit a seemingly endless variety of food possibilities and other resources.

11.7 Summary of Evolution

The phenomenon of evolution is possibly one of the more important aspects of systems science. As we have seen throughout the book so far systems may be organized, unified wholes, but they may also be subject to changes in their internal structures and functions even while remaining unified. There is, perhaps, a philosophical point arguing that the modified system is not the same as the pre-modified system. For example, biologists use the categorical construct of species to differentiate two supposedly different *kinds* of organisms. But, as we have seen from the above section on Descent with Modification, there is continuity between the original system (a species) and the evolved system (a new species), with the latter being derived from the original by a sufficiently significant alteration in morphology and/or behavior so as to cause a reproductive affinity split in the lines. The species are different but their genus is not; it just has more members in the set.

A more subtle problem comes from looking at the human brain as it learns new concepts and even more so when it learns concepts that are contrary to previously known ones (e.g., conflicting beliefs). Learning is an evolutionary process in that the brain is structurally altered at the level of cortical neural networks and their interconnection strengths even though there are no new macro-level structures that magically emerge. These new or altered concepts result in possibly altered behaviors by the possessor. Then the question "is this the same system?" becomes a good deal more difficult to answer. Just because you learn something new does that mean you are a different person? Most people would probably insist not.

This points to the fact that if we are to understand systems in the context of evolutionary processes, we must be extremely precise in how we define the system. Moreover, as we will see in Chap. 12, Systems Analysis, it definitely complicates the task of "understanding" systems that evolve new capabilities or lose old ones. These systems change even while under observation. Think of a forest ecosystem that has been stable for many hundreds or thousands of years. What happens when an invasive, but non-deleterious species comes into the system and establishes itself into the existing food web? The system is different but it is also the same. What do systems analysts working on analyzing a business process for future automation in a company do when a manager, without notification, changes something in the existing process? When the software is delivered and it does not incorporate the change, there is a problem. Was the process the same, only different?

Evolution is an ongoing process that generally involves systems becoming more complex over time. Those chance new combinations of subsystems that prove beneficial are favored in the long run, whether by reproductive fitness or by business success or ecosystem stability. Those combinations that don't work out so well can bring down the whole system.

We can offer no definitive guidance on when to say that an evolving system is a different system per se. What appears to be a unity may change and perform new functions. It may change and cease performing old functions. Or it may change internally and have exactly the same behavior but do it in internally hidden new ways. This will always be a problem. The real point is that systems science, to be complete, must always be attuned to the fact that the universe produces non-stationary environments to those systems of interest to us. And that being the case, those systems will be subject to evolutionary pressures. Our suggestion for systems scientists is to simply always be on the lookout for evolutionary effects and pay especial attention to time scales. Otherwise the system you think you understand may not be the one you originally understood.

Finally, it is interesting to note how much of that non-stationarity actually comes from other systems in the environment that are themselves evolving. In other words, as we have seen, systems really co-evolve, in recurrent feedback loops of change begetting change.

Bibliography and Further Reading

Bourke AFG (2011) Social evolution. Oxford University Press, Oxford

Carroll SB (2005) Endless forms most beautiful: the new science of Evo Devo. W. W. Norton & Company, New York, NY

Carroll SB (2006) The making of the fittest: DNA and the ultimate forensic record of evolution. W.W. Norton & Company, New York, NY

Darwin C (1860) On the origin of species, American edition, New York, NY: D. Appleton and Company. Available at: http://publicliterature.org/books/origin_of_species/1. Accessed Aug 24, 2013

Dawkins R (1976) The selfish gene. Oxford University Press, New York, NY

Dawkins R (1987) The blind watchmaker: why the evidence of evolution reveals a universe without design. W.W. Norton & Company, New York, NY

Dennett DC (1995) Darwin's dangerous idea: evolution and the meanings of life. Simon & Schuster, New York, NY

Dobzhansky T (1973) Nothing in biology makes sense except in the light of evolution. Am Biol Teach 35:125

Hawken P (1993) A declaration of sustainability. Utne Read 59:54–61

Jablonka E, Lamb MJ (2006) Evolution in four dimensions: genetics, epigenetics, behavioral, and symbolic variation in the history of life. The MIT Press, Cambridge, MA

Kauffman S (1995) At home in the universe: the search for the laws of self-organization and complexity. Oxford University Press, New York, NY

Kauffman S (2000) Investigations. Oxford University Press, New York, NY

Lovins A et al (1999) Natural capitalism: creating the next industrial revolution. Rocky Mountain Institute, Boulder, CO

Margulis L, Sagan D (2000) What is life? University of California Press, Los Angeles, CA

Morowitz HJ (1992) The beginnings of cellular life: metabolism recapitulates biogenesis. Yale University Press, New Haven, CT

Norgaard R (1994) Development betrayed. Routledge, London

Nowak MA (2012) Why we help. Scientific American, July Issue, pp 34–39

Schneider ED, Sagan D (2006) Into the cool: energy flow, thermodynamics, and life. University of Chicago Press, Chicago, IL

Sober E, Wilson DS (1998) Unto others: the evolution and psychology of unselfish behavior. Harvard University Press, Cambridge, MA

Striedter GF (2005) Principles of brain evolution. Sinauer Associates, Inc., Sunderland, MA

Tainter J (1988) The collapse of complex societies. Cambridge University Press, Cambridge, UK

von Neumann J (1966) Theory of self-reproducing automata, University of Illinois Press, Urbana, IL. Available at: http://web.archive.org/web/20080306035741/http://www.walenz.org/vonNeumann/index.html

Weiner J (1994) The beak of the finch: a story of evolution in our time. Alfred A. Knopf, Inc., New York, NY

Wilson EO (2013) The social conquest of earth. Liveright, New York, NY

Part V
Methodological Aspects

1.1 Working with Systems

The following chapters present an overview of the methods of systems science with their application to systems engineering. Chapter 12 deals with the analysis of systems, either existing systems or desired ones. Analysis is typically thought of as a deconstructive process or a reduction of a system to its component parts. While basically true, this is only part of the story. Whole system analysis looks not just at the components but how they interact with other components in the normal functioning of the system. Modern systems analysis is informed by the principles covered in Part II so that analysts know to look for complexity, networked relations, dynamics and so on as they proceed from a black box view toward a hierarchical white box view. The objective of analysis is to understand a system not just in terms of what it is made of, but how it works (and sometimes how it doesn't work) and how it might evolve over time.

Chapter 13 examines one of the main tools in the armamentarium of systems science, the use of modeling to represent systems in ways that allow scientists and engineers to test various hypotheses about a system without actually changing anything in the system or exposing the real system to forces that might be disruptive. The chapter will survey a number of modeling methods but will go into greater detail in one method, called systems dynamics modeling, in order to give the reader a better understanding of how modeling is done and what can be done with a model.

Chapter 14 considers the constructive approach to systems design and engineering, that is, the synthesis of complex systems. For the human-built world, functions that need to be performed are implemented in systems. Every modern tool or machine is complex enough to warrant the title of system. We will provide an overview of the systems design and engineering process with some examples of modern systems development.

Chapter 12
Systems Analysis

"Any fool can know. The point is to understand."

Albert Einstein (attributed)

"The improvement of understanding is for two ends: first, our own increase of knowledge; secondly, to enable us to deliver that knowledge to others."

John Locke

Abstract In order to come to understand a system, we must start by analyzing it in terms of its components and their relations to one another. But then we can treat every component as a subsystem, if necessary, and analyze them each in the same fashion. Analysis is the first step in developing a complete understanding. It is necessary but not sufficient. We cover the use of modeling systems in the next chapter. Here we examine the process of systems analysis which is used in one guise or another in all scientific (or systematic) inquiry. We include some specific tools and techniques that are used to guide the analysis of any system of any kind. Specific sciences require specialized instruments for the medium in which they work, but the process is fundamentally universal.

12.1 Introduction: Metascience Methodology

Systems analysis (SA) is essentially a methodology employed in many diverse fields in one form or another, even when it is not called systems analysis. Essentially every discipline that attempts to explain the systems that comprise their domain of interest uses some form of systems analysis. Some are elaborate and highly formal, involving complex tools, procedures, and documentation. Others may be looser and involve mere accepted "practices." Regardless the general procedures are basically the same. In this chapter we are going to outline the general methods of doing systems analysis. We will present it in a semiformal format; we do not want to get too hung up in procedural details; otherwise we wouldn't be able to cover the topic sufficiently in a single chapter.

© Springer Science+Business Media New York 2015
G.E. Mobus, M.C. Kalton, *Principles of Systems Science*, Understanding Complex Systems, DOI 10.1007/978-1-4939-1920-8_12

Indeed there are many books and journal papers covering various formal methods, especially in the fields of computing systems and systems engineering (the latter to be introduced in Chap. 14). There have been many different "flavors" of SA just within the computerized systems field alone. Most of these have been motivated by the need to produce high-quality software. That has been problematic because the advances in hardware have always outpaced those in software. We'll provide an example of SA applied to a software system later in the chapter.

We start with the purpose of systems analysis. In the many fields that employ some form of SA and which are motivated by immediate economic rewards, the deep purpose of SA is sometimes lost, leading to a loss of effectiveness in obtaining veridical results. In these fields there has been an almost endless search for the "right" method. The deep purpose is what we will cover first.

12.2   Gaining Understanding

Principle 11: Systems Can Be Understood
The purpose of systems analysis is to gain an *understanding* of the system. Understanding is not just a matter of knowing some facts about components and relations. It is an ability to construct hypotheses about how the system works and eventually proposing a *theory* of the system. The word theory, as used here and throughout science in general, means that given some set of environmental circumstances, one can, with confidence, predict how the system will behave. At that point you can say that you understand the system.

In this chapter we explore the basic concepts of analysis of systems so as to produce understanding that can be shared broadly among those who seek such an understanding. Our approach is to use all of the knowledge we have covered so far about systemness to explicate a general procedure for analysis, which will be applicable in any specific domain and that if followed rigorously will produce understanding of the system(s) of interest in that domain. It should not surprise the reader to realize that such a procedure has a great deal in common with the general science process (which includes the scientific procedure). What is different in this approach to understanding is the complete framing of it in terms of systems science and the 12 principles we enumerated in Chap. 1.

For example, consider that not too long ago the word "analysis" chiefly meant breaking something down into its constituent parts (as opposed to synthesis, which meant construct something from parts). It was associated with scientific reductionism, or explaining the object in terms of its components. Analysis, as we use it here, certainly involves something called "decomposition." But we will show that decomposition alone is not enough for understanding. In addition to breaking a system into its constituent parts and identifying the behaviors and interrelations, we go a step further. We gain better understanding if we can build a functional model of the system and use that model to test ideas (hypotheses) about the system. In effect, we combine decomposition and synthesis (of a model) into one analytical process.

In the following chapter we will cover the general concepts and methods of building and using models to test our understanding. In this chapter we will show how to decompose a system in a systematic way that provides the starting point for constructing a model.

We return to the principles of Chap. 1. Here is a brief outline of how we can use those principles to guide our gaining of understanding. The methods of systems analysis are based on the active obtaining of information and the formation of knowledge, as described in Chap. 7. We use computation (Chap. 8) to accomplish this and to use the knowledge to construct models (Chap. 13), which, in turn, demonstrate the clarity (or not) of our understanding.

12.2.1   Understanding Organization

The process of decomposition provides a way to obtain details of how a system is organized in terms of its internal structure but also its connections with its environment. We use network theory in combination with hierarchy theory to guide our analysis of that structure. We will see how we identify and analyze boundaries and boundary conditions.

12.2.2   Understanding Complexity

Our capacity to understand systems is directly affected by the complexity of those systems. Complexity is a potential "threat" to understanding. However, now that we have an understanding of complexity itself, we should be in a position to understand how to manage it in our analysis. An important aspect to grasp is that there is probably no such thing as *complete or perfect understanding*. However, that doesn't mean we can't have *practical* understanding of very complex systems. We will revisit the notion of black boxes and the concept of abstraction to see that there are ways of managing complexity by not getting lost in unnecessary details.

Question Box 12.1
Understanding has many levels. Think of something you do not consider yourself to understand at all. It would probably have to be something you do not deal with at all. Next, what would be something about which you have considerable practical understanding but no theoretical understanding? What makes it practical, that is, you are able to deal with it in daily life? What more would it take to move to some degree of theoretical understanding? What can you do with the theoretical understanding you could not do with just the practical understanding?

12.2.3   Understanding Behaviors (Especially Nonlinear)

In exactly the same way that we will deal with complexity, we will show how to incorporate nonlinear behaviors. The most interesting systems are those with the most complex behaviors. This, of course, not only means describing a system's overt behavior but also what internally motivates its behaviors. For example, when it comes to societies and organizations, we need to grapple with the internal agents (e.g., humans) and how they behave.

12.2.4   Understanding Adaptability

We have seen that the most "interesting" systems are those that have the capacity to adapt to changing environments. By changing we mean that environments have dynamic properties but are basically stationary when it comes to the values that interactive forces might take on. Adaptability is important for a system to maintain stability and resilience.

In Part III of this book, we looked at the nature of information and knowledge and how these operate within systems that are able to use information to make knowledge and change their behaviors as a result. Both computing capability and a cybernetic framework provide this important capability. As we analyze adaptive systems, we show how to identify and model these aspects.

12.2.5   Understanding Persistence

Systems that are adaptive are able to behave stably for longer time scales. The complete hierarchical cybernetic system gives a system stability, resilience, and sustainability. The latter may actually be based on making substantive changes within the system in order to meet new and persisting demands from its environment.

Our analysis of systems includes life history information. We need to look at systems not as static things, but dynamic, and generally long-lived entities. We are alerted to look for signs of adaptability but also evolvability as it pertains to that longevity.

Question Box 12.2
It is now thought that modern birds evolved from the dinosaurs. In what sense would this mean birds and dinosaurs are a single evolvable system? What kind of features might be taken to indicate this?

12.2.6   Understanding Forming and Evolving Systems

Systems form and come into existence. The history of system evolution and development is as important as its current configuration. We will see, for example, that some of the most important systems that are a part of our lives are part of an evolutionary process. Looked at from this perspective, we find that we should not consider the world as we find it as the way things will be forever.

Systems analysis is the approach we use to gain understanding of how the world and all of the subsystems within it work. Humans are voracious informavores. We have always sought understanding of everything we see, and many things we could not see but realized were there (like atoms). Long ago when humans were first developing consciousness of their own consciousness, they wondered in theoretical ignorance (they needed practical understanding to even survive!) of why the world worked as they observed. They developed language to *describe* the phenomena they observed. Over time they discovered they could examine these phenomena more closely and use causal relations to begin to explain them. During the last several centuries, this has culminated in the formal definition of the scientific process.

Unfortunately the "plain vanilla" scientific process developed into a reductionist paradigm, analysis by decomposition of layer after layer of component systems. Scientists tended to become more specialized within a relatively narrow domain of interest. Even while the sciences were thus delving deeper and deeper into the mechanisms of phenomena, the scientists were also tending to pull apart from one another in interests and methodologies. Very different and specialized languages were evolved that reinforced the separation. What got lost in this process was the systemic connections among all these subsystems and an understanding of how new behaviors could emerge as components formed new relational wholes.

But within the last several decades, reductionism is being increasingly complemented with more holistic considerations. Scientists and their specific disciplines have discovered that further progress in developing knowledge depends on understanding the systemness of the phenomena in which they are interested. They are also discovering how interrelated all phenomena truly are.

Today, the scientific process has matured considerably. One piece of evidence of this and the consilience toward common understanding is the way in which systems analysis has risen as a common approach to understanding systems at any and every scale of time and space. This process is what we want to examine in this chapter.

12.2.7   Cautions and Pitfalls

It is not our intention to present systems analysis as a fully formed guaranteed procedure for gaining understanding in *every* domain of knowledge. Nor is it guaranteed to be practically useful in every domain. This is not because conceptually systems analysis lacks anything, but because in many domains of interest, in which we talk about systems, the sciences of those domains have yet to develop the tools for exploration needed to conduct successful decompositions. When those tools for understanding emerge, they will amount to systems analysis tools aimed at some sort of understanding in the general categories of systems analysis we have presented. See the discussion below about "microscopes." The steps of systems analysis we will be presenting may seem concrete, but their actual implementation requires tools appropriate to the system of interest, so practitioners in various domains must use some caution when tackling problems using systems analysis.

One of the most difficult problems faced in doing systems analysis on very complex adaptive and evolvable systems is that too often the system changes before an analysis is completed. There has been very little research on this problem, but it has been well recognized in several fields. The area that has long used systems analysis and design approaches has been in the automation of business processes. A typical problem in very large commercial systems projects is that by the time the analysis is completed and a design implementation is under way, the business itself has changed and the design (based on the analysis) is no longer correct. If this isn't recognized in time, the delivered system won't do what the business needed. An ongoing challenge for the business community and the software development industry in general is to continue to check the needs requirements and find ways that the changes can be readily incorporated into the ongoing design and implementation. Modular programming and object-oriented (OO) programming languages have been advanced in support of maintaining some kind of flexibility in developing the software. Some gains have been made in many instances in the field, but the scientific understanding of analysis and design of evolvable systems is still a major problem.

Yet another example of analysis of evolvable systems leading to questionable results is that of the human mind. Human brains are able to evolve, that is, they can learn new ideas. This can lead to changed or new behaviors. All of us, in our social lives, are always trying to figure out how our friends, family, and enemies (should we have any) are thinking and trying to anticipate what they will do in different situations. We are doing systems analysis in an intuitive way, with the systems being the other people in our lives. And how often do people we think we know well still manage to surprise us?

Another common pitfall involves a misunderstanding of the nature of the result of applying systems analysis. Systems analysis is often practically applied to organizational systems with the intent of engineering and constructing a "better" system. For example, the most common use of systems analysis has been to automate (computerize) various business processes that involve large data sets and frequent report generation. This is certainly a perfectly legitimate use of systems analysis since a business is a system, but all too often the real purpose of analysis gets lost in the desire to get a system running quickly—a business decision. *The purpose of systems analysis is gaining understanding.* As we will see this dictates the depth and breadth of the analysis, including embedding the system of interest in its environment. When used properly, with this core intention, it can be very successful in terms of producing all that is needed to design a computationally based system. But if the analysts are motivated to take shortcuts in the interest of speeding a "solution" to gain profit leverage (a common condition in the business arena), the outcome all too often is a nonoptimal system performance with concomitant null gain of profits.

A final caution involves the pitfall of "digging too deep in the wrong subsystems." As we will shortly see, systems decomposition produces a plethora of subsystem objects that can be further decomposed. Some of these may, in turn, have considerable complexity depth. At the same time, the problem being tackled may not require as much breadth and depth in decomposing all of the subsystems. Unfortunately there is no specific rule for selecting which subsystems to go further with. Nor is it always possible to determine how deep to go. In this sense, systems analysis is still very much as much an art as a scientific procedure. Novices often find themselves bogged down in useless details while missing some important ones. It takes practice (especially domain-specific practice) and experience to become wise in making judgments of where to apply the tools of systems analysis.

12.3   Decomposing a System

Take a moment and take another look at Figs. 3.2 and 3.14. In Chap. 3 we established the organizational structure of systems and discussed how a system is comprised of subsystems that interact within it. In turn we showed that systems of interest (where our focus lay) are really subsystems in some much larger metasystem: systems belong to yet larger systems. You can always rise to a higher level where you can see this.

Later in Chap. 3, Sect. 3.3.3.3, we discussed how this gives rise to a hierarchical conceptual structure. You might want to review that section as it is the basis of what follows.

And what follows is the general procedure for explicating this structure. The procedure is called decomposition. It will be used to delve deeper into the system, level of organization by level, using black and white box analysis to find out what is there, what is connected to what, and how. And we use it to tease out the dynamics of the subsystems we find.

At the end of the process (and we will explain what we mean by "end"), you will have an astonishing product. You will have a functional and structural map of the system from which you can construct a model. At that point you are in a position to build and run the model to test hypotheses and make predictions. How to build the model will actually be explained in the next chapter. Here we will cover the decomposition process as the necessary first step and the uses of functional models to complete our understanding.

Question Box 12.3

If you already have a functional and structural map of the system, why do you still need to construct a model and test how it behaves? What does that add to understanding?

12.3.1   Language of System Decomposition

Throughout the book and especially in Chap. 6—Behavior—we have used figures showing systems with boundaries and flow arrows of various kinds. We have been hinting about the language of systems throughout. We will now explicate the language as a set of lexical elements needed to describe a system. This language will be used here to analyze and document the parts of a system explicitly. In the following chapter we will be using this language in the process of constructing models of the systems we've analyzed.

Let's now take an explicit look at these elements or language "objects." Once we understand what these objects are and what they do in a system, we will show how to use them in the process of decomposition. Figure 12.1 contains a set of lexical elements or objects. These are used to determine functionality and can be used to provide a graphical view of the system and subsystems as the analysis proceeds.

12.3.1.1   Lexical Elements

12.3.1.1.1   Sources and Sinks

The "parenthesis-like" objects in Fig. 12.1 (upper left) represent what we call un-modeled sources of flows and sinks to which a flow goes. They are un-modeled in the sense that they are objects in the environment of the system of interest that interact with the system but whose internals are unknown. Throughout the book we have shown lots of diagrams of processes/systems receiving flows from sources and sending flows out to sinks.

12.3.1.1.2   Process/System

We have covered this subject extensively so will leave it to the reader to review the concept from prior chapters.

12.3.1.1.3   Flows

The right-side, upper part of Fig. 12.1 shows a set of flows. We have broken them down into types of "stuff" flowing through. These are the following:

12.3.1.1.3.1   Organized Material Flows

Here organized means that the material already has a structure that was the result of prior work being done. For example, we've written about products and components needed to make them. Both products and components are organized

Fig. 12.1 The analysis
lexicon and its symbols are
the "objects" we use to
describe systems

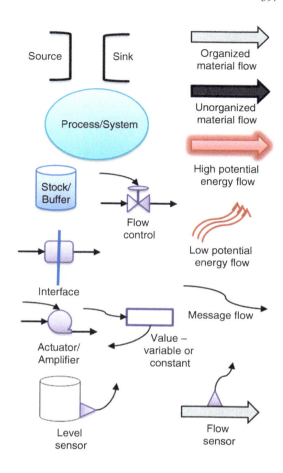

sufficiently to be "useful" in the sense that some process can use them, doing additional work, to produce a highly organized output. Biomass is a great example of organized material.

12.3.1.1.3.2   *Unorganized Material Flows*

Waste products, breakdown products, or any material that is being extruded from a process, which will require future additional work performed on it in order to make it "useful" again. This is a little tricky in that in the biological world some waste products from one kind of organism may be a resource to another kind. Going back to our notions of levels of organization from Chap. 3, we can see that there are, in fact, gradations of organization. In general, for any system we can show the outflows of unorganized material as being unusable by that system and treat it as waste.

Question Box 12.4
Conditions of a system's environment determine whether or not its outflows
are a form of pollution. Ants pasture aphids on tasty plants because they eat
the "honeydew" excreted by the aphids. Algae in streams flourish on the nitrogen runoff from fertilized fields, yet this is regarded as a serious form of
pollution. What's the difference?

12.3.1.1.3.3 *Energy Flows*

Energy flows from a source of "high" potential to a sink of "low" potential. For
example, heat flows from a high temperature object to a low temperature one. It can
be the case of warm air melting an ice cube or an automobile engine exhausting hot,
but unusable gasses out the tail pipe into the cooler air. We have seen already how
energy flow through a system drives the work processes that create greater
organization.

12.3.1.1.3.4 *Message Flow*

From Chaps. 7–9 we should have a very good notion of messages and what they do
in systems. Messages convey information when the receiving system receives a
message state that it wasn't particularly expecting. If it does, that information will
cause some change to occur in the system. We account for message flows explicitly
particularly because they are essential elements of cybernetic systems.

12.3.1.1.4   Stocks or Buffers

All of the various flows mentioned above are sourced from or flow into stocks.
External sources and sinks are actually types of stocks but generally considered as
infinite reservoirs. Stocks are internal storage containers within a process that are
used to temporarily hold whatever is in the flow stream. Inventories, batteries, reservoirs, and, in biological systems, livers and fat tissues are just a few examples of
storage containers. Most often these containers are used as buffers to resolve difference in flow timing between some source and an ultimate sink. Other kinds of stocks
can be viewed as repositories. Bank vaults and databases are examples of stocks of
this kind. The latter example, however, is unique in that drawing data out of a database does not actually remove it from storage. As we saw in Chaps. 7 and 8, data in
memories can be copied an endless number of times without depleting.

12.3.1.1.5 Flow Controls

These devices were explored in Chap. 9. Control systems need several kinds of actuators to produce physical results in the processes. Flow controls, also sometimes referred to as "valves," are mechanisms that can reduce a flow down from some maximum value per unit time, usually down to zero. They are actuated by a message flow as shown in the figure.

12.3.1.1.6 Interfaces

We haven't delved much into this type of object, but these are the components that explicitly act to transfer inputs and outputs across boundaries. Interfaces establish mechanisms that do the transfer or can also computationally translate inputs into a form usable by subsystems inside the system of interest. A biological example of an interface is a protein channel embedded in a cell membrane. These channels are activated by different mechanisms, for example, a neurotransmitter attaches to a sodium channel ligand site in a postsynaptic membrane, signals the opening of the channel letting sodium ions flow into the cell interior, and triggers a depolarization event. Another interface, quite common in nearly everyone's experience, is the USB (Universal Serial Bus) ports on your computer. This device has self-contained within it the ability to translate electronic signals from very different devices into a common form for getting data into and out of the computer.

12.3.1.1.7 Values Slots

These are relatively specialized forms of a stock or buffer that acts as an internal source of a value that is used, generally, in the control of flows (above) and actuators (below).

12.3.1.1.8 Sensors (Level or Flow)

As in Chap. 9 we will see many examples of systems that are able to sense externally and internally as they attempt to adjust to changes in their environments. There are basically two kinds of sensors: one type to sense the level in a stock and the other to sense the rate of a flow. These devices transduce a physical property (e.g., water depth) into a message, which, as seen in Chap. 9, is used by the controller process (a computation, Chap. 8) to make decisions based on the current state of the system being monitored.

12.3.1.1.9 Actuators

The final piece is, in some ways, the antithesis of a sensor. An actuator is any device that amplifies or increases a flow pressure, thus possibly increasing the flow rate. A pump, for example, is a device that transfers high powered mechanical kinetic energy into increased flow rates. A transistor is a form of amplifier that modulates the flow of a higher level of voltage given a low-level input signal (a message). Muscles are another important kind of actuator. They are based on biochemical reactions that result in the shortening of fibers and exerting a pulling force to create motion of biological tissue.

12.3.1.2 Uses in Decomposition

It may seem ambitious to make such a claim, but we do assert that any complex system can be described using these lexical elements with the addition of augmenting data. For example, we may find in our system of interest a subsystem (process) which takes in a few input types (flows) and outputs a processed type. One important piece of data to augment this with is an identifier, a name that indicates an item's uniqueness and what role it plays in the system. So, for example, we find a product manufacturing process as part of a larger business enterprise. It receives parts from an inventory stock, energy from the internal power grid, labor from the labor pool, management decisions from the logistics level coordinator, etc. and it produces finished products that are sent to the product inventory for later shipment to customers. If there were several similar product manufacturing units, we would want to call this one the "Widget_manufacturing" process as opposed to, say, the "Whatchamacallit_manufacturing" process. We will see more of this labeling later.

Important attributes of flows include what is flowing (what kind of material), by name, how much is flowing, by units per unit time, and possibly other quality measures such as density, temperature ranges, etc. Similarly stocks are labeled according to their contents, and such.

These attributes are called state variables and any measurement that describes the state of the object per unit time is specified. Figure 12.2 shows the kinds of labels and attributes that augment some of the objects in a system. Note that the ID numbers can be used to encode the level of decomposition as well as which higher-level object "owns" the specific object (see discussion of the composition hierarchy below).

As a system is decomposed, the analyst looks for these items at each level, labels them, and provides the augmented state data in the form of acceptable or likely ranges. For example, a stock might have a fixed maximum level (like a water tank or a battery) and a fixed zero minimum. But it might be necessary to define the range

Fig. 12.2 The augmentation data kept with representative types of objects found in the decomposition of a system. See the text for explanations

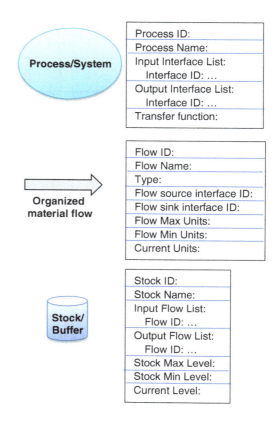

of acceptable levels. Later we have to determine what happens when the measurement of level gets outside that range.

In short, decomposition proceeds level by level (see Fig. 12.3 below) with a delineation of all of the subsystems, flows, flow controls, etc. until we have a consistent picture of what the parts are and how they are connected.

In the figure both the process object and the stock object include input and output flow lists. These are lists of all of the inputs and outputs associated with that particular object. The interface IDs are listed. Some interfaces receive and some supply the flows that enter and leave processes. Note that the interface IDs of sources and sinks are included in the flow object data as well. Thus, it would be possible to determine the interfaces by just knowing the flow ID, but that would take more time to search through the data.

The process object includes a data item called "Transfer function." This is a formula or equation that gives the output(s) as a function of inputs. Since there are multiple inputs and possibly multiple outputs, this is a more complex equation than many readers are used to seeing. Actually in most cases this will provide a set of equations with single outputs that have to be solved simultaneously using linear algebra techniques. One of the hardest parts of analyzing a system is determining

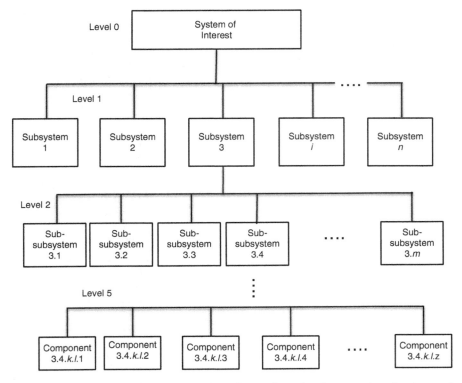

Fig. 12.3 A formal, hierarchical decomposition diagram is used to document the subsystems and their sub-subsystems, down to the component level. In this example subsystem 3 has been further decomposed to a second level. The *dots* represent additional subsystems that are not shown. Subsystem 3.4 is shown as having been further decomposed for three more levels (levels 3 and 4 represented by *vertical dots*). For this particular decomposition level 5 is the lowest level and the entities are now labeled as components, signifying no further decomposition is needed. Other hierarchies may go to deeper or shallower levels. See text for explanation of the numbering system

the transfer function, or what is called "system identification." The methods are beyond the scope of this book, but many books have been written about the procedures and methods that are appropriate. The transfer function will be used in modeling to simulate the dynamics of the system and its subsystems.

Question Box 12.6
Although the language of this sort of analysis reflects its roots in the world of manufactured processing, its potential is not limited to such systems. For example, try applying it to the analysis of strengths, limits, and implications of two different teaching styles, the instructor simply lecturing students versus emphasis being put on student preparation and participation in discussion.

12.3.2   A Top-Down Process

Decomposition, in general, is the taking apart of a system or breaking it down into its constituent parts. This statement will require a fair amount of unpacking.

We start with the system of interest as our focal point. Along with the system, which at this point is treated as a black box, we identify the inputs and outputs, the various flows that penetrate the boundary of the system. Sources of inputs are treated as un-modeled objects as described in the lexicon above. The same goes for sinks. Once all of the inputs and outputs have been identified and labeled, the process of decomposition begins. One special note: the boundary is not necessarily explicitly described, though in some situations this is desirable. Rather, the boundary is implied and will be represented by the set of interfaces that provide the "portals" through the boundary for the flows of inputs and outputs. These for many purposes sufficiently define a systemic "inside" and "outside."

12.3.2.1   Tools for Decomposition: Microscopes

The generic tool for converting a black box into a white box is what we will call a "microscope." This is a term we apply to any method that we have for looking at the internals of the system. What we are going to do is dissect the system and look at the component parts as if we were dissecting an organism and looking at the internals through a magnifying lens. In business process analysis this is done by an analyst examining documents and asking the participants questions. The microscope used in examining an ecosystem would be field collection of samples and lab analyses. Different realms have different kinds of systems that require different tools and methods of teasing apart the pieces. In all cases, though, the intent is to observe increasingly finer details as we proceed down the composition hierarchy. The real objective of many of these microscopes is to not require destruction of the system being observed. Ideally the system can be observed "in vivo" or actually operating without causing any disruption to its workings.

With the dramatic lowering of costs of computation and sensing devices, an increasingly important tool and method for gathering detailed functional data on the workings of systems is the use of distributed intelligent sensor networks. These consist of sensor arrays where individual sensors are distributed at key points inside a system; generally multiple sensor types are clustered together. A microcontroller (an embedded computer) samples the sensors on regular intervals and forwards the data collected, usually via wireless networks, to central computers where the data is put into databases for analysis. Some authors refer to these as "macroscopes" rather than microscopes in the sense that they are used to get the big picture. However, we prefer to stick to microscopes since they are designed to gather detailed data on the structure and dynamics of systems without being particularly intrusive.

These kinds of systems are being deployed to gather data on everything from what's going on in human bodies to the physical attributes of cityscapes. They are used to monitor building efficiencies (e.g., heat and lighting), and the analyses of this data can be used in altering the controls of the heating, venting, and air-conditioning systems.

Another very important nondestructive microscope being used in neuroscience and psychology today is the functional magnetic resonance imaging (fMRI) scanner. This device can take real-time pictures of what is going on in terms of activity in various parts of the brain while a subject is actively engaged in cognitive or affective activities. This device has proven invaluable in helping scientist begin to understand how the brain works to produce the myriad forms of behavior in animals and humans. We will provide an example of systems analysis of the human brain later in the chapter.

12.3.2.2   Scale, Accuracy, and Precision of Measurements

The microscopes used are subject to some important rules in terms of what measurements are made, what units are used, and the accuracy and precision of measurements at particular scales. Since decomposition produces smaller and smaller scales of attention, the measurements taken must also become more refined as one moves down the levels of the hierarchy. Since we are ultimately interested in the dynamics of systems as well as their structures, we need to pay particular attention to the units per unit of time. For example, flows that would ordinarily be measured in, say, kilograms per minute at one level might need to be considered at grams per second at the next level down. Thus, decomposition of measurements should be thought of as a form of taking derivatives using smaller and smaller units, whereas going the opposite direction involves the integration over longer time scales. The analysts need always be mindful of these measurement technicalities when performing the decompositions.

12.3.3   Composition Hierarchy

Figure 12.3 shows a structural result of system decomposition. As we have repeatedly pointed out in the previous chapters, complex systems are structured hierarchically. Here we see another view of what Simon called "near decomposability." Actually, most systems are theoretically completely decomposable. But practically they tend to be only partially so. Analysts often have to infer lower level structures and functions from a black box analysis of a subsystem. In many cases we find that part of the problem is the lack of appropriate microscope tools at the level of analysis desired. At other times it simply boils down to a matter of not having enough time.

Knowing that we are looking for the composition (also called structural) hierarchy, here is how we proceed.

The system of interest, as a whole, is assigned to level 0. This includes identifying all of the relevant inputs and outputs (see, e.g., Fig. 12.4) and their sources and

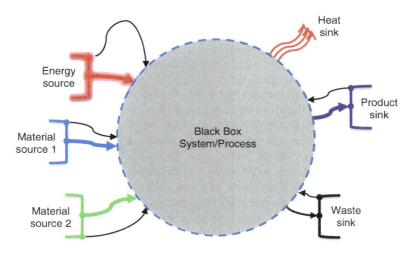

Fig. 12.4 Level 0, the system of interest showing the mapping of inputs and outputs is approached as a black box. *Thick arrows* indicate flows of materials and energy (the *wavy red lines* represent the flow of unusable heat energy being dissipated into the environment). The *thin black arrows* represent message flows. Here we only show messages from external sources and sinks that are received by the system. In real systems there can be many two-way communications with material/energy sources and sinks. In addition a complex adaptive system can be observing many additional objects in the environment that are not directly interacting with the system (see Chap. 8, Strategic Management)

sinks. At this stage nothing is known of the internals of the system. Observational analysis (using the microscope) should provide the formal functional relations between the inputs and outputs as described in Chap. 6. After finding the dynamics of the whole system, its behavior under varying input conditions, the next step is to increase the resolution of the microscope and start looking inside the system.

What the analyst now looks for is subsystems, stocks, and flows. The input flows will map from the boundary interfaces to either stocks or processes internally. In some cases special processes that include actuators are actively regulating the flows. Some of those same processes are responsible for distributing the inputs, for example, to various other processes. Below we will go through a detailed description of discovering and mapping the internals of the system—the procedure of decomposition.

In terms of our structural hierarchy as in Fig. 12.3, the internals are delineated as level 1 of the top-down tree structure. In the figure we have taken the liberty to use the terms subsystem sub-subsystem to cover the range of objects including stocks and flows. Here we are only interested in the structural arrangement of subsystems relative to the holding system. Sub-subsystems are really just subsystems of the level object above them. We use this terminology just to indicate that we are delving lower into the hierarchy. The component level is where the analysis will halt further decomposition. A component, in this sense, is a whole object that need not be further decomposed as its function is well known. For electronic equipment, for example, this would be at the resistor, capacitor, transistor, etc. level.

For an organization we might decide to stop at the individual participant level unless we thought that different personalities had a relevance to further analysis. Again, the analyst has to have developed a deep understanding of the different kinds of systems within their domain of expertise to exercise judgment about how deep the decomposition needs to go.

In this relatively straightforward hierarchy, we have adopted a numbering scheme to provide unique identifications for each object. The form is called a "dotted" number. This format is often used in outlines (which are clearly hierarchical in structure), in technical documents for easy section referral, and, as you have surely noticed, in this and other similar books, again for easy reference.

The leading number is the index number of the object (subsystem) at level 1, the first decomposition level. Every item is given a unique number. As you will see below, the number is actually preceded by a descriptor character, such as P for process, or S for stock. We have condensed these into the boxes representing subsystems to keep the figure from being busier than it is!

The next number represents the subsystem index within the first number. So S3.4, for example, is the fourth index object found in decomposition of subsystem 3 at level 2. The number sequence between dots is the level in the hierarchy. The figure shows us skipping the details from S3.4 to S3.m. There are m objects comprising the subsystem S3. Again this is just to show the general format of the hierarchy.

We use a similar method to show the decomposition down to level five for the branch rooted at S3.4. Levels 3 and 4 are represented by the vertical ellipses. The numbers in the series of objects at this level are indicating the following: k is the index of the object at level 3, and l is the index of the object at level 4, that is, a subsystem of object S3.4.k. Then all of the components are indexed as before.

These identification codes allow us to rapidly place an object at any level in the hierarchy. They are going to be used, however, as we document the decomposition in a special data repository for later use in analysis.

12.3.4 Structural and Functional Decomposition

We are now ready to decompose a representative system. The process of decomposition will expose both structural and functional relations between objects as it proceeds. The starting place is the whole system of interest embedded within its environment. That environment is, by definition, not part of the system, and the various sources and sinks are un-modeled except to be given names and identifications for record-keeping purposes. The flows, however, are specified. We do this by analyzing the boundary conditions of the system. Imagine examining the entire boundary of the system and locating inflows and outflows of all types. When a flow is discovered, it is characterized and measurements of the rates are noted.

Once the totality of flows into and out of a system has been characterized, it is feasible, at least in principle, to identify the whole system function, that is, the states of outputs given the inputs. We say "in principle" because if all inputs and outputs are accounted for over a very long period of time, a deterministic transfer function exists. The downside is that part about observing over a long enough period of time. This is hard to establish in practice for most complex systems, and for adaptive or evolvable systems, it is nearly impossible. At first this might seem like a show stopper, but in practice we can never really know all there is to know about any system. Our knowledge of systems, itself, evolves over time. The best that can be done is to make some judgments about the length of time of observation needed to obtain reasonable accuracy based on short-term observations of the fluctuations in flow levels. Again this is a matter of experience in the domain of interest. In later examples we will provide some heuristics that can be employed in various domains.

Once the flows, sources, and sinks have been identified and characterized, we are ready to begin.

12.3.4.1 The System of Interest: Starting the Process

Figure 12.4 shows the initial system as a black box. The inputs and outputs have been identified and characterized sufficiently well that we can begin to decompose the system to discover its internals.

Applying the appropriate "microscope(s)," we can begin to explore the various subsystems within.

12.3.4.2 Decomposing Level 0

The first step is to identify the various subsystems. The input and output flows give hints as to what to look for. Every such flow will be found to have an associated actuator process, processes to receive inputs and processes to push out outputs. These processes will be found in tight association with the whole system boundary so are generally easy to locate. Other internal processes may be more difficult to locate, but we will employ a method for finding them.

Figure 12.5 shows internal processes that have been identified. Note the numbering system used to identify the objects found. Also note that we have omitted interfaces for now to simplify the picture. In this figure we do show the internal processes and stocks because they are "obvious" to casual observation. In other cases they must be discovered.

All input processes send their flows on to either other internal processes or to stocks. Similarly output processes have to get their flows from somewhere in the interior. So in both cases, following the flows, forward from the input processes or backward from the output processes will lead to either interior processes or stocks. Figure 12.7 shows the mapping of flows of energy and material through the system.

Even though we have indicated that the mapping of processes and stocks and the mapping of flows are two stages, in reality there is often an iterative process of

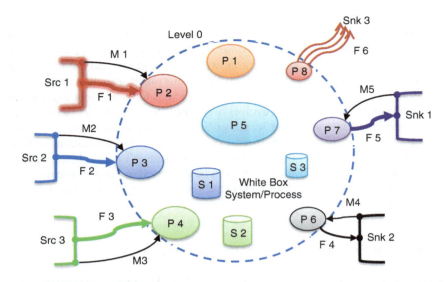

Fig. 12.5 The first stage of decomposition involves discovery of all of the internal processes and stocks. All objects relevant to level 0 are labeled accordingly. Src stands for source, Snk for sink, P for process (or subsystem), S for stock, F for flow (energy or matter), and M for message

refinement of these maps, where the analyst may need to use the flow maps to better identify the processes or stocks.

Stage two requires the accurate mapping of flows internally. For all processes that have been identified, we treat each process as the system of interest. That is we isolate each process and do a boundary condition analysis just as we did for the original whole system. Figure 12.6 shows the treatment of process P2, the energy input process for the whole system. The figure shows that a single input of energy is output as eight flows, through an interface. Also shown are a number of message flows both in and out as well as their interfaces. The message arrows from/to I9 are two headed indicating a two-way communication. The implication of these flows is that there are eight processes to which flows of energy and messages go. The two message arrows from/to I8 are part of the control system we will investigate later. The incoming message arrow to I7 is the original message input from the external energy source shown in Fig. 12.5.

The same treatment is given to each of the input and output processes in order to identify and catalog all of the flows that need to be accounted for within the system. The next step is to map the outputs of the input processes to the other, possibly yet not discovered, internal process. Note that during this step, the output of energy from P2 will match up with inputs to the other input/output processes, each of which will be found to have an energy input (used to power those processes).

Figure 12.7 shows the mapping of energy flows into P2 and then out to all of the other processes, both input/output and internal. The figure includes a similar mapping of material flows. Message flows will be examined shortly, but were left out of

Fig. 12.6 Doing the same
kind of boundary analysis as
was done for the original
system for process P2 reveals
the various inputs and outputs
that cross the boundary

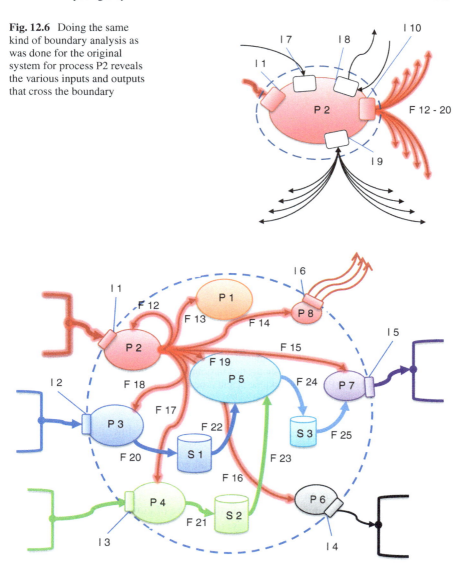

Fig. 12.7 In stage 2 of decomposition, we map all of the flows between internal processes and
stocks. In this figure we show only the high potential energy and material flows. Note the numbering system used to provide unique identifiers for each object. We have also added the interfaces
between the external flows and the internal processes. Other internal interfaces have been omitted
for simplicity

this figure to keep it uncluttered. The same is the case for interfaces. We can get by
with this because it is safe to assume there will be an interface associated with every
flow of every kind at the boundary of any process. However, what we do here to keep
a figure uncluttered should not be an excuse to not pay attention to real interfaces!

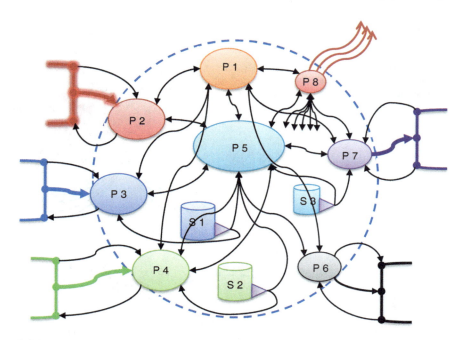

Fig. 12.8 Communications channels in the system are represented by *thin black arrows. Two headed arrows* represent two-way communications. Note the *purple sensors* on the stocks. These are measuring levels in the stocks and supply that information to both the coordinator (P1) and the material input processes (P3 and P4) and the output processor (P7)

We will show a few examples of analysis of interfaces to show the importance of doing so.

Figure 12.7 only shows the flows of material and energy into the processes and stocks. It does not show the waste material or heat flows that are handled by P6 and P8. We will leave that as an exercise for the reader.

However, in order to impress upon you the nature of complexity even for such a seemingly simple system, we will show you some of the communications (message flows) that go on internally and externally to accomplish logistic and tactical controls. Figure 12.8 should convince you that information is what makes the world go round. And the thin black arrows in the figure are merely representations of the channels of message flows that are needed to convey cooperation and coordination in this simple system.

Think of this example as a small manufacturing company that gets exactly two kinds of components (parts) from outside vendors. It gets electricity from the local utility (red source). It sells its product, which it produces in process P5 to a single customer. And it generates both waste materials, which it must pay the garbage company to take away, and heat; process P8 is the air conditioner.

We have not explicitly shown the message flow ID numbers in this figure because, yes, it would get too cluttered. You can start at the upper left corner of the figure with what was labeled as M1 in Fig. 12.5 and start labeling the message flow arrows not already given IDs. Then start in the same corner but inside the boundary giving IDs to the arrows. Don't forget to give two IDs to the double-headed arrows! The double-headed arrows are just a convenience to collapse many likely channels into one item for diagramming purposes. We've also left off interfaces as before.

At this point, and we are still just at level 0, you are probably convinced that we are producing a substantial amount of data with all of these objects and their attributes. And you are right. Systems analysis can be excruciatingly detailed and complex. Finding the objects and recording their attributes are a time-consuming process. Keeping track of all of that data and being able to use it to develop our *UNDERSTANDING* are, arguably, an even bigger challenge.

12.3.5 System Knowledge Base

At this point you will undoubtedly see the rapid explosion of detail that needs to be tracked as the decomposition proceeds. Recall the augmented data that was needed for every object we find. In order to keep track of all of this data, we introduce a special database, sometimes called a system knowledge base. The system knowledge base captures all of the detailed data associated with every object and has built in procedures for checking consistency, e.g., checking that an outflow from one process is accounted for as an inflow to another process or a stock. The analysis of complex systems would be impossible without this tool. It will be found in one form or another in every field. Huge databases of data cataloging everything from stars and galaxies to genes and their variant alleles have been constructed. These databases are queried for specific relations between objects cataloged. Data mining, or finding unsuspected patterns in the data, is another mechanism being used to aid analysis of systems.

All of this is accomplished with computers of course.

12.3.6 The Structural Hierarchy (So Far)

Figure 12.9 shows the structural hierarchy of the system after the decomposition from level 0. The objects identified, cataloged, and described with data are the starting points for decomposition at level 1. The tree structure here has been drawn in a peculiar way in order to show as much of the structure as possible. Note that we used vertical lines to connect the families of objects to the main horizontal line. In truth, of course, the structure should be strung out horizontally with each object at the same level in the tree.

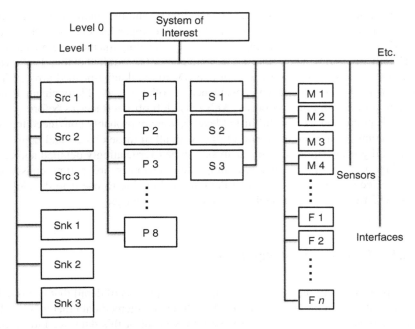

Fig. 12.9 The decomposition of the original system of interest produces a structural hierarchy (tree). Each *labeled box* contains all of the augmented data associated with each object. This is how the data are organized in the system knowledge base

12.3.7 Specifics Regarding Flows, Interfaces, and the Objects of Interest

One detail we haven't gone into that probably needs some further explanation is the role of interfaces in getting flows out of and into various other objects, namely, stocks and processes. We will not be concerned with sources and sinks since they are un-modeled objects; we assume they have the appropriate interface.

Interfaces make it possible for flows to connect to the various other objects. Figure 12.10 shows a typical arrangement, but in particular, one involving two boundaries—the outer containing boundary of the meta-system and the boundary of the process that actually receives the flow. In this case, the interface is first recognized at the whole system boundary and given an ID accordingly. However, the interface is also associated with the process (P3 is responsible for obtaining the parts and getting them into inventory (S1). When we do the decomposition of P3, we will have to show that the interface is valid for both levels.

Here we trace the flow of a material—a part—from the vendor (Src 2) into the system via process P3. P3 is a special kind of process, a resource obtainer, that actively gets the parts needed and transfers them to inventory (stock S1). In the figure we see three interfaces along with their names and inferring their "jobs."

Fig. 12.10 Interfaces can be as simple as a parts receiving dock or involve internal processes themselves

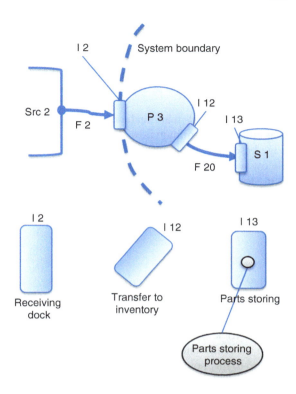

When we decompose P3 and I2 at level 1, we will see how the receiving dock works to get material into the system and how P3 works to get it into inventory. At this point we only need to note that interface I2 serves a single purpose but is recognized at two levels of the analysis. Also, anticipating further decomposition of I13, note that this interface involves an internal process of its own. Stocks are not considered active in the way a process is. Somebody (or a robot) has to physically move the parts from the shelf where incoming inventory is deposited by P3 (the shelf being part of the interface) and put it in an appropriate bin in the inventory cage or room. We show an internal process inside the interface to handle this small detail.

12.3.8   Where We Are Now

At this point we have two products of the decomposition. The structural hierarchy as shown in Fig. 12.9 shows how the objects discovered from level 0 to level 1 are related. The nature of the links in this tree diagram can be thought of as "belongs to" or "is part of" from level 1 to level 0.

The second product is the functional map. All of the figures shown with ovals and arrows are representative of this map. The figures we have shown lack the details that a complete map would have. However, if all of the data structures in the system knowledge base are completed, e.g., flows identify both source and sink interfaces, we actually have all the information necessary to construct a complete map. Computer programs then can search through the system knowledge base and generate graphic maps at any level of resolution needed. There will be more about this later.

The functional map includes detailed data on the flow metrics, including maximum and minimum flow rates. So the map (as represented in the system knowledge base) has all the information needed to describe the dynamics of the system. Below we will give a preliminary description of modeling in which the computer, using the functional map and a simulation program, simulates the dynamics of the system. That will be fully covered in the next chapter.

At this juncture we know the overall structure and function(s) at level 1 for the whole system.

12.3.9 Recursive Decomposition

Now that we have covered the basics of decomposition and its products, we perform a magic trick. We don't yet have enough detailed knowledge about the system to claim we really understand it. It is time to decompose the objects we have discovered up to this point. And the magic is to simply do the same operations we did for level 0 to all of the decomposable elements in the system knowledge base at level 1.

As an example, Fig. 12.11 shows the starting point for decomposing process P1, which is actually the management process for the whole system. We've already started the decomposition to show the internal processes (or some of them anyway); management consists of a logistics coordination process, a tactical coordination process, a strategic management process, and an energy import and distribution process. Shown are just four of the sources/sinks that interact with P1. P2 is the main energy importer and distributor for the whole system as in Fig. 12.8. Here it, along with P3 (a material importer as in Fig. 12.10) and P5 (the main manufacturing process), and P8 (the waste heat remover) are shown as sources and sinks as if they were un-modeled. Insofar as our focus is now on P1, it is legitimate to proceed as if they were un-modeled because in a similar decomposition of these objects, they will be modeled and added to the system knowledge base.

The decomposition of P1 follows exactly the same procedure as with the whole system at level 1. We therefore produce a level 2 map of everything that is relevant inside P1. This reapplication of the same procedure, but at a lower level in the structural hierarchy, is called "recursive" decomposition. Once the full map of P1 is exposed, we would start with process P1.1 (the strategic management process) and perform the procedure over again. What we end up with is a tree structure as shown in Fig. 12.12.

The tree in this figure resembles that in Fig. 12.9 because we have applied a recursive method to extend the decomposition to the next level down. The self-similar structure is a signature of recursion. The same procedure is repeated for all of the

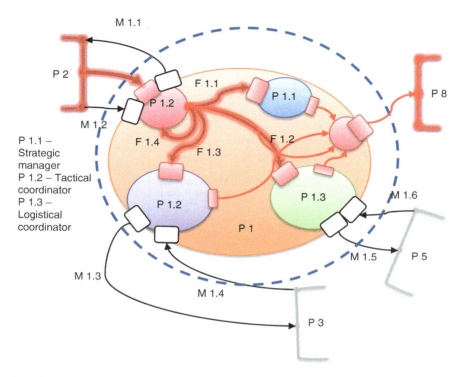

P 1.1 –
Strategic
manager
P 1.2 – Tactical
coordinator
P 1.3 –
Logistical
coordinator

Fig. 12.11 Process P1 is now ready to be decomposed. Here we show other processes from level 1 as if they are un-modeled sources and sinks. The mapping shows message flows between P1 and three other processes. Internally we show the start of decomposition to discover internal processes and flows of energy

level 1 processes (and interfaces as necessary). The tree starts to get very "broad"; the number of nodes at each level of the tree increases. Clearly this is not a tree we can easily draw to show all details. Thus, some kind of abstraction is needed to represent the tree visually, leaving the details to be handled by the system knowledge base. Figure 12.13 shows the use of abstraction by encapsulating the full details of a process inside the process. The result is a process hierarchy.

Decomposition can be done in either a depth-first manner or a breadth-first manner. Or the analyst can alternate as seems necessary. Depth-first means that decomposition proceeds to the deepest (lowest) level of the tree structure before decomposing the sister processes at a given level. Figure 12.14 shows this approach. The brown lines in the tree represent the path of decomposition.

Breadth-first decomposition goes level by level first decomposing all processes at level 1, then starting from the left-hand-most process in level 2, it decomposes to level 3, and so on.

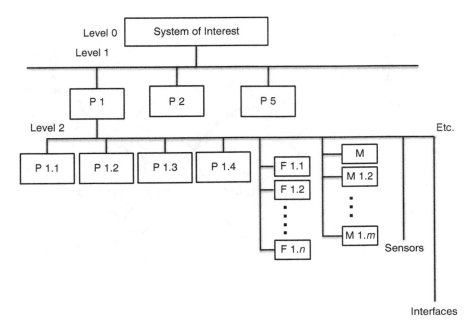

Fig. 12.12 Decomposing process P1 produces a structural hierarchy that resembles the tree rooted at the system of interest (Fig. 12.9)

Both of these approaches produce the same tree structure, and there are operational arguments for doing either at some particular time during the analysis.

12.3.9.1   When to Stop Decomposition

Regardless of whether one takes a breadth-first or depth-first approach (or a combination) ultimately the question arises, "Where does this stop?" Since all systems are physical structures, it is conceivable that we could drive down to the atomic level or lower. That is, we could adopt a reductionist attitude and try to explain the system in terms of its most fundamental elements. But in addition to the fact that the deeper levels would not capture properties that emerge through the more complex relationality at higher levels, for most questions there is a level beyond which deeper and more detailed analysis adds little of relevance. Clearly there is a point beyond which decomposition is neither necessary nor practical.

In Fig. 12.3 we introduced the terminology of "component" as the lowest nodes on a branch of a structural hierarchy without elaboration. We use the word component in several different ways. It could mean a subsystem that is discovered within a higher-level system, when it is first identified as an object. Or it could mean (as suggested in the figure) a simple object that needs no further decomposition. Here we address what this means and how one answers the question, "Is this a component, or do I need to decompose it further?"

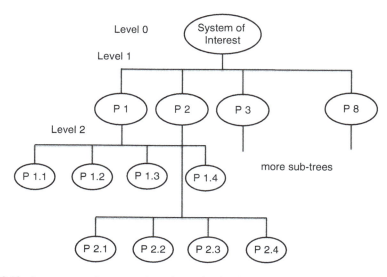

Fig. 12.13   A process tree is a convenient abstraction for the entire decomposition tree. Each process (now represented by a simple *oval* to avoid confusion with a decomposition tree) contains all of the details regarding sub-processes and flows. Here we show P1 and P2 down to level 2. The sub-processes of P2 are shown lower only to avoid clutter. As indicated by their P numbers, they are at the same level as the P1 sub-processes

Fig. 12.14   A depth-first decomposition proceeds along a path as shown here. The *brown line* shows the progress along the *leftmost* branch of the tree. ETC. stands for any further depth that might be discovered by further decomposition

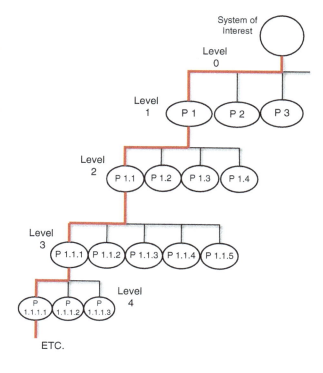

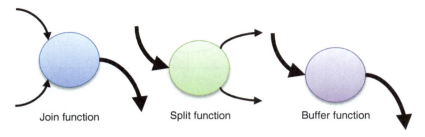

Join function   Split function   Buffer function

Fig. 12.15 Simple functions such as these can be used to make a decision to stop decomposition and treat these objects as components. The flows are materials or energy. In the case of material flows, there will also be an energy flow to support the function that is not shown

Unfortunately there are several problems (see below) with this question. A lot depends on the kind of system one is decomposing. Components, such as in an electronic circuit, are easy to identify, and one who "reverse engineers" circuits—one kind of decomposition—knows that once identified as a part, no further decomposition is needed. Indeed, standard modular designs, for example, in digital circuits, make it possible to stop decomposing at a fairly high level. The same cannot be said for decomposition in biological systems. For example, in analyzing the workings of a single cell, we can decompose the subsystems and find objects like mitochondria which are still quite complex and whose function depends on sub-subsystems within it. The question is even more difficult when analyzing organizations or ecosystems.

One heuristic for deciding that an object need not be further decomposed is shown in Fig. 12.15. This is the "simple function" heuristic or stopping rule. It is based on the black box analysis of simple subsystems that perform these simple functions.

The joining of two inflows of the same type to produce a single outflow implies that the internal work is simple and needs no further decomposition. In the case of material flows, there will also be an energy input and a heat output. We ignore these energy flows because they are effectively *givens* and the black box analysis is sufficient to specify them functionally. In the case of energy joining (e.g., an amplifier or actuator) one of the two inputs is the source of work power and only the heat output is considered. Similarly splitting an input flow into two outputs is considered simple, with the same consideration for energy flows. If three or more inflows are joined or three or more outflows are split, then further decomposition might be needed to determine whether or not these operations are done in a single process or more.

The buffer function is actually a material or energy stock in the disguise of a process. It too will have energy inputs and heat outflows that can be handled by the black box analysis, but we know from the fact that there are simply inflows and outflows of exactly the same type that all that happens in this object is a temporary residence of the stuff that is flowing (e.g., a water tank or a battery). If the buffer function is for messages, however, then we have a slightly different problem. The process itself is a computational one and the output messages may not be the same as the input messages. It is usually a good idea to further decompose such a buffer since computations have to be dealt with more rigorously than say the storage of water or electricity. It is important to know the algorithm(s) employed. Simple black box analysis cannot always be used to reliably infer what the computation involves.

This heuristic is generally a pretty good guide to when to stop decomposition on any branch of the structural tree. It is, however, not foolproof so when in doubt it doesn't hurt to at least start a white box analysis to see what you find. If it turns out that the internals are really as simple as implied by the black box analysis, then little harm is done in starting an analysis and abandoning it when the determination is made. A little time may be lost, but knowledge is always gained.

> **Question Box 12.7** Analysis of organizations often stops at some high level of system function such as accounting, research, marketing, etc. What kind of questions would prompt further decomposition of these units? What sort of question might call for a decomposition of a unit's workforce into individual personality types? Can you think of a question where further decomposition into individual biological or metabolic traits would be relevant? How often would such information be relevant to the kind of questions that prompt a map of flows from accounting to research and marketing and the like?

12.3.9.2 Tree Balance (or Not)

Decompositions of very complex systems rarely give rise to a balanced tree structure, that is, one in which all components are on the same level. In fact the situation as shown in Fig. 12.16 is more common. This shows a root process node deep in a complex system structural hierarchy. Some of the children levels have both process nodes and component nodes. The process nodes require further decomposition. The component nodes do not.

12.3.10 Open Issues, Challenges, and Practice

Systems analysis is a work in progress. There are many parts of the process that are as much art as science. In this section we mention a few "caveats" that make analysis something less than routine work. In several of these areas, there are open research questions that systems scientists are pursuing.

Fig. 12.16 Some process nodes will be composed of both subsystem processes and components that need no further decomposition. This gives rise to a fairly unbalanced tree structure

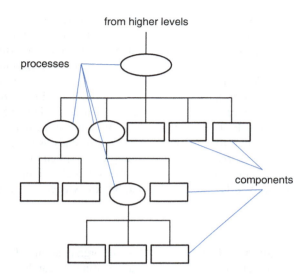

12.3.10.1   Recognizing Boundaries for Subsystems

Recall from Chap. 3 that system boundaries can be tricky. Physical boundaries may exist yet be difficult to locate. For example, the boundary may be the result of having stronger coupling links between components on the boundary and the internal components of the system (Fig. 3.5, Chap. 3). When decomposing a system, it may take some time to recognize portions of the internals as bounded subsystems. One approach that seems promising is to use network theory and our understanding of clusters and hubs (Chap. 4, Figs. 4.5 and 4.6) to identify likely candidate subsystems. If you recall from that chapter, clusters, for example, are formed when a group of nodes are more strongly connected to one another than to other nodes in the network. A cluster can be found in several different ways, but network analysis using what are called clique-finding algorithms is one approach.

This is an area in which more research is needed to find reliable "microscopes" for identifying boundaries and, hence, subsystems. In the meantime there is nothing like experience and meta-knowledge of the "medium" of the type of systems one works with to accomplish this task.

12.3.10.2   Adaptable and Evolvable Systems

Decomposition can work relatively well in discerning the dynamics of a system, both its external and internal behaviors. In many kinds of systems, where the right kinds of "microscopes" are available, it is possible to carry on decomposition even in a functioning (dynamic) system. But what about cases where systems are undergoing adaptation to changes in the environment? Should these situations just be treated as another kind of dynamics? In fact, how would the analyst recognize a condition in which a system is adapting or has adapted to a changed environment? This question comes up repeatedly in, for example, laboratory experiments on animals' behaviors. The difference between wild-type behavior and experimentally induced behavior, if such a difference exists, might be missed. The animals being tested may behave differently in a lab setting than in their natural setting for reasons the experimenter does not know. The animal has adapted itself to a new environment. How does the experimenter factor that into her understanding of behavior?

In the world of enterprise (organizational) systems analysis, a great deal of work has been done to accommodate the nature of adaptable systems. Indeed, adaptation or learning has been recognized as a necessary part of any successful organization because they must operate in changing environments.[1] Learning organizations can be more than simply adaptable however. When what an organization learns leads to major, permanent changes, then the system is evolving. And evolvable systems are arguably the very hardest to analyze.

Part of the reason for this is that evolution cannot be predicted. Adaptations of a system to changing environments are, to a degree, predictable because the nature of the adaptation is inherent in the system to begin with. Recall from Chap. 10 that we put restrictions on the term adaptation. An adaptive system is one that has the capacity already built in to adapt to a change. Adaptation is just the temporary shift in resource allocation internally in order to meet a change in demands on the system's existing response processes. Evolution, on the other hand, involves permanent structural and functional changes to the system that are then tested by selection forces from the environment. They may or may not succeed in providing sustainability for the system.

Where there exists a body of background knowledge about systems in a particular category, e.g., living systems, it is feasible to recognize adaptability and take it into account while a particular decomposition is under way. For example, in a living system, we know in advance about homeostasis and autopoiesis, and our "microscopes" have the ability to discern aspects like the ranges of tolerance for specific environmental influences. This is possible only because many years (decades actually) have been spent studying the adaptability of living systems. Biologists' background knowledge is significant, and the newer tools used to decompose living systems, even while alive and behaving, are able to provide the kind of information the biologists can use to successfully grasp the range of adaptivity of the systems they study.

Decomposing a system that is evolvable, and, indeed, undergoing some kind of selection, is inherently hard at best. The analyst has to have considerable background knowledge of the environment of the system to understand or recognize what features of the environment are changing in such a way as to increase or decrease some particular selection force. Organizations can do this to some degree;

---

[1] See Senge (2006) for an example of systems thinking and understanding of the way in which organizations learn (and when they don't).

recall from Chap. 11 our discussion of strategic management and its attention to the larger environment and to the future. Successful organizations tend to do strategic management well. They can intentionally alter their internal workings and external behaviors to meet future challenges.

In the field of ecology, there are some very interesting problems that stem from attempts to understand an evolvable system. One is what happens to an ecosystem with the invasion of a foreign species (so-called invasive species). Ecologists need to understand how that species may upset the balance of an ecosystem as they study it. Another problem is one of habitat loss and its effects on the survivability of a species. Yet a third involves understanding how changing climate conditions will affect a species or whole ecosystem. The problems associated with decomposing the system while it is undergoing evolutionary changes are immense. We'll discuss some of them and how new instruments may give ecologists better "microscopes" to work with.

In any case, systems analysis that works to decompose an intact, operating system that is either adapting or evolving (and usually a mixture of both) cannot be done in one swoop down the structural hierarchy. One can get down to a level of detail and not be aware of changes that have occurred further up that structure, or finish a decomposition on one branch of tree, go to another branch (as in a depth-first analysis), and then have a change occur in the branch just completed. One of the single biggest mistakes that systems analysts who work in information systems, for example, run into is that they assume the analyzed system is static in structure, complete their work, and specify the automated design. When the automated system is delivered, they find out that the business has changed in ways that make their software systems obsolete even before they are installed! Not recognizing that a system is evolvable and that evolution could take place even while a systems analysis is under way is a surefire way to produce a bad result. There are methods and practices in enterprise systems analysis to try to minimize this kind of problem. Mostly they involve an iterative rechecking of previous results against current conditions in selected branches of the tree; a kind of double check to make sure nothing has changed. Unfortunately this approach is quite expensive and tends to slow the analysis process down—not something to be encouraged in the business world. This is another area ripe for more research.

12.3.11   The Final Products of Decomposition

At the end of the decomposition process, there are three basic products that will be used in the next phase. These are really just different perspectives on the system based on all of the data captured during decomposition. The first, the one that is most important to the next phase, is the system knowledge base. Literally all of the specific data about every object in the system should have been captured in this product. Because this knowledge base contains all of the linkages needed to reconstruct the full network of objects, it can be used to generate the other two products.

Software designed to provide visualizations can generate the other two products from this knowledge base data.

The other two products, a tree and a map of flows and processes, are really means for humans to visualize the system systematically. It is useful to be able to "see" the system from the perspective of a structural hierarchy—a tree—in order to make inferences regarding the system's complexity, for example. The depth of the tree is related to the metric we covered in Chap. 5. The structural hierarchies we have been showing are models of what this product looks like. Much like some "browsers" in software packages most of us use, the user can generate a picture of all or part of a tree and by "clicking" on a node in the tree bring up a data definition box showing all of the attributes and their values.

It is equally useful to be able to see the map of flows and processes that conveys understanding of the system's dynamics. This map can also be generated from the system knowledge base by special software. And as with the tree, the user/analyst can click on an object, such as processes, flows, controls, etc. and bring up the same kind of data definition box appropriate to that object.

These visualization tools work on the same knowledge base to show static pictures of the system as analyzed through decomposition. The next phase of analysis, however, will also use this knowledge base to "breathe life" into the data. We now describe this modeling phase in which the objective is to see the system in action, at least to see a model of the system in action.

12.4   Life Cycle Analysis

Decomposition is the method used to understand the organization and function of complex systems as they are observed. It does not, however, address another important aspect of understanding, and that is what happens to the system over the course of its lifetime. Here we assume that a system has a life cycle. There is an origin, a development, a useful life, and death and decommissioning at the end. Species go extinct. Individuals die and decompose (under natural conditions). Corporations either eventually fail or they morph into some new kind of company, or they are absorbed by a different company. Civilizations rise and fall. Even the Earth as a planet coalesced from dust, heated, cooled, supports life, and will likely be burned to a cinder when the Sun morphs into a brown dwarf star.

Cycles of birth, maturation, operations, and death are applicable to every system we can think of. And this means the system we decompose may not be the system that started out or will be the system that eventually dies. Thus, another analysis framework is needed to answer questions about the life history of systems.

Biologists are always interested in the life histories of living systems for obvious reasons. Many historians and sociologists are interested in how nations and organizations come into existence, last for a while, and then melt away. But it turns out that life cycle analysis (LCA) has become extremely important in the world of human artifacts as well. For example, a nuclear reactor-based power station goes through the distinct phases of design (including site environmental impact analysis), building, operating life, and decommissioning when it is no longer safe to operate (for whatever reason). Why is LCA applied? It is because the power station is to produce electricity and it should produce far more energy over its working life than it takes to work through all of those phases. For example, if it turned out to take a huge amount of energy to process, haul away, and store safely the spent fuel modules, the net energy to society to run its electric motors may end up being much lower than the cost of building and operating the station in the first place warranted.

LCA adds some important new dimensions to systems analysis, duration and cost being chief. The system in question has to have a long enough working life to produce sufficient product or service to offset the investment and operating costs associated with it. This is relatively easy to see in the case of something like a nuclear power station. It is much less easy to see with something like a wetland ecosystem. Yet the concept of energy costs versus service provided (e.g., water purification) is relevant to many policy decisions regarding such environmental issues.

Question Box 12.8
Change and innovation are sometimes regarded as desirable in themselves. What kind of "speed limit" on innovation would LCA suggest when it comes to introducing new models of products such as cell phones? Can you find analogous speed limits on change in other systemic areas?

12.5   Modeling a System

In the next chapter we will be covering the methods of modeling in more detail. In this section we want to describe the way in which modeling interacts with decomposition to produce understanding of the system. Up until now we may have left the reader with the impression that decomposition is fully completed before we try to execute a model. This is generally not the case. In fact we can build and execute models of systems at any level of decomposition. That is we can use the system knowledge base at any level to construct a "rough" model of the system. This is related to black box analysis in that we don't actually know how the various subsystems at that level might work in detail. But if our transfer functions are approximately correct, then we have an approximate model that should give reasonable results for that level of "resolution."

Start with the system pictured in Fig. 12.4. This is the "whole" system along with the parts of its environment with which it directly interacts. Recall that doing a black box analysis means gathering data about the inflows and outflows in order to estimate a transfer function—determining the outputs given the state of the inputs at any time tick. This is hard to do for any really complex system, especially if you are not decomposing destructively; nevertheless, this is the starting place for all analysis.

At this point the transfer function is most likely only a rough approximation of what the system does in response to changes in the environment. Even so, it is possible to build a model of the system as a whole and check the veracity of your transfer function. By comparing the outputs of the model with the actual behavior of the system, the analyst gets clues about how to modify the transfer function. More importantly she gets clues about what may be found at the next level of decomposition. Thus, decomposition and modeling go hand in hand in systems analysis. Both are needed to ferret out the understanding we seek.

12.5.1 Modeling Engine

A modeling engine is the software that takes a representation of all of the data in the system knowledge base and iterates over all of the objects, updating their states as functions of objects upstream of them in the network. We will provide more details in the next chapter. Here we provide a basic idea of what is involved in modeling and what the outputs tell us about the system. And we will describe how we can use these results to guide our further analysis and to produce understanding at whatever level of the analysis that has been completed.

A model is an abstract representation of the real system. A computerized model allows us to take a static representation (the data in the system knowledge base) and simulate the behavior of the system over time. Just as the picture in Fig. 12.4 shows "inputs" to the system, the modeling engine has to be able to modify the states of the inputs to the model as they might happen to the actual system.

Figure 12.17 depicts a typical modeling environment showing the modeling engine and the various other components that are needed to build and run a simulation.

In a fully integrated environment, the user interface can coordinate the uses of the decomposition tools to construct the system knowledge base (purple stock). When the user/analyst reaches a point in decomposition where it is feasible to try a model run, they can direct the model constructor to take the appropriate state data out of the system knowledge base, along with using the metadata regarding object linkages, and set up the model representation (data). The user/analyst has to also set up the input database. This is a file of data elements that represent the time series of inputs to the model. The user/analyst designates how many time steps the model is to run and pushes the "run" button.

As the engine runs, on each iteration its first job is to read in the input data and put it into an array that matches the input processors of the model. It then goes through each object in the model, starting with the input processes and computes their new state variables (which basically means levels in stocks and flows based on the control requirements). The newly computed values represent the new states of the whole system and are stored back to the model data replacing the previous values. They will be used as the "current" values in the next time step.

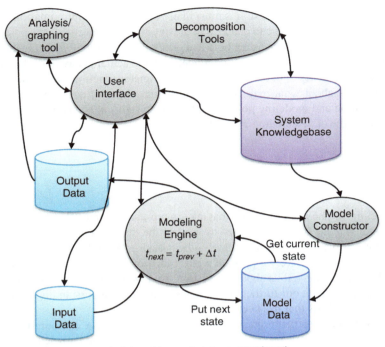

next state = f (current state, current input)

Fig. 12.17 A digital simulation of the system employs several data and software tools. The modeling engine uses the data representing the state(s) of the model objects, computes the states for the next time step, and replaces them into the model data in preparations for the next iteration

The engine then writes the current values that had been specified by the user to the output database. This data will be used in various ways to analyze the system's dynamics (next chapter).

The engine runs like this, through each time step, until it completes the time span the user requested. At the end of the run the user can use the output data to graph the behavior of the system.

Question Box 12.9
How do the components and flows of the modeling process presented in Fig.12.17 compare with the elements of conscious modeling as used, for example, when we cross a busy street in the middle of the block? Are there any shortcuts, or do we need all that complexity?

12.5.1.1 System Representation

The system knowledge base includes a large amount of data about each object in the system. Some of this data are the state variables, the values of flows and stocks that change during the system's simulation (and in the actual system as it behaves). Some of the data is called metadata because it is data about the relations between objects, e.g., the source and sink interfaces for any flow object. The former data is represented in the computer memory, but the latter kind of data is only used at start-up to set up how that state data is organized for efficient computation of new states. Generally speaking the organization looks like multiple multidimensioned arrays in computer memory, with pointers from source objects to sink objects. The simulation algorithm starts from the input matrix and follows the pointers to the objects that are receivers of that data, computing the new state variables in those objects. It then follows pointers from those objects to the next set of objects downstream. At the end of the chain of pointers, the algorithm updates the final output values, which are then written to the output data file.

12.5.1.2 Time Steps

The type of simulation being described here is called discrete time modeling (we will describe several other kinds of modeling approaches in the next chapter). The engine iterates over time steps that are representative of a given duration. For example, a model may simulate changes of state every model second, or every minute, or every year depending on the scale and complexity of the system being modeled. For example, when modeling living neurons in brains, the time step might be measured in milliseconds or even microseconds if the temporal resolution for specific behaviors needs to be that small a unit of time. Generally, all of the time steps are of equal duration, usually designated as Δt. This is a model run parameter set by the user. What is going on in the computer, depending on the size of the model (measured in data elements) is that each iteration over the model might take a few milliseconds (for a large model). Each pass through the computations represents one of these designated time durations. So if the model is large and the time step is supposed to represent, say, a hundred microseconds of real time, then it takes longer than real time to run the model. On the other hand, if the time increment represents, say, a week of real time, then the model can be run many times faster than real time.

Models that run slower than real time can only be used to verify our understanding of systems. Models that run faster than real time, however, have the potential to be used to generate predictive scenarios. It is these latter kinds of models and their ability to be run faster than the real systems they represent that were used extensively in Chap. 9 when discussing the use of models in control structures.

12.5.1.3 Input Data

The user/analyst has to construct a set of input data that represent a scenario of environmental inputs to the system. This is a set of data elements that represent a time series of the various inputs in the same time step intervals as the Δt. However, not all inputs will vary much or at all in every time step. There are "tricks" that can be used to simplify the input data to just represent the changes that occur in a particular input stream when they happen so that the system doesn't have to store a huge amount of the exact same data for many time ticks. For example, when running a simulation, it is often the case that the user will want to run an "experiment," varying one or a few inputs over time while holding all other inputs at some nominal constant value.

12.5.1.4 Instrumentation and Data Output Recording

The user/analyst has to specify the state variables that they wish to track during any run of the simulation. They may only be interested in the levels of key stock values, for example. Since the engine has access to every variable in the model, it is relatively easy to record the values of those selected variables either each time step or at integer multiples of the time step (as a kind of summary data). This is called instrumenting the model to capture key data. It is effectively the same as placing real sensors inside the real system and recording the data directly. Indeed, the deployment of remote sensors in the "field," i.e., inside the system of interest, is just one of the kinds of "microscopes" that we've alluded to previously. Such sensor networks are currently being employed to collect operational data on a wide variety of eco- and human-built environments (descriptions below in the Sect. 12.6). These sensor networks and their generated data are proving invaluable for doing much better decomposition on very complex systems of these types. See Chap. 6 and Figs. 6.5 and 6.7 to refresh your understanding of "instrumenting" a system.

12.5.1.5 Graphing the Results

As discussed in Chap. 6 (Behavior), the end product of gathering data from the run of a simulation is an ability to visualize the dynamics of the system. Graphic representations such as shown in Fig. 6.1 help analysts understand the behavior of systems and their component objects over time under different input conditions (also review Graphs 9.1 through 9.3 in Chap. 9, Cybernetics). The general modeling environment, as shown in the above figure, therefore includes a graphing tool, with other analysis tools such as curve fitting algorithms. These tools give the user/analyst a visual aid in understanding what the system does under differing input conditions.

12.5.2   The System Knowledge Base Is the Model!

The astute student may have noticed something rather remarkable about the products of decomposition. Namely, the system knowledge base actually contains all of the knowledge needed to construct the model. That is shown in Fig. 12.17 as well. The runtime model is taken directly out of the knowledge base. This fact puts extra importance on the rigor with which the decomposition phase is done. In the next chapter we are going to do what amounts to a decomposition and construction of a model "by hand" to show what goes on in this process. The reader will start to appreciate the value of good software tools to support these activities.

12.5.3   Top-Down Model Runs and Decomposition

In the introductory paragraphs of this whole section (Sect. 12.5), it was mentioned that models could be constructed and simulated at any level in a decomposition as long as the level was complete. In fact this is a common practice. It turns out that running a model at higher levels in the structural hierarchy tree can help with decisions that need to be made about the next level of decomposition. For example, from a model run at level 1, say, it may become clear that process P1.3 shows the most activity in terms of flows in and out. Or it turns out to have the highest leverage in controlling other processes. This would suggest that this would be a more important process to decompose next, especially in terms of where to spend your time and energy resources. Indeed, the decision on whether to switch to depth-first decomposition may be influenced by such findings.

Another aspect of model simulation while decomposing is still in process is the ability to run models of sub-processes as if they were whole systems. This is possible because, of course, they are whole systems in their own right. In the above example of finding process P3 very important and possibly demanding a closer look, that process could be isolated as if it were the top level system and simulated independently to see how it behaves under different input conditions. This, of course, is part of the black box analysis leading to a better transfer function definition and eventually to a more informed white box analysis.

In fact, due to computational limitation (size and time), it is sometimes necessary to simulate only parts of very complex systems at a time, using the output data as input data to other parts. This is inherently dangerous if any mistakes in one simulation lead to false results in other, otherwise, correct simulation models. We'll revisit this problem in the last chapter of the book on Systems Engineering.

12.6   Examples

In the balance of this chapter, we want to provide several examples of the use of systems analysis in a number of different domains. The term "systems analysis" might be credited to IBM in the 1970s when they were starting to provide business computing solutions to large corporations. One of the first analytic frameworks for enterprise systems was called Hierarchical-Input-Process-Output (HIPO) Modeling.[2] This brought the term "system" to the forefront of popular thinking. But the notions of analyzing complex objects as systems were laid down in the 1950s and 1960s in a diverse number of fields before IBM's analysts went about analyzing businesses. For example, in the 1960s Howard T. Odum[3] was already analyzing ecosystems as systems. He is considered the father of systems ecology and developed a graphic language similar to what we've shown above.

Our first example comes from cell biology. The cell is a complex adaptive system but it is not evolvable in itself. The other four examples are evolvable. Enterprise systems, economic systems, and ecosystems are potentially subject to major alterations even while the analysis is under way. The human brain is evolvable in terms of learning but not in terms of genetic evolution. In all these cases the relevant sciences are conducting systems analyses.

12.6.1   Cells and Organisms

Ever since the assertion of the "cell theory" of biology (early 1800s), biological science has been based on cell systems and meta-systems comprised of cells (e.g., multicellular organisms). The cell theory[4] basically states that a unit of organization, the cell, is the basic system upon which all of life is dependent. The core of modern biology is the drive to understand life by examining how cells work. Though most traditional biologists might not readily refer to what they have been doing as systems analysis, in fact cellular biology fits the definition and description quite well.

The reason we adopted the word "microscope" to describe the instruments for decomposing systems is that it was the microscope,[5] invented by Roger Bacon sometime in the 1200s, that has been the main and most important tool for decomposing cellular structures. Combined with various chemical dyes and light filtering methods, the compound light microscope has been the single most useful tool for analyzing the nature of cells. Today there is a substantial armamentarium of tools

---

[2] See HIPO, http://en.wikipedia.org/wiki/HIPO.

[3] See Howard T. Odum, http://en.wikipedia.org/wiki/Howard_T._Odum. Also, Odum (2007).

[4] See Cell Theory, http://en.wikipedia.org/wiki/Cell_theory. Also, Harold (2001).

[5] See Microscope, http://en.wikipedia.org/wiki/Microscope.

used by cellular biologists to functionally and structurally decompose cells and discover just how they work. These include relatively recent advances in genetic sequencing abilities coupled with genetic engineering techniques that allow biologists to modify genes responsible for the construction of specific proteins. Then they can see what the biochemical and morphological changes wrought inside the cell lead to as compared with "normal" cell structures.

All cells have membrane boundaries that are porous but regulated with respect to flows into and out of the cell. They have elaborate internal structures. Bacterial cells are somewhat less complicated; eukaryotic cells have organelles that are encapsulated by membranes, and most importantly, a membrane encapsulated nucleus. They are the quintessence of a complex adaptive system.

By themselves cells are not evolvable. Obviously species of bacteria and single-celled organisms are evolvable, but our interest for the moment is in the cell as it exists today.[6]

Open any general biology textbook and you will find elaborate drawings and micrographs (photographs of cells through a microscope). They are labeled with the names of the objects, subsystems, found and visible within the various membrane compartments. Look at the metabolism chapter(s) and you will find the flows of biochemical moieties and their reactions. Many of these, like the production of ATP (adenosine triphosphate), the basic energy distribution molecule, are associated with specific subsystems (mitochondria in the case of ATP), and they flow in well-defined pathways to other subsystems. As an example, take a look at Fig. 12.18. It has a lot going on but you should be able to pick out the subsystem processes and various kinds of flows.

The two subsystems shown are at different size scales (the mitochondria are much smaller than ribosomes). Many mitochondria are pumping out ATP molecules which flow to many different organelles in the cytoplasm. These are the main energy sources for the work of all these organelles. The mitochondrion subsystem is shown as a process where different sub-processes are responsible for taking in raw energy (carbohydrate molecules) and oxygen molecules and through a series of biochemical reactions (different colored ovals inside the mitochondrion) mediated by various enzymes (catalysts) that are recycled, and outputting the high energy molecules of ATP. At the synthesis site on the ribosome, the energy is used to join the incoming amino acid to the outgoing amino acid chain (the polypeptide).

As an exercise for the student, try drawing a structural hierarchy tree from the above "map" of processes and flows.

---

[6] See Morowitz (1992) for a vision of the evolution of cells from the beginning of life as protocellular systems.

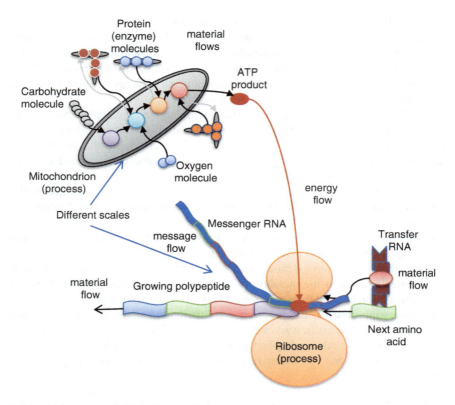

Fig. 12.18 As an example of subsystems (processes) and flows in a living cell, we consider the processes for making energy available to the cell and a subsystem for producing polypeptide molecules (the precursors of proteins). The mitochondrial process produces adenosine triphosphate, ATP, which carries energy to the ribosome process. The latter manufactures polypeptide string from raw materials, amino acids. The amino acid molecule is attached to a transfer RNA molecule with an exposed code (three nucleotides) that the ribosome recognizes as matching the messenger RNA "program" tape under the "read head"

12.6.2 Business Process

As mentioned above, business enterprises were among the first organizations to undergo systematic analysis for the purpose of automating previously paper-based subsystems. The early work started with one of the most highly defined subsystems in all of business—accounting. The processes of recording transaction data and using it to periodically produce a "picture" of the health of a company had been in practice for over 400 years in essentially the form we see it today. Accounting practices, considered the most important in a business in terms of making sure the organization is fulfilling its objectives (mostly making profits), had been well worked

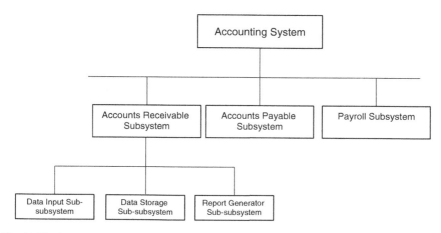

Fig. 12.19 A structural hierarchy was sufficient to represent the logical relations between subsystems and the whole "system" in the early days

out and codified in a way that made its analysis fairly straightforward. In the early days the problem was expressed in terms of a system for working with data and was not concerned with matter or energy flows—it was an information system. Each of the sub-subsystems, e.g., accounts payable or accounts receivable, could be analyzed in terms of procedures for handling data which basically amounted to building programs for obtaining the transaction data and classifying it according to which accounts they affected. That was the input, the I in HIPO. Figure 12.19 shows a "typical" structural diagram of parts of an accounting system.

At the time such structural diagrams were the main documentation at the analysis stage. Each subsystem, sub-subsystem, etc. could be represented in this modular form without any need to represent processes or functional maps. Processes were readily defined by the existing accounting procedures. The system knowledge base was paper based.

It was not long, however, before businesses themselves started to change and adopt more rigor in other operations. Eventually every department in an organization demanded automation of its procedures even if those procedures were not as well defined as was the accounting function. Most were not what we would call stable, that is, the procedures did not remain the same over long periods of time.

At first there were inroads made into some of the departments that had direct interfaces with the accounting system: finance, purchasing, shipping, and inventory management were among the first non-accounting subsystems to be automated. Production floor management (for manufacturing), marketing, sales, etc. were rapidly tackled. Lots of computers were sold. The development of minicomputers in the late 1970s and later desktop computers in the late 1980s introduced some flexibility, allowing departments to automate certain functions on their own. However, they needed to employ systems analysts separate from the main data processing center's personnel and things got messy!

While automation could certainly be shown to improve productivity where it was done well, unfortunately more often than not, it was not done right. The trade literature of the time had article after article bemoaning the "software development crisis," which basically meant that large-scale software systems were not being successfully completed on-time and on-budget. And they were becoming progressively more costly. It is not a stretch to claim the crisis continues to this day. If one surveys the successes and the failures, asking what is a common feature shared by all within one or the other category, one finds that the successes (as defined by the acceptance and level of satisfaction by the user department or client) tend to either be relatively simple and stable or (if complex) put much more effort into performing a systematic analysis at the beginning and going back from time to time to make sure nothing has changed. Contrariwise, those that are failures, or for which the client is dissatisfied, tend to be systems that were not stable or simple, and in spite of their complexity, little effort was put into an adequate up-front analysis. This latter group has one other common characteristic of the environment of the systems development process: the clients or upper management is too impatient to wait for a sufficient systems analysis to be completed. A common observation in case studies of these failures is that upper management rarely understood anything about software development other than that programmers were supposed to be programming. Analysts drawing complicated pictures did not strike them as being productive. A famous saying emerged from this short-term thinking: "Why is it we never have time to do it right the first time, but always have time to do it over?"

It wasn't for lack of adequate procedures for systems analysis that these kinds of failures kept coming up. In the 1980s various authors/researchers began developing and promoting "structured analysis,"[7] which included many tools and techniques for doing real systems analysis. Some of those techniques include the same kind of decomposition that we've covered here.

As we argued at the beginning of this chapter, the point of systems analysis is *understanding*. The latter may or may not bring profits in the sense of a business' purpose. Thus, in spite of the existence of methodologies that were designed exactly to obtain understanding of the business processes that were to be "improved" so that the implementers would succeed in that objective, the rush to profit-making too often prevented the very process that could have brought that success.

On the other hand, enterprise systems implementation as subsystems of businesses has learned a great deal over the decades, and today there are increasingly more successes than failures. More and more the efforts of business systems analysts are being directed toward business expansion, that is, the evolution of new functions in the framework suggested by Senge (2006). Corporations are employing systems analysis, including both decomposition and modeling, to explore new business opportunities. They have learned how to do strategic management.

---

[7] See, for example: Structured Analysis, http://en.wikipedia.org/wiki/Structured_analysis. Also, DeMarco (1979, 2001a, b).

12.6.3 Biophysical Economics

The economy is a system. It is the system for obtaining natural resources, converting them into useful artifacts, distributing them to customers, and ensuring that necessary nonmanufacturing services get done. The underlying "philosophies" of how to best to accomplish this have been endlessly debated and many experiments performed over the centuries. Today, it seems the dominant philosophy is the free market, capitalism backed by a democratic governance system. At least that is the theory.

Along with this dominant philosophy, the dominant model of economics emerged. It was basically derived in the 1800s during the takeoff of the Industrial Revolution as the successes of that particular philosophy seemed to prevail. It is called the neoclassical economics model.

In recent decades the neoclassical model has come under increasing scrutiny and criticism on many fronts. For example, the standard model assumes that all buyers and sellers in a market are rational utility maximizers (called, somewhat tongue-in-cheek, *Homo economicus*). They are always calculating what is most profitable to them in any transaction. This assumption is necessary in order to show how markets function to resolve all problems of supply, demand, price, and distribution. The model only works if this is the case. Unfortunately social psychology has discovered that humans are anything but rational agents, at least as defined by economists. That is an extremely interesting story but beyond the scope of this book, so we will refer the reader to some good sources in the bibliography (see: Gilovich et al. 2002; Kahneman 2011).

Another very important hole in the neoclassical model has to do with natural resource depletion and especially energy flows. As we have demonstrated throughout this book, systems depend on these aspects as inputs. We have also stressed that outputs, if they are not "products" to be used by some other system, are wastes that are essentially dumped into the system's environment. Neoclassical economics treats the economy as essentially a closed system in which the actors are "firms" and "households" (producers and consumers) and the transactions are mediated by money. Resource depletion and sink capacities are treated as "externalities" that will somehow take care of themselves outside the economic system.

Criticisms include the increasingly poor performance record of neoclassical economics to either explain adequately or alleviate the economic stresses that have plagued the globe for many decades. They cannot make sufficiently accurate predictions that would ordinarily be considered the basis of policy decisions in government. Many of those criticisms are coming from professionals who routinely work in systems analysis such as systems ecologists, biologists, and systems engineers. The root of these criticisms is the recognition that the neoclassical model is based on a set of assumptions that simply do not work in the real world. Another example of an unrealistic premise (assumption) of the model is that as the price of a commodity (resource) goes up due to declining availability (depletion), customers will simply find substitutes. Unfortunately, given the complexities of our technologies and tremendous reliance on the high power provided by fossil fuels, it turns out there may not be many economical substitutes for many of these resources.

The list of other assumptions that are being shown to be contrary to reality includes the notion of unlimited growth (say as measured by gross domestic product, or GDP[8]) and that money is independent of any physical quantity so it can be created as needed.

But recently a group of systems scientists from various disciplines, including some economists who naturally think systemically, have begun to reconsider economics in light of systems principles. Much of the thinking along these lines can be attributed to Howard T. Odum, who has been mentioned previously. Noting how the economic system was very similar in dynamics to an ecological system, Odum (2007) began constructing various ways to think about the economy in the same way he thought about natural ecosystems. Many of these ideas gave rise to several approaches to economics that opened the model up to more realistic bases, for example, ecological economics.[9] One of Odum's PhD students, Charles Hall, professor emeritus at SUNY-ESF in systems ecology, went on to develop these ideas further, creating a new field called biophysical economics. Unlike other versions of neoclassical economics, Hall followed Odum's lead and based the study of economic systems on energy flows. The reason is that energy flow and physical work is what is at the base of all wealth creation (Hall and Klitgaard 2011).

One of us (Mobus) has applied systems analysis to the study of biophysical economics. Figure 12.20 shows a first-level, conceptual decomposition and a few representative processes and flows.

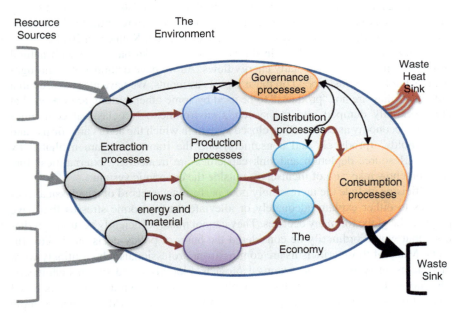

Fig. 12.20 The economy studied as a system is comprised mainly of extractive processes, production (conversion) processes, distribution processes, and consumption processes. A governance process, based on any number of philosophical models, e.g., capitalism/democracy, helps to regulate the otherwise market (cooperative) flows of messages (e.g., money) that determine what work is to be done

[8] See Gross Domestic Product, http://en.wikipedia.org/wiki/GDP.

[9] See Costanza et al. (1997), Daly and Townsend (1993).

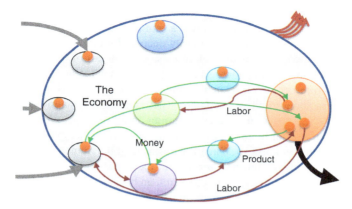

Fig. 12.21 The *orange circles* represent adaptive agents as described in Chap. 9. The consumer agents (*orange circles* in far right consumption process *oval*) decide what products (and services) they want and how much they want them. They then "send" money as a message to the distributor, who, in turn, sends money to producers, and they send money to the extractors (or suppliers). The *red arrows* represent embedded energy (value added) or labor energy. The same agents that are consumers also supply labor to the other processes. In this case the management agent (*orange circles* inside all the other processes) decides how much labor is needed and how valuable it is. This is a basic market system in which decision agents communicate directly with one another, through price, to settle on how the flow of energy will move through the entire economy. The governance process has not been included here—consider this a primitive economy that could get by with cooperation alone

Figure 12.21. shows some detail regarding the fact that adaptive agents (people) are both consumers and sources of labor (energy flow).

This analysis is aimed at understanding better the role of money in communications between consumers and firms. As well, we seek to better understand the general flows of matter, energy, and other communications (e.g., financial markets and regulatory subsystems).

Question Box 12.10

In an economic model, it is important to account for costs/energy. Where in Fig. 12.20 would you locate the costs for resource depletion? How about sink problems such as pollution?

12.6.4 Human Brain and Mind

One of the oldest and most profound questions humans have asked is: "What is the mind?" The very subjective experience of thinking, along with feelings and emotions, has mystified those minds that have reflected upon their own workings.

Even the ancients had a sense that the brain had something to do with the experience of mind. Today we realize that whatever a mind is, it is a result of what the brain does. Clearly this subject is extraordinarily deep and broad, so we are not going to try to "answer" any questions or explore even a small part of the subject.[10] Rather, we are interested in showing how recent developments in technologies have made it possible to do more complete systems analysis on the brain and help resolve some age-old puzzles about mind and behavior. These technologies allow nondestructive monitoring of brain functions as they are going on in living people.

Arguably the human brain is one of the most complex, adaptive, and evolvable (learning) systems we know of. One of the aspects of the human brain that makes it so complex is not just its internal wiring, but the fact that it cannot exist as an isolated system, but must interact with other human brains in order to function properly.[11] That is, we cannot talk about the brain as an isolated system, but must embed it in the larger social and physical framework that defines human experience.

Up until about two decades ago, the only means for performing functional and structural decomposition on the human brain was postmortem dissections and correlations between behavioral/cognitive/emotional deficits in patients during their lives and brain lesions found after their deaths. Neuroscientists were forced to extrapolate from brain damage and behavioral problems to what functions those parts of the brain were responsible for. For example, neurologists had recognized that damage to an area of the left brain, known as Broca's area,[12] was related to speech deficits. Large lesions due to strokes, for example, could leave patients speechless.

Microscopic examination of various regions of the brain revealed an astounding plethora of cell types, neurons and glia cells. Neuroscientists were able to map numerous tissue regions based on the complements of cells in those regions and how they were connected to one another and/or sent branching connections out to other areas.[13]

The human brain is the result of hundreds of millions of years of animal evolution, with many lower-level species making genetic contributions to the complex organ

---

[10] As we think the brain as a system is so incredibly illustrative of every aspect of systems science, we will provide an extensive bibliography on the subject at the end of the chapter including brain, consciousness, and behavior as well as how these arts thought to have evolved.

[11] Of course it must also have a functioning body to support its existence. In this short description, we are focusing on the brain rather than its interactions with the whole body system.

[12] A good, readable description of Broca's area can be found in Carter (1999), in particular page 138. Also: Broca's area, http://en.wikipedia.org/wiki/Brocca_area.

[13] For example, the neocortex has been mapped into what are called Brodmann areas, after Korbinian Brodmann, a German neuroanatomist. The mapping into areas was based on the cell types contained and the wiring of those cells, called the cytoarchitecture. See: http://en.wikipedia.org/wiki/Brodmann_area.

we find today.[14] The brain is roughly divided into three major "modules" based on that evolution. The most "primitive" region is popularly called the "reptilian" brain. It is the oldest vertebrate brain and could just have easily been called the "fish" brain. The reptilian label comes from the association with tetrapods (legged). This brain is responsible for all of the low-level operational body controls and instinctive behaviors. When the first primitive mammals evolved, the brain added a new layer of tissues overlaying the older brain based on a "flat" layout, called a cortex. This type of tissue is now recognized as the basis for learning associations between environmental contingencies and consequences to the organism. The "paleocortex" added a much more flexible behavior mode and much better tactical control (as in Chap. 9). It is the basis for the bulk of adaptability in behavior. Parts of the reptilian brain along with this newer cortical brain are referred to as the "limbic" centers.[15] Finally, with the rise of higher mammals, particularly carnivores and primates, we see the emergence of yet a more complex cortical tissue overlaying the paleocortex. This is the neocortex, the gray matter that allows animals to build more elaborate mental models and exploit much more flexible behaviors. Moreover, in carnivores and primates, the frontal most part of the brain, the prefrontal cortex, expanded considerably in proportion to the rest of the brain. The prefrontal cortex is now recognized as the executive center, associated with thinking and reasoning, as well as conscious experiences of those thoughts and of emotions.

How have neuroscientists come to this conclusion? How do they know, for example, that another region, deeper in the limbic brain, called the amygdala, is responsible for fear responses?[16]

For most of the history of the neurosciences, investigators relied on the above-mentioned correlation techniques along with experimental probes in living animal models.[17] Along with very course grained maps of brain activity from encephalographic techniques, they struggled to piece together a coherent picture of gross brain functions.

Keep in mind that the brain covers many levels of complexity. Not only were scientists trying to grapple with the high-level functions of brain regions, they had to consider the "wiring diagram" of interconnecting regions and modules. On top of

---

[14] For a complete grasp of how brains evolved, you will need to tackle Striedter (2005). And if you like that and want to really go into the topic deeper, you will want to take a look at Geary (2005).

[15] See Cater (1999), p. 16. And Limbic System, http://en.wikipedia.org/wiki/Limbic_system.

[16] Joseph LeDoux (2002) has spent many years studying the amygdala and its relationship to fear responses.

[17] We should also mention the use of electrical probes of living human brains during surgeries. Wilder Penfield (1891–1976), a Canadian surgeon, used electrical stimulation to regions of the brain of patients who were undergoing brain surgery. There are no pain receptors in the brain, and so these patients were awake and reported their mental images or thoughts resulting from these probes. See: Wilder Penfield, http://en.wikipedia.org/wiki/Wilder_Penfield.

that was that plethora of neuron types mentioned above. And at the sub-neuron level, the methods by which cells communicate with one another via sending electrochemical signals (action potentials) down long axons, how they received messages at their synapses, how those synapses changed their dynamic behavior as a function of those communications, and the underlying biochemical/genetic reactions that supported it all were being tackled as best possible with the tools at hand.

And those tools have been undergoing substantial improvement in recent decades. For example, microelectrodes can now be inserted into living brain tissue in order to record or stimulate the activity of single neurons! The new imaging technologies, such as functional magnetic resonance imaging (fMRI), allow neuroscientists to watch in real time the activity of a specific brain module in response to conscious and subconscious stimulations. The resolution of these images, in both time and space, has been getting better over the years such that the researchers are able to pinpoint some amazing phenomena associated with thinking and emotions.

The tools for decomposing the living human brain without destroying it in the process have come a long way. Former president George H. W. Bush declared the decade 1990–1999 as the Decade of the Brain.[18] A great deal of funding was directed toward human brain studies, much of it going to improving the analytical tools.

Mirroring the impressive improvements in microscopes for the brain, there is currently a project (in early stages) to use computer models to further understand the functioning of the brain and its subsystems. In addition, the Obama administration recently announced an initiative to understand the brain in a manner not unlike the Human Genome Project. Called the BRAIN Initiative,[19] the plan is an aggressive one to map all aspects of brain activity, especially with an eye to understanding (and eventually treating) a wide array of brain disorders, such as Parkinson's and Alzheimer's diseases.

The Human Genome Project was a multidisciplinary one involving scientists from numerous fields who had to learn to speak the same language. This brain initiative will be even more interdisciplinary requiring people from the behavioral sciences as well as neuroscientists, and even bioethicists. They will all need to speak the same language. Guess what that language is? If you said the language of systems, you would be correct. Systems thinking is the epitome of interdisciplinarity.

Think Box 12 How the Brain Analyzes the World and Itself
Analysis is a process whereby one pattern is compared with a second pattern where differences and similarities are noted and used for some purpose (to be described later). If we have two patterns that could be said to be in the same

(continued)

---

[18] See Project on the Decade of the Brain (official government document), http://www.loc.gov/loc/brain. Accessed 18 July 2013

[19] See BRAIN Initiative, http://en.wikipedia.org/wiki/BRAIN_Initiative.

Think Box 12 (continued)

category, we can find some structural homomorphism that allows us to do analysis using one point in one pattern where we can find its homologous point in the second pattern. Then it is possible to use topological differences to decide if there needs to be a new instance or even a new category. The structural forms of the patterns, in the brain, are encoded in the concepts as we discussed in Think Boxes 5 and 6. When two concepts share a large number of sensory features, then they are likely to be considered as in the same category or class.

Remember the Chihuahua example in Think Box 7? In that case there were enough similarities between the Chihuahua and the category of dog-ness that your brain decided not to create a totally new category at that level. It created a new subcategory labeled Chihuahua-ness and perhaps a specific instance memory of your friend's pet.

If, on the other hand, your friend had gotten a ferret as a pet (and assuming you had never seen one before), your brain would analyze the two mappings (concepts) of dog-ness (the usual sort of pet) and this new thing. It would most generally find that there were too many differences between the two animals for it to subcategorize the ferret as a kind of dog. Being told it was a ferret, your brain would simply create a new category for any animal that fits that same pattern (like a mink). At first it would only have the one instance as a memory, but over time, with encounters with more ferrets and ferret-like creatures, the category would be modified to generalize the concept.

The comparison is carried out by brain modules that monitor when multiple concept clusters at the same level of complexity are excited from sensory inputs from below and concept inputs from above. In our dog versus ferret example, we have a case of a brand new cluster being excited by sensory input, the ferret, and a cluster is being excited from above—the higher concept of dog-ness. The dog-ness concept then sends excitatory signals back down to lower levels of sensory clusters that represent the various percepts and features of dog-ness. The comparator module can "sense" that the sensory-driven lower clusters are not the same as the concept-driven lower clusters. These two groups of perceptual inputs, with a few exceptions like HAS-HAIR (in other words mammal-ness), are not sufficiently the same. The comparator then outputs a signal telling the neocortex to set up the new cluster as a potentially permanent new concept and to get the higher concept (mammal-ness) to include the new concept as a sub-concept.

(continued)

Think Box 12 (continued)

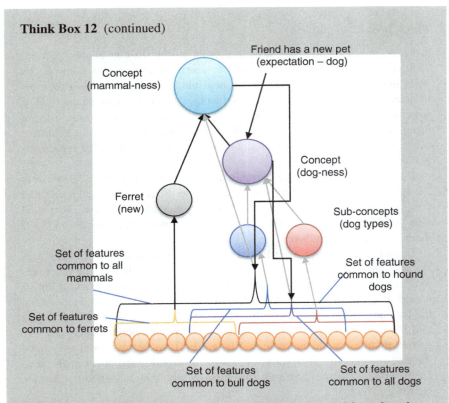

Fig. TB 12.1 The higher-order concept, dog-ness, has been activated from above by an expectation that a friend's new pet is a dog and sends excitatory signals downward to activate the set of features associated with it, the *purple set* of features (that would drive it if they were in the sensory field). But the pet is a ferret which sets off a different set of features (*yellow set*), only a few in common with dog-ness. A comparator module (not pictured) detects that there are features in the set of ferret that are firing strongly that aren't in the set of dog-ness features so it elevates the concept of ferret to ferret-ness, at least provisionally, allowing for future encounters to strengthen it as a concept

The amount of difference between the ferret and a dog is the information (Chap. 7) conveyed. Expectations that the pet would be a dog and the fact that it is quite different from a dog trigger the brain to start learning—processing information to create knowledge. In this case it creates a new potential category concept that can be strengthened in the future by more encounters with ferrets.

As we saw in Think Box 8, the brain doesn't actually do math in the way we normally think of it. The comparator module does not add up the number of features in the ferret set and the dog-ness set and then subtract one from the other to come up with a numerical difference. Rather it performs something

(continued)

Think Box 12 (continued)

akin to estimation of differences between the activation levels of the two sets and uses that estimation (in the form of some specific subcluster being activated above its ground state) to signal the need for learning back in the neocortex module that has been activated for ferret-ness. Note that this process, internal to the brain, is not at all unlike the process that naturalists go through when analyzing the difference in features between species and then assigning animals to different categories in the phylogenetic tree. The former is implicitly happening, while the latter is explicitly going on using language.

12.7   Summary of Systems Analysis

All sciences have been doing systems analysis in some form or another. We humans have been doing systems analysis for much of our history. It wasn't until the late twentieth century, however, with the rapid developments in computer and communications technologies, that the formal picture of systems and systems analysis began to be recognized. SA is applicable to everything that is a system, and since, as we have argued, everything is a system in one sense or another, everything can be analyzed in this systematic way.

That doesn't mean that we always have all of the tools we need to perform adequate analysis. We might not have the right microscopes with which to nondestructively decompose systems. Even when we have some rather sophisticated microscopes, such as with modern neuroscience, they still might not have the level of resolution needed to adequately discern the fine details of deeply complex systems. And we might not have the computing power necessary to do justice to building models and running simulations. Our global climate models, for example, are still basically rough approximations of possible future scenarios rather than predictions in which we can have confidence.

So even though we now have a much clearer understanding of the process of systems analysis and its relation to gaining understanding of systems, we are still in a formative stage in terms of being able to apply the method to everything in which we are interested. Some of our most important and relevant systems remain black boxes insofar as understanding them well enough to exploit that knowledge for the benefit of humanity and the natural world. We are beginning to get a glimpse of how societies work and how humans think, feel, and act. There is hope that we can understand human institutions like the economy such that we can work toward more functional designs. Even then it will be interesting to see if we can actually intervene in a system like the economy, with better designs, say for more equitable distribution of wealth than seems to be in place today. It may be that only an evolutionary process will produce that kind of result. Either way, the more we understand "the system," the better off we will be in the end.

Bibliography and Further Reading

Alkon DL (1987) Memory traces in the brain. Cambridge University Press, Cambridge, MA

Arbib MA (1964) Brains, machines, and mathematics, 2nd edn. Springer, New York

Baars BJ, Gage NM (2007) Cognition, brain, and consciousness: introduction to cognitive neuro-science. Elsevier, New York

Calvin WH (1996) How brains think: evolving intelligence, then and now. Weidenfeld & Nicolson, London

Carter R (1999) Mapping the mind. University of California Press, Berkeley, CA

Costanza R et al (1997) Introduction to ecological economics. CRC Press, Boca Rotan, FL

Daly HE, Townsend KN (1993) Valuing the earth. The MIT Press, Cambridge, MA

Damasio AR (1994) Descartes' error: emotion, reason, and the human brain. Penguin Putnam, New York

Deacon TW (1997) The symbolic species: the co-evolution of language and the brain. W. W. Norton & Company, New York

DeMarco T (1979) Structured analysis and system specification. Prentice Hall, New York

DeMarco T (2001) Structured analysis: beginnings of a new discipline. In: Broy M, Denert E (eds) Software development & management conference, 2001. Software Pioneers, Springer 2002.

DeMarco T (2001b) A mind so rare: the evolution of human consciousness. W. W. Norton & Company, New York

Fuster JM (1999) Memory in the cerebral cortex. The MIT Press, Cambridge, MA

Geary DC (2005) The origin of mind: evolution of brain, cognition, and general intelligence. American Psychological Association, Washington, DC

Gilovich T et al (eds) (2002) Heuristics and biases: the psychology of intuitive judgment. Cambridge University Press, Cambridge, UK

Hall C, Klitgaard K (2011) Energy and the wealth of nations: understanding the biophysical economy. Springer, New York

Harold FM (2001) The way of the cell: molecules, organisms, and the order of life. Oxford University Press, Oxford

Kahneman D (2011) Thinking fast and slow. Farrar, Straus and Giroux, New York

LeDoux J (1998) The emotional brain: the mysterious underpinnings of emotional life. Simon and Schuster, New York

LeDoux J (2002) Synaptic self: how our brains become who we are. Viking, New York

Merlin D (1991) Origins of the modern mind. Harvard University Press, Cambridge, MA

Morowitz HJ (1992) Beginnings of cellular life: metabolism recapitulates biogenesis. Yale University Press, New Haven, CT

Odum HT (2007) Environment, power, and society for the twenty-first century. Columbia University Press, New York

Pinker S (1997) How the mind works. W. W. Norton & Company, New York

Pinker S (2007) The stuff of thought: language as a window into human nature. Viking, New York

Senge PM (2006) The fifth discipline: the art & practice of the learning organization. Doubleday, New York

Striedter GF (2005) Principles of brain evolution. Sinauer Associates, Inc., Sunderland, MA

Swanson LW (2003) Brain architecture: understanding the basic plan. Oxford University Press, Oxford

Chapter 13
Systems Modeling

Students should learn that all decisions are made on the basis of models.
Most models are in our heads. Mental models are not true and accurate images of our surroundings, but are only sets of assumptions and observations gained from experience.

However, mental models have serious shortcomings. Partly, the weaknesses in mental models arise from incompleteness and internal contradictions. But more serious is our mental inability to draw correct dynamic conclusions from the structural and policy information in our mental models.
System dynamic computer simulation goes a long way toward compensating for deficiencies in mental models.

Jay W. Forrester, 1994

Abstract Analysis of a system (as described in the last chapter) is not sufficient to foster understanding of that system. One must be able to show behavior and functions of the system as well. The primary tool for grasping how systems actually work is modeling the system. This is an integrative (as opposed to analytical) process in which the modeler will attempt to reconstruct the system from its components. Modern computer methods for modeling allow investigators to test their understanding of system functions by making predictions from running their models to simulate the "real" system and then observing the actual physical system. Where there is correspondence, it suggests that the model did capture some essential qualities of how the system functions. If the model fails to predict (or post-dict) the behavior of the physical system, then the modeler seeks to improve the model in various ways. We present a survey of modeling techniques and uses. Several examples of kinds of models are given.

13.1 Introduction: Coming to a Better Understanding

Systems thinking is about understanding how the world, or at least some portion of it, works. The way we do this is to build models of the system based on how we think it works and what we already know. The model is then used to answer questions that

© Springer Science+Business Media New York 2015
G.E. Mobus, M.C. Kalton, *Principles of Systems Science*, Understanding
Complex Systems, DOI 10.1007/978-1-4939-1920-8_13

we do not understand. What is truly amazing about the nature of models and systems is that they represent one another in complex and interesting ways. A model is a simplified version of a modeled system, in a medium (e.g., a computer programs or brains) in which tests can be performed (questions asked). But a model is also a system. Even more amazing is that systems can contain models of other systems with which they interact. Having such a model facilitates interactions.

The function and potentials of models thus engage three of the systemic principles we introduced in Chap. 1. We will begin by a brief discussion of each to frame the broader meaning of modeling in systems science.

Principle 9: Systems Can Contain Models of Other Systems
It may seem a stretch to claim that this principle applies to all systems at every scale of complexity and spatiotemporal dimensions, but looked at carefully it really is true. For example, the valence electrons in atoms provide an interface in exchanges between atoms when forming (in this case) covalent bonds. Built into the structure of the atoms is knowledge of other entities, atoms of different elements, with which atoms "know" how to interact. We don't generally think of this structural configuration as knowledge because it wasn't exactly "learned." But as we saw in Chap. 7, knowledge is really any lasting configuration of physical structures that allow a system to exchange flows or make connections. In this sense, we say that an atom contains, in its outer electron shell, a model of other entities in its environment.

The most interesting application of this principle comes from looking at complex adaptive systems where behaviors in their world depend on having more elaborate and even learned models of that world. Living systems, of course, are the most clear example of systems containing models of other systems. So that is what we will consider for the most part. These models are necessary for the systems (organisms or populations) to succeed in surviving and reproducing. Some aspects of models are acquired over evolutionary time, and we see their influences in instinctive behaviors. Others, particularly in more complex animals such as mammals and birds, are acquired through experience and learning. They come about through the modification of their brains. All such models are "computational" in that they process information input to produce decisions and behavioral outputs. Animal brains contain models of other systems in their world.

Principle 10: Sufficiently Complex, Adaptive Systems Can Contain Models of Themselves
Once the construction of models within a living system became possible, another interesting possibility emerged. At a sufficient level of complexity in the machinery of mental computation, brains could come to build models of themselves as well as of other similar entities. This is the realm of consciousness in humans and some animals (at least suspected). In particular the capacity in the human brain to construct a model of its own behaviors is what gives us self-awareness—the kind that allows us to see ourselves in a mirror and recognize that it is us we are seeing.[1]

---

[1] There is considerable evidence that chimpanzees and even elephants can self-recognize. For example they will attempt to remove a paint mark placed on their foreheads when they see their reflections in a mirror. Dogs and cats do not show a similar interest in mirror images of themselves.

Unfortunately, the scope of this book does not allow us to delve deeply into this extraordinarily interesting topic. The issues of what consciousness is and how it manifests in human experience are still hotly contested. However, in fields such as social psychology and sociobiology, the evidence that social creatures interact with one another on the basis of some limited form of self-awareness and other awareness makes it clear that the brains of these animals must have at least primitive models of themselves. This is probably one of the most exciting realms of research today.

Principle 11: Systems Can Be Understood
And this brings us to a fundamental principle—that systems can be understood, at least *in principle*. In this chapter, we are going to look at the way in which this can be accomplished by the use of models. Here we emphasize the construction of manipulatable models, that is, the kind that we can modify and test in various configurations as we attempt to gain understanding of the system in question. The great advantage of models of this form is that they can be "run" in fast-forward to produce virtual results that can predict the future behavior of the system in question given the current configuration of the model. For example, we can predict the weather 5 days from now using computer models of climate patterns and current weather conditions. If our predictions consistently turn out to be wrong, then we can investigate the reason (not a complete or accurate model) and modify the computer code to make the model better in the sense of making predictions. The fact that a model fails to predict the future of the system properly is a spur to look deeper into the real system and, thus, cause us to better understand that system. This chapter is about how humans construct and use models to better understand the world.

13.1.1   Models Contained in Systems

For the last 50 years or more, we have been building models of systems of interest using computer programming to represent the systems and simulate their actions. The computer itself is a system. And in this sense, one system can contain a model of another system. Indeed, we can take this to the point where the one system, the computer, can be used to control the actions of the other system by virtue of containing that model (Chap. 8). The acceleration computer controlling your automobile engine contains a model of the fuel injector as well as environmental conditions and reacts to a real-time accelerator input to compute just the right mix of oxygen and fuel and the right volume to inject.

Your brain is a system. And it contains many, many models of systems in the world, including a model of your own self. In fact, you probably have many conceptual models of the same systems in the world. And, of interest to psychologists, some of those models may even be inconsistent with one another! We're still working on how that can be the case. But the point is that everything you know about the world and yourself is based on conceptual models composed of neural networks and memory traces that can be activated to simulate a world inside your head (e.g., daydreaming!)

A model is a system, and a system can be modeled. This is what led us to principles 9 and 10.

13.1.2   What Is a Model?

Throughout this book, we have shown a number of *models* of systems. These have been conceptual and visual models, for the most part. In the last chapter, we introduced the use of computer-based models as part of the process of analysis of systems. In this chapter, we will more completely describe various kinds of computer-based models, the modeling process, and some examples of the use of models to complete the analysis. We will also introduce the use of models as a prelude to systems engineering (next chapter), designing new solutions to complex problems.

In the previous chapter, we presented the notion of coming to understand a system by virtue of systems analysis. We described how the process of analysis of an existing system, if conducted through functional and structural decomposition, actually produces the system knowledge base used to construct a computer-based model. We also introduced the basic computer programming environment, the modeling engine that "runs" these models so that analysts can see the dynamics of the system as a whole as well as subsystems within. In this chapter, we will complete the description of coming to understand a system by using the model. We will amplify these ideas and provide some examples.

There are so many good books and other references on how to actually *DO* modeling in various forms. Our purpose in this chapter is not to teach you how to do modeling so much as what role modeling plays in the process of understanding systems. We will take a survey of several different approaches to building models in a way that can simulate the behavior of the real system using computers. With this basic knowledge, you should be able to quickly orient yourself to any specific modeling approach, subject to the appropriate mathematical skills you will need. See our Quant Boxes for examples of the latter.

The term "model" is used extensively in many domains and in many different ways. For our purposes, we will generally use the term to describe a computational object that *simulates* a real system. In most uses of the term, a model is a representation of a system that is more abstract and reduced in form. Models are implemented in a medium that is quite different from the real system. The model includes only the essential elements that represent the important aspects of the system. Of course, deciding what those elements are is not exactly straightforward at times, so some aspects of model building can be "tweaking" what is included and what is left out. We'll show some examples of this in the chapter.

Throughout the book, in the various examples of systems we have used, we've seen that there are different representations used in modeling. For example, when we were working on network theory, we saw that an abstract object called a "graph" in mathematics could be used to very succinctly represent any kind of network. The graph, and its representation in a computer memory, is then used to explore questions about structure and functions that are cogent to network dynamics. One of those questions we saw was: How do subnetworks coalesce when a large network (like the Internet) is growing and evolving? A graph model does not necessarily need to be concerned with, for example, energy flows, when the questions are about structural matters. On the other hand, some network representations (using flow graphs) may be used to explore questions of function.

Table 13.1 shows some representative system/model types, the kinds of questions researchers might be interested in asking, and what tools might be applied to build the models. We will provide a brief description of modeling tools later.

Table 13.1 Here are some examples of system/model types that are used to simulate the real system under different input assumptions

System type (example)	Representative question	Model type	Modeling tool
Simple physical—airfoil	How does turbulence affect the lift?	Scale model of the system	Wind tunnel using smoke and force sensors[a]
Complex physical (mechanical)—space shuttle	How will the various subsystems interact and behave under stressful conditions?	Coupled subsystem simulations	Ad hoc computer codes, tied to mechanical simulators
Complex nonadaptive dynamic system—chemical processing plant	How do we establish and maintain stability in the chemical reactions of interest?	System dynamics	Computerized SD languages
Complex nonlinear systems—climate	What will the average annual global temperature be in 50 years?	Coupled cellular (grid) models	Global climate model (general circulation model)[b]
Complex nonlinear adaptive systems—ecosystems	How will the system respond to the introduction of an invasive species?	System dynamics and agent based	SD and AB languages
Network evolution—the World Wide Web	What pattern of connectivity emerges over time?	Evolvable graph	Graph theoretical tools still being developed
Non-evolving economic system	How will the price of a finite commodity be affected as a result of depletion?	System dynamics and agent based	SD and AB languages
Evolutionary—artificial life	How do systems become more complex over time?	Multiple ad hoc methods	Various

Note the first example is a physical model of a physical system that has almost ceased to be used anymore

[a]These days, this is actually not the way new designs are tested. There are now computers powerful enough to be able to build digital models to test the systems

[b]See global climate model, http://en.wikipedia.org/wiki/Global_climate

13.1.3 Deeper Understanding

Building and running models in simulation is the final step in coming to an understanding of systems of interest. If we get a model to produce the same behavior as the real-life system, then we have a great deal of confidence that we really understand the system. It changes our knowledge base of the system from a mere collection of facts about components and interconnection relations into a deeper understanding of how the system works and, sometimes, why it works the way it does.

Modeling a system successfully fulfills principle 11. Systems can be understood, but with a caveat. All models are incomplete reductions of the real system so can never really represent any kind of ultimate knowledge. Understanding is not some kind of *absolute* property. It is a question of relative understanding. When we analyze more deeply and build more detailed models that provide more information, then we can say we are "coming" to understand the system more deeply. The question that systems thinkers have to answer is: "How deeply do we need to understand a system?" The question is contingent on context, unfortunately. It depends on what kinds of problems we are trying to solve with respect to the system. For example, researchers in nuclear fusion power have discovered that very much deeper understanding of the fusion process was needed when their early designs ran into unpredicted problems containing the reaction. Deeper analyses (particle and fusion dynamics) were needed, and models based on those deeper facts revealed why the prior experiments didn't work.

Systems need to be understood because doing so offers some kind of advantage to the understander. The deepness of understanding needed depends on some practical limits. For example, there is always a point where the information returns on analysis effort diminish and deeper understanding brings no further benefit. There is also the practical aspect of what kinds of microscopes (analysis tools) are available and whether it is thought that investment in more powerful microscopes would really have a payoff. For example, the next instrument needed to dig deeper into fusion reactions, the ITER[2] project, carries an extremely huge price tag, and many are questioning if that much money will really buy understanding of fusion reactions sufficient to build practical power generators using it.

Insofar as we are guided by the understanding embodied in them, models mediate our action upon, interaction with, and participation in various types of system. This means that the relation between model/understanding and the system itself is a two-way process, a dynamic feedback loop in which systems may be modified by the way they are understood, and the model is changed as we experience its effects. This is especially relevant when it is a matter of modeling aspects of complex systems in which we ourselves participate, such as social relations, the environment, or a market economy. As we ourselves are components in a dynamic relational process, the process itself continually shifts and adjusts in response to the way we

---

[2] For the International Thermonuclear Experimental Reactor (ITER), see http://en.wikipedia.org/wiki/Iter.

understand and act within it. In this respect, deeper understanding commonly includes the understanding that this is an open-ended process in which there is always something more to be understood.

13.2   General Technical Issues

There are several issues that are relevant regardless of the kind of model or modeling procedure we are using. We cover those here so they will be in the back of the reader's mind as we look at specific examples of modeling approaches and models.

13.2.1   Resolution

How fine a detail should the model provide? How small a time increment should we be able to see? These are questions related to the resolution of the model. For example, the temperature of a gas in a container is a very simple number that gives the average momentum of the aggregate of gas molecules. The alternative would be to somehow "clock" the velocities of each molecule at the same moment and then find the average. The latter is an extraordinary resolution, but it is not only impossible to do in practice, and it doesn't actually provide any more information than does simply taking the temperature.

Model builders need to be very careful not to try to get too much detail into their models. Nor should they try to get too fine a resolution in time. That is, they should not choose time deltas too small. The cost of both of these mistakes is computational time eaten up without gaining any more information.

Another way to look at this relates back to the complexity hierarchy of Chap. 5 and the depth of analysis in the last chapter. Depending on what kinds of questions we are trying to answer, we could find that detailing a model down to, say, level 2 would be sufficient. Going further than that, even just to level 3, buys you no more information but costs you greatly in running time.

Therefore, one of the primary technical issues that need to be resolved early in the modeling process is what questions you are trying to answer and peg the level of resolution of the model to those. For example, later we will provide an example of modeling more realistic biological neurons by one of us (Mobus). The questions being asked revolved around the dynamic behavior of the synaptic junctions over very long time scales and in response to stimulating inputs. The big question involved how synaptic junctions could store memory traces. These dynamics are dependent on a large set of biochemical reactions each of which runs with different time constants ranging from milliseconds to hours. Would it be necessary to build a model of each of these reactions at the molecular level (with hundreds of different kinds of molecules)? Or could many of them be aggregated in a way similar to what we do with temperature? Had the model attempted to detail out every single chemical reaction, it would have taken a supercomputer and years of time to do even a small run. Instead the various reactions that had very similar time constants were lumped together and the average time constants used. The resulting four time domains (from real time in milliseconds, to minutes for short-term memory, to hours for intermediate-term memory, to very long-term memory encoding in days) were used and produced results that were close enough to real synaptic dynamics to simulate learning long-term associations in robot brains (Mobus 1994).

Question Box 13.1
There is a price for too little and a price for too much resolution. What kind of problems go with too little resolution? What are the problems of too much? If one is feeling their way to the "just right" resolution, is it generally best to "play it safe" in one direction or the other?

13.2.2 Accuracy and Precision

Related to the issue of resolution are accuracy and precision. The model builder needs to consider these up front. How close to the real system should the various model output variables come? How precisely do those numbers need to be tested?

Accuracy is a measure of how much deviation there is between the model output and the system as measured in the field given that both receive the same inputs. Deviation can be simple deterministic such as: the system produces measure y with input x, but the model produces $y + \Delta y$ with the same value of x, where Δ is always the same. More often, however, both the system and the model are stochastic[3] in nature so that the Δ is an average and has a standard deviation.

Precision is closer to the issue of resolution. Regardless of accuracy, the precision issue asks how many digits are required to represent the value(s). The more digits it takes, the more computational overhead you impose on the modeling engine. Sometimes scaling factors can be used to reduce the number of digits, but can introduce errors. The modeler has to be careful in choosing. For example, suppose a model of population could be built that accounts for every individual. The integer size needed to represent the US human population today would be eight digits (decimal to tens of millions). That is, no problem for a modern computer. Even slightly older personal computers can work with 9 digits of precision (and their newer brothers can handle 19 digits) without fancy software emulation of

---

[3] A stochastic model is called a Monte Carlo simulation. The model includes one or several kinds of noise injected in order to simulate the real stochastic system. Monte Carlo techniques (like rolling dice or spinning a roulette wheel) require multiple model runs and using the same statistical analysis of the collection of runs as are used to directly analyze the real system.

what is known as infinite precision arithmetic. But suppose you are modeling the population of bacteria in a petri dish culture? Clearly you would use some kind of estimation or approximation method based on a scaling factor, a one to a billion ratio might be appropriate!

13.2.3   Temporal Issues

As mentioned above, one temporal issue has to do with the time step size one chooses.[4] In general, and based on the capabilities of most of the modeling languages available, the step size is determined by the smallest time constant needed at the lowest level of resolution. A time step is the amount of time between successive measurable changes in a variable's value at the chosen precision. For example, in the population models of humans and bacteria, a time step of 1 month for a precision of eight digits would work for the human model, but a time step of 1 s would probably be more appropriate for the bacteria.

Another issue with time is that some models might require level modularity. That is, some of the subsystems at any level might need to be resolved down to a lower level, or selectively resolved down at critical points in a model run. These would have much smaller time constants than the higher-level processes. Currently, most languages do not support multiple time scale processing, so the time constant of the lowest level is chosen. In this case, the higher-level inputs need only be read into the model engine using time step sizes (integrals of the smallest time step) much larger than the lower level. The same can be said for recording result data.

13.2.4   Verification and Validation

Models of complex processes are notoriously complex themselves. And the construction of the model can be prone to errors. If a model is to be used to generate likely scenarios or make predictions, and especially if the model is going to be used to make policy recommendations, the modeler should be sure the model accurately reflects the actual system it is supposed to represent.[5]

Verification and validation are operations for the quality assurance of modeling. Verification is the process whereby the modeler or an independent party checks and rechecks that the model structure is a true reflection of the systems knowledge base

---

[4] Here we are assuming the model is to be run in a computer simulation that is iterated each time step.

[5] Often times there are other documents besides the analysis knowledge base. These could include requirements, performance, and test specifications. We will cover these documents in the last chapter on engineering. Modeling may be done in an engineering context, but we will wait to that chapter to discuss these.

as derived by systems analysis. Just as typos can show up in a written document, errors can be introduced in the process of encoding the knowledge base into the modeling language. The implemented version of the model (the program) needs to be verified as a true implementation of the analyzed model. This may seem only common sense, but in modeling complex systems such as the economy or the weather, simplification is inevitable, and how the simplifications affect the "reality" of the model can be a critical and much disputed issue.

Validation is a form of testing. That is, the test is to compare the model outputs under defined inputs to those of the real system (assuming that one exists). If the model performs the same behaviors as the real system given the same inputs and environmental conditions, then there is a high likelihood that the model is valid. Of course the model must be tested under varying conditions or combinations of inputs and environmental conditions. This is a form of empirical testing of the model given observations of how the real system performs. Its strength is only to the degree that there is adequate data on the real system under a wide variety of conditions. If there is a strong correspondence between the model outputs and the system outputs under a wide range of combinations, then our confidence in the model's validity is increased.

Verification and validation of relatively high-resolution models of complex systems is time-consuming but absolutely essential if our confidence in predictions made by the model is to be high.

13.2.5   Incremental Development

Rarely are analysis and model development done in one iteration. As we saw with analysis, we may have to reanalyze a system or subsystem to determine if (1) something has changed, (2) we were right the first time, or (3) new evidence suggests we need to dive deeper into the hierarchy in a particular area. Typically, the first and third situations result in multiple and increasingly refined versions of the knowledge base. The same is true for model building. Here we start from the top level, building a model at level 0 and testing it. Then, we can build the model at level 1 and check to see if the results are consistent with the top-level model. In other words, we incrementally increase the resolution of the model as we seek to answer questions that involve the lower-level dynamics of the system.

13.3   A Survey of Models

In this section, we will take a quick look at the range of modeling methods that are used, the kinds of systems that are investigated, and the questions that the models are intended to answer. This will be a survey that expands in Table 13.1.

13.3.1 Kinds of Systems and Their Models

We look at different modeling methods from the perspective of the kinds of systems that we are interested in and from the kinds of questions we ask in attempts to understand them. With each kind, we supply an example of the type of question that might be asked when using the modeling approach.

13.3.1.1 Physical

This is a general category of systems in which one might be interested in its behavior in a very dynamic environment, such as how a particular wing design might react to turbulence. The system is a physical object, and the environment has to be simulated in some fashion in order to measure the behavior of that object. In the "old days," for example, a scale model physical wing would be built and put into a wind tunnel for testing. Force meters attached to the wing could measure stresses on the airfoil as the tunnel simulated various kinds of air flows. The measurements were generally analog and required additional mathematical analysis in order to derive answers.

Today many such physical systems can be simulated digitally, meaning that both the system itself and its environment can be modeled in software with the output data directly available for analysis using numerical methods. Modern airplane designs, for example, are now done almost entirely in digital form and subjected to testing directly from the designs. This very much shortens the time between a design specification and the test results.

Example Question 13.1
At what wind speed will turbulence develop in the air flow over the wing?

13.3.1.2 Mathematical

Some system dynamics can be modeled entirely in mathematical equations that have what are called "closed-form" solutions. These could be algebraic or use the calculus to describe the behavior. Their solution is just a matter of plugging in the independent variable values and a time value, solving for the dependent variable, and voila, the answer tells you what the characteristics of the system will be at some future point in time. An example would be the logistical function discussed in Quant Box 6.1. Once the values of the parameters (constants obtained from empirical observations and curve fitting as in Quant Box 6.1) are given, the equation can be solved for any future time, t, in one calculation.

Mathematical models are extremely powerful just because they can be computed inexpensively and quickly. But they only provide answers about the "state" of a system, say at some future time, and cannot tell you how or why the system got to that state. They are the ultimate abstraction, working very well to make specific predictions for (primarily strictly deterministic) systems where exact formulas have been developed. Their uses are therefore limited to situations where only state information in needed. Any causal relations that are embedded in the model are implicit rather than explicit. As a consequence, they may provide quick solutions regarding the state of the system, but do not directly lead to understanding of the system, other than by analogy to another system for which the same set of equations are known to work.

Example Question 13.2
With reference to Quant Box 6.1, what will the population size be at time $t = 300$?

13.3.1.3 Statistical

This class of models is also mathematical but is used for stochastic processes where a certain amount of noise might mask the behavior of the system. Statistical modeling relies on methods of inference of behavior, so, once again, cannot provide causal relations. These models are generally used to make predictions of dependent variable state (values) given, typically, a set of independent variables and their states at a particular instance. Since noise is involved, the methods attempt to find a best fitting model that can then be used to predict the *most likely* state of the dependent variable for any relevant combination of states for the independent variables.

Regression analysis seeks to find a best-fit curve in multidimensional space from empirical data (again as in Quant Box 6.1). That then becomes the model. Various powerful statistical methods are used to put points on the curve into reasonable confidence intervals. The latter can be used to assess things like risk of making a wrong prediction.

Statistical modeling has been the tool of choice for many researchers in the social sciences where the idea of controlled experiments, as are often done in the natural sciences, is not really feasible. The scientists collect data from the field from the system as it is operating and then use statistical modeling to try to find strong correlations between factors. Those correlations can then be used to make predictions such as, "If we see these conditions arise in the system, then it is likely that the system will enter such-and-such a state."

It is important to recognize a major caveat with this kind of approach. Correlations do not necessarily mean causations. They can tell you that several factors vary together reliably, but they cannot tell you what causes what. One way to infer causation, however, involves time-shifted correlations. That is, if one or more independent factors are seen to lead in variation, and the dependent factor follows *AND* the time lags are consistent (though possibly stochastic), then one can infer that the leading factors have a causal role in changes in the following factor (see the discussion in Chap. 9, Sect. 9.6.3.2—Anticipatory systems).

Question Box 13.3 Every autumn the days get shorter, school starts, the weather is cooler, birds migrate, trees turn color, plants drop their seeds, and the sale of digital cameras increases. How might you go about figuring out causal connections in this scenario?

This is still a weak form of causal inference as compared with what we will discuss below. Very sophisticated methods for determining time-shifted covariances (even multidimensional ones) have been developed that have proven useful and successful in strengthening the causal inference. But if one has to be very certain of the causal relations in the behavior of a system, then a modeling method that directly tests those relations is needed.

Example Question 13.3
What is the likelihood that the system will be in state X given input M at time t?

13.3.1.4   Computerized (Iterated Solutions)

Below we will describe in greater detail several modeling methods that explicitly examine the causal relations between components and subsystems in order to obtain a deep understanding of the system in question. The above modeling methods, of course, use computation to solve the equations and analyze data. By "computerized," here we mean that the system is built from the knowledge base described as a product of systems analysis as in the prior chapter. The model includes details of all of the subsystems and components and their interrelations with one another. The model is then "run" in the manner described in the last chapter in section [Modeling a System]. All of the explicit state variables of the system are updated on subsequent time steps so that the modeler can track the evolution of the system dynamics over time. Rather than simply solving an equation for the final state, these models provide a graphic representation of the changes in state. This allows the modeler to verify (or modify as needed) the causal relations. Most importantly, it allows the modeler to more deeply understand the internal workings of the system than the previous types did.

Below, in our Survey of Systems Modeling Approaches, we will cover three basic computer-intensive categories of models. The first, system dynamics, provides the most direct approach to modeling complex systems. Using the knowledge base of systems analysis, it is possible to build highly detailed models that provide a lot of information about the actual workings of a system. Most of our behavior (fixed and adaptive), emergence, and evolutionary systems have been described in the line of system dynamics.

The second, agent-based modeling, is explicitly used to describe and simulate systems based on an aggregate or population of essentially similar (homogeneous) decision agents. Its main contribution is to show emergent group (e.g., social) behaviors and structures often based on simple decision rules. We will describe some of the simpler ones. But we will also introduce some of the developing ideas in building models of heterogeneous and adaptive (learning) agents that are being developed to study human social systems.

The third, operations research, is actually a very large body of mathematical methods for tackling all kinds of problems associated with finding optimal behavior or conditions in complex systems. We will describe a method that is used to find an optimal end state through a process of progressive constraint satisfaction among multiple interacting constraints. The ideas presented in Chap. 10 on auto-organization are of this sort. A system with high potential complexity is pushed (by energy flow) to explore a large space of possible configurations but is guided by the constraints imposed by selection forces and interaction potentials.

Example Question 13.4
How long, in time units, will it take for the system to enter state X with the stream of inputs in vector Y?

13.3.2   Uses of Models

Here we summarize the various uses of models. We have mentioned these several times in various contexts so it would be good to review them here in one place as applied to all models.

13.3.2.1   Prediction of Behavior

Presumably if we understand how a system works given various input scenarios, we should be able to predict how the system will behave under specific scenarios, even ones not actually witnessed in the real system. For example (and a critical one at that), global climate models are used to predict the climate changes we should expect to ensue with increasing average temperature due to greenhouse gas forcing. Climate scientists are spending a great deal of effort trying to refine their models as they try to understand what effects global warming will have on the future climate. This is critical because food production, water cycles, and so much else that we depend on are affected by climate, and if there are going to be detrimental impacts on these, we need to know about it and use that knowledge to prepare.

Naturally the predictions coming from models can only be as good as the quality of the models themselves. For model predictions of behaviors that have never been observed (as in the case of global climate change), it is hard to gauge just how good those predictions will be. Therefore, quite a lot of extra energy and care are needed in the system analysis phase.

13.3.2.2   Scenario Testing

Related to prediction, but not quite the same thing, scenarios are possible outcomes based on different constellations of system inputs. This is particularly important for systems that have nonlinear internal subsystems (which is basically all interesting systems!). Scenarios allow us to consider outcomes that may vary with slight tweaking of one or more inputs. The modeler varies the combination of values for inputs for a sequence of model runs and graphs the states and outputs to see how the system might respond to those combinations.

Scenario testing can be particularly useful if the modeler is seeking some kind of control over the future of the system. For example, if a company has a good model of sales of various products and their profits from those sales (with different profit margins on different products), they can plan the mix of product volumes that will maximize overall profits. They can then tailor their sales efforts and marketing to put more resources into higher-profit products.

> **Question Box 13.6**
> We use our imaginations in conjunction with remembered experience constantly to do scenario testing as we advance into anticipated futures. What features cause us to be cautious about imagined scenarios? How does this compare with computer-produced scenarios? Are they more "objective?" What features of systems/computers might enter into how confident we are in these computer-generated scenarios?

13.3.2.3   Verification of Understanding

The sciences are more interested in understanding systems than using predictions or scenarios for purposes of exploitation. Model building and simulations are a way to test hypotheses about the internal components and relations that produce overall behavior. Model run data is compared with data collected from real systems and their environments to verify that the model at least behaves like the real system. This then lends weight to the hypothesized structure and function of the system's internals. Such verification is related to the black/white box analyses where when the model system, operating in the model environment, behaves as the real system in its real environment, the scientists who built the model can then have greater confidence in the validity of their model and claim greater understanding of the system.

One commonly practiced method is called post-diction (as opposed to prediction). In this approach, the model system/environment is set up in an initial condition similar to a historical condition of the real system. The scientists already know the outcome of the real system, so if the model system arrives at the same state as the real system did, then this is taken as evidence that the model is correct. Post-diction has been used with climate models, for example, to see if given the conditions of the planet and its prehistoric climate (derived from geological and ice core sampling), the model evolves to a new state that replicates what actually happened on Earth. The deviations from the actual outcomes can be used to tune the model, which is a form of better understanding the system. When the model outcomes match the real prehistoric outcomes, then the climate model is more likely to be good at predicting future climate conditions because it captures the mechanisms of the Earth's climate system quite well.

13.3.2.4 Design Testing

In the next chapter, we will discuss the final principle and the process of systems design and engineering. Briefly the process of engineering involves developing systems that will fulfill specific functions with specific performance criteria. Systems engineering implies that the system to be designed is sufficiently complex that we cannot just draw specifications on paper and then build the system. Many forms of testing must be done, but the earliest form involves building computerized models of components and of the whole system to test parameters of the design. These days not only are airfoils tested digitally, but whole airplanes are designed and simulated long before the first piece of metal, plastic, or wire is produced. We will have more to say about this in that chapter.

13.3.2.5 Embedded Control Systems

As we saw in Chap. 9, controllers use models of the system, and they are controlling to determine what control signals best match the current condition of the process and the objective. Today microcontrollers, microprocessors along with ancillary circuits needed to interface them to the environment through sensors and motor outputs, are ubiquitous. If you have a smartphone in your pocket, you are carrying a very sophisticated embedded controller that handles the transmission protocols for several different wireless options and works cooperatively with a more standard microprocessor that handles the user interface. The automobile example mentioned at the start of the chapter is another example. Modern thermostats are yet one more example. In fact almost every electronic gizmo, from entertainment systems to refrigerators, and even some fancy toasters contain microcontrollers. The programs in these devices are the model of their environments, including users and the process they are controlling.

13.4 A Survey of Systems Modeling Approaches

Now let's take a look at a few modeling approaches that are used to gain that deeper understanding of systems. This will be brief and only look at a few examples since the topic is broad and deep, we can't do more than whet the students' appetite for what is possible. We cover three basic approaches that are widely used system dynamics (which will look familiar from Chap. 12), agent-based modeling (both typical and a glance at a more desirable way), operations research (optimization), and evolutionary modeling.

13.4.1 System Dynamics

13.4.1.1 Background

Systems dynamics[6] modeling started with the work of Jay Forrester[7] of the Massachusetts Institute of Technology School of Management. Professor Forrester had a background in science and engineering that he brought to bear in thinking about the very complex issues of managing human organizations. What Forrester did was see how one could formalize key aspects of how all systems seemed to work. He devised a "language" of systems, the semantics of which could be described as "stocks and flows with feedback controls." The beauty of this approach is that one could develop visualization tools (graphics) that captured the semantics and made it possible to construct pictures of systems (most of the graphics you have seen so far in this book have been in the same vein).

Forrester invented the DYNAMO[8] programming language that allowed a model builder to express the visual elements of a system diagram—the conceptual model—in a programming language having the same semantics. Thus, one could first construct a visual conceptual model that could then be coded into a computer program for actual execution.

Figures 13.1 and 13.2 show two forms of diagramming that are used to develop models in systems dynamics. The first simply captures the major factors that are at play in the model and determines the flow of influence or causality from one factor to another. A plus sign next to an arrow head means the influence causes the factor to increase in whatever units are appropriate. The minus sign means it causes a decrease. There is no attempt to necessarily identify the proper units nor is the implication that one kind of unit "flows" from one factor to another. The intent of the diagramming method is just to outline the factors and their interaction influences

---

[6] See System Dynamics, http://en.wikipedia.org/wiki/System_Dynamics.

[7] See Jay Wright Forrester, http://en.wikipedia.org/wiki/Jay_Wright_Forrester.

[8] See DYNAMO, http://en.wikipedia.org/wiki/DYNAMO_%28programming_language.

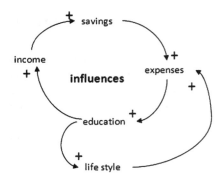

Fig. 13.1 An influence diagram shows what factors influence or cause changes in what other factors in a closed-loop model. In this system, income feeds the savings of an individual, while expenses (from living) reduce the amount in savings. One of the expenses, however, is for education that, in general, can cause an increase in income. At the same time, a more educated person, with a larger income, might be tempted to raise their standard of living which then increases the expenses. This model is not meant to assert any claims about the economics of life, just illustrate the way in which variables can affect other variables in feedback loops

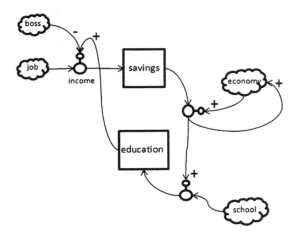

Fig. 13.2 This figure shows a structural/operational model of the same system shown in Fig. 13.1. This model is considered operational because the various flows and stocks (money, knowledge) can be parameterized and, at least in principle, measured. This kind of model can then be converted to a computerized version that can be run to see how the long-term effects of continuing education affect the wealth of the individual. See if you can conceptualize the savings of the individual who keeps getting more degrees from college. If only it really worked this way!

on one another. The second purpose of this kind of diagram is to identify loops, positive and negative, where feedback will cause "interesting" dynamic behavior.

The second figure, 13.2, takes the same model and operationalizes it or casts it into actual stocks, flows, and controls. The clouds represent un-modeled sources and sinks, just as we used open square brackets in the last chapter. The valves are rate controls. The arrows terminating on the valve handles show the influence of changing the flow rate (again, positive and negative).

Note that this particular example model is tremendously oversimplified and fails to actually capture reality very well. These diagrams are only meant to convey the basic ideas in how systems dynamic models can be constructed. In the last chapter, we saw how to analyze a system so as to produce these kinds of objects and their relations. Now we see how we use that process to begin building simulation models.

The stock-and-flow model is, in one sense, extremely simple. Every substance is represented in one of these two forms. A stock is a reservoir that holds something (matter, energy, information), like potential energy or an inventory. A flow is a transfer of something from, say, an input source to a stock within the system of interest, like kinetic energy charging a battery or material receiving bringing goods into the inventory. Those things can then flow out of the one stock and into another. The various flow rates can be controlled by forces (like a pump or gravity) and constraints (like a valve). It is possible, with some ingenuity, to break any system down into its constituent stocks and flows.

An important feature of the stock-and-flow model is the ability to show how a stock, say, in a downstream location, can influence the rates of flow of that or other stuff (as in Fig. 13.1 and see the figure below). This is the principle of feedback, and it is very important for the dynamics of complex systems. Representing feedback dynamics in dynamic systems theory turns out to be a bit harder, requiring complex systems of partial differential equations. In many cases, it turns out, also, that these equations admit to no closed-form solution, which is part of the advantage of taking the dynamic systems approach in the first place. But in DYNAMO, Forrester chose instead to represent flows and stocks through difference equations that allow for much easier expression in this instance. The computational load of iteratively processing large sets of these difference equations can be quite large, but as computers have become faster and more efficient, that disadvantage is less noticed. Figure 13.3 shows a message feedback system used to control the flow from one stock (or reservoir) to another based on the "level" in the second, receiving stock.

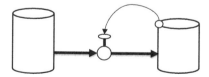

Fig. 13.3 The level in a downstream stock can be sensed (measured), and the information fed back to control the rate of flow from one stock to the downstream stock. The valve is used to control the rate of flow, and the small oval on the corner of the second stock (*right hand*) represents a measurement of the level of the stock. When the stock level is too low, the control valve opens to allow more flow-through. When it is over a limit, it closes the valve down. This is just a partial view of a larger system in which inflows and outflows from sources and sinks have been omitted

13.4.1.2 Strengths of System Dynamics

By focusing on the stocks, flows, and feedback controls, SD captures all of the essential aspects needed to demonstrate dynamic behavior. One has to say it is exquisitely simple, and therefore, easy to use by noncomputer scientists/programmers. In essence, everything important about a system's behavior is represented by these relatively simple constructs, stocks of stuff, flows of stuff, and feedback loops that help regulate the flows and levels. Very little more need be identified in order to build models that provide information about the system's dynamics. As is the case for many mathematical and formal language constructions, the simpler ones are often the most expressive. They are the most flexible in terms of being able to express any complex system using just a few lexical elements.

The reason this is the case is that stocks, flows, and controls apply to every system. Thus, in principle, any system can be expressed in this language. Derivatives of DYNAMO, such as STELLA,[9] Simile,[10] and Vensim,[11] are all based on this fundamental semantics. They have different graphic interfaces (see Ford 2010, Appendix C for a comparison) with different icons representing the various lexical elements. Nevertheless, they all work on the basis of modeling stocks and flows, providing graphical outputs to visualize dynamic behavior.

13.4.1.3 Limitations of Stock and Flow

The original approach to systems dynamics modeling has stood the test of time, and the stock-and-flow semantics are still in use today. It has been very powerful in solving many complex systems models. For example, a language called STELLA can take a purely visual representation in the stock-and-flow format, as above, and directly translate it to a DYNAMO-like program making its use extremely friendly.

But the stock-and-flow semantics can be very cumbersome when it comes to expressing some kinds of complexity in systems models. These semantics are almost too general (which usually has the advantage of increasing the expressiveness of a language) in that anything imaginable can be defined as a flow (substance). But it turns out that the overall architecture of a language is just as important, if not more so, to the efficiency with which concepts can be expressed. This is where the process perspective comes in. In languages like DYNAMO, the representation of processes has to be expressed strictly in terms of stocks and flows with their attendant controls. While possible to do, it can lead to some awkward constructs (see Fig. 13.4). As it happens since all of the substances in the real world are matter, energy, or messages, it is possible to restrict definitions of flows/stocks to being of

---

[9] See STELLA, http://www.iseesystems.com/ web site for description.

[10] See Simile, http://www.simulistics.com/ web site for description.

[11] See Vensim, http://vensim.com/ web site for description.

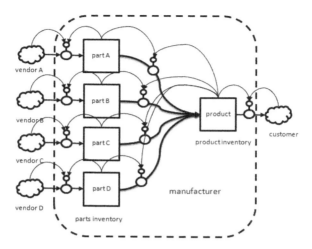

Fig 13.4 A model of manufacturing using the stock-and-flow semantics leaves out a lot of detail. This is a minimal model that assumes the rates of flows of parts from vendors to parts inventory (stocks) are arbitrated between the inventory management (un-modeled) and the vendors. It similarly assumes that the flow controls from parts to product model the combinations of parts that go to make up the product. Arbitration between the parts inventory management and product inventory management (somehow) regulate those flows. Nowhere in this model do we see the actual process of combining parts to produce the product. In theory, this model could be further decomposed into sub-stocks and sub-flows and combined with more explicit management decision clouds to make it more representative of the manufacturing process. In practice such decomposition becomes tedious and highly complex. A process semantics-based language should accomplish two things. It should simplify the model construction while incorporating a lot more details of the process based on accounting rules and the laws of nature. See Figs. 5.7 and 5.8 for comparison

these three types. Furthermore, in nature the ways these three substances interact are quite well defined and follow specific laws. By using these ideas to constrain the framework, we might get a great deal of expressiveness at a lower cost (in conceptualization effort—remember Chap. 1!) in terms of the complexity of the model.

Taking a lesson from the world of object-oriented computer programming,[12] a semantics based on an object, which is a *process*, provides a more natural way to express dynamic systems as we did in the last chapter.

---

[12] Object-oriented programming (OOP) is a paradigm that was invented expressly for building models and realized in languages like Smalltalk, C++, and Java. In these languages, every "module" that performs actions (functions) is defined as an object. Objects have internal state (variables) and behaviors. They can provide services to other modules by virtue of those modules communicating with one another. See Wikipedia—http://en.wikipedia.org/wiki/Object-oriented_programming.

13.4.2 Agent-Based Modeling

13.4.2.1 Background

Agent-based modeling has been used extensively to investigate the phenomenon of emergence for both structures and behaviors in aggregates of similar "agents." At the low end of the approach, investigators were interested in how simple, rule-based decision entities would interact with each other given the parameters of their environment. We saw this case in Chap. 10, Sect. 10.3, in the form of auto-organization. At the high end, researchers are interested in human societies, the formation of social groups, economies, and other emergent phenomena that arise from the interactions of "autonomous" decision entities. The latter are called agents, but this term should probably apply to a broader range of entities than just humans. The rule-based entities may not be autonomous (see definitions below) in the conventional sense, but when you include the stochastic processes that are involved in jostling their interactions (chance meetings), they can, at times, appear to be autonomous. At least they can produce collective surprising actions, which is what we mean by emergence.

13.4.2.2 Modeling Framework

The general framework for agent-based modeling involves the generation of a large population of initially similar agent objects. These agents have intercommunications capabilities, individual internal memories, decision heuristics, action outputs, and possibly variations on motivations for actions. As they are each generated from a common template, their internal parameters may be initialized variously (e.g., randomly) to start them with varying personalities. Computationally these agents are implemented using a combination of object-oriented structures and distributed artificial intelligence processing, as shown in Fig. 13.5.

In the language of systems decomposition we presented in the last chapter, agents are complex decision processes. There are two ways that agents can be incorporated into models. We saw an example of this in Fig. 12.21 in the last chapter regarding adaptive agents making decisions in the context of a larger dynamic system, i.e., an economy. The second way, used to investigate emergence directly, is to simply put a lot of agents into an initially non-organized arena, let them interact according to their individual personalities, and see what comes out of it. The output of the former might be to discover nonoptimal decisions or interaction dynamics, as when customers decide to pay too much for a product leading to overproduction relative to other economic goods. The output of the latter would be to discover how agents cluster or show preferred interactions with one another as an indication of some emergent behavior.

Figure 13.6 shows a general structure that is used for the latter kind of model. The "interaction matrix" is a switching mechanism that allows any agent to interact with any other agent in the population. The database contains a time series recording

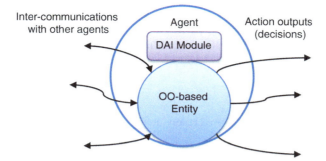

Fig. 13.5 Agents are implemented using distributed artificial intelligence (DAI) and object-oriented (OO)-based processing. The *two-headed arrows* represent communications between agents and the *single-headed* (output) *arrows* represent actions or decisions that agents take

Fig. 13.6 A typical model of agents interacting to see what emerges can be implemented like this

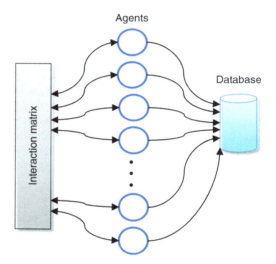

of agent decisions that can then be analyzed for patterns of interactions and decisions. The simulation proceeds as described in the last chapter as iteration over the array of agents for each time step.

The actual patterns of interactions between agents depend entirely on assumptions made about what sorts of potential interactions there are and the kinds of decisions each agent can make. For example, suppose a researcher is interested in understanding what is involved in any agent deciding to be friends with any other agent. They would have to develop a hypothesis about how those kinds of decision are made and then program the AI part to execute that hypothetical mechanism as individual agents get information from their environments of other agents.

At the end of a run, the researcher looks at the strength of interaction connections that evolved as an indication of who is friends with whom and takes a close look at exactly which kinds of interactions produced that result.

13.4.2.3 Definitions

Let's take a moment to look more closely at definitions of terms we've used.

13.4.2.3.1 Decision Processes

In various parts of the book so far, we have discussed decision processes, especially in computational terms (Chaps. 8 and 9). In the case of autonomous agents, we have to recognize that the decision processes they engage in are extremely complex and the actual details can change as the agents "learn" from their experiences.

13.4.2.3.1.1 Decision Trees

Decision processes can be represented in the form of a tree in which each level (going down) represents a decision point going forward in time. The usual interpretation is that each node represents a current state of affairs, which includes both the state of the agent and the state of the environment. At a particular node, the actual state then dictates a path to the next level down. Figure 13.7 shows the general structure of a portion of such a tree.

> **Question Box 13.7**
> Regarding decision nodes in Fig. 13.7, why does each node need to represent both the agent and the environment?

Generally, the agent is seeking a goal state, but that state may actually just represent an intermediate state and the start of a new tree. Decision trees are used in game theory where each node is the state of the game, the goal is to win, and each level represents a move (alternate levels being moves for opponents), say in a game of chess.

A decision tree can also represent a map of decision paths taken even when there is no rule as to which path to take from any one state. In such a case, the tree is a "history" of decisions and states experienced. There might not be an explicit goal other than to not get destroyed along the way! The phylogenetic tree, the record of evolution, represents this kind of tree. Technically, evolution isn't making decisions, certainly not explicitly. But the actions of selection on traits effectively make a kind of choice. And, as they say, the rest is history.

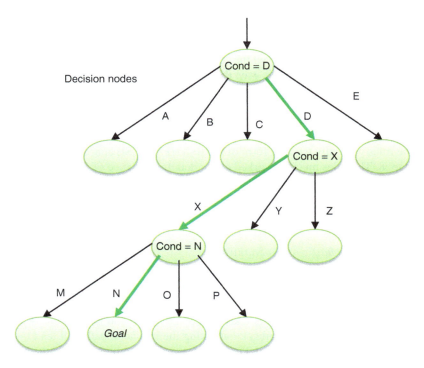

Fig. 13.7 A decision tree is constructed from a priori knowledge of which path to take in any given node given the conditions that are found at that juncture. Each node represents a state of affairs for the system (and environment). When the agent enters the top node, a particular state, if the condition found is "D," then the agent must choose to follow path "D." Nodes may also represent actions taken given each state condition found. The bottom row represents final states, one of which is the "goal" of the system

Decision trees are useful to agents if there are some means of choosing a path based on the current state and, perhaps, some history of prior experiences—a memory. We will look at three examples of decision tree structures that are used in artificial and real agents as they make choices. The first is what we will call absolutely deterministic. It really isn't much in the way of intelligence, but it is instructive in how decision trees might be constructed with intelligence. The second one is somewhat typical of AI agents. It is a probabilistic structure in which the choices are poorly known a priori, but can be refined with experience. Its use is when the same kinds of decision states recur over time. The last is closer to how we think humans actually process information to make decisions. Researchers are thinking about how this might be incorporated into automata, but none have emerged from the research labs as yet. Interestingly the approach has a name that is a tongue-in-cheek play on artificial intelligence. It is sometimes called artificial *stupidity* because the decision agent is imperfect and can make stupid decisions (like some human beings we all know!).

13.4.2.3.1.2 Rule Based

A rule-based tree is just what the name implies. At each node, there exists a rule that determines the choice made given the conditions found at that node. Figure 13.7, as shown, can be interpreted in this way. Rule-based decision processes can be used by agents when all of the possible conditions at each level in the tree can be known in advance. Examples are diagnostic processes for deterministic situations, e.g., diagnosing a problem with an automobile. This is applicable where the conditions at the nodes are fully known in advance and can be sensed unambiguously.

The rules generally take the form of the IF-THEN-ELSE construct we examined in Chap. 8. The rules must be worked out in detail in advance of running the agents through their paces. The applications for such trees are limited. Think about a tic-tac-toe playing program. Not very exciting, and it can always win if the opponent goes first.

13.4.2.3.1.3 Stochastic or Probabilistic

Decisions get more interesting if the pathways have some uncertainty associated with them. Even if the state evaluation is certain at any node, the pathways to the next level down may only be assessed probabilistically. For example, in the figure above, given that the condition is taken to be "D" in the top node, there is only a numerical likelihood (a probability) that taking the indicated path will lead to state "X," which is on a path to the goal. Put another way, that path might have the highest probability of leading to "X" of the choices available. The rule then becomes IF goal is "X," THEN choose the path with highest probability of reaching it. In the case where multiple pathways have equal and highest probabilities then choose one at random!

Probabilistic choices certainly look a lot more realistic when thinking about real life. However, they are not too different from their deterministic rule-based cousins above. The probabilities of each choice leading to a favorable next state have to come from somewhere. But this does lead to interesting possibilities for AI. As long as there is a way for the program to evaluate the relative "goodness" of a choice made, compared with its ideal state, then it can assign weights to the various choices a posteriori. The assumption is that the agent will go through this kind of decision again in the future, so it needs to learn the probabilities as it goes and use what it learns in the future when it comes to the same place in the tree. There are a number of machine learning methods that associate the current state pattern of a node with the pathway choices made over many iterations of going through the tree. They can even be programmed to be exploratory and try new paths from time to time as they build up their table of probabilities (see Chap. 7 regarding how knowledge is built from information gained in past experiences).

Question Box 13.8
Getting a car repaired and getting a medical condition fixed seem like similar decision tree diagnostic processes. But usually the car mechanic sounds rule based and the doctor frustratingly probabilistic. Is the difference just a matter of medical ignorance, that is, should the ideal for the decision tree be the same? Why/why not?

The key to this, however, is that the agent cannot be permanently killed by making a bad choice. Think here of computer/console shooter games. If the protagonist doesn't manage to shoot all of the bad guys, he or she may wind up being shot and loose energy points. Loose enough and he/she dies. But, lo and behold, the player can simply start over in some manner and go through the game again. If he/she learned anything from failing the prior time, then they can use that to prevent being killed again. Obviously this doesn't work in real life where major damage or death is on the line (e.g., extreme sports!)

13.4.2.3.1.4   Judgment Based

Agents that make decisions of the prior forms are not terribly intelligent. These are sometimes referred to as "reactive" agents. Many forms of agent-based modeling can be based on rule-based and heuristic-guided agents. To model human or even animal social systems, however, requires agents that are far more intelligent and make decisions based on current information and stored knowledge. These are called "cognitive" agents.

In this final example, we look at what we think the nature of cognitive decision processing looks like and give a brief review of the differences. One of the first differences is that the tree metaphor might not really work for real-life decisions. A tree is just a special case of a directed graph, one without cycles. For real-life conditions, the better metaphor is a complex web of states and multiple connections between nodes.

Indeed, a better graph structure might be that of a complex of underground chambers with tunnels leading between them. Figure 13.8 gives a picture of this metaphor. The problem is framed as a search by the agent for some goal contained in one or more of the chambers. Thus, the problem can be couched in graph theoretical terms as a traversal from a starting node to a goal node. In this case, the search agent has to be pretty clever and use information local to the chamber and memory of past searches to conduct an, hopefully, efficient search through the tunnels.

This is a maze-like structure. Each chamber represents a state (again for both the agent and the environment). From each there is a plethora of tunnels leading out to other nodes. Here the graph is nondirected which means that an agent can go either way through a tunnel and indeed get stuck going in a circle. Moreover, note that given the nature of the tunnels, there is no hint from just the structure of the connections

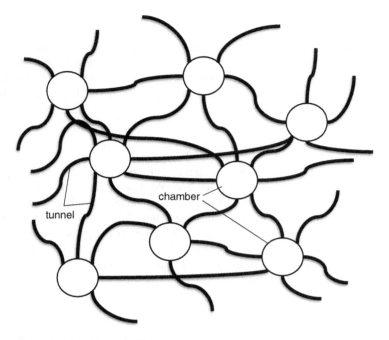

Fig. 13.8 Real-life decisions might seem more like a huge set of state nodes (chambers) and multiple links (tunnels) that form a complex web of choices

how to proceed. Imagine our subterranean agent going through a tunnel and coming out into a chamber. It looks around the chamber and discovers multiple other tunnels leading to other chambers. Which tunnel should the agent choose if the current chamber does not contain the goal (like a treasure chest!)? This is a very hard problem. And this is precisely where real intelligence comes into play.

If our agent has an ability to learn to associate state knowledge with pathways, then it is possible to build up a knowledge base of which pathways out of a chamber have proven most efficacious in prior experiences. Not only must the agent evaluate the current state of affairs but must have a memory of which pathways (in the past) have led to the most favorable subsequent state. We will allow that inside each chamber, on the wall just next to an outgoing tunnel, is some set of tokens that are actually connected with the contents of the chamber to which the tunnel connects. Moreover, we allow that next to each tunnel is a chalk board on which the agent may write any symbol it chooses on the board next to the tunnel it selects so that if it is ever in exactly this same chamber again it will recognize that it had been this way before. The tokens and symbols decorating a chamber are used by the agent to make decisions about where to go next.

It turns out that this representation is not just metaphorical. It comes into play in a real-life *foraging search* where the animal is situated in a real environmental position and has to decide which direction to follow in order to find, say, food (Mobus 1999).

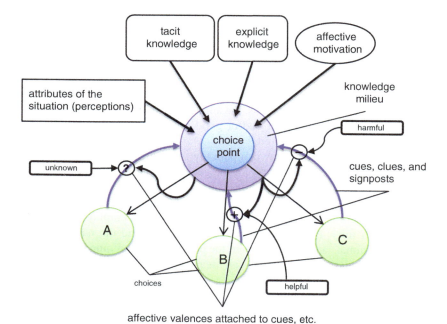

Fig. 13.9 An intelligent decision graph would include nodes that are augmented with additional knowledge and information along with motivational constraints. See text for explanation

Natural intelligence is a product of the evolution of animal brains as new species entered more complex environments in which the finding of resources (e.g., food) and mates became increasingly difficult. Much greater computational power was needed.

In Fig. 13.9, we show a mental construct of a choice point that, essentially, corresponds with the kind of graph structure above. Each choice point represents a "chamber" or position in the larger environment. Using local information, knowledge from memories, and motivations (e.g., attraction or repulsion), an intelligent agent evaluates the possible choices and selects the "tunnel" that will most likely help it find its goal.

In the figure, the choice point (in the graph) is surrounded by a *knowledge milieu*. Perceptual information from the environment (all of the tokens and symbols next to the tunnels), what we call cues, clues, and signposts, is combined with the *tacit knowledge* held in long-term memory as well as *explicit knowledge* in working memory and *affective motivation* (e.g., hunger) all of which act to modulate the decision process.

The most primitive decision processing in animals is based on affective valences attached to the clues, etc. by past experience.[13] Valence means that either a clue (or

---

[13] Damasio (1994). Damasio calls these valence tags "somatic markers," page 184 in paperback edition.

pattern of them) is tagged as negative (something bad happened sometime in the past when I made this choice), positive (something good happened), or unknown (not marked). These tags are neural markers for punishment and reward in the brain, and once an engram for bad or good is associated with a pattern of clues and cues, it is hard to undo. Fundamentally, given no other basis for choice, an animal would choose the positive valence (or any positive valence)-tagged pathway and avoid a negatively tagged one. For unknowns, the animal might choose it to be exploratory. There might be something either bad or good down that path, but it will remain unknown until that path is explored.

As with the decision trees above, a path toward a goal is taken by going to the next node ("chamber"), looking first for the goal, but if not finding it looking at the choices presented and then combining all information and knowledge to determine the best next move.

At present, there are no artificial intelligent agents that employ this judgment-based approach fully. But, as we have come to understand the nature of mental processing in animals and humans better, the possibility of building such systems in models seems at least feasible. This is an area of intense interest and much current research.[14]

Question Box 13.9
Based on the above, what is involved in making a judgment? What is difficult about a clear rule to discriminate between better and worse judgments?

13.4.2.3.2 Autonomy

Autonomy is often taken to mean that a system can take decisions and actions without any external control affecting it. Insofar as agents are already set up with some kind of defined reaction potentials to environmental conditions, this notion of autonomy needs further reflection, though sometimes this extreme version is identified with what in humans is called "free will." We won't explore the issue of whether or not humans truly have free will in this sense. Instead we will look at a systems perspective on what autonomous decision making could mean.

From that perspective, we recognize that all systems react to environmental inputs, and in certain cases, such as the impact of forces or ill effects of chemical substances, the system has little choice in how it reacts. Think of a reflex response. If you put your hand on a very hot object without realizing it, then without thinking you will instantly retract your hand. Your nervous system is wired to do this.

---

[14] Cf. the *Blue Brain Project* at École polytechnique fédérale de Lausanne (EPFL). http://bluebrain. epfl.ch/. Accessed, August 10, 2013

Of course, if you are pre-aware of the fact that a surface is hot and for some reason other factors in your environment compel you to do so, you can override this response and keep your hand on the surface, even to the point of getting burned. So it seems you, as an autonomous being, have some choice, but notice the sleight of hand we just pulled—we said that something else in your environment, your circumstances, compelled you to go against your body's wisdom. Would it make a difference if we said "motivated" rather than "compelled?"

Determinism implicitly takes a Newtonian physics model[15] for all causal processes: every cause, including combinations of causes, has in principle a determined result. Can systems that respond with internal activation guided by information have more than a single possibility when the complex of contradictory and complementary informed motives is summed up? That must be left up to the reader; no one as yet even knows how to do such a calculation. For our purposes, we will take autonomy to be a condition where a very complex, adaptive system is capable of taking decisions in the sense described above (e.g., Fig. 13.9) in a very complex situation based on internal motivations and knowledge along with clues from the environment. Since we include internal motivations (such as a desire to not be eaten) that are programmed into the system, we have to accept that under this definition systems are not completely free to choose any course of action. Nor do they randomly choose. That is, choices are motivated, and a systems version of autonomy does not imply otherwise.

There are, then, degrees of autonomy as we apply it to agents (below). At one extreme, the agent obeys fixed rules. A little less rigid is an ability to evaluate preferences for multiple possibilities and choose the most attractive one. There are still heuristic rules being followed, but there is an appearance of greater autonomy to an outside observer watching the agent. At the other extreme, as far as we are able to gauge, are agents having a plethora of options to choose from at each and every decision point. Furthermore, they may not have the requisite knowledge needed to adequately evaluate the situation and thus might have to "guess" what to do. This is where something like "judgment" comes into play—using past experiences to build "suggestions" into the decision process.

One final consideration for what might constitute autonomy of agents in a computer simulation is the absence of a centralized controller. There is no central decision center that gives instructions to the agents. The free market of economists' dreams is an example. According to the theory, every agent acts in its own self-interest and makes decisions that maximize its well-being.

---

[15] Much of the so-called quantum weirdness has to do with the nondeterminacy of simultaneously possible or "superposed" states of unobserved particles. This may have nothing to do with questions of choice, except that it seems to challenge the scope of the simple deterministic model of causality assumed in Newtonian physics.

13.4.2.3.3   Agents

In the context of modeling using agents, these are independent objects with their own local memory. Each interacts with other agents and the general environment in such a way that they develop idiosyncratic memory values.

More generally, agents are entities that have some degree of autonomy (as described above) and act either in their own behalves or on the behalf of clients (e.g., a travel agent searching for a great vacation package). Intelligent and adaptive agents have been simulated in software to varying degrees, and this is still an extremely intense area of research.

13.4.2.4   Emergence of Macrostructures and Behaviors

In Chap. 10, we explored the meaning of auto-organization and emergence. With agent-based modeling, it is possible to simulate these phenomena. Unlike working with an existing system that we model, say using system dynamic methods, we are interested here in exploring the results of auto-organization as agents interact over time. We are looking for what sorts of structures of mass behaviors emerge given our starting assumptions about agent decision-making capabilities and other "personality" traits. We can start with an essentially amorphous structure—an aggregate of agents—and see how they organize into groups or other subsystems as they interact with one another. The only a priori structure in this approach is the initial agent(s). A very wide set of questions regarding evolving system dynamics can thus be explored in a "bottom-up" approach.

What researchers are attempting to verify is that when they design interaction potentials for their agents and they run a simulation, do the agents behave the same way the real agents appear to? Do the kinds of social structures and behaviors emerge that are observed in real life? When the answer is yes, the researcher builds confidence that their understanding of the real agent(s) is(are) correct. Social psychologists are using this approach to test theories about human interactions and the emergence of social structures based on their understanding of human psychology. Below we will give an example of another kind of social structure that arises from the interactions of agents somewhat simpler than human beings. We will show how the behaviors of ant colonies emerge from the actions of a few different kinds of agents.

13.4.2.5   Strengths of Agent-Based Modeling

Emergent behavior can arise from interacting similar agents, even as simple as represented by binary decision rules. The fact that emergence of complex behaviors from simple agent systems is, in itself, a valuable insight that we have gotten from these models. More complex agent-based models, that is, where the agents are more complex and their decision processes are more autonomous, are just now starting to be explored with greater vigor and already show some surprising behaviors. It is clear that social systems models must involve populations of interacting agents and we look forward to more advanced developments in this field.

Large populations of agents can be readily simulated on computers. The space of possible model designs is enormous so there are many possibilities for exploring this fertile approach.

13.4.2.6 Limitations of Agent-Based Modeling

On the other hand, there are some remaining challenges. Most of our interests will likely be in the societies of human agents. That is, we are most interested in what we humans might do in social contexts. Current models of agents are very limited in terms of their autonomy. Some approaches seek to provide more degrees of freedom to agents by using Monte Carlo techniques (injecting randomness). But this isn't really what humans do. One example of the problems with agent-based models where unrealistic assumptions about human decision processes have been made is the neoclassical economics model of *Homo economicus*.[16] Numerous psychological investigations have now shown that the basic assumptions about how humans make economic decisions are quite wrong—mere conveniences to make the modeling simpler.[17]

Agent-based modeling is promising but, perhaps, still in its infancy. Agents that can make the kinds of judgments portrayed above in Sect. 13.4.2.3.1.4 are still problematic to model. However, research is active and vigorous in this arena, so we should keep our eyes open to the possibilities.

13.4.3   Operations Research: An Overview

In Chap. 9, we discussed the nature of logistical control in a hierarchical cybernetic system. Logistics, you may remember, is a decision process in which the operations of multiple operational units have to be coordinated, and it usually requires that the coordination produces an optimal final output from the system being managed. Optimality is essentially the situation when the qualitative measure of the output is maximal subject to the constraints imposed by inputs and the internal structure/ functionality of the system as a whole. There are costs associated with trying to boost the behavior of any given subsystem, or for acquiring more or better inputs from the environmental sources. Those costs figure into decisions about how best to produce the best possible output (which is associated with an income). Businesses,

---

[16] See http://en.wikipedia.org/wiki/Homo_economicus for a description of the assumptions used to build agent-based models of economic systems.

[17] Gilovich et al. (2002) provide a large collection of papers on just how unlike we humans are from Homo economicus!

for example, use profit maximization as their quality measure but balance that against constraints such as input costs (like labor) and requirements such as quality standards for the products or customer satisfaction.

This is not an easy task in an extremely complex system.

"Operations research" (OR) is essentially a branch of applied mathematics which uses a number of methods from other mathematical realms to build models for optimization. Most optimization problems cannot be solved directly but require a computer program that iterates over a matrix of equations making incremental changes and looking for those changes that move the current solution closer to a maximum. Of course, by definition, the maximum cannot be known in advance, so most of the methods simply look for improvement with each iteration, going further if an improvement is found or backtracking if the solution was less. Quant Box 13.1 provides an example of the kind of problem that can be solved by a method known as *linear programming*.

As in the example in the Quant Box, OR practitioners are effectively building models of complex systems in a computer memory, and then the processing operates to find the solution by iterating over the system of equations. These systems of equations are quite similar to those used in system dynamic models and are based on detailed information about how all of the relevant parts of a system interrelate.

OR techniques are used extensively in both logistical and tactical management in industry and the military for budgeting, moving assets to points of greatest effectiveness (e.g., shipping from plants to distribution points), and financial planning.

Quant Box 13.1 Linear Programming
Solving for an Optimal Value in a Complex Set of Requirements and Constraints
In complex dynamic systems, we often find that the interactions between various parts of the system can produce a suboptimal overall performance (function). Linear programming is a mathematical approach to finding a global optimal solution that is also a "feasible" solution by iteratively taking into account all of those interactions. Starting with a linear objective function and subject to various linear constraints, the method computes a feasible region, a convex polyhedron, and finds the point where the objective function is either a minimum (say for costs) or a maximum (for profits).

A classic optimization problem is solving for a maximum profit obtained from the production of a mix of products where the capacity to produce any one product can be utilized at the expense of production of any of the other products. Each product generates revenues but also costs that vary with production rates chosen. So the problem is to find the right mix of products given their revenue potentials and cost requirements that give the largest profit (total revenues less total costs).

(continued)

Quant Box 13.1 (continued)
Standard form of the problem
The objective is to maximize the value of a function of the form

$$f(x_1 x_2 \ldots x_n) = c_1 x_1 + c_2 x_2 + \ldots + c_n x_n$$

where x_1, x_2,..., x_n are variables that need to be solved for and c_1, etc. are constants.

But the solution is subject to a set of linear constraints of the form

$$a_{11} x_1 + a_{12} x_2 + \ldots + a_{1n} x_n \leq b_1$$
$$a_{21} x_1 + a_{22} x_2 + \ldots + a_{2n} x_n \leq b_2$$
$$\vdots$$
$$a_{n1} x_1 + a_{n2} x_2 + \ldots + a_{nn} x_n \leq b_n$$

where xs are the variables and the as and bs are constraint requirement constants. Another requirement is that the variables cannot be negative values.

An example
The Wikipedia article on linear programming (http://en.wikipedia.org/wiki/Linear_programming) has a simple example that is based on a farmer's desire to maximize the profits he will realize from planting a fixed size field with some mix of two crops (or just one or the other). He has to solve for the fraction of the field that will go to each of the crops that will produce the maximum sales value. One crop sells for S_1 dollars per square area, the other sells for S_2. The objective function would then look like

$$\max S_t = S_1 x_1 + S_2 x_2$$

or find the values of x_1 and x_2 that maximize the total dollars. But there are constraints. First, the amount of land in the field is fixed. That is,

$$x_1 + x_2 \leq L$$

where L is the total amount of land available.

Each crop requires different amounts of fertilizer and insecticide to produce maximum yield. And the farmer only has F kilograms of fertilizer and P kilograms of insecticide to work with. So he has to subject the objective function to constraints such as

(continued)

Quant Box 13.1 (continued)

$$F_1 x_1 + F_2 x_2 \leq F$$

$$P_1 x_1 + P_2 x_2 \leq P$$

and no amount of land can be negative!

$$x_1 \geq 0, \qquad x_2 \geq 0$$

Solving

It is beyond the scope of this example to go through the solution process. The above set of inequalities can be readily represented in matrix forms, and then an algorithm known as the "simplex method" is used to find the optimum (maximum in this example) if such exists. See Wikipedia article: simplex algorithm, http://en.wikipedia.org/wiki/Simplex_algorithm.

Linear programming has numerous uses for solving many kinds of optimization problems. It is a way to construct a model showing the relations between multiple components (variables) based on multiple independent constraints. Once set up, as above, it is just a matter of running the model (e.g., the simplex method) to determine the outcome.

13.4.3.1 Strengths of OR

There are many well-known optimization problems that are amenable to OR models, and their uses, especially in areas such as logistics management, are very well developed. Models, such as the linear programming example in Quant Box 13.1, have been used quite effectively in many such areas.

13.4.3.2 Weaknesses of OR

Sometimes it is not easy to identify all of the requirements and constraints needed to adequately characterize a problem to set it up for an OR approach. Without a very thorough systems analysis, it is sometimes easy to assume certain conditions to be true (such as that all relations in a system are linear) when they are not. It happens that people do tend to treat some problems that have a superficial resemblance to a classical OR form as such, apply the tools, and then get very surprised when the real system does not perform optimally. This is called "the law of the instrument," best articulated by Abraham Maslow (the psychologist) as: "If the only tool you have is a hammer you are likely to treat every problem as a nail!"[18]

This isn't so much a weakness of OR per se as it is of the systems analyst/engineer who applies it inappropriately. When the problem legitimately fits the conditions of optimization, it is a very powerful tool indeed.

13.4.4   Evolutionary Models

There have been numerous approaches to building models for exploring evolution. Whereas agent-based models have become somewhat "standardized" for exploring emergence, there are many different approaches used to understand evolution through modeling. Since this is itself an emerging field, we will only mention a few approaches and describe how they attempt to simulate evolutionary phenomena.

The common feature of these models is the use of environmental selection forces to eliminate poor outcomes.

13.4.4.1   Evolutionary Programming/Genetic Algorithms

This approach is related to OR techniques in that the final product is hoped to be an optimal structure/function. The starting premise is that there exists a program or algorithm that would solve a complex problem, but it isn't clear what that algorithm entails. The objective is to do the same kind of gradient following as we saw in OR as the program iterates over "generations" of solution attempts.

For example, suppose there must exist an analog circuit that could most efficiently compute the solution to a specific instance of a very hard problem. The entities making up a population of solvers are circuit models where component parts (resistors, etc.) are wired in essentially random fashion. Note that the solution of the problem itself is already known. What we are looking for is to evolve a circuit that can solve the problem directly. Thus, the selection criterion is how close does the generated circuit come to the solution. Many circuits are generated in a single generation. All are tested, and only one or a few that produce the "best" solution are kept as the base for the next iteration (generation).

Now, this next part is meant to emulate the novelty generation in evolving living systems. The circuits are represented by a "genetic code" in such a way that a randomizing process can cause mutations to occur in the wiring diagram and thus give rise to mutated circuits. Lots of circuits with different mutations are generated in each generation. And then the selection test is applied to all of them as before. This cycle is repeated as often as necessary until the circuit solves the problem as close to perfection as some arbitrary tolerance allows.

Sometimes, the process also includes such features as chromosomal crossover (another kind of mutation) and a form of sexual mating where half of the gene set of

---

[18] See http://en.wikipedia.org/wiki/Law_of_the_instrument.

any one circuit is mixed with half a set from another circuit (both of which are among the best performers, of course).

These models have been applied to a number of different problems that are inherently hard to solve directly algorithmically. Researchers have used the method to design mechanical devices as well as discover computer programs that solve problems (like circuits but in digital form).

Question Box 13.9
Evolution in the natural world is an open-ended process in which the metric is the moving target of fit with an ever-changing environment. Computer programs are good at a recursive selective feedback loop on fit with the goal set by the programmer. What factors differentiate this from the open-ended evolution of the natural world? What would it take to bridge the gap, or is that just flat out impossible?

13.4.4.2   Artificial Life

An area that has been popular with complexity theorists studying emergence has been *artificial life* (AL). Like its close cousin, artificial intelligence (AI), AL attempts to replicate life processes such as metabolism (Bagley and Farmer 1992) and ant swarms.[19] Many of the methods used are some form of rule-based agent modeling where the overall behavior of a meta-system emerges from the interactions of simple agents.

13.5   Examples

13.5.1   Modeling Population Dynamics with System Dynamics

Numerous models of populations (of humans and other species) have been explored using system dynamics. The population of living beings is a stock, births are inflows, and deaths are outflows. This seems straightforward enough, but when we do a closer systems analysis of populations as systems of interest, we discover many complications.

It turns out that demographic factors complicate the picture. Take just the fact that a population is actually an aggregate of subpopulations, say of age groups. Births are only inputs to the youngest age group, but deaths affect all age groups, just at different rates. Here we will take a look at a relatively simple population model and demonstrate the ways in which the model is constructed and simulated. We will look at the results in terms of population levels plotted as a function of time to see how the population grows.

---

[19] See Swarm Intelligence, Ant-based routing, http://en.wikipedia.org/wiki/Swarm_intelligence#Ant-based_routing.

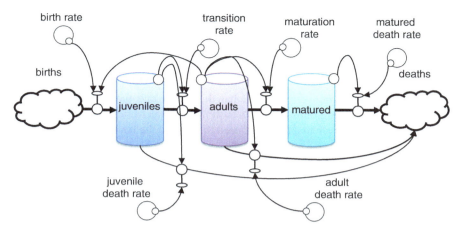

Fig. 13.10 A simple population model might use three stocks to represent different age groups based on their contribution to birth and death rates. The *circles* with *smaller circles* represent external constants that control valve opening. *Circles attached to stocks* are sensors that measure the level in the stock. The model leaves out some details of how multiple control factors influence a valve. *Clouds* are the essentially infinite source and sink. Following conventions in Meadows (2008)

13.5.1.1 The Model Diagram

Let's start with a simple version of the population model shown in Fig. 13.10. In this model, we will consider the population to be represented by three age groups based on their contribution to births. We define contributors as "adults." Noncontributors include the group called "juveniles" and (euphemistically) "matured." Since individuals may enter the adults and matured stocks at different ages (e.g., when a 16-year-old girl gives birth, she becomes classified as an adult), we supply separate transfer rates that represent average ages of transfer rather than strict brackets.

We also have to account for premature deaths from the juvenile and adult categories. Note in the figure that births depend on the inherent birth rate, a function of the number of women in the adult population able to get pregnant. Similarly note that the number of deaths in all categories depends not just on an inherent death rate for that group but also how many individuals are in the group.

Individuals move from one population stock to the next based on both a rate constant and the level within that stock, i.e., more people pass from one stock to the next if there are more of them in the first stock.

13.5.1.2 Converting the Diagram to Computer Code

There are any number of off-the-shelf packages that could be used to transform the diagram (using what is called a drag-and-drop graphic front end) to a program that the computer understands. Or, in this case, we can implement the model using a spreadsheet program. The columns of the spreadsheet basically represent the stocks, and the formulas in the cells represent the flow functions. Rate constants (see below) are special reference cells. The rows of the spreadsheet represent the time steps. We start at 0 time where the stocks are initialized. After that we let the spreadsheet software do its work.

For models that are more complicated than this simple one, we would switch to one of those off-the-shelf programs or write a customized object-oriented program. All of these methods are equivalent and can be used to produce the results seen below.

13.5.1.3 Getting the Output Graphed

Graph 13.1 shows a typical population growth rate using the constants and starting conditions in Table 13.2. The graph shows the growth of the population in each of the three groups and the total for all three together. The graph is produced by the same spreadsheet software. Off-the-shelf programs generally have graphing capabilities, or you can export the output data to a spreadsheet for graphing.

Table 13.2 These are the parameters used to produce the simulation results shown in Graph 13.1

Juveniles		Adults		Matured	
Initial: 100k		Initial: 200k		Initial: 80k	
Birth rate	Death rate	Transition rate	Death rate	Maturation rate	Death rate
0.040	0.005	0.020	0.004	0.020	0.018

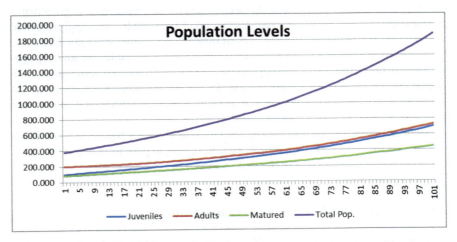

Graph 13.1 Given the values in Table 13.2, the model in Fig. 13.10 produces this exponential growth over 100 time units = 10 years

13.5.1.4 Discussion

This model is unrealistic in several ways. Among them is the fact that most populations, regardless of kind of plant or animal, face an upper boundary called the "carrying capacity," which refers to the number of individuals that a given environment can support with renewable resources, like food and water. The exponential growth shown in Graph 13.1 is not sustainable in the long run.

Though the details are a bit complex for explanation in a book of this scope, we modified the model to incorporate a hard limit carrying capacity. Basically, this modification means that the birth and death rates are no longer strictly constants. They are affected by the total population itself. We observe in many kinds of populations a reduction in birth rates and increases in death rates as the population increases beyond the carrying capacity.

Graph 13.2 shows an oscillatory behavior of the population as a result of going over the limit, rapidly declining as a result, and then once below the limit rising exponentially again. Note that the oscillations occur around the carrying capacity level (10,000k) with a secondary periodicity that is actually an artifact of oversimplification, which illustrates how a relatively small change to a model can change its dynamics. In this case, the population appears to be, on average, stable at the carrying capacity and does not grow exponentially forever. We simulated 200 time units in this run just to make sure.

Even this model is still too simple to demonstrate the natural dynamics of real populations, but it does illustrate how models can generate some important information. Note that in this case the model provides estimates of the subpopulations for

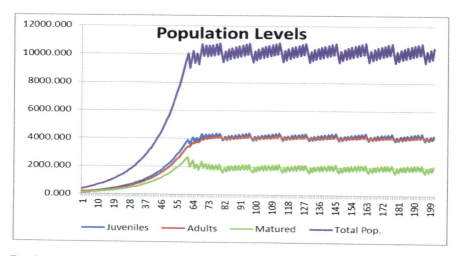

Graph 13.2 More interesting dynamics arise in a model of the population growth model that includes a hard carrying capacity at 10,000k. See text for explanation

juveniles, adults, and matured individuals. If this is the human population in a given society, then these numbers might be useful in planning public policy. For example, the predictions of how many adults would be available at any time to be productive workers capable of supporting the other two populations would have impacts on economic and health-care policies.

For many more details about building system dynamic models and to see examples of simulation results for many different kinds of systems, see Ford (2010), Meadows (2008), and Meadows et al. (2004).

13.5.2   Modeling Social Insect Collective Intelligence

A number of researchers have been interested in the way the eusocial insects such as ant cooperate to achieve many complex and seemingly intelligent behaviors. Ant colonies contain a variety of castes of nonreproducing individuals and at least one reproducing queen (fertilized once by short-lived male drones). There are general workers that take care of construction of tunnels and chambers, keeping the nest clean. There are caregivers that nourish and manage pupae. There are soldiers that are responsible for protecting the nest. And there are workers that specialize in finding and bringing back food for the colony.

One area that has gained a lot of attention in AI/AL circles is the way in which scout ants search for food and then signal workers as to where the food is located. The workers then organize to march out to the food source, clip bits and pieces, and carry them back to the colony nest for processing. Biologists have determined that these insects have an elaborate chemical-based signaling system. The "scout" ants, when they find a food source, find their way back to the nest by the most efficient path, and as they go, they lay down a chemical (pheromone) trail. This trail can then be followed by workers back to the food. The chemical will evaporate over time; the trail is a kind of short-term memory (compare this with the adaptrode model discussed below) that will be lost if not reinforced. And that is exactly what the workers do as they carry food back to the nest they also release the pheromone, thus increasing the time that the trail is "active." When the food source is exhausted, the workers start returning to the nest with less and less food, and eventually none. As the food diminishes, the ants lay down less and less pheromone so that the trail starts to fade, signaling other workers that there is no reason to go back to the food source—the memory fades.

Agent-based models of this behavior have been developed both to better understand the dynamics of the ant colony and to possibly apply to technological problems (Dorigo and Stützle 2004). Artificial ants have been used to compute an optimal routing path through a congested Internet packet-switching network for packets attempting to get from a source to a destination. They used a digital analog of the pheromone depositing method!

13.5.3 Biological Neurons: A Hybrid Agent-Based and System Dynamic Model

Many researchers in neurobiology and artificial intelligence (AI) have long been interested in how brains process data to make decisions that guide actions. The way neurons communicate with one another and process those communications has been the focus of these interests. Computer scientists and engineers have investigated the computational prospects of massively parallel processing using artificial (modeled) neurons.[20] Neural processing has many important attributes that AI researchers would love to emulate in their computers. The model discussed here was developed by one of us (Mobus) as a component for the autonomous control of a mobile robot, which necessarily involved a memory/learning for the purpose of controlling an autonomous mobile robot.[21]

Neurons can be thought of as adaptive decision agents that form networks. The communications are carried through a long process called the axon outward from the neuronal body. The axon can divide numerous times so that the signal is sent to multiple other neurons where they make contact with the receiving neurons through special connections (interfaces) called synapses.[22] In Chap. 8, our main concern was to demonstrate the nature of biological computation. But we provided the model developed by Mobus of biologically more realistic neurons and neural networks.

Classical artificial neural networks have been based on what Mobus felt was an overly simplified model of neurons, particularly as it relates to how learning takes place. Now, here, we demonstrate the "adaptrode" model of plastic synaptic junctions that encode memory traces in multiple time domains. In Chap. 9, Graphs 9.2 and 9.3, we showed the physiological costs associated with responding to a stimulus based on adaptrode learning. Graph 9.2 showed a single short-term memory trace that caused the response mechanism to activate slightly faster due to having some memory of the stimulus. The costs were reduced because the response came sooner than the simple S-R model in Graph 9.1. Graph 9.3 showed the effects of a multiple time domain adaptrode with associative learning that resulted in substantial reductions in total costs of responding to a stimulus.

The kind of learning that the adaptrode emulates is very close to the actual plasticity dynamics in real synapses. In Chap. 8, computation, in the section on neural computation, we discussed the kind of stimulus-response learning that occurs in classical conditioning and showed how the unconditioned stimulus, and if it arrives shortly after the conditionable stimulus input has arrived (as close sequences of action potentials), then the conditionable input synapse will be "strengthened" such

---

[20] See references in the bibliography marked N.

[21] Mobus and Fisher (1994). You can watch a "promotional" video of one of Mobus' mobile robotics class at the University of Washington, Tacoma at http://faculty.washington.edu/gmobus/AutoRoboDemo.m4v.

[22] Consider this as a crash course in neural processing, but recall the figure in Chap. 3, Fig. 3.21. Also we covered neural net computation in Chap. 8. See Fig. 8.7.

that subsequent inputs to that synapse will cause stronger responses, at least for a while after the first signals. This ability for a synapse to retain a memory of correlated inputs is called *potentiation*. It is time and activity dependent. If the frequency of input signals is high and the unconditioned signal arrives in time, the potentiation will be slightly higher than for a low-frequency burst. If there is a long duration pause between one set of input spikes and a second, then the potentiation can have time to decay away—a sort of forgetting. In Quant Box 13.1, we provide a stock-and-flow model of a three-level adaptrode. The adaptrode is not implemented in one of the standard system dynamic modeling languages, but the treatment in the Quant Box shows how systems models can be equivalent.

In the following graphs, we show a small set of results of the adaptrode model showing how it emulates the response of synapses under slightly different conditions. The first graph shows a single time domain memory trace. The response (red trace) to a single action potential spike (blue trace) is rapid but then begins exponential decay. A second spike a short time later results in a slightly elevated response compared with the first one due to the fact that the first trace has not quite decayed away completely (actually to a base line). A third AP spike much later produces essentially the same response as the first one since the memory trace has decayed away almost completely.

Real synapses (as explained in Chap. 8) have a different memory trace dynamic. This is due to additional time domain physiological changes in the postsynaptic compartment. The surface-level response is like an action potential; it is a depolarization event that travels from the synaptic receiver to the cell body where it is temporally integrated (summed) with other correlated depolarization events from other synapses. If the sum of these events exceeds a threshold for the neuron, it fires off an action potential down its axon (review the section in Chap. 8). Graph 13.4 shows an adaptrode with three time domains. The green trace models a biochemical reaction that reinforces the primary one (red trace). It acts as a floor below which the response trace cannot go, so as it rises it keeps the memory trace alive longer than what we saw in Graph 13.3.

The purple trace is a third time domain biochemical reaction that takes longer to activate and also much longer to decay after the stimulus stops. It supports the second time domain trace which cannot decay faster. This type of synapse is non-associative. It potentiates simply as a function of input frequencies within action potential bursts.[23]

Graph 13.5 shows a gated three-level adaptrode that is used to encode a long-term trace only if temporally correlated with the unconditioned input signal (opens

---

[23] The rise in W2 in this model represents a phenomenon called long-term potentiation (LTP) in real synapses. It is far beyond the scope of this book to elaborate, especially in a chapter on modeling(!), but for readers interested in going deeper into the neuroscience of this phenomenon, see http://en.wikipedia.org/wiki/Long-term_potentiation and LeDoux (2002), pages 139–61 for a basic explanation.

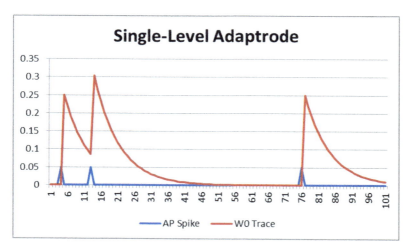

Graph 13.3 The adaptrode model emulates the postsynaptic response, the contribution of a receiving neuron membrane after an action potential (*blue*). This graph shows the increase in responsiveness with a short interval between the first two spikes. The memory trace decays exponentially fast so that after a longer duration (between second and third spikes) the responsiveness has decayed back to the base level

the gate). In the graph, you can see that the green trace does not rise simply in response to the red trace as in Graph 13.2. Rather a secondary signal arrives sometime after the primary signal and opens the gate after which the green trace starts to rise. But because it got a later start than in Graph 13.2, it doesn't rise as much, so the long-term trace is not as potentiated. This means that the association must be reinforced over multiple trials in order for the memory to be as strong. In this way, the memory trace of an association can only be strengthened if it represents a real, repeating phenomenon and not just a spurious relation.

The adaptrode model leaves out quite a lot of biochemical details that are found in real living synapses. Nevertheless, its dynamic behavior is sufficiently close to what we see in real neurons that it can be used to build model brains (of snails) that can actually learn, via Pavlovian association, how to negotiate a complex environment filled with both rewarding and punishing stimuli.

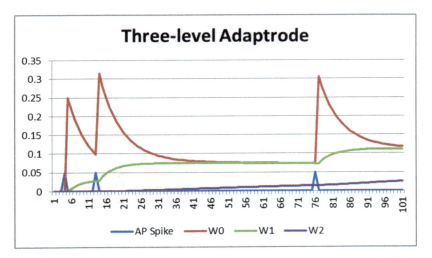

Graph 13.4 Real synapses can maintain a longer-term memory of the spiking history. A three-level adaptrode emulates the responses to the same action potential trace. The memory of two fast spikes (one and two) is maintained much longer. The *green trace* creates a floor below which the *red trace* (responsiveness) cannot go. See text for full explanation

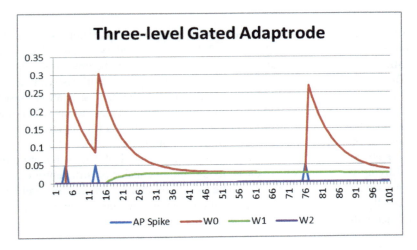

Graph 13.5 A gated adaptrode emulates the memory trace of an associative synapse in real neurons. The second-level "weight" rises only after a secondary signal opens the gate. See Chap. 8, Sect. 8.2.5.1 for more complete explanation. Also see Quant Box 13.1 for the model and the equations

Quant Box 13.2 The Adaptrode Model

Figure QB 13.2.1 (summarizing Fig. 8.8) shows a model of an adaptrode-based neuron or what we call a "neuromimic" processor. The term neuromimic means mimicking a neuron. It behaves dynamically in a manner similar to a real neuron. The adaptrodes mimic postsynaptic compartments in the neuron. The single flat-headed arrow represents a non-learning input, and the two-headed arrows represent learning (plastic) adaptrodes capable of encoding long-term memory traces.

A three-level adaptrode, the output of which is shown in the graphs above, is shown in Fig. QB 13.2.2. The explanation follows. The equations that govern adaptrode memory trace encoding is given below.

The three stocks represent the levels of potentiation within an adaptrode. The red oval is an output processor. We have not shown the details since the graphs above only trace the values of the stocks, but the output is based on both the level in the w^0 stock and the pulsed input (described below). By tradition in the neural network modeling community, the stocks are called "weights" and are labeled w^0, w^1, and w^2 (as in the graphs).[24] The stocks can be considered

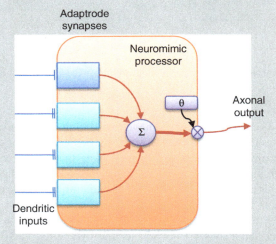

Fig. QB 13.2.1 Adaptrodes, mimicking synaptic processing, encode memory traces (as per Fig. 8.8). The summation, Σ, of outputs from all of the adaptrodes is compared to a threshold value θ. If the value of Σ exceeds the threshold, then the neuron outputs a signal through the axon to other neurons, to their dendritic inputs

---

[24] In conventional artificial neurons, there is only one weight associated with each synapse. The weight value changes according to some "learning rule." The dynamics of a single weight variable do not even come close to mimicking a real synapse, however. See, especially, Rumelhart and McClelland (1986) for many descriptions of single weighted synapses and their learning rules.

(continued)

Quant Box 13.2 (continued)

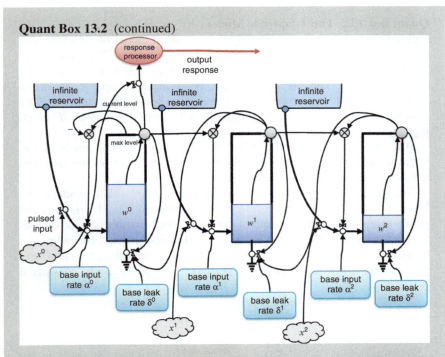

Fig. QB 13.2.2 This is a system dynamic model of a three-level adaptrode. See text for explanations

similar to fluid vessels that receive inflows from infinite reservoirs. The rates of inflows, however, are controlled by a somewhat complex set of controls. The thick black arrows represent those inflows. Note that each stock also has a leakage to "ground." That is, the levels in the stocks are constantly leaking or decaying.[25] This is an important aspect of potentiation and is related to "forgetting" as memory traces can fade over time.

The clouds represent the inputs from other neurons. The x^0 input is the direct dendritic input shown as two-headed arrows in Fig. QB 13.2.1. It is a pulsed input—the action potential arriving from another neuron's axon. It opens the valve fully letting "fluid" flow to the rate limited valve leading into the w^0 stock when it receives a pulse (see the graphs).

The other clouds provide what are called "neuromodulator" inputs. The x^1 input, for example, is the first associative enabling input discussed in Chap. 8.

---

[25] There is an alternative way to model this kind of system. Electronic circuits can be designed which are analogs to these "mechanical" versions. For example, the stock can be modeled using a capacitor which is another kind of leaky integrator.

(continued)

Quant Box 13.2 (continued)

These inputs come from other areas of the brain that are processing temporally correlated information and act to gate the potentiation of the next stock.

The blue rounded rectangles contain rate constants for both inflow and leakage rates (α, δ). The decay constants are much smaller than the inflow constants, and the decay rate for w^2, for example, is very much smaller than for w^0 or w^1. This is what accounts for the multiple time scales for increasing and decaying. w^2, in this instance, represents a long-term memory potentiation.

We will describe the basic operation of the w^0 level and the effect of w^0 on the input to w^1. We will leave it to the reader to work out the details of the other two levels.

The valve closest to the input to the w^0 stock is controlled by base input rate constant, α^0, and the difference between the maximum level possible (saturation) and the actual current level when the signal arrives from x^0. The closer to max that the current level is, the less "fluid" is allowed into the stock. The leakage from the stock is a product of the base decay rate, the level in the stock, and the level in the w^1 stock. The latter has the effect of slowing the decay as the level in w^1 rises. Effectively the level in w^1 acts as a floor value for w^0, preventing it from decaying to a value less than w^1. Since the decay base rate is much less than the input flow base rate, the stock will fill up at a rate proportional to the frequency of pulses coming in through x^0 and its current level.

In a similar fashion, the level in w^0 has an impact on the rate of filling for w^0. The higher the value is in w^0 the faster w^1 will tend to fill.

Now for the math! Here is the general form equation for updating each "weight" stock:

$$w^l_{(t+1)} = w^l_{(t)} + \alpha^l x^l_{(t)}\left(w^{l-1}_{(t)} - w^l_{(t)}\right) - \delta^l\left(w^l_{(t)} - w^{l+1}_{(t)}\right) \tag{QB 13.2.1}$$

where

$$x^l_t = \begin{cases} 1 & \text{on condition A,B,C, etc.} \\ 0 & \text{otherwise} \end{cases} \tag{QB 13.2.2}$$

$$A, B, C = if\left(\left(w^l_{(t)} - w^{l+1}_{(t)}\right) > \sigma^l\right) \tag{QB 13.2.3}$$

A, B, C, etc. are gates that are opened given the condition in Eq. (QB 13.2.3). Sigma is an empirically derived threshold constant appropriate to the time scale of the level l.

Also, when $l=0$ then w^{l-1} is replaced by w^{max}. This equation can be used for any number of levels, but practically four levels seem to be sufficient to capture the essential character of memory trace encoding from short term to effectively permanent.

Think Box Mental Models of the World

In the last Think Box, we gave a brief explanation of how the brain builds concepts and categories by analyzing differences in perceptual features between objects that share some features and not others. When there are sufficient differences, then a new category is encoded, and as more encounters with objects that better fit into that category occur, it is strengthened as a memory. In this chapter, we show how a memory trace is formed at neural synapses and how it can be strengthened over time by more reinforcing encounters (the Adaptrode model, Sect. 13.5.3 and Quant Box 13.2). Thus, memory "engrams" are developed that represent concepts at all levels of abstraction. At this point, however, we have mostly been talking about the formation of "thing" concepts—nouns like Fido (my dog), dog (-ness), mammal (-ness), etc. What about verbs (actions) and relations?

As we saw in this chapter, models are dynamic representations of not only the things but also the relations between things and how those change over time (dynamics). The brains of all animals that have brains have evolved to be models of the particular kind of world the animals occupy. Animals low on the phylogenetic tree have models largely "learned" through evolutionary processes, and they are constructed in the brain structures by genetic endowment. These are called instinctive behaviors, and all animals at all levels in the tree have inherited some of them from earlier species. But models in animals that have more capacity to learn through acquired neural representations gave rise to more expanded options in behaviors and, therefore, greater success in life (Think Box 11). In mammals, with their large cerebral neocortex (and in birds with an analogous structure), the ability to learn elaborate models of their worlds has not merely augmented built-in instincts, and it has allowed the animal to modify and even override some instinctive behaviors giving them far more adaptive behavioral options.

How are actions and relations coded into the brain? It turns out that human language gives us a clue. Words are names for the representations we are able to model. So, for example, the word, "move" is a name for an action. And the word "on" is a name of a relation. We already know that the brain encodes namable things as concepts so it should be no surprise that these are encoded in the very same way that regular nouns are encoded. Thus, all of the dynamic and relational attributes we seek to capture in a model are present in neural representations. What we need is the programming that associates these representations in temporal ordering. That is what we recognize as a universal grammar. The formation of sentences that describe, for example, a subject (noun), an action (verb), and sometimes an object (another noun), along with some subsidiary modifiers, etc., provides just such an ordering. The temporal sequencing of regular sentences is the running of a model! It encodes causal relations, as in "A did such-and-such to B." And, the chain of sentences that tell a story produce the sequence of such relations.

(continued)

Think Box (continued)

Telling a story is, in essence, running a model in the brain which outputs, for humans anyway, the verbalization (silently or aloud) that attends language. This is what we recognize as thinking. It is thought that other mammals (and maybe birds) actually have this same capability to represent things, actions, and relations (ever doubt that your pet dog or cat know you and that you are the source of their sustenance?), but they do not have the brain structures that allow encoding of associated abstract sounds (words), so likely do not think in an explicit language.

As it turns out, you do not do most of your thinking, that is, running your models of the world, in explicit words either. We now realize that most of the brain's thinking activity goes on below the level of consciousness, that is, in the subconscious. There, it seems that multiple stories are running in parallel—what Daniel Dennett calls "multiple drafts" (1991 pp. 111 ff)—and being edited (analyzed for various attributes) before some portion of just one gains such saliency that it comes into focus in conscious thinking. It could be that the subconscious versions of your stories are indeed wordless just as they are in other animals and that bringing them into consciousness is when the associated words and sentence structures get attached and used in the narration process.

Mental models have one more very important function. Since the mind can construct stories, it can build a historical record (memories of the past), a narrative of the present (formation of new memories), and, most useful of all, fiction stories (of the possible future). You'll recall from the discussion in this chapter on the uses of models, Sect. 13.3.2.2, models can be used to predict different outcomes if given variations on the inputs, what we called scenario testing. This is also called "what-if" testing. With mental models we can not only change the values of standard inputs but change what even counts as an input. We can also construct variations within the model itself. We call this imagination, being able to construct a memory of something different from reality using bits and pieces of existing models of reality put together in new ways. This is what we call creativity, and it has given humans incredible powers to explore possible alternative worlds in which consequences of change can be anticipated in thoughts rather than simply happen in a trial-and-error manner.

13.6 Summary of Modeling

13.6.1 Completing Our Understanding

When we analyze a system, we come to understand its composition, structure, and subfunctions. This is a better understanding to possess compared with a mere black box viewpoint. But as this chapter has asserted, our understanding is not solidified until we build a model of the system from what we learned and simulate the whole to see if we got everything right. Most often we don't accomplish that on the first pass. For sufficiently complex systems, we may have to go through hundreds, even thousands of iterations. Our analysis of human brains and societies continues and will for the foreseeable future. At the end of each iteration, though, we do understand better.

Modeling is essential to getting a completion of understanding. And in this chapter, we have surveyed a number of approaches used to accomplish the simulation of systems. What are we trying to get out of these approaches?

- Behavior of the Whole
  Model simulations should tell us how the whole system behaves in varying environments. We saw how system dynamics is suited to this need. We saw how a stepwise decomposition of a system can be modeled at each stage to compare the behavior.
- Behavior of the Parts
  We are interested in the dynamics of the parts, but we are also interested in how the parts interact to produce something not readily predictable from merely knowing the dynamics of each part. Agent-based modeling can show us how new levels of organization emerge from those dynamics.
- Looking at the Past
  Models can be used to post-dict past conditions. We have seen how this capability is becoming increasingly important in understanding climate change and what its impacts on the planet have been in the distant past. This is critical for the next objective.
- Looking at the Future
  The ultimate value of good models is that they can suggest what conditions in the future might be given various sets of starting conditions. Models can be used to predict but also to generate various scenarios, such as best case, most likely case, and worst case, so that we can plan and take actions in the present.

Modeling brings us full circle in the quest for understanding. One particular complex adaptive and evolvable system is the human brain, and a population of these systems is attempting to understand other systems, as well as themselves.

13.6.2 Postscript: An Ideal Modeling Approach

System dynamics, adaptive agent-based, optimization, and evolutionary modeling approaches have been used to ask questions about systems from different perspectives. Sometimes those questions are particular to the modeling approach, for example, the agent-based approach, we have seen, helps us understand processes of emergence of new structures and functions. System dynamics helps us understand the whole system behavior (dynamics). But if there is one message we hope has been conveyed in this book, it is that real systems in the world need to be understood from all perspectives. In particular, what approach should we use to understand complex adaptive and evolvable systems? The answer has to be "all of the above."

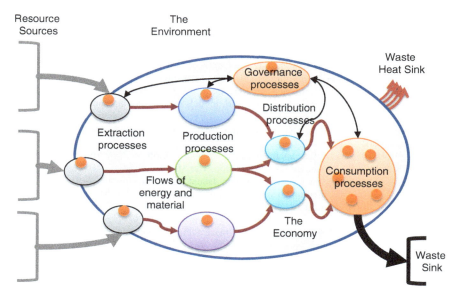

Fig. 13.11 Putting it all together in a single modeling environment would allow the building of complex adaptive and evolvable systems such as the economy. The *small orange circles* represent adaptive agents managing the processes

Currently, there is no single development tool/environment that allows such systems to be modeled. This may prove to be an impediment to the understanding of significant complex systems like the environment or the economy.

An ideal modeling environment should be able to combine aspects of all of the above approaches into one tool. Imagine a society of adaptive agents organized into process structures that interact with one another in an economy. Figure 13.11, here, replicates Fig. 12.20 and adds symbols representing adaptive agents managing the various processes (as in Fig. 12.21). This system employs system dynamics, agents, hierarchical cybernetics, optimization, and essentially everything that we have presented in this chapter.

Such a modeling environment would be a massive software development undertaking. On the other hand, if the principles of systems science that we have been elucidating in these chapters are taken seriously, then the possibility of building such a program (or more likely a cluster of programs running on a distributed platform!) should be feasible. We assert that it is highly desirable to do so. We even propose a name: complex adaptive and evolvable systems, CAES!

In the next chapter, we cover how we use systems science principles to design and build such complex artifacts as CAES. Perhaps some ambitious computer science graduate students will follow the hint.

Bibliography and Further Reading

Note: Entries marked with an * are references for artificial neural network modeling.

Bagley R, Farmer DJ (1992). Spontaneous emergence of a metabolism. In: Langton et al (eds) Artificial life II, a proceedings volume in the Santa Fe Institute Studies in the sciences of complexity. Addison-Wesley Publishing Co., Redwood City. pp 93–140

*Commons ML et al (eds) (1991) Neural network models of conditioning and action. Lawrence Erlbaum Associates, Publishers, Mahwah

Damasio AR (1994) Descartes' ERROR: emotion, reason, and the human brain. G.P. Putnam's Sons, New York

Dennett D (1991) Consciousness explained. Little, Brown & Company, New York

Dorigo M, Stützle T (2004) Ant colony optimization. The MIT, Cambridge

Ford A (2010) Modeling the environment, 2nd edn. Island Press, Washington

Forrester JW (1994). Learning through system dynamics as preparation for the 21st century. http://clexchange.org/ftp/documents/whyk12sd/Y_2009-02LearningThroughSD.pdf. Accessed 18 Feb 2014

Gilovich T et al (eds) (2002) Heruistics and biases: the psychology of intuitive judgment. Cambridge University Press, New York

Langton CG et al (eds) (1992) Artificial life II, a proceedings volume in the santa fe institute studies in the sciences of complexity. Addison-Wesley Publishing Co., Redwood City

*Levine DS, Aparicio IV M (1994) Neural networks for knowledge representation and inference. Lawrence Erlbaum Associates, Mahwah

*Levine DS et al (eds) (2000) Oscillations in neural systems. Lawrence Erlbaum Associates, Publishers, Mahwah

Meadows DH (2008) Thinking in systems: a primer. Chelsea Green Publishing, White River Junction

Meadows DH et al (2004) Limits to growth: the 30-year update. Chelsea Green Publishing Company, White River Junction

*Mobus GE (1994) Toward a theory of learning and representing causal inferences in neural networks. In: Levine DS, Aparicio M (eds). Neural networks for knowledge representation and inference. Lawrence Erlbaum Associates, Mahwah. pp 339–374

*Mobus GE (1999) Foraging search: prototypical intelligence. In: Dubois D (ed) Third international conference on computing anticipatory systems. Center for Hyperincursion and Anticipation in Ordered Systems, Institute of Mathematics, University of Liege, HEC Liege

*Mobus GE, Fisher P (1994) MAVRIC's brain. In: Proceedings of the seventh international conference on industrial and engineering applications of artificial intelligence and expert systems, Association for Computing Machinery, 31 May to 3 June 1994, Austin, Texas. pp 315–320. http://faculty.washington.edu/gmobus/Mavric/MAVRICS_Brain_rev.html

LeDoux J (2002) Synaptic self: how our brains become who we are. Viking, New York

*Rumelhart DE et al (eds) (1986) Parallel distributed processing: explorations in the microstructure of cognition, vols. 1 and 2, The MIT, Cambridge

Sawyer RK (2005) Social emergence: societies as complex systems. Cambridge University Press, New York

*Sutton RS, Barto AG (1998) Reinforcement learning: an introduction. The MIT, Cambridge

Chapter 14
Systems Engineering

"The noblest pleasure is the joy of understanding."

Leonardo da Vinci

"Engineering refers to the practice of organizing the design and construction [and, I would add operation] of any artifice which transforms the physical world around us to meet some recognized need."

G.F.C. Rogers, 1983

Abstract Human-built systems are engineered. The process is one of first determining the desired function of the system and then working backwards, down the analytical tree produced in systems analysis, through the subfunctions needed, to design the connections between components that will produce those functions. Engineering is a mixture of intentional process and exploration of possibilities. Thus, it is also subject to evolutionary forces. Artifacts evolve in a way similar to how biological systems evolve, with human tastes and needs acting as the selective environment. Systems engineering is a form of meta-engineering that involves the integration of complex subsystems that are often designed by different subdisciplines, such as in the case of an airplane, the parts of which are designed by mechanical, electrical, avionics, and computer engineers (to name a few). All of the parts they design have to be brought together in a coherent construction of the whole system.

14.1 Introduction: Crafting Artifacts to Solve Problems

Principle 12: Systems Can Be Improved
Humans make things, not only physical objects, machines, building, bridges, etc., but they create procedures, institutions, policies, organizations, and all of the components of culture that permeate our social systems. Of all of the primate species alive today, humans are the only ones that make compound tools, that is, tools made from multiple parts. And humans are the only species that has evolved a symbol-manipulating cognitive capacity along with the language to express thoughts.

© Springer Science+Business Media New York 2015
G.E. Mobus, M.C. Kalton, *Principles of Systems Science*, Understanding
Complex Systems, DOI 10.1007/978-1-4939-1920-8_14

All of these components come together in the human ability to adapt to every conceivable environment on the planet and, as it turns out, to adapt the environment to our desires.

In this final chapter, we will examine how humans have been able to craft more complex artifacts to solve more complex problems in an attempt to improve the system we call civilization. Systems engineering is the means by which these complex artifacts can be developed. It goes beyond traditional engineering practices (which are still required within the framework of systems engineering) by considering the whole system framework in which artifacts are to be used. It is informed by systems science, and we will show how the principles that have been the subject of the prior chapters play a role in guiding a more holistic approach to engineering.

14.1.1 Problems to Be Solved

What do we mean by a "problem?" It is a very general term (and we will offer a precise definition below) that relates to a condition in which one or more people want to achieve some outcome (a desired goal) but cannot do so by any simple actions on their part. For example, if you want to get to a specific location that is within walking distance, then that is not a problem. On the other hand, if you want to get to a town on the other side of the continent, then that could be a problem. Problems require that some extra work be done to create a tool that can be used to achieve the goal. People developed trains and airplanes to solve the problem of long-distance travel.

Every artifact that humans have created from the origins of the genus has been meant to solve problems. For prehistoric humans, the main problems revolved around getting adequate food and finding shelter. Fire starter kits, stone blades for butchering meat, and digging sticks for getting roots greatly increased the efficiency in solving those problems.

In this chapter, we examine the systems approach to solving problems by creating or improving artifacts. In the context of systems science, systems engineering isn't just about the artifacts themselves, but about the system in which the artifacts are to operate. In other words, systems engineering takes into account the meta-system in which the artifact is a subsystem or component. Engineers in every domain seek to take into account the "usability" of their artifacts, but all too often, the defining of that usability is a rough approximation taken from "user's" statements of what they want. Too often we forget that the user is really part of a larger system that provides context and motivation for the need for the artifact. That is where a systems approach helps to better define the usage of the artifact.

Systems engineering also applies to the design and construction of the artifact itself. Today so many new technologies are really systems composed of components and subsystems that have to work together and are designed in several different traditional domains, e.g., electrical or mechanical engineering. Systems engineering is involved in making certain that these different subsystems integrate properly to produce the complete artifact.

There are several aspects about human cognition that give us the ability to recognize problems and develop solutions. We will take a quick look at some of those here.

14.1.2 Affordance

One of man's cognitive competencies is the ability to recognize the capacity for an object (or concept) to be used in a way, not originally part of its "design," to achieve an objective. William Gibson (1904–1979), the American psychologist called this capacity "affordance."[1] It turns out to be a critical element in the process of invention. An example of affordance would be using a low table as a casual seat (as if it were a chair) because it has properties such as flat top and supported by legs from below, which afford it the ability to perform as a seat. But note that it is the human mind that accomplishes this not the table. Nor did the original designer think of the table as a seat. But it conveniently works for that purpose under the right circumstances (educators can often sit on a table at the front of a class while lecturing!)

Affordance works as well for natural objects as for human-built ones. Imagine a hiker in the woods who gets tired and decides to sit down on a convenient rock or fallen tree. Nature didn't design either as seats, but humans find such things convenient to use as such. One can imagine a long-ago *Homo habilis* creature[2] seeing a stone that had accidently been broken leaving a sharp edge recognizing it as something that might more readily assist in the butchering of a carcass. Thus, was born the idea of a knife! It was possibly a relatively small jump to the discovery that such stone implements could be intentionally made by chipping at an appropriate rock. This is a small jump but it may have taken many thousands of years.

Every human, to one extent or another, is capable of using what is at hand to accomplish a goal whether what is at hand was intended for that purpose or not. Two of the greatest inventions of modern times have been bailing wire and duct tape!

14.1.3 Invention

Invention is the intentional shaping and combining of components to form a tool *system*[3] that did not previously exist. This does not only mean physical implements, such as a telescope or stethoscope, but extends to creating procedures and methods as well as forming clubs and corporations. All of these are tool systems in the sense that they help us solve problems. They help us achieve what we could not have formerly achieved, and with luck, they do this with efficiency.

---

[1] Gibson (1977). Also see: http://en.wikipedia.org/wiki/Affordance.

[2] One of the first recognized species in the genus *Homo*. See: http://en.wikipedia.org/wiki/Homo_habilis.

[3] A tool system refers to a complex of different objects that are combined to provide the functionality of a tool. A bow and arrow is an example of an early form of tool system. So is clothing and constructed shelters.

Much invention has been the result of tinkering, a trial-and-error method of putting things together and then trying to solve the problem to see if that combination will work. Edison did this when he invented the first practical incandescent light bulb. He tried many different materials as filaments before settling on carbonized fibers. While this approach is considered valid, a method of doing "research" to find a solution, as knowledge from the sciences has built up through the eighteenth, nineteenth, and early twentieth centuries, invention has become less an exploratory process and more one of engineering. That is, an artifact that would likely solve a problem is first conceived and then the details are developed through engineering. Using the quantitatively based knowledge of the sciences, engineers create new tool systems through intentional design.

14.1.4 Abstract Thinking

In Chap. 8, Sect. 8.2.5.2.1.3, we discussed how concepts are formed and maintained in the brain's cortical layers. Concepts are encoded in neuronal networks and are activated when the correlated firing of neurons in that network generate an output. Concepts are abstractions of actual things, actions, and events in the physical world. And when we think, we are manipulating these abstractions in a purposeful manner. Our brains have the capacity to represent many different concepts but also to combine and even copy and modify concepts on a trial basis (imagination).

The human brain has the capacity to translate these abstract representations into several external forms for the purpose of communication. Language, or the translation of concepts and associations of concepts into words, is a primary way to share those concepts between people. We invented writing to encode those words in external forms that could be saved and reused. Closely related to writing (and preceding it historically), we are also able to draw pictures and give dimensions in terms of numbers. On top of that we are able to use numerical-based abstractions to derive mathematical relations that provide precise descriptions of our concepts. And, finally, we are able to manipulate those relations according to rules so as to derive more complex relations.

All of these capacities are brought to bear in science and engineering in highly disciplined fashion.

14.1.5 Crafting by Using Language, Art, and Mathematical Relations

Humans communicate abstract concepts through language, artistic rendering, and, more precisely, through mathematics. This facility allows us to share ideas and specifications between members of a design group and even to ourselves across time. Without these abilities, we could never convey complex ideas (for complex artifacts) to others or to ourselves. Throughout this book, we have seen examples of all three modes, descriptions, figures, and quant boxes.

Engineering is an intentional process of crafting artifacts using mathematics and knowledge from numerous disciplines. The word "discipline" connotes a methodological approach to a subject. Engineering is a discipline. It sometimes still includes a bit of "research" when there are gaps in our knowledge. Sometimes affordance is required to see a new way to use an old thing in the process. But even when these "arts" are part of the process, ultimately the engineer will test the design not by merely seeing if a prototype works, but calculating (or computing) its abilities to do the intended job.

Engineers must communicate their ideas very precisely so that other engineers understand and those who will actually build the artifact will have clear instructions to follow. The engineering disciplines have developed clear and concise ways to express complex ideas in abstractions in language (descriptions and specifications), drawing (diagrams and blueprints), and mathematics (also part of specifications).

14.1.5.1   Engineering and Science: Relations

Strictly speaking science and engineering are different enterprises. The former is an exploratory quest for knowledge and understanding. The latter is the disciplined use of that knowledge to design and construct artifacts that serve a human/social purpose. Engineering is using knowledge for practical purposes. So why do we have a chapter in a book on systems science devoted to engineering? The answer is that these two enterprises are deeply intertwined and sometimes hard to distinguish in practice. Engineers often do explorations of their own when they have no guidance from science. They can experiment, for example, testing the capabilities of a new design when there is no operative theory to provide an analytic solution.

Scientists, on the other hand, are continually faced with the problem of measurement, which means they have to design some kind of instrument in order to conduct their research. Particle physicists have become as much tool makers as interpreters of experimental data. They design and build the huge particle accelerators and detectors just so they can do their experiments.

Thus, no treatment of systems science would be complete without some treatment of the engineering of systems. Of course, there are numerous flavors of engineering such as electrical, mechanical, and chemical. There are offshoots and variations on the traditional engineering disciplines, such as biomedical and genetic engineering. These exist because there are differences in the problems being solved and the substrates of the artifacts needed to solve them. Even so there are common practices and theories that are shared across all of these domains.

And that is where systems engineering comes into the picture. Where mechanical engineering takes the knowledge of physics (along with economics and a few other areas) and uses it to design machines, systems engineers apply the knowledge of systems science to the crafting of complex systems that involve components from many different domains. Think of a jumbo jet airplane. It includes mechanical, electrical, chemical, avionics (airplane electronics), and many more engineering disciplines that have to bring their expertise together to produce a single very complex machine. Systems engineers are employed to help make sure all of the parts fit properly and that everything needed to accomplish the ultimate objective of the system is accounted for.

Today so many of our technological and scientific advances involve complex systems of tools that have to work together seamlessly and efficiently that they cannot be engineered by expecting practitioners from the various disciplines to somehow get along. Systems engineers are increasingly important members of engineering projects. We will see how this works.

14.1.5.2   Mathematics in Engineering

Engineering is a discipline based on mathematics. It seeks accuracy and precision in dealing with its subject matter (physical, chemical, etc.). The discipline requires sophisticated mathematical tools in order to solve complex relational problems when designing an artifact.

Systems engineers also require a fair amount of mathematics even though they do not directly do the same kind of design work as done by the domain experts (e.g., including, say, financial planners!). Nevertheless, they do use several mathematical tools, such as matrix algebra and linear systems (remember OR in the last chapter?). They also need to have a basic grasp of the mathematics employed by the domain experts so as to understand their language. Remember, one of the chief jobs of a systems engineer is coordination between sometimes disparate disciplines, so a healthy knowledge of various engineering mathematical tool kits is advisable.

14.2   Problem Solving

The raison d'être for engineering is to solve complex problems most effectively and efficiently. What are some examples of problems and the invention of the tool systems used to solve them?

Getting back to our early ancestors, this time in the Mesolithic era, the problems basically came down to finding adequate food, water, and shelter for the family/tribe and protecting themselves from becoming food for some nasty predators. They had long known the effectiveness of stone cutting instruments for butchering meat and cutting fibers (for clothing) handed down from the Paleolithic era. The crafting methods for shaping just the right stone implement had been well worked out—a kind of science of cutting edges in stones. Similarly the wooden-tipped spear had a long history of making the hunts more successful. The jump to a stone-tipped spear that could be used even more effectively for both hunting and protection against an attacker required an early engineer to purposely cut a notch in the end of a spear, put a new kind of spear point cutter into the notch, and bind the two pieces together with a cord or sinew. The cave person who managed this first imagined it in their mind and set about to make it happen. It was no accident. And it must have worked spectacularly for its intended purpose because human soon began to hunt some mega-fauna, such as wooly mammoths to extinction.

There were other factors involved in hunting these large creatures besides blade-tipped spears. They also learned to use grass fires to "herd" the animals into an ambush. They invented the hunting *methodology*. And the problem of how to get enough food was effectively solved. Agriculture was basically the same thing. The problem was how to improve the odds of finding enough grains, vegetables, and fruits over what they were with the pure hunter-gatherer lifestyle. Planting and animal husbandry were the solutions.

14.2.1 Defining "Problem"

In the spirit of working in a disciplined way, the first step in understanding engineering as disciplined problem solving would be to define precisely what we mean by the word "problem." We distinguish a problem from other "puzzlements" such as situations or predicaments, which tend to be more elusive when it comes time to completely characterize what the needs are. In order to solve a problem, we need to be able to characterize it because we need to be able to tell when we have found a solution.

14.2.1.1 Definition

The first thing we have to do is establish the context or domain in which problems that can be solved by engineering exist. We generally don't think that complex social issues can be solved by engineering, even systems engineering. But in the domains of economics, technology, medicine, and the like, we can use this definition of a problem and then use it to clarify and characterize the problem so that it can be solved.

A problem exists when one or more persons want to accomplish a *legitimate* goal but are prevented from doing so by the *infeasibility* of using existing artifacts. A problem may arise because someone sees a new need that would, if fulfilled, produce a profit (to be explained) or something that people had been doing is threatened by a loss of one or more resource inputs and a new way to do the old job has to be found.

The two highlighted terms are important. Legitimate means in both the sense of the accomplishment is of something that is within the laws of nature and within the laws of the land. For example, if someone wants to accomplish building a machine that produces more energy than it consumes (called a perpetual motion machine), then they can puzzle around all they want, even use higher math, but they will be trying to break the First Law of Thermodynamics and that isn't a legitimate problem. Similarly if someone wants to invent a method for entering a bank vault undetected so they can steal the money, then they are not really solving a problem of importance. Indeed if they succeed, they will be creating bigger problems for other people.

Infeasibility is a key to understanding why problems are complex. It could mean that the current artifacts are incompetent for the task or that they are too expensive to use on the task. There would not be a profit payoff given what capital would need to be invested. Profit, here, means that society and not just the immediate user is better off. The returns can be monetary, but they can also be in the form of other well-being factors like enjoyment, better education, better health, etc.

14.2.2   Modern Problems

Today it is hard for people living in the developed world to imagine that getting enough food is a problem. Most of the readers of this book will simply go to a grocery store and purchase affordable foods. Yet for nearly half of the current population of 7+ billion people, getting food is a primary problem and over 1 billion people are either undernourished or malnourished. They are hungry most of the time. This is a complex problem involving many aspects, not just food production. We'll return to this and see what role systems engineering might play in solving it.

For now consider some more common "problems" in the developed world.

Consider the problem of traffic congestion on freeways during "rush hours." The problem is that people want to get to and from work expeditiously, but everyone drives automobiles at roughly the same time of day to commute. The question is posed, "How can we reduce traffic slowdowns due to high rates of single-occupant automobiles on the roads during rush hour?" The desire is to minimize delays so that people are not inconvenienced (and turn a little grumpy from the experience). It is a complex problem that needs to be understood as a system. Too often, however, the conventional "solution" will end up being chosen. The planners will simply call for widening the freeways to accommodate more traffic, but the temporary easing of congestion often simply leads to more people moving to the suburbs, lured by less costly housing, more open space, and not-too-bad commute times (now that the highway is wider). That, in turn, leads to more traffic during rush hours and the system is right back to where it had been. Nobody is happy.

Most of our modern problems, especially urban and suburban living, large-scale food production, transportation, infrastructure maintenance, and many more are of this complex nature. They are systemic, meaning that solving one small part will have impacts on connected parts and those impacts cannot always be predicted. How often do we hear the words "unintended consequences" uttered when looking back at the larger-scale results of solving some local problem? It is the nature of a complex interconnected society, like those in the modern developed world, to have systemic problems that should be understood in the systems perspective (but too frequently aren't). Systems engineering based on systems science includes not just treating the solution to a problem (like a specific product, process, or organization) as a system, but recognizing that it is also a subsystem in a larger meta-system and will have potentially unforeseen consequences on the whole system unless the whole system is taken into account.

Such an approach is possible in principle, but unfortunately it might be infeasible in practice due to the extra costs involved in analyzing the larger system. Other problems, such as limits on access to information regarding the other components in a larger system, can make the job nearly impossible. For example, when studying economic systems in a private enterprise larger system, it is difficult to obtain information on other companies because their data is proprietary and a competitive marketplace favors keeping it private.

Thus, most systems engineering tends to be somewhat limited in terms of meta-systems analysis. The practice is starting to take an interest in, and works to understand, the way in which a solution (a system of interest) might impact the natural environment. For example, systems engineering now includes working to understand the waste streams created by new systems, to minimize them, and, hopefully, to see where those streams could feed into other processes not as wastes, but as resources. The city in which the authors work, Tacoma, Washington, has started processing sludge from its sewer treatment plants into a high-grade soil supplement called "Tagro" which it sells to the public for nonfood gardening purposes.[4] Many coal-fired power generation plants extract sulfur from their combustion fumes and are able to sell it to commercial processes that use sulfur as an input. Systems engineers can seek similar connections for the systems they design.

In the commercial world, we are witnessing the increasing cooperative interactions between separate companies in terms of supply chains for parts and services. These chains are made possible by computing and communications technologies that also allow companies to share more data so that their cooperation makes them all more competitive in the larger marketplace. However, the chains are really a kind of mesh network so that, in principle, more and more companies are finding it advantageous, in terms of cost savings, to approach their operations much more systemically. It may be that one day systems engineers will be able to take a more comprehensive approach to meta-systems analysis (at least in the economic systems) as a result of companies discovering the advantages of the systems approach.

14.2.3   Enter the Engineering of Systems

Modern problems are nothing if not complex and involve many disciplinary practitioners to craft solutions. This is where systems engineering[5] comes into the picture. It is the job of a systems engineer to apply the knowledge of systems science to

---

[4] See, http://www.cityoftacoma.org/cms/one.aspx?objectId=16884.

[5] For a fairly good background, including history of systems engineering, see http://en.wikipedia.org/wiki/Systems_engineering.

integrate and manage the activities of a large and diverse group of domain experts who are brought together in multidisciplinary teams to analyze a systemic problem, design a solution in the form of artifact systems, build, test, and deploy the system and monitor its operations over its life cycle.

14.2.3.1   Role of the Systems Engineer

All engineering activities have the above-described attributes in common. Systems engineering, however, has a special role to play in the development of complex artifact systems. Just as systems science is a meta-science that weaves a common theme throughout all of the sciences and exposes the common perspective in all the sciences, systems engineering plays a similar role in a systems development project. A systems project is a large undertaking that will used the capabilities of a large number of people from various disciplines. The systems engineer (usually working with a project manager) does not directly design pieces of the system. That is left to domain expert engineers.

14.2.3.1.1   Domain Expert Engineers

We use this term to designate the talents and knowledge needed by people who design and construct the various component parts of the larger system. In very large complex systems, people in domains that are not ordinarily thought of as engineering are needed to design and build those parts of a system that are not hardware per se. For example, designing the management structures that will be used requires the talents of someone who has specialized in management decision hierarchies. They can work with, for example, software engineers in the development of a management decision support subsystem (computers and communications). They supply the business knowledge that guides the software engineers in designing the programs and databases needed. Systems projects are inherently multidisciplinary. We have lumped all contributors to this single category of domain experts and called them engineers because, regardless of the medium in which they work, they are still crafting artifacts that meet specifications. Domain experts will be engaged in all stages of a systems project.

14.2.3.1.2   The Systems Engineer

The work of a large and diverse team working on multidisciplinary projects requires an extraordinary level of integration that, generally, no one discipline can bring to bear. This is where the systems engineer comes into the picture. Their role is to provide the whole system integration throughout the life cycle of the system and particularly during the project process.

You might say that the systems engineer's main job is to make sure all of the parties to a project are "talking" to each other. Domain experts know the language of their own domain, but do not always understand that of other domains. The systems engineer has to make sure that these experts succeed in communicating with one another. Not surprisingly the use of systems language acts as a Rosetta Stone-like mechanism to aid translations from one domain to another. The interfaces at a subsystem boundary, discussed in both Chaps. 12 and 13, are a key element in making this work. The systems engineer is constantly monitoring how these interfaces are developing as the project proceeds.

Because domain experts are involved in all stages of a project, the systems engineer's job is to keep track of the big picture of the whole system and where things stand as the stages develop. This in no babysitting or hand-holding job. The systems engineer has to have a fair understanding of aspects of all of the domains and be able to facilitate the completion of specific tasks within each stage, relative to the contributions of the domain experts. So one way to characterize this job is that it is all about facilitation and integration.

14.3   The System Life Cycle

All systems are "born" at some point in time. All systems "grow," "develop," and "mature." And all systems eventually "die." That is, all complex systems can be thought of as emerging from components, developing as time passes, maturing, operating for some lifetime, and then either being upgraded (evolved) or replaced (decommissioned and put to rest). This is what we call the *systems life cycle*. The systems engineering process is involved with every phase of the life cycle, not just the birthing process. Even though the engineers don't actually operate the system, they do need to monitor its operational performance as feedback to the design.[6] Though most people don't think about it this way, real systems do age. Many artifact systems become less useful as the environment in which they act changes in ways in which they cannot be adapted economically. Economic decisions regarding just how long a given system should be operational before any substantive intervention, or decommissioning of the system, are often reviewed periodically. For example, suppose a given system has been in operation for some time but a very new technology that becomes available would make the system obsolete in terms of its economic viability (the new technology might deliver more results at cheaper costs). Good management decisions involve doing the economic assessments and determining when a system should be replaced.

---

[6] Typically the engineers instrument the system to record operational performance data that can be analyzed later for indications of "aging." This information can give the next round of engineers ideas about what needs to be fixed or altered in a new version.

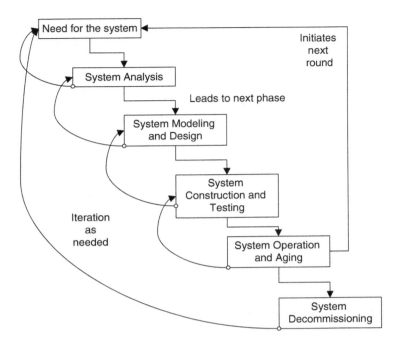

Fig. 14.1 The artifact system life cycle runs from the time there is a recognized need for a system to solve a particular problem through to the final decommissioning of a "spent" system. The latter might simply involve an evolutionary upgrade of the system or replacement by a completely new system to solve an evolved version of the original problem. The decision to decommission and upgrade or replace starts the process of needs analysis over again

Here we describe the life cycle in "naturalistic" terms. Even human artifact systems resemble naturally developing systems in many aspects. This should not be surprising in that even though the mechanism of human intention and exercise of intelligence is the operative factor in creating these systems, that process is also natural (as opposed to supernatural) and so artifact systems obey the same life cycle dynamics as every other complex system. Also see Fig. 14.1 to see an outline of the life cycle concept mapped to the engineering process.

14.3.1 Prenatal Development and Birth

Whereas living embryos are created (through fertilization) and develop within an egg structure (like birds) or a womb (like mammals), artifact systems are created and develop within the minds of humans with the aid of symbolic languages, external storage media, and electronic mental assistants (computers). Even the fertilization of ideas is analogous to fertilization of an egg to form a zygote in that new ideas most often are combinations of older ideas. It is probably no accident that we often refer to inventors as having fertile minds.

The initial impetus for creating a new artifact system is recognition of a need for a system that will perform a useful function. Then the act of creating the idea for the artifact starts a process of development that involves systems analysis (Chap. 12), modeling (Chap. 13), and engineering (the current topic).

The birth of the system is when the first prototype or product is delivered into the hands of the person or group that first defined the need (client). The clients and engineers then bear a certain resemblance to "parents" who have just had a baby! They proceed to nurture the prototype through early development.

14.3.2   Early Development

Most people have heard the term "beta testing" and even without knowing the specific meaning of the term have an idea that it involves the using and testing of a product that is still not completely "perfect." Indeed, beta testing is one of several formal testing stages in which a produced artifact is "exercised" to determine any latent defects that somehow eluded the engineers during design. The prototype is still a "child," and the engineers and clients, together, are in the process of "educating" that child. Depending on the circumstances (and especially the specific kind of artifact, e.g., software system, for a customer, or weapons system, for the military), there can be many phases of testing in which all "stakeholders" participate. We will provide a few examples below.

Eventually the system is ready for deployment and fulfilling its purpose. It is delivered, users are trained in its use, and it goes "live" in the social context of its working environment.

14.3.3   Useful Life: Maturing

Like most of us, a majority of the useful life of an artifact system is spent doing its job. It is producing its functional output. And like us when we get a cold of the flu or have other life crises, the system can at times falter and perform suboptimally. The human components of such systems (e.g., operators, managers, even customers of its product or service) may make mistakes that cause the system to malfunction. Repairs need to be made. As a complex adaptive system with built-in resilience, this should be doable within economic constraints.

A well-designed system will endure over what we call its economically useful life. In the for-profit business world, this means the use of the system continues to produce profit for the organization. In the nonprofit world, this could simply mean the system is remaining effective; it is doing its job correctly and serving the needs of the beneficiaries. The systems engineers who created and designed the system considered the potential life history of the system in the context of its operating environment. For example, take the case of a large solar photovoltaic electricity-generating system (an example we have mentioned in prior chapters). Let's say this system is supposed to operate in a dessert environment with dust, sand, and winds posing threats to the operational efficiency of the array of collectors. Furthermore, let's say the system has been determined to only be economically viable if it operates with minimal repairs for at least 30 years. The engineers should consider this harsh environment and design low-cost cleaning subsystems that can keep the system operating at what are called design specification (so many kilowatts per hour output per unit). If that requirement is not considered in the design (and this has happened in the real world!), then the useful life of the system may be much shorter than its economic investment required. The system will have been a failure overall.

Even so, as we have seen in prior chapters, in different guises, systems do suffer internal decay, and as they age, they require increasing maintenance. They are like us. The older we get, the more health-care we generally require. No system stays youthful forever.

14.3.4   Senescence and Obsolescence

Systems can enter the end game (so to speak) through one of two pathways. One of these is straight out obsolescence, or becoming nonuseful because the environment has changed radically. New technology, new fashions, and changes in attitudes, there are many number of reasons why a system might no longer be useful (see comments below about Detroit and the automobile manufacturing industry). In the electronic entertainment, communications, and computer platform arenas, we have seen incredible rates of obsolescence compared with other technologies. Some people have gotten into the habit of replacing their cell phones just because a new model has been released. Obsolescence is not easy to predict because it is a result of changes in the environment. Thus, the response of a merely adaptive system depends on just how adaptive it is. When the environmental changes are extreme, then the system either is capable of evolving in place (e.g., a company redesigning a product using the new technology) or will give rise to evolved new species having the same basic functions but being more fit to its environment.

But in the more routine case, systems get old. When a system has aged over the period of its useful life, subsystems start to falter and fail. Repairs are costly and so toward the end of a system's life, it is necessary to determine whether there is an economic benefit in making the repairs. The breakdown of components in older systems is just a natural phenomenon associated with what are called entropic processes or, more correctly, decay. Every physical object is subject to decay. Complex systems with many, many subsystems will experience decay and breakdown of different subsystems at different time. Take, for example, the automobile. As everyone knows after so many years or so many miles driven, parts start to break and need replacement. At first it may be a rare failure of a part here and there. But over the later years, more parts start to wear out and the cost of replacements starts to climb. At some point, it is time to trade in the car or give it to charity (or scrap it).

Senescence is the phase of a life cycle when the system is still operating but losing its ability to do so economically.

A good systems engineering job includes planning for senescence. It includes designing the system for decommissioning processes that, themselves, will be economically feasible.

All too often many of our systems (products and real properties) were not considered economically decommissionable. Some of the photos emerging from the city of Detroit Michigan at the time of this writing (late August, 2013) are dismal at best. They show derelict buildings and vehicles abandoned and decaying where they sit. These structures and machines were not decommissioned; they were simply left to rot. Detroit was once an economically booming town due to the automobile industry and its spinoffs. Today auto manufacturing has been moved out of Detroit, and with its departure, the economic base of the city is devastated. Hence, there is no money with which to reclaim the materials from which those buildings and machines were made.[7]

There can be an economic incentive to keep a system operating beyond its design life if it can be done economically and the system still produces quality functionality. Doing so makes it possible to put off replacing it, which is also economically beneficial. But when it is determined that a system is no longer producing the intended functions at an economic price, then it is time to decommission it.

14.3.5   Death (Decommissioning)

Basically this is the dismantling and salvaging or recycling the parts of the system so that waste output is minimized. Recycling is not something we humans have done very well in the past, and it really wasn't recognized as a necessary part of the system life cycle until we had a much better understanding of systems in general. In the past, we were ignorant of and therefore did not consider what are called externalized costs for not recycling. These are costs that are not accounted for because nobody knew they existed until their very real impacts started to accumulate. They arise from both depleting natural resources on the input side, which eventually is recognized in higher extraction costs, and dumping wastes freely into landfills, streams, and air. The latter is the result of not finding economic uses for the wastes. For example, until recently we did not recycle old computers and associated equipment. These were simply shipped to developing world countries where they were dumped, sometimes right next to water supplies of villages. It turns out these devices contain many toxic substances that leached out of them and poisoned the drinking water. Moreover, they also contain somewhat valuable metals like gold and copper,

---

[7] However, scavengers have managed to extract most of the copper wiring that has a market value worth taking the risk.

if you are willing to pick them apart by hand. The poor villagers started dismantling the machines, burning the wiring to get rid of the insulation so as to retrieve the copper. They could sell this metal for cash so it seemed worth it to them to do so. The problem was that the burning insulation gave off toxic fumes that caused sickness in some of the villagers. The external cost here is in the lives of innocent people. In the Western world, our trash problem was solved but at the cost of lives in the underdeveloped world.

Systems engineers, today, know that they have to design for several factors that reduce the cost of maturation, senescence, and decommissioning. The intent is to encourage the safe recycling of material components. This kind of design is still being learned and developed. But more and more product and even whole community systems are being designed for decommissioning should it become necessary.[8]

Systems engineering, then, is a way to use systems science as a guide to engineering whole systems (CASs) and that entails engineering for the complete system life cycle such that wastes are minimized. Except for energy resources consumed, the inputs to the system can ultimately come from recycled materials. For energy the engineering effort focuses on efficiency, but not just the internal efficiency of the system per se. It looks at the external energy consumption needed to produce the inputs; it takes into account the global efficiency of the system.

14.4   The Systems Engineering Process

In this section, we will outline the general process by which systems engineers work from a perceived need for a system to do a specific (albeit generally very complex) function to the delivery of that system and to the monitoring and analysis of upgrade requirements. The systems engineering process matches the needs of the system life cycle described above.

Be aware that what we present here is a very generic framework. Many organizations and commercial companies have developed variations on the main framework. Their details and terminologies will vary from place to place, but if you examine them carefully, you will find they follow the general framework we cover here.

Systems engineering has come under some criticism of late owing to the fact that a little too often, practitioners deviate from the process or get hung up in over-formal methods. The latter refers to the fact that some "formalized" processes that have been "specified" have paid more attention to rigid procedures and documentation that too often fail to capture the real "spirit" of the system being engineered. Engineering, in general, can sometimes suffer from this *zeal* to be rigorous. After all engineers are expected to measure, compute, and specify numerical values. Systems engineering, however, has a very broad scope that includes human influences (e.g., "users") that cannot always be specified or even identified. The human factors are

---

[8] See Braungart and McDonough (2002) and McKibben, Bill (2007).

not just a matter of, say, a client's desires, but include how the humans who work within the system will behave, how the system impacts the larger social setting, and many others. Just the communications links between human participants can require greater insights from the engineers than merely specifying the equipment at each end of the communications channel. In other words, systems engineering is much more than machine engineering. It is engineering so that all the components, including the humans, work smoothly together as intended. In this sense, systems engineering is somewhat akin to architecture. An architect cannot only be concerned with designing a building without taking into account how it is going to be used and perceived.

What we present here is an *adaptable and evolvable* version of systems engineering. That is, the process is a system in the full sense that we have developed throughout the book.

Figure 14.1 provides a flow diagram or "roadmap" of the systems engineering process that outlines correspondence with the system life cycle . It is shown as a kind of cascading flow from the top down. Starting with a "needs analysis," the process moves down, the completion of each stage leading to the next lower stage. However the process is rarely (some would say never) linear. As indicated in the figure, there are feedback loops from a lower stage back to the higher stage. These result from discoveries made at a subsequent stage that indicate that the prior stage did not cover some important aspect. For example, systems analysis might discover a hidden factor that was not taken into account in the needs assessment and that might materially impact the definition of the needs. Similarly systems analysis might miss something that is discovered when trying to build a model. These pathways allow the whole process to self-correct or, should the flaw be fatal to the project, allow a proper review and decisions to be made. The figure only shows immediate feedback pathways from lower to next higher stages. In fact pathways from much lower stages can be used in those cases where a design problem somehow remains hidden until much later in the process but needs to be fixed before going forward.

The process starts with a needs assessment carried out with (and usually driven by) a client. After the client and engineer agree that they have a good description of the need, the system engineer proceeds to analyze the system. This may involve analyzing an existing system that needs upgrading, or it could be the analysis of a new system by analyzing the problem and the environment (as in Chap. 12). A successful analysis leads to modeling (Chap. 13) and the beginning of designing. Once a design is complete, the system (or a prototype) is constructed and tested extensively. At the completion of all functional and operational tests, the system is deployed or delivered to the client for use. Eventually, as described above, the life cycle of the system approaches its end and the system will be decommissioned, possibly replaced by an upgrade or something new entirely. Below we will provide some more details of what goes on in each of these stages.

Figure 14.2 shows a somewhat similar flow chart that is more detailed with respect to systems engineering tasks. This is called the systems engineering process life cycle (to differentiate it from the system life cycle). Here the emphasis is more on what the engineer must do. For example, the needs assessment includes a process we call "problem identification." This is actually what the engineer is doing when

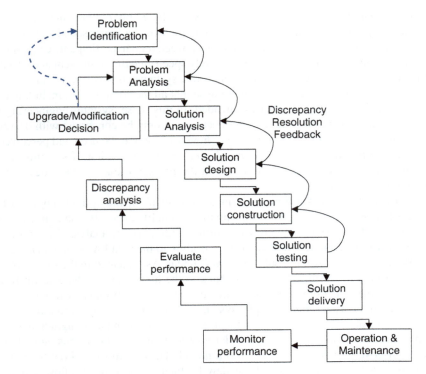

Fig. 14.2 The engineering process life cycle can follow a model such as this. Note that it is really a cycle that corresponds with the system life cycle. During the analysis, design, and construction/ testing stages, any discrepancies or problems encountered at any stage can result in feedback and resolution at a higher stage. Thus, much of this process is iterative as a final design is developed

trying to help the client assess their needs. You can see from the identifiers on each stage what the engineer is working on in progression. As with the above flow chart, there are feedback pathways that allow a reanalysis or redesign if something doesn't quite work right in a lower stage. This is shown again in the below figure.

14.4.1   Needs Assessment: The Client Role

People observe and recognize the need for systems. The general description is that they are experiencing a problem. Perhaps they want to accomplish something that no current system can perform. Perhaps they want to do what they are already doing but do it less expensively; the current system is too costly to operate at a profit. Occasionally an inventor just wants to see if they can create something new. Finally there is the motivation just to gain knowledge. The scientist who needs to do a complex experiment will need a complex instrument as well as other equipment and people in order to explore the unknown.

We will refer to this generic condition as problem solving. But before a problem can be solved, it has to be understood. Systems engineering begins with consultation with the person or people (called the client) who are experiencing the problem and have a need to solve it. The initial needs assessment is driven by the client. They understand their situation better than anyone else so they are in the best position to describe the problem as they perceive it. It is, however, up to the systems engineer to recognize that there are times when the client's perceptions might be incomplete. It is sometimes prudent to make sure you understand what the client "needs" and not just accept what the client says they "want." Of course that requires tact!

After an initial, and generally informal, discussion and description of the perceived problem, where the systems engineer listens and ask questions but does not rush to suggest solutions,[9] the process of needs assessment gets under way.

Actually a lot goes on during this phase. One very important dynamic that needs to get established is the communications modes between client and engineer. The engineer needs to establish in the client a sense of confidence in their competence. A very successful engineer knows that this doesn't come from impressing the client with their vast knowledge (even if they have it). It comes, ironically, from the engineer asking the right questions at the right moments and carefully listening to the answers. That does more to convey the impression of expertise and rapport than anything else the engineer can do. It is a behavior vital to success in the coming projects.

The objective of needs assessment is to start outlining the various parameters that will condition the development of a system. These include the client's description of what the system should do, the economic constraints, and the economic and operational risks that might be involved (e.g., liability insurance rates for life-critical systems), finding out who all the stakeholders are (client, customers, employees, stockholders, and also the environment!), and other relevant issues. This is what we call the "big picture." The engineer is trying to get a really broad view of everything that will impact the project and the life cycle of the system. From the engineer's point of view, the central work is called problem identification (as in Fig. 14.2).

There are various kinds of documentation that derive from this initial phase. Some may be as simple as a short whitepaper describing the system and how it will solve the problem if it is feasible to build. But more generally the documents will include general descriptions and an initial feasibility assessment and budget considerations. It should *NOT* include a description of the solution, but it should provide a preliminary statement of what all parties view as the problem.

Some systems engineering organizations have formal documents (forms) that capture the results of the needs assessment. As long as these documents don't try to get the cart in front of the horse (suggest solutions prematurely), they can be a great help as a kind of checklist of things to do in accomplishing the assessment. When properly completed, they measure success at this first stage and can be reviewed by

---

[9] One of the most common points of failure in this process is when either the client is sure they know exactly what they need or the system engineer jumps to a possible solution without going through analysis. For example, the engineer might be reminded of a similar problem that had a particular solution and conclude, prematurely, that the same kind of conclusion might work here. Very dangerous.

all parties as a prelude to continuing the work. This kind of "stop and see where we are and are we making progress" method is used at every phase in systems engineering. If followed rigorously, it will prevent building a useless system (we know this from historical experience!).

14.4.2   Systems Analysis for Artifacts to be Developed

14.4.2.1   Problem Identification

The needs analysis helps to identify the larger issues in the problem the client needs solved. But this is just a method of arriving at a first approximation of the problem and its scope. Once the client and engineer agree to a mutual understanding of the client's needs, the real work begins.

The output from a well-done problem identification and analysis process is a specification of what the solution would be. That is, it is not a specific solution design; that comes much later. Rather it is a specification of what the solution should do to solve the problem and how you will know it is successfully doing so. The engineer must establish the criteria of merit that determines when a specific design will have succeeded in solving the problem within the technical and economic constraint and according to performance standards.

Figure 14.3 shows a more detailed flow diagram for this process.

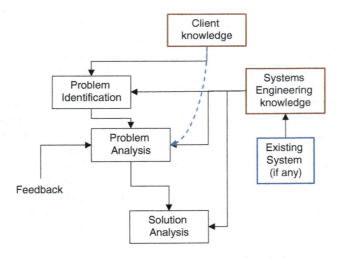

Fig. 14.3 The problem identification and analysis involve inputs from both systems engineers and clients. If there is an existing system that is being upgraded or replaced, it will be used to help understand the problem. Problem identification is as much driven by the client as by the systems engineer. Problem analysis, however, requires the disciplined methodologies covered in Chap. 12

Problem identification is often a natural extension of needs assessment. But it involves more directed question asking by the systems engineer and involvement by the client.

As an example, suppose a certain city is experiencing traffic snarls in a major thoroughfare through the downtown area and the city planners are convinced that the problem will be resolved by widening the street (similar to the freeway expansion discussed above). However, the city council has heard that systems engineering might be useful in finding possible alternative means for resolving the problem. For one thing, suppose the widening of the street has a negative impact on storefront businesses. They would be anxious to avoid such negative results just to solve this problem. So they hire you as a systems engineering consultant.

You sit down with the planners (assuming they are friendly to your being hired!) and start probing for their opinions and ideas about the problem being experienced. You discover, among other factors, that traffic patterns are constrained by existing one-way streets, no real alternate route options, and the effected street acting as a connector to two separated hubs of commercial activity, a banking district uptown, and a major retail district downtown. Using network analysis, you map out the city using a flow network and ask for data regarding traffic loads at different time of day. Already systems science is coming into play to help you define the problem. Note that this is analysis, but it is still fairly high level and is being used just to get a formal description of the problem that you will document and present to the client.

In this case, what you are trying to determine is the scope of the problem, that is, how bad is it at the worst times, and who is being negatively affected by it, and so on. Later you will extend this into an even more formal analysis, which we call the problem analysis (Fig. 14.4).

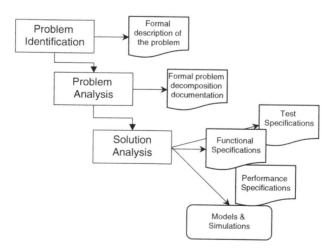

Fig. 14.4 Documentation is required at each stage of the process to ensure completeness, consistency, and agreement among all parties. The nature of these documents is covered in the text

Even though we called the first stage problem identification and the second stage problem analysis, you should recognize that you, as the engineer, are always engaged in analytical thinking. You always have to be asking questions of the client, of the problem, and of your own thinking about the problem.

In this example, suppose that you, as the engineer, also interview shop owners, pedestrians, and drivers who are routinely caught in the snarls. You also think to interview employers at both hubs to see if there is a problem from their point of view. From the city's perspective, they see the traffic jams and believe this is hurting business and aggravating drivers. It turns out that the retail merchants downtown do not perceive any harm to their business due to rush hour traffic. In fact several mention that they think more people spend more time in the stores just to avoid the traffic! Clearly this is a multidimensional problem and being a good systems engineer, you have identified more aspects than originally assumed.

Most professional systems engineering operations require the engineer to produce a formal document at the end of the identification stage that details what has been identified. This document provides the basis for clients and other stakeholders to agree with the engineer that what is documented is indeed the problem that needs to be solved. Once that agreement is in place, the process can proceed with some confidence that we know what it is we are trying to do!

14.4.2.2   Problem Analysis

The problem identification found the main parameters of the problem, the different facets of it, and the scope (who/what was affected). Now it is time to put some numbers on these and start a system decomposition to get to the details. Following decomposition procedures as outlined in Chap. 12, the engineer uses the scope information from the identification phase as a first approximation to putting a boundary on the system of interest. In the case of the traffic problem, this might be seen as the two business hubs and the street between them. Inputs and outputs would be automobiles (and possibly pedestrians) from side streets and any other streets leading into and out of the main street. Starting with a preliminary definition of the system, it is then possible to start decomposing things like the street itself (how many lanes, condition, etc.) trying to get values for flow rates. An important subsystem of the street turns out to be the traffic lights, which regulate flows in both directions and entries and exists of traffic.

The systems engineer gathers data on volumes, establishes sample rates, notes fluctuations, and records anything that can be used to better understand the system as it exists. The result is the set of documents as described in Chap. 12, maps, hierarchical structures, and knowledge bases. Note in Fig. 14.3 that the client has a role in this stage. They are expected to review the documents from the analysis as it proceeds and also at the end to make sure they still have an understanding of the problem and the current system. This is often a weak point for some engineering efforts. Too often the kinds of documents they use are arcane insofar as the non-systems engineering (lay) public is concerned, full of acronyms and abbreviations, even mathematics that are off-putting to most clients. The astute engineer (and the engineering company) has developed formats for documents that are clear and easily understood by clients. This is not really hard, but it is seemingly easier for some organizations to use the more "technical" formats. Indeed there are some organizations that revel in showing the clients how incredibly intelligent and knowledgeable they are by showing them technical documents. Clients may very well agree with their conclusions just so as not to look dumb. But most of the later project failures occur right at that point. No systems analysis process is perfect. And it is usually the case that the client can spot some deviation from their original shared understanding of the problem from the identification stage. If the clients show straightforward documentation, they are very likely to grasp the results of the analysis and be able to verify that the engineer has got it right, or point out the deviations that need review.

Multimillion dollar projects have failed miserably because the engineers did not sufficiently involve the client in this critical stage. Remember, the outcome of systems analysis is not just a bunch of mathematical relations; it is the *understanding* of the system. And that should be accessible to all parties/stakeholders.

14.4.2.3 Solution Analysis

Once a problem has been identified and analyzed, it is time to consider solutions. At this point, it is also time to bring in specialist engineers to start looking for a specific subsystem according to their expertise. For example, with the city congestion problem, a team of specialists from traffic engineers to business consultants may need to be assembled. The job of the systems engineer now starts to deviate from what we normally think of as engineering. It is not so much to develop solutions as to coordinate the efforts of people who can develop subsystem solutions and make sure that all of the subsystem requirements are covered and the various experts are proceeding with their piece of the puzzle apace with everyone else.

Systems engineering is not so much about inventing the whole solution as it is about making sure the solution is whole! Using systems science principles, systems engineers pay attention to the boundaries between subsystems and especially the communications and flows between subsystems to ensure that they are being properly integrated.

During this stage, the systems engineer becomes one of those logistical and tactical level managers we discussed in Chap. 9.

14.4.2.3.1   What Is a Solution?

When we talk about solutions at this stage, we are not actually talking about completed designs or the built artifact system. Those come later. What we mean by a solution is a determined *specification* of what the artifact system to be developed should do (function) and how well it should do it under specified constraints (performance). This applies to the system as a whole, but also to each of the subsystems as well. The method for working these specifications out is to model the whole and each subsystem (independently) just as we described in Chap. 13. As much as anything, this stage is about describing what a solution to the problem looks like, how it will solve the problem, and what its parameters will be so that it is known what a realized solution—the artifact—will be like. The documents from this stage, the functional and performance specifications along with the models used to develop and test them, provide a complete description of what the solution is.

14.4.2.3.2   Feasibility

In generating solutions to problems, in terms of the design and construction of an artifact system, it is important to establish the feasibility of the solution proposed. Feasibility covers several considerations. Can the system be designed and constructed economically (that is within budget constraints)? Can the system perform as expected under conditions expected? And so on. Before launching a major effort to design, build, and deploy a system, the engineer must analyze these kinds of factors and document that analysis.

14.4.2.3.3   Sub-solutions

Remember the system hierarchical structure of Chap. 12? The system analyzed is comprised of subsystems, and they of sub-subsystems, etc. Thus, it is not surprising to see that the solution system will be comprised of sub-solution systems. So just as we applied a recursive process to decompose the system in analysis, here we follow that structure and analyze solutions by subsystems. In this case, the analysis can go upward as a process of integration. When subsystem solutions are proposed, they must be evaluated in the context of their higher level meta-system.

14.4.2.3.4   Modeling Sub-solutions

As we discussed in Chap. 13, models can be built by building submodels and then integrating them. The type of model that is used will depend on the kind of subsystem being analyzed. However, this can raise some issues when a specific subsystem is best modeled using one approach, e.g., a particular component might be modeled with a dynamic systems (differential equations) approach, while the larger meta-system is best modeled using a system dynamic framework. At this date, there are no easy solutions to integrating these seemingly disparate approaches. However, a top-down analysis might suggest that a decomposed system dynamics framework should be used even on the component so that the proper dimensions, time constants, and other factors are already integrated. Some engineers might protest that that approach is inefficient vis-à-vis the component and they would be correct. However, the amount of time needed to translate from one model language to another to resolve the parameters might be greater than whatever is lost in modeling the component in discrete time versus continuous.

Ideally the meta-model of the solution system will readily incorporate all sub-models at, however, many levels are required. We say "ideally" because this is usually a very difficult problem. Perhaps someday there will exist the appropriate modeling tools that will expedite such integration and allow engineers to "zoom in" on lower level models or "zoom out" to see the big picture.

14.4.2.3.5   Specifications

The output products of this stage are called "specifications" and essentially describe in great detail every aspect of the solution system in terms of function, performance, and how the system will be tested (see Fig. 14.4). These specification documents are organized in the same hierarchical fashion as the system. That is, there are whole system functional specifications as well as functional specifications for every subsystem all the way down the hierarchy to the lowest components. Oftentimes the numbering scheme used to identify these documents correspond with that we showed in Chap. 12, reflecting the system hierarchy.

Functional specifications are just what the name implies. They detail the function(s) that the system performs. These are tied directly to the solution of the problem. For example, the function of the city traffic control problem is to maximize the flow through subject to the arriving volume of traffic at either end of the corridor. The performance specification tells how well the functions are performed. A performance criterion for the city traffic might be something like "no automobile should be on this street for more than ten minutes." This specifically puts requirements into measurable units. This is important for later monitoring of the system to make sure it is performing as expected and required (see Sect. 14.4.2.8 below).

The test specification details what testing is required for each subsystem (sometimes called "unit tests") from the bottom of the hierarchy up to the system as a whole. The kind of testing that can be used, just like the kind of modeling, depends on the nature of the subsystem. For example, one of the subsystems of the traffic flow system is the "intelligent traffic light" timing. It might be possible to construct a test bed somewhere out in the country where engineers can simulate several city blocks and test real light timing algorithms. That is probably a fairly costly approach (although it turns out there are such test beds already built that can be "rented" for doing this kind of testing!). Contrast that with what it takes to test the operational characteristics of the computer controls for the lights. For example, the performance specification may require the switching response time for changing the lights along with responses to the triggering events (called a real-time response requirement). This can easily be tested in an electronics laboratory so the test specification would call for the kinds of real-time response tests that would be required for the hardware once it had been determined.

At the end of each solution analysis for each component, sub-subsystem, subsystem, and the system as a whole, the systems engineer's job is to verify the completeness of all of these documents. This includes making sure that the inputs to a subsystem are accounted for as outputs from another subsystem or interface. This is where the completeness of the systems analysis in Chap. 12 is particularly important. If that work was done well and completely, the placeholders for all of the flows (inputs and outputs) are already in place and just need to be filled in as specifications. Unfortunately this is perhaps the most tedious part of systems engineering; there is a lot of paperwork to deal with! However, tools for aiding this are in the works and a few have already been developed. The ideal tool would assist the engineer to do the analysis, produce the placeholder slots in the knowledgebase, generate the models, test the integration, and produce forms the domain experts can fill out directly. Perhaps a systems engineer can get to work on producing that kind of tool—it is a system after all.

14.4.2.4 Solution Design

Once a specification for a subsystem has been determined, it is time to get to work on designing the physical aspects of the solution. This is, again, where domain expert engineers are required to do the detail work. The specifications developed during the solution analysis now pass to those experts for their work (Fig. 14.5).

This is where things get really messy! Up to this point, the work undertaken by the systems engineer has been very generic. Even the solution analysis process is essentially the same regardless of the nature of the system solution. But because the design stage involves much more specific work within domains of expertise, there is very little we can say about how this stage proceeds.

The work of the systems engineer may become that of project management, keeping track of manpower resources, schedules, etc., the scope of which is beyond this book. The methods of project (or sometimes called program) management are covered in detail in many fine textbooks and trade books. So we will effectively wave our hands and proclaim "a miracle happens!"

Out of the design stage, a set of design documents appropriate to the nature of the components/subsystems and the system as a whole emerges to guide the actual construction or implementation of the system.

14.4.2.5 Solution Construction

At this point, we should mention the fact that sometimes systems will be built as prototypes versus the final product. Prototypes are often needed because there is still some uncertainty in how the design will work in practice. Some systems cannot be prototyped, for example, the city traffic flow system is just going to be constructed in place. Models of such systems might be used instead of prototypes to test as much as possible. Regardless of whether the system to be constructed is a prototype or the final system, the process is fundamentally the same.

As with design, the work of construction is given to domain experts who know how to build them. The job of the systems engineer is, again, one of coordination and monitoring progress of all of the parts. And again this is largely a paperwork process. If all of the specifications and models were of high quality, the systems engineer has a lot of background documentation to guide their work and decisions.

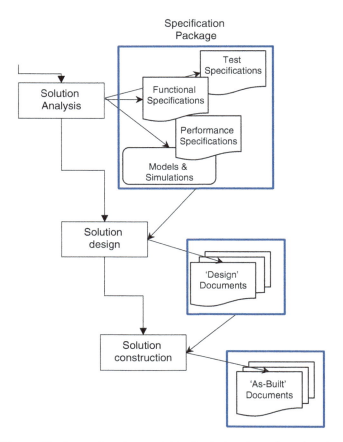

Fig. 14.5 The specification package is passed into the design stage where domain expert engineers practice their trade creating designs that will fulfill the specifications. They also have access to the models to assist in understanding the dynamics of the system. The output is a set of design documents like blueprints and schematics—whatever form is needed to communicate the design to those who will construct the system. The "as-built" documents provide information about how actual physical construction may have had to deviate from the design because of various factors

One interesting aspect of actual system construction is the fact that all too often, there will be latent discrepancies still in the designs that the design engineers were unaware of. For example where an engineer may have called out a component part with which they were familiar and which, according to the catalog, had the functional and performance specifications needed, the company that manufactured that component may have a new model or even discontinued the part. The construction worker usually has some discretion in substituting another component in the latter case or using the upgraded component if the price isn't too much greater. In all such cases, the constructors mark copies of the design documents with the altered specifications. These documents are often called "as-built" because they reflect what was actually built. As a rule, when these documents are generated, the systems engineer is responsible for making sure they find their way back to the design engineers for confirmation or redesign. If they confirm that the as-built system is OK, they will sign off and the original design documents may be modified to reflect the new configuration.

14.4.2.6   Solution Testing

At this point, we should probably mention that while what we have so far described as a linear, cascading process (as in Fig. 14.2), the reality is that there is a lot of up and down, back and forth between many of these stages. As we just described, where the design engineers had to verify an as-built design before the project proceeds, throughout the various stages, discrepancies between what was determined in the prior stage and the current work turn up. That is why in Fig. 14.2 we showed some recurrent arrows back from lower stages back up to the previous stage. In fact it can be much messier than that. A problem discovered in a lower stage might end up propagating further up to a very early stage. The point of having check points along the way is to catch these discrepancies as they occur and get them fixed as soon as possible so that the whole project is not disrupted.

Testing at every level is a good example. Remember the test specifications written in the solution analysis stage? Once a unit (component or subsystem, final, or prototype) is built, it is tested according to the specification.

14.4.2.7   Solution Delivery (Deployment)

Eventually, and hopefully on time and budget, the solution is delivered to the client. The systems engineer is still involved in this process because for complex systems, the client's employees (users) need to be trained. The systems engineer may not be directly involved in the training but is involved in making sure the training documentation is properly prepared and may even manage the training process.

In addition, deployment of the artifact system is not just a matter of turning on the equipment and running it. Most very complex systems require a "staged" start-up, meaning that parts are put into place and put into limited operations, checked for the validity of operations (users trained, etc.), and then certified by both clients and engineers. Systems engineers guide the system of installation and start-up just as they managed the systems of analysis and design. Their overall role is to monitor this process and double-check results against expectations to ensure the client is satisfied with the project results.

14.4.2.8   Monitor Performance

A complete system life cycle process requires that the systems engineers continue to collect data on the artifact system's on-going performance. They can identify problems that might develop and intervene when necessary. The kind of data they collect is based on the specifics of the system so there isn't much to say about specifics. But the general idea is to collect the data to minimize any performance degradation and have the data on hand for a later evaluation process that should take place before the next generation of the system, or its replacement, is started. Performance histories of existing systems feed back into the analysis and design of newer systems in the future. Unfortunately, aside from ongoing managerial analysis for operating costs, the monitoring of systems for purposes of future design is often considered too costly and therefore not followed through on. Detailed performance data are generally not collected and that will have a negative impact when it comes time to upgrade or replace the system (which is inevitable). We say unfortunate because it turns out that if the data on performance is collected and saved, it will cut costs of later analysis.

14.4.2.9   Evaluate Performance

In Fig. 14.2, performance monitoring feeds into an evaluation process. As just mentioned, the only performance analysis that usually takes place is the business case, i.e., cost of operations versus benefits (e.g., profits). Operations and financial managers look at the dollars and make decisions about the value of the system as it operates. Performance evaluation, however, has more to do with the criteria of performance of subsystems as well as the system as a whole. The data that we suggested above to be collected involves physical as well as financial performance measurements. If the former are collected and periodically analyzed, it is possible to identify problems before they hit the financial performance. But even more important is looking to the future. Assuming the client will still be in business (in the generic sense of that word) when it comes time to upgrade or replace the system, the ongoing performance data will prove valuable.

Evaluated periodically, that data can help in determining whether or not the time is coming to start considering such upgrading or replacement. In the former case, the systems engineer is looking for performance issues that imply adaptation intervention; they may need to modify select parts of the system to bring it back into compliance with its long-term performance requirements. In the latter case, we may be talking about an evolutionary change.

14.4.2.10   Performance Discrepancy Analysis

We treat this as somewhat different from performance evaluation in that the systems engineer is looking closely at the data and comparing the results with long-term expected performance. Recall that all systems exist in nonstationary environments. Changes in the environment may actually call for changes in the performance requirements of the system. The systems engineer has to look for discrepancies between environments over time and system performance over time.

14.4.2.11 Upgrade/Modification Decision

Finally, the system is subject to modification (adaptation) or upgrade (evolution) when sufficient discrepancies between actual performance and desired performance (which can include those due to new environmental factors) reach a critical point. The client and the systems engineer consult over the results of ongoing monitoring and analysis to come to that decision. For the client, it comes down to the cost/benefit analysis as well as future projections for operations (e.g., would replacement cause major disruptions to other operations). The system may be reaching the end of its planned life cycle. More often than not, in this modern world of rapid change, it may simply be a case of the environment changing unpredictably and the larger meta-system, say a company, needing to change its ways of doing things to continue to operate in a relevant manner.

Whatever the case, we find ourselves back to the beginning of the system development cycle. This is where that data collection and analysis we mentioned above can really be a saving. In Fig. 14.2, we show two arrows out of the decision box, one leading back to problem analysis and one, dashed, leading back up to problem identification. The need to go back to problem identification is often the result of not doing a good job of monitoring the system during its operating life. In the second case, the problem is already well known and the jump to problem analysis will save time and money. Moreover, the actual amount of analysis that would need to be done may be foreshortened because most of the problem is understood from the performance analysis steps.

Think Box The Brain as Tool Maker
We come to the end of our thinking inside the boxes with a recognition of how our human brains, having mental models with which to ask "what-if" questions about the world, are able to generate stories that combine different elements together in new ways. We can try out combinations of things, actions, and relations and use our powerful analytical abilities to test the worthiness of the new story by comparing it with our current models of reality. As noted in Think Box 13, this is imagination.

The current chapter is about the notion of systems improving through intentional design, what we call engineering. People are forever exploring their environments in search of resources. Using affordance (as described above), people can "see" possible ways in which those resources can be extracted or exploited. What they can see is opportunity but also they see a "problem." In order to actualize the exploitation of a resource, they have to also have a "tool." It is one thing to dig up a root vegetable with your hands or catch a rabbit by running it down. But to really get production going requires some kind of digging tool or a bow and arrow. In other words, it takes tools to obtain more useful work than can be done by hand alone.

(continued)

> **Think Box** (continued)
> The brain commands our muscles to perform actions that are feasible for human bodies to do. But when the actions needed are not within the abilities of the body, there needs to be some instrument devised that can do the work. Tools are instruments that extend and amplify the capacity of the human body to do work. With the ability of the brain to build models of possible world states, to see how natural objects can be shaped or put together to achieve a purpose, and to grasp the rewards to be reaped by investing time and energy in the tool construction process, humans have become consummate tool makers. In fact an argument could be made that that ability is the basis of our culture. Even art can be considered a tool for exploiting enjoyment!
> Engineering is actually the art of creating tools or making tools better. A craftsperson is one who makes the tools with skill. But the engineering mind conceives of the tool and how it can be used to do other work. This whole process is the epitome of the cybernetic hierarchy of the brain, with the strategic level actively planning the design and construction of the first tool for its purpose.

14.4.3 Process Summary

Throughout the systems engineering process, as it is coupled with the system life cycle, we have seen a continuing role for the systems engineers to be involved in all stages. Primarily the systems engineers act as "big picture" overseers to make sure that all the parts of both the physical system and of the processes in each stage fit and are completed properly (and on time and within budget). Unlike say a computer engineer who might be involved only in developing, testing, and delivering one component in the system, the systems engineer lives with the system for its entire life, all the way through development, etc., and then through its ongoing operations. The latter is because evolution is invariably the dominant long-term force that shapes artifact systems. And those systems are evolvable, not just adaptable. Look at the modern automobile as an example of an artifact system. Not only has the system improved in performance criteria over the decades, it has had to evolve to meet changing demands from its environment. That environment includes the physical infrastructure (the roads and fuel supply systems), customer preferences, manufacturing processes, new technologies, and many more factors. The fundamental function of the automobile has not changed particularly (in spite of many predictions, it has not really morphed into a transformer-like private airplane or boat on demand). It is still a personal/family ground transportation unit. But how it performs its functions has been very influenced by the environment.

Among many changes that have taken place in the automobile industry over the years is how cars are designed and engineered. Today new models are developed by teams of designers/engineers who, collectively, act as systems engineers. That is, when you look at the process through which new autos are created, you will find the elements of systems engineering even when no one person is necessarily designated with that title. Indeed, look at almost any industry that produces very complex systems products or services, and you will find that they involve every element of the systems engineering process. Some industries, for example, the large passenger aircraft industry, has officially adopted the role of systems engineers because they recognize that what they are doing is systems engineering. Having a titled group of people not only recognizes that function but also effectively makes the organization "cognizant" of it.

In the coming years, we expect to see more industries and also governmental and nongovernmental agencies that undertake very complex systems projects to turn to "official" systems engineering to improve their successes.

14.5 Systems Engineering in the Real World

Systems engineering has been employed by many companies that build extremely complex artifact systems such as large airplanes (Boeing Company) and government-based organizations that have complex missions (NASA—putting men on the moon). Many other organizations, business, government, civil social, and others that have very complex missions and need to organize around artifact systems (both physical objects and operations procedures) are already practicing systems engineering, either explicitly or, more generally, implicitly. Complex missions require complex artifacts to solve problems, and these organizations have no choice but to use many of the tools of systems science and engineering to tackle these problems.

As systems engineering becomes more recognized for what it can accomplish, it is likely that more organizations will adopt formal methods. Several organizations have formed to develop standards and methodologies for a formal approach. The International Council on Systems Engineering (INCOSE),[10] formed in 1990, is dedicated to advancing the practice of systems engineering in all arenas. The Institute of Electrical and Electronic Engineers (IEEE) has a standard setting process and has worked out a number of standards related to systems engineering. For example, "15288-2004—IEEE Standard for Systems engineering—System life cycle processes" provides guidance for just what its name implies. From the web page:
...establishes a common framework for describing the life cycle of systems created by humans. It defines a set of processes and associated terminology. These processes can be applied at any level in the hierarchy of a system's structure.[11]

NASA (National Aeronautics and Space Administration) also has produced a set of standards for its organization and suppliers to use when working on complex missions.[12] Not all applications of systems engineering are necessarily about

---

[10] See http://en.wikipedia.org/wiki/INCOSE.

[11] See http://standards.ieee.org/findstds/standard/15288-2004.html.

[12] You can download one of the documents, NASA (2013): nodis3.gsfc.nasa.gov/npg_img/N_PR_7123_001B_/N_PR_7123_001B_.doc.

complex products like airplanes, complex technical missions like putting people on the moon, or information systems to support a supply chain operation. All of us live in complex systems. And those systems are likely to show stresses if they are not well designed.

More and more of the "problems" that our civilization face are inherently complex and, hence, systemic. Their solution, if they can be solved, will involve systems engineering. We will need the application of systems engineering for what we often take for granted—sustainable living systems in a world beset by extraordinarily complex problems. Our current society is based on the availability of fossil fuels (over 80 % of our energy comes from fossil fuels) and those fuels are a finite and diminishing resource. The burning of fossil fuels has dumped an amazing amount of carbon into the atmosphere and oceans (CO_2) and has contributed to heating the planet and changing the basic climate system. The change in climate and human uses of freshwater is already changing the availability of water for drinking and agriculture. There are other resource threats that we have to consider.

Bibliography and Further Reading

Gibson JJ (1977) The theory of affordances. In: Shaw R, Bransford J (eds) Perceiving, acting, and knowing. Lawrence Erlbaum, New York

McDonough W, Braungart M (2002) Cradle to cradle: remaking the way we make things. North Point Press, Berkeley, CA

McKibben B (2007) Deep economy: the wealth of communities and the durable future. Henry Holt & Company, New York

NASA (2013) NASA systems engineering processes and requirements NPR 7123.1B, April 2013 NASA procedural requirements. nodis3.gsfc.nasa.gov/npg_img/N_PR_7123_001B_/N_PR_7123_001B_.doc. Accessed 9 May 2013

Rogers GFC (1983) The nature of engineering—a philosophy of knowledge. The MacMillan Press Ltd, London

Index

© Springer Science+Business Media New York 2015
G.E. Mobus, M.C. Kalton, *Principles of Systems Science*, Understanding
Complex Systems, DOI 10.1007/978-1-4939-1920-8

CPSIA information can be obtained at www.ICGtesting.com
Printed in the USA
LVOW02*1142071214

417619LV00001B/1/P